THE WAY THINGS WORK

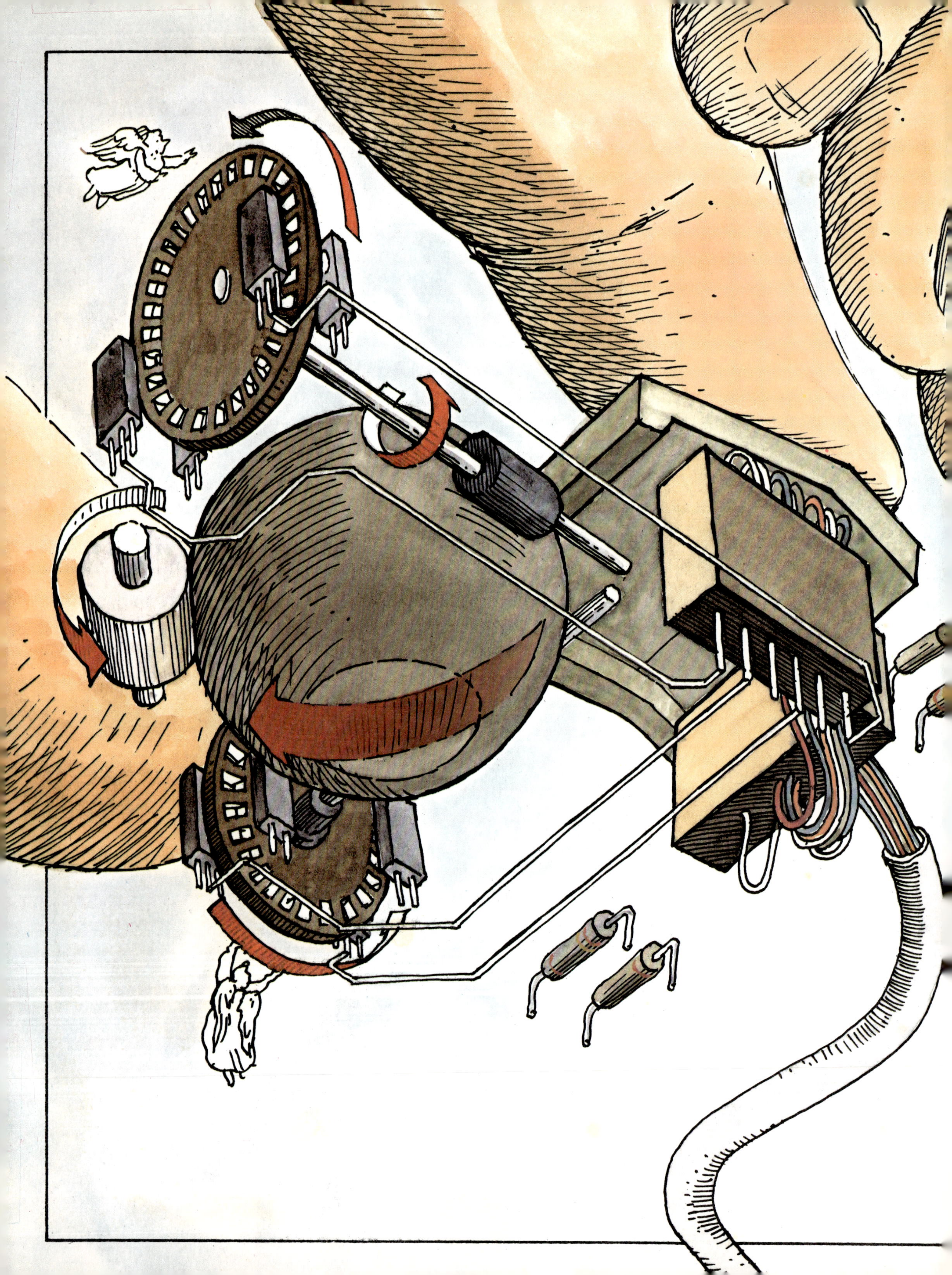

THE WAY THINGS WORK

DAVID MACAULAY

A Scholastic Book Club Edition

HOUGHTON MIFFLIN COMPANY
BOSTON

THE WAY THINGS WORK has taken almost three years to create. It could easily have taken another three years without the help and encouragement of several people. My gratitude, first and foremost, goes to Neil Ardley, who provided all the technical text and who shared his vast knowledge of the subject with tremendous patience and enthusiasm. Another member of the team I would like to thank is David Burnie, whose editorial input, organizational skills and unfailing sense of the book from beginning to end were matched only by the illusion of calmness he maintained as each deadline came and went. Designer Peter Luff both provided a framework which ensured a visual consistency and was also an invaluable critic throughout the making of the pictures. I also wish to thank Christopher Davis who used the old ruse of the after-dinner chat to get me into this whole thing in the first place, and also Linda, Ben and Sam Davis who provided me with a second home during my many meetings in London. Finally, my thanks go to my wife Ruth, who has alternately encouraged and tolerated this project and all its demands. To her, with love, this book is dedicated.

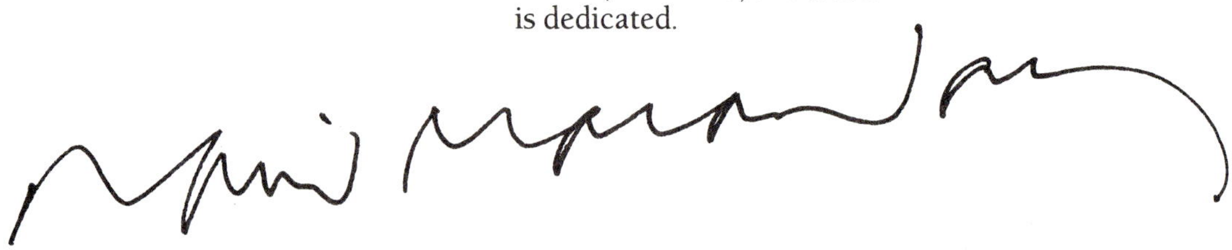

Library of Congress Cataloging-in-Publication Data

Macaulay, David.
The way things work/David Macaulay.
p. cm.
Includes index
ISBN 0-590-42989-2
1 Technology—Popular works. I. Title.
T47.M18 1988
600-dc 19 88-11270
CIP

Published in the United States by Houghton Mifflin Company
Published in Great Britain by Dorling Kindersley Limited

Printed in the United States of America

J 10 9 8 7

Contents

EARLY·WORK ON THE·ROTATION OF·THE·EARTH

PART 1

THE MECHANICS OF MOVEMENT

CONTENTS

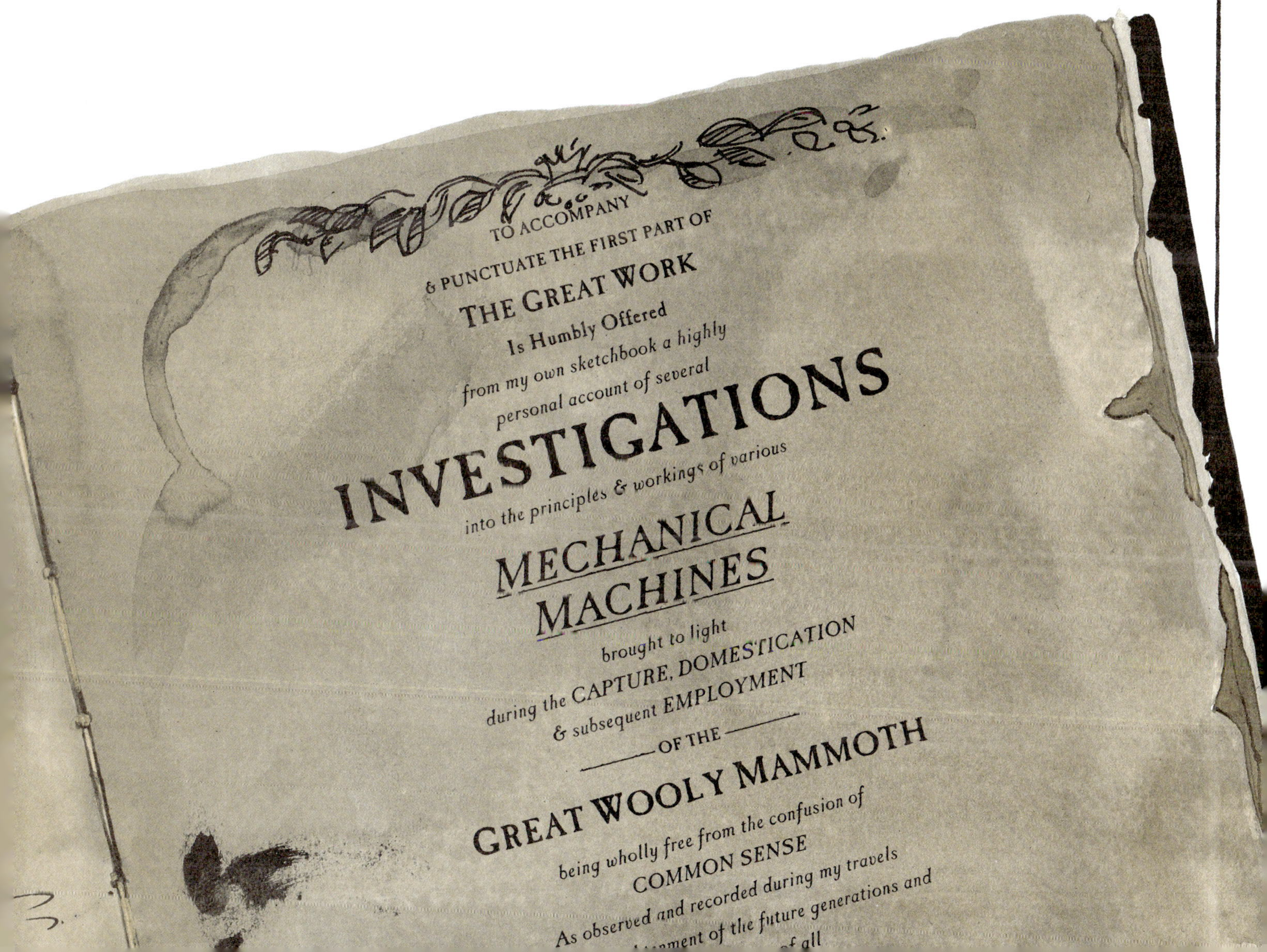

TO ACCOMPANY
& PUNCTUATE THE FIRST PART OF
THE GREAT WORK
Is Humbly Offered
from my own sketchbook a highly
personal account of several
INVESTIGATIONS
into the principles & workings of various
MECHANICAL
MACHINES
brought to light
during the CAPTURE, DOMESTICATION
& subsequent EMPLOYMENT
OF THE
GREAT WOOLY MAMMOTH
being wholly free from the confusion of
COMMON SENSE
As observed and recorded during my travels

INTRODUCTION

TO ANY MACHINE, work is a matter of principle, because everything a machine does is in accordance with a set of principles or scientific laws. To see the way a machine works, you can take the covers off and look inside. But to understand what goes on, you need to get to know the principles that govern its actions.

The machines in this and the following parts of *The Way Things Work* are therefore grouped by their principles rather than by their uses. This produces some interesting neighbors: the plow rubs shoulders with the zipper, for example, and the hydroelectric power station with the dentist's drill. They may look very different, be vastly different in scale, and have different purposes, but when seen in terms of principles, they work in the same way.

MACHINERY IN MOTION

Mechanical machines work with parts that move. These parts include levers, gears, belts, wheels, cams, cranks and springs, and they are often interconnected in complex linkages, some large enough to move mountains and others almost invisible. Their movement can be so fast that it disappears in a blur of spinning axles and whirling gears, or it can be so slow that nothing seems to be moving at all.

But whatever their nature, all machines that use mechanical parts are built with the same single aim: to ensure that exactly the right amount of force produces just the right amount of movement precisely where it is needed.

MOVEMENT AND FORCE

Many mechanical machines exist to convert one form of movement into another. This movement may be in a straight line (in which case it is often backward-and-forward, as in the shuttling of a piston-rod) or it may be in a circle. Many machines convert linear movement into circular or rotary movement and vice-versa, often because the power source driving the machine moves in one way and the machine in another. But whether direction is altered or not, the mechanical parts move to change the force applied into one – either larger or smaller – that is appropriate for the task to be tackled.

Mechanical machines all deal with forces. In one way they are just like people when it comes to getting them on the move: it always takes some effort. Movement does not simply occur of its own accord, even when you drop something. It needs a driving force – the push of a motor, the pull of muscles or gravity, for example. In a machine, this driving force must then be conveyed to the right place in the right amount.

There is a lot of ingenuity involved in transmitting force from one place to another, and ensuring that it arrives in just the right quantity. When you squeeze and twist the handles of a can opener, the blade cuts easily through the lid of the can. This device makes light work of something that would otherwise be impossible. It does this not by giving you strength that you have not got, but by converting the force that your wrist produces into the most useful form for the job – in this case, by increasing it – and applying it where it is needed.

HOLDING MATTER TOGETHER

Every object on Earth is held together and in place by three basic kinds of force; virtually all machines make use of only two of them.

The first kind of force is gravity, which pulls any two pieces of matter together. Gravity may seem to be a very strong force, but in fact it is by far the weakest of the three. Its effects are noticeable only because it depends on the masses of the two pieces of matter involved, and because one of these pieces of matter – the whole Earth – is enormous.

The second force is the electrical force that exists between atoms. This is responsible for electricity, a subject explored in Part 4 of this book. Electrical force binds the atoms which make up all materials, and it holds them together with tremendous strength. Movement in machines is transmitted – unless the parts break – only because the atoms and molecules (groups of atoms) in these parts are held together by electrical force. So all mechanical machines use this force indirectly. In addition, some machines, such as springs and friction devices, use it directly, both to produce movement and to prevent it.

The third force is the nuclear force that binds particles in the nuclei of atoms. This force is the strongest of all, and is released only by machines that produce nuclear power.

THE CONSERVATION OF ENERGY

Underlying the actions of all machines is one principle which encompasses all the others – the principle of conservation of energy. This is not about saving energy, but about what happens to energy when it is used. It holds that

you can only get as much energy out of a machine as you put into it in the first place – no more and no less.

As a motor or muscles move to supply force to a machine, they give it energy; more force or more movement provides more energy. Movement is a particular form of energy called kinetic energy. It is produced by converting other forms of energy, such as the potential energy stored in a spring, the heat energy in a gasoline engine, the electric energy in an electric motor, or the chemical energy in muscles.

When a machine transmits force and applies it, it can only expend the same amount of energy as that put into it to get things moving. If the force the machine applies is to be greater, then the movement produced must be correspondingly smaller, and vice-versa. Overall, the total energy always remains the same. The principle of conservation of energy governs all actions. Springs may store energy, and friction will convert energy to heat, but when everything is taken into account, no energy is created and none destroyed.

If the principle of conservation of energy were suddenly to be dropped from the rule-book that governs machines, then nothing would work. If energy were destroyed as machines worked then, no matter how powerful those machines might be, they would slow down and stop. And if the workings of machines created energy, then all machines would get faster and faster in an energy build-up of titanic proportions. Either way, the world would end – with a whimper in one case and a bang in the other. But the principle of conservation of energy holds good and all machines obey. Or nearly all. Nuclear machines are an exception – but that is a story for the second part of this book.

THE INCLINED PLANE

ON CAPTURING A MAMMOTH

In the spring of that year, I was invited to the land of the much sought-after wooly mammoth, a land dotted by the now familiar high wooden towers of the mammoth captors. In ancient times the mammoth had been hunted simply for its meat. But its subsequent usefulness in industry and growing popularity as a pet had brought about the development of a more sophisticated and less terminal means of apprehension.

Each unsuspecting beast was lured to the base of a tower from which a boulder of reasonable dimensions was then dropped from a humanitarian height onto its thick skull. Once stunned, a mammoth could easily be lead to the paddock where an ice pack and fresh swamp grass would quickly overcome hurt feelings and innate distrust.

THE PRINCIPLE OF THE INCLINED PLANE

The laws of physics decree that raising an object, such as a mammoth-stunning boulder, to a particular height requires a certain amount of work. Those same laws also decree that no way can ever be found to reduce that amount. The ramp makes life easier not by altering the amount of work that is needed, but by altering the *way* in which the work is done.

Work has two aspects to it: the effort that you put in, and the distance over which you maintain the effort. If the effort increases, the distance must decrease, and vice versa.

This is easiest to understand by looking at two extremes. Climbing a hill by the steepest route requires the most effort, but the distance that you have to cover is shortest. Climbing up the gentlest slope requires the least effort, but the distance is greatest. The work you do is the same in either case, and equals the effort (the force you exert) multiplied by the distance over which you maintain the effort.

So what you gain in effort, you pay in distance. This is a basic rule that is obeyed by many mechanical devices, and it is the reason why the ramp works: it reduces the effort needed to raise an object by increasing the distance that it moves.

The ramp is an example of an inclined plane. The principle behind the inclined plane was made use of in ancient times. Ramps enabled the Egyptians to build their pyramids and temples. Since then, the inclined plane has been put to work in a whole host of devices from locks and cutters to plows and zippers, as well as in all the many machines that make use of the screw.

While the process was more or less successful, it had a couple of major drawbacks. The biggest problem was that of simply getting a heavy boulder to the right height. This required an almost Herculean effort, and Hercules was not due to be born for several centuries yet. The second problem was that the mammoth, once hit, would invariably crash into the tower, either hurling his captors to the ground, or at least seriously damaging the structure.

After making a few calculations, I informed my hosts that both problems could be solved simultaneously by building earth ramps rather than wooden towers. The inherent sturdiness of the ramp would make it virtually indestructible should a mammoth fall against it. And now, rather than trying to hoist the boulder straight up, it could be rolled gradually to the required height, therefore needing far less effort.

At first, the simplicity of my solution was greeted with understandable scepticism. "What do we do with the towers?" they asked. I made a few more calculations and then suggested commercial and retail development on the lower levels and luxury apartments above.

HOW EFFORT AND DISTANCE ARE LINKED

The sloping face of this ramp is twice as long as its vertical face. The effort that is needed to move a load up the sloping face is therefore half that needed to raise it up the vertical face.

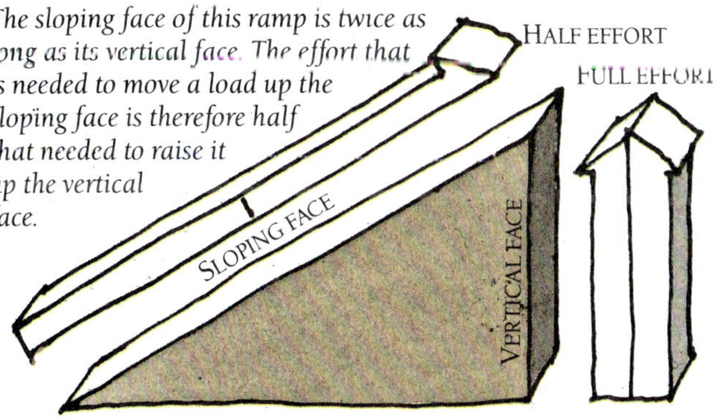

THE WEDGE

In most of the machines that make use of the inclined plane, it appears in the form of a wedge. A door wedge is a simple application; you push the sharp end of the wedge under the door and it moves in to jam the door open.

The wedge acts as a moving inclined plane. Instead of having an object move up an inclined plane, the plane itself can move to raise the object. As the plane moves a greater distance than the object, it raises the object with a greater force. The door wedge works in this way. As it jams under the door, the wedge raises the door slightly and exerts a strong force on it. The door in turn forces the wedge hard against the floor, and friction (see pp. 86-7) with the floor makes the wedge grip the floor so that it holds the door open.

LOCKS AND KEYS

Here's a puzzle not unconnected with locks: how to separate two blocks held together by five two-part pins. The gap in each pin is at a different height. In order to separate the blocks, the pins must be raised so that the gaps line up.

Knowing the principle of the inclined plane, we insert a wedge. It pushes up the pins easily enough, but by the wrong distances.

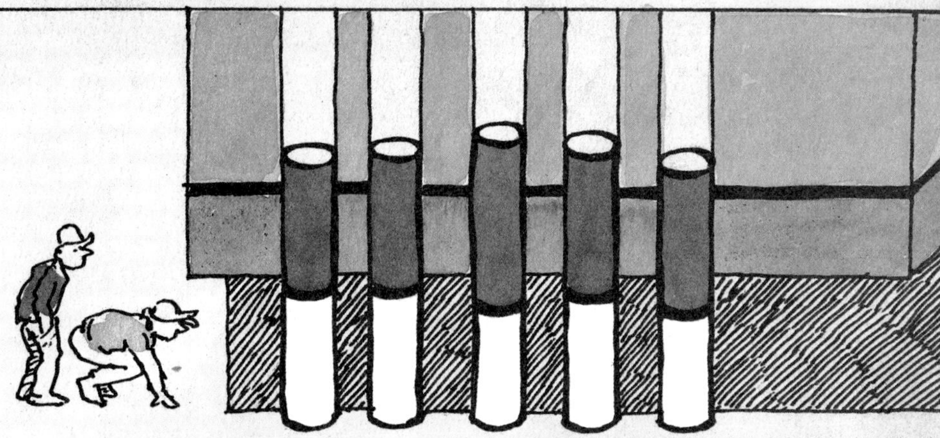

More thought suggests five wedges – one for each pin. This raises the pins so that the gaps line up, freeing the two halves of the block. However, the wedges themselves are now stuck fast in the lower half.

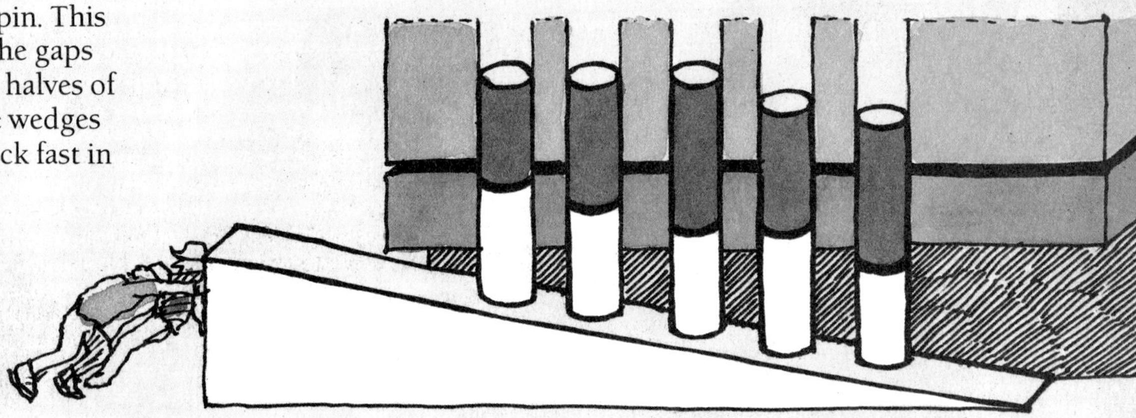

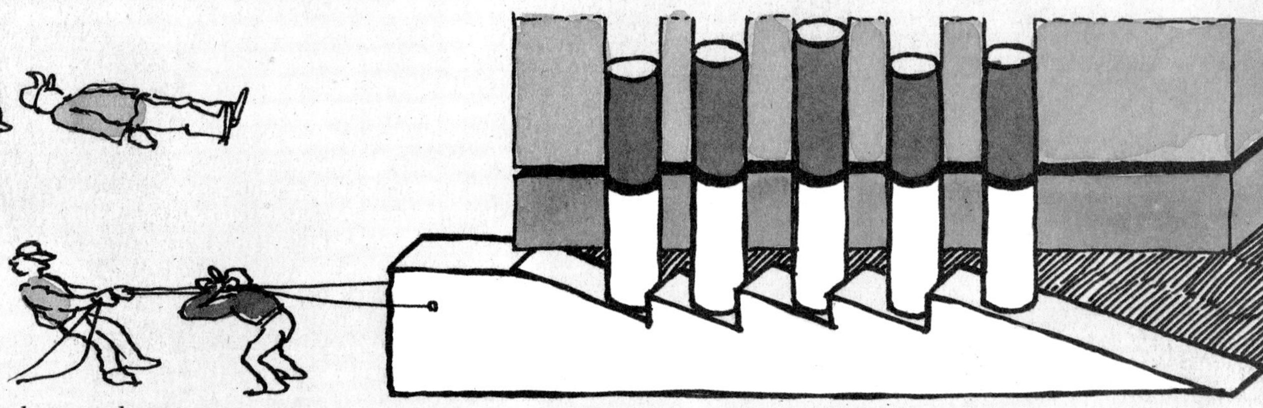

The key to the puzzle is just that – a key – because the block is a simplified cylinder lock. The serrated edge of the key acts as a series of wedges that raises the pins to free the lock. Because the serrations on the key are double sided, the key can be removed after use. The springs will then push the pins firmly back into position, closing the lock.

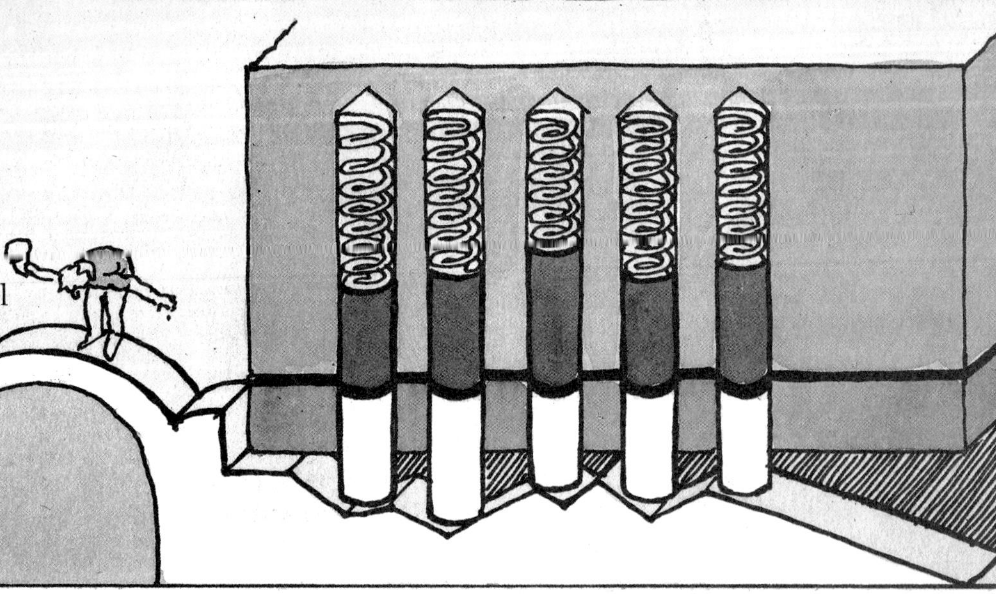

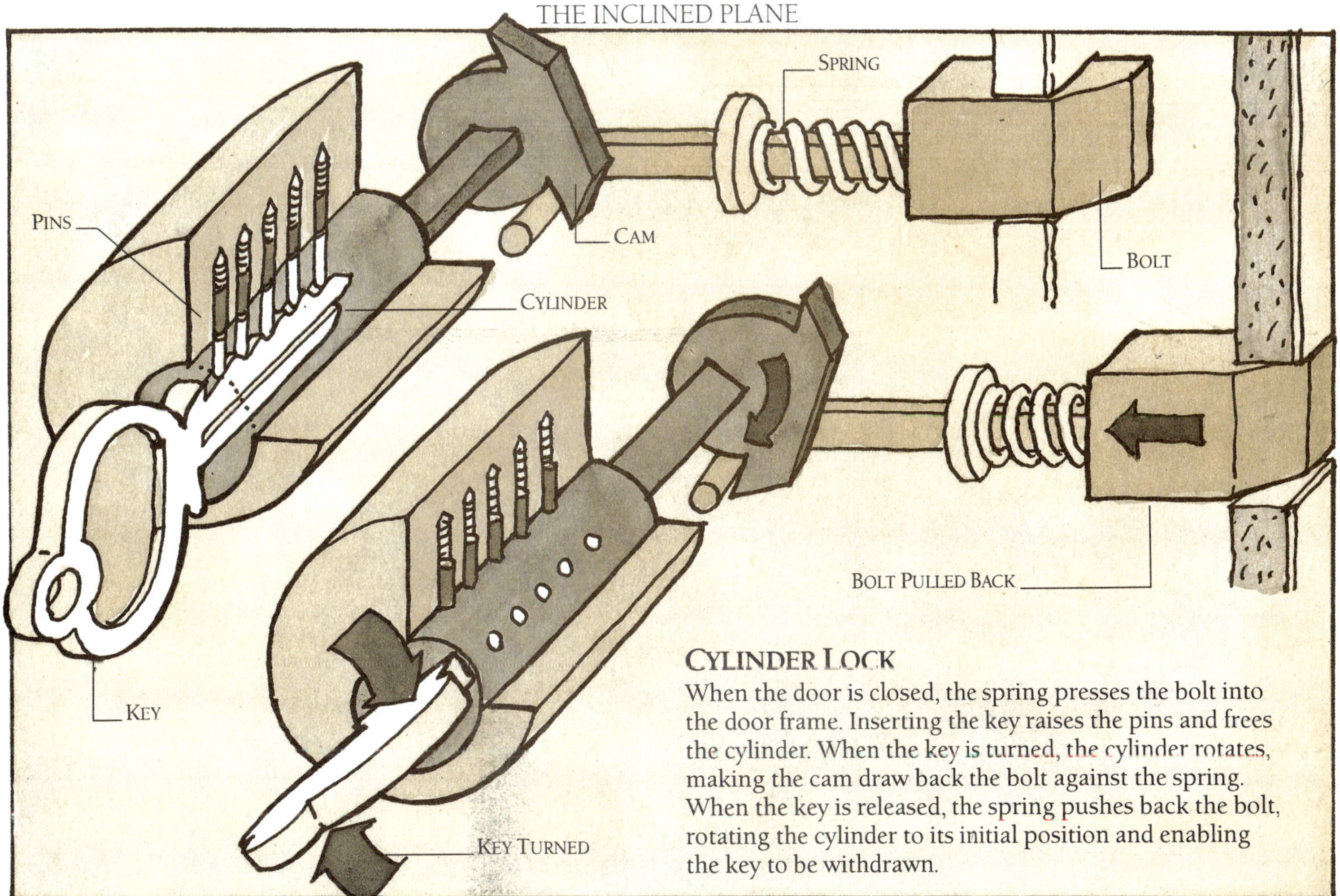

CYLINDER LOCK

When the door is closed, the spring presses the bolt into the door frame. Inserting the key raises the pins and frees the cylinder. When the key is turned, the cylinder rotates, making the cam draw back the bolt against the spring. When the key is released, the spring pushes back the bolt, rotating the cylinder to its initial position and enabling the key to be withdrawn.

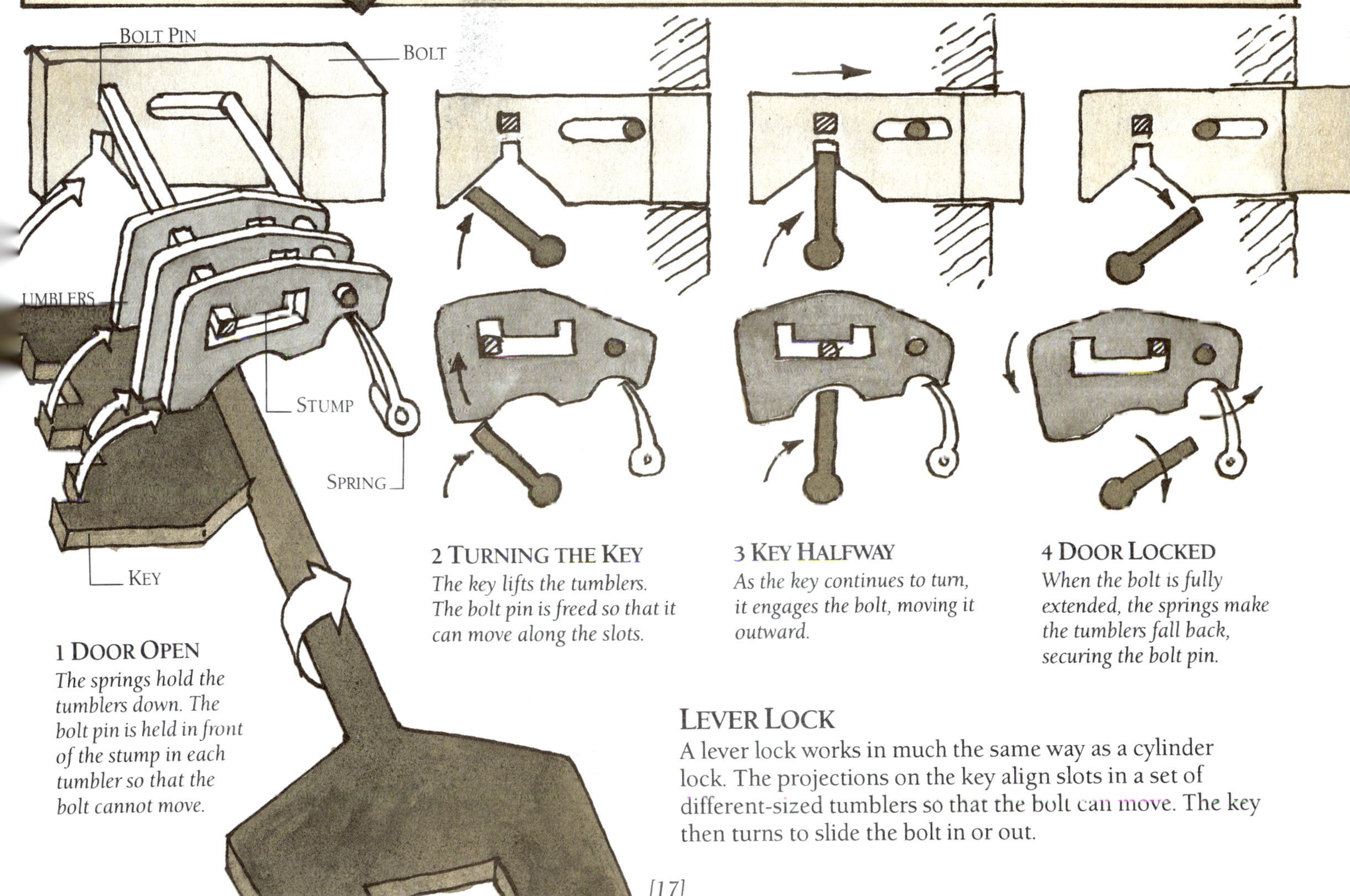

1 DOOR OPEN
The springs hold the tumblers down. The bolt pin is held in front of the stump in each tumbler so that the bolt cannot move.

2 TURNING THE KEY
The key lifts the tumblers. The bolt pin is freed so that it can move along the slots.

3 KEY HALFWAY
As the key continues to turn, it engages the bolt, moving it outward.

4 DOOR LOCKED
When the bolt is fully extended, the springs make the tumblers fall back, securing the bolt pin.

LEVER LOCK

A lever lock works in much the same way as a cylinder lock. The projections on the key align slots in a set of different-sized tumblers so that the bolt can move. The key then turns to slide the bolt in or out.

CUTTING MACHINES

Nearly all cutting machines make use of the wedge, a form of inclined plane. A wedge-shaped blade converts a forward movement into a parting movement that acts at right angles to the blade.

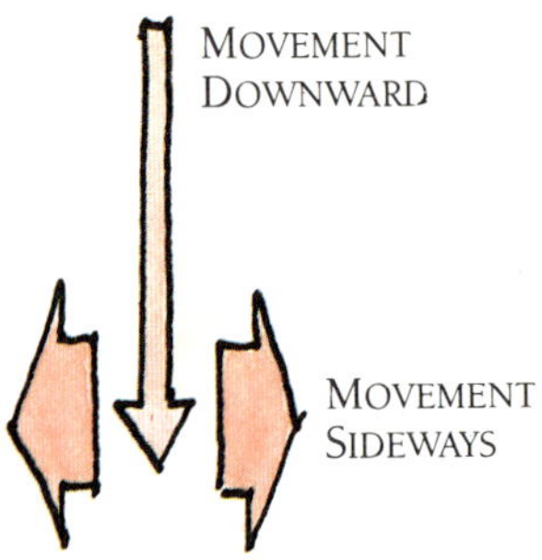

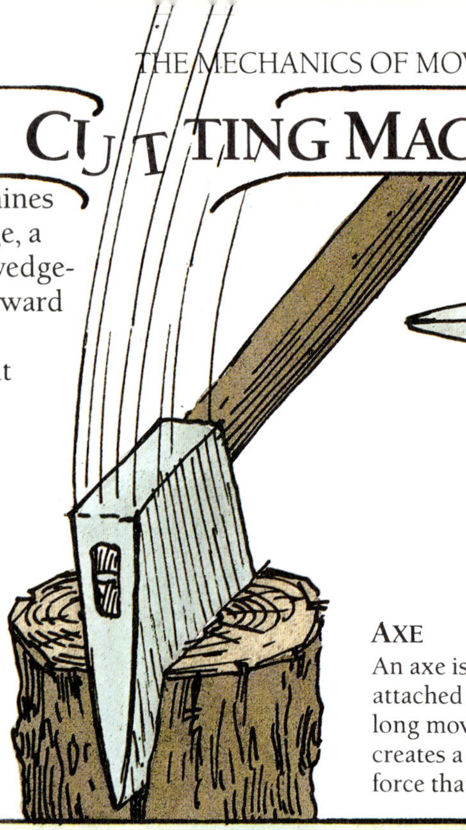

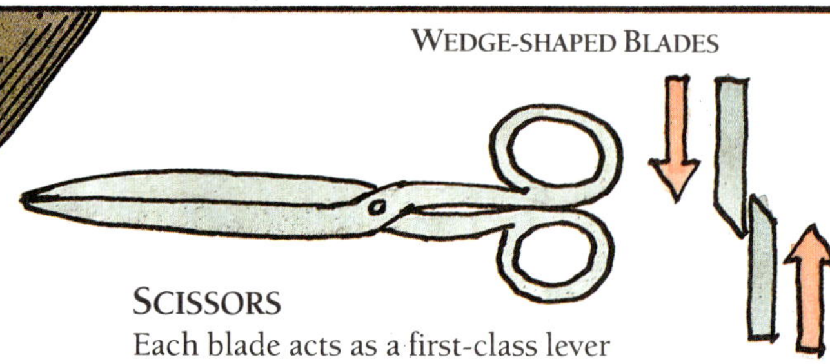

SCISSORS

Each blade acts as a first-class lever (see p.23). The sharpened edges of the blades form two wedges that cut with great force into a material from opposite directions. As they meet, they part the material sideways.

AXE

An axe is simply a wedge attached to a shaft. The axe's long movement downward creates a powerful sideways force that splits open the wood.

The axe has another built-in wedge: a sliver of metal is driven into the top of the shaft, and this jams the shaft tightly into the socket in the axe's head.

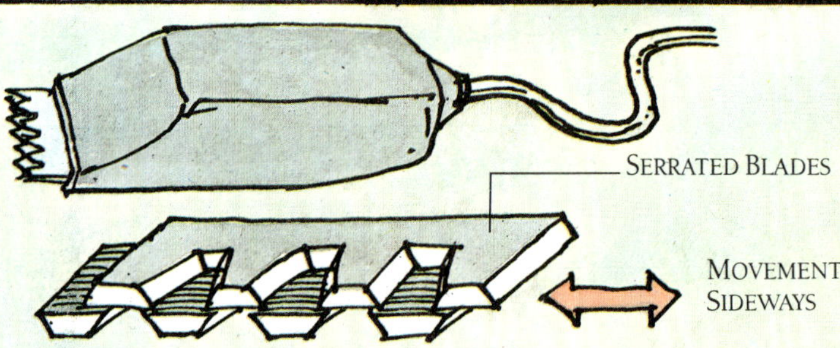

ELECTRIC TRIMMER

An electric trimmer contains two serrated blades driven by a crank mechanism (see pp.52-3) The blades move to and fro over each other. As gaps open between the serrations, stems or hairs enter to be trapped and then sliced as the blades cross. The trimmer's blades act as paired wedges like the blades of scissors.

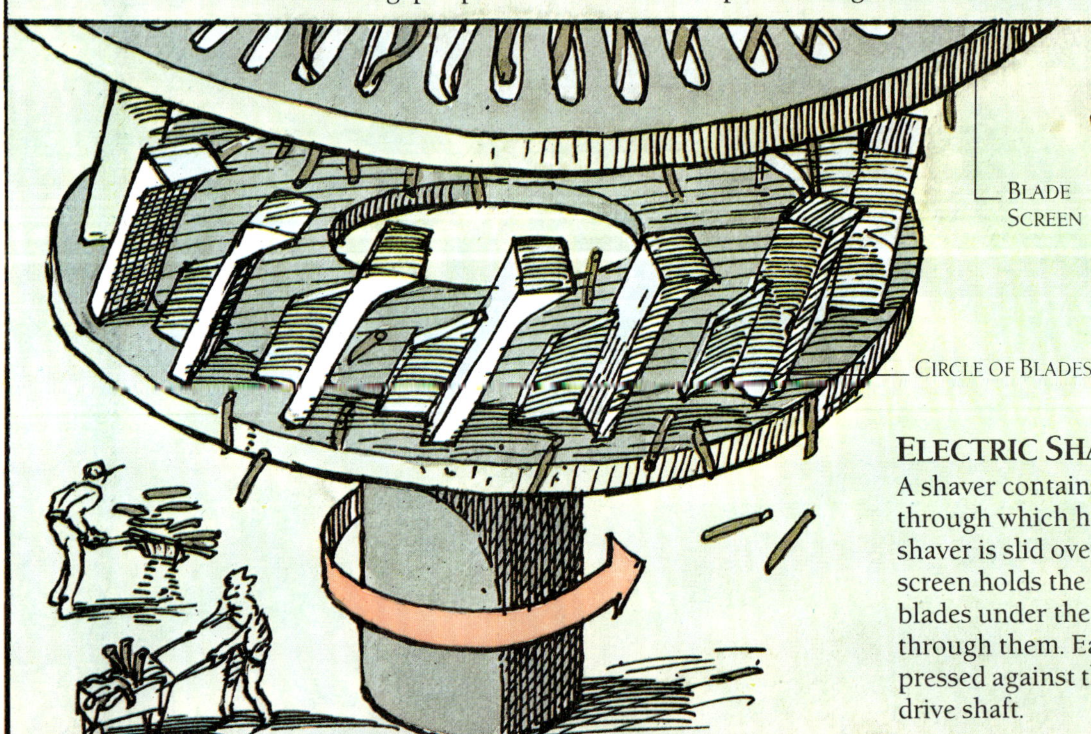

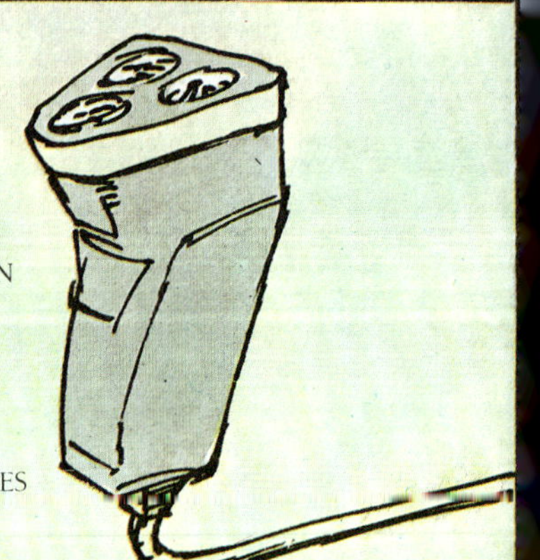

ELECTRIC SHAVER

A shaver contains a fine screen through which hairs protrude as the shaver is slid over the skin. The screen holds the hairs so that cutting blades under the screen can slice through them. Each circle of blades is pressed against the screen by a sprung drive shaft.

THE CAN OPENER

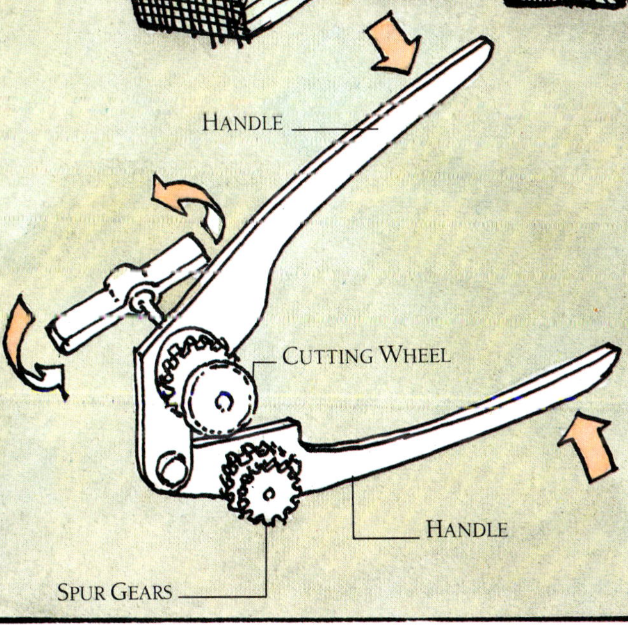

A can opener has a sharp-edged cutting blade or wheel that slices into the lid. A toothed wheel fits beneath the lip of the can, and rotates the can so that the cutting wheel is forced into the lid. Two further toothed wheels – one above the other – form a pair of spur gears (see p.41) to transmit the turning force from the handle.

THE PLOW

A plow is a wedge that is dragged through the ground either by a draft animal or a tractor. It cuts away the top layer of soil, and then lifts and turns over the layer. In this way, the soil is broken up for planting crops. In addition, vegetation growing in or lying on the soil is buried so that it rots and provides nutrients for the crops. The plow is one of mankind's oldest machines. Wooden plows have been in use for about five thousand years, although metal plows date back less than two centuries.

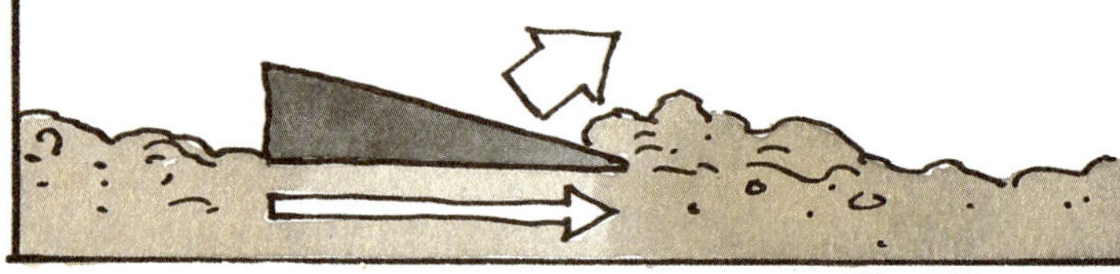

THE MAIN PARTS OF A PLOW

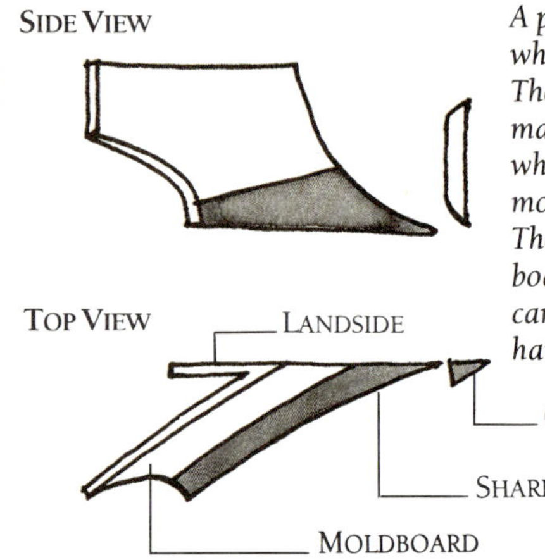

A plow has four main parts, which are all made of steel. The coulter precedes the main body of the plow, which consists of the share, moldboard and landside. The coulter, share and moldboard all act as wedges, and can exert great force to plow hard and heavy soil.

THE PLOWING SEQUENCE

THE COULTER *slices a furrow in the soil.*

The coulter creates a furrow by making a vertical cut in the soil. With animal plows, the coulter is a knife-like blade. Tractor-drawn plows normally have disk coulters, which are sharp-edged wheels that spin freely as the plow is drawn forward.

The share follows the coulter, making a horizontal cut and freeing the top layer of soil. Attached to the share is the moldboard, which lifts and turns the layer. The landside is fixed to the side of the moldboard and slides along the vertical wall of the furrow. It thrusts the moldboard outward to move the layer of soil.

THE SHARE *cuts loose the top layer of soil.*

THE MOLDBOARD *lifts and turns the layer of soil.*

THE ZIP

The zipper cleverly exploits the principle of the inclined plane to join or separate two rows of interlocking teeth. The zipper's slide contains wedges that turn the little effort with which you pull it into a strong force that opens and closes the fastener. The teeth are designed so that they can only be opened or closed one after the other. Without using the slide, it is practically impossible to free the teeth or make them mesh together.

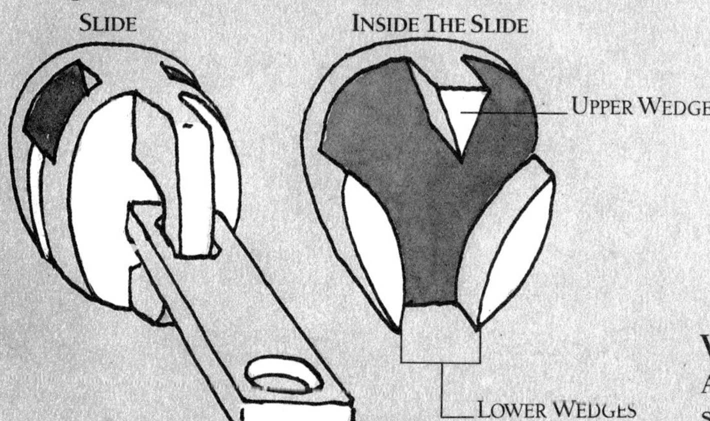

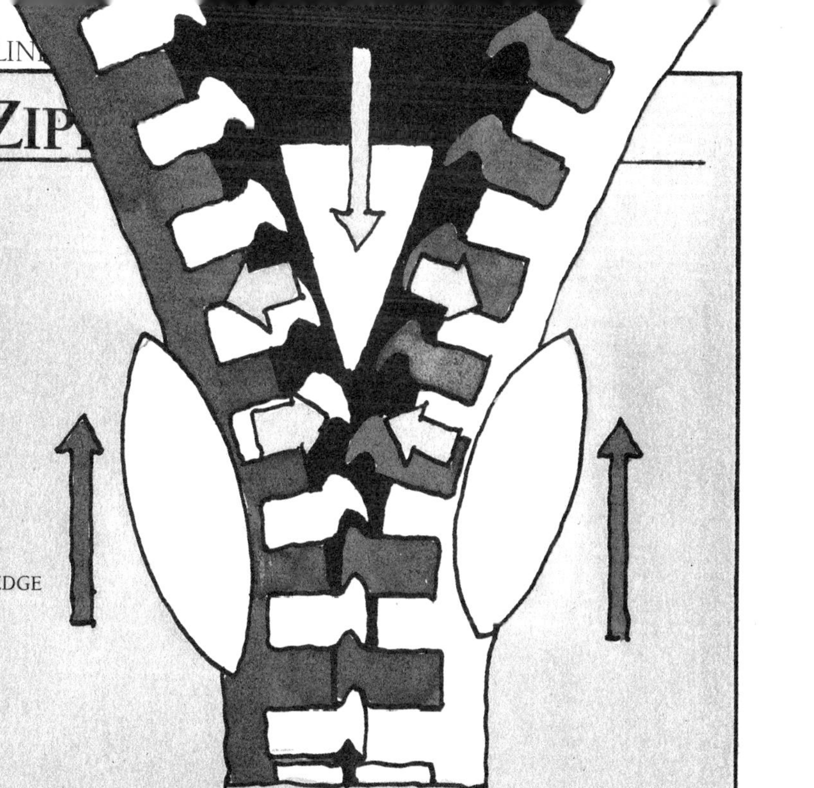

WEDGES AT WORK

As you open a zipper, the triangular upper wedge in the slide detaches the teeth and forces them apart. On closing, the two lower wedges (which are often the curved sides of the slide) force the teeth back together so that they intermesh. Plastic zippers contain two intermeshing spirals instead of two rows of teeth.

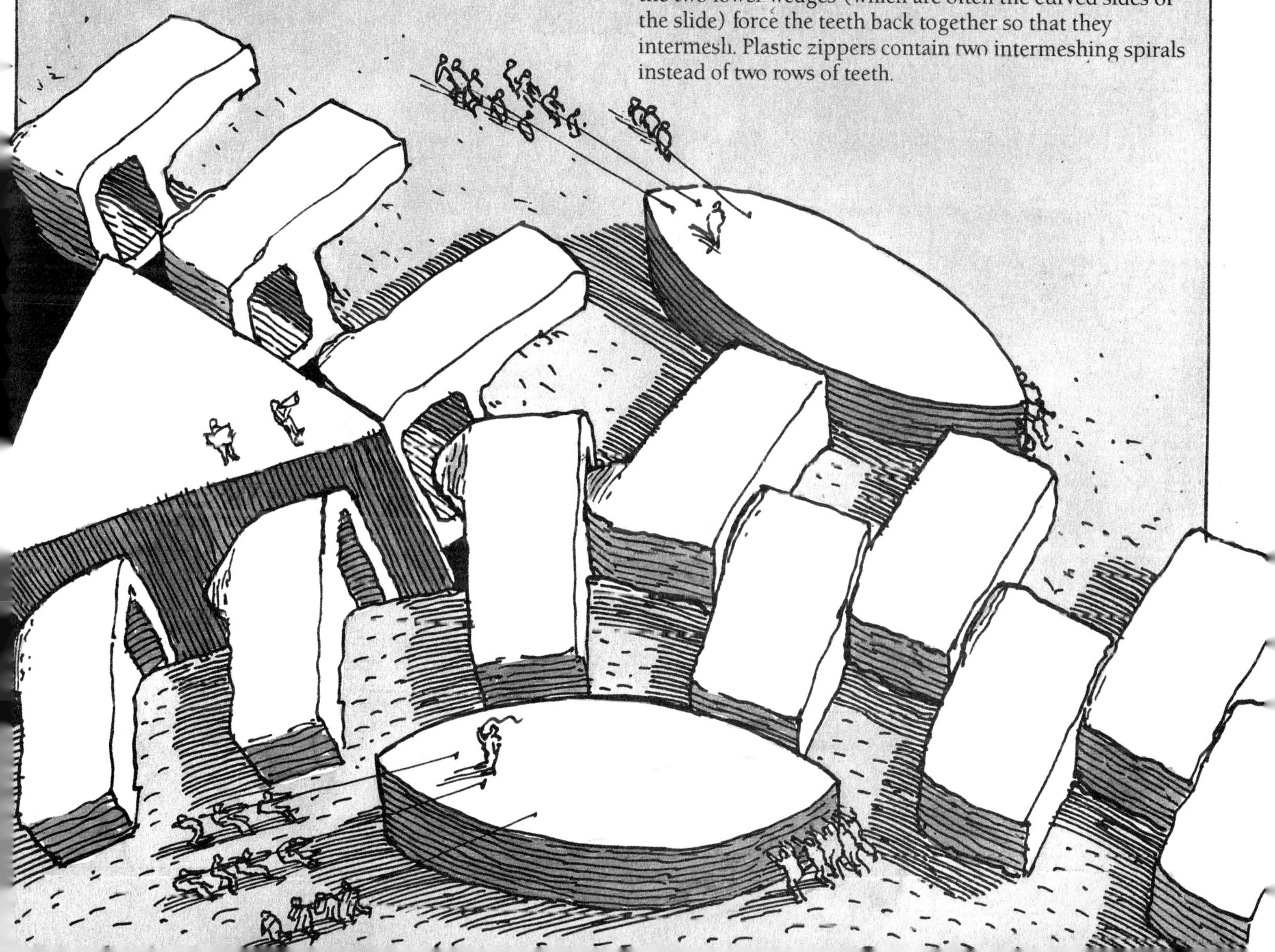

LEVERS

ON WEIGHING A MAMMOTH

Before being shipped to its final destination, a mammoth must be weighed. I was fortunate enough in one village to witness the procedure at first hand. The center of a tree trunk was placed directly on a boulder. One end of the trunk was then pulled down and the mammoth encouraged to sit on it. No sooner did the beast seem reasonably comfortable than a number of villagers scrambled onto the other end of the trunk. Slowly their end sank and, as it did, the startled mammoth rose into the air. I was told that when the trunk reached a horizontal position, the combined weight of the people would equal that of the mammoth. This seemed reasonable enough.

THE PRINCIPLE OF LEVERS

The tree trunk is acting as a lever, which is simply a bar or rod that tilts on a pivot, or fulcrum. If you apply a force by pushing or pulling on one part of the lever, the lever swings about the fulcrum to produce a useful action at another point. The force that you apply is called the *effort*, and the lever moves at another point to raise a weight or overcome a resistance, both of which are called the *load*.

Where you move a lever is just as important as the amount of effort you apply to it. Less effort can move the same load, provided that it is applied further from the fulcrum; however, the effort has to move a greater distance to shift the load. As with the inclined plane, you gain in force what you pay in distance. Some levers reverse this effect to produce a gain in distance moved at the expense of force.

With levers, the distances moved by the effort and load depend on how far they are from the fulcrum. The principle of levers, which relates the effort and load, states that the effort times its distance from the fulcrum equals the load times its distance from the fulcrum.

FULCRUM IN CENTER

The effort and load are the same distance from the fulcrum. In this situation, the load and effort are equal, and both move the same distance up or down as the lever balances.

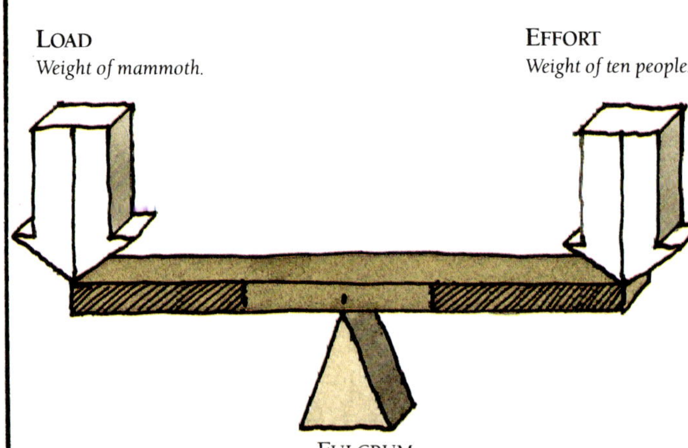

It was then that I noticed across the square a second equally large mammoth about to be weighed, this time using far fewer people as a counterbalance. As I watched, anticipating disaster, the boulder was rolled closer to what would be the mammoth's end of the tree trunk. Once the mammoth was in place, a mere handful of people climbed onto the other end. To my amazement the tree trunk gently assumed a horizontal position. I was then informed that the length of the tree trunk from the people to the boulder multiplied by their combined weight would equal the length of the trunk from the boulder to the mammoth multiplied by its weight.

I was in the midst of calculations to verify this most unlikely theory when I heard a scream. Apparently, all the villagers had not disembarked from the trunk simultaneously, thereby causing a lad to be dramatically launched. I made a note, thinking that one day this might be useful.

FIRST-CLASS LEVERS

There are three different basic kinds of levers. All the levers on these two pages are first-class levers. First-class levers aren't superior to other-class levers; they are just levers in which the fulcrum is always placed between the effort and the load.

If the fulcrum is placed in the center – as in the diagram on the left – the effort and load are at the same distance from it and are equal. The weight of the people is the same as the weight of the mammoth.

However, if the people are placed twice as far from the fulcrum as the mammoth – as in the diagram on the right – only half the number of people is needed to raise the mammoth. And if the people were three times as far from the fulcrum than the mammoth, only a third would be needed, and so on, because the lever magnifies the force applied to it.

These mammoth-weighing levers balance in order to measure weight, which is why this kind of weighing machine is called a balance. When the lever comes to rest, the force of the effort balances the force of the load, which is its weight. Many other kinds of levers work to produce movement.

FULCRUM OFF-CENTER

The effort is twice as far from the fulcrum as the load. Here, the effort moves twice as far as the load, but is only half its amount.

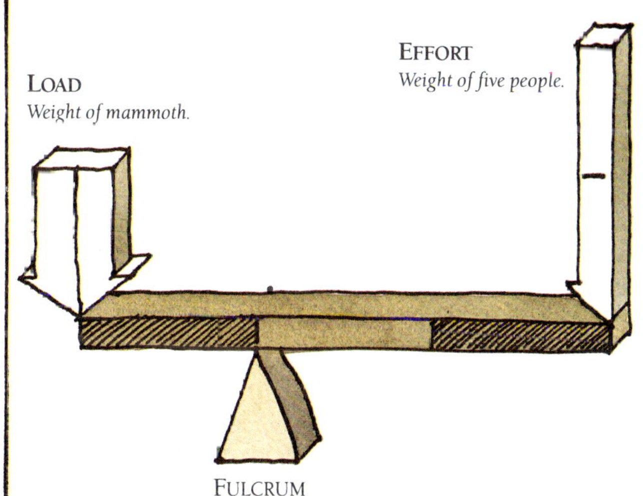

ON MAMMOTH HYGIENE

Early in my researches I discovered that mammoths smell and so unavoidably do their immediate surroundings. I was therefore gratified to observe that in order to minimize any unpleasant odor, the staff of the mammoth paddock train their animals to sit on mats. These they change with some regularity.

Since mammoths often refuse to budge once settled, the keepers have learned how to remove mats while in use. One end of a trimmed tree trunk is carefully slipped under the mammoth. By raising the other end a fair distance the mammoth is lifted just enough to release the fetid fabric.

I found the ease with which the keepers could raise their oversize charges quite astounding. I also noted that wheelbarrows are an invaluable asset during clean-up sessions.

SECOND-CLASS LEVERS

Both the mammoth-lifter and the wheelbarrow are examples of second-class levers. Here, the fulcrum is at one end of the bar or rod and the effort is applied to the other end. The load to be raised or overcome lies between them.

With this kind of lever, the effort is always further from the fulcrum than the load. As a result, the load cannot move as far as the effort, but the force with which it moves is always greater than the effort. The closer the load is to the fulcrum, the more the force is increased, and the easier it becomes to raise the load. A second-class lever always magnifies force but decreases the distance moved.

A wheelbarrow works in the same way as the mammoth-lifter, allowing one to lift and shift a heavy load with the wheel as a fulcrum. Levers can also act to press on objects with great force rather than to lift them. In this case, the load is the resistance that the object makes to the pressing force. Scissors and nutcrackers (see p.26) are first-class and second-class examples. These devices are compound levers, which are pairs of levers hinged at the fulcrum.

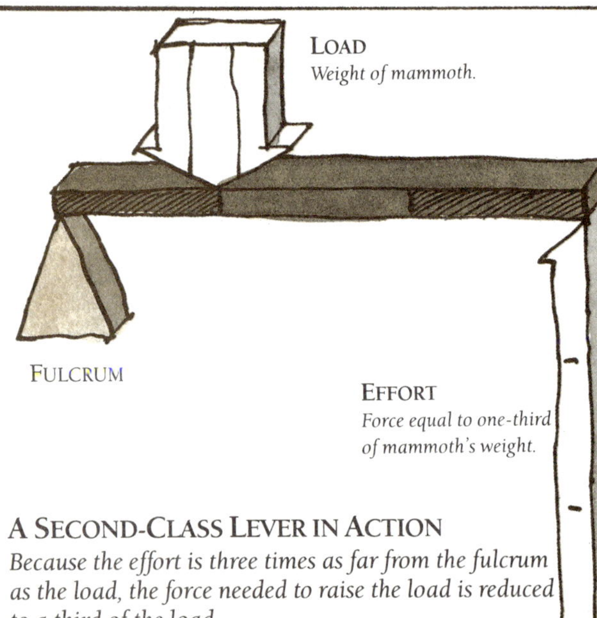

A SECOND-CLASS LEVER IN ACTION
Because the effort is three times as far from the fulcrum as the load, the force needed to raise the load is reduced to a third of the load.

On Tusk Trimming and Attendant Problems

I watched with great curiosity a mammoth that was having its tusks trimmed as a precaution prior to being shipped. The beast was clearly not happy as the workers sawed, chipped and filed away. No sooner had I recalled the old adage that second only to the fury of a woman scorned is the wrath of a disgruntled mammoth, when suddenly the enraged creature wrapped its trunk around the end of a nearby weighing log and began swinging it from side to side.

I was able to note during the ensuing collapse of the trimming stand that although the mammoth's head pivoted only a short distance, the free end of the log was swinging much further thereby achieving a stunning velocity.

Third-Class Levers

In extending its trunk with that of a tree, the mammoth now has something in common with such innocuous devices as a fishing rod and a pair of tweezers. It has become a giant-sized third-class lever.

Here, the fulcrum is again at one end of the lever but this time the positions of the load and effort are reversed. The load to be raised or overcome is furthest away from the fulcrum, while the effort is applied between the fulcrum and the load. As the load is furthest out, it always moves with less force than the effort, but it travels a proportionately greater distance. A third-class lever therefore always magnifies the distance moved but reduces the force.

The mammoth's neck is the fulcrum, and the end of the log moves a greater distance than the trunk gripping it. The force with which the log strikes the people is less than the effort of the trunk, but still enough to overcome their weight and scatter them far and wide. The end of the log moves faster than the trunk and builds up quite a speed to get the people moving.

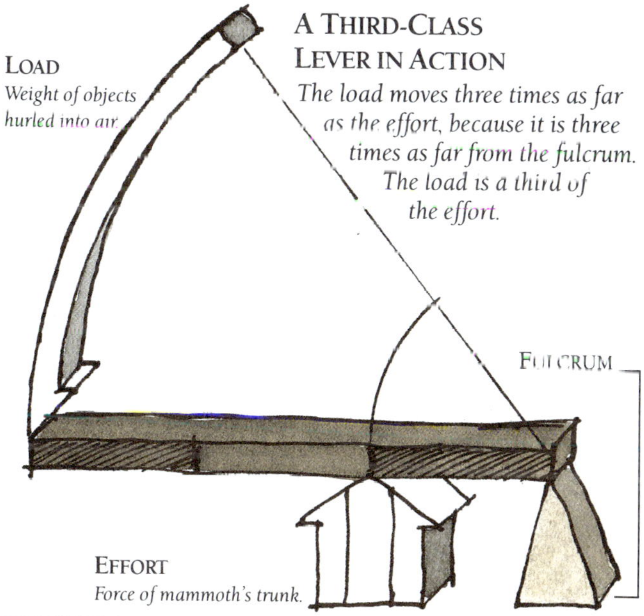

A Third-Class Lever in Action
The load moves three times as far as the effort, because it is three times as far from the fulcrum. The load is a third of the effort.

LEVERS IN ACTION

FIRST-CLASS LEVERS

BALANCE

The object to be weighed is the load and the weights make up the effort. The two are equal, being at the same distance from the fulcrum.

BEAM SCALE

The fulcrum is off-center, and the weight is moved along the bar until it balances the object being weighed.

NAIL EXTRACTOR

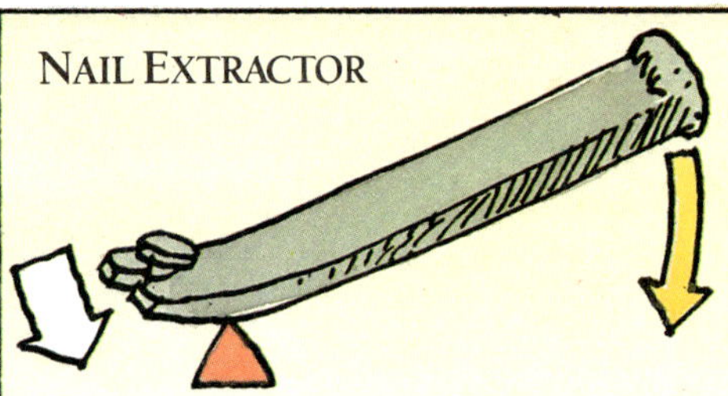

The effort of the hand is magnified to pull out a nail. The load is the resistance of a nail to extraction.

HAND CART

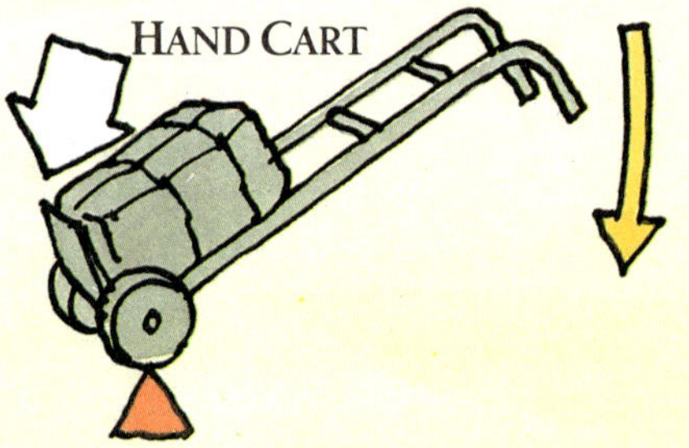

Tipping the handle of the hand cart with a light effort raises a heavy load.

PLIERS

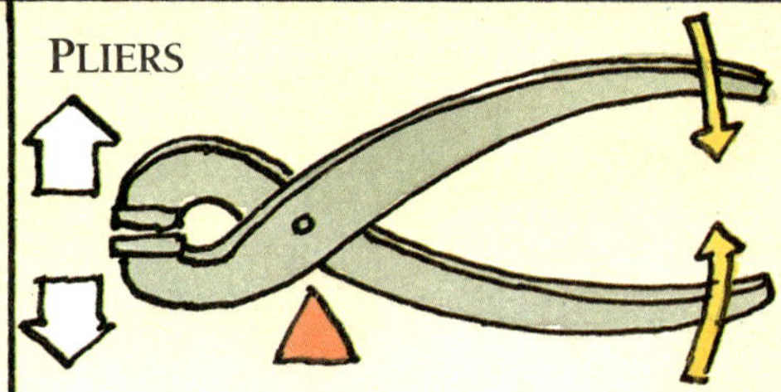

A pair of pliers is a compound lever (a pair of levers hinged at the fulcrum). The load is the resistance of an object to the grip of the pliers.

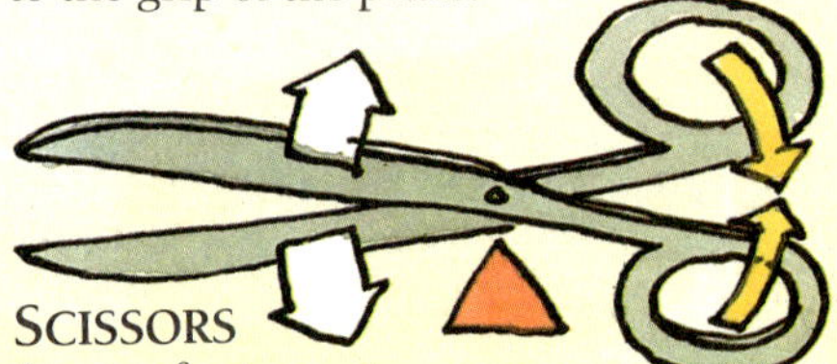

SCISSORS

A pair of scissors is a compound first-class lever. It produces a strong cutting action very near the hinge. The load is the resistance of the fabric to the cutting blades.

SECOND-CLASS LEVERS

WHEELBARROW

Lifting the handles with a light effort raises a heavy load nearer the wheel.

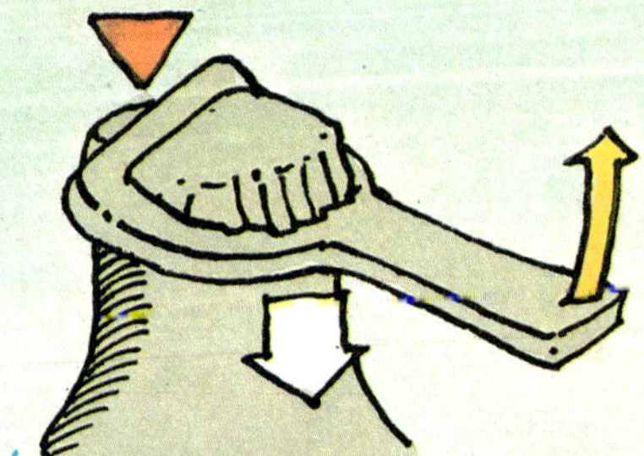

BOTTLE OPENER

Pushing up the handle overcomes the strong resistance of a bottle cap.

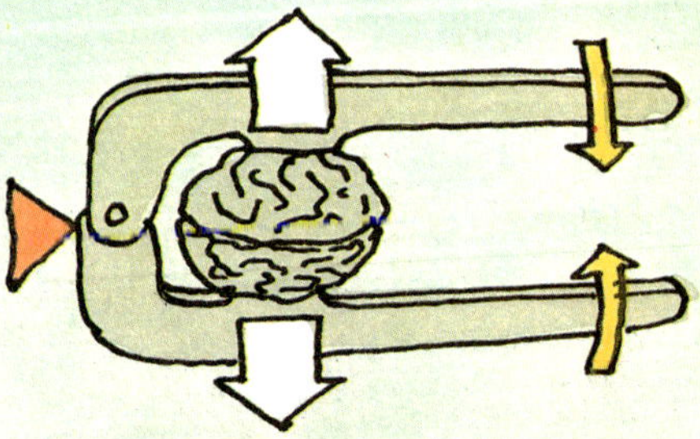

NUTCRACKERS

A pair of nutcrackers is a compound second-class lever. The load is the resistance of a shell to cracking.

Third-Class Levers

Hammer

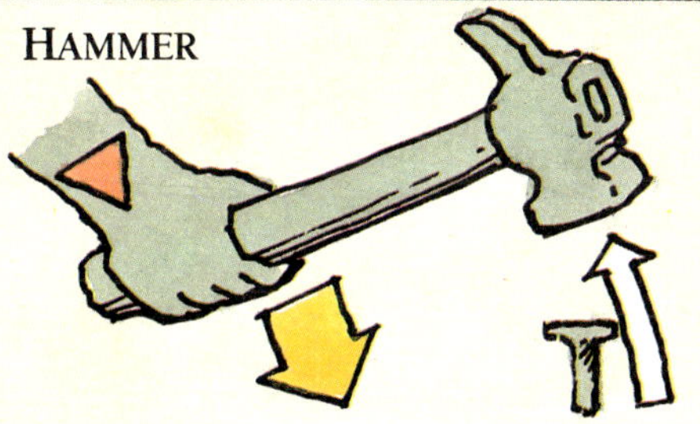

A hammer acts as a third-class lever when it is used to drive in a nail. The fulcrum is the wrist, and the load is the resistance of the wood. The hammer head moves faster than the hand to strike the nail

Fishing Rod

One hand supplies the effort to move the rod, while the other hand acts as the fulcrum. The load is the weight of the fish, which is raised a long distance with a short movement of the hand.

Tweezers

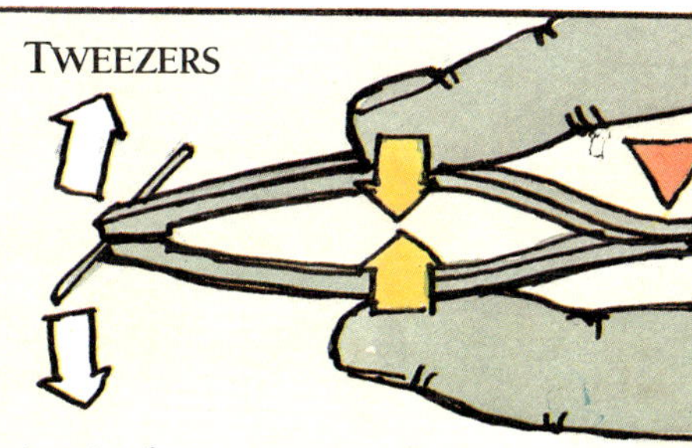

A pair of tweezers is a compound third-class lever. A small movement of the fingers produces a longer movement of the tweezer tips in order to grip a hair. The load is the resistance of the hair.

Multiple Levers

Excavator

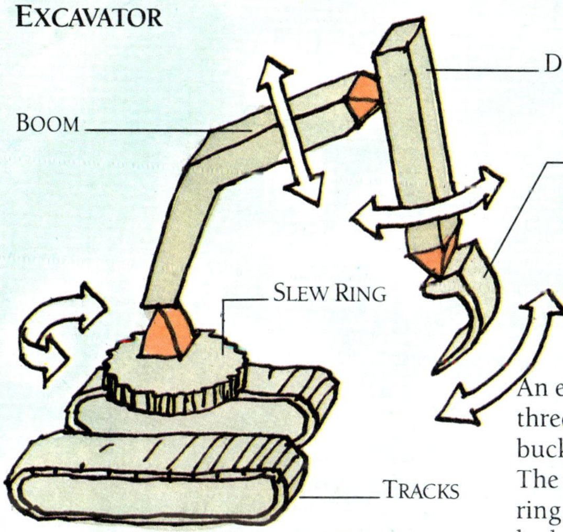

An excavator is a rotating assembly of three levers – the boom, dipper and the bucket – mounted on caterpillar tracks. The assembly swings round on the slew ring while the three levers, powered by hydraulic rams (see pp.136-7), combine to place the bucket in any position. The boom is a third-class lever that raises or lowers the dipper. The dipper is a first-class lever that moves the bucket in and out. The bucket is itself another first-class lever that tilts to dig a hole and empty its load.

Nail Clippers

Nail clippers are a neat combination of two levers that produce a strong cutting action while at the same time being easy to control. The handle is a second-class lever that presses the cutting blades together. It produces a strong effort on the blades, which form a compound third-class lever. The cutting edges move a short distance to overcome the tough resistance of the nail as they slice through it.

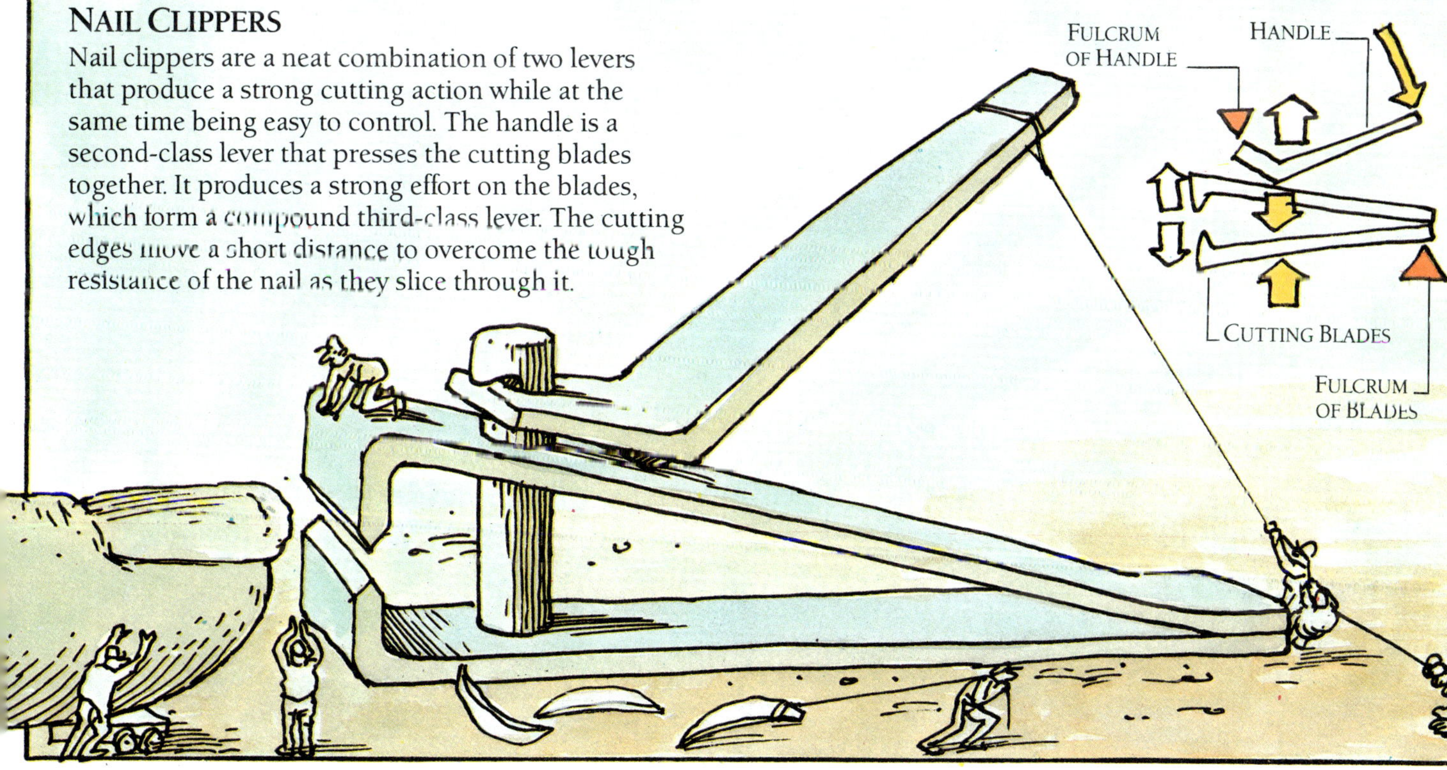

WEIGHING MACHINES

THE ROBERVAL ENIGMA

This simple kitchen balance or scales is based on the Roberval enigma, a linkage of parallel levers devised by the French mathematician Roberval in 1669. This allows the pans to remain horizontal, and also allows objects and weights to be placed at any position in the pans without affecting the accuracy of the scales. Their weight acts at the support of each pan. (At first sight, this seems to defy the principle of levers – hence the enigma).

A balance is a first-class lever. As the centers of the two pans are the same distance from the central pivot, they balance when the weight on one pan equals that on the other. Balances with suspended pans work in the same way.

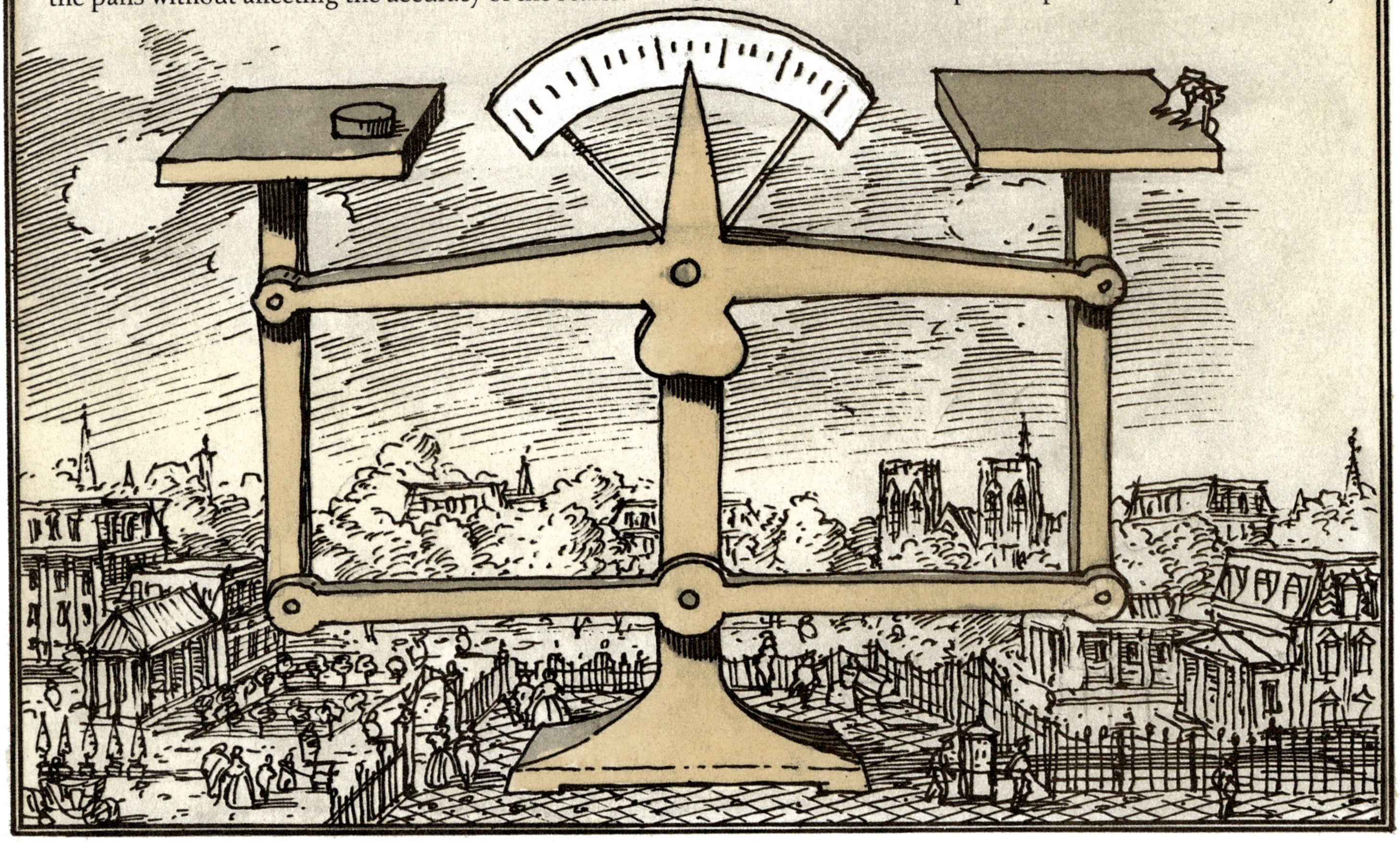

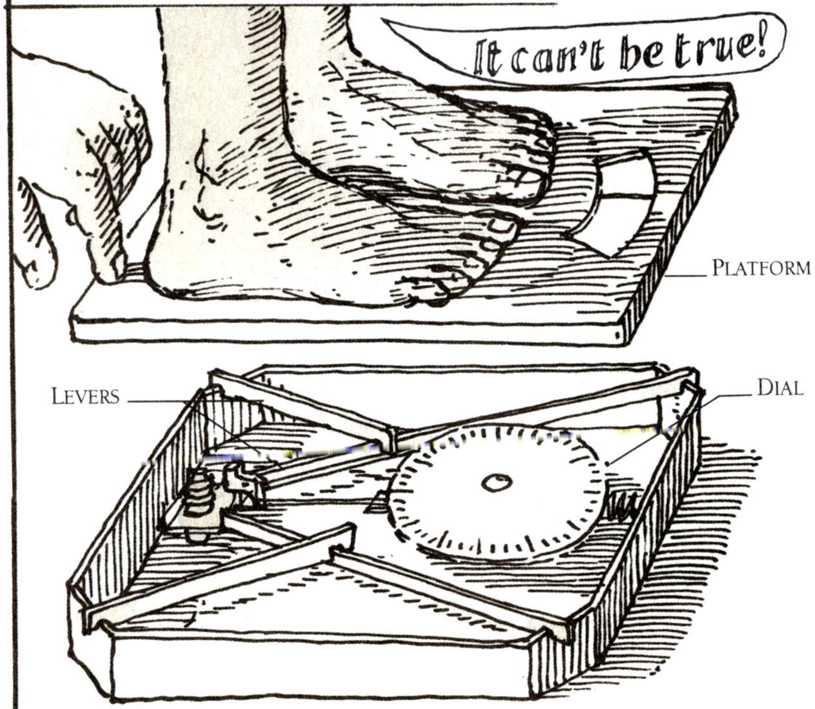

BATHROOM SCALES

For safety, your bathroom scales will barely move as you step onto the platform, no matter how heavy you are. The mechanism inside magnifies this tiny movement considerably, turning the dial sufficiently to register your weight.

A system of third-class levers pivoting on the case beneath the platform transmits its movement to the calibrating plate, which is attached to the powerful main spring. The levers force the plate down, extending the spring by an amount in exact proportion to your weight, one of the key properties of springs (see pp.82-3). The crank – a first-class lever – turns, pulled by another spring attached to the dial mechanism. This contains a rack and pinion gear (see p.41) which turns the dial, showing your weight through a window in the platform.

On stepping off the platform, the main spring retracts. It raises the platform and turns the crank to return the dial to zero.

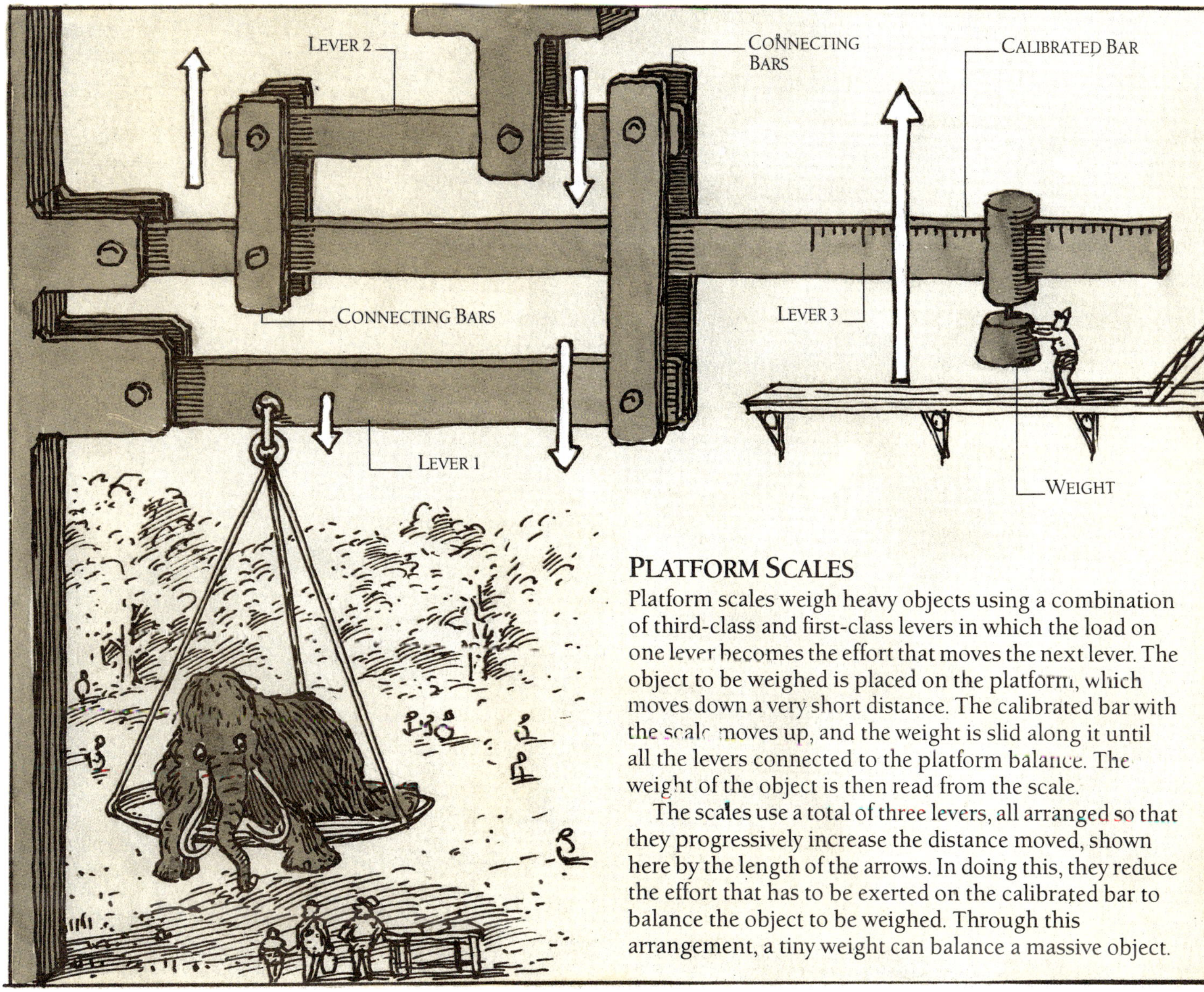

PLATFORM SCALES

Platform scales weigh heavy objects using a combination of third-class and first-class levers in which the load on one lever becomes the effort that moves the next lever. The object to be weighed is placed on the platform, which moves down a very short distance. The calibrated bar with the scale moves up, and the weight is slid along it until all the levers connected to the platform balance. The weight of the object is then read from the scale.

The scales use a total of three levers, all arranged so that they progressively increase the distance moved, shown here by the length of the arrows. In doing this, they reduce the effort that has to be exerted on the calibrated bar to balance the object to be weighed. Through this arrangement, a tiny weight can balance a massive object.

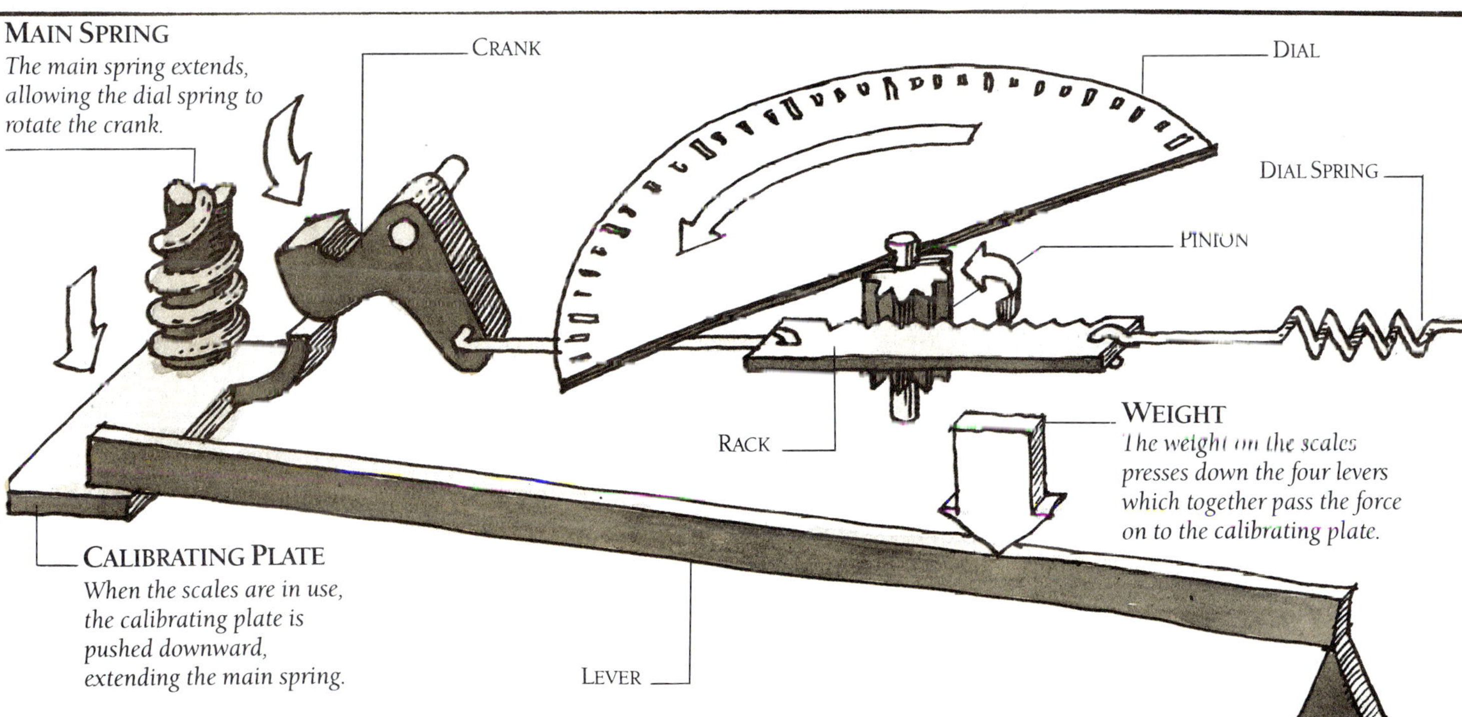

GRAND PIANO

Each key of a piano is linked to a complex system of levers called the action. Overall, the levers transmit the movement of the fingertip to the felt-tipped hammer that strikes the taut piano wire and sounds a note. The action magnifies movement so that the hammer moves a greater distance than the player's fingertip. The system of levers is very responsive, allowing the pianist to play quickly and produce a wide range of volume.

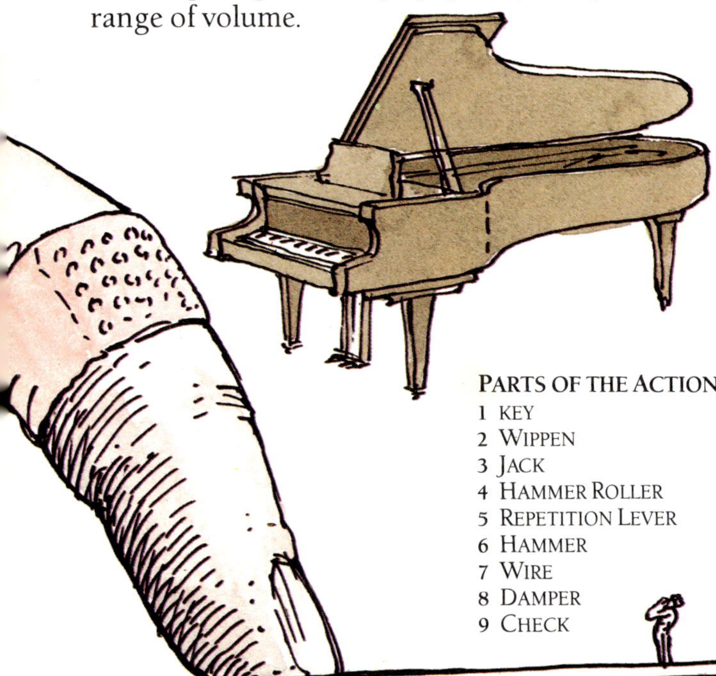

THE ACTION IN ACTION

The key raises the wippen, which forces up the jack against the hammer roller and lifts the lever carrying the hammer. The key also raises the damper and immediately after striking the wire, the hammer drops back, allowing the wire to sound. On releasing the key, the damper drops back onto the wire, cutting off the sound.

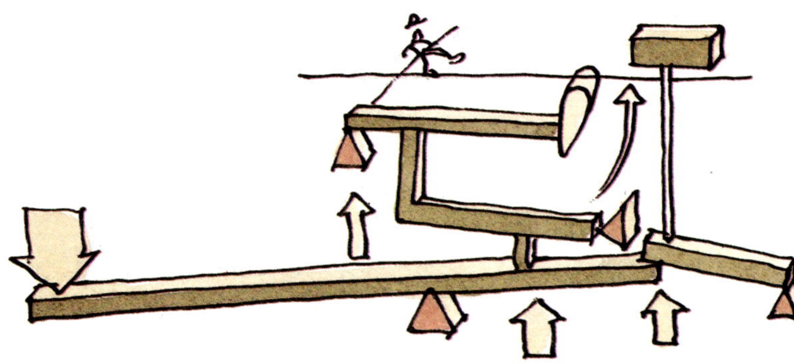

PARTS OF THE ACTION

1 KEY
2 WIPPEN
3 JACK
4 HAMMER ROLLER
5 REPETITION LEVER
6 HAMMER
7 WIRE
8 DAMPER
9 CHECK

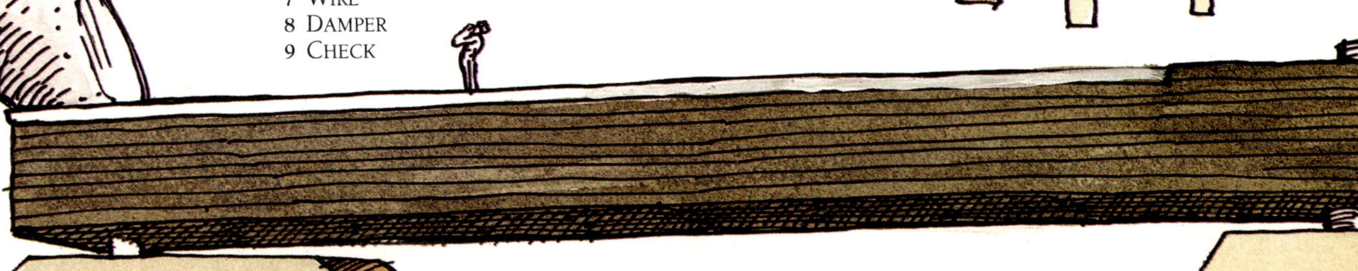

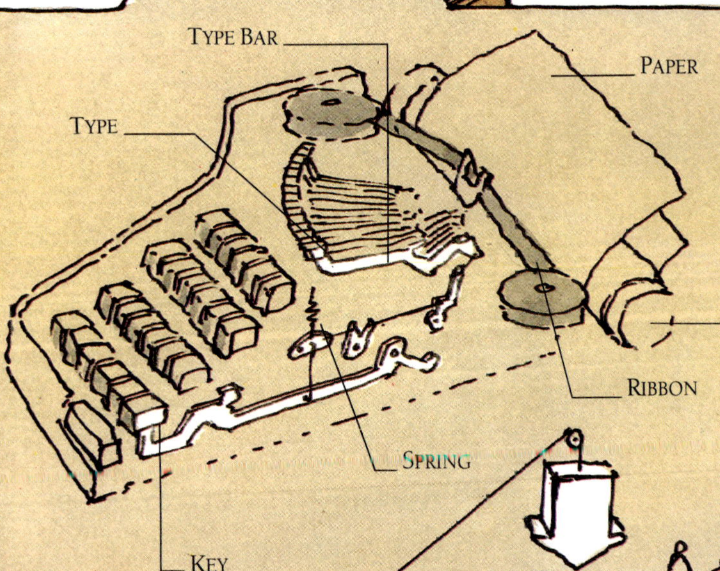

TYPE BARS

The type bars are arranged in a semicircle so that each strikes the center of the machine, the paper and ribbon moving on between each strike.

UPPER AND LOWER CASE

Each type bar bears both upper- and lower-case letters. Pressing the shift key lowers the type bar so that the upper-case letter strikes the ribbon.

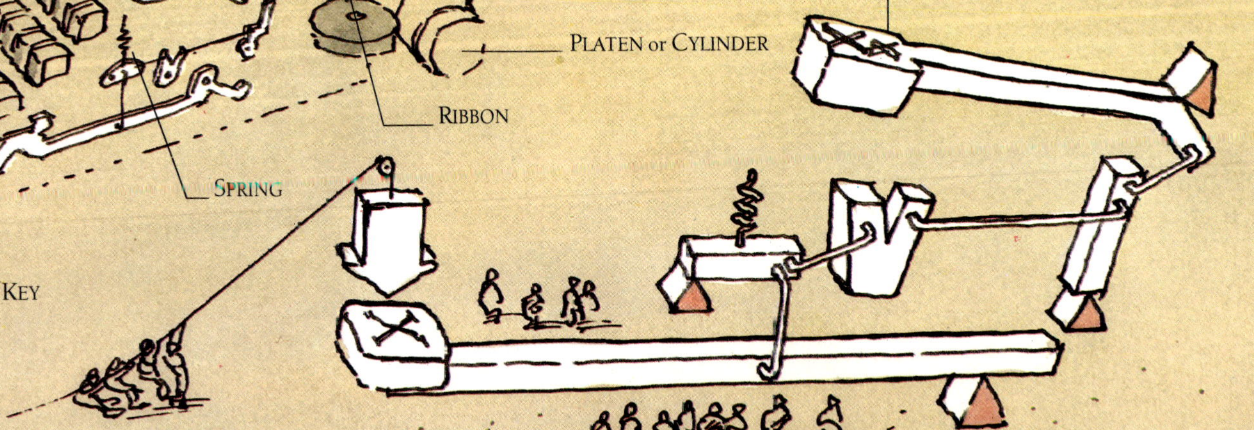

Repeating a Note

The hammer drops back after striking the wire. If the key is not released, the fall of the hammer is arrested by the check and repetition lever. The hammer is held in this position so that it is ready to strike the wire rapidly if the key is immediately pressed again.

Manual Typewriter

Like the piano, the typewriter also contains a system of levers that converts the small movement of a fingertip on a key into a long movement – in this case the movement of the raised type on the end of the type bar. As the typewriter is always played *fortissimo*, a simple system of levers suffices to connect the key to the type. Most manual typewriters use at least five levers between key and type bar.

THE PARKING METER

One of the most ingenious combinations of levers does nothing more than buy time; it is the mechanical parking meter. The meter has to cope with coins of different sizes and values, which trigger the mechanism to move the pointer to the right amount of time. Clockwork powers the levers and moves the pointer to show the amount of time remaining. It also pops up a penalty flag when the time expires.

BUYING TWO HOURS

One mammoth, a coin of modest dimensions but high value, buys two hours – the maximum time that the meter can indicate. The coin trips levers that move the pointer as far as it can go.

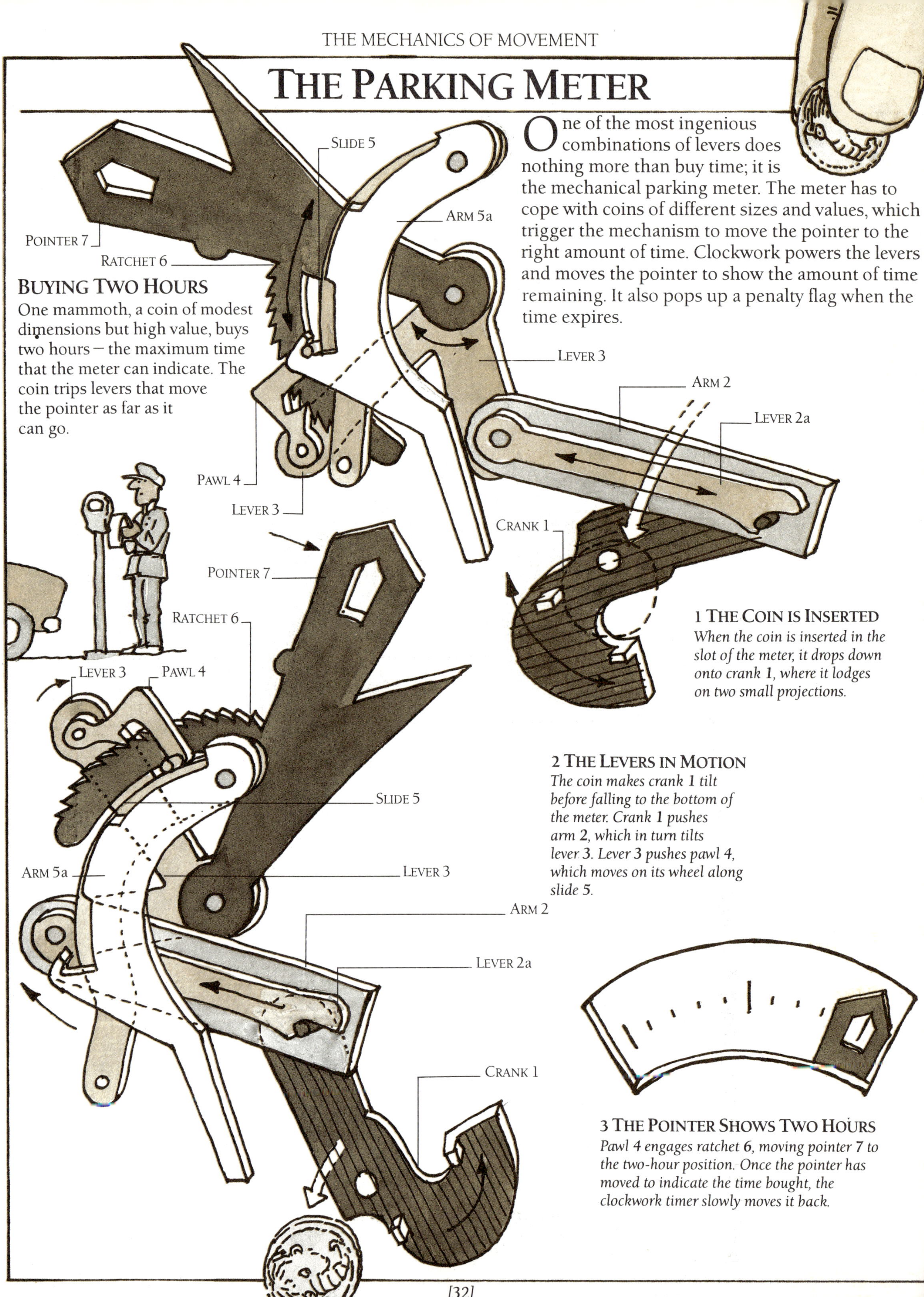

1 THE COIN IS INSERTED

When the coin is inserted in the slot of the meter, it drops down onto crank 1, where it lodges on two small projections.

2 THE LEVERS IN MOTION

The coin makes crank 1 tilt before falling to the bottom of the meter. Crank 1 pushes arm 2, which in turn tilts lever 3. Lever 3 pushes pawl 4, which moves on its wheel along slide 5.

3 THE POINTER SHOWS TWO HOURS

Pawl 4 engages ratchet 6, moving pointer 7 to the two-hour position. Once the pointer has moved to indicate the time bought, the clockwork timer slowly moves it back.

BUYING AN HOUR

A hog is a large coin, but has only half the value of a mammoth and so buys an hour. The size of this coin makes another lever – 1b – come into operation. This sets off a sequence of movements that moves the pointer to the one-hour position.

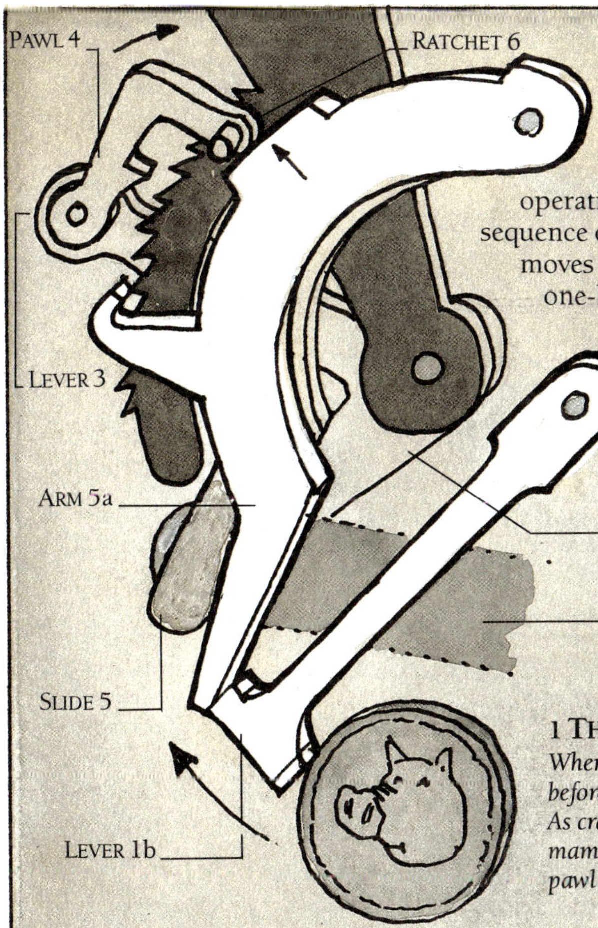

1 THE COIN IS INSERTED

When the coin strikes crank 1, it pushes aside lever 1b before it tilts crank 1. Lever 1b lifts arm 5a on slide 5. As crank 1 tilts, the same action occurs as with the mammoth coin to move pawl 4. However, arm 5a lifts pawl 4 above ratchet 6 for half of its movement.

2 THE POINTER SHOWS ONE HOUR

After half its movement, pawl 4 descends to engage ratchet 6 and move pointer 7 to the one-hour position.

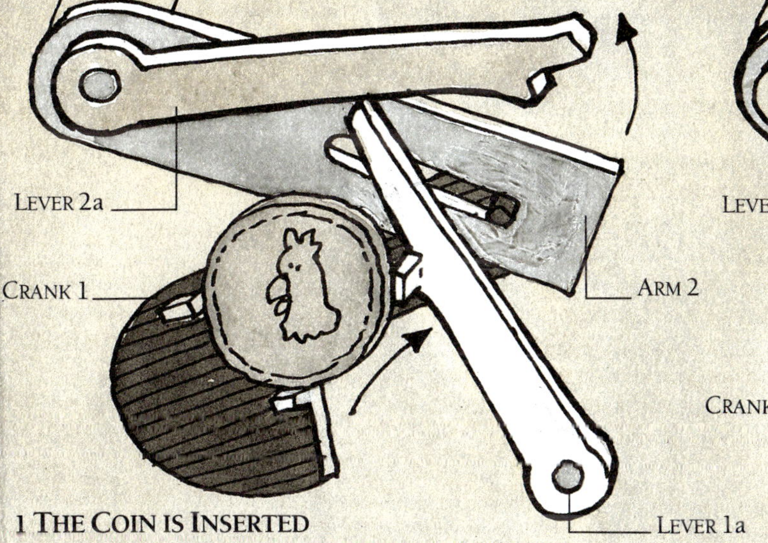

1 THE COIN IS INSERTED

When the chicken strikes crank 1, it is big enough to lift lever 1a, which in turn raises lever 2a on arm 2. As crank 1 tilts, the pin on crank 1 slides along the slot in arm 2.

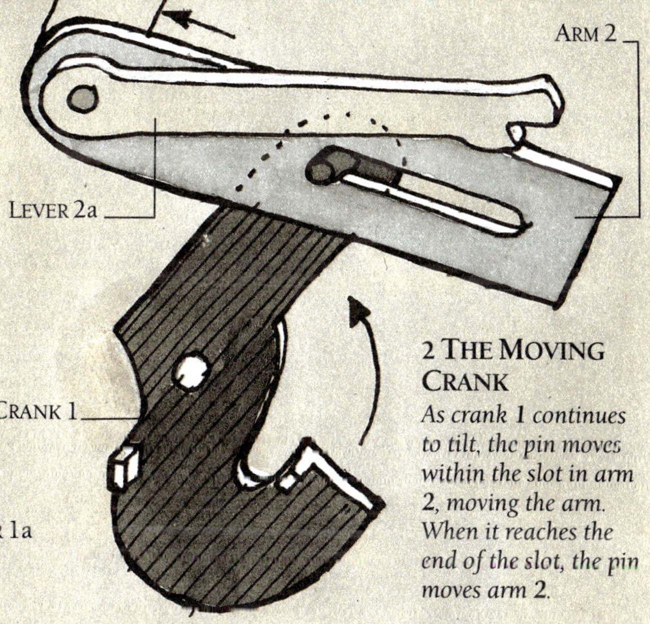

2 THE MOVING CRANK

As crank 1 continues to tilt, the pin moves within the slot in arm 2, moving the arm. When it reaches the end of the slot, the pin moves arm 2.

BUYING 12 MINUTES

A chicken is lowly in value and buys only 12 minutes, but it happens to be a medium-sized coin. For the chicken, lever 2a now comes into action to move the pointer to the correct position.

3 THE POINTER SHOWS 12 MINUTES

The movement of arm 2 pushes pointer 7 as with the mammoth coin, but the pointer reaches only the 12-minute position.

THE WHEEL AND AXLE

ON THE GROOMING OF MAMMOTHS

The problem with washing a mammoth, assuming that you can get close enough with the water (a point I will address further on), is the length of time it takes for the creature's hair to dry. The problem is greatly aggravated when steady sunshine is unavailable.

Recalling the incident between the mammoth and the tusk trimmers, and particularly the motion of the free end of the log, I invented a mechanical drier. It was composed of feathers secured to the ends of long spokes which radiated from one end of a sturdy shaft. At the other end of the shaft radiated a set of short boards. The entire machine was powered by a continuous line of sprightly workers who leaped one by one from a raised platform onto the projecting boards. Their weight against the boards turned the shaft. Because the spokes at the opposite end of the shaft were considerably longer than the boards, their feathered ends naturally turned much faster thereby producing the steady wind required for speedy drying.

A colleague once suggested I replace manpower with a constant stream of water. I left him in no doubt as to my views on this ludicrous proposal.

In the very same village where I built my first feather drier there began the strange – albeit fashionable – practice of tusk modification. A blindfolded mammoth was drawn up against a fixed post or tree by ropes secured to its tusks. The other ends of the ropes were fastened around the drum of a powerful winch – a most ingenious device. As workers turned the drum with handles which projected from it, they were able to straighten the tusks.

Keeping the tusks straight would of course require frequent visits and have been quite lucrative. However, since the process not only made movement through doors impossible, but also affected a mammoth's breathing, it had to be abandoned.

Wheels as Levers

While many machines work with parts that move up and down or in and out, most depend on rotary motion. These machines contain wheels, but not only wheels that roll on roads. Just as important are a class of devices known as the wheel and axle, which are used to transmit force. Some of these devices look like wheels with axles, while others do not. However, they all rotate around a fixed point to act as a rotating lever.

The center of the wheel and axle is the fulcrum of the rotating lever. The wheel is the outer part of the lever, and the axle is the inner part near the center. In the mammoth drier, the feathers form the wheel and the boards are the axle. In the winch, the drum is the axle and the handles form the wheel.

As the device rotates, the wheel moves a greater distance than the axle but turns with less force. Effort applied to the wheel, as in the winch, causes the axle to turn with a greater force than the wheel. Many machines use the wheel and axle to increase force in this way. Turning the axle, as in the mammoth drier, makes the wheel move at a greater speed than the axle.

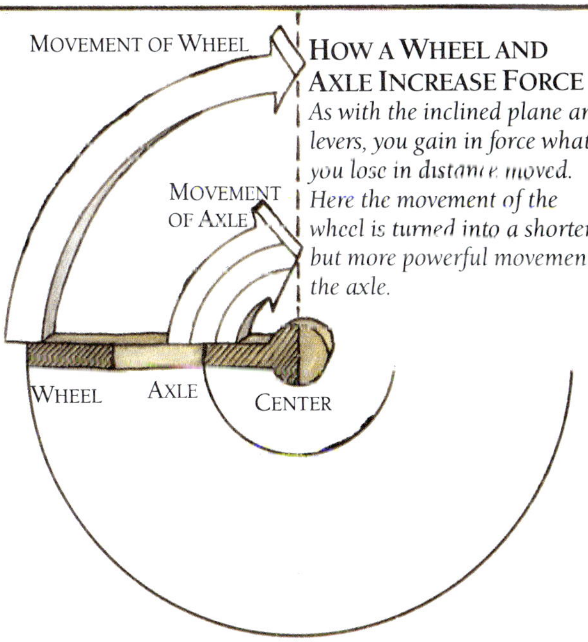

How a Wheel and Axle Increase Force
As with the inclined plane and levers, you gain in force what you lose in distance moved. Here the movement of the wheel is turned into a shorter but more powerful movement at the axle.

THE WHEEL AND AXLE AT WORK

SCREWDRIVER

The handle of a screwdriver does more than enable you to hold it. It amplifies the force with which you turn it to drive the screw home.

SARDINE CAN

The key of a sardine can exerts a strong force to pull the metal sealing band away from the can.

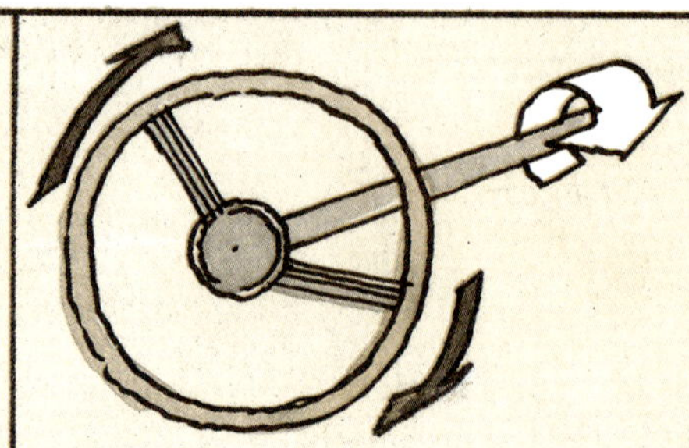

STEERING WHEEL

The force of the driver's hands is magnified to turn the shaft, producing sufficient force to operate the steering mechanism.

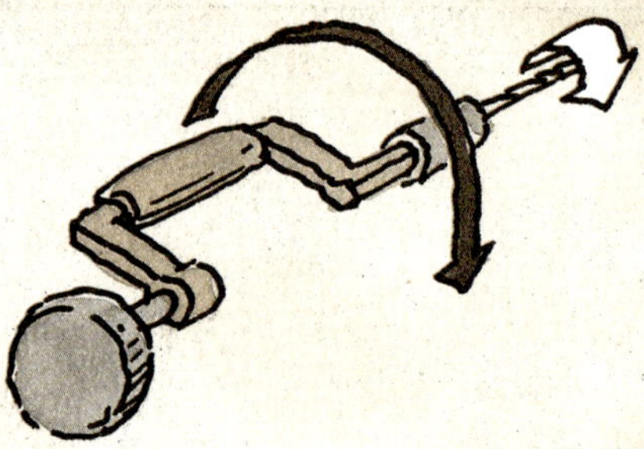

BRACE AND BIT

The handle of the brace moves a greater distance than the drill bit at the center, so the bit turns with a stronger force than the handle.

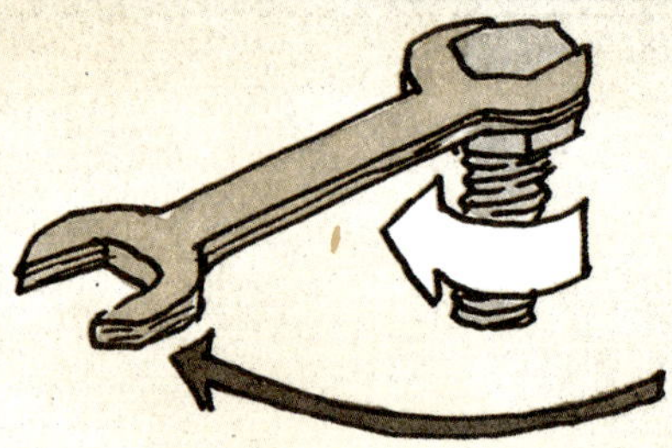

WRENCH

Pulling one end of the wrench exerts a powerful force on the bolt at the other end, screwing it tight.

FAUCET

The handle of a faucet magnifies the force of the hand to screw down the washer inside firmly and prevent the faucet dripping.

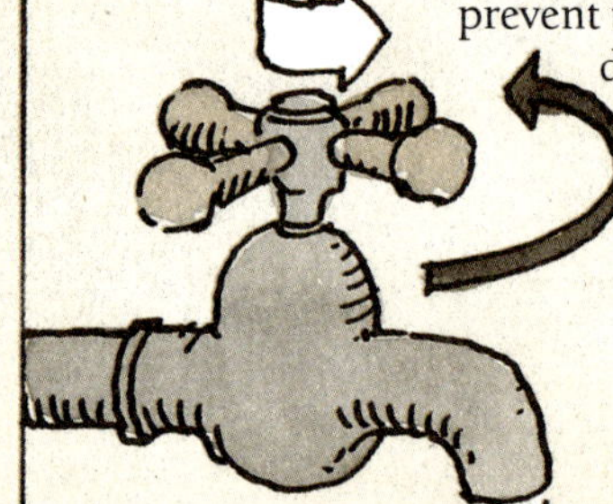

WATERWHEEL

The earliest waterwheel – the Greek mill of the first century BC – had a horizontal wheel. It was superseded by vertical wheels, which could be built to a larger size and thus developed greater power. Waterwheels obey the principle of the wheel and axle, with the force of the water on the paddles at the rim producing a strong driving force at the central shaft.

GREEK MILL

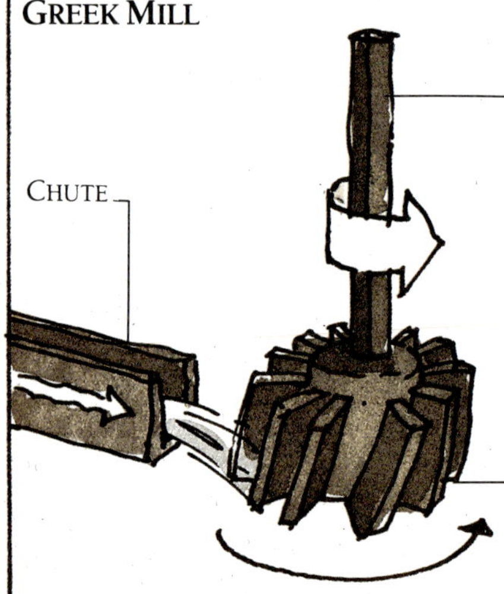

OVERSHOT WATERWHEEL

HYDROELECTRIC TURBINE

Hydroelectric power stations contain water turbines that are direct descendants of waterwheels. An efficient turbine extracts as much energy from the water as possible, reducing a powerful intake flow to a relatively weak outlet flow. Modern turbines, such as the Francis turbine shown here, are carefully designed so that the water is guided onto the blades with the minimum of energy-wasting turbulence.

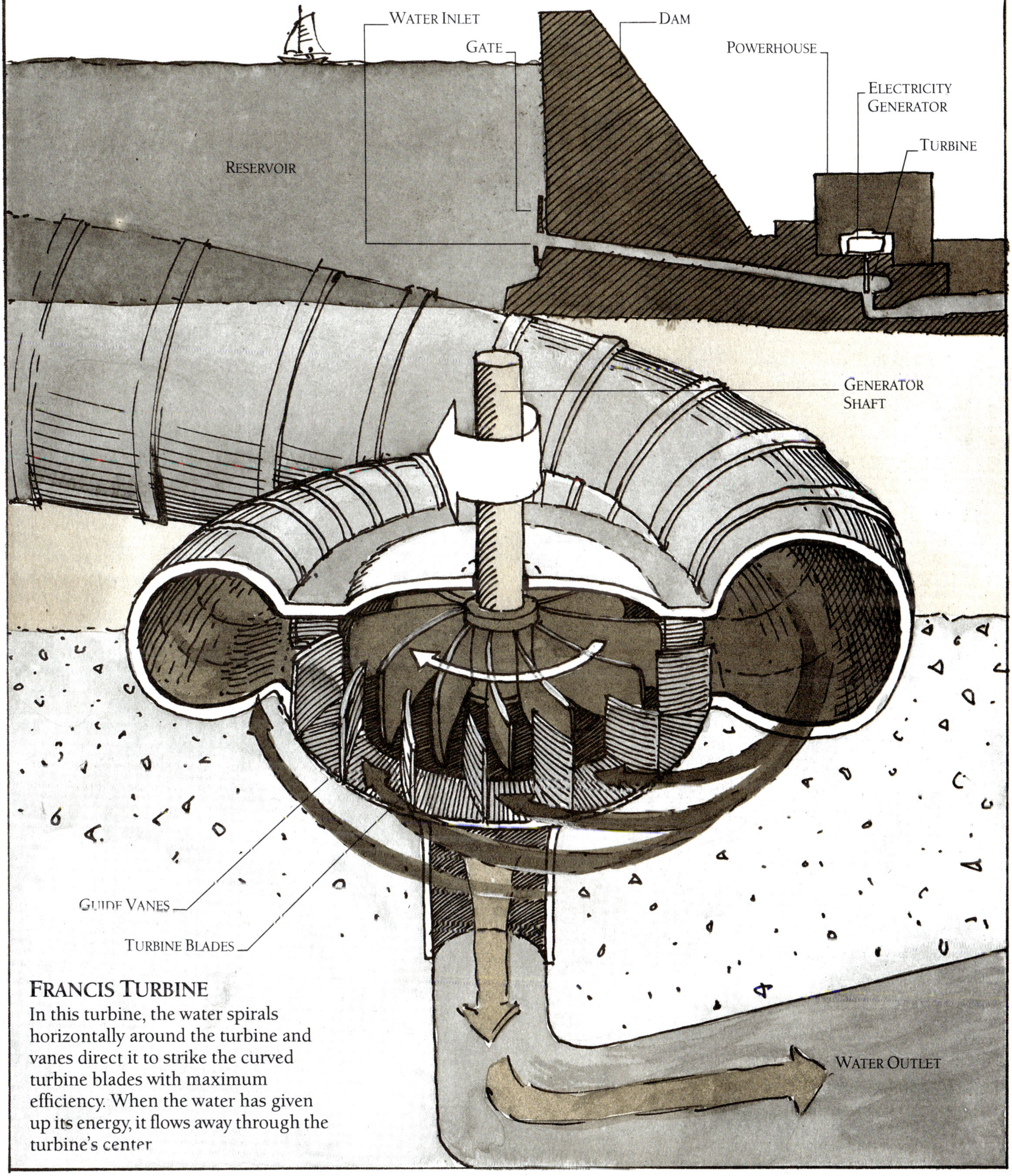

FRANCIS TURBINE

In this turbine, the water spirals horizontally around the turbine and vanes direct it to strike the curved turbine blades with maximum efficiency. When the water has given up its energy, it flows away through the turbine's center

THE WINDMILL

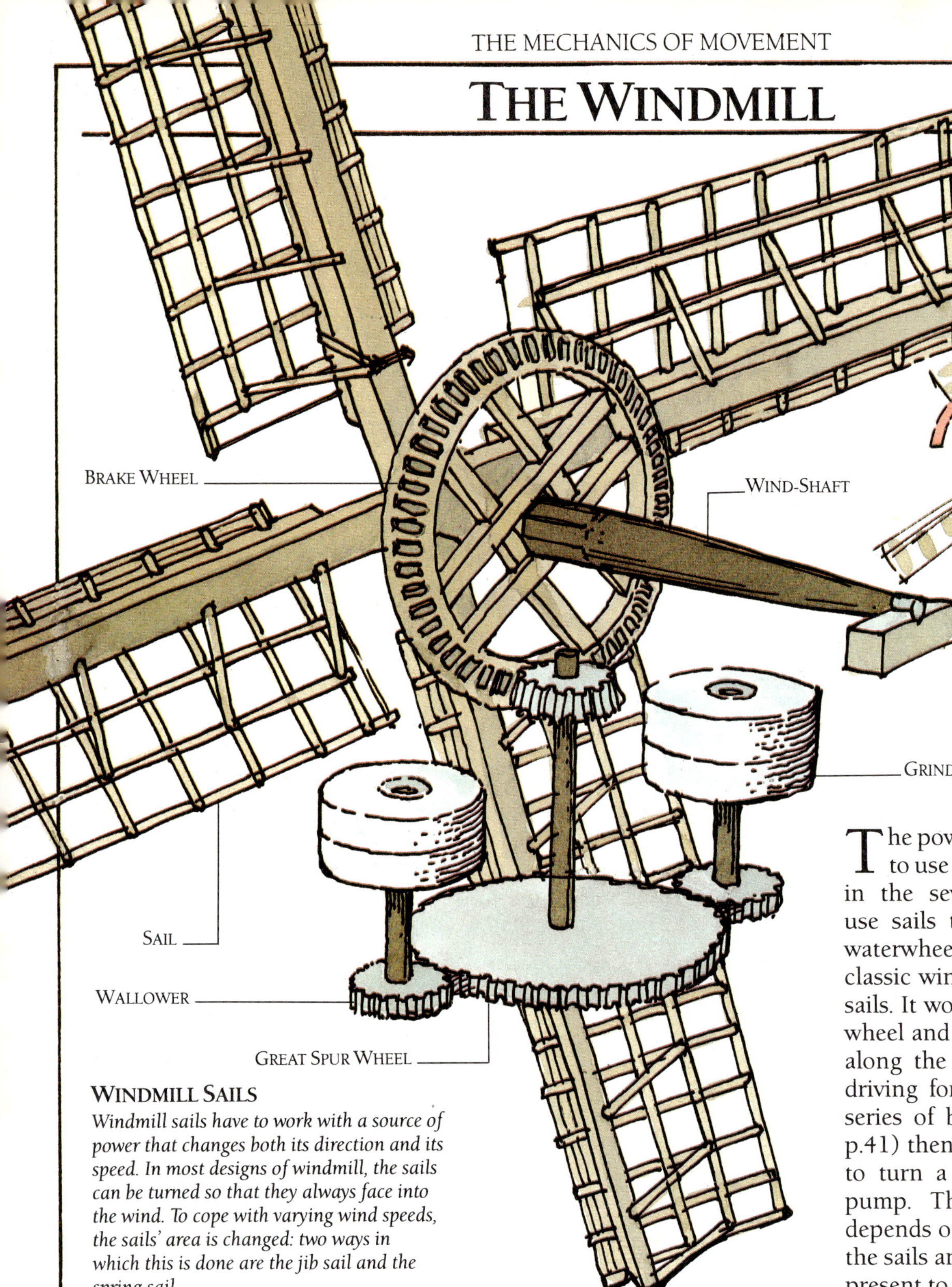

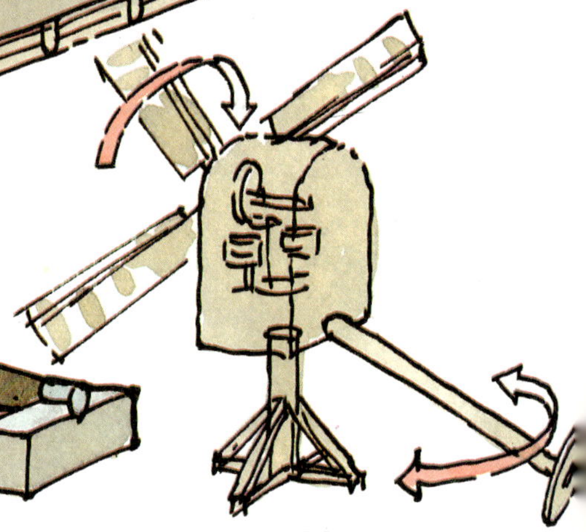

POST MILL

The power of the wind was first put to use in a windmill built in Persia in the seventh century. Windmills use sails to develop power, just as waterwheels employ paddles. The classic windmill has four big vertical sails. It works by the principle of the wheel and axle: the force of the wind along the sails produces a stronger driving force at the central shaft. A series of bevel and spur gears (see p.41) then transmit this force, usually to turn a grindstone or to drive a pump. The power of a windmill depends on the speed of the wind on the sails and on the area that the sails present to the wind.

WINDMILL SAILS

Windmill sails have to work with a source of power that changes both its direction and its speed. In most designs of windmill, the sails can be turned so that they always face into the wind. To cope with varying wind speeds, the sails' area is changed: two ways in which this is done are the jib sail and the spring sail.

JIB SAIL

Around the Mediterranean Sea from Portugal to Turkey, windmills can still be seen with jib sails – simple triangular cloth sails like the sails of boats. To deal with a change in wind speed, the miller just furls or unfurls each sail.

SPRING SAIL

Sails composed of rows of wooden shutters replaced cloth sails in the late 1700s. The shutters pivot against a spring, opening when the wind gusts and closing when it drops. In this way, a constant wind force is maintained on the sails.

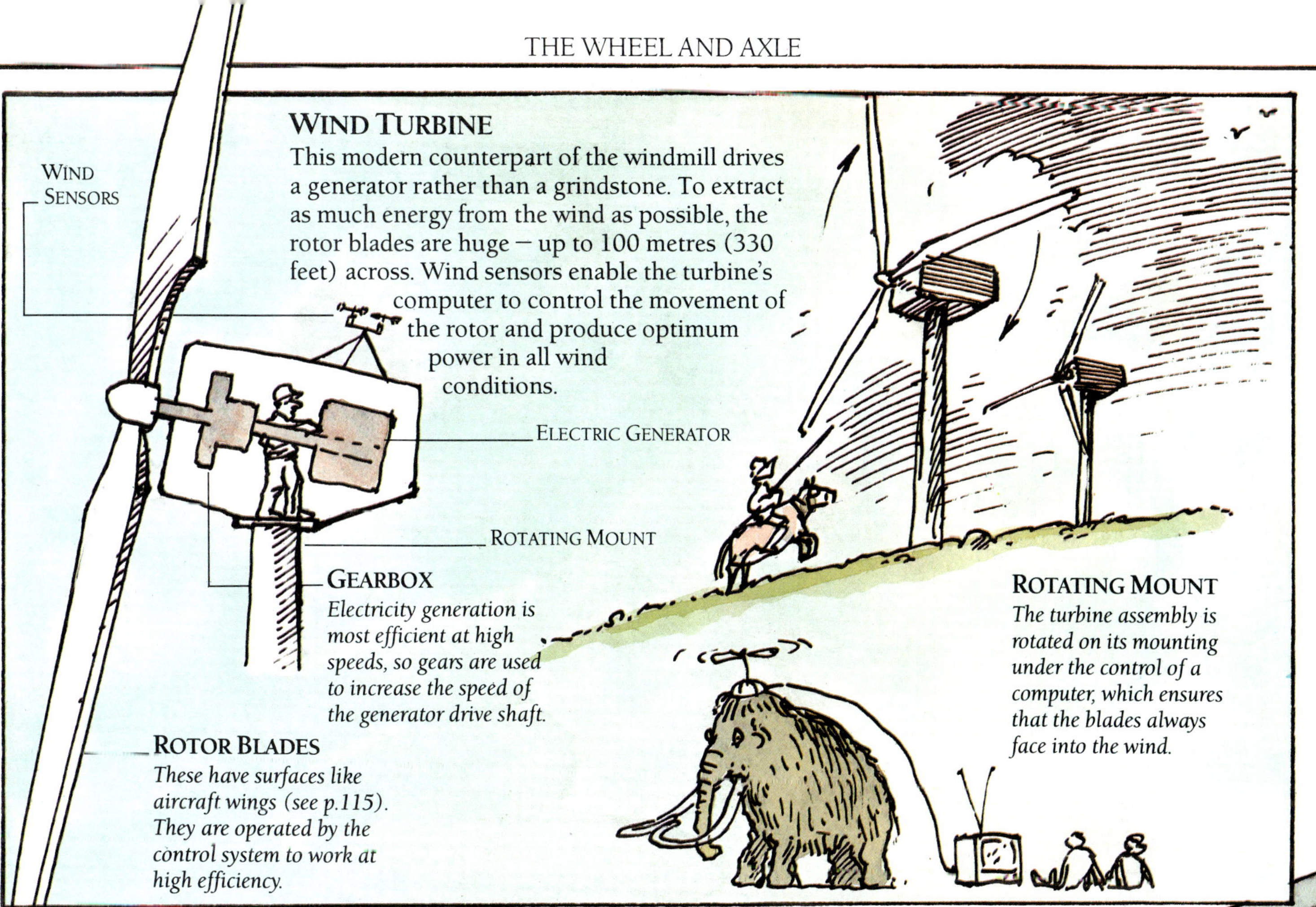

WIND TURBINE

This modern counterpart of the windmill drives a generator rather than a grindstone. To extract as much energy from the wind as possible, the rotor blades are huge – up to 100 metres (330 feet) across. Wind sensors enable the turbine's computer to control the movement of the rotor and produce optimum power in all wind conditions.

GEARBOX

Electricity generation is most efficient at high speeds, so gears are used to increase the speed of the generator drive shaft.

ROTOR BLADES

These have surfaces like aircraft wings (see p.115). They are operated by the control system to work at high efficiency.

ROTATING MOUNT

The turbine assembly is rotated on its mounting under the control of a computer, which ensures that the blades always face into the wind.

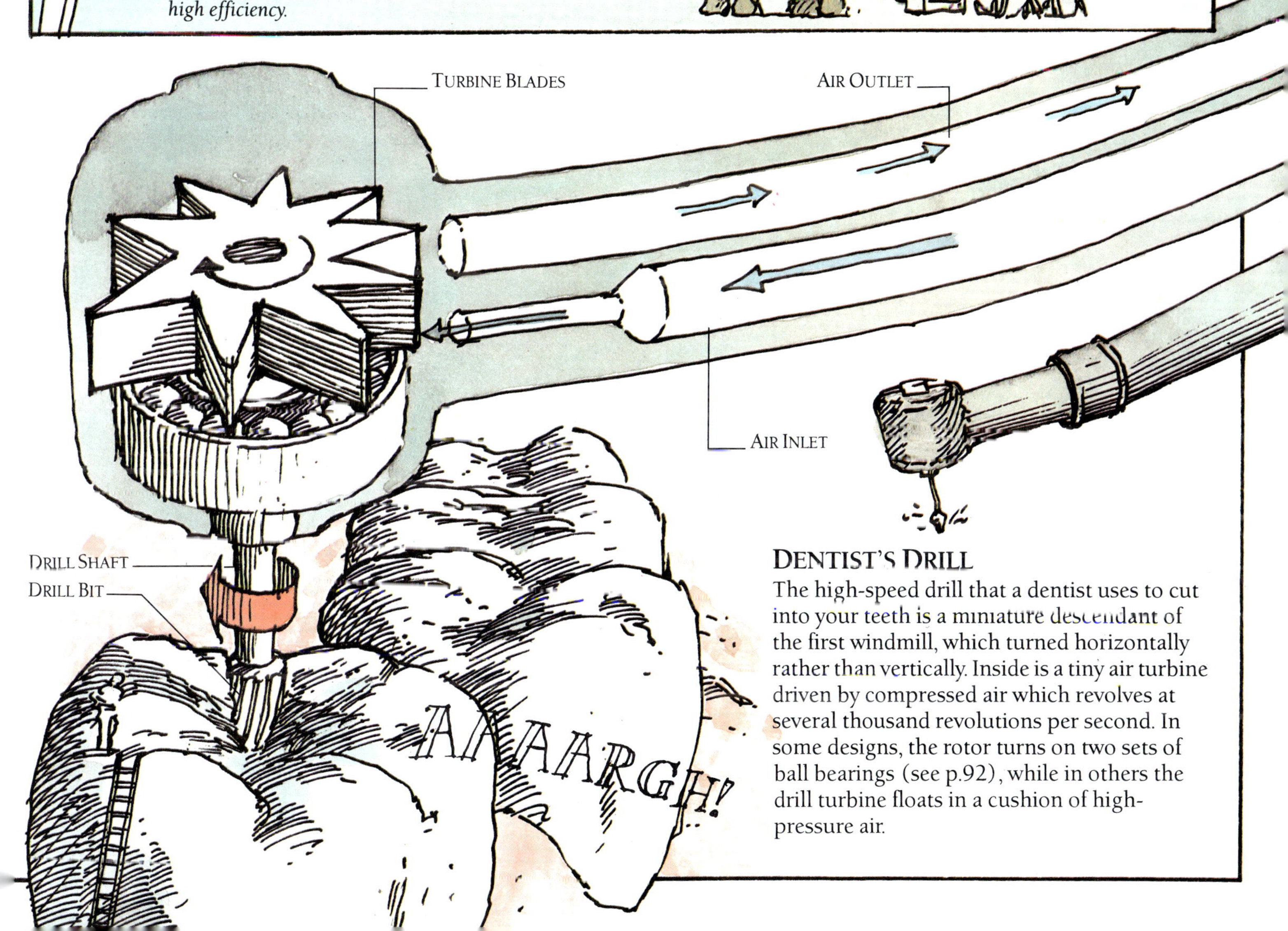

DENTIST'S DRILL

The high-speed drill that a dentist uses to cut into your teeth is a miniature descendant of the first windmill, which turned horizontally rather than vertically. Inside is a tiny air turbine driven by compressed air which revolves at several thousand revolutions per second. In some designs, the rotor turns on two sets of ball bearings (see p.92), while in others the drill turbine floats in a cushion of high-pressure air.

GEARS AND BELTS

ON EARLY MAMMOTH POWER

As far as I can ascertain, the first use of mammoths in industry was to provide power for the famous merry-go-round experiment. The equipment consisted of two wheels, one large and one small, placed edge to edge so that when the mammoths turned one wheel, the other would turn automatically. At first seats were hung from the small wheel which was driven by the large wheel. The result was a hair-raising ride. When the wheels were reversed, the ride was far too sedate. Eventually belts connected to drive wheels of different sizes operated two rides simultaneously, one fast and one gentle. Carrot consumption during the experiment was astronomical.

Types of Gears

Gears come in a variety of sizes with their teeth straight or curved and inclined at a variety of angles. They are connected together in various ways to transmit motion and force in machines. However, there are only four basic types of gears. They all act so that one gear wheel turns faster or slower than the other, or moves in a different direction. A difference in speed between two gears produces a change in the force transmitted.

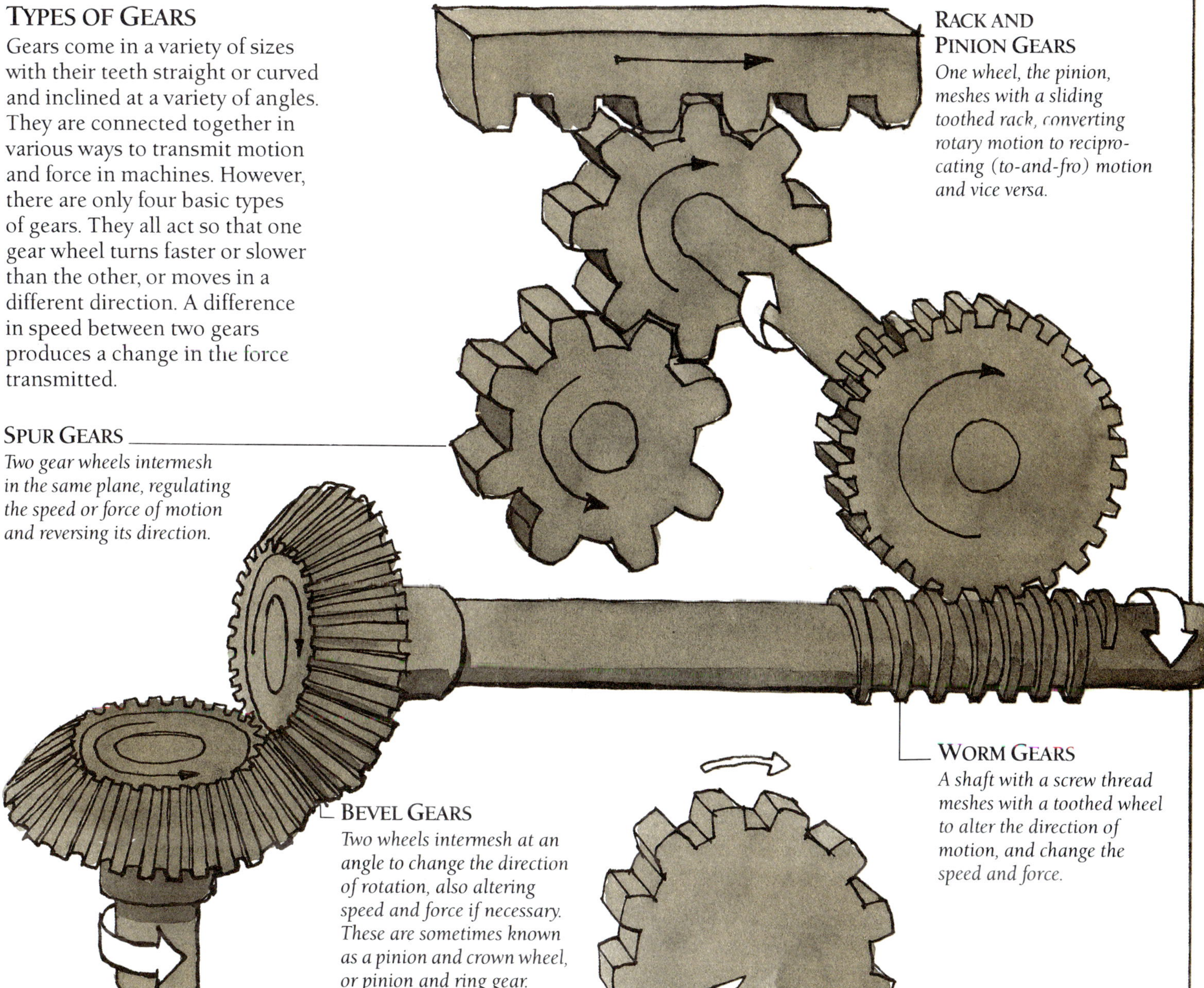

Rack and Pinion Gears

One wheel, the pinion, meshes with a sliding toothed rack, converting rotary motion to reciprocating (to-and-fro) motion and vice versa.

Spur Gears

Two gear wheels intermesh in the same plane, regulating the speed or force of motion and reversing its direction.

Worm Gears

A shaft with a screw thread meshes with a toothed wheel to alter the direction of motion, and change the speed and force.

Bevel Gears

Two wheels intermesh at an angle to change the direction of rotation, also altering speed and force if necessary. These are sometimes known as a pinion and crown wheel, or pinion and ring gear.

Gears

The big wheel has twice the number of teeth, and twice the circumference, of the small wheel. It rotates with twice the force and half the speed in the opposite direction.

How Gears and Belts Work

The way gears and belts control movement depends entirely on the sizes of the two connecting wheels. In any pair of wheels, the larger wheel will rotate more slowly than the smaller wheel, but it will rotate with greater force. The bigger the difference in size between the two wheels, the bigger will be the difference in speed and force.

Wheels connected by belts or chains work in just the same way as gears, the only difference being in the direction that the wheels rotate.

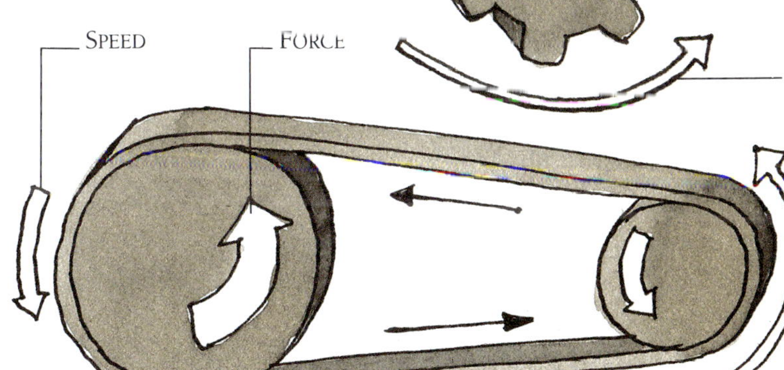

Belts

The big wheel has twice the circumference of the small wheel. It also rotates with twice the force and half the speed, but in the same direction.

SPUR GEARS

DERAILLEUR GEARS

The chain connecting the pedals of a bicycle to the rear wheel acts as a belt to make the wheel turn faster than the feet. To ride on the level or downhill, the rear-wheel sprocket needs to be small for high speed. But to climb hills, it needs to be large so that the rear wheel turns with less speed but more force.

Derailleur gears solve the problem by having rear-wheel sprockets of different sizes. A gear-changing mechanism transfers the chain from one sprocket to the next.

DRIVING GEAR

The driving gear is turned by one tooth every time the wheel rotates. The counter records how many revolutions take place, converting this figure into the distance traveled.

DRUMS

The drums bear twenty teeth – two for each numeral – on their right sides. On the left side of each drum is a gap by the numeral 2 and two projections on either side of the gap.

GEAR WHEELS

The wheels intermesh with each drum except the one at the right-hand end.

BICYCLE DISTANCE COUNTER

The counter is mounted on the front axle, and driven by a small peg fixed to a front wheel spoke. A reduction gear makes the right-hand numbered drum revolve once every mile or kilometer. As it makes a complete revolution, it makes the next drum to the left move by one-tenth of a revolution, and so on. The movement of the drums is produced by the row of small gear wheels beneath the drums. The wheels have teeth that are alternately wide and narrow, a feature which enables them to lock adjoining drums together when one drum completes a revolution.

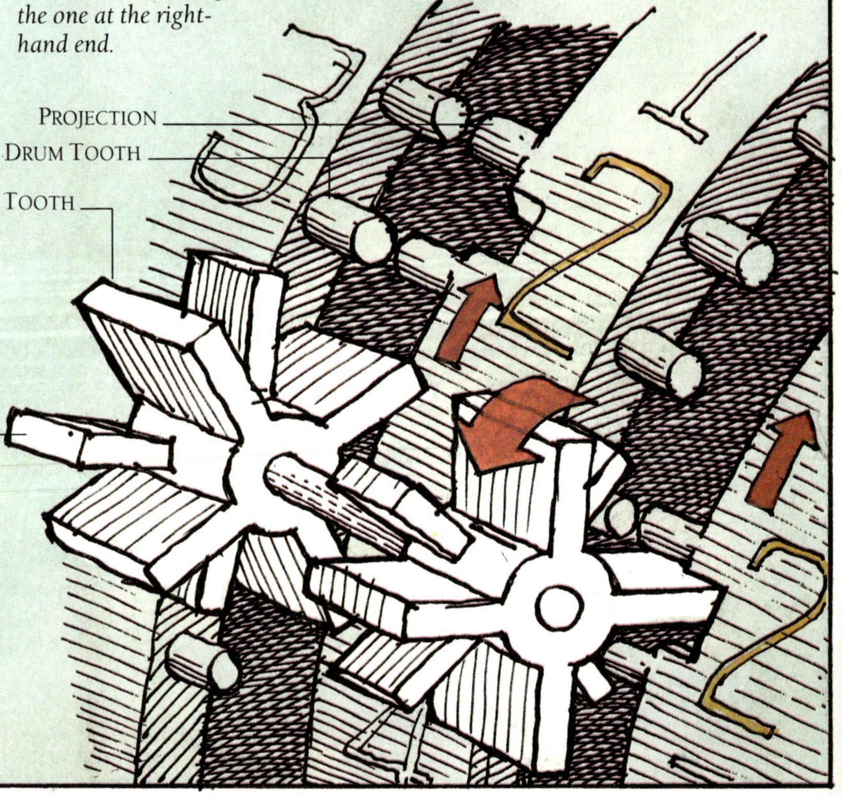

HOW ADJOINING DRUMS LOCK

Normally, a narrow wheel tooth fits between two drum teeth and the drums do not move. When a 9 moves up into the viewing window at the top of the drums, the projection on that drum catches the narrow tooth on the gear wheel, making it rotate. The next wide tooth fits into the gap by the 2 and locks the drum and the next left drum together. As the 9 changes to 0, the gear wheel rotates and the next drum also moves up one numeral.

Salad Spinner

A salad spinner rotates at high speed, throwing off water by centrifugal force (see p.75). The drive mechanism consists of an epicyclic or planetary gear – a system of spur gears in which an outer gear ring turns an inner planet gear that drives a small central sun gear. Epicyclic gears can achieve a high speed magnification yet are simple and compact.

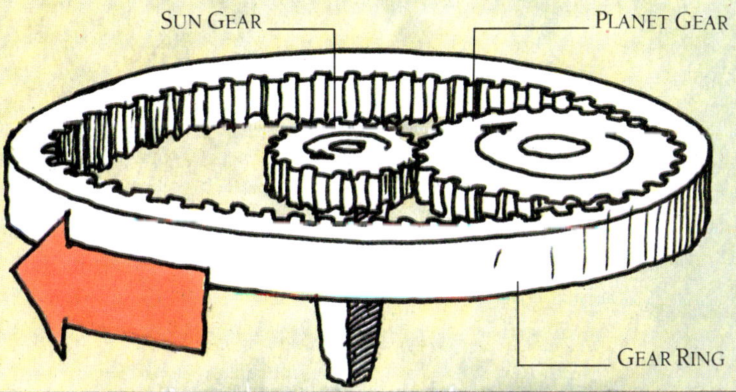

Car Window Winder

The handle in a car door turns a small cog that moves a toothed quadrant (a section of a large spur gear), which in turn raises or lowers levers supporting the car window. Electrically operated windows work on the same principle, but more gears are required because the speed of the motor has to be stepped down to provide a small but powerful movement.

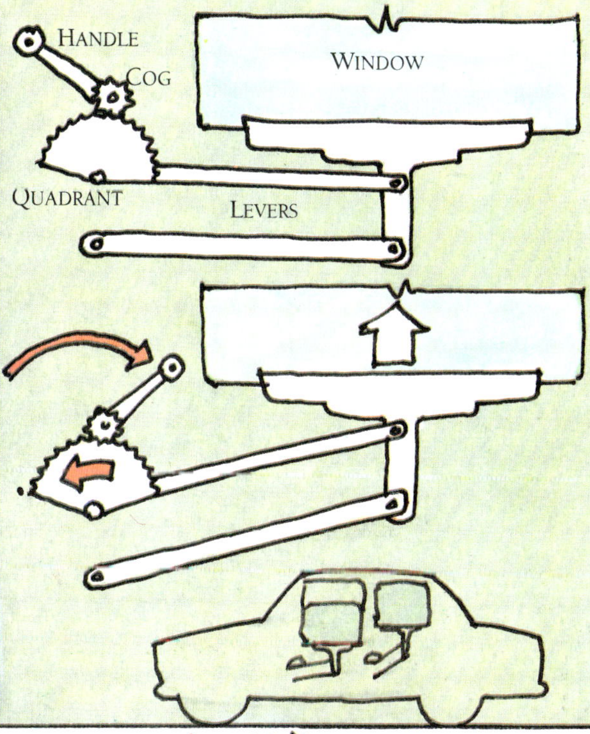

Lawn Mower

A push-mower contains an internal pinion turned by a gear ring inside the wheel. The pinion turns the cylinder of cutting blades as the wheels move, and the grass is trapped between the spinning blades and a fixed cutting blade. The blades spin much faster than the wheels so that each blade slices only a small amount of grass, mowing the lawn neatly at a single pass.

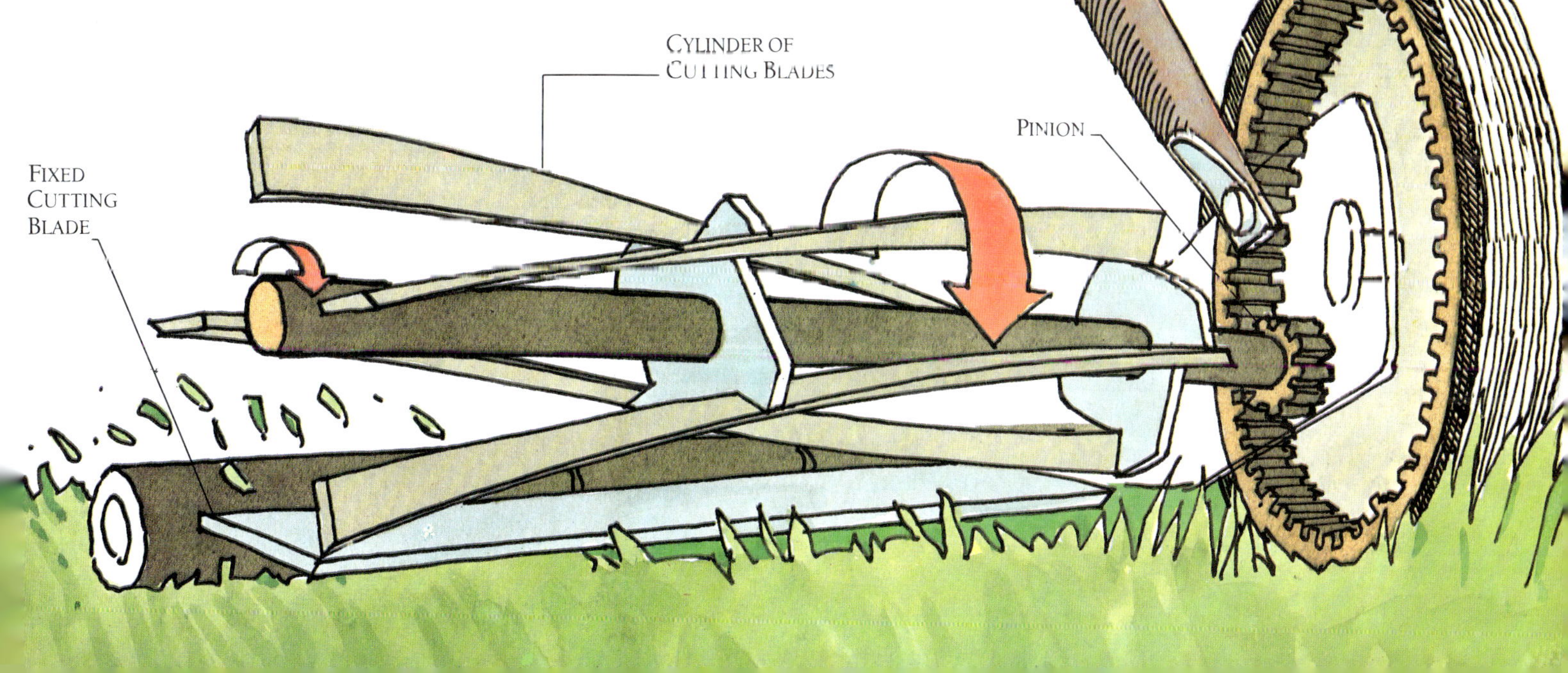

THE GEARBOX

All gasoline engines work best if they run at a high but limited rate of revolutions. The job of the gearbox is to keep the engine running at its most efficient rate while allowing the car to travel at a large range of speed.

The crankshaft always turns faster than the wheels – from about twelve times as fast in first gear to about four times as fast in top gear. The differential (see p.49) reduces engine speed by four times. The rest of the reduction takes place in the gearbox.

The gearbox lies between the clutch (see p.88) and the differential. The gears can only be changed when the clutch has disengaged the engine. Operating the gear lever of a manual gearbox brings a different train of spur gears into play for each gear, except fourth gear. In fourth gear, no gear wheels are engaged and transmission goes directly through the gearbox from the clutch to the differential.

The different ratios of teeth on the gears involved produce different speeds. Selecting reverse gear simply introduces an extra gear wheel which reverses the rotation of the transmission shaft.

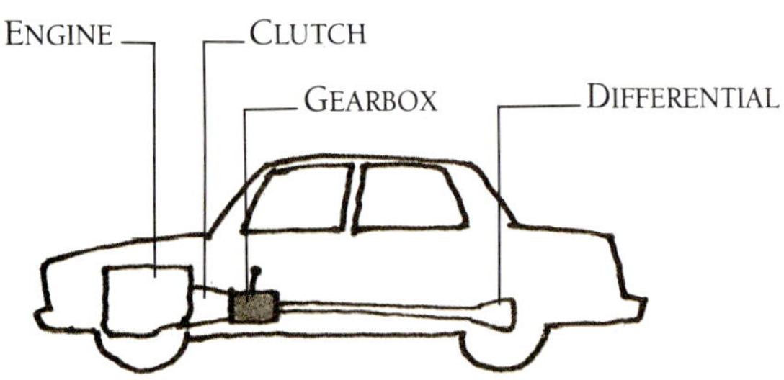

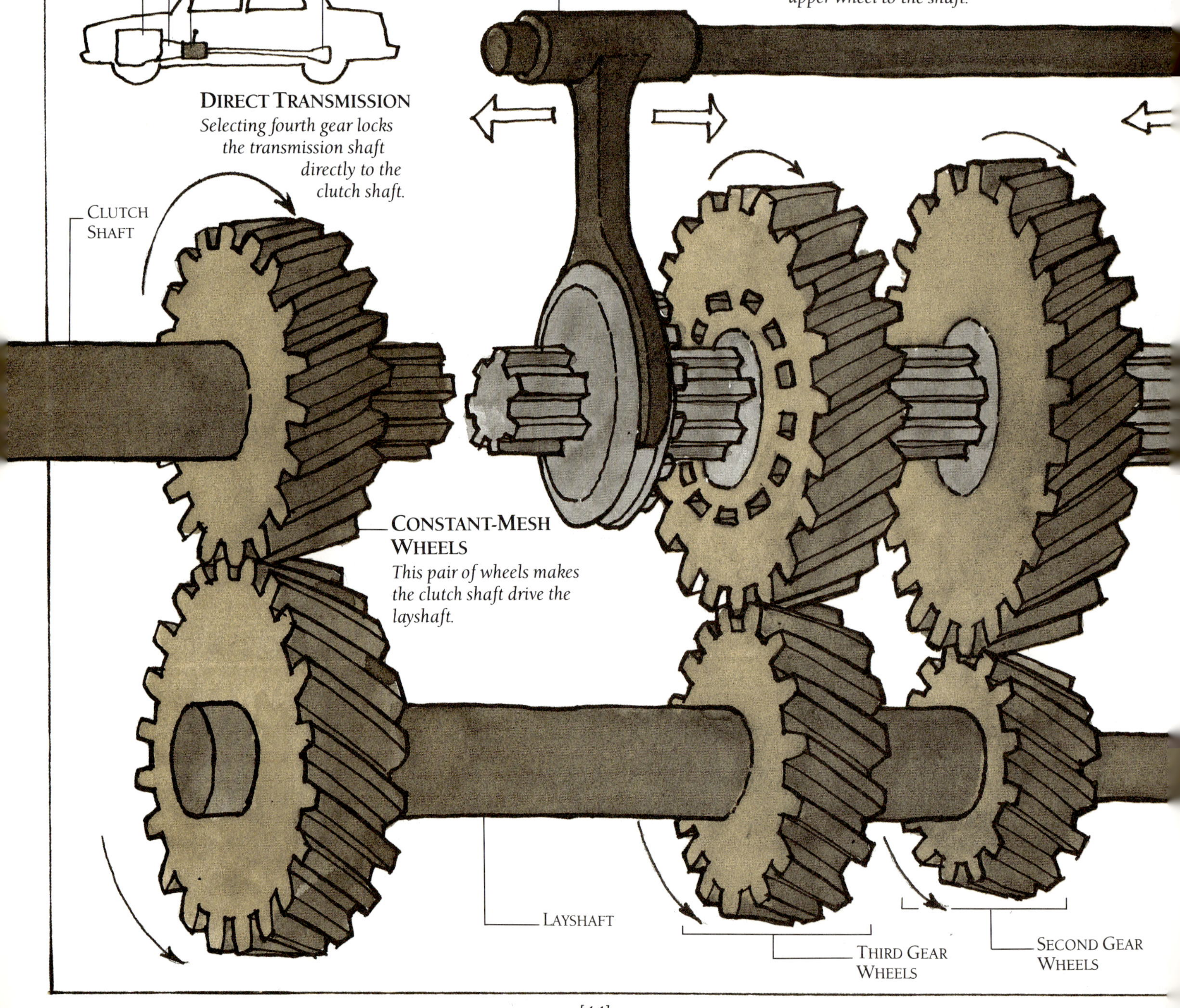

SELECTING A GEAR

Moving the gear lever tilts it so that it pushes or pulls one of the three selector rods. Except in reverse gear, the selector fork then shifts a collar that makes the dog teeth lock the required gear wheel to the transmission shaft. The speeds of the rotating parts are matched by the synchromesh (see p.89). In reverse gear, the fork engages the idler wheel.

This illustration (right) shows first gear being selected, in which the transmission goes from the clutch shaft via the layshaft and first gear wheels to the transmission shaft.

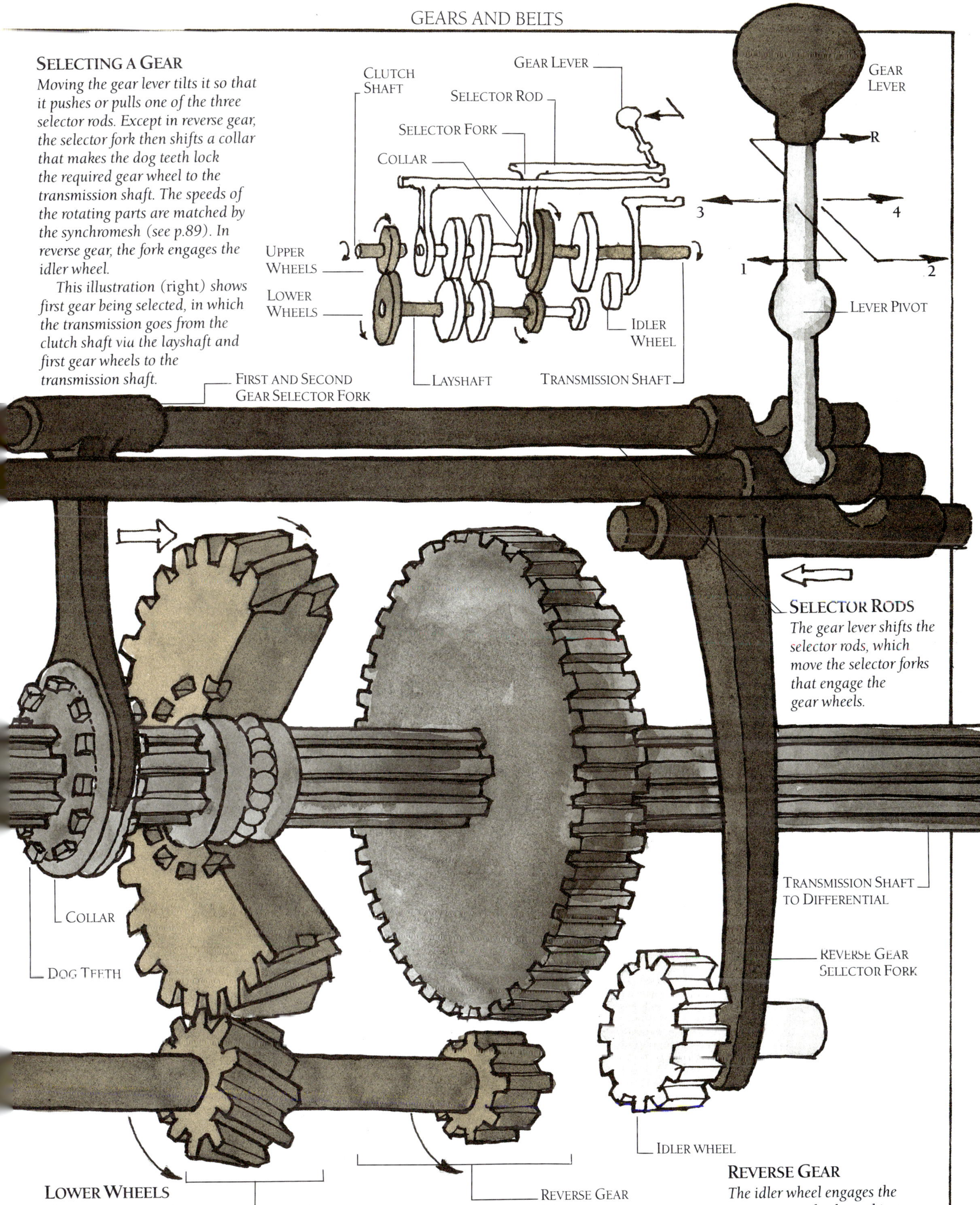

SELECTOR RODS

The gear lever shifts the selector rods, which move the selector forks that engage the gear wheels.

LOWER WHEELS

The lower gear wheels are turned by the layshaft and drive the upper wheels.

REVERSE GEAR

The idler wheel engages the reverse gear wheels, making the transmission shaft turn in the opposite direction.

MECHANICAL CLOCKS AND WATCHES

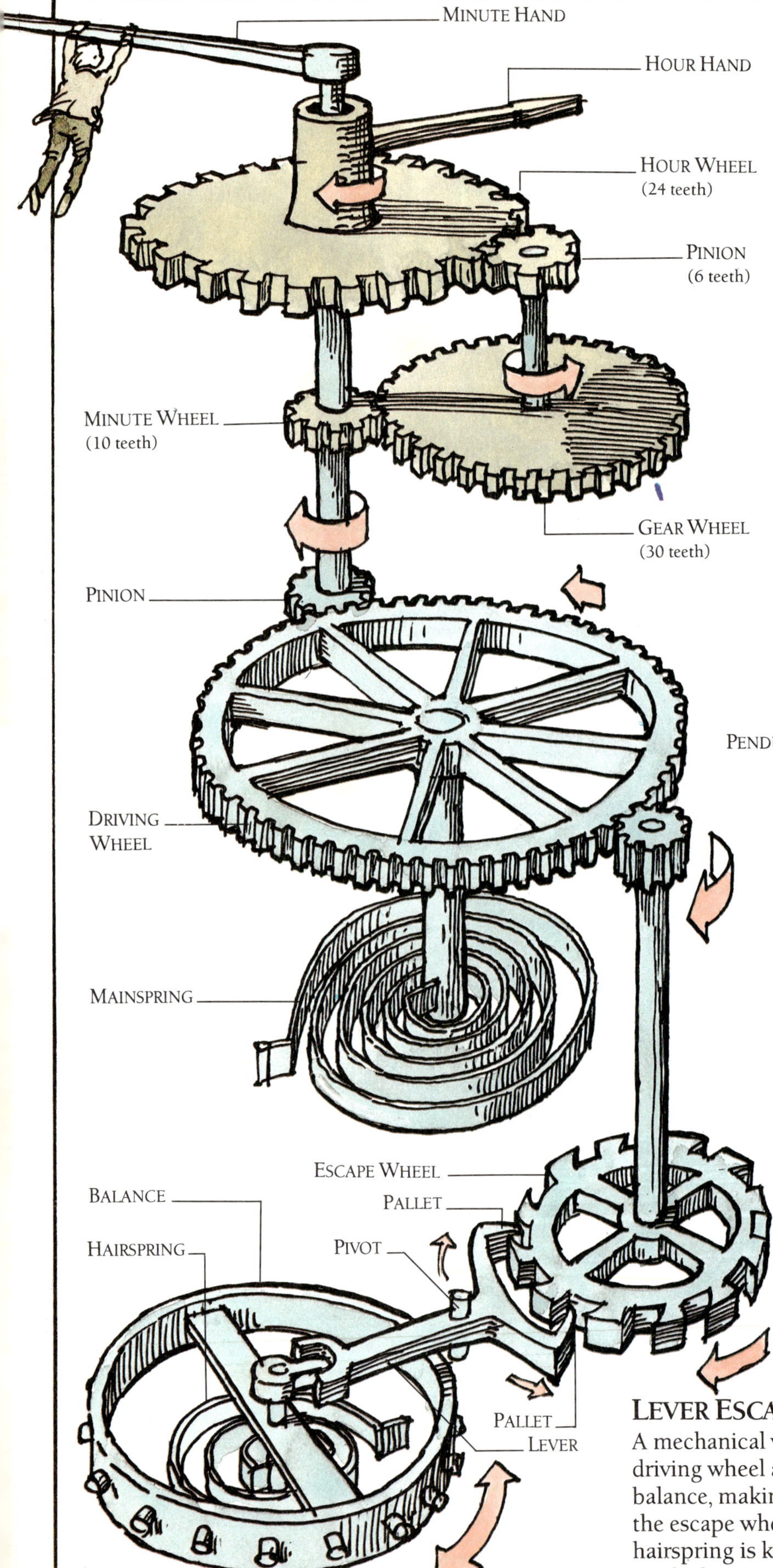

Spur gears lie at the heart of mechanical timepieces. Powered by a falling weight or an unwinding spring, they turn the two hands, ensuring that the minute hand moves exactly twelve times around the dial for every revolution of the hour hand.

The hands are turned by the driving wheel through a pinion that is geared to rotate once an hour. The pinion drives the minute hand directly. The hour hand is driven through two sets of spur gears which together reduce its speed to one-twelfth that of the hour hand.

A further train of gears controls the speed at which the driving wheel rotates by connecting it to the escapement, which is the heart of the time-keeping mechanism.

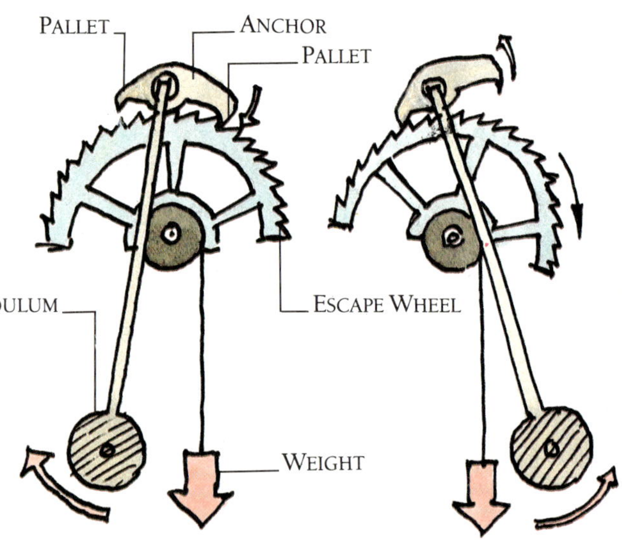

ANCHOR ESCAPEMENT

Many pendulum clocks are powered by a weight that turns the escape wheel, which is itself connected through gear trains to the hands. The escape wheel moves in precise steps. The swinging pendulum rocks the anchor so that the pallets alternately engage the teeth on the escape wheel. Each swing releases the escape wheel for a short interval to allow it to move on by one tooth. As the teeth of the escape wheel move, they push the anchor to keep the pendulum swinging.

LEVER ESCAPEMENT

A mechanical watch is powered by its mainspring, which turns the driving wheel and escape wheel. The hairspring oscillates in the balance, making the lever rock to and fro so that the pallets release the escape wheel in the same way as an anchor escapement. The hairspring is kept moving by the pressure of the escape wheel teeth on the lever.

THE RACK AND PINION

CAR STEERING

In rack and pinion steering, the steering column turns a pinion that shifts a rack to the right or left. Each end of the rack moves a track rod linked to a steering arm that turns the axle of each front wheel. Overall, a wheel and axle (in the steering wheel), a rack and pinion and a lever combine to multiply the force of the hands and turn the wheels.

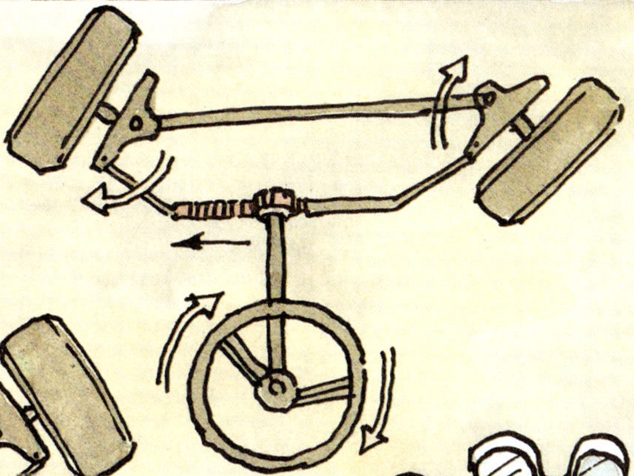

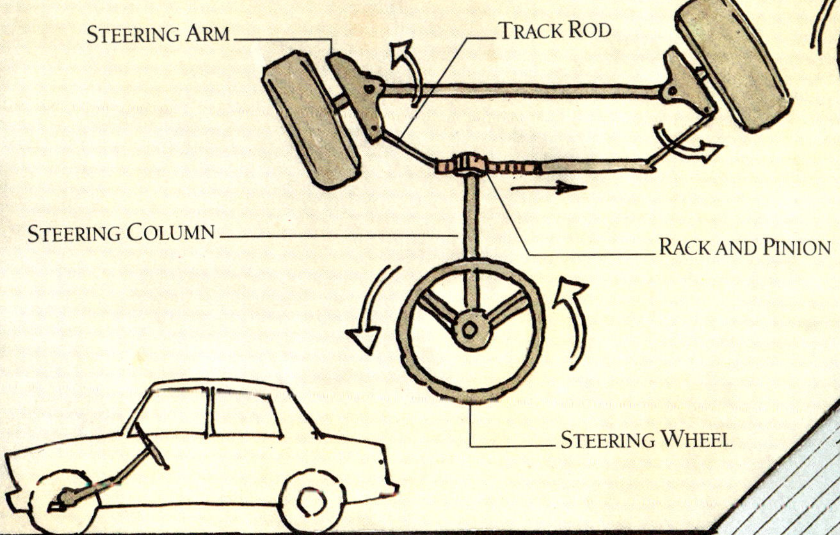

CORKSCREW

One good design of corkscrew makes use of the screw (see pp.66-7) and the rack and pinion to pull a cork from a bottle. The long handles ending in pinions produce considerable leverage on the rack, enabling the cork to be extracted without having to pull it out.

REMOVING A CORK

The corkscrew is first placed in position (1), and as it is screwed in, the handles rise (2). When the screw is fully inserted, the handles are pushed down (3), so that the pinions force up the rack, and with it, the cork.

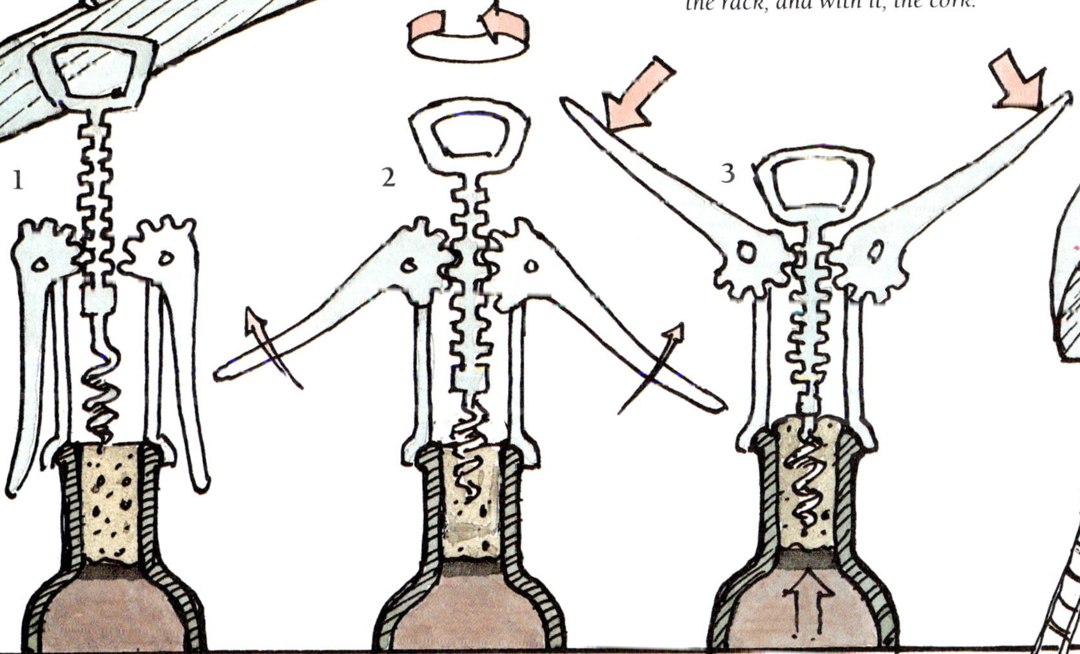

BEVEL GEARS

EGG BEATER

In bevel gears, the gear wheels are often of very different sizes. This difference serves to change either the force that is applied to one of the gears, or to increase or decrease the speed of motion. An egg beater converts a slow rotation into two much faster rotations that work in opposite directions. Its handle turns a large double-sided crown wheel, which in turn drives two bevel pinions to spin the beaters. The great increase in speed is produced by the much larger diameter of the crown wheel compared to the bevel pinions.

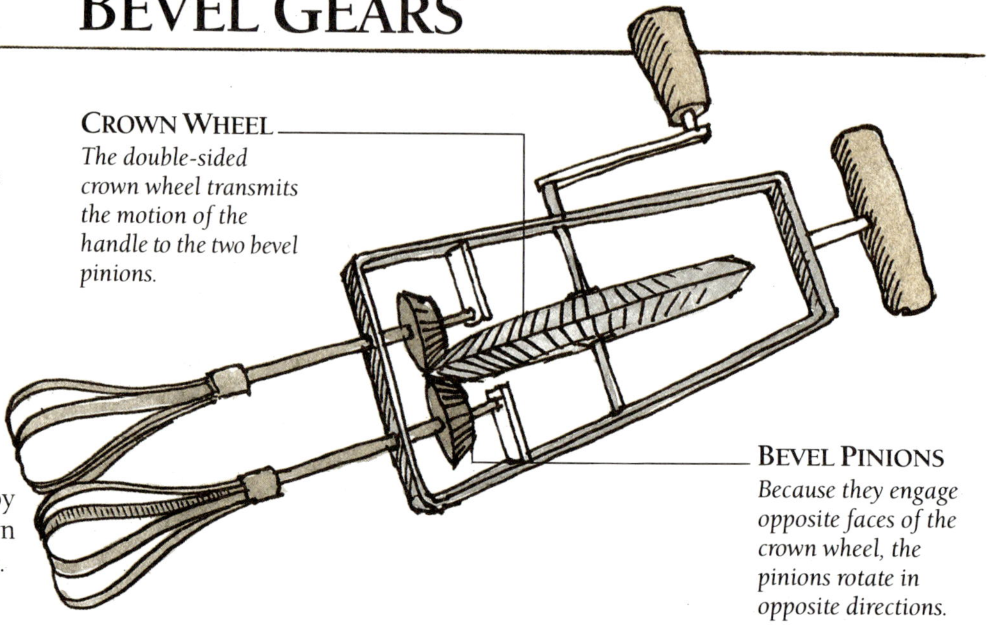

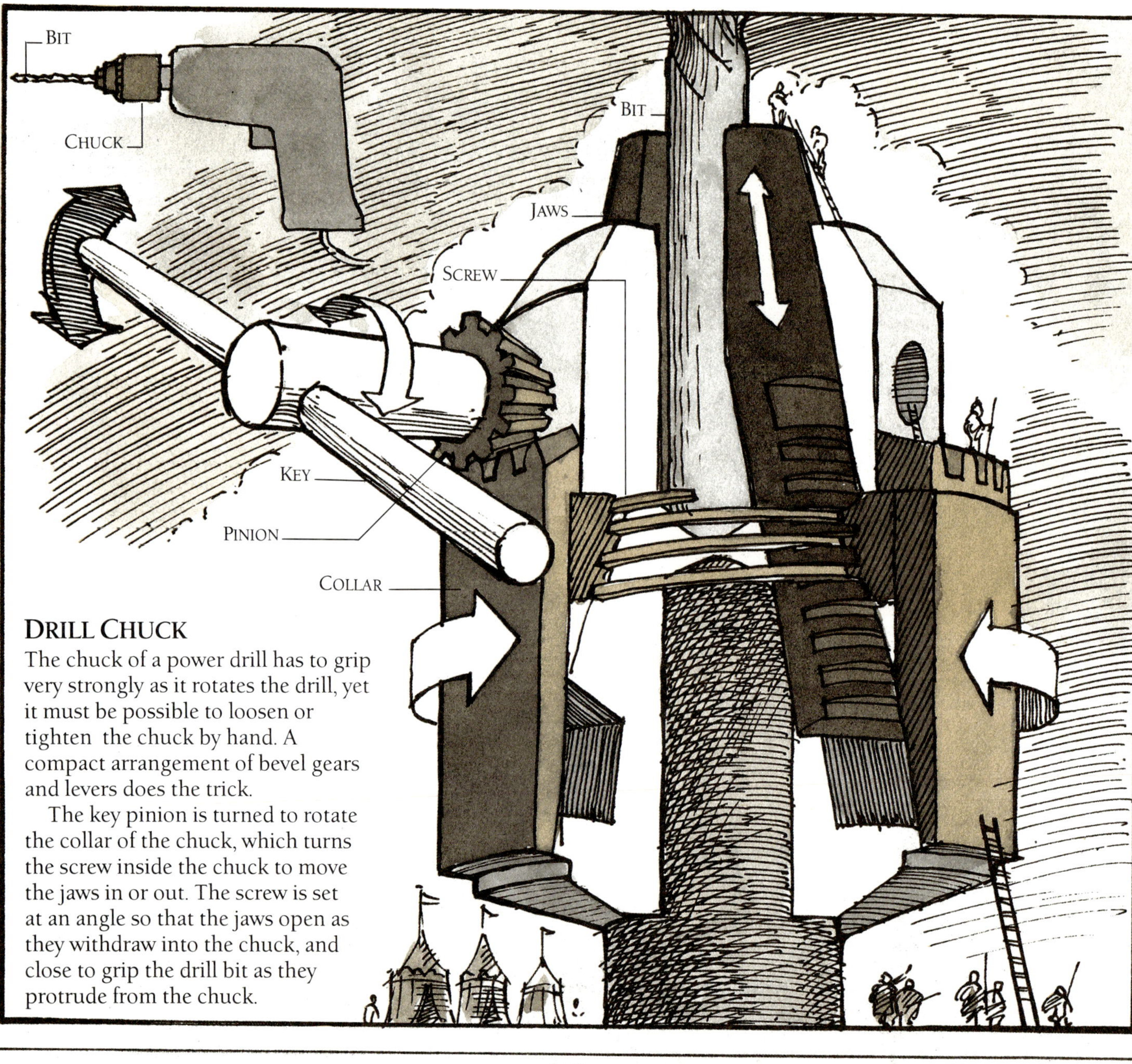

DRILL CHUCK

The chuck of a power drill has to grip very strongly as it rotates the drill, yet it must be possible to loosen or tighten the chuck by hand. A compact arrangement of bevel gears and levers does the trick.

The key pinion is turned to rotate the collar of the chuck, which turns the screw inside the chuck to move the jaws in or out. The screw is set at an angle so that the jaws open as they withdraw into the chuck, and close to grip the drill bit as they protrude from the chuck.

THE DIFFERENTIAL

When a car goes around a corner, the outer driving wheel must be turned at a greater speed than the inner one. This is achieved through the differential. It lies midway between the two driving wheels, linked to each wheel by a half-shaft turned through a bevel gear.

The half-shafts have sun gears connected by free-wheeling planet pinions. When traveling straight, the planet pinions do not spin and drive both half-shafts at the same speed. As the car corners, the planet pinions do spin, driving the sun gears and half-shafts at different speeds.

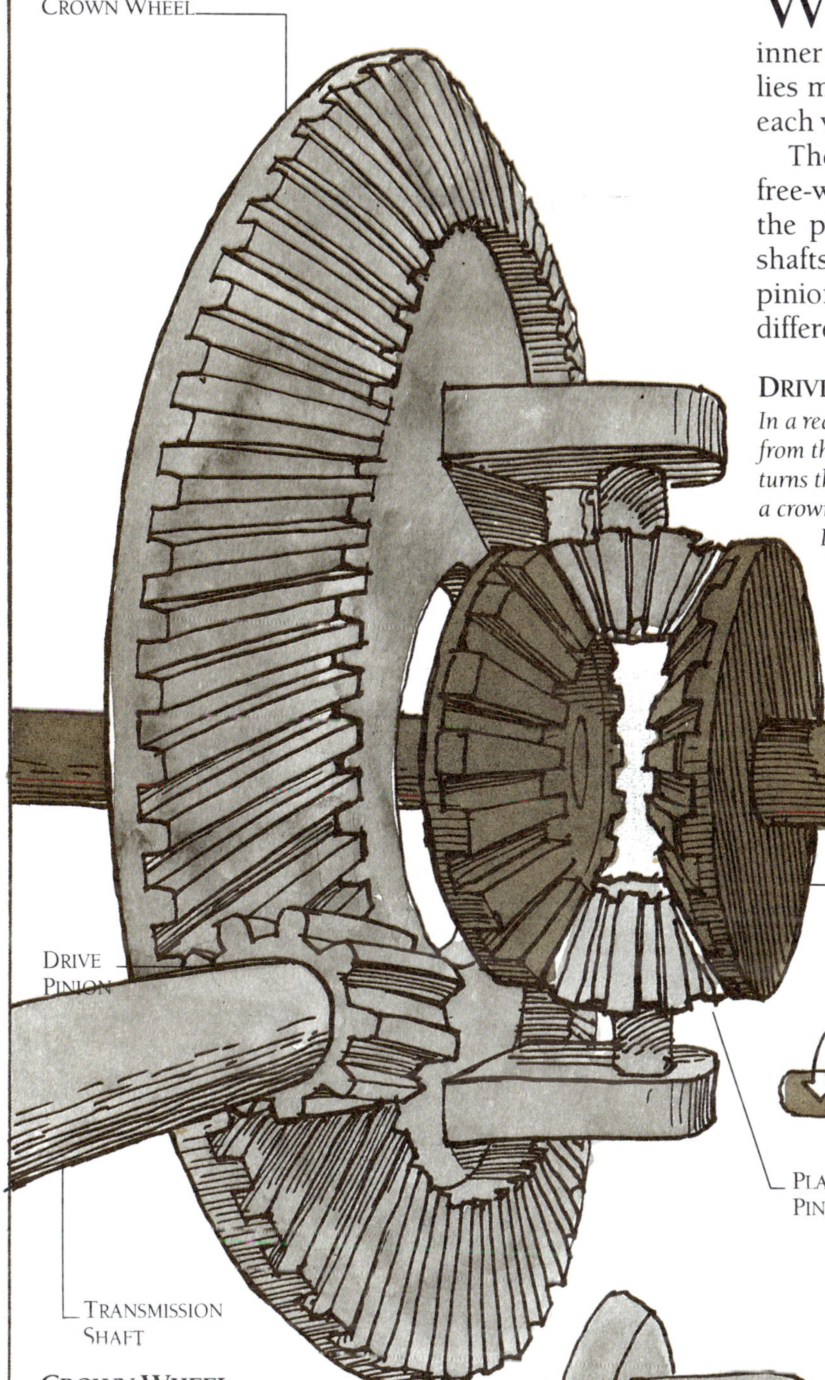

DRIVING THE DIFFERENTIAL

In a rear-wheel drive car, the transmission shaft from the gearbox (see pp.44-5) turns the differential through a crown wheel and pinion. In a front-wheel drive car, the gearbox may drive the differential directly through a pair of spur gears. Four-wheel drive cars have two differentials, one for each pair of wheels.

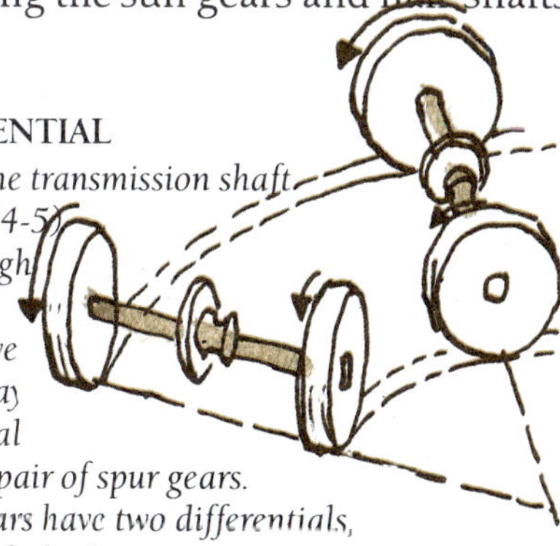

TRAVELING STRAIGHT

The planet pinions circle around within the differential without spinning. They drive both the half-shafts at the same speed.

CROWN WHEEL

The teeth on the crown wheel and the drive pinion are helical, or curved. This allows the transmission shaft to rise up and down slightly if the road surface is uneven.

TURNING A CORNER

The planet pinions both circle around within the differential and spin. The half-shafts now rotate at different speeds.

WORM GEARS

ELECTRIC MIXER

An electric mixer has a pair of contra-rotating beaters, just like an egg beater. However, electric motors rotate at very high speeds and develop heat. The speed therefore has to be reduced when the motor is put to work, rather than increased as in the hand-powered egg beater. A worm gear is used to drive the beater shafts, and a fan attached to the motor shaft blows air over the motor to cool it.

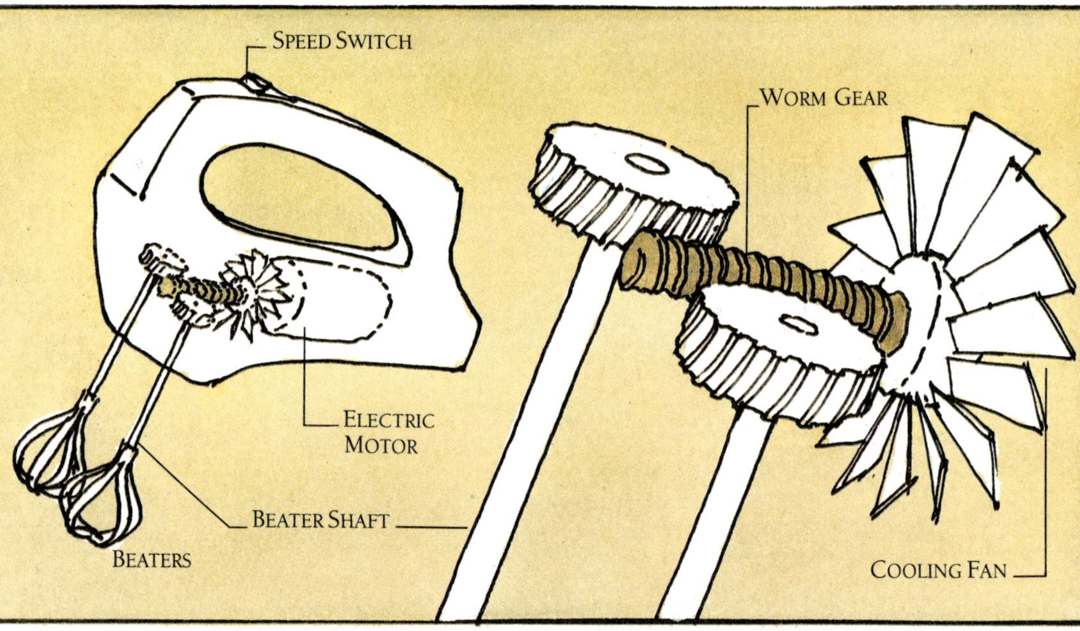

SPEEDOMETER

A car's speedometer uses worm gears to produce an enormous reduction in speed. The final drum of numbers in the distance counter turns just once every hundred thousand miles or kilometers, while the transmission shaft that drives it turns several hundred million times.

The speedometer is driven by a flexible cable. This contains a rotating wire connected to a small drive wheel which is rotated by a large worm on a shaft that drives the wheels. Inside the speedometer, the wire drives the speed indicator through electromagnetic induction (see pp.304-5). Its speed is further reduced by another worm gear to turn the distance counter, which itself contains reducing gears so that each numeral drum rotates at a tenth the speed of its neighbor.

SPEED INDICATOR

The shaft rotates a magnet inside a drag cup.

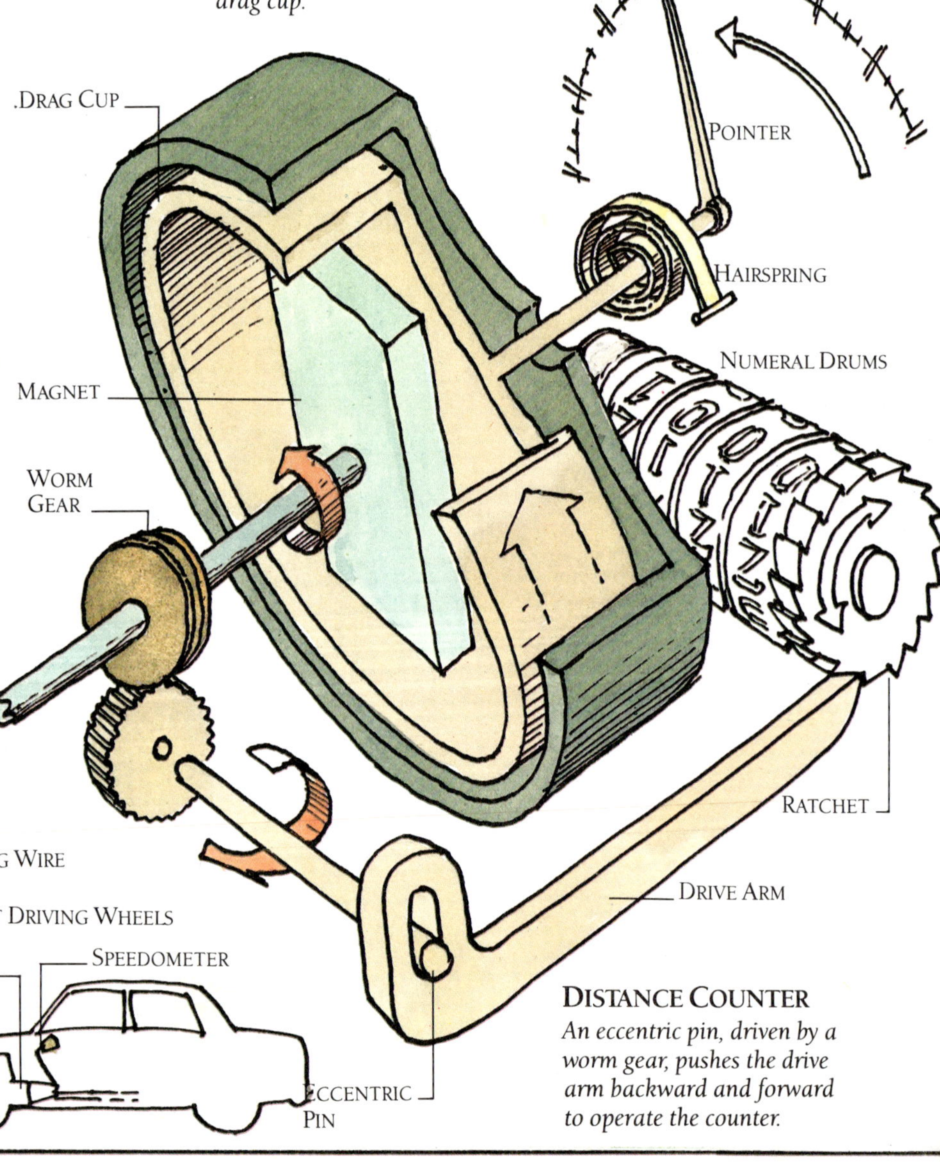

CABLE CONNECTION

The speedometer cable is attached to the gearbox output shaft, transmission shaft or differential, all of which rotate at a speed that depends on the road speed.

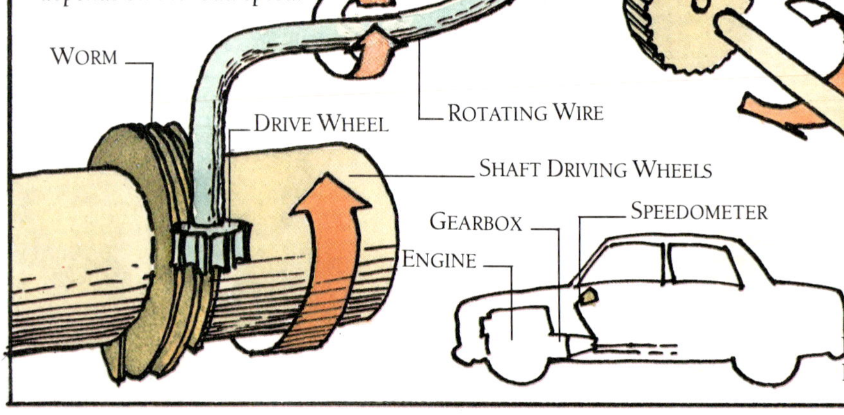

DISTANCE COUNTER

An eccentric pin, driven by a worm gear, pushes the drive arm backward and forward to operate the counter.

LAWN SPRINKLER

A good sprinkler not only produces a fine spray of water but also swings the spray to and fro to water a wide area of grass. No extra source of power is needed, because the mechanism is driven by the movement of the water through the sprinkler, using a system of worm gears.

As the water enters the sprinkler, it drives a turbine at high speed and then rushes to the spray tube. The turbine drives two worm gears that reduce the speed of the turbine to turn a crank at low speed. The crank moves the spray tube slowly to and fro.

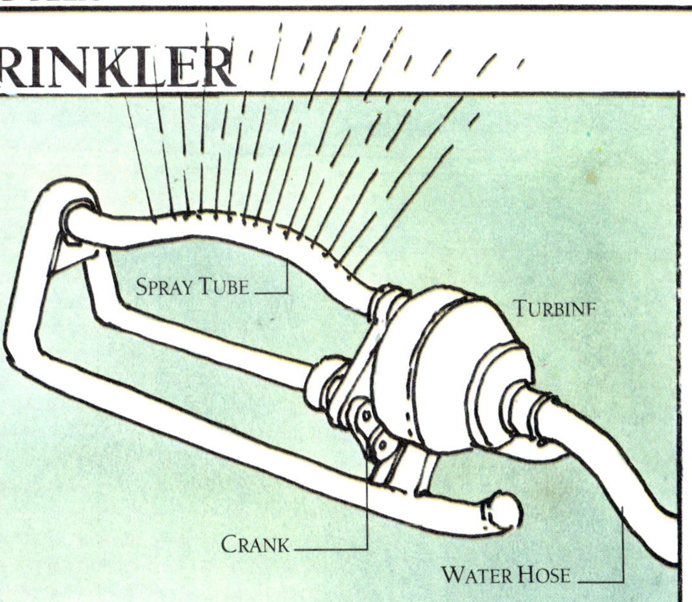

CAMS AND CRANKS

ON AN ANCIENT MACHINE

I have recently come across the remains of an extraordinary machine, the operation of which is here depicted. I believe that the machine was designed to crack the eggs of some huge and now extinct beast. Each egg was shattered by a mammoth-powered hammer, and the broken shell pushed out of the way by a shovel. My discovery prompts two observations: (a) the mammoth merry-go-round may not have been the first industrial use of mammoths after all, and (b) there must have been considerable demand for omelettes of prodigious proportions.

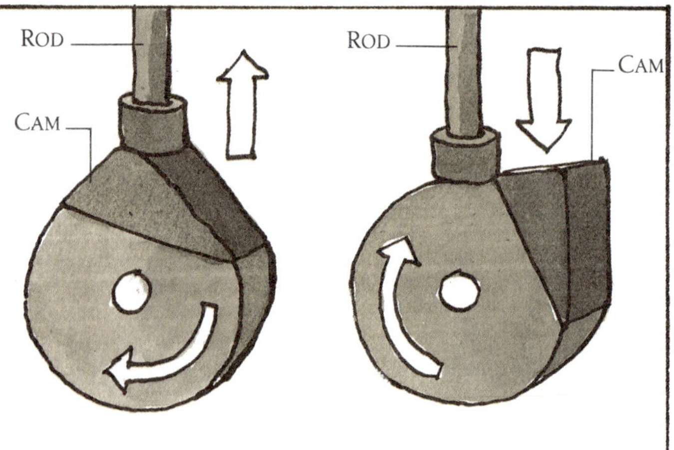

THE CAM

The egg-cracker uses a cam, a device which in its most basic form is simply a fixed wheel with one or more projections. A rod is pressed against the wheel, and as the wheel rotates, the rod moves out and in as the projection passes.

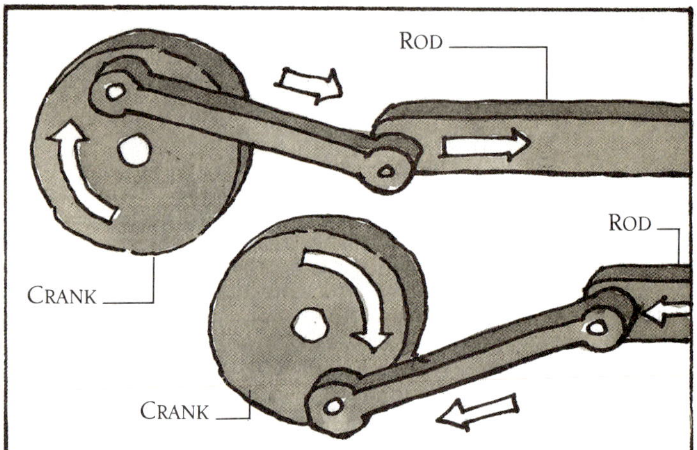

THE CRANK

The shovel is moved by a crank. This is a wheel with a pivot to which a rod is attached. The other end of the rod is hinged so that the rod moves backward and forward as the wheel rotates. Unlike cams, cranks may work in reverse, with the rod making the wheel rotate.

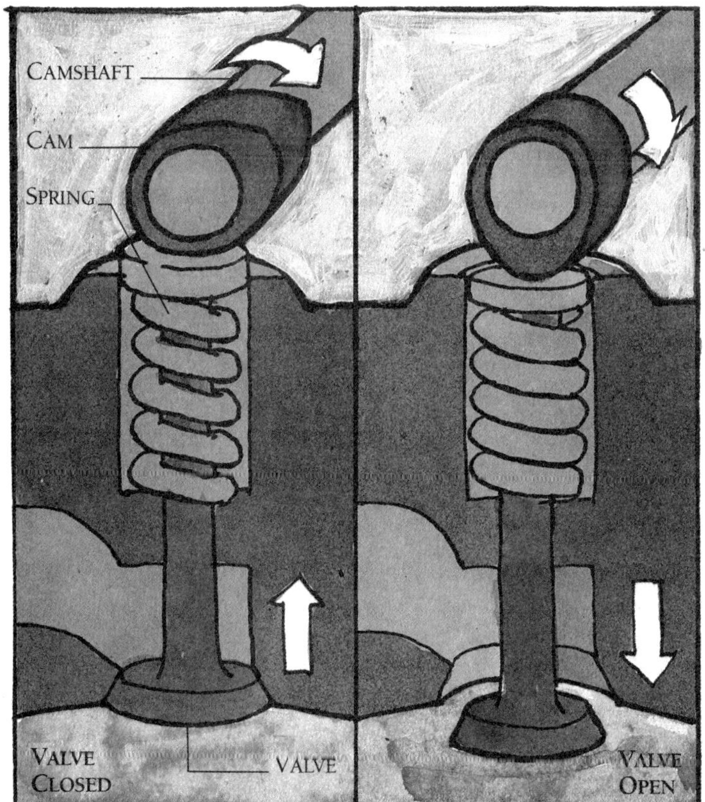

CAR ENGINE CAMSHAFT

Each cylinder of a car engine contains valves that admit the fuel or expel the exhaust gases. Each valve is operated by a cam attached to a rotating camshaft. The cam opens the valve by forcing it down against a spring. The spring then closes the valve until the cam comes around again. The cam may operate the valve directly, as here, or through levers, as shown on the next page.

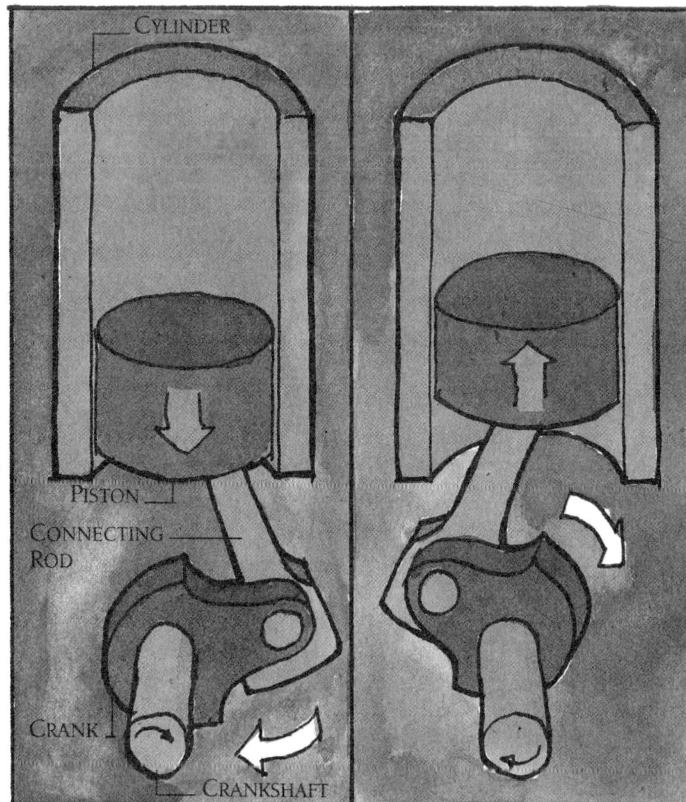

CAR ENGINE CRANKSHAFT

Powered by the explosion of the fuel, a piston moves down inside each cylinder of a car engine. A connecting rod links the piston to a crank on the crankshaft. The rod turns the crank, which then continues to rotate and drives the piston back up the cylinder. In this way, the crankshaft converts the movement of the pistons into rotary power.

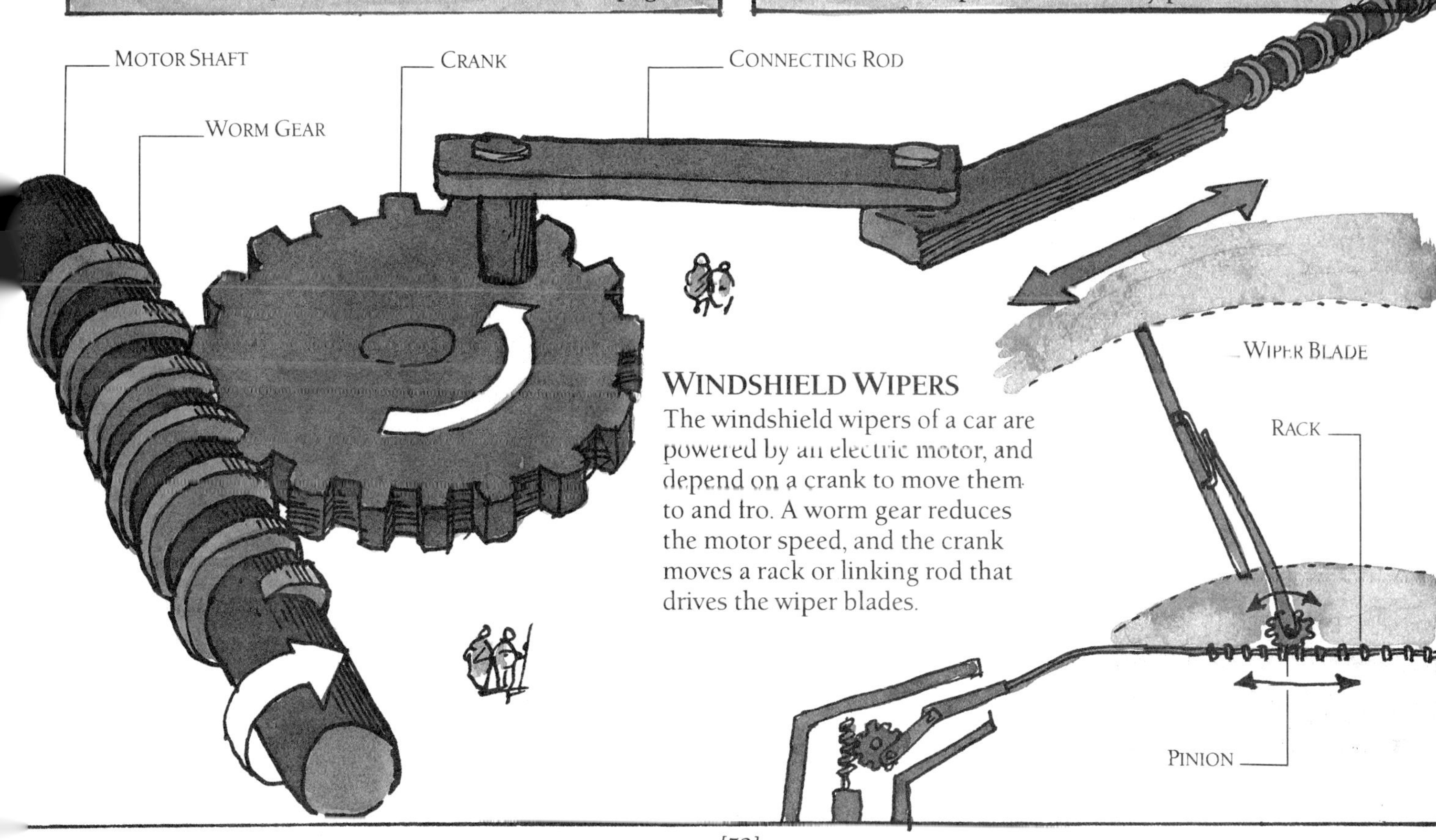

WINDSHIELD WIPERS

The windshield wipers of a car are powered by an electric motor, and depend on a crank to move them to and fro. A worm gear reduces the motor speed, and the crank moves a rack or linking rod that drives the wiper blades.

CAMS AND CRANKS IN THE CAR

Split-second timing is essential for smooth and powerful running in a car engine. It is achieved by the engine's camshaft and crankshaft working in concert.

As the pistons move up and down in the cylinders, they drive the crankshaft which turns the flywheel and, ultimately, the wheels. But, through a chain linkage, the crankshaft also turns the camshaft. As the camshaft rotates, the cams operate the cylinder valves. In an overhead camshaft engine, the cams lie over the valves and move the valves directly. Here, the camshaft is to one side, and it operates the valves through push-rods and rockers. The cams and cranks involved open and close the valves in step with the movements of the pistons to follow the four-stroke cycle (see pp.164-5).

The camshaft and crankshaft may also drive other parts of the engine. A gear wheel on the camshaft, for example, drives the oil pump (see p.132) and the distributor (see p.309).

The Sewing Machine

The sewing machine is a marvel of mechanical ingenuity. Its source of power is the simple rotary movement of an electric motor. The machine converts this into a complex sequence of movements that makes each stitch and shifts the fabric between stitches. Cams and cranks play an important part in the mechanism. A crank drives the needle up and down, while two trains of cams and cranks move the serrated feed-dog that shifts the fabric.

In order to make a stitch, the sewing machine has to loop one thread around another. The first thread passes through the eye of the needle and the second thread is beneath the fabric. As the needle moves up and down, a curved hook rotates to loop the thread and form a stitch. When a stitch has been completed, the feed-dog repositions the fabric so that the next can be made. The amount of fabric moved by the feed-dog can be altered to produce long or short stitches.

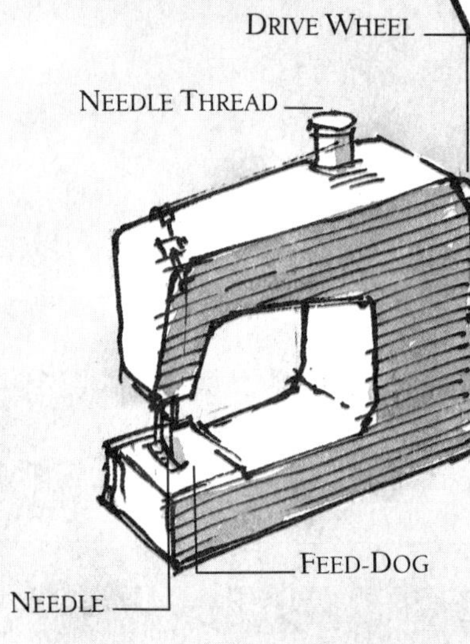

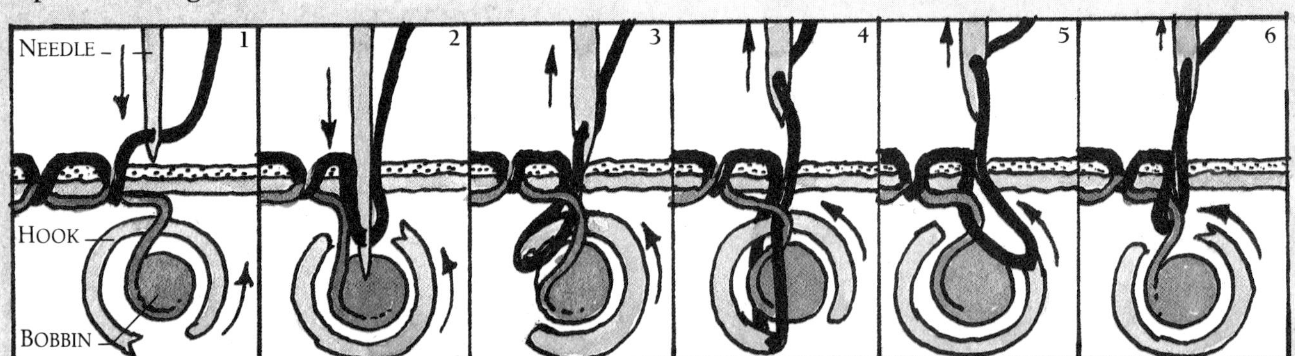

FORMING THE LOOP
The needle, carrying one thread, moves down (1). The other thread is wound on the bobbin in the rotary shuttle below the fabric. The needle pierces the fabric and then moves up, leaving a loop of thread beneath the fabric (2).

HOOKING THE LOOP
The hook on the shuttle catches the loop of needle thread (3). It then pulls the loop around the bobbin and around the bobbin thread (4). The bobbin thread is effectively put through the loop of needle thread.

COMPLETING THE STITCH
The hook continues to turn (5). The loop then slips off the hook as the needle rises above the fabric (6). The needle thread is then pulled tight by a lever on the sewing machine to form the stitch.

THE FEED-DOG

This moves the fabric forward. One train of cams and cranks moves the feed-dog forward and backward, while the other makes it rise and fall. Both are powered by a wheel driven by the electric motor, synchronizing their movements. The feed-dog rises and moves forward between stitches to shift the fabric and then dips and moves back.

PULLEYS

ON MILKING A MAMMOTH

Although it has a rather strong flavor, mammoth milk is rich in minerals and vitamins. I have passed through countless villages of white-toothed, strong-boned folk all of whom attribute their remarkable health to a life of drinking this exceptionally nutritious fluid. The only problem in milking these creatures, besides obtaining enough buckets (they produce an unbelievable amount of milk), is the animals' great reluctance to be touched. It is necessary therefore to raise the mammoth far enough above the ground to deny it any traction. The milker is only safe when the milkee is dangling helplessly.

In many villages, I observed mammoths being lifted in a harness using a number of wheels. These wheels, around which a strong rope traveled, were hung in a given order from a very stout framework. Although the weight to be lifted was often tremendous, a system of wheels greatly reduced the effort required. I noticed that the more wheels the villagers used, the easier it was to lift the weight, but by the same token, it was also necessary to pull in much more rope to get the mammoth up to a sufficient height.

PULLEY POWER

For some, lifting a heavy weight while climbing a ladder poses no problems. For most of us, however, pulling something down is a lot easier than lifting it up.

This change of direction can be arranged with no more than a wheel and a rope. The wheel is fixed to a support and the rope is run over the wheel to the load. A pull downward on the rope can lift the load as high as the support. And because the puller's body weight works downward, it now becomes a help rather than a hindrance. A wheel used in this way is a pulley and the lifting system it makes up is a simple crane.

Single pulleys are used in machines where the direction of a movement must be changed, as for example in an elevator (see p.65) where the upward movement of the elevator must be linked to the downward movement of a counterweight.

In an ideal pulley, the effort with which the rope is pulled is equal to the weight of the load. In practice, the effort is always slightly more than the load because it has to overcome the force of friction (see pp.86-7) in the pulley wheel as well as raise the load. Friction reduces the efficiency of all machines in this way.

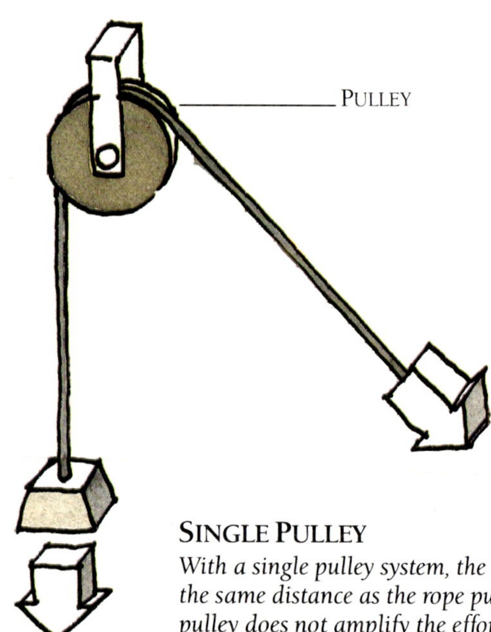

SINGLE PULLEY
With a single pulley system, the load moves the same distance as the rope pulled in. The pulley does not amplify the effort with which the rope is pulled – it just allows the pulling to be done in a downward direction.

One wise old milker informed me that to avoid unnecessary delay and expenditure of energy, the gender of the mammoth should be checked before *attaching the harness.*

CONNECTED PULLEYS

As well as changing a pulling force's direction, pulleys can also be used to amplify it, just like levers. Connecting pulley wheels together to make a compound pulley enables one person to raise loads many times their own weight.

In a system with two pulleys, one pulley is attached to the load and the other to the support. The rope runs over the upper pulley, down and around the lower pulley and back up to the upper pulley, where it is fixed. The lower pulley is free to move and as the rope is pulled, it raises the load. This arrangement of pulleys causes the load to move only half as far as the free end of the rope. But in return, the force raising the load is doubled. As with levers, the distance moved is traded off against force – much to the puller's advantage.

The amount by which a compound pulley amplifies the pull or effort to raise a load depends on how many wheels it has. Ideally, the amplification is equal to the number of sections of rope that raise the lower set of pulleys attached to the load. In practice, the effort has to overcome friction in all the pulleys and raise the weight of the lower set of pulleys as well as the load. This reduces the amplification of the effort.

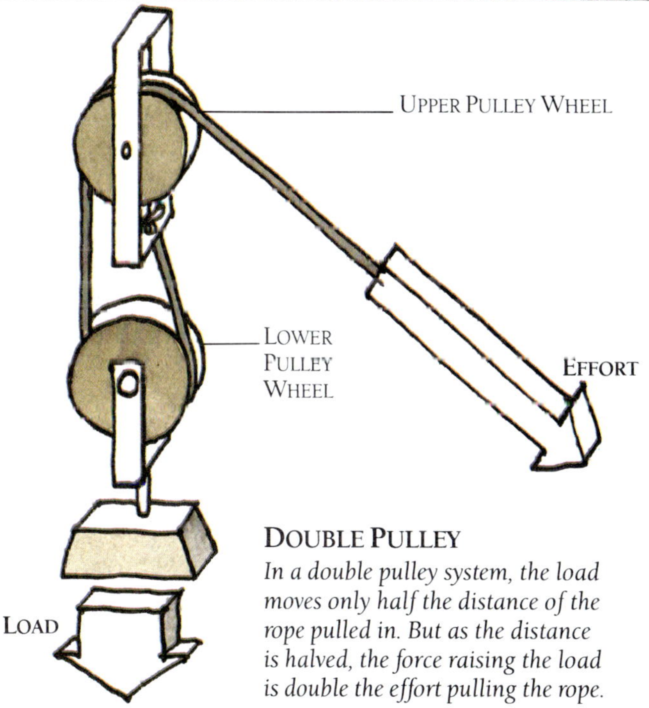

DOUBLE PULLEY

In a double pulley system, the load moves only half the distance of the rope pulled in. But as the distance is halved, the force raising the load is double the effort pulling the rope.

CHAIN HOIST

The chain hoist consists of an endless chain looped around three pulleys. The upper two pulleys are fixed together, while the load hangs from a lower pulley, which is supported by a loop of chain. The load remains still unless the chain is moved. Just how much effort is needed to move the load depends on the difference in diameter between the two upper pulleys.

RAISING AND LOWERING THE HOIST

When the chain is pulled so that the paired pulleys rotate counterclockwise (left), the larger wheel pulls in more chain than the smaller wheel lets out, magnifying the pull exerted and raising the load a shorter distance. When the chain moves in the reverse direction (below), the load is lowered.

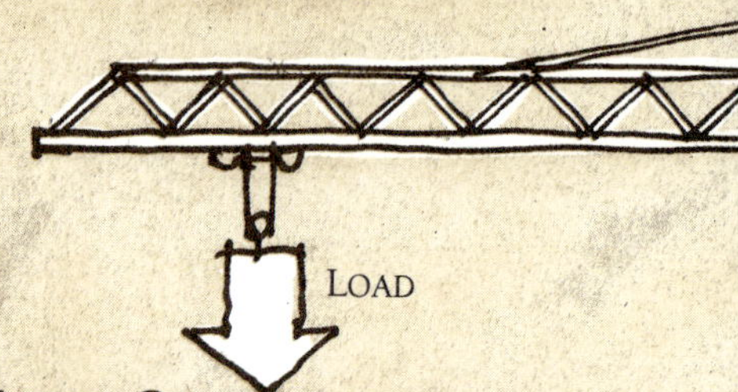

TOWER CRANE

The tower crane is a modern equivalent of the shadoof, using a counterweight to balance its load in the same way.

SHADOOF

This water-raising machine, invented in antiquity, has a counterweight at one end of a pivoted beam which balances a container of water at the other end. When full, the container can be raised with little more than a light touch.

FORK-LIFT TRUCK

The heavy counterweight at the rear of a fork-lift truck helps raise a load high into the air by preventing the truck from toppling forward.

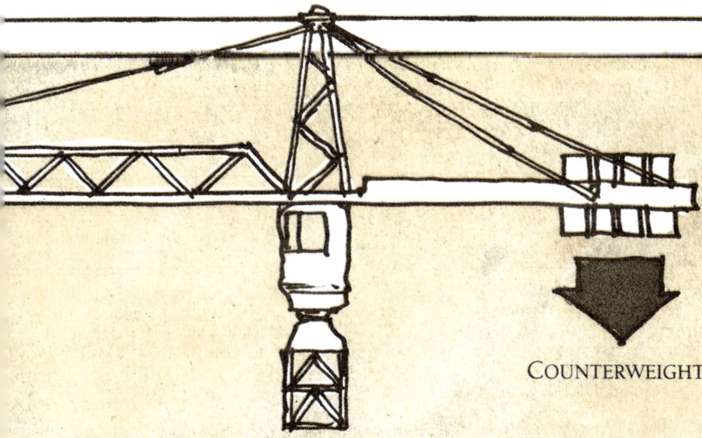

COUNTERWEIGHTS

Cranes and other lifting machines often make use of counterweights in raising loads. The counterweight balances the weight of the load so that the machine's motor has only to move the load and not to support it. The counterweight may also stop the machine tipping over as the load leaves the ground. In accordance with the principle of levers (see p.22), a heavy counterweight placed near the fulcrum of a machine such as a crane has the same effect as a lighter counterweight positioned further away.

MOBILE CRANE

Once on site, a mobile crane uses outrigger beams and hydraulic jacks to relieve the suspension of the strain during lifting. The telescopic boom with its block and tackle can swivel around and extend far outward, secured by the counterweight at the base of the boom.

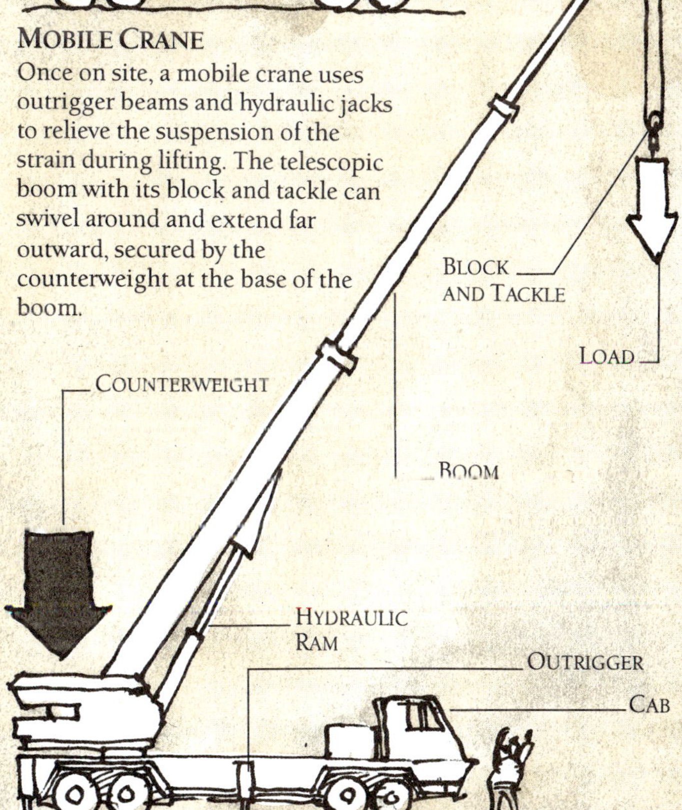

BLOCK AND TACKLE

The block and tackle is a system of pulleys that is compact yet able to raise substantial loads. It is commonly used at the end of the boom of a crane to increase the force of the crane's motor in lifting a load.

The system contains one rope wound around two separate sets of pulleys. The pulleys in each set are free to rotate independently on the same axle. The upper set is fixed to a support such as a boom, while the lower set is attached to the load. Pulling the rope raises the lower set of pulleys. The magnification of the force that the block and tackle produces is equal to the number of pulley wheels it contains.

This block and tackle contains five pulleys in each set plus a guide wheel above. The load is raised by ten pulleys, so the block and tackle increases the force applied to it by ten times.

TOWER CRANE

The aptly-named tower crane uses several sets of pulleys for precise lifting work over a wide area.

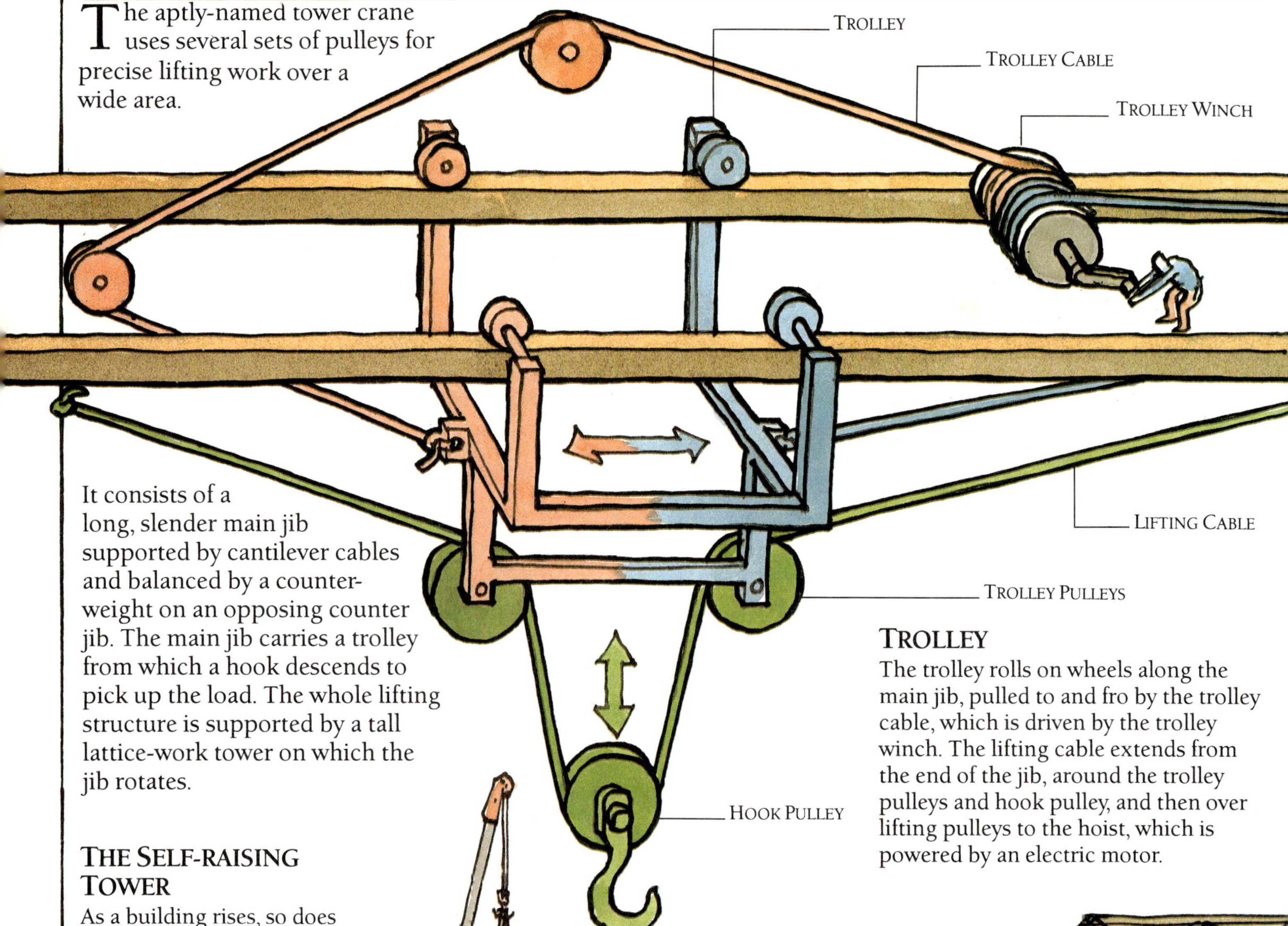

It consists of a long, slender main jib supported by cantilever cables and balanced by a counter-weight on an opposing counter jib. The main jib carries a trolley from which a hook descends to pick up the load. The whole lifting structure is supported by a tall lattice-work tower on which the jib rotates.

TROLLEY

The trolley rolls on wheels along the main jib, pulled to and fro by the trolley cable, which is driven by the trolley winch. The lifting cable extends from the end of the jib, around the trolley pulleys and hook pulley, and then over lifting pulleys to the hoist, which is powered by an electric motor.

THE SELF-RAISING TOWER

As a building rises, so does the crane that helps in its construction. Tower cranes do not expand telescopically like mobile cranes; instead, they extend themselves section by section. They do this by using a hydraulically operated climbing frame which raises the cab to make room for additional sections.

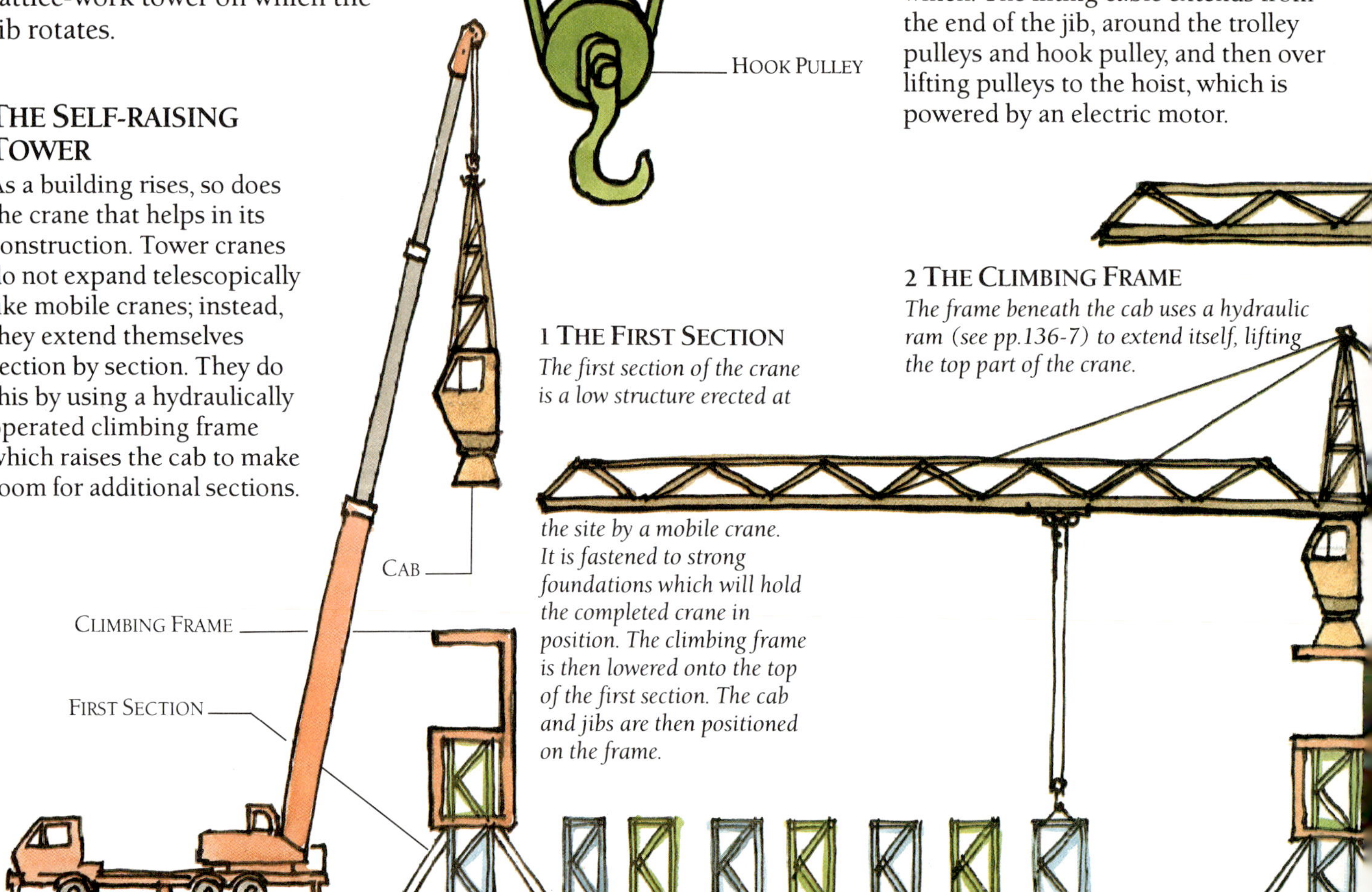

1 THE FIRST SECTION

The first section of the crane is a low structure erected at the site by a mobile crane. It is fastened to strong foundations which will hold the completed crane in position. The climbing frame is then lowered onto the top of the first section. The cab and jibs are then positioned on the frame.

2 THE CLIMBING FRAME

The frame beneath the cab uses a hydraulic ram (see pp.136-7) to extend itself, lifting the top part of the crane.

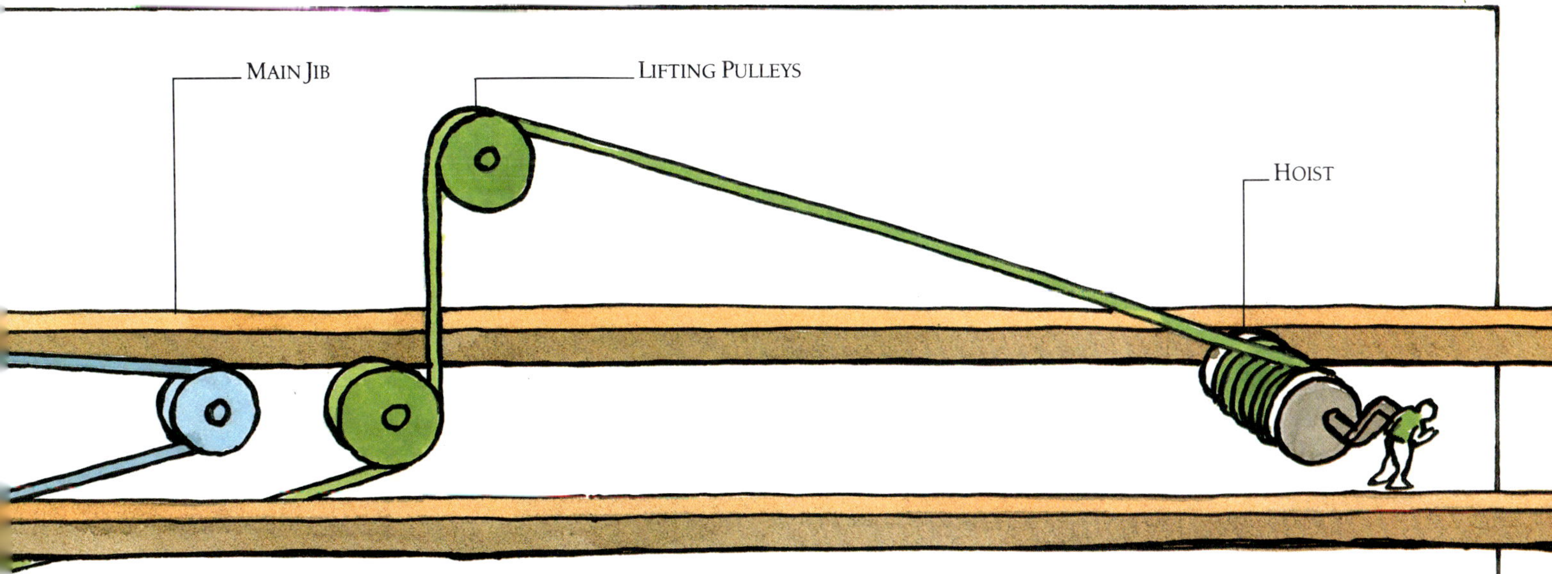

THE HOIST

The hoist winds the lifting cable in and out, raising and lowering the hook. The trolley pulleys and hook pulley together double the force exerted by the hoist by doubling the length of cable used. Extra pulleys may be incorporated to increase the force still further by quadrupling the length of cable.

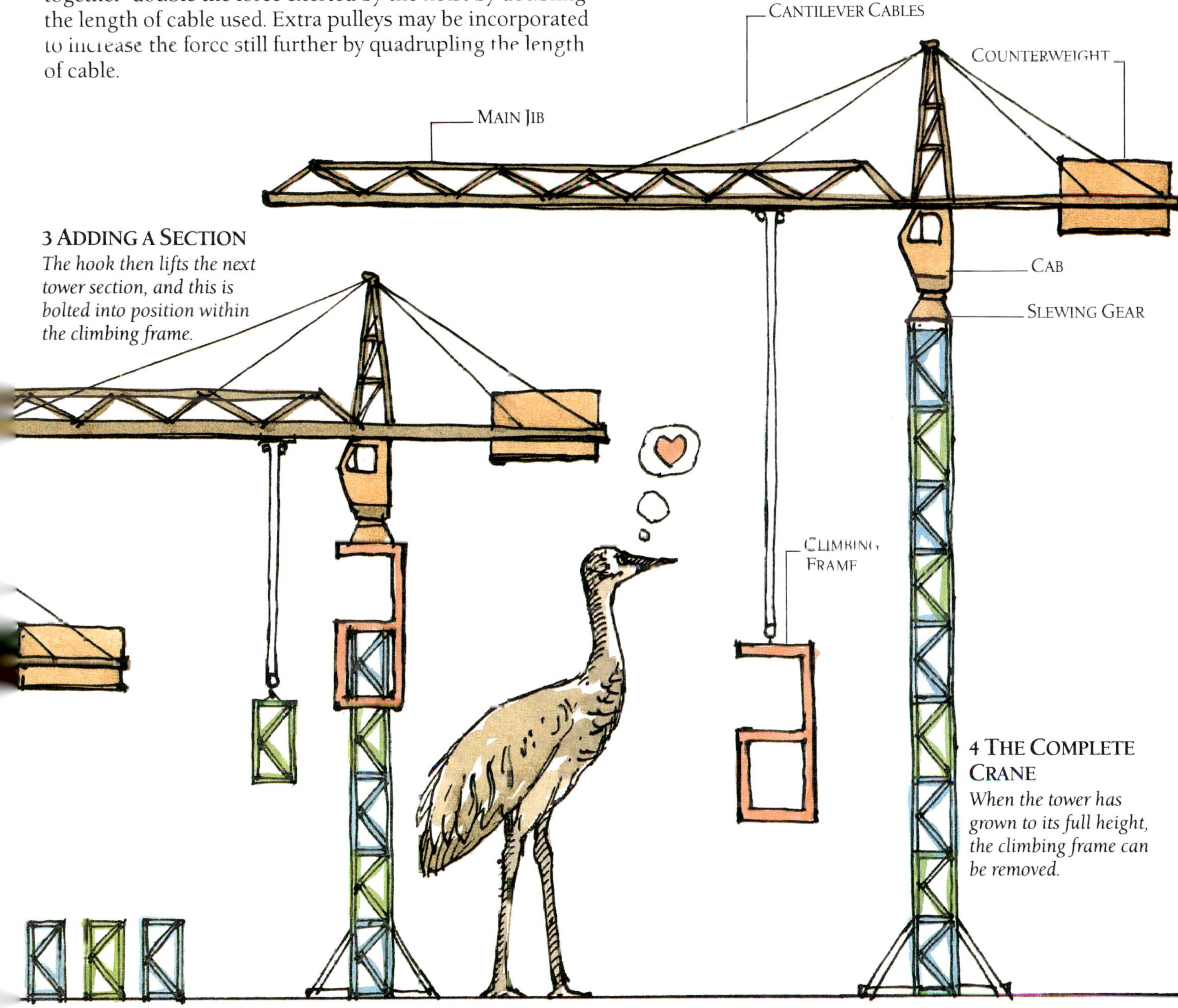

3 ADDING A SECTION

The hook then lifts the next tower section, and this is bolted into position within the climbing frame.

4 THE COMPLETE CRANE

When the tower has grown to its full height, the climbing frame can be removed.

ESCALATOR AND ELEVATOR

Escalators and elevators are both lifting machines that make use of pulleys and counterweights. This is obvious in the elevator, where the cable supporting the elevator car runs over a pulley to a counterweight. The pulley also drives the cable. Although it is not so immediately apparent, the escalator works in a similar way. A drive wheel moves a chain attached to the stairs, while the returning stairs act as a counterweight.

ESCALATOR

Escalator stairs are connected to an endless chain that runs around a drive wheel. The wheel is powered by an electric motor at the top of the escalator. The descending half of the stairs acts as a counterweight to the ascending half, so that the motor moves only the weight of the people riding.

Every stair has a pair of wheels on each side and each pair runs on two rails beneath the stair. The rails are in line except at the top and the bottom of the escalator. Here, the inner rail goes beneath the outer rail so that each stair moves to the level of the next stair. In this way, the stairs fold flat for people to get on and off.

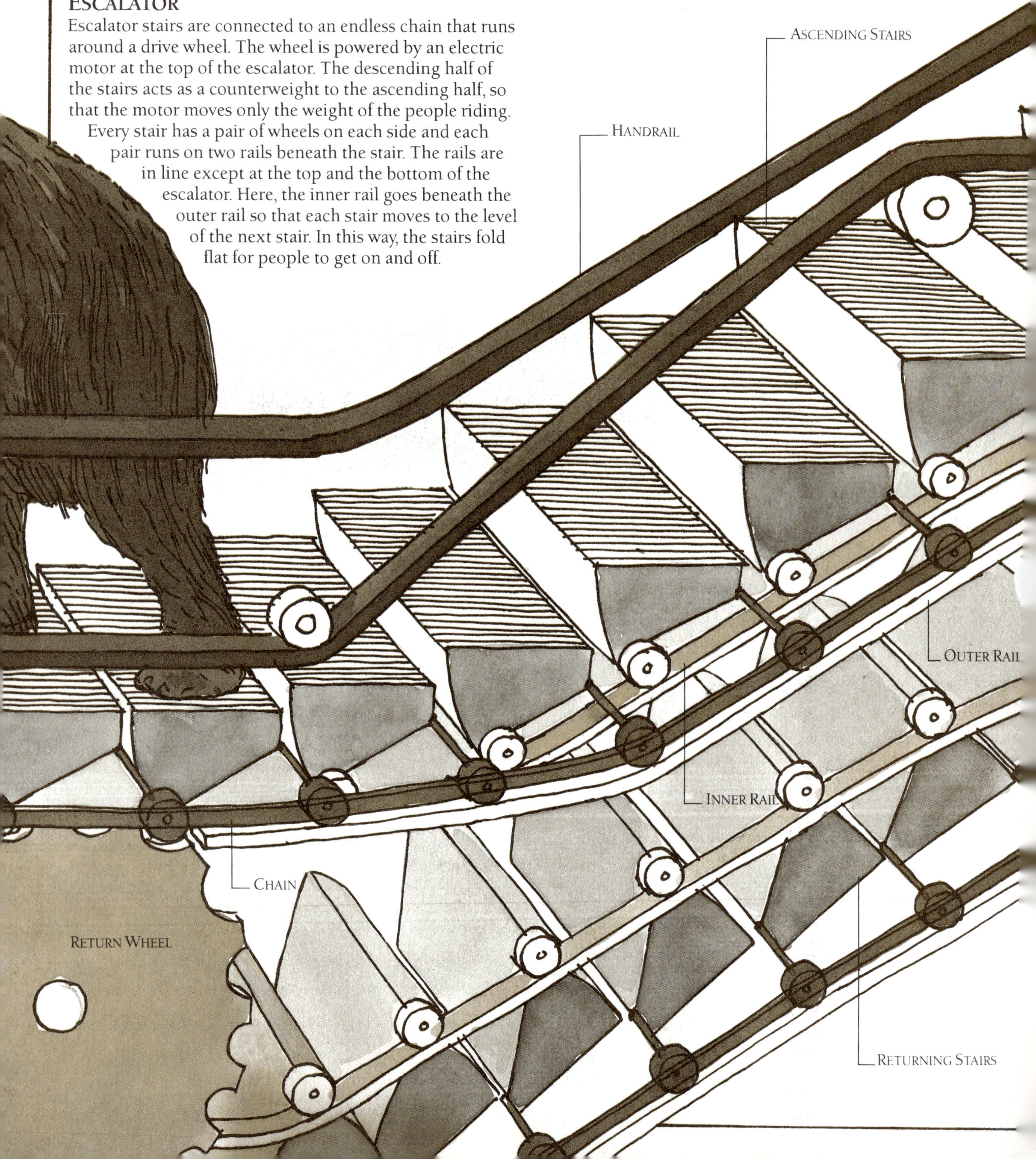

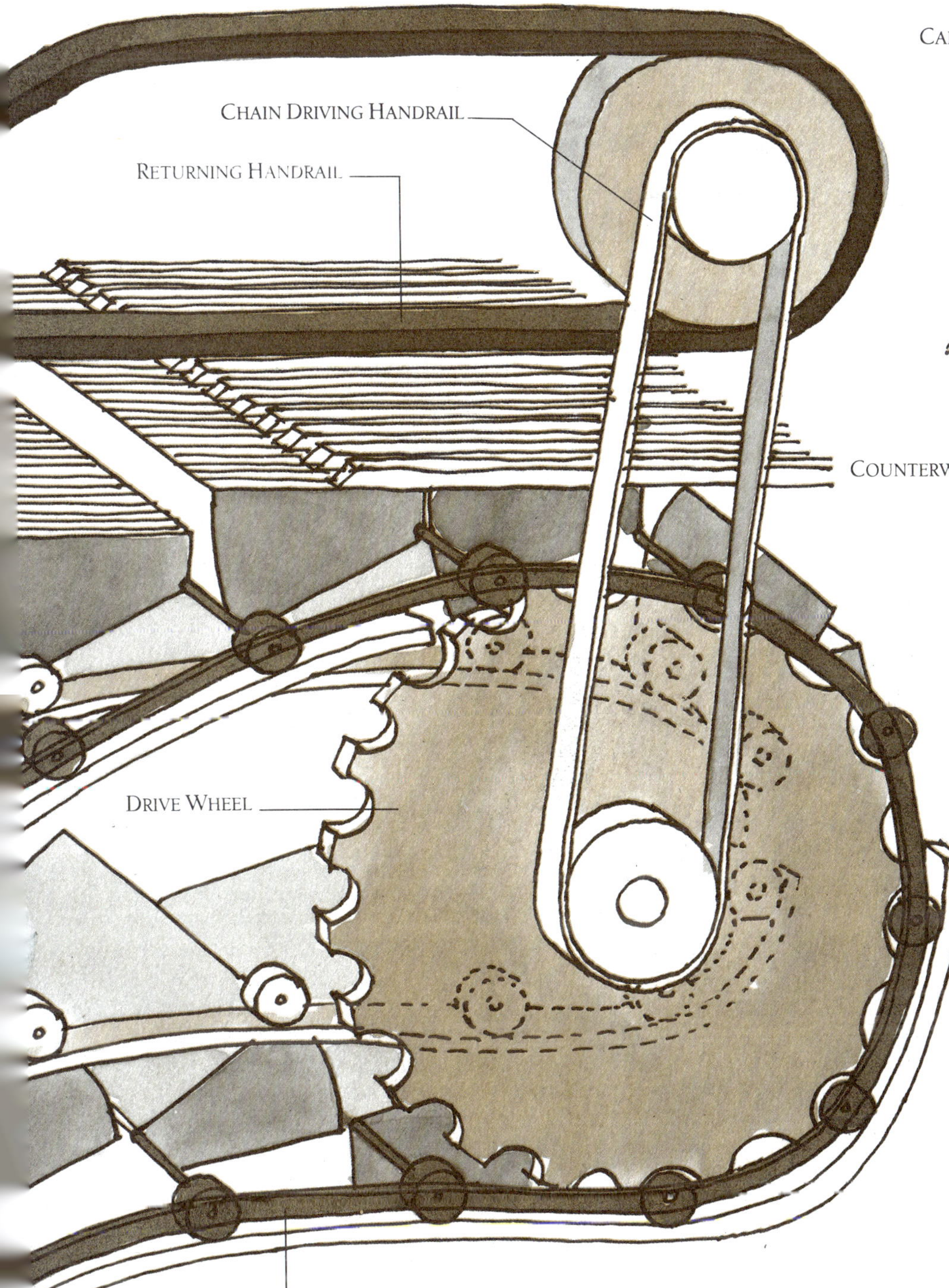

ELEVATOR

An elevator is a single pulley lifting machine. The car is raised or lowered by a cable running over a pulley at the top of the elevator shaft. At the other end of the cable is a counterweight that balances the weight of the elevator car plus an average number of passengers. Both car and counterweight run up and down the shaft on guide rails. An electric motor drives the pulley to move the car, needing only enough power to raise the difference in weight between car and its passengers and the counterweight.

BELOW THE ESCALATOR

The weight of the stairs returning to the foot of the escalator offsets the weight of the stairs traveling back up to the top. All the motor has to lift is the weight of the passengers.

SCREWS

ON THE INTELLIGENCE OF MAMMOTHS

I have recently unearthed a document which, I believe, proves beyond doubt the much-debated intellectual capacity of mammoths. One day, the document records, while seeking some good to do, a knight and his mammoth came upon a damsel imprisoned at the top of a stone tower. The tower contained no doors and only tiny windows. The knight attempted to rescue the damsel with a short ladder but his armor was so heavy that he found the climb impossible. Next it appears that he built a long ramp by tying several planks together. Unfortunately, the knight was no good at tying knots.

NUTS AND BOLTS

The screw is a heavily disguised form of inclined plane, one which is wrapped around a cylinder – just as the knight's ramp encircles the tower. As we have already seen on p.14, inclined planes alter force and distance. When something moves along a screw thread, like a nut on a bolt, it has to turn several times to move forward a short distance. As in a linear inclined plane, when distance decreases, force increases. A nut therefore moves along the bolt with a much greater force than the effort used to turn it.

A nut and bolt hold objects together because they grip the object with great force. Friction (see pp.86-7) stops the nut working loose.

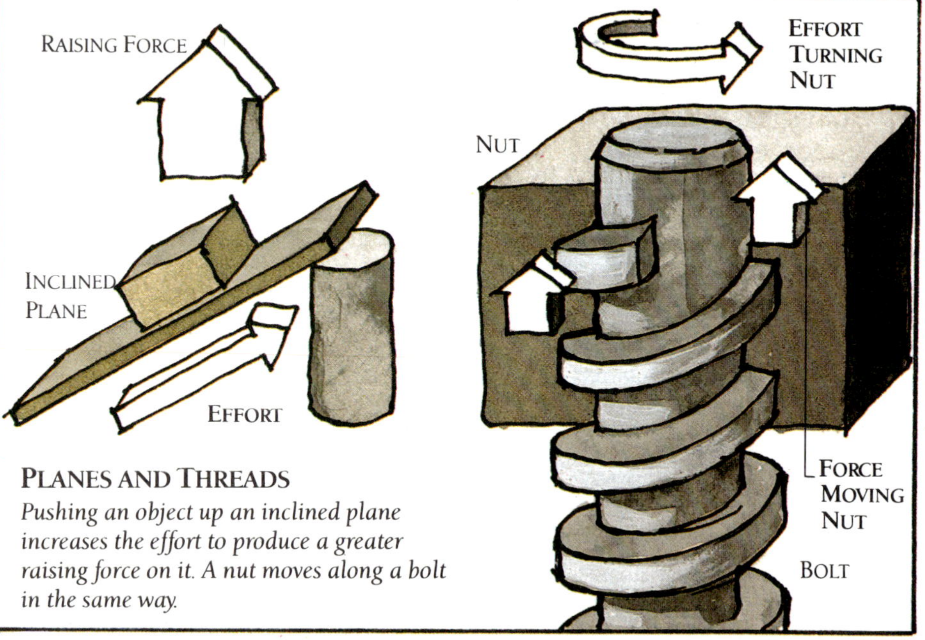

PLANES AND THREADS
Pushing an object up an inclined plane increases the effort to produce a greater raising force on it. A nut moves along a bolt in the same way.

The knight's next idea was to assemble the planks into another ramp, and to fix it in a spiral around the tower. But the ramp was not long enough to reach the damsel.

At this point, the trusty mammoth acted. He picked up a nearby tree trunk, inserted it into one of the windows and turned the entire tower. Uncertain of what was going on, the knight joined in. To his amazement, the end of the ramp started to dig into the soil. By turning the tower many times they slowly screwed it into the ground. Soon the top of the tower was within easy reach of the ladder, and the dizzy damsel skipped to freedom.

SCREWS

Straight inclined planes are often used as wedges, in which the plane moves to force a load upward. Spiral inclined planes can work like wedges too. In most kinds of screws, the screw turns and moves itself into the material – like the damsel's tower. As with the nut and bolt, the turning effort is magnified so that the screw moves forward with an increased force. The force acts on the material to drive the screw into it.

As in the case of the nut and bolt, friction acts to hold the screw in the material. The friction occurs between the spiral thread and the material around it. It is strong because the spiral thread is long and the force between the thread and material is powerful.

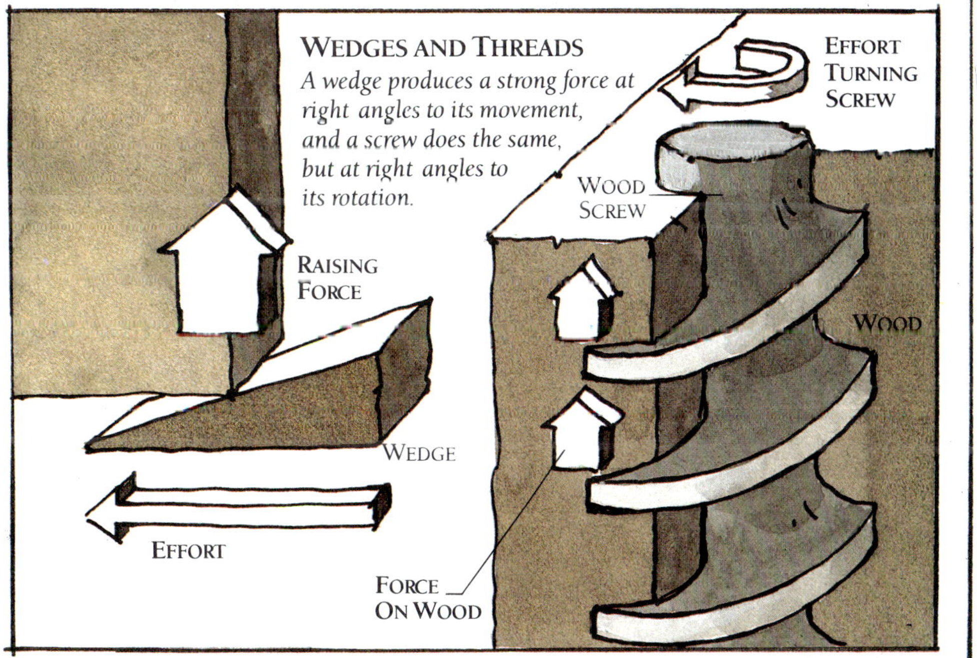

The Screw at Work

Wood Screw
The thread of a wood screw pushes strongly against the wood as it turns and drives itself into the wood. The screwdriver helps to increase the driving force even more.

Nut and Bolt
The thread forces a nut and bolt together. The turning force is increased by the leverage of a wrench.

Screw Jack
A screw jack uses a screw mechanism to lift a car. The handle may move fifty times further than the car, so the force on the car is fifty times greater than the effort on the handle.

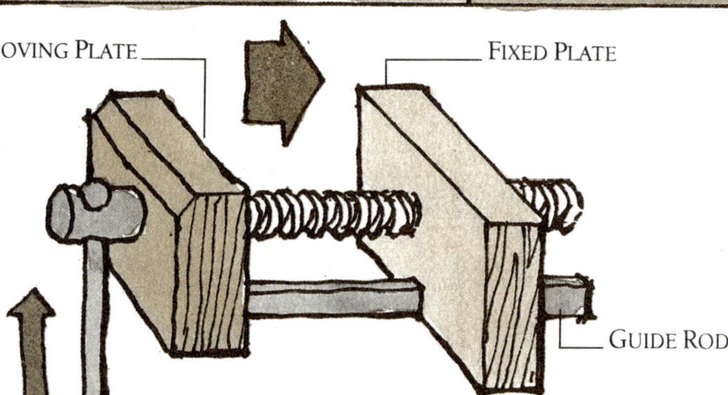

Vice
The vice uses a screw to grip an object tightly on a work surface.

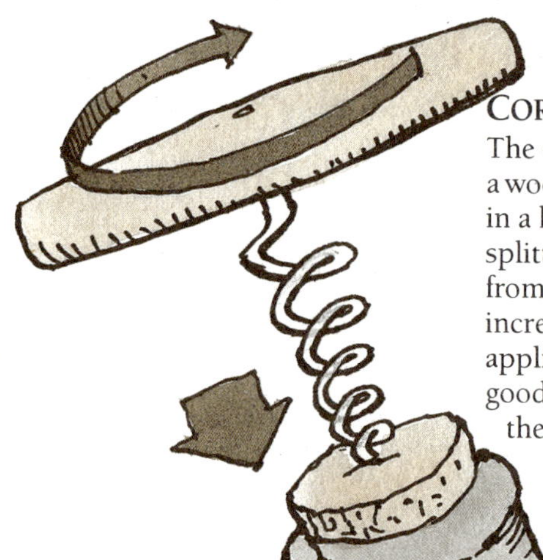

Corkscrew
The corkscrew works like a wood screw, but is shaped in a helix to stop the cork splitting when it is pulled from the bottle. The handle increases the turning force applied, and provides a good grip for extracting the cork.

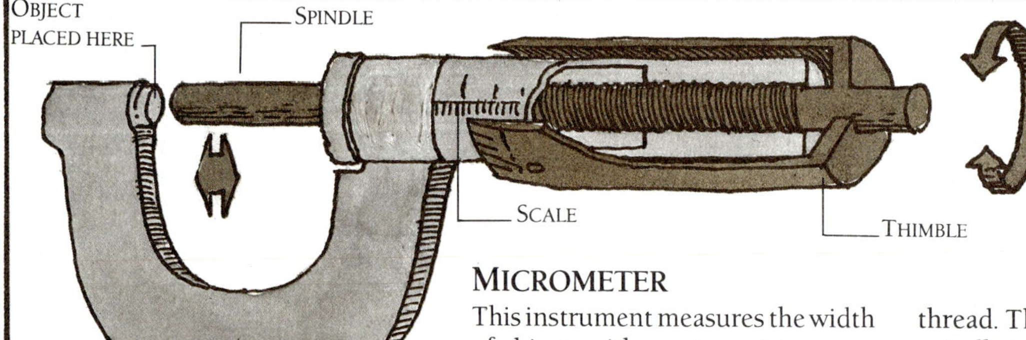

Thimble
The thimble turns on a ratchet mechanism. The ratchet stops the spindle moving forward when it touches the object.

Micrometer
This instrument measures the width of objects with great precision. The object is placed in the micrometer and the thimble turned until the spindle touches the object. The spindle and thimble gradually move along a screw thread. The movement of the spindle is read on a scale, while the graduations on the thimble itself show small fractions of a revolution. Added together, the two figures give a highly accurate measurement.

E FAUCET

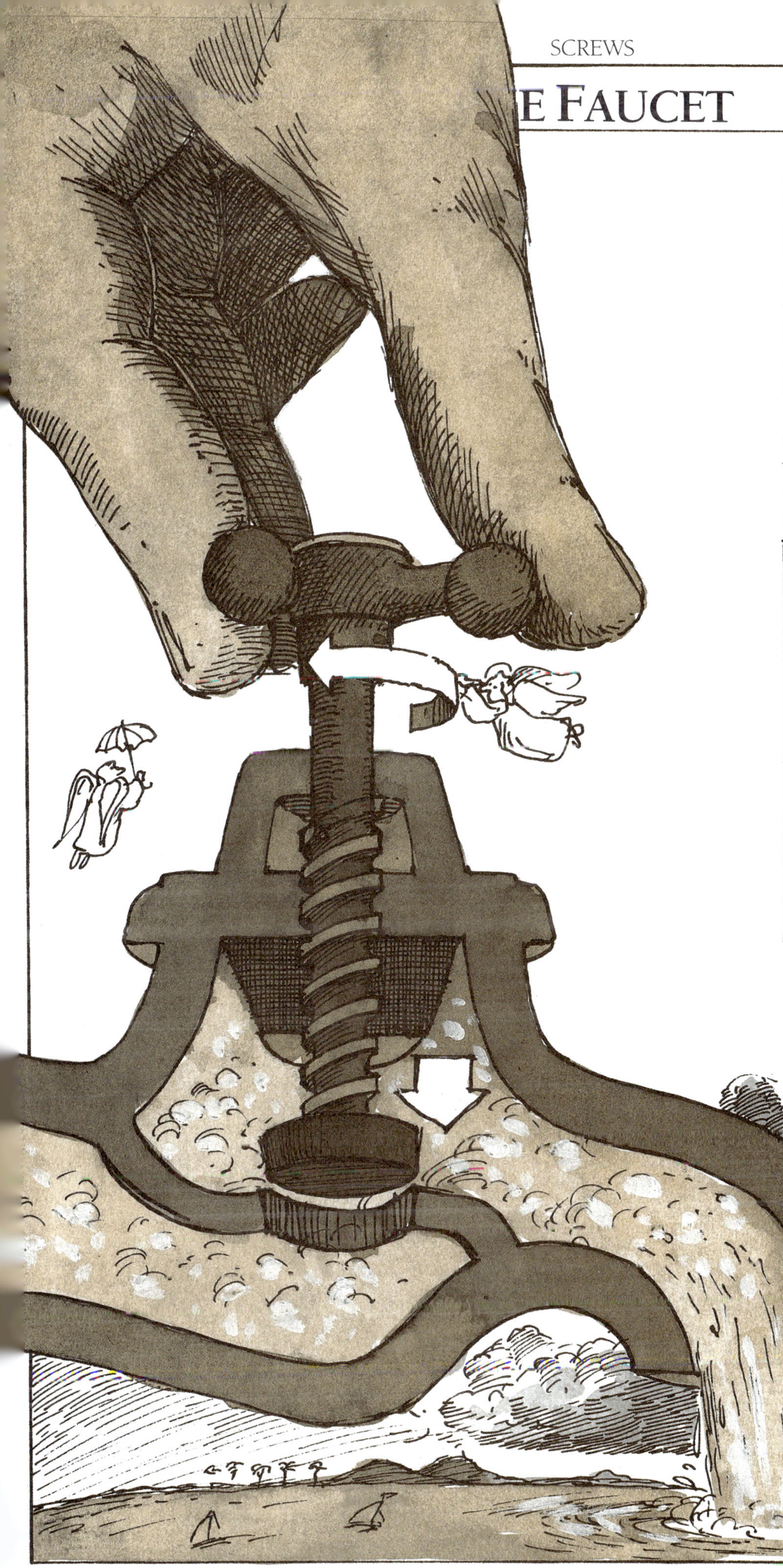

If you've ever tried to stop water flowing from a faucet with a finger, you'll know just how much pressure the water can exert. But a faucet controls the flow with little effort, using a screw (aided by the use of the wheel and axle in the handle) to drive the washer down against the water flow with great force. Once tightened, friction (see pp.86-7) acts on a screw thread to prevent the screw working loose. A steep pitch on the screw thread minimizes the turning needed to work the faucet.

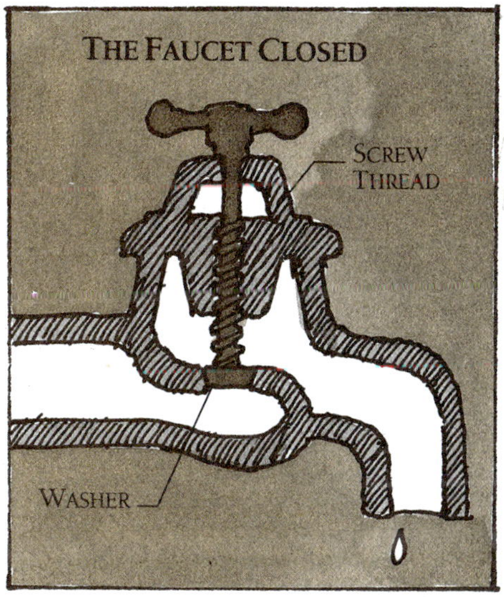

DRILLS AND AUGERS

In drills and augers, the screw is used as a means of carrying loose material. As a drill cuts forward into a material with its sharp point, it also channels waste away backward along along its screw-shaped grooves. In large-diameter drills, the grooves that remove waste material are more pronounced and these give the drill a corkscrew shape.

BRACE AND BIT

When a lot of force is needed – for example, in drilling a wide-diameter hole – an ordinary hand drill will grind to a halt. The answer is a brace and bit. The bowed handle enables the bit to be turned with great leverage.

HAND DRILL

A hand drill uses a bevel gear (see p.41) to step up the speed at which the bit rotates. One bevel gear transmits the turning force, while the other freewheels. Hand drills are fast, but not very powerful.

POWER DRILL

An electric power drill has gears to drive the bit at high speed. It may also have an impact mechanism that hammers the drill bit through a tough material.

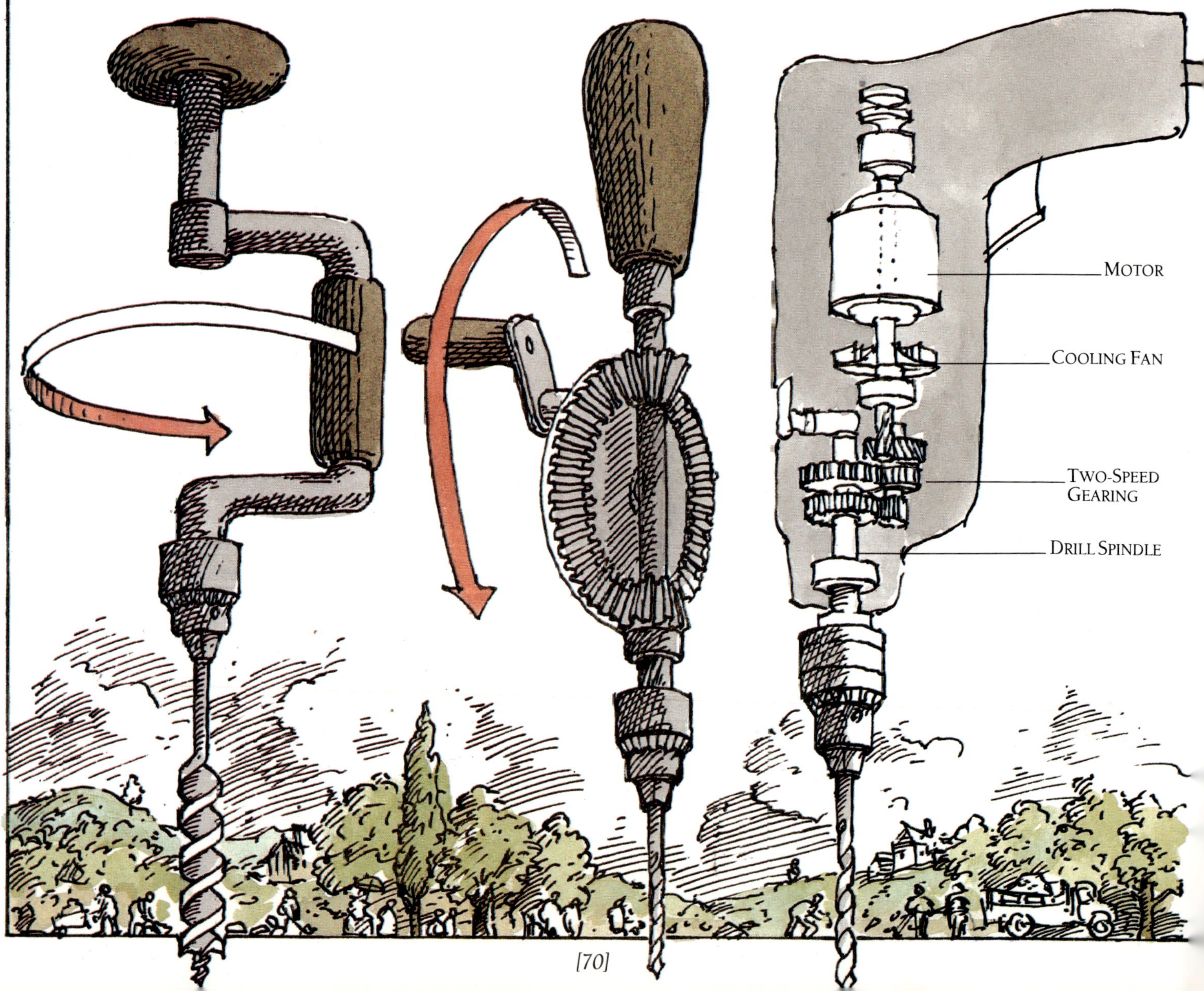

MEAT GRINDER

As anyone who has trapped their finger in one will know, a kitchen grinder can reduce even the toughest chunks of meat to shreds. Turning the handle turns the cutting blades and also an auger which forces the meat into the cutters. The wheel and axle action of the handle combines with the action of the auger to magnify the turning force, moving and cutting the meat with tremendous force.

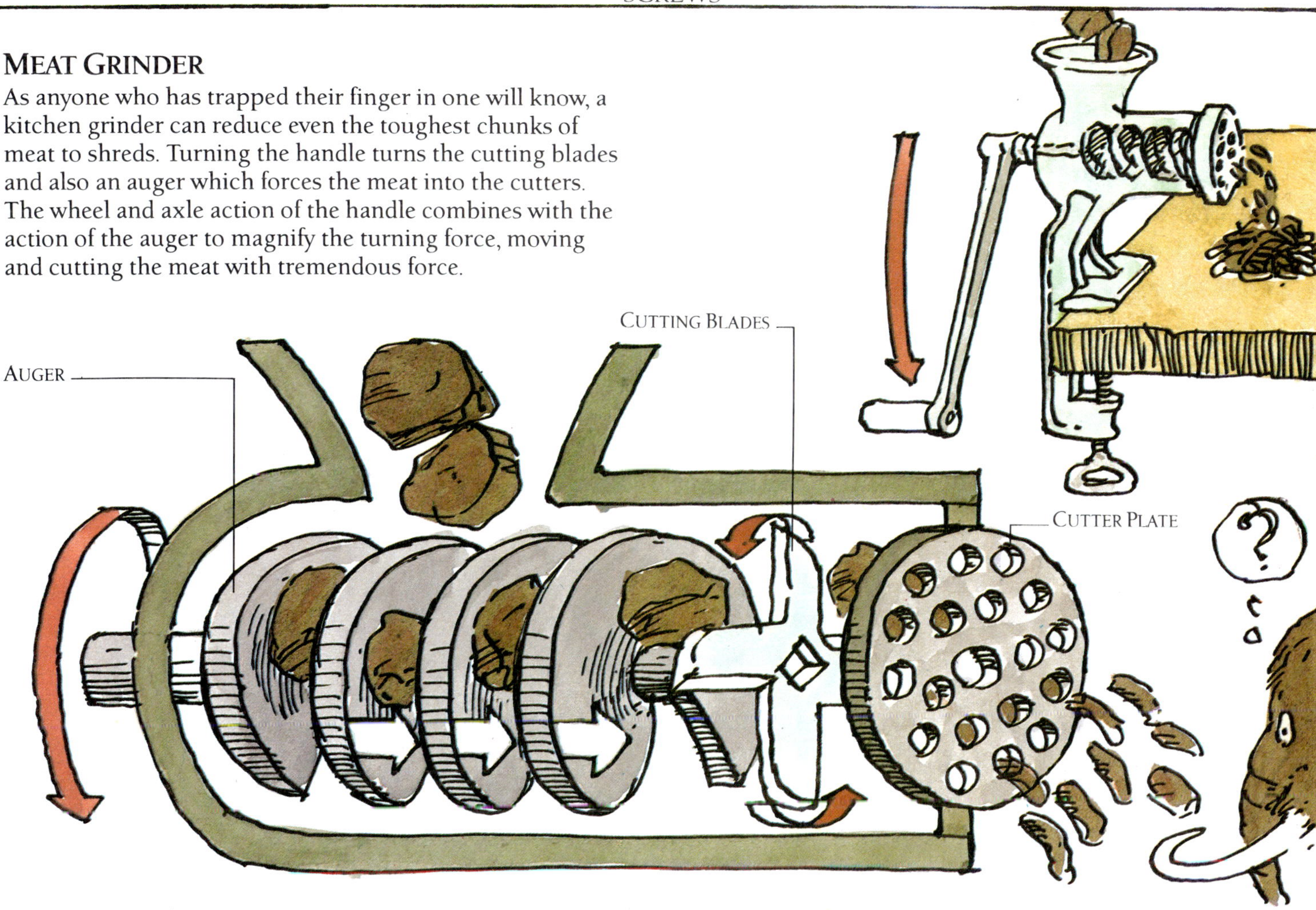

CONSTRUCTION AUGER

Augers are used to drill holes in soft ground for the piers of large buildings. As the auger rotates, it becomes filled with soil. It is then lifted to the surface where the soil is removed, after which the auger is lowered again. In this way, an auger of limited length can excavate deep holes.

MECHANICAL MOLE

The mechanical mole is a tunneling machine able to burrow its way through soil or soft rock. The cutting blades scour away at the workface, and as the mole advances, the tunnel behind it is lined to prevent collapsing.

The waste produced by the cutting blades is passed to one or more augers which transport it away from the workface.

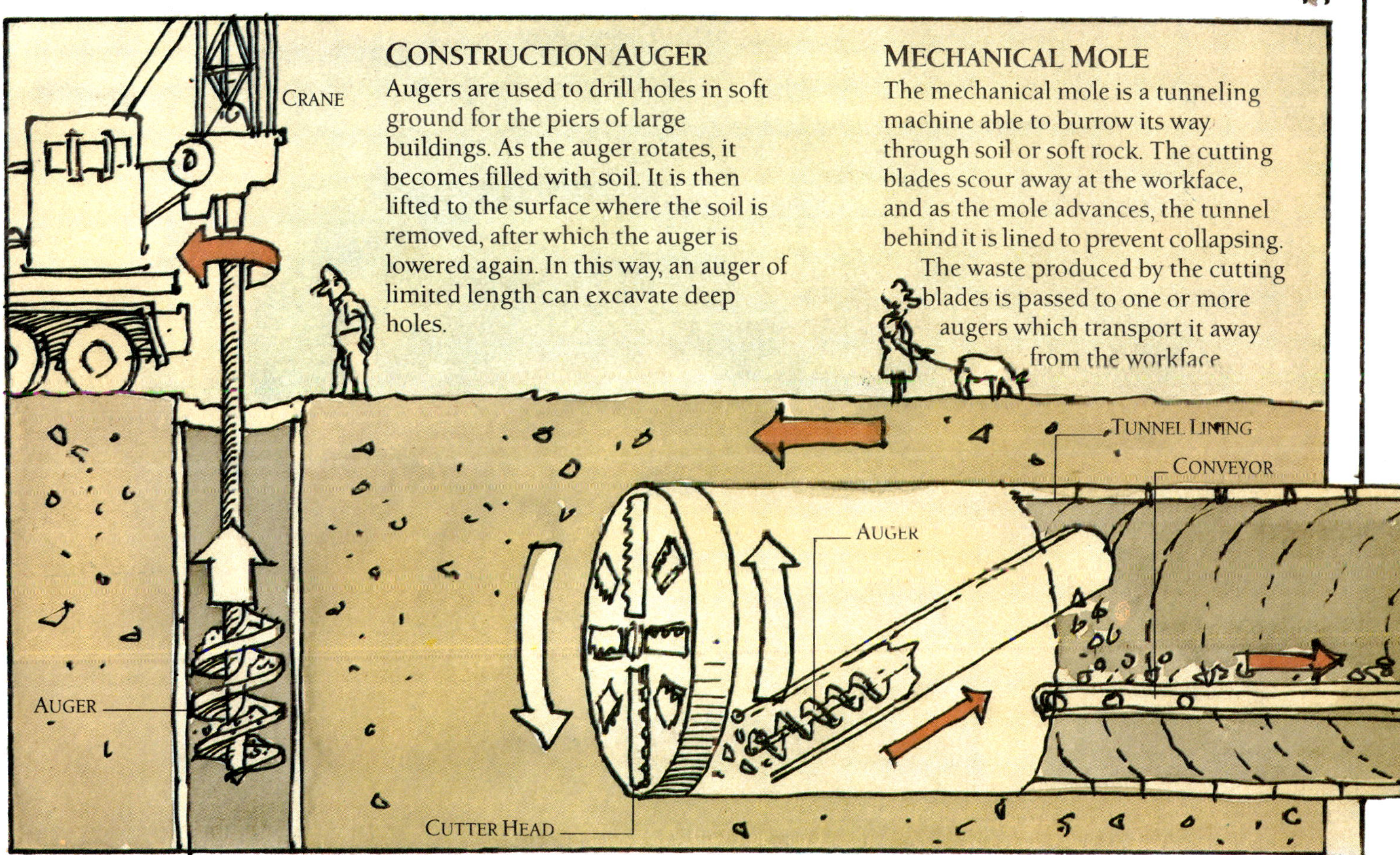

THE COMBINE HARVESTER

The combine harvester gets its name because it combines the two basic harvesting activities of reaping (cutting the crop) and threshing (separating out the grain). It may also bale the straw so that large fields can be harvested and cleared in one quick and tidy operation. Combine harvesters feature a number of screw mechanisms to transport the grain within the machine. Harvesters for seed crops other than grain work in similar ways.

KEY TO PARTS

1 REEL
The reel sweeps the stalks of the crop into the cutter bar.

2 CUTTER BAR
The bar contains a knife that moves to and fro between the prongs, slicing the stalks near ground level.

3 STALK AUGER
This transports the stalks to the elevator.

4 ELEVATOR
The elevator carries the stalks up to the threshing cylinder.

5 THRESHING CYLINDER
This contains a set of bars that rotates at high speed. The grain is separated from the heads and falls through the concave to the grain pan.

6 REAR BEATER
As this rotates, the straw (the threshed stalks) is moved to the straw walkers.

7 STRAW WALKERS
These carry the straw to the rear of the harvester, where it drops to the ground or is packed into bales.

8 GRAIN PAN
The vibrating surface of the pan transports the grain to the sieves.

9 SIEVES
The grain, unthreshed heads and chaff fall onto vibrating sieves. Air blows the chaff out of the rear of the harvester, while the sieves retain the unthreshed heads. The grain falls through the sieves to the base of the harvester.

10 TAILINGS ELEVATOR
This returns the unthreshed heads blown from the sieves to the threshing cylinder.

11 GRAIN AUGER AND ELEVATOR
The grain is carried by the auger and elevator to the grain tank.

12 UNLOADING AUGER
This transports the grain from the tank to a trailer or into bags.

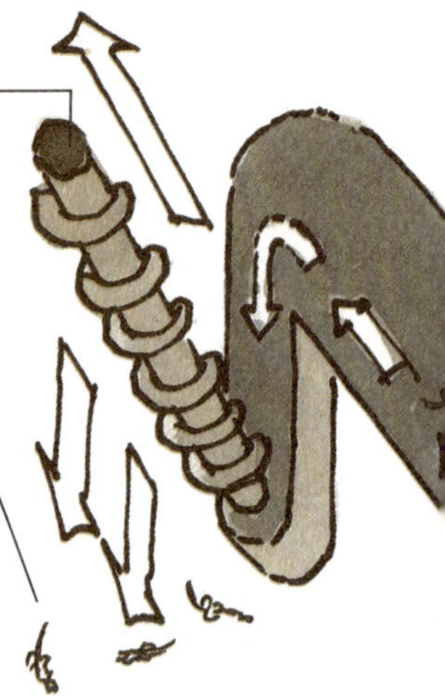

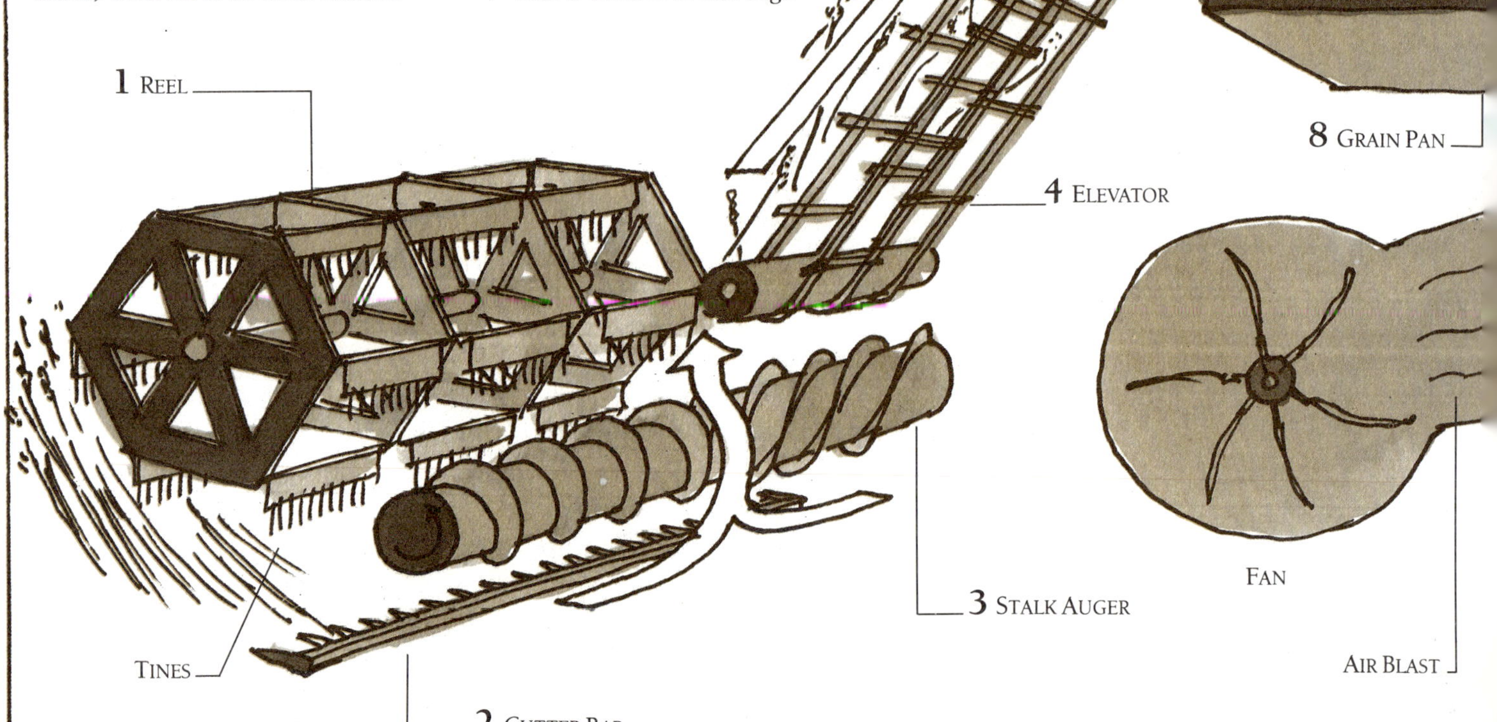

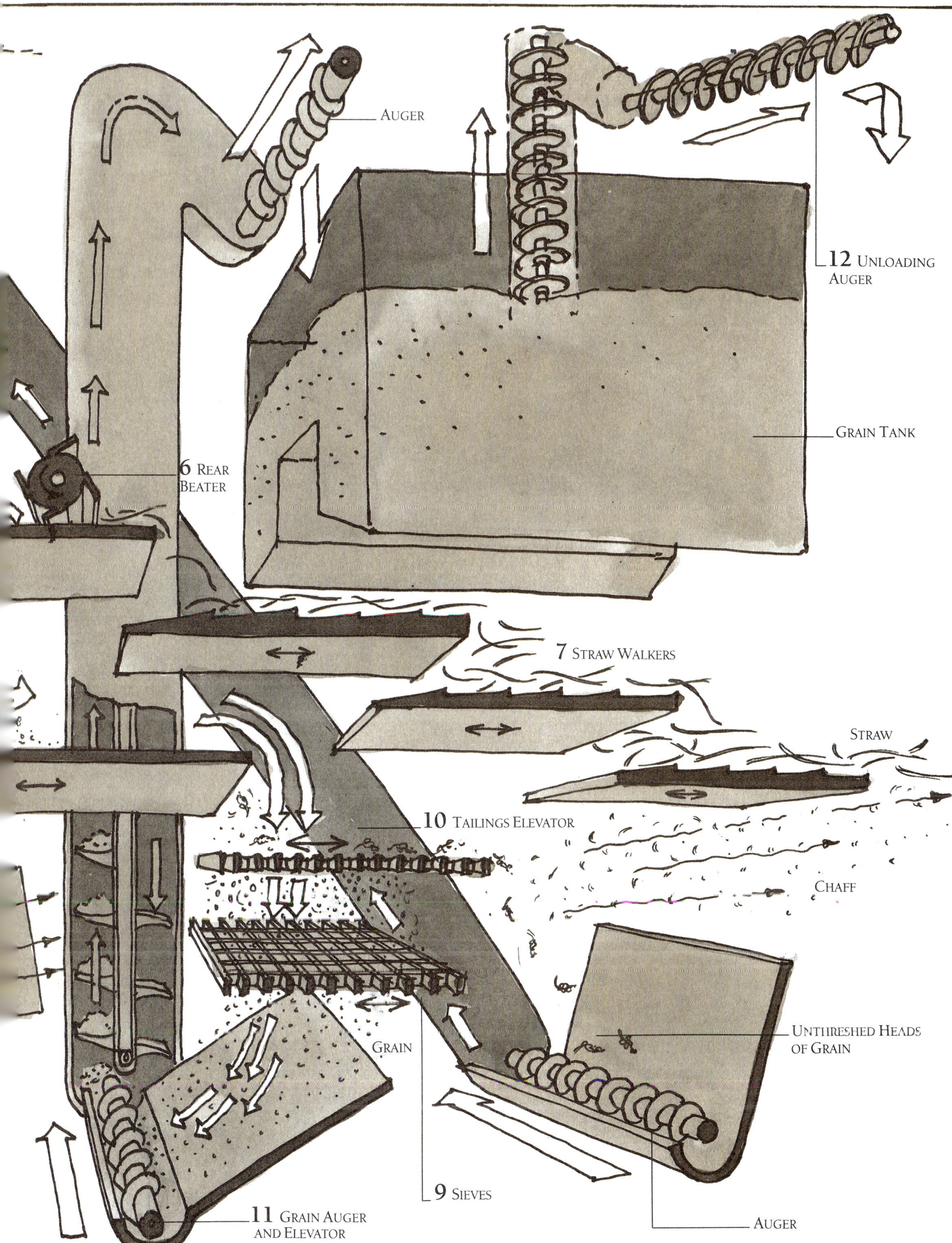
Auger
12 Unloading Auger
Grain Tank
6 Rear Beater
7 Straw Walkers
Straw
10 Tailings Elevator
Chaff
Unthreshed Heads of Grain
Grain
9 Sieves
11 Grain Auger and Elevator
Auger

ROTATING WHEELS

ON LEARNING FROM MAMMOTH ADVERSITY

I once made the mistake of leaving my unicycle unattended in the presence of a young mammoth. Being innately curious, the mischievous creature promptly took to the road. Even as I shouted, I could not help noting the extraordinary stability of the rotating wheel which allowed the novice cyclist to make good its escape.

Although the mammoth soon lost interest in the undertaking, the wheel—now rolling along at full tilt—seemed reluctant to stop. By the time the unicycle had reached the top of a small hill, its terrified rider was being carried helplessly forward. Everything in their path was promptly and unceremoniously flattened.

PRECESSION

Precession is a strange kind of motion that occurs in wheels and other rotating objects. You can feel its effects for yourself if you hold a spinning bicycle wheel by its axle. When you try to turn it, you will find that the wheel won't turn in the way you intend it to. Instead, it will "precess", so that the axle actually turns at right angles to the direction you expect.

Precession makes a wheel rolling on its own stay upright, and it enables a cyclist (or unicyclist) to ride. We use precession instinctively by slightly swiveling the front wheel. Each swivel brings precession into play to correct tilting and the bicycle remains upright.

The force of precession increases with speed. Conversely, it decreases as a wheel slows down. This is why it is difficult to ride a bicycle that is moving slowly. Remaining upright on a stationary cycle is purely a feat of balance, and does not involve precession.

INERTIA

You'll have experienced the effects of inertia if you've ever had to push a car in order to start it. It takes a lot of effort to get a car moving, but once it is going, it will carry on for some distance without further pushing and, with luck, will start itself.

Inertia accounts for all the pushing and shoving. It is the resistance of objects to any change in their speed, even if the speed is zero. Everything has inertia, and the amount depends on mass. The greater an object's mass, the more inertia it has.

In a rotating wheel, inertia also depends on how the mass is distributed. A wheel has more inertia if its mass is concentrated near the rim than if it is concentrated around the center. This means that two wheels of the same mass can have different inertia. Wheels designed to exploit inertia in machines often have heavy or thickened rims to provide the maximum resistance to any change in speed.

As I raced down the hill and over the wreckage, I wondered about my insurance coverage. Then I noticed the pond and its stunned occupant. The mammoth's little adventure had ended, but it was several minutes before I could approach my vehicle. Although upside down in the mud, the wheel was spinning rapidly and as it did, it flung everything attached to it a considerable distance.

CENTRIFUGAL FORCE

When an object moves in a circle, it is also always changing direction. Its inertia resists any change in direction as well as speed, and will make the object move straight on if it is free to leave the circle.

So, relative to the circle, the object is always trying to move away from the center under an apparent outward-acting force. This is known as centrifugal force, and anything moving in a circle – like the mud on the unicycle – experiences it. The faster an object is traveling, the stronger the force is.

Centrifugal force is used in machines to throw something outward. The simplest example is probably the spin drier, in which a spinning drum holds clothes while the water in them is forced outward through holes in the drum. Other machines use the centrifugal force that is generated by a sudden movement to activate catches and ratchets.

INERTIA AT WORK

POTTER'S WHEEL

The potter's wheel is a heavy disk with an axle. It is usually turned either by kicking the axle around or by operating a treadle. The wheel has considerable inertia, and this keeps it turning between kicks or presses of the treadle.

FRICTION-DRIVE TOY

Friction-drive toys store up energy in a flywheel. When you push the toy along the floor, the flywheel is set spinning by the wheels. Its inertia keeps it spinning so that when the toy is put down, it scoots across the floor.

TURNTABLE

The turntable of a record player has to rotate at a very constant speed. To do this, it has a heavy rim so that most of its mass is concentrated in the part that moves fastest, thereby raising its inertia. The inertia of the turntable cancels out any slight variation in speed that occurs in the turntable motor.

STARTER MOT

Inertia comes into play both in starting a car and in producing a smooth ride. A car's starter motor turns the engine by meshing with the teeth of the flywheel. An ingenious use of inertia allows the starter motor to engage and disengage the flywheel through a simple spring and screw system. Once the engine has started, the inertia of the heavy flywheel smooths out the jerky movement of the pistons.

1 STARTING UP

When the ignition key is turned, the starter motor rotates rapidly. The motor shaft turns more quickly than the pinion, which is slowed by inertia. The pinion therefore moves along the screw thread.

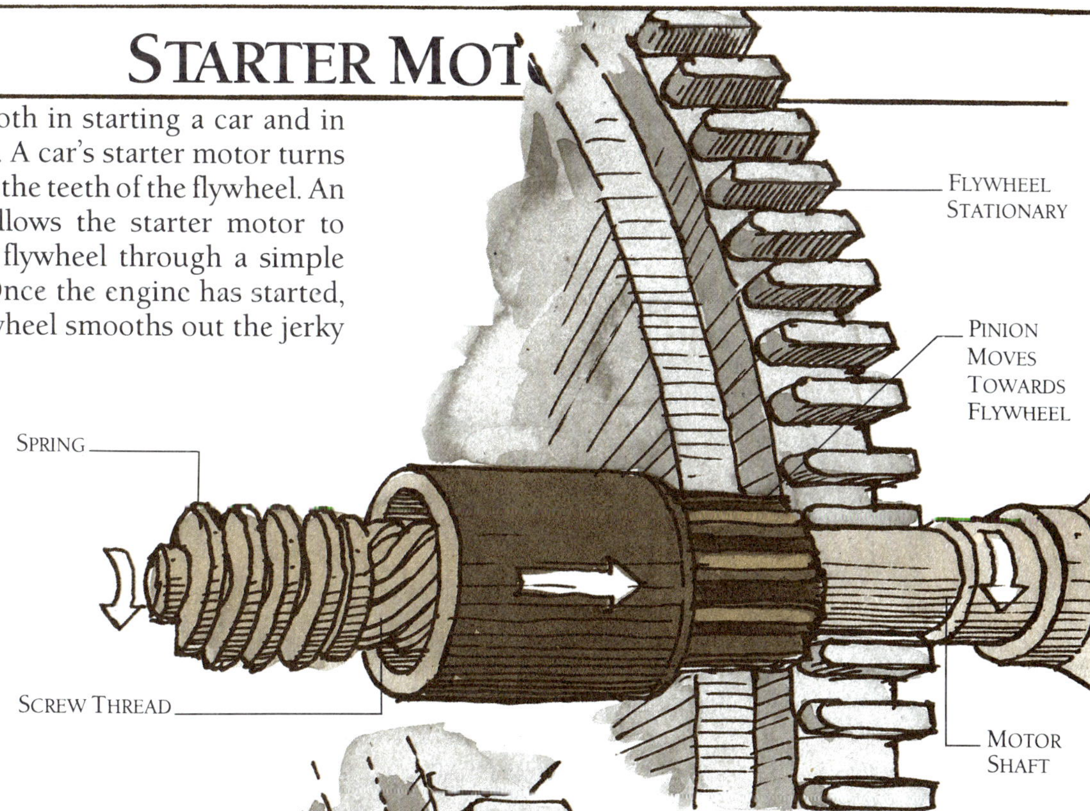

2 ENGINE RUNNING

The teeth of the pinion engage the flywheel, and through its contact with the flywheel, the starter motor turns the crankshaft.

3 STARTER DISENGAGED

When the engine starts, the pinion now begins to rotate faster than the starter motor shaft, and so it moves back along the screw thread, disengaging the flywheel.

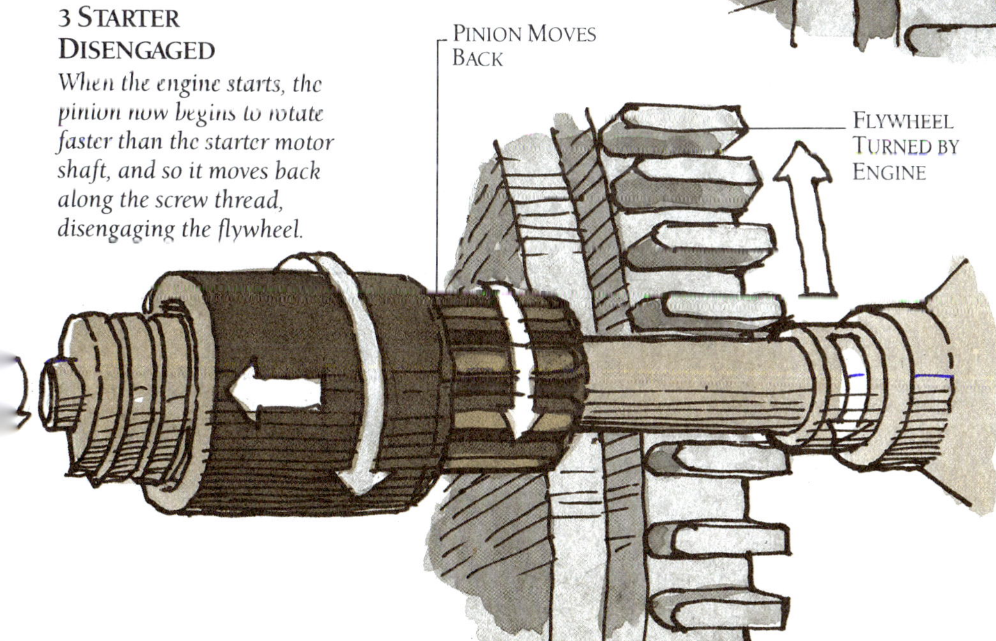

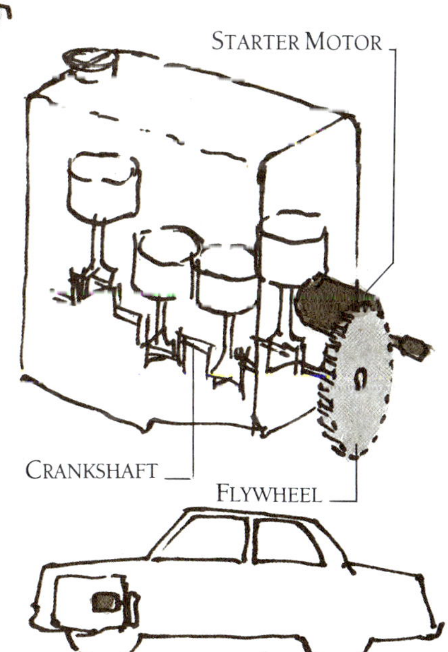

WINDOW SHADE

A window shade is lowered simply by pulling it down; the shade unrolls and remains in any position. To raise it, all that is needed is a sharp tug and the whole shade will roll up. But how can it tell a gentle pull from a sharp tug?

The shaft on which the shade is rolled contains a powerful spring. This winds up as the shade is lowered. A locking mechanism – a simple ratchet – prevents the spring unwinding if it is released gently. But when the shade is pulled suddenly, the ratchet no longer holds the shade in position. The motion makes a centrifugal device in the locking mechanism release the spring: the spring unwinds, releasing the energy that it has stored, and up goes the shade.

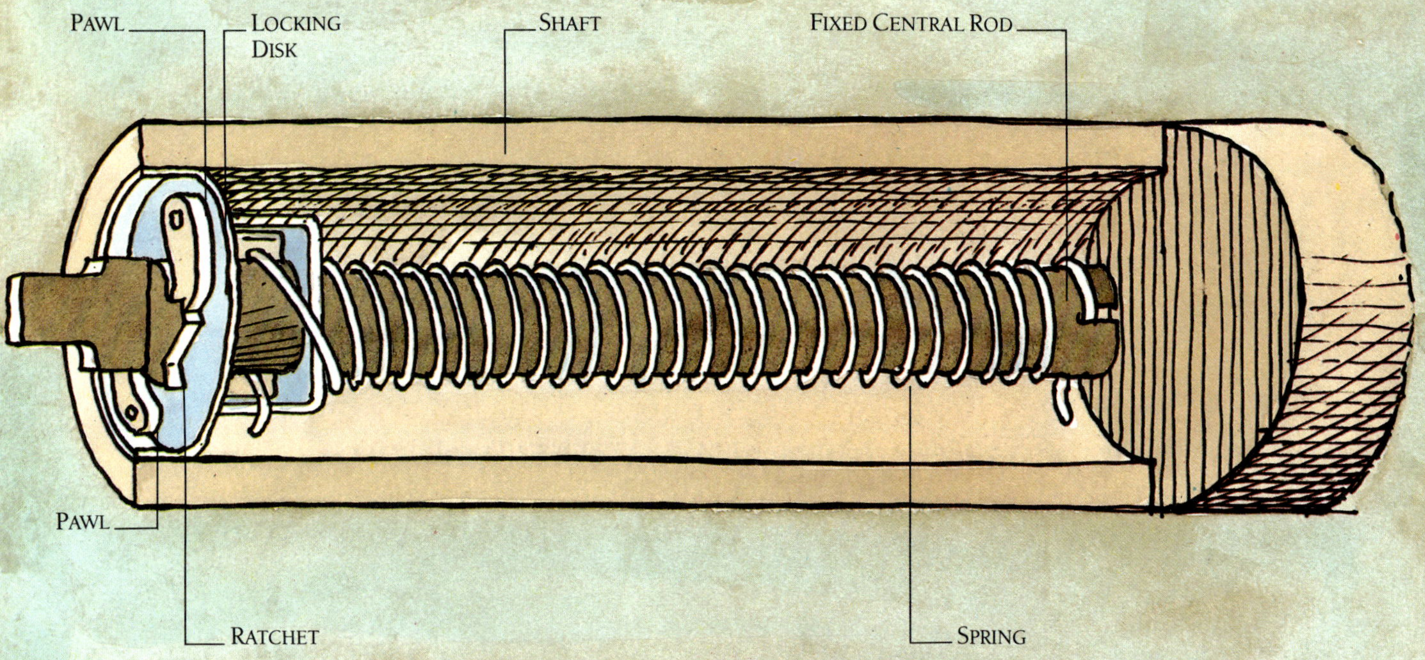

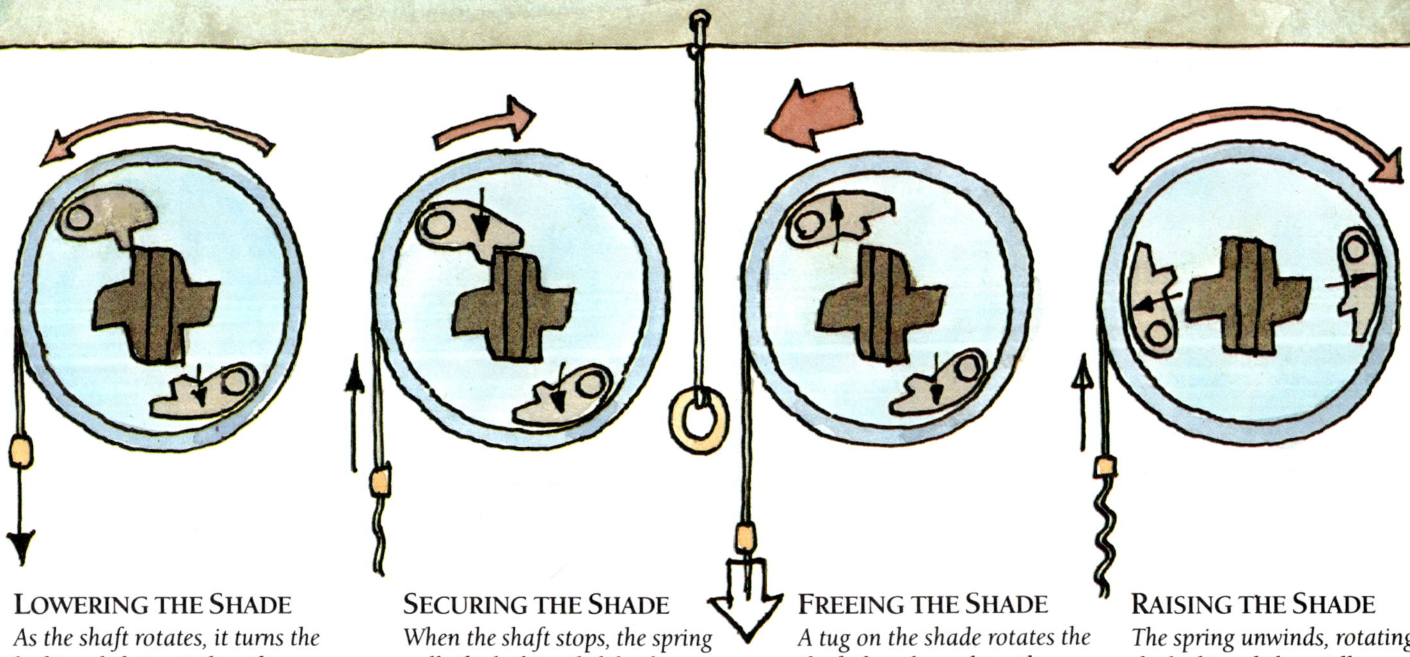

LOWERING THE SHADE
As the shaft rotates, it turns the locking disk to wind up the spring. The pawls are hinged and move over the ratchet, which is fixed to the central rod and does not move.

SECURING THE SHADE
When the shaft stops, the spring pulls the locking disk back slightly. One of the pawls falls to engage the ratchet, securing the locking disk.

FREEING THE SHADE
A tug on the shade rotates the shaft sharply, making the locking pawl move back and disengage the ratchet. The locking disk is now free to move.

RAISING THE SHADE
The spring unwinds, rotating the locking disk rapidly. Centrifugal force holds the pawls away from the ratchet, and the shade rolls up.

CAR SEAT BELT

A car seat belt works in the reverse way to the window shade. Instead of locking when the belt is pulled gently, it locks when the belt is given a sharp tug of the kind that would occur in a crash, and so secures the driver or passenger. The belt remains unlocked when pulled slowly, allowing normal movement in the seat. At the heart of the seat belt is a centrifugal clutch.

1 THE BELT MOVES FREELY

During normal use, the toothed plate is not in contact with the clutch and so the plate, and therefore the belt shaft, are free to rotate slowly.

2 THE CLUTCH ENGAGES

A sudden movement makes the toothed plate rotate quickly within the clutch. Centrifugal force makes it slide outward to engage the inner teeth of the clutch.

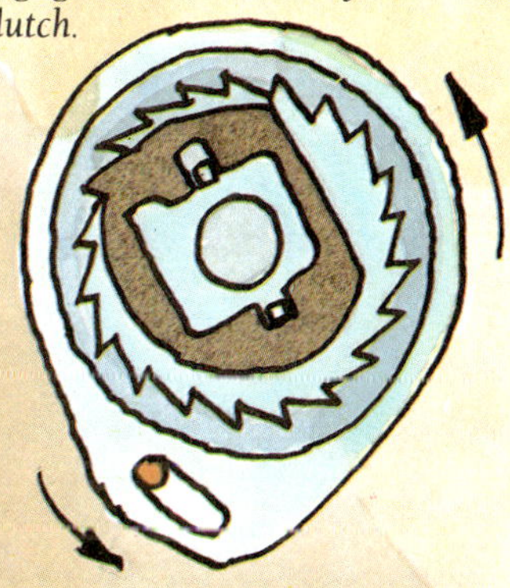

3 THE BELT LOCKS

Once the clutch has engaged, it rotates to move a pawl which in turn engages the ratchet. The pawl is fixed to the car body, while the ratchet is attached to the belt shaft. The pawl prevents the ratchet turning, so locking the belt. When the belt slackens, springs return the parts to their initial positions and free the belt.

GYROSCOPE

A spinning gyroscope can balance on a pivot, defying gravity by remaining horizontal while resting just on the tip of its axle. Instead of falling off the pivot, the gyroscope circles around it. The explanation for this amazing feat lies in the effects of precession. Like all other objects, the rotating wheel of the gyroscope is subjected to gravity. However, as long as the gyroscope spins, precession overcomes gravity by transforming it into a force that causes the gyroscope to circle instead of falling.

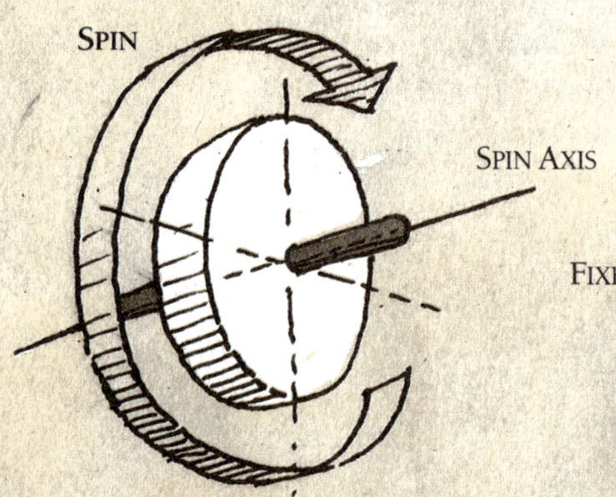

1 THE GYROSCOPE STARTS SPINNING

The gyroscope is set spinning so that its axle is horizontal and the wheel is vertical. The whole gyroscope rotates around the spin axis, which runs along the axle.

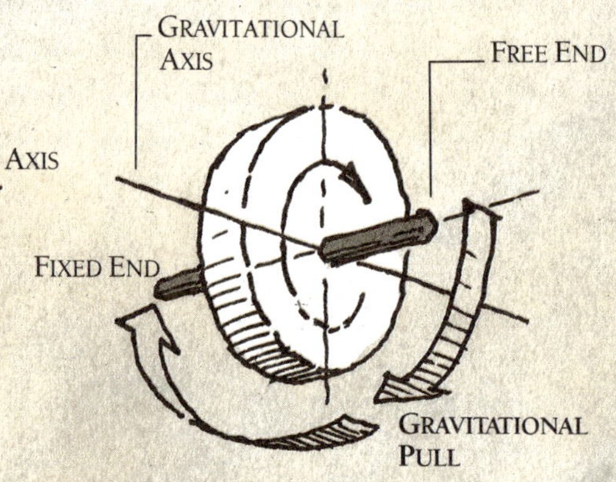

2 GRAVITY BEGINS TO ACT

The gyroscope is now placed so that one end of the axle is free to move. Gravity tries to pull this end downward, rotating the gyroscope around a second axis, the gravitational axis.

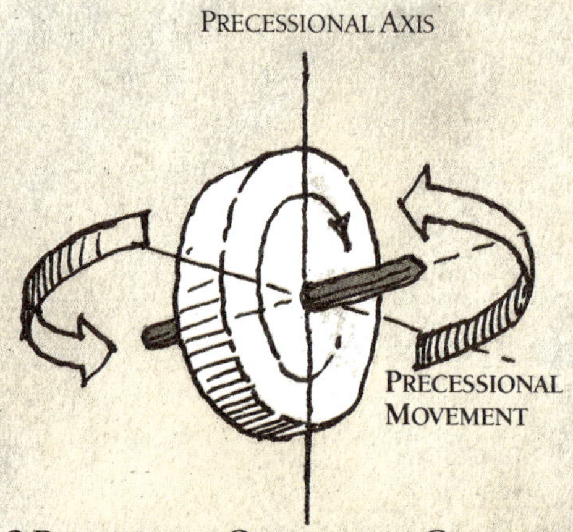

3 PRECESSION OVERCOMES GRAVITY

At this point, precession occurs. Instead of obeying the pull of gravity, precession makes the gyroscope move in a horizontal circle—in effect rotating it about a third axis, a precessional axis.

CHANGING DIRECTION

If gravity operates to turn the gyroscope in the opposite direction then it precesses in the opposite direction too.

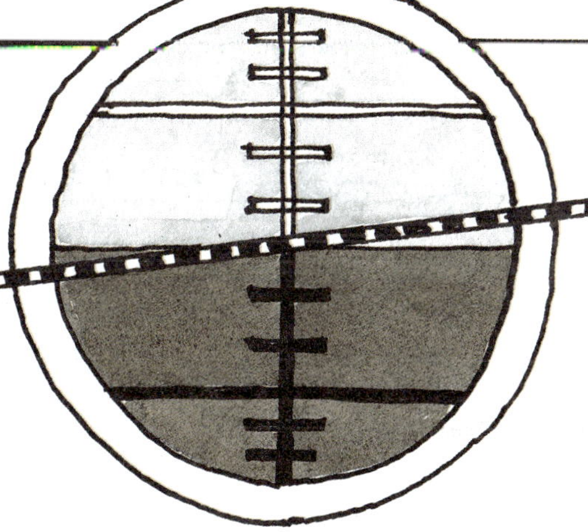

Artificial Horizon

Gyroscopes are very important in navigation. A spinning gyroscope possesses gyroscopic inertia, which makes it resist any change in its direction. The axle of the gyroscope remains pointing in the initial direction to which it is set. In the artificial horizon – an instrument which indicates the angle at which an aircraft banks – a gyroscope controls an indicator. Gimbals allow the gyroscope axle to remain horizontal. As the aircraft banks, the indicator also remains horizontal and shows the angle of the aircraft.

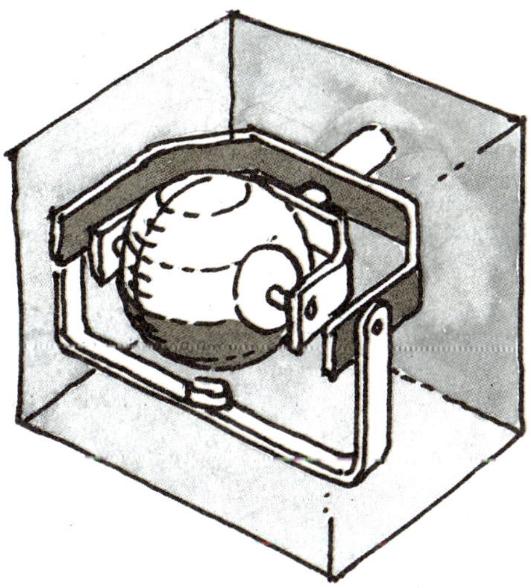

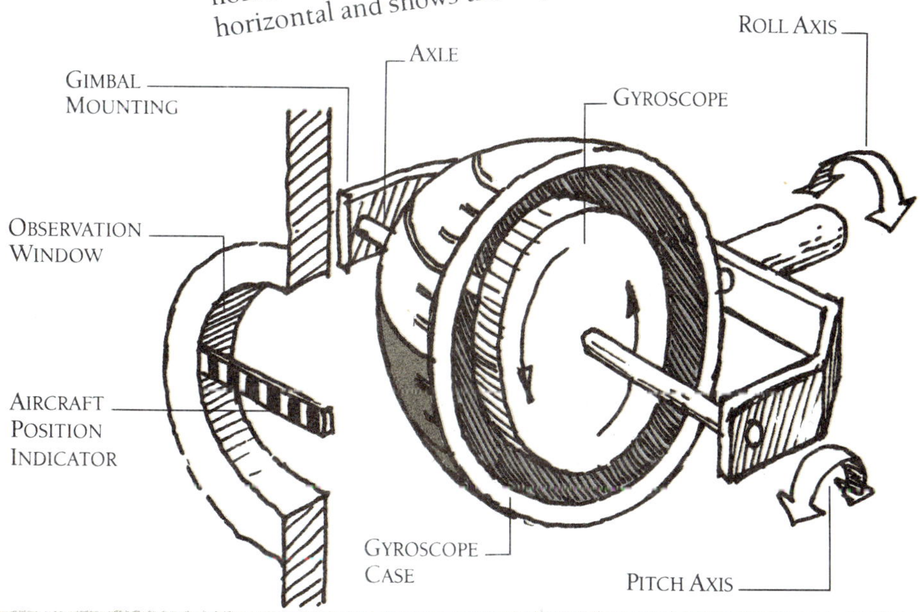

Gyrocompass

A gyrocompass makes use of the gyroscope to indicate direction. The axis of the gyroscope rotor is set in a north-south direction and the rotor is set spinning. The gyroscope is connected to an indicator so that as the ship or aircraft carrying the compass turns, the gyroscope keeps the indicator pointing north.

However, just as in the toy gyroscope, friction in the gyrocompass can cause it to drift out of true, and this may have to be corrected. In some gyrocompasses, this is done automatically by using the Earth's gravity. The gyroscope is connected to a weight, such as a tube of mercury, that acts as a pendulum. If the gyrocompass begins to point away from north, the pendulum tilts the axis of the rotor. Precession then occurs to bring the axis back to true north.

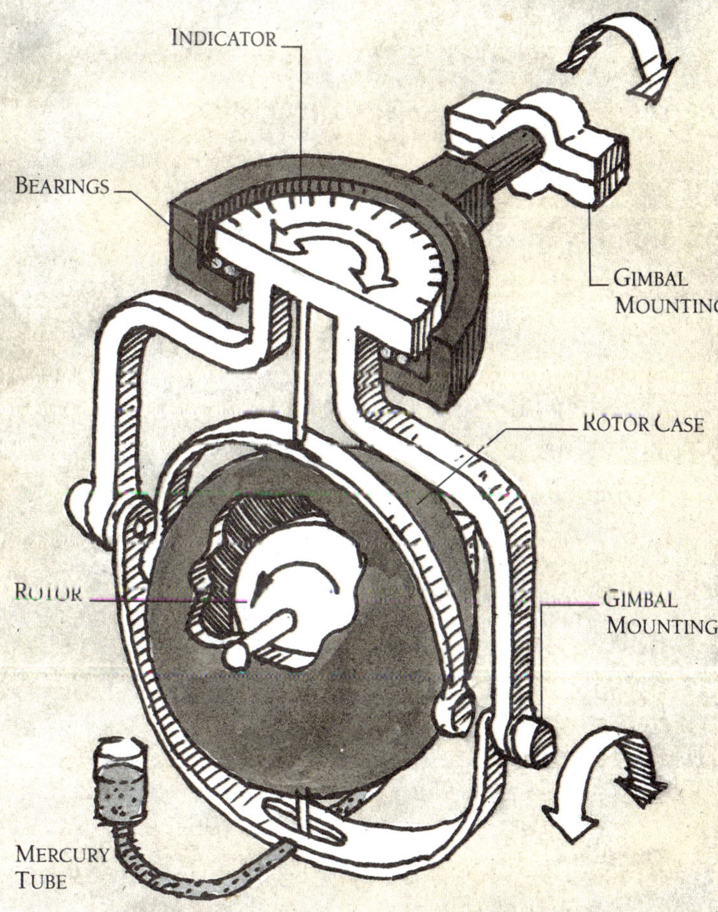

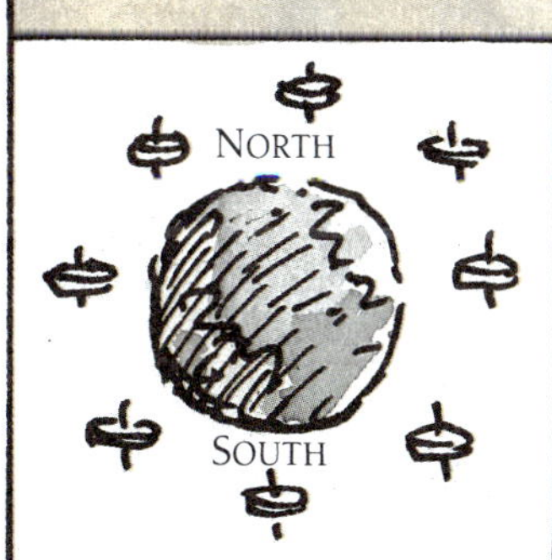

The Non-Magnetic Compass

A magnetic compass points to the north magnetic pole, which is away from true north, so correction is needed. Because gyrocompasses do not use magnetism, they always point to true north.

SPRINGS

ON A MAMMOTH HARVEST

A great many mammoths, in spite of their generally placid temperament, are ill-suited to inside work. Their preference for the outdoors combined with their tremendous strength makes them marvelous helpers in the field. I well recall seeing mammoths assisting eagerly during a particularly heavy coconut harvest. Instead of climbing each tree and simply dropping the coconuts, which could damage the shells, the farmer used his mammoth to bring the coconuts within reach of a ladder for effortless picking.

SPRINGS THAT REGAIN THEIR SHAPE

Springs have two basic forms – either a coil or a bending bar – and they have three main uses in machines. The first is simply to return something to its previous position. A door return spring, for example, contracts after being stretched, while the valve springs of a car engine expand after being compressed.

SPRINGS THAT MEASURE FORCE

The second use of springs depends on the amount by which springs change shape when they are subjected to a force. This is exactly proportional to the strength of the force exerted on the spring – the more you pull a spring, the more it stretches. Many weighing machines use springs in this way.

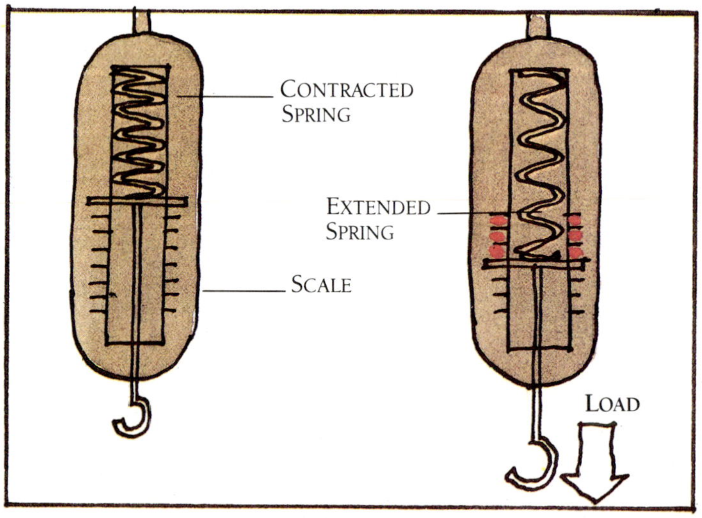

But as I mused on this harmonious partnership between man and mammoth, disaster struck. The unexpected appearance of a mouse so deranged the mammoth that it released the rope. The tree then obeyed its natural desire to return to its original configuration, thereby dispensing the coconuts – and the farmer – far and wide.

SPRINGS THAT STORE ENERGY

The third main use of springs is to store energy. When you stretch or compress a spring, you use up energy to make it move. This energy can be released immediately, as in a door spring, but if not, the energy remains stored. When the spring is released, it gives up the energy. Spring-driven clocks work by releasing the energy stored in springs.

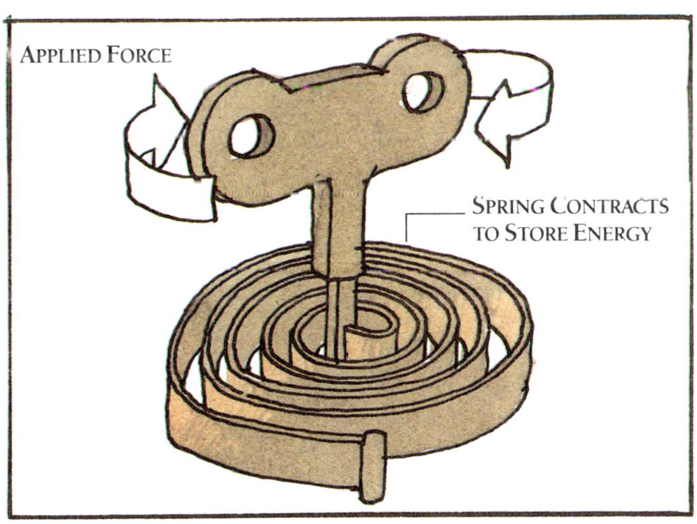

ELASTICITY

The special property of springs, their elasticity, is conferred on them by the way their molecules interact. Two main kinds of force operate on the molecules in a material – an attracting force that pulls molecules together, and a repelling force that pushes them apart. Normally these balance so the molecules stay a certain distance apart.

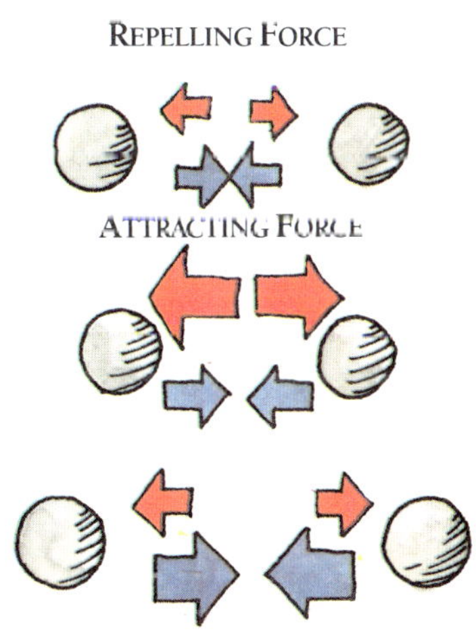

SPRING AT REST
The attracting and repelling forces are balanced.

SQUEEZED SPRING
Squeezing builds up the repelling force. When released, the force pushes the molecules apart again.

STRETCHED SPRING
Stretching builds up the attracting force. When released, this pulls the molecules back together.

THE STAPLER

A stapler is an everyday device that conceals an ingenious arrangement of springs. It uses both a coil spring and a leaf spring, which feed the staples along the magazine and return the stapler to its original position once it has been used. Pushing down the stapler causes the blade to descend into the magazine, forcing the front staple through the papers. The anvil bends the ends of the staple to clip the papers together. The return spring then raises the magazine and blade, allowing the magazine spring to advance the next staple into position.

CAR SUSPENSION

The suspension of a car allows it to drive smoothly over a bumpy road. The wheels may jolt up and down, but springs between the wheel axles and the body of the car flex and take up the force of the jolts. This ensures that the force of the bumping is not transferred to the car. Springs alone produce a bouncing motion, so the suspension also contains dampers, commonly known as shock absorbers. These slow the movement of the springs to prevent the car and its occupants bouncing up and down.

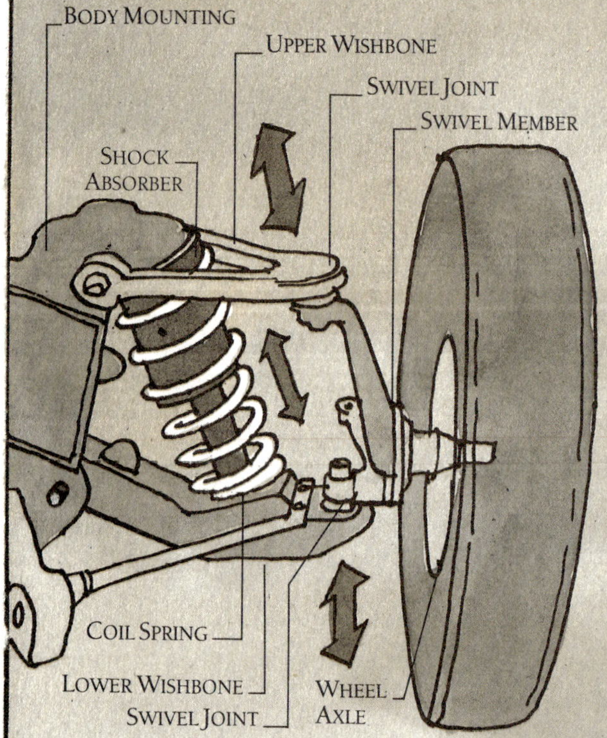

SHOCK ABSORBER

A shock absorber is fixed between the wheel axle and car body. Its piston moves up or down as the suspension spring flexes. As it does so, the oil is squeezed through channels in the piston, slowing the piston's movement.

COIL SPRING

Smaller vehicles have a coil spring and shock absorber attached to each wheel. The axle of the wheel is attached to hinged struts so that it can move up and down. The spring and shock absorber are fixed between the car body and the struts, or "wishbones".

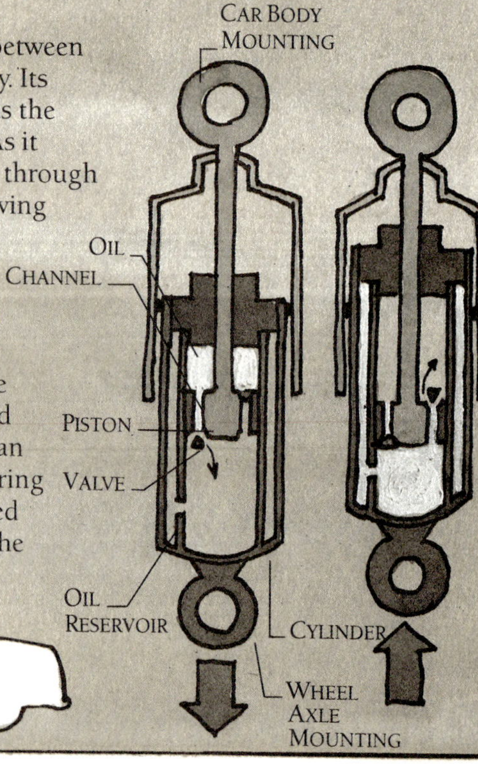

LEAF SPRING

Larger vehicles have heavy-duty leaf springs and shock absorbers to cushion the ride. The leaf spring is a stack of steel strips slightly curved so that the spring straightens when the vehicle is loaded. The axle is attached at or near the center of the leaf spring, and the ends of the spring are fixed to the body. The shock absorber is fixed between the axle and body.

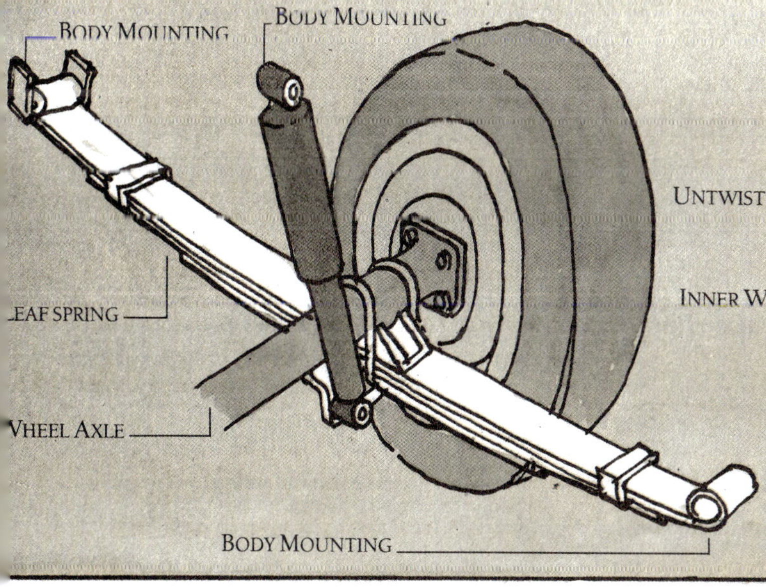

TORSION BAR

A torsion bar is a steel rod that acts like a spring to take up a twisting force. If the bar is forced to twist in one direction, it resists the movement and then twists back when the force is removed. Many cars contain an anti-roll bar fixed between the front axles. This rotates as the wheels go up and down. If the car begins to roll over on a tight corner, the anti-roll bar prevents the roll from increasing.

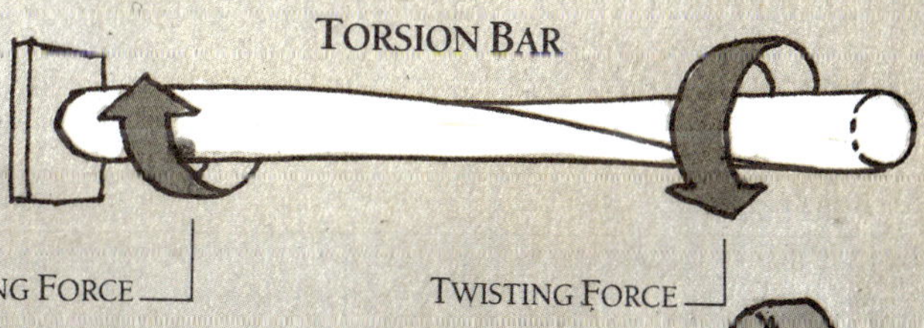

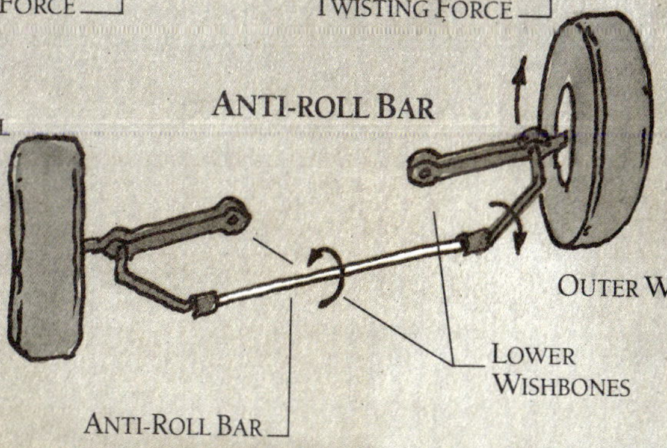

FRICTION

ON MAMMOTHS AND BATHING

Like children, mammoths in a domestic situation must be bathed with some regularity. Also like children, they tend to see bathing both as an annoying interruption and a needless indignity. Frequent bathing is virtually impossible, but when it must be done the most difficult part of the process is just getting the beast near the tub.

GETTING A GRIP

Friction is a force that appears whenever one surface rubs against another, or when an object moves through water, air, or any other liquid or gas. It always opposes motion. Friction happens because two surfaces in close contact grip each other. The harder they press together, the stronger the grip. The same molecular forces are at work as in springs. Forces between the molecules in the surfaces pull the surfaces together. The closer the molecules get, the stronger the force of friction.

The bathing team to have contend with the mammoth's superior weight which gives it the better grip on the ground. Only by reducing friction with the soap and marbles – a lubricant and bearings – can they move it.

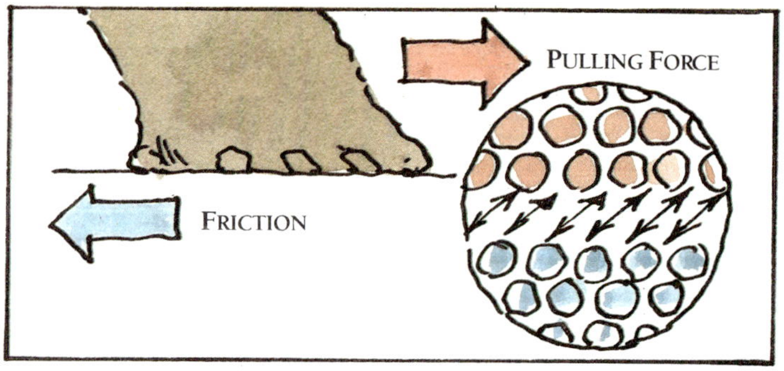

You can never get the same amount of useful work from a mechanical device as you put into it; friction will always rob you of some of the energy that is transmitted through the machine. Instead of useful motion, this lost energy appears as heat and sound. Excessive heat and strange noises coming from a machine are sure signs that it is not performing well.

Designers and engineers strive to overcome friction and make machines as efficient as possible. But paradoxically, many machines depend on friction. If it were suddenly to be banished, cars would slide out of control with wheels spinning helplessly. Brakes, which depend on friction, would be of no use, and neither would the clutch. Grinding machines would not make even a scratch, while parachutes would plummet from the sky.

The bathing scene I remember most vividly was not unlike the weighing of a large mammoth in its communal atmosphere. A large sneaker-clad crowd gathered on one side of a bath filled with soap suds. A dirty mammoth sat defiantly on the other. It should be noted that a mammoth's weight is its greatest defense and that just by standing or sitting still, it is able to resist all but the most determined efforts to move it.

Once ropes had been attached to the animal, they were pulled tight. Meanwhile another team used a technique which I had not previously encountered in my researches. First, they employed second-class levers to raise the beast slightly. Just when I had concluded that they intended to lever it all the way to the tub, some of their number poured a mixture of liquid soap and marbles between the protesting creature and the floor.

The result was astonishing: the animal's resistance was suddenly reduced, and despite its struggles it was hauled inexorably toward the water. Working simultaneously from both ends, it took little more than half an hour to get the mammoth close enough to the foam-filled tub for a good scrub behind the ears.

CAR TIRE

Car tires use friction to provide traction and steering: they grip the road so that the force of the engine is converted into forces that propel and turn the car. Tires must grip the road surface in all conditions, including wet weather. If a film of water becomes sandwiched between the tire and the road, then friction – and with it traction and steering – is lost.

The raised tread on the surface of a tire is designed to maintain friction on a wet road by dispersing the water.

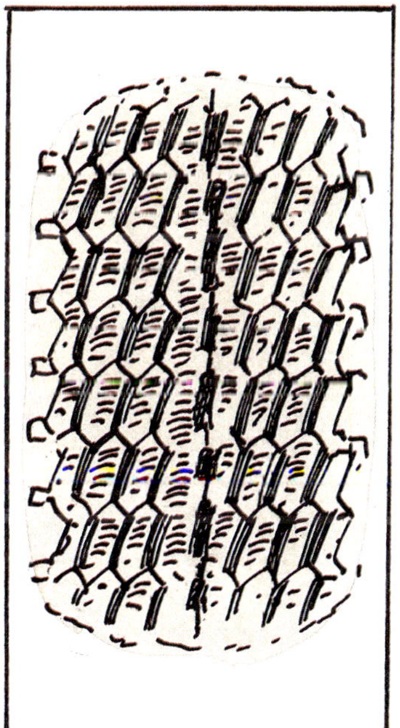

PARACHUTE

As a parachute opens, it develops a large force of friction with the air because it is moving rapidly. Friction is initially greater than gravity so the parachutist slows down. As the speed of the parachute lessens, friction decreases until it equals the force of gravity. At this point, there is no overall force acting on the parachutist; so he or she continues to descend without speeding up or slowing down.

THE CLUTCH

In a car, the clutch makes use of friction to transmit the rotation of the engine crankshaft to the gearbox, and then to the wheels. It can take up the rotation slowly so that the car moves smoothly away.

In a car with a manual gearbox, the clutch is disengaged when the clutch pedal is pressed down. The pedal operates the thrust pad, which presses on levers at the center of the rotating clutch cover. This raises the pressure plate away from the clutch plate, disconnecting the flywheel, which is turned by the crankshaft, from the transmission shaft. When the clutch pedal is lifted, the springs force the pressure plate and clutch plate against the flywheel. Friction linings on the clutch plate allow the plate to slide before it becomes fully engaged, which prevents jerking.

Synchromesh

The synchromesh is a mechanism in a car's gearbox (see pp.44-5) that enables the driver to change gear easily. It prevents gear wheels inside the gearbox from engaging at different speeds and crunching together. Before any forward gear is selected, gear wheels driven by the engine freewheel on the transmission shaft. For a gear to be engaged, the wheel and shaft need to be brought to the same speed and locked together. The synchromesh uses friction to do this smoothly and quietly.

Pushed by the selector fork, the collar slides along the transmission shaft, rotating with it. The collar fits over a cone on the gear wheel, making the wheel speed up or slow down until both are moving at the same speed. The outer toothed ring on the collar then engages the dog teeth on the cone, locking the collar to the gear wheel.

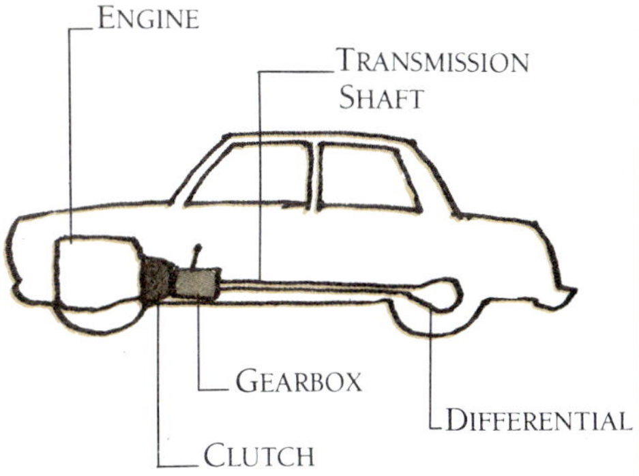

CLUTCH ENGAGED

Releasing the pedal allows the clutch springs to force the clutch plate and flywheel together, so the flywheel drives the transmission shaft.

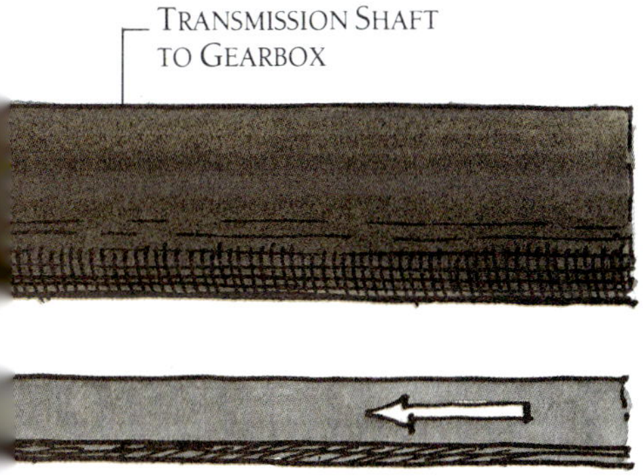

CLUTCH DISENGAGED

Pressing the pedal pushes in the thrust pad, which in turn pulls back the pressure plate. The flywheel and transmission shaft are now disconnected, so the engine cannot turn the wheels.

SYNCHROMESH DISENGAGED

When the synchromesh is disengaged, the collar and gear wheel are not connected, and the gear wheel freewheels on the transmission shaft.

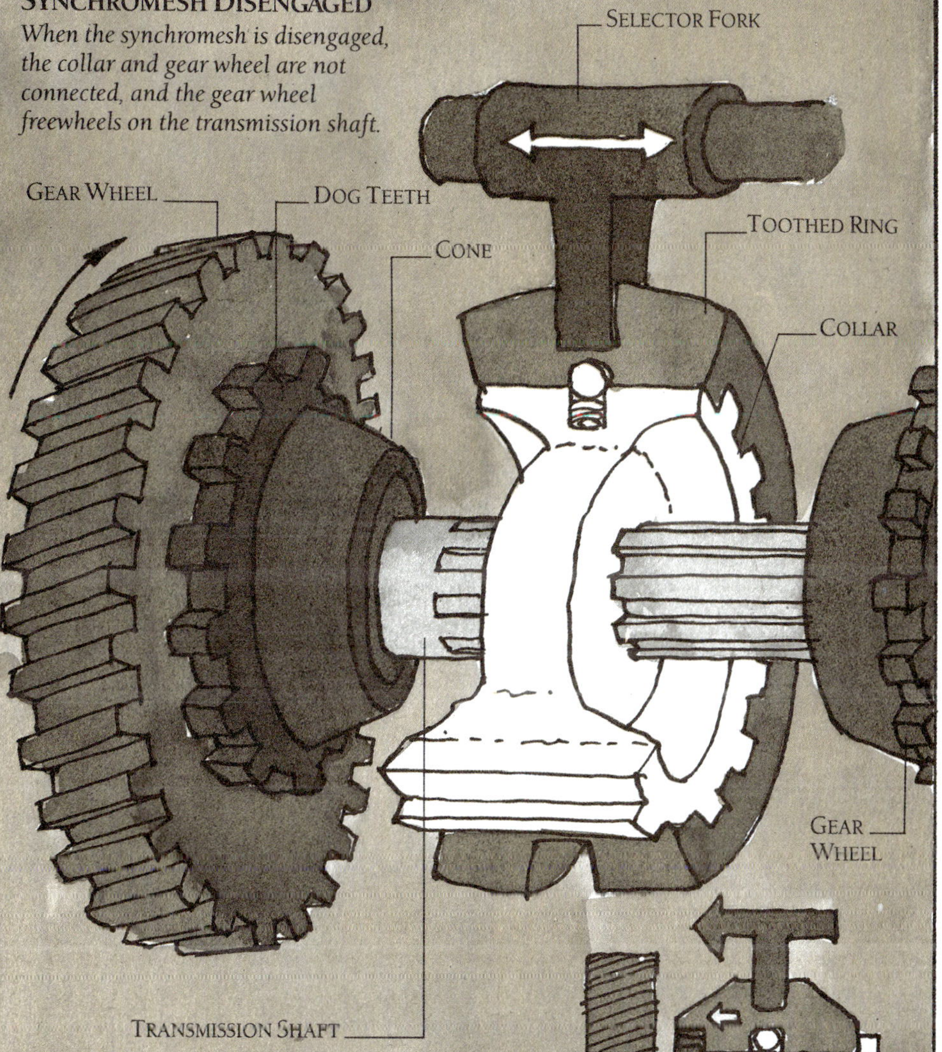

SYNCHROMESH ENGAGED

The collar makes contact with the cone, and friction between them brings them to the same speed. The teeth mesh together. The gear wheel is now locked to the transmission shaft and so can transmit power to it from the engine, turning the wheels.

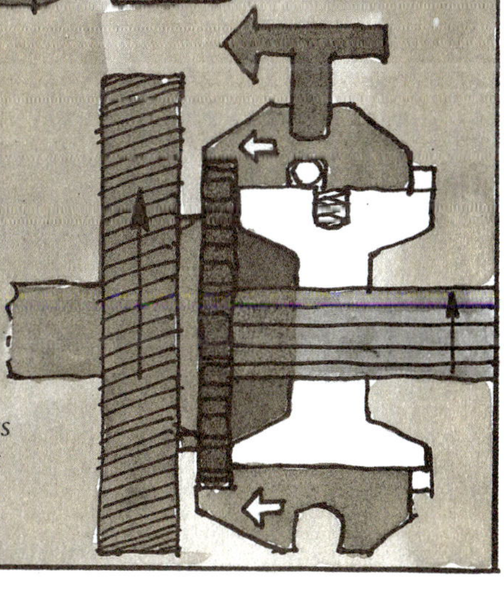

CAR BRAKES

To bring a fast-moving car and its passengers to a halt in a few seconds, car brakes must create a greater force than the engine does. Yet this force is produced by friction between surfaces with a total area only about the size of your hands.

Brakes are powerful because the brake pad or shoe and the brake disk or drum are pushed together with great force. In unassisted brakes, the force of the driver's foot is amplified by the hydraulics in the braking system (see p.137). In power brakes, the brakes are kept disengaged by a partial vacuum which is created and maintained by the engine. Pressing the brake pedal lets in air, and atmospheric pressure pushes the brakes into the engaged position.

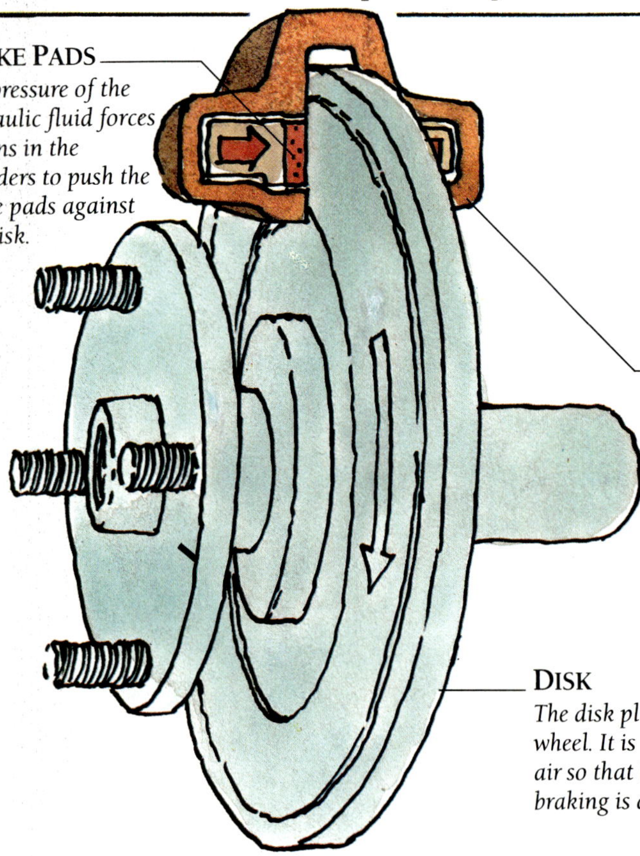

BRAKE PADS
The pressure of the hydraulic fluid forces pistons in the cylinders to push the brake pads against the disk.

CALIPER
The caliper fits around the disk and houses the brake pads and the hydraulic cylinders.

DISK
The disk plate is fixed to the wheel. It is exposed to the air so that heat generated by braking is dissipated.

DISK BRAKES

In disk brakes, friction is applied to both sides of a spinning disk by the brake pads. Much heat can be generated without affecting performance, giving great braking power. This is because the heat is removed by air flowing over the disk. Disk brakes are fitted to the front wheels of a car, where more braking power is needed, or to all wheels.

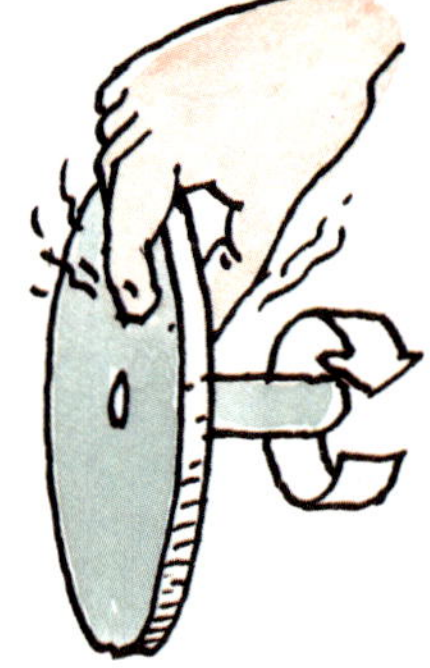

DRUM BRAKES

In drum brakes, friction is applied to the inside of a spinning drum by the brake shoes. Heat build-up tends to reduce friction, causing drum brakes to "fade" and give less braking power. Drum brakes are fitted to the rear wheels of many cars. The handbrake or parking brake often operates the rear brakes via a mechanical linkage.

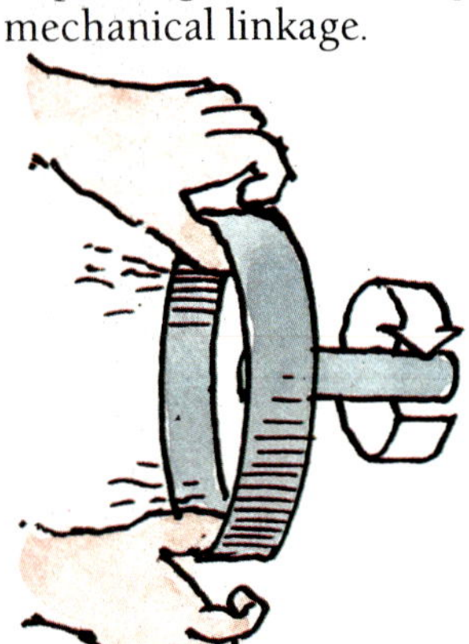

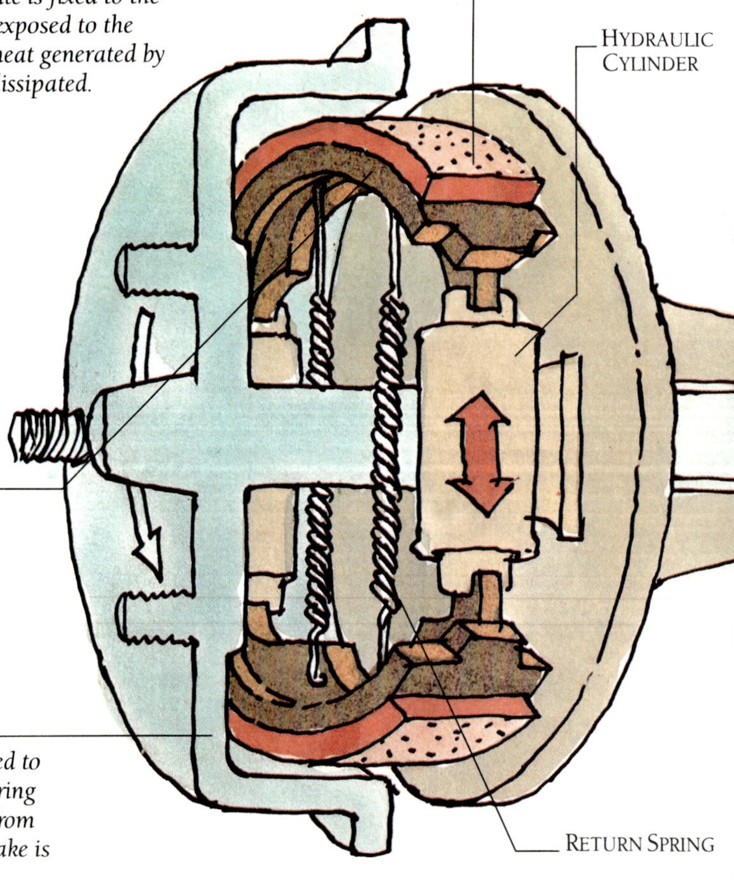

BRAKE SHOES
The brake shoes are either hinged at one end or moved by two hydraulic cylinders. The linings on the shoes come into contact with the brake drum.

BRAKE DRUM
The brake drum is fixed to the wheel. A return spring pulls the shoes away from the drum when the brake is released.

OIL RIG

Drilling rigs often have to penetrate deep into hard rock. The drill bit grinds its way into the ground, breaking the rock up into small pieces. Grinding is an extreme form of friction; it develops great heat, which is removed by a cooling fluid mud that is pumped down the shaft. Oil rigs are set up above a deposit of oil or gas, which may be found under land or the seabed. Offshore rigs either stand on the seabed or on long legs, or float at the surface anchored in position.

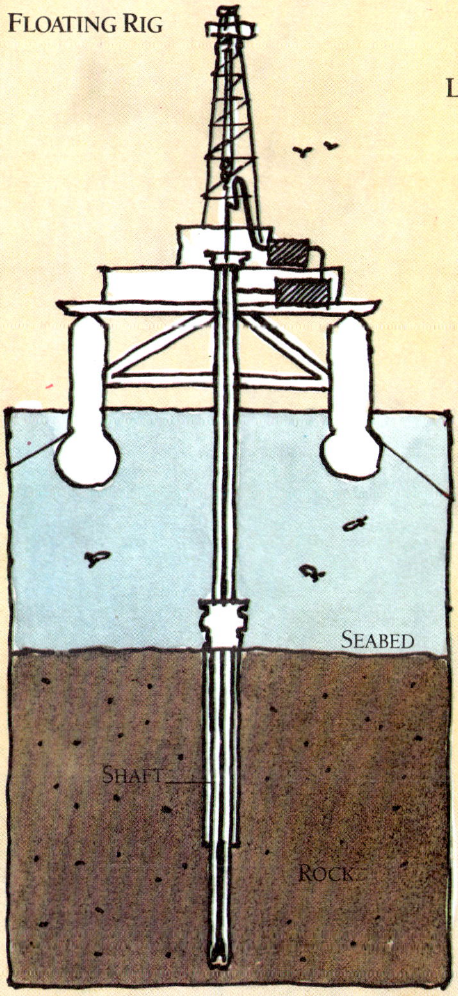

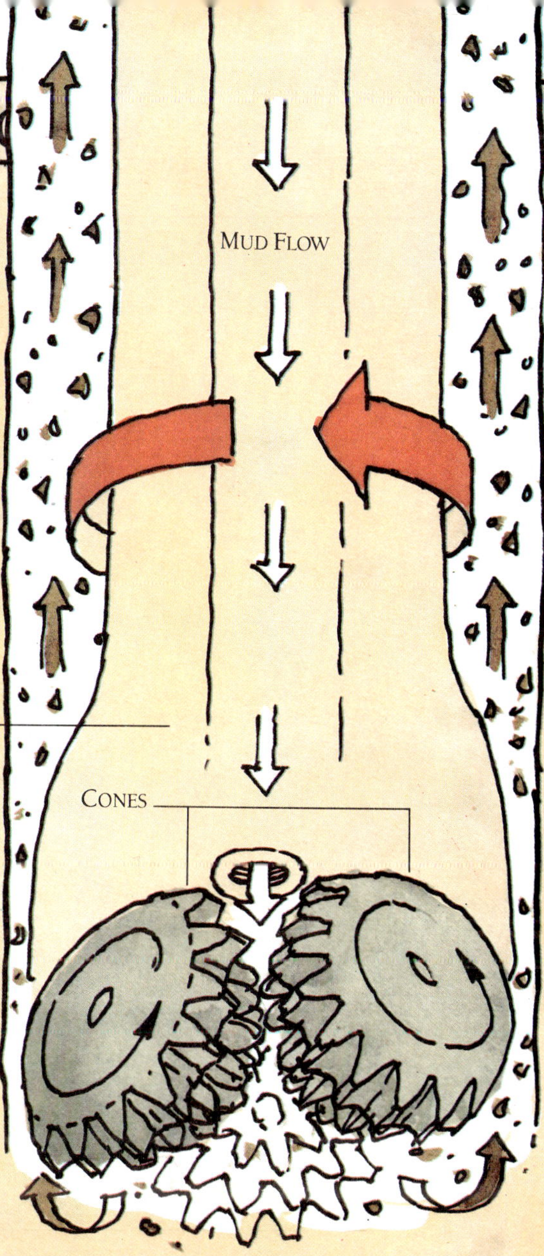

ROTARY BIT

The bit that drills the shaft is mounted on the end of a long drill pipe, which is rotated by an engine in the rig above. A tricone rotary bit has three cones studded with teeth that turn as the drill pipe rotates. The weight of the pipe on the bit helps it to crush and grind the rock.

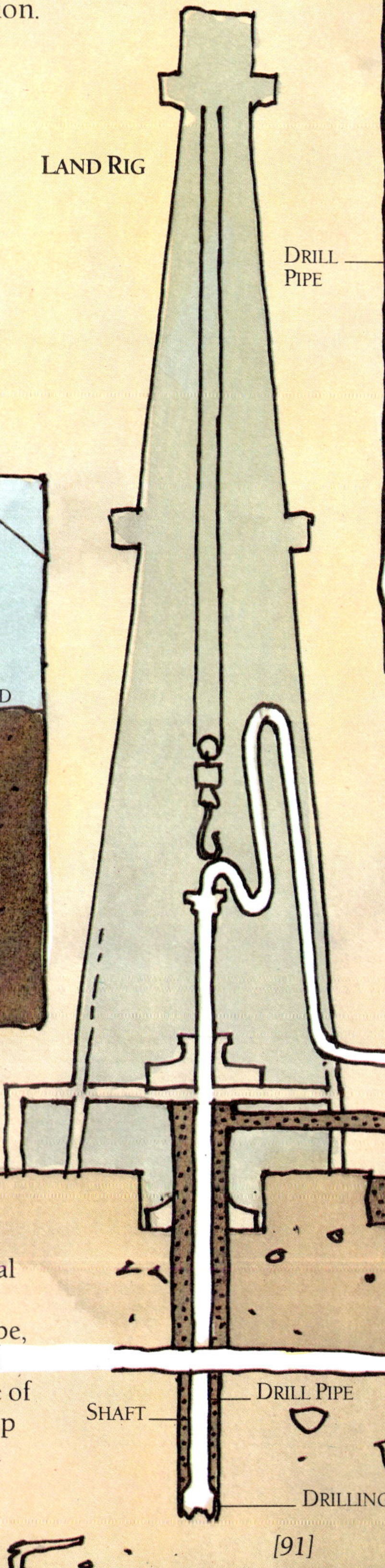

DRILLING MUD

The mud used on oil rigs is a special liquid developed for drilling. It is pumped into the top of the drill pipe, and from there it flows down to the drilling bit and then up the outside of the pipe back to the rig, bringing up the ground rock, before it is filtered and recycled.

FREEDOM FROM FRICTIO

Machines that move themselves or that create movement are limited by friction. In the moving parts of an engine, for example, friction lowers performance and may produce overheating. Reducing friction reduces energy needs and so improves efficiency. This reduction is achieved by minimizing the frictional contact through bearings, streamlining and lubrication.

BALL BEARING

In a ball bearing, the area of contact between the balls and the moving parts is very small, and friction is therefore very low. Roller bearings contain cylindrical rollers instead of balls but work in the same way.

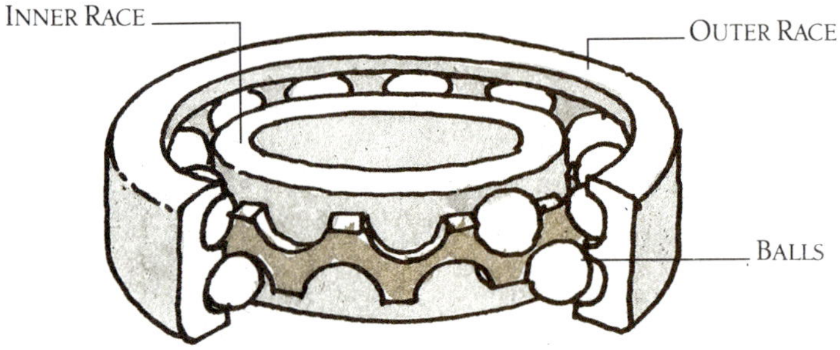

CAR LUBRICATION

A car has several sections with moving parts and good lubrication is essential. In the suspension, steering, gearbox and differential, filling with oil or grease is sufficient. The engine, however, needs a special lubrication system to get oil to its components as they work.

Oil is contained in the sump, which is a chamber at the base of the engine. A pump (see p.132) forces oil up from the sump through the oil filter, which removes dirt particles, and then to all the bearings and other moving parts of the engine, such as the pistons. The parts contain narrow channels that lead the oil to the moving surfaces. The oil then returns to the sump to be recirculated.

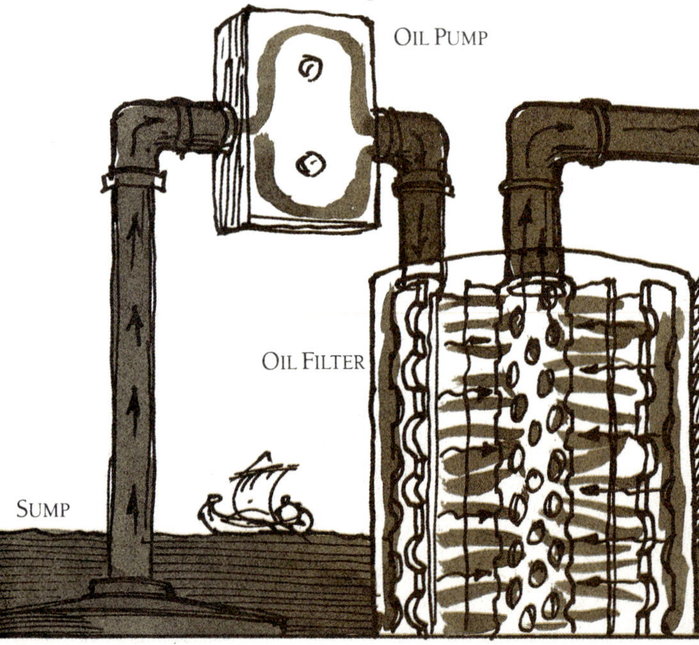

PERPETUAL MOTION

Even with the very best bearings, lubricants and streamlining, a little friction still remains. Without a continual supply of fuel or electricity, friction gradually consumes a machine's kinetic energy (its energy of movement) and the machine slows down and stops. This is why the goal of inventors throughout the ages to build a machine capable of perpetual motion can never be achieved...at least, on Earth.

In space, matters are different. No air exists to cause friction and slow a spacecraft. Once launched into space, a spacecraft is freed from friction. It can continue to move in perpetuity without ever firing its engine again. Thus, in the space probes voyaging outward toward the stars, we have achieved perpetual motion, a pure movement governed only by the celestial mechanics of gravity.

VOLUNTEER MOLECULES
ENTER·THE·FIRST
WHOOPEE·CUSHION

PART 2

HARNESSING THE ELEMENTS

CONTENTS

THE SECOND PART
OF
THE GREAT WORK
Is Here Presented
in which the principles
& workings of various
STARTLING
PHENOMENA
including acts of TRANSPORTING,
SQUIRTING & the supply of USEFUL
ENERGY are unwittingly
demonstrated
BY THE
GREAT WOOLY MAMMOTH
remaining free from the intrusion of
COMMON SENSE
recorded during my travels

INTRODUCTION

It was the ancient Greeks who first had the idea that everything is made up of elements. They conjured up just four of them – earth, fire, air and water. As it has turned out, the idea was right but the elements wrong. Modern elements are less evocative but more numerous; they make up just over one hundred basic substances. Some are commonplace, like hydrogen, oxygen, iron and carbon; others are rare and precious, such as mercury, uranium and gold.

Purely by the power of reason, the Greeks also made another fundamental discovery, which is that all things consist of particles called atoms. Elements are substances that contain only one kind of atom. All other substances are compounds of two or more elements in which the atoms group together to form molecules.

The way molecules behave governs the workings of many machines, in particular those which are covered in this part of *The Way Things Work* – machines such as ships, airplanes, pumps, refrigerators and combustion engines, all of which harness the ancient elements and set molecules to work.

MORE ABOUT MOLECULES

The idea that everything is made of particles takes some imagination to understand. For example, as you read this, molecules of oxygen and nitrogen traveling at supersonic speed are bombarding you from all directions. The reason that you are unaware of this is that the molecules (which, along with those of other gases, make up the air) are on the

small side. You could get about 400 million million million of them into an empty matchbox. In fact, it would be truer to say that you could get all those millions of molecules *out* of the matchbox, because the molecules of gases are so hyperactive that they will fill any space open to them. Like five-year-olds, they dash about in all directions with unflagging energy, crashing into any obstacle they meet. In liquids, the molecules are less energetic and go haphazardly about in small groups, rather like drunken dancers prone to colliding with the walls of the dance hall. The molecules in solids are the least energetic; they just huddle together like a flock of sheep shuffling around in a field.

However invisible molecules might be, their existence does explain the properties and behavior of materials that are put to use in machines. In a solid, the molecular bonds are strong and hold the molecules firmly together so that the solid is hard and rigid. The weaker bonds between liquid molecules pull them together to give the liquid a set volume, but the bonds are sufficiently weak to allow the liquid to flow. The bonds between gas molecules are weaker still, and they enable the molecules to move apart so the gas expands and fills any space.

In all materials, the molecules' urge to stick together or spread apart is very strong, and it is put to use in machines and devices as different as the rocket, the toilet tank and the aqualung.

STRENGTH IN NUMBERS

Because molecules in liquids and in gases are always on the move, they have power. Each one of them may not have much, but together they become a force to be reckoned with. A liner floats because billions upon billions of moving

water molecules support the hull, while a jumbo jet can fly thanks to countless air molecules clustering under its wings and holding them aloft.

Molecules continually bombard any surface they encounter. Each collision produces a little force as the molecule hits the surface and bounces back, trampoline-style. Over the whole surface, a large force builds up – this is known as the pressure of the liquid or gas. If you squeeze more molecules into the same space, you get more pressure as more molecules strike the surface. The weight of all the liquid or gas molecules above a surface pushes down to increase pressure too.

The pressure produced by this restless movement of molecules is put to work in many ways. Machines such as the dishwasher work by producing pressure while others, such as the pneumatic drill, are powered by it.

SPEEDING THINGS UP

There is another way of strongly increasing pressure that does not involve physical effort, one that drives cars, trains, aircraft, spacecraft even. This is heat, which is a form of energy.

We feel heat, or the lack of it, as a change in temperature. But on a molecular level, heat is just movement. If you touch something cold, the molecules in your fingers slow down as they lose heat; if you touch something hot, they gain heat and speed up. That's all there is to it.

Whenever molecules are heated, they respond by moving faster. The pressure increases unless the molecules get further apart, in which case the material expands. If the

molecules are made to move fast enough, the bonds between them start to give way: a solid melts into a liquid and a liquid into a gas.

If an object is cooled down the speed of its molecules slows. The material loses pressure or contracts. As the bonds between molecules reassert themselves, a gas may condense to a liquid and a liquid freeze to a solid. There is a point at which all heat vanishes, although no one can quite achieve it. If a material were cooled to −273°C (−459°F), the molecular motion would cease altogether, making this – absolute zero – the lowest temperature possible.

Machines that make or use heat all get molecules on the move. The extra motion can strain relations within the families of atoms inside molecules, making them change partners and form new molecules. Fire and explosions are some of the possible results, but so too are the making of steel and toast.

BREAKING THE BONDS

The atoms of elements are made up of even smaller particles – electrons, which form the outer shells of each atom, and protons and neutrons, which make up its core, or nucleus. We tap the energy of electrons – in the form of electrical heat – in everyday devices from hairdriers to heaters. However, breaking the bonds that hold together the nucleus of an atom is a more serious business altogether.

As we shall see in the last section of *Harnessing the Elements*, these bonds are the strongest of all forces. Breaking them unleashes the most powerful and potentially dangerous source of energy known.

FLOATING

ON TRANSPORTING A MAMMOTH

Once, when waiting to board a ferry, I observed further downstream a rival operator attempting to shunt a particularly large mammoth onto a sizeable raft. No sooner had the craft and its protesting cargo been launched, than both quickly sank.

Taken aback by this turn of events, I relinquished my position in line to inquire whether I might not be of assistance. My offer was quickly accepted by the soggy pair. After interviewing those involved and making some hasty calculations, I deduced that the spirit of the water, clearly afraid of the raft's imposing cargo, had simply moved out of the way as it approached. This left nothing below the raft and so it sank.

Clearly, a little subterfuge was necessary to keep the cargo afloat. The mammoth, I suggested, must be hidden from the spirit of the water.

RAFTS AND BOATS

Although characteristically wayward, the inventor's explanation of the mammoth's adventures contains an element of truth. Water does move out of the way when anything enters it. But rather than leaving nothing below an immersed object, the water around it pushes back and tries to support the object. If the water succeeds, the object floats.

Take the case of the raft before the mammoth is on board. Its weight pulls it down into the water. But the water pushes back, supporting the raft with a force called upthrust. The amount of upthrust depends on how much water the raft displaces, or pushes aside, as it enters the water. Upthrust increases as more and more of the raft settles in the water. At some point, the upthrust becomes equal to the weight of the raft and the raft floats.

Now let's load the mammoth. The extra weight makes the raft settle deeper. Although the upthrust increases, it cannot become great enough to equal the weight of the raft and the mammoth because not enough water gets displaced. The raft and its load sink to the bottom.

The boat is a different matter. Because it is hollow, it can settle deeper in the water and displace enough water to provide the necessary upthrust to support the weight of the boat and the mammoth.

Things can also float in a gas and, like the inflated mammoth, a balloon floats in air for the same reason that a boat floats on water. In this case, the upthrust is equal to the weight of air that is displaced. If the weight of the balloon, the air that it contains and the occupants is less than the upthrust, the balloon will rise. If it is greater, the balloon will sink.

THE EFFECT OF DENSITY

Why should a heavy wooden raft float while a pin sinks? And if a steel pin sinks, why does a steel boat float? The answer is density. This factor, rather than weight, determines whether things float or sink.

The density of an object is equal to its weight divided by its volume. Every substance, including water, has its own particular density at a given temperature (density varies as a substance gets hotter or colder). Any solid less dense than water floats, while one that is more dense sinks. However, a hollow object such as a boat floats if its *overall* density – its total weight divided by its total volume – is less than the density of water.

A wooden wall was therefore built around the top of the raft. To everyone's amazement, except of course my own, both raft and cargo floated safely across.

In deference to the beast's obvious displeasure at even the risk of a second dunking, I had suggested that it be clothed in a rubber diving suit. I confess that I am still unable to fully explain what happened shortly after landing. The mammoth was tied up near the dock, wearing the suit and getting some sun. The suit began to expand and to my astonishment the enormous beast rose into the air. Quite why this happened is completely baffling. Perhaps it had something to do with the spirit of the air? We still have so much to learn.

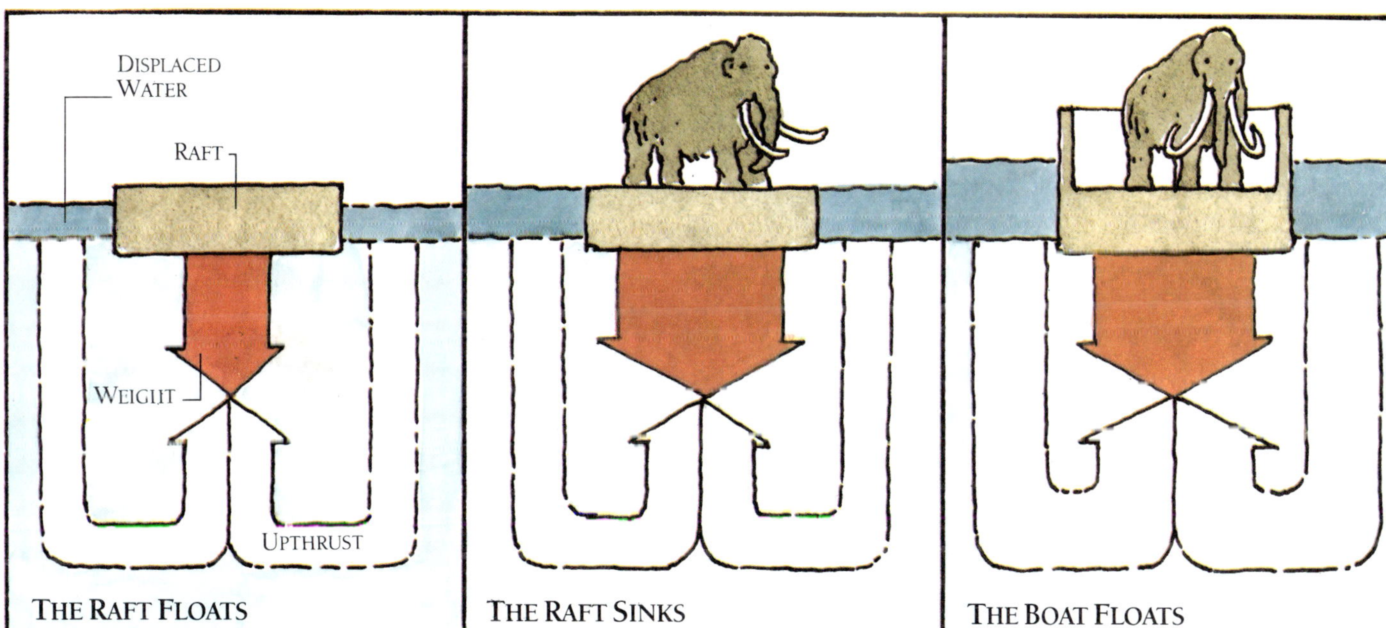

THE RAFT FLOATS
The force of the upthrust created by the displaced water equals the weight of the raft, supporting the raft so that it floats.

THE RAFT SINKS
The weight of the raft and mammoth exceeds the upthrust because little extra water has been displaced. The raft sinks.

THE BOAT FLOATS
The boat displaces more water, producing sufficient upthrust to support its weight plus that of the mammoth, so it floats.

THE SUBMERSIBLE

Submersibles are designed for use at great depths. They need to be able to sink, to rise and also to float underwater. They do this by altering their weight with a system of ballast tanks which can hold either air or water. If a craft's ballast tanks are flooded with water, the craft's weight increases. If the water is then expelled by compressed air, the weight decreases. By adjusting the amount of water in the tanks, the craft's weight and buoyancy can be precisely regulated.

Submersibles are designed to perform delicate tasks deep underwater, and are therefore designed to withstand high pressure and to be highly maneuverable. They do not need to move at speed and therefore, unlike submarines, they are not streamlined.

THE SUBMARINE

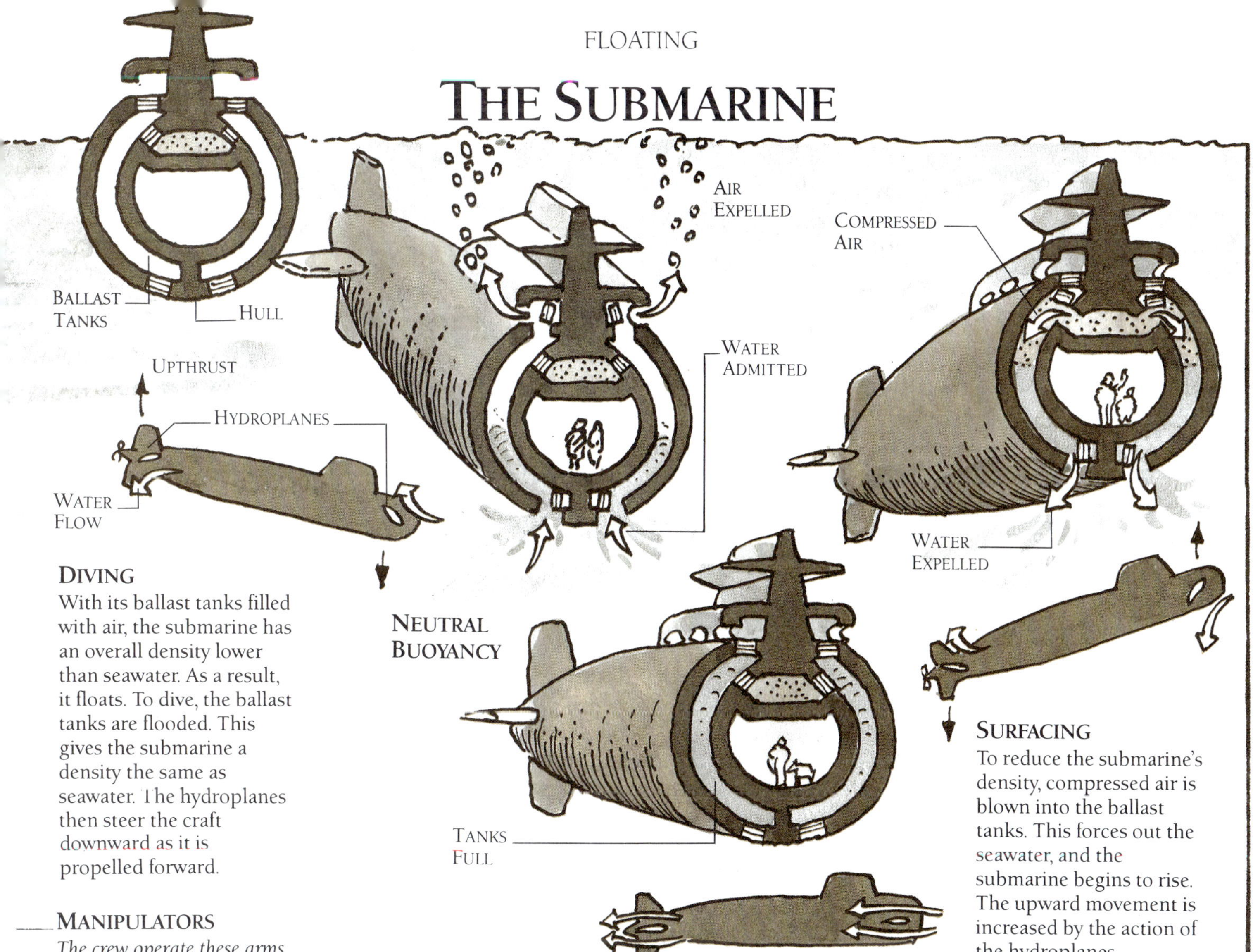

DIVING
With its ballast tanks filled with air, the submarine has an overall density lower than seawater. As a result, it floats. To dive, the ballast tanks are flooded. This gives the submarine a density the same as seawater. The hydroplanes then steer the craft downward as it is propelled forward.

SURFACING
To reduce the submarine's density, compressed air is blown into the ballast tanks. This forces out the seawater, and the submarine begins to rise. The upward movement is increased by the action of the hydroplanes.

A submarine works in much the same way as a submersible, with the exception that it is able to use the force driving it forward to control its depth. Fins on either side called hydroplanes swivel to deflect the flow of water around the hull. This lifts or drops the nose so that the submarine can ascend or descend under the power of its propellers. As in the submersible, buoyancy is controlled by ballast tanks. These are flooded when diving; when surfacing, the water in the tanks is expelled by compressed air.

MANIPULATORS
The crew operate these arms, which are equipped with lights and gripping claws, from the crew compartment.

PASSENGER BOAT

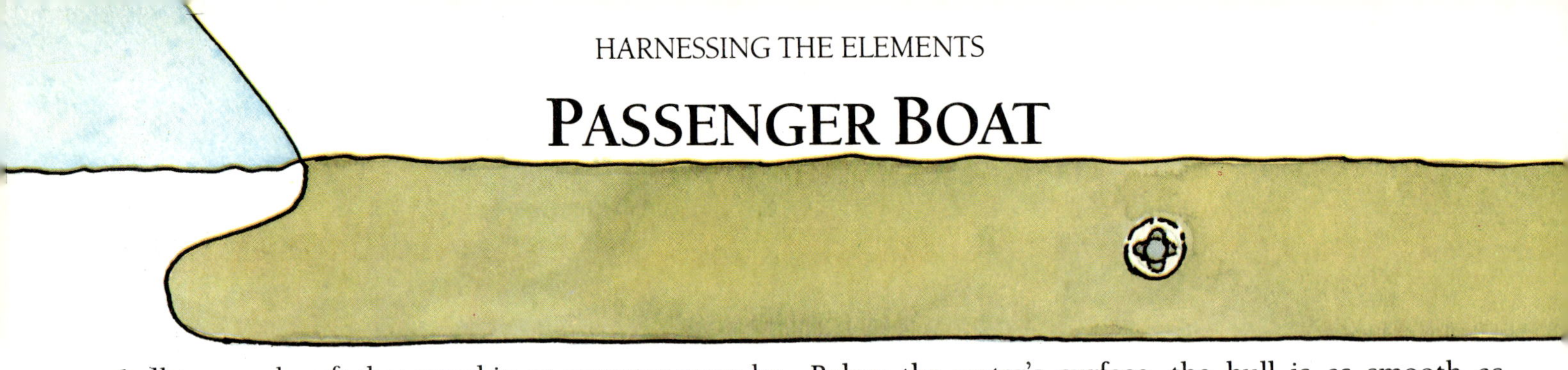

All powered craft that travel in or on water move by imparting movement to the water or air around them, and they steer by altering the direction in which the water or air flows. In a large ship, power is provided by the propellers, and the direction is governed by a rudder. But large ships also need to be able to control their movement sideways when docking, and their roll during heavy seas. They do this with bow thrusters and stabilizers, two devices that act in the same way as the main propellers and the rudder.

Below the water's surface, the hull is as smooth as possible to reduce the ship's water resistance and increase efficiency. The bow thrusters are recessed, and therefore do not disturb the water flow. The stabilizers are retractable, folding away inside hatches when not in use. At the bow, the hull may project forward beneath the water in a huge bulb. This bulb reduces the bow wave that the ship makes as it slices through the water. The water resistance of the ship is lessened, and this raises the speed or saves fuel.

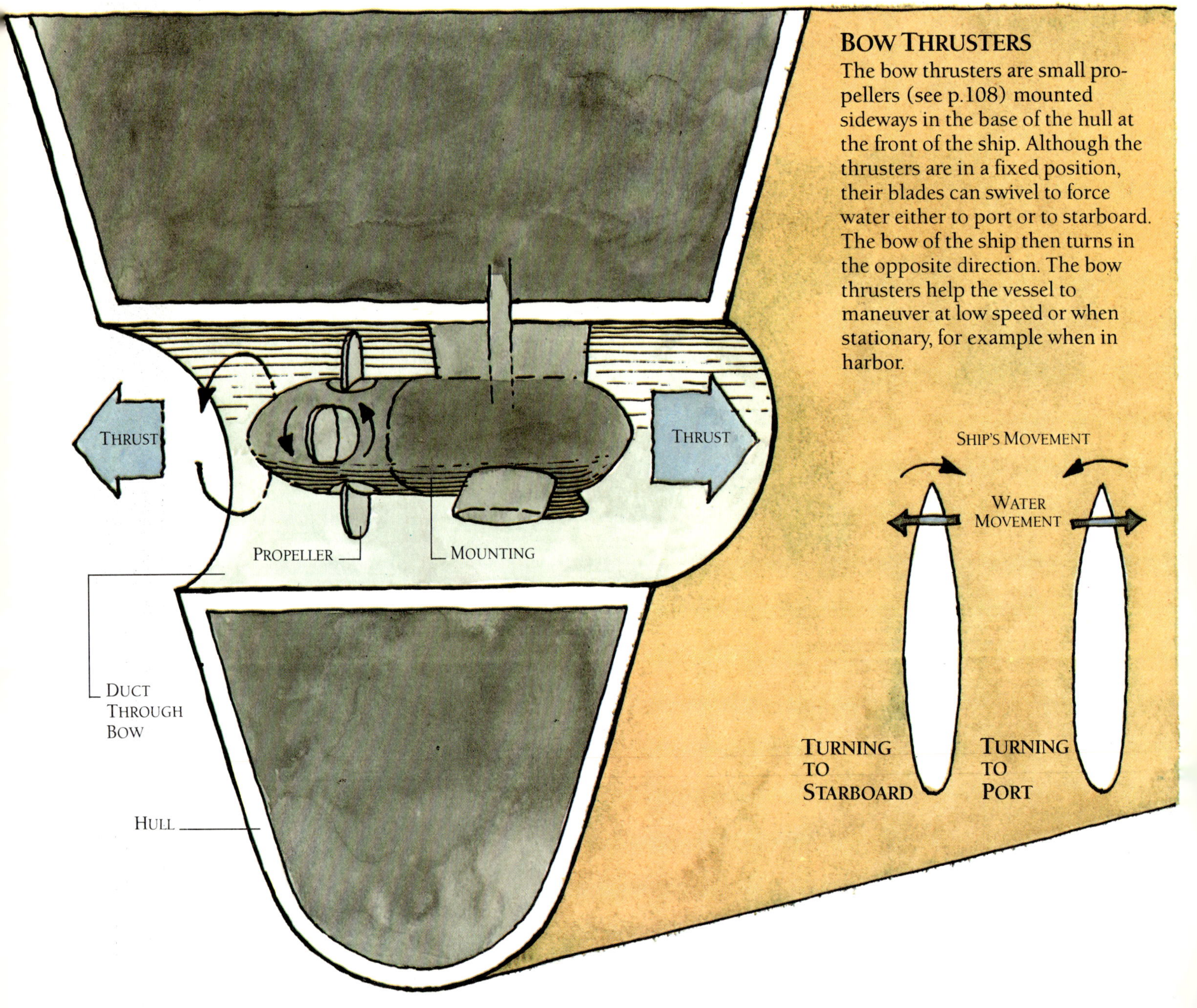

BOW THRUSTERS

The bow thrusters are small propellers (see p.108) mounted sideways in the base of the hull at the front of the ship. Although the thrusters are in a fixed position, their blades can swivel to force water either to port or to starboard. The bow of the ship then turns in the opposite direction. The bow thrusters help the vessel to maneuver at low speed or when stationary, for example when in harbor.

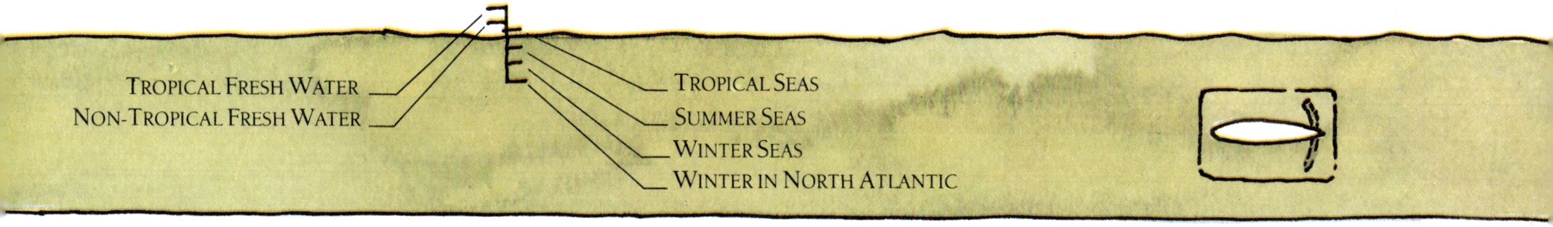

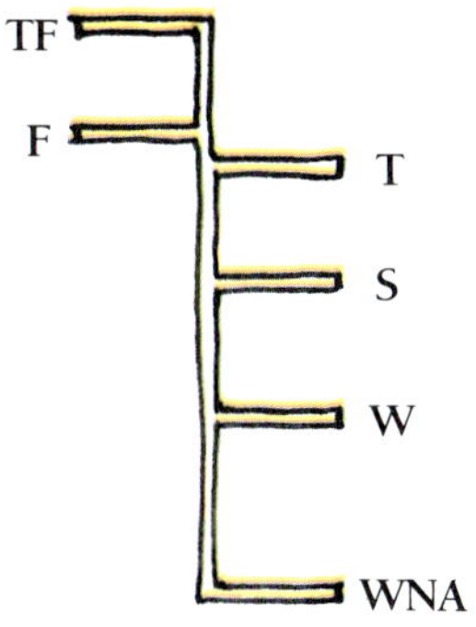

Plimsoll Line or Load Line

The loading of a ship is regulated by marks on the side of the hull. The lines indicate loading limits for a variety of seas and seasons. As the ship is loaded, it settles deeper in the water. For safety, it must not be loaded so that the relevant mark goes below the water.

The different levels are due to differences in the density of water, and therefore the upthrust it produces. Salt water is denser than fresh water, and cold water denser than warm water.

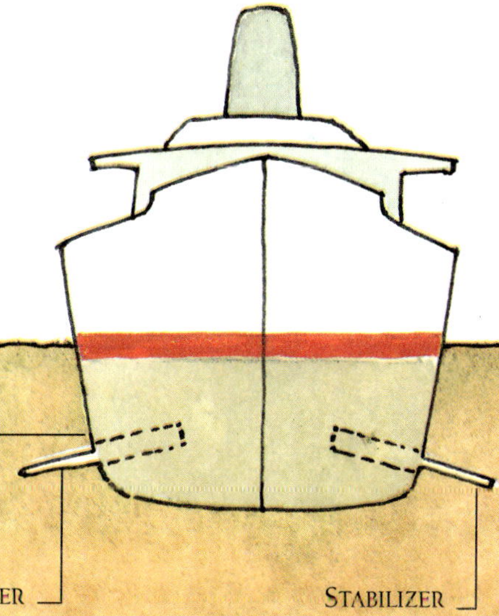

Stabilizers

Ships roll from side to side as they encounter high waves. To reduce rolling, they have stabilizers. These are two large fins that extend from the hull. The fins swivel as the hull begins to roll, acting like horizontal rudders (see p.109) to produce upward or downward forces that counteract the roll. The stabilizers are often controlled by a gyroscope (see p.80) that senses the ship's motion. The fins can reduce rolling movement by 90 per cent.

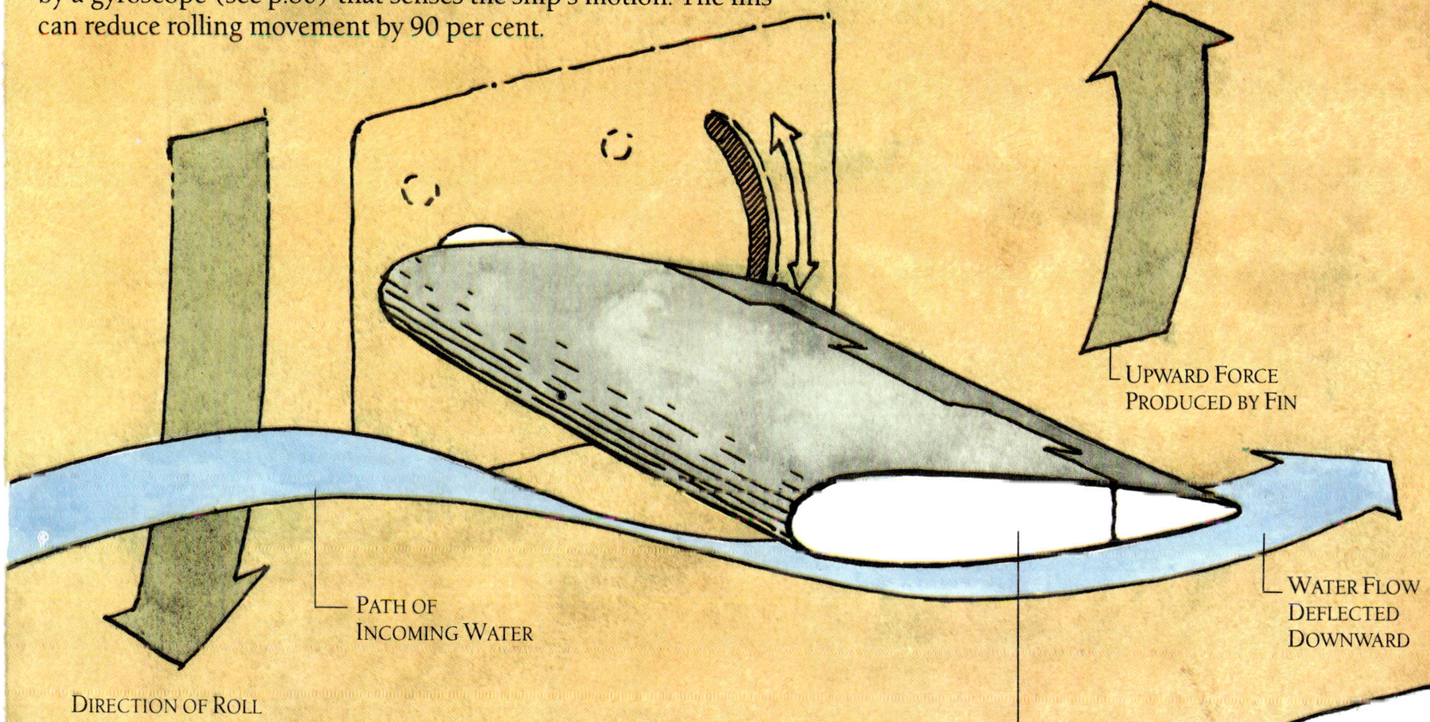

Extended Fin

When the hull rolls down, the front edge of the fin tilts up to deflect the water flow downward. This produces an upward force on the fin, which stops the roll. Tilting the fin in the opposite direction stops an upward roll.

Passenger Boat

Most craft that travel on water need a source of power to propel them forward and also a means of steering. These requirements are met by propellers and rudders, two devices which work by the same pair of principles.

The first principle is action and reaction. As the blades of a propeller spin, they strike the water and make it move toward the rear of the vessel. The force with which the blades move the water is called the action. The water pushes back on the blades as it begins to move, producing an equal force called the reaction which drives the propeller forward.

The second principle is called suction. The surface of each propeller blade is curved so that the blade has the shape of an airfoil (see p.115). Water flows around the blade as it rotates, moving faster over the front surface. The faster motion lowers the water pressure at the front surface, and the blade is sucked forward.

Overall, a combination of reaction and suction drives the spinning propeller through the water.

A rudder affects the water flowing around it in the same way. Reaction and suction produce a turning force that changes the boat's direction.

Propellers drive most surface vessels as well as submarines and submersibles. They also work in air, powering airships and many aircraft. Virtually all forms of water transport and most forms of air transport steer with rudders.

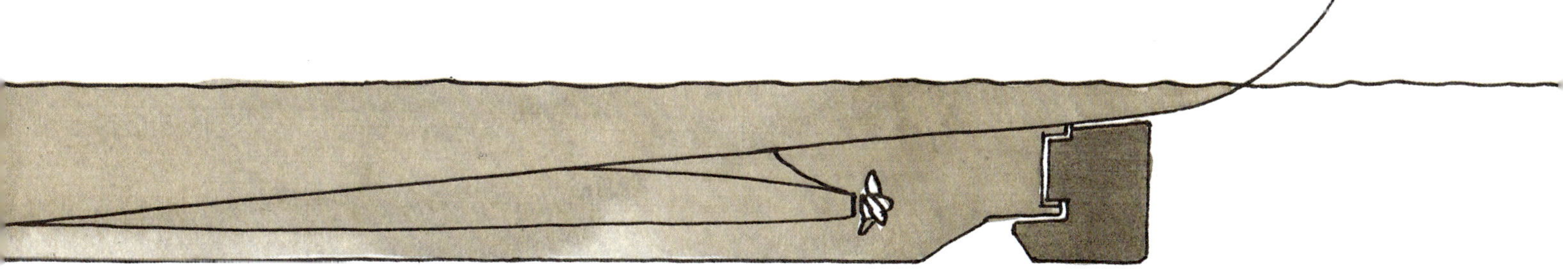

PROPELLER

The blades of a ship's propeller are broad and curved like scimitars to slash strongly through the water. The propeller does not turn rapidly, but, having broad blades, it moves a large amount of water to produce a powerful reaction as well as strong suction. Small high-speed vessels have fast-spinning propellers with narrow blades that move less water but which give high suction. At very high speeds the propeller may make water vaporize, causing a loss of power.

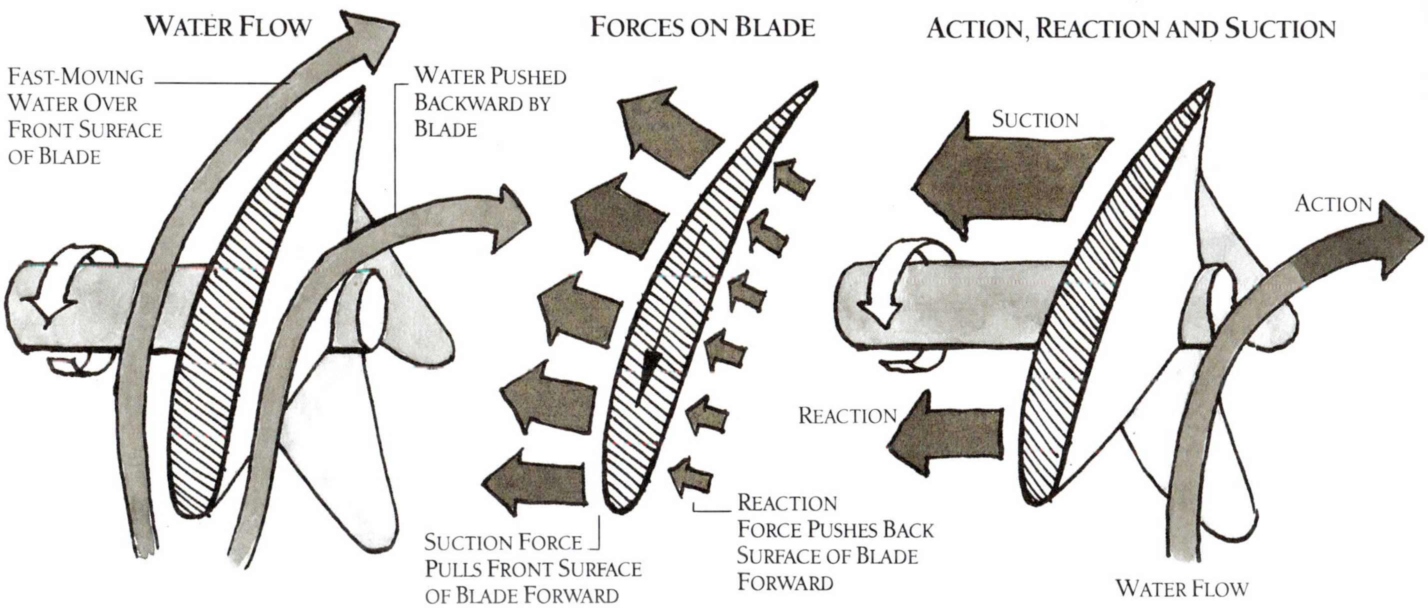

RUDDER

The rudder acts on the water flowing past the vessel and the backward flow generated by the propeller. The rudder blade swivels to deflect this flow. As the water changes direction, it pushes back with a reaction force and the blade moves in the opposite direction. Suction produced by water flowing around the blade assists reaction. These forces move the stern of the boat and the whole vessel turns around its center so that the bow points in a new direction.

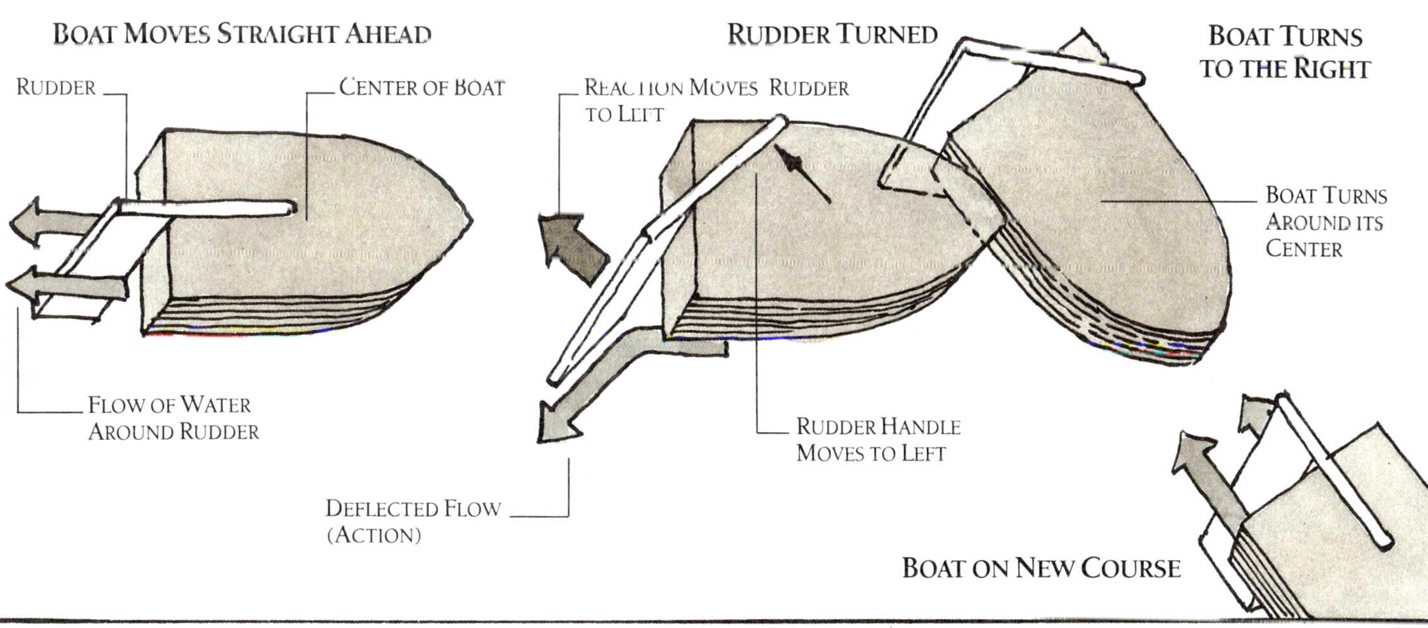

THE WINDSURFER

Modern sailing craft from the windsurfer to the racing yacht can use the power of the wind to propel them in any direction, no matter which quarter the wind may blow from.

This versatility is achieved with a triangular sail that can be shifted around the boat's mast to engage the wind at various angles. The sail is able to propel the boat at any angle to the wind, except head-on. However, sailing boats are able to make progress into the wind, although in an indirect way. They do this by "tacking", or following a zig-zagging course which keeps the sail at an angle to the wind, and so enables it to provide power.

The windsurfer is the simplest craft with a movable sail. It is basically a raft with a sail on a tilting mast and a small keel beneath. The person aboard the windsurfer grips a curved bar to move the sail in any direction to take advantage of the wind. The sail not only drives the windsurfer forward but also steers it.

SAILING BEFORE THE WIND

When the wind is directly behind the windsurfer, the sail is held at right angles to the wind. The force of the wind pushing the sail drives the board forward.

SAILING ACROSS THE WIND

The sail is still held at right angles to the wind, but water resistance on the keel prevents it moving sideways. The force of the wind is split into thrust driving the board forward and a heeling force acting on the sail.

SAILING INTO THE WIND

The sail is held edge-on to the wind so that the wind blows around it. The wind inflates the sail, curving it so that the sail becomes an airfoil (see p.115). The air flow produces a suction force that pulls the sail at right angles to the wind. This pull is split into thrust and heeling force to move the windsurfer forward.

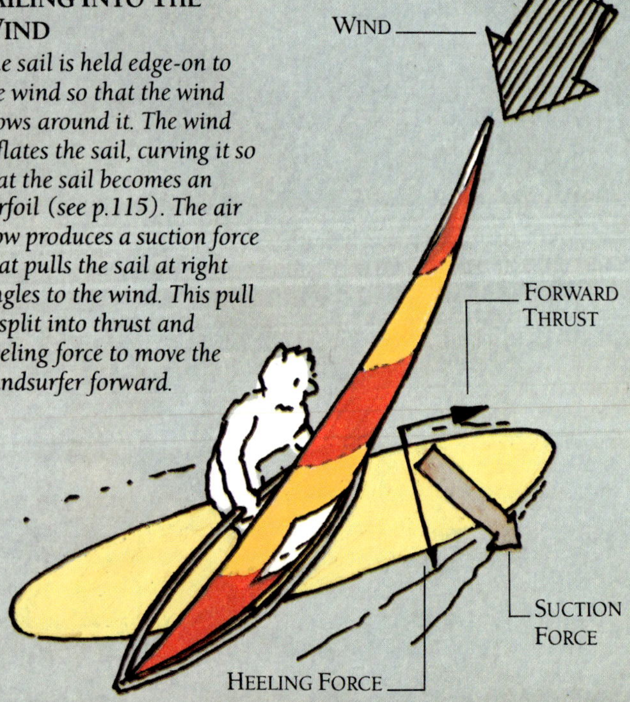

TURNING AWAY FROM THE WIND

If the mast is tipped forward, the heeling force on the sail moves in front of the keel. The water resistance on the keel and the heeling force combine to turn the board away from the wind.

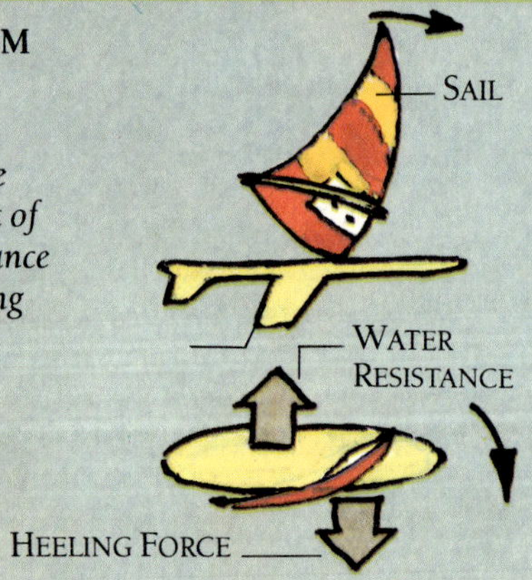

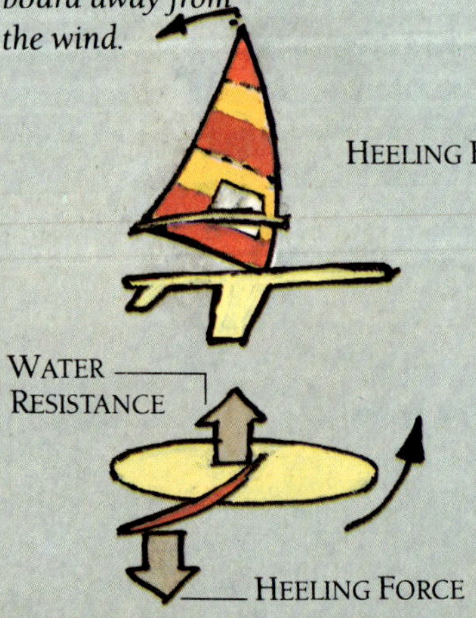

TURNING INTO THE WIND

Tipping the mast backward makes the windsurfer turn into the wind. The heeling force moves behind the keel. The water resistance on the keel and the heeling force on the sail combine to turn the board into the wind.

THE YACHT

A yacht usually has two triangular sails – the mainsail and the jib. The sails propel the yacht before, across or into the wind in the same way as the windsurfer. When sailing into the wind, the two sails combine to act as one large airfoil with a slot in the center. The slot channels air over both sails, producing a powerful suction force. This force splits into two separate forces – thrust that propels the yacht forward and a heeling force that tilts it over. Water resistance on the hull and keel prevent the yacht moving sideways.

A yacht is steered with a rudder (see p.109), which deflects the flow of water that passes the hull to turn the yacht in the required direction. As the yacht turns, the crew let out or pull in the sails so that they take up the best angle to the wind. A balloon-like spinnaker sail may be used when the yacht is sailing before the wind.

THE AIRSHIP

An airship has a vast envelope that creates a powerful upthrust to lift the substantial weight of the cabin, engine, fans and passengers. The bulk of the envelope contains helium, a light gas which reduces the weight of the airship so that it is equal to the upthrust, thereby producing neutral buoyancy. Inside the envelope are compartments of air called ballonets. Pumping air out of or into the ballonets decreases or increases the airship's weight and it ascends or descends. The airship also has propellers called ducted fans that drive it through the air and which swivel to maneuver the airship at take-off or landing. Tail fins and a rudder can tilt or turn the whole craft as it floats through the sky. In this way, the airship travels from place to place like an airborne combination of submarine and submersible.

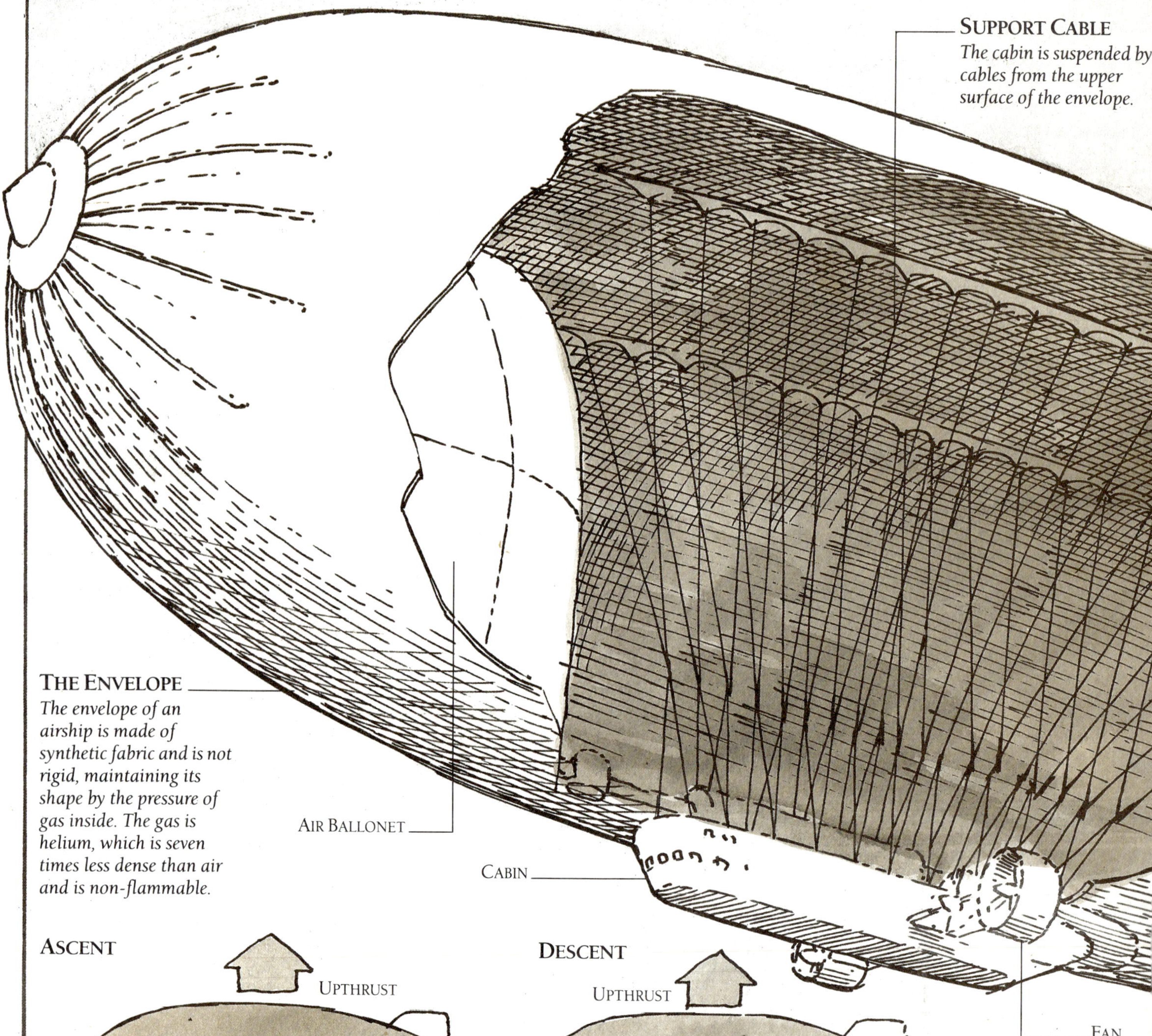

ASCENT

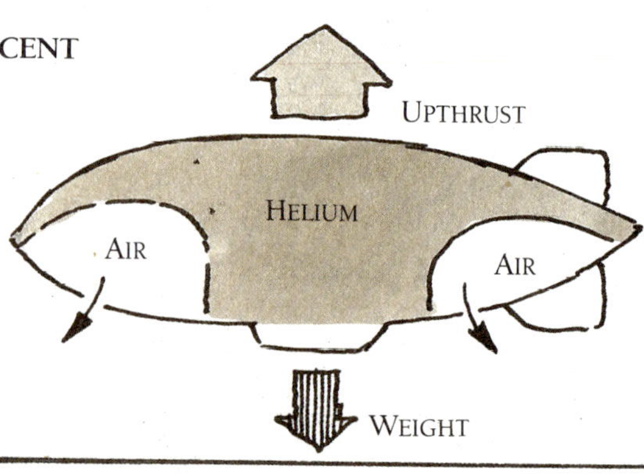

DESCENT

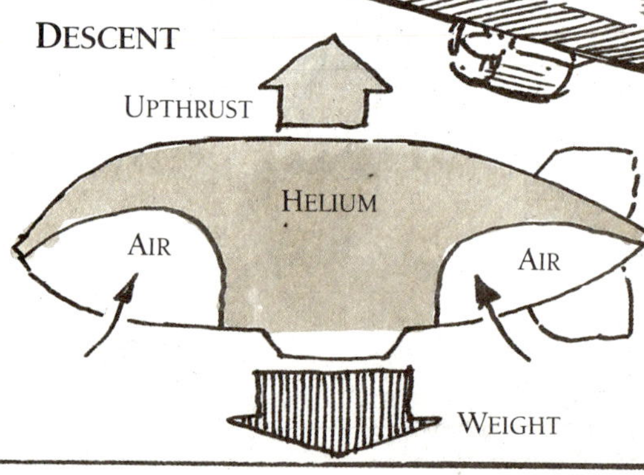

THE HOT-AIR BALL

The envelope of a hot-air balloon has to be big so that it can displace a large amount of air, thereby creating sufficient upthrust to float the basket and its occupants through the air. The balloon works like an underwater craft in reverse. Operating the burner heats the air in the envelope; the air expands and some escapes from the envelope. The overall weight decreases, and the upthrust carries the balloon upward. When the burner cuts out, the air in the envelope cools and contracts. Air now enters the envelope, increasing the balloon's weight and causing it to descend. Fast descent can be achieved by opening a port in the top of the envelope. This partially deflates the envelope to reduce the upthrust.

A hot-air balloon has no means of propulsion and drifts with the wind. Intermittent blasts of the burner enable the balloon to stay at a constant height.

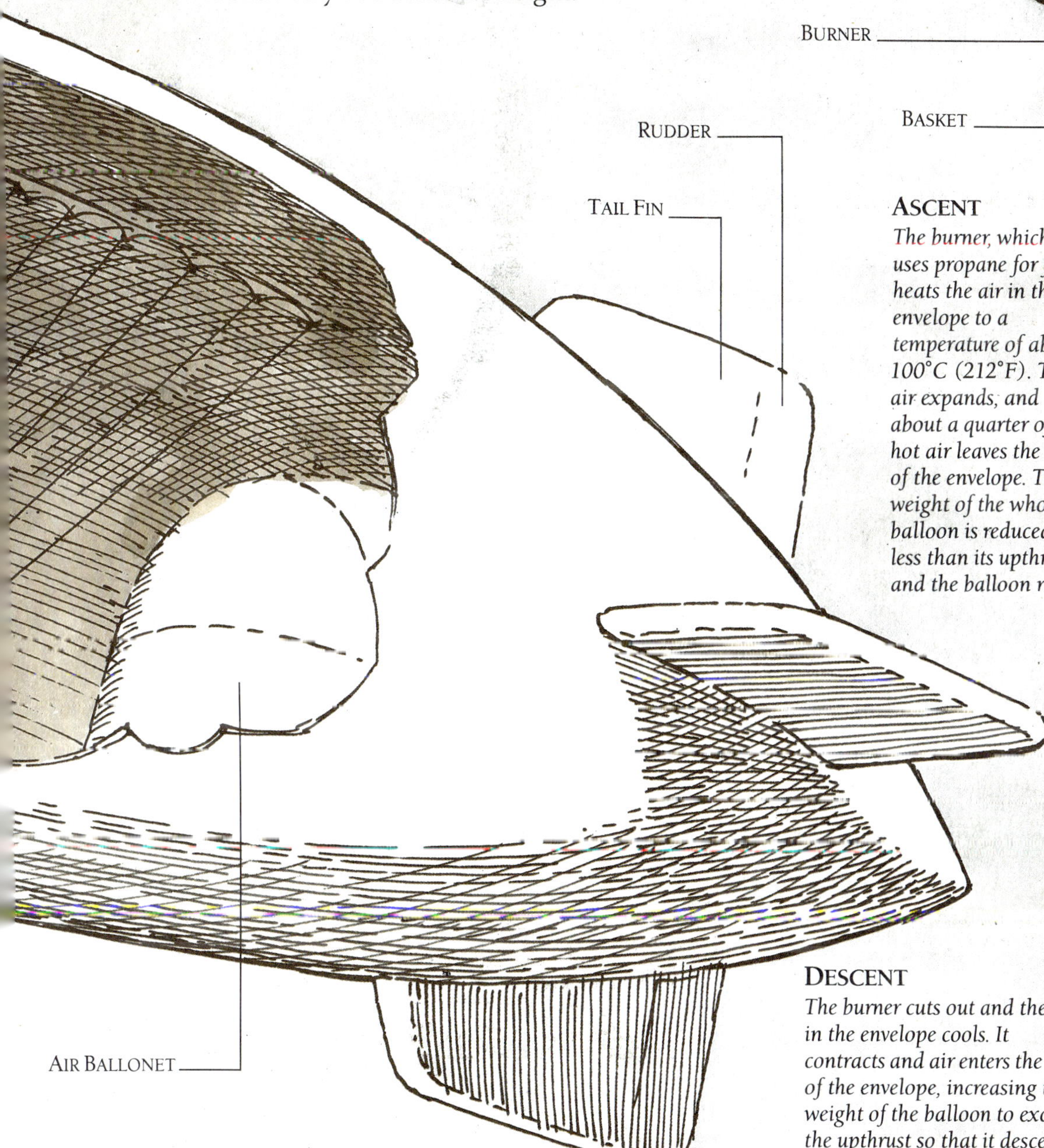

ASCENT

The burner, which uses propane for fuel, heats the air in the envelope to a temperature of about 100°C (212°F). The air expands, and about a quarter of the hot air leaves the base of the envelope. The weight of the whole balloon is reduced to less than its upthrust, and the balloon rises.

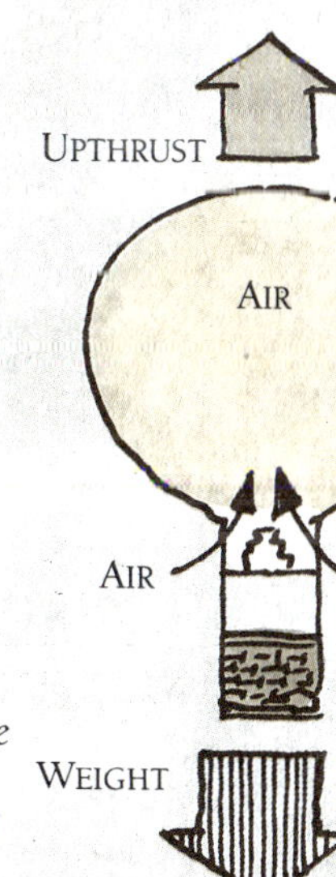

DESCENT

The burner cuts out and the air in the envelope cools. It contracts and air enters the base of the envelope, increasing the weight of the balloon to exceed the upthrust so that it descends.

FLYING

ON THE ADVENT OF AIRFREIGHT

One day I chanced upon a delivery mammoth from a local awning manufacturer sighing under the weight of a large wooden frame over which was stretched a piece of canvas. Apparently waiting for its driver, the mammoth was tethered to a tree with the awning firmly secured to its back. Suddenly the wind picked up, lifting the startled beast dramatically into the sky. I noticed that as long as the wind blew and the rope between tree and mammoth held, the creature remained airborne. . .

. . .but when the wind abruptly died, the mammoth returned to the ground without ceremony, destroying not only the awning but also the manufacturer's entire premises.

HEAVIER-THAN-AIR FLIGHT

In the struggle to overcome its not inconsiderable weight and launch itself into the air, the mammoth becomes in turn a kite, a glider and finally a powered aircraft. These are three quite different ways by which an object that is heavier than air can be made to fly.

Like balloons and airships, heavier-than-air machines achieve flight by generating a force that overcomes their weight and which supports them in the air But because they cannot float in air, they work in different ways to balloons.

Kites employ the power of the wind to keep them aloft, while all winged aircraft, including gliders and helicopters, make use of the airfoil and its power of lift. Vertical take-off aircraft direct the power of their jet engines downward and heave themselves off the ground by brute force.

The two principles that govern heavier-than-air flight are the same as those that propel powered vessels – action and reaction, and suction (see pp.108-9). When applied to flight, suction is known as lift.

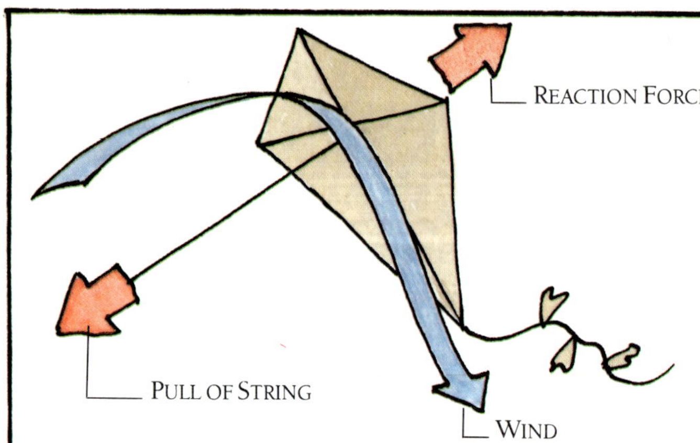

KITE

A kite flies only in a wind, and it is held by its string so that it deflects the wind downward. The wind provides the force for flight. It exerts a reaction force that equals the pull of the string and supports the kite in the air.

During my own experiments with awning delivery, I discovered that by securing a slightly curved awning to a volunteer mammoth's back, the danger and considerable expense of crash landings could be greatly reduced. Should the wind speed drop or the rope break, the mammoth would usually glide back to Earth in a gentle spiral. I planned one further improvement in which friction-reducing foot-gear would enable the mammoth to leave the ground simply by blowing backward with its trunk.

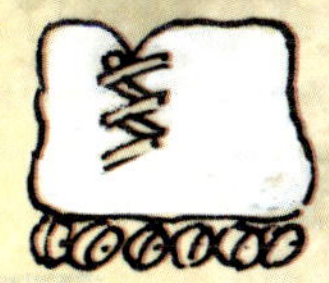

However, despite repeated attempts, the mammoth never got far enough off the ground to make this novel form of delivery a practical procedure. Even with the specially designed foot-gear in place, landings remained somewhat unpredictable.

I recall one most unfortunate incident in which a mammoth had to be completely bandaged after an unusually clumsy four-point landing. This resulted in the rather interesting streamlined form depicted here. It is not one that I feel could ever leave the ground.

AIRFOIL

The cross-section of a wing has a shape called an airfoil. As the wing moves through the air, the air divides to pass around the wing. The airfoil is curved so that air passing above the wing moves faster than air passing beneath. Fast-moving air has a lower pressure than slow-moving air. The pressure of the air is therefore greater beneath the wing than above it. This difference in air pressure forces the wing upward. The force is called lift.

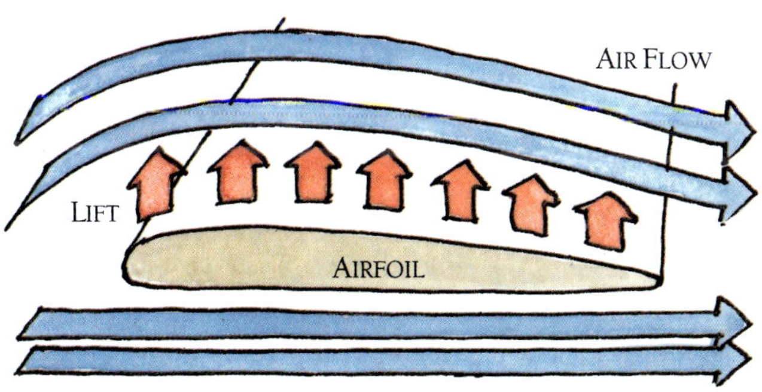

GLIDER

A glider is the simplest kind of winged aircraft. It is first pulled along the ground until it is moving fast enough for the lift generated by the wings to exceed its weight. The glider then rises into the air and flies. After release, the glider continues to move forward as it drops slowly, pulled by a thrust force due to gravity. Friction with the air produces a force called drag that acts to hold the glider back. These two pairs of opposing forces – lift and weight, thrust and drag – act on all aircraft.

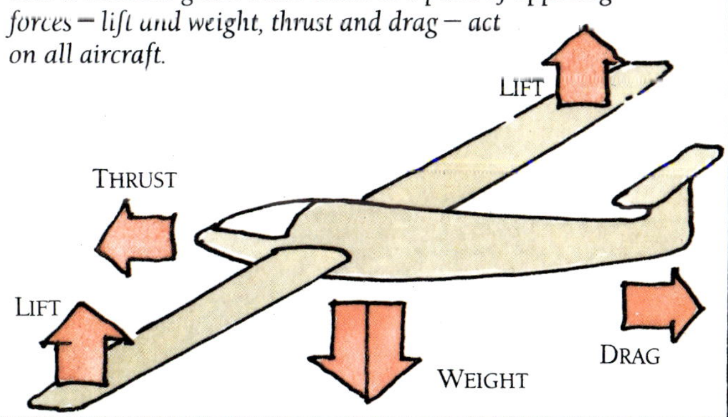

THE AIRPLANE

Adding an engine to a flying machine gives it the power to dispense with winds and air currents that govern the flight of unpowered craft such as balloons and gliders. In order to steer an airplane, a system of flaps is used. These act just like the rudder of a boat (see p.109). They deflect the air flow and turn or tilt the airplane so that it rotates around its center of gravity, which in all airplanes lies between the wings.

Airplanes usually have one pair of wings to provide lift, and the wings and tail have flaps that turn or tilt the aircraft in flight. Power is provided by a propeller (see p.108) mounted on the nose, or by several propellers on the wings, or by jet engines (see pp.168-9) mounted on the wings, tail, or inside the fuselage.

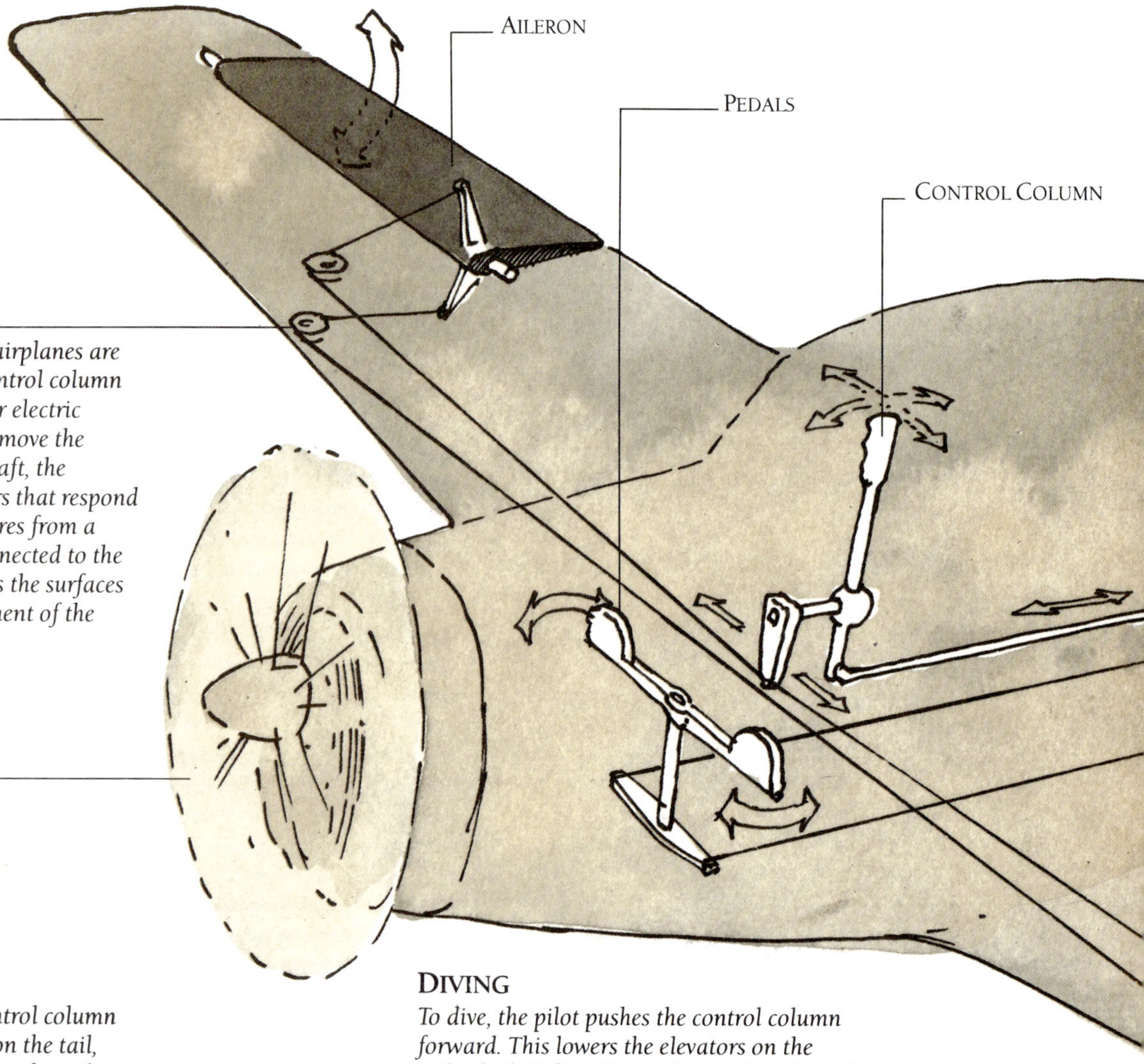

CONTROL CABLES

The control surfaces of many airplanes are physically connected to the control column by cables. Hydraulic systems or electric motors operated by the cables move the surfaces. In "fly-by-wire" aircraft, the surfaces are operated by motors that respond to control signals fed along wires from a computer. The computer is connected to the control column, and it controls the surfaces to produce the required movement of the plane.

CLIMBING

To climb, the pilot pulls the control column back. This raises the elevators on the tail, which deflect the air flow so that the tail drops. The nose rises and the aircraft climbs.

DIVING

To dive, the pilot pushes the control column forward. This lowers the elevators on the tail, which deflect the air flow so that the tail rises. The nose drops and the aircraft dives.

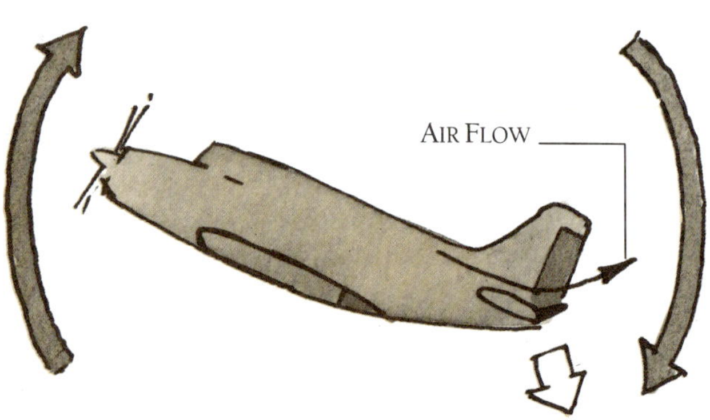

TURNING

To turn to the right or left, the pilot presses the pedals to swivel the rudder on the tail and moves the control column to the side to raise or lower the ailerons on the wings. The rudder turns the aircraft, and one aileron goes up while the other goes down to bank the aircraft as it turns.

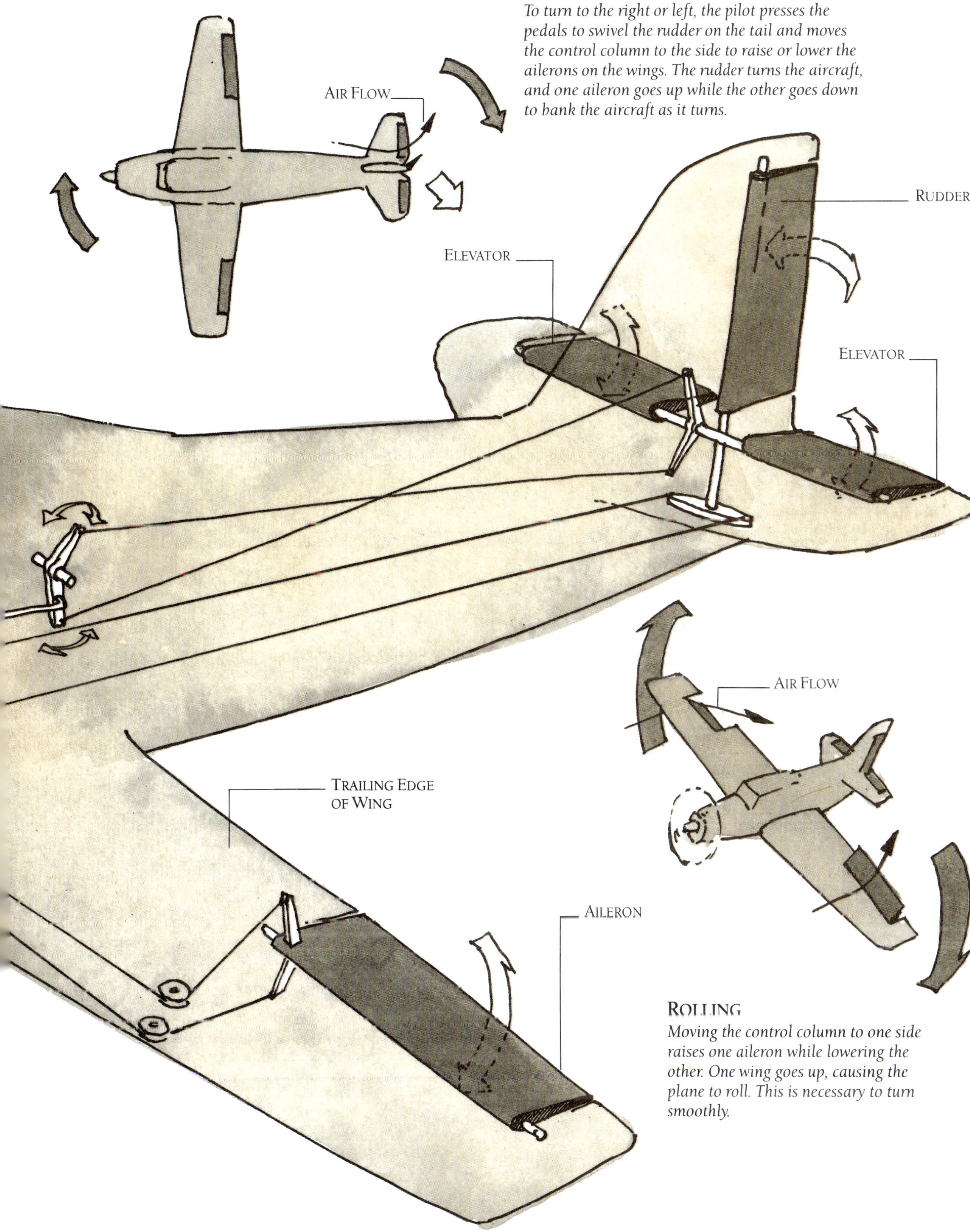

ROLLING

Moving the control column to one side raises one aileron while lowering the other. One wing goes up, causing the plane to roll. This is necessary to turn smoothly.

FLYING MACHINES

Many different flying machines now fill our skies. They range from solo sports and airobatic planes to wide-bodied and supersonic jet airliners which carry hundreds of passengers. Some, such as pedal-powered planes, lumber along just above the ground, while others, such as reconnaissance aircraft, streak at three times the speed of sound at a height three times that of Mount Everest.

There are also unpowered gliders, of which the returning space shuttles are the largest and hang gliders the simplest. Development in other directions has led to helicopters and vertical take-off aircraft which are capable of rising vertically and hovering in the air. There are also kites of all shapes and sizes, some large enough to carry a person.

Machines also fly through water. Hydrofoils flying through the waves employ exactly the same principles that keep winged airplanes aloft.

GLIDER
Being unpowered, a glider cannot travel fast and so has long straight wings that produce high lift at very low speed.

SPACE SHUTTLE
The space shuttle re-enters the atmosphere at very high speed, and so has a delta wing like a supersonic airliner. It then glides to a high-speed landing.

LIGHT AIRCRAFT
Short straight wings produce good lift and low drag at medium speed. Propellers or jet engines provide the power that produces the lift.

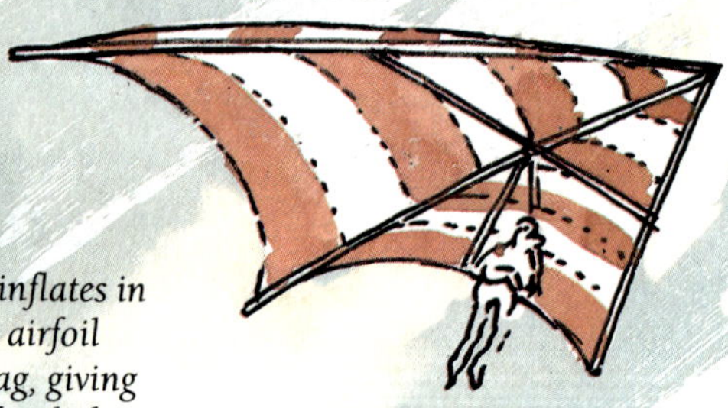

HANG GLIDER
The A-shaped wing inflates in flight to produce an airfoil with low lift and drag, giving low-speed flight with a light load.

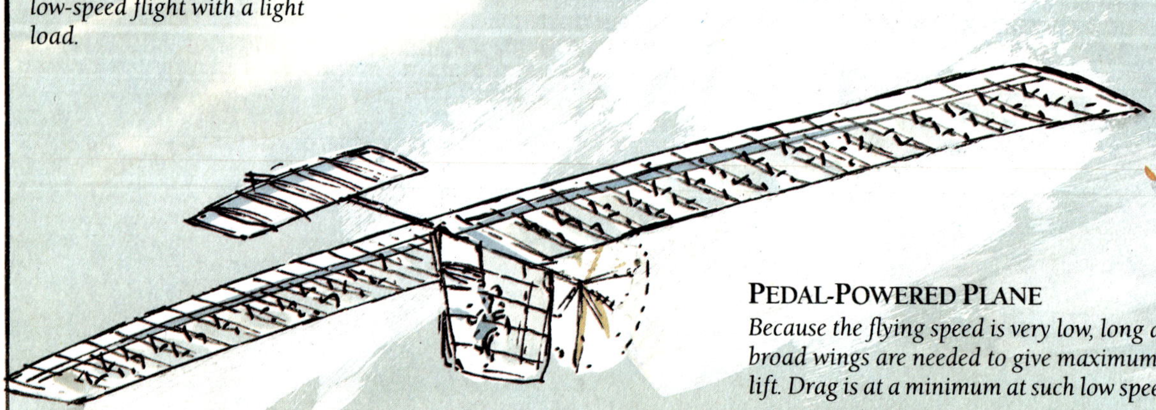

PEDAL-POWERED PLANE
Because the flying speed is very low, long and broad wings are needed to give maximum lift. Drag is at a minimum at such low speeds.

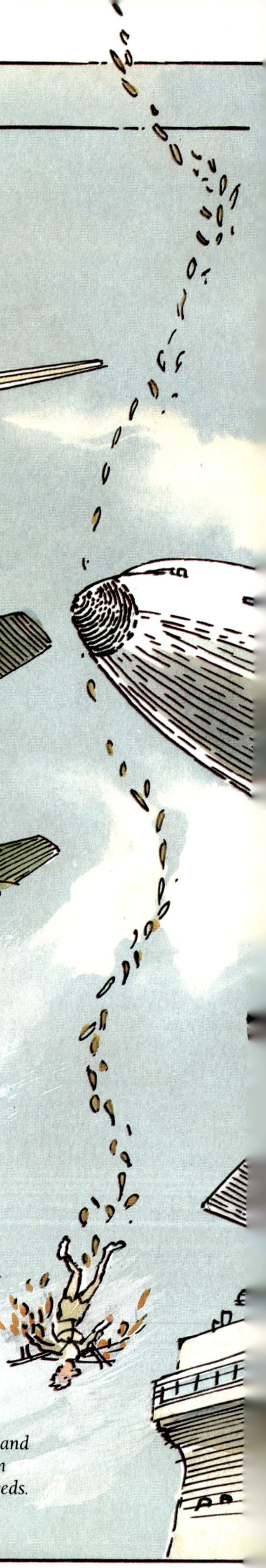

Forward-Swept Wings
This experimental design gives high lift and low drag to produce good maneuverability at high speed. Two small forward wings called canards aid control.

Swing-Wing Aircraft
The wings are straight at take-off and landing to increase lift so that take-off and landing speeds are low. In flight, the wings swing back to reduce drag and enable high-speed flight.

Supersonic Airliner
Aircraft that fly faster than the speed of sound often have dart-shaped delta wings. This is because a shock wave forms in the air around the aircraft, and the wings stay inside the shock wave so that control of the aircraft is retained at supersonic speed. Take-off and landing speeds are very high as lift is low.

Airliner
Swept-back wings are needed to minimize drag at high speed. However, lift is also reduced, requiring high take-off and landing speeds.

Flapping Wings
This is a highly efficient wing design that you should look out for, particularly in places where bird feeding is encouraged.

AIRLINER WING

On a small airplane, the wings need little more than simple hinged ailerons to control flight. An airliner wing, however, experiences enormous and varying forces both in the air and on the ground. To cope with these, it uses an array of complex flaps that change the wing's shape.

During take-off and landing, the wing shape needs to be very different to that needed for cruising. By adjusting the area of the flaps presented to the air, and their angle to it, a pilot is able to vary the amount of lift and drag generated by the wing to suit different phases of the flight.

There are four basic kinds of flaps. Leading-edge flaps line the front edge of the wing, while trailing-edge flaps take up part of the rear edge. These flaps extend to increase the area of the wing, producing more lift and also drag. Spoilers are flaps on top of the wing that rise to reduce lift and increase drag. Ailerons are flaps at the rear edge that are raised or lowered to roll the aircraft in a turn.

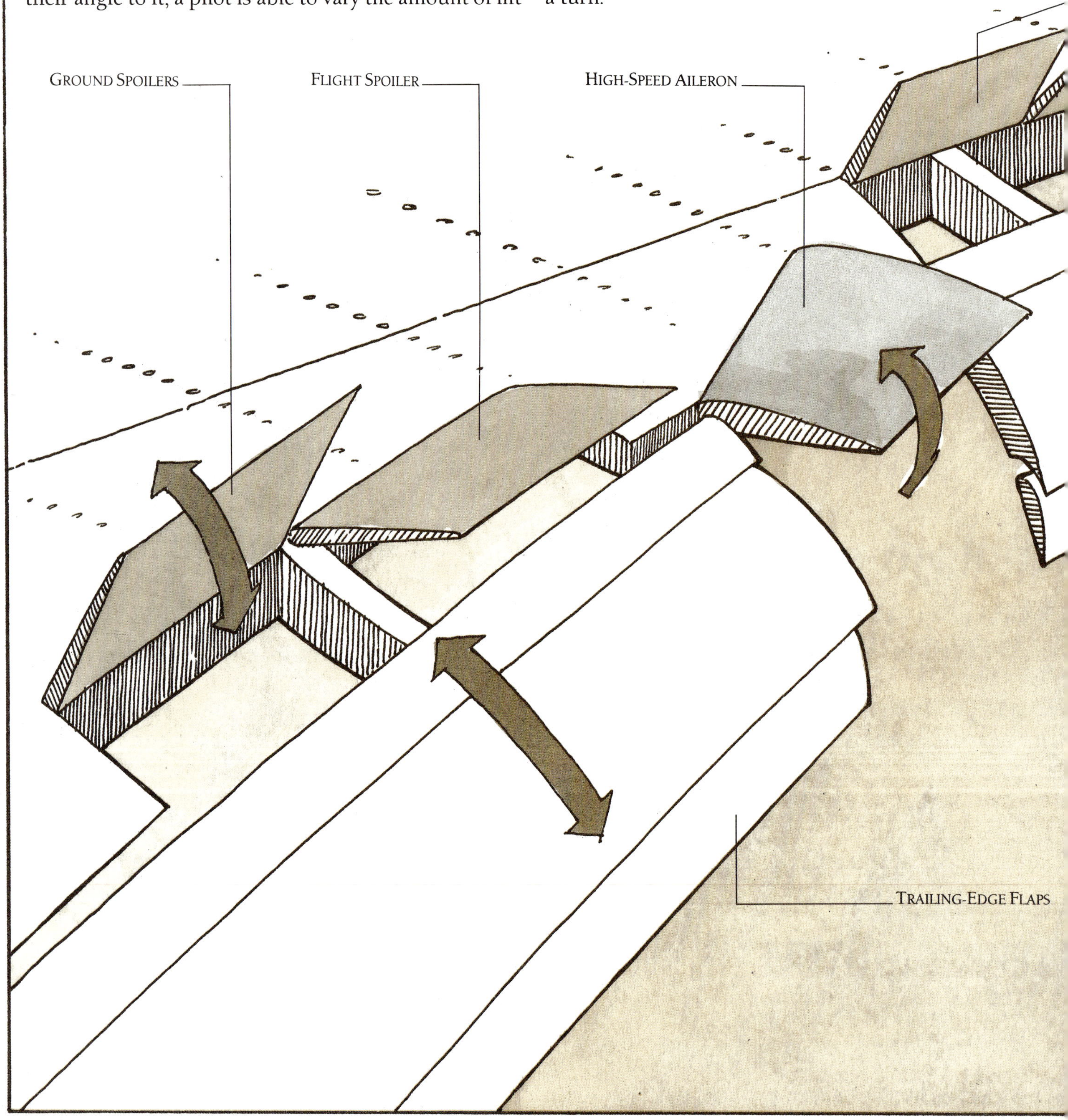

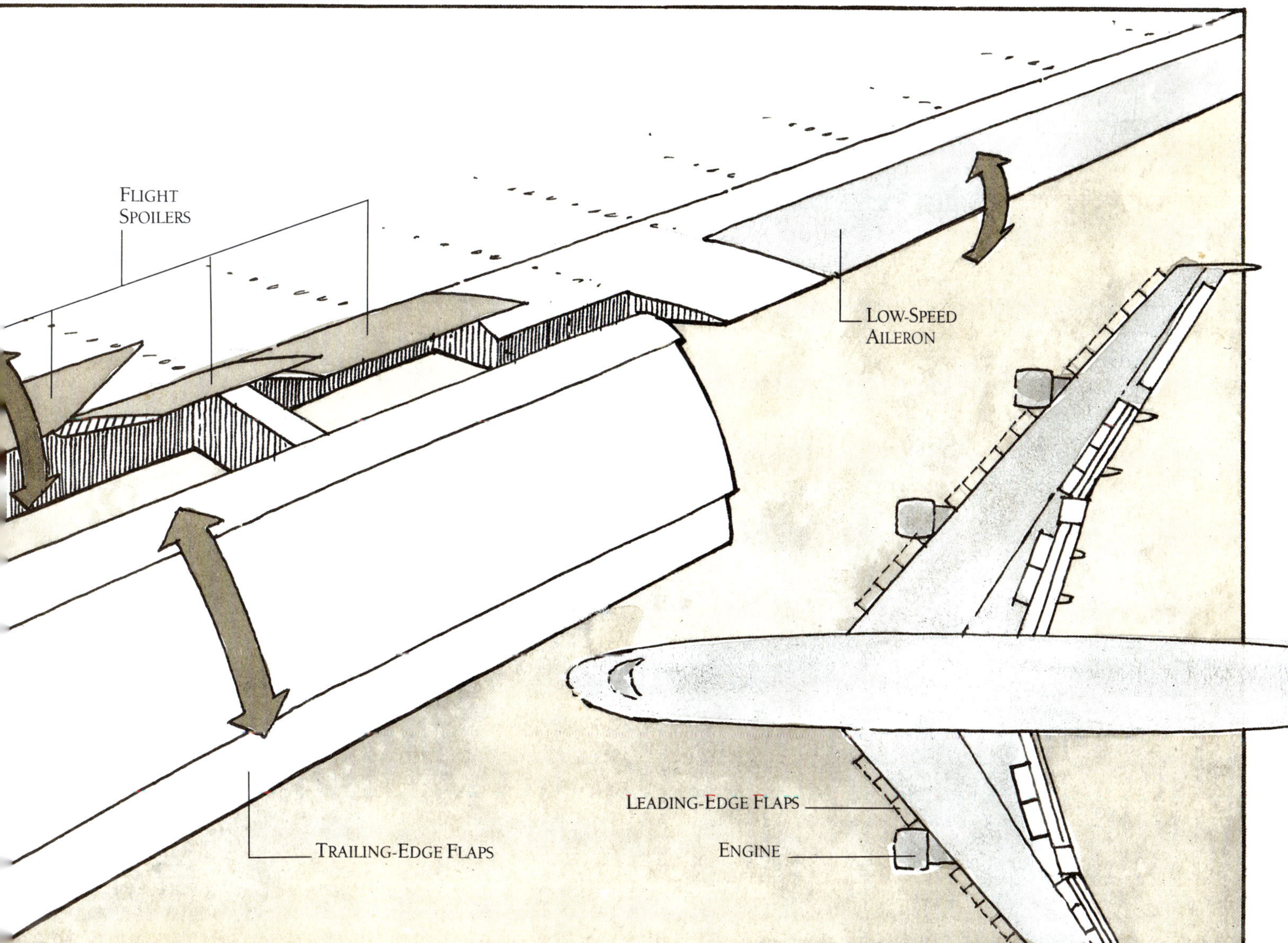

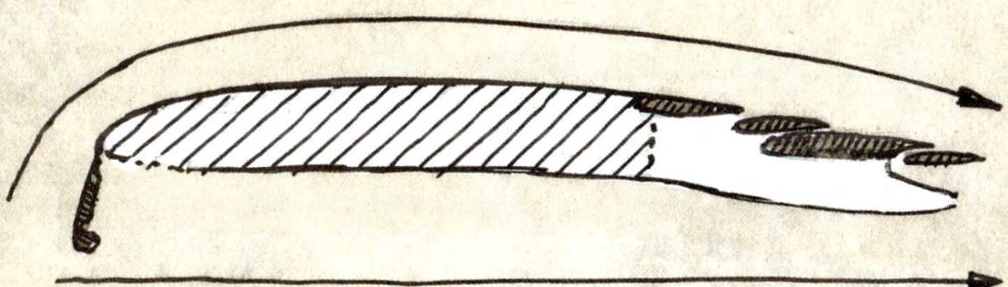

TAKE-OFF

The leading edge flaps extend and the trailing-edge flaps are raised to increase the area of the wing. This improves lift at low speed without incurring much extra drag, so that take-off speed is not high and the take-off run not prolonged.

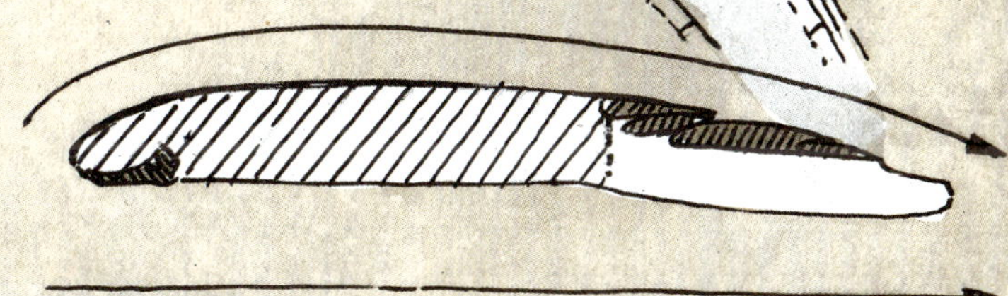

CRUISING

Leading-edge and trailing-edge flaps are both retracted for minimum drag, so the wing presents the minimum area to the oncoming air. The ailerons operate to control the flight, and may be assisted by the spoilers.

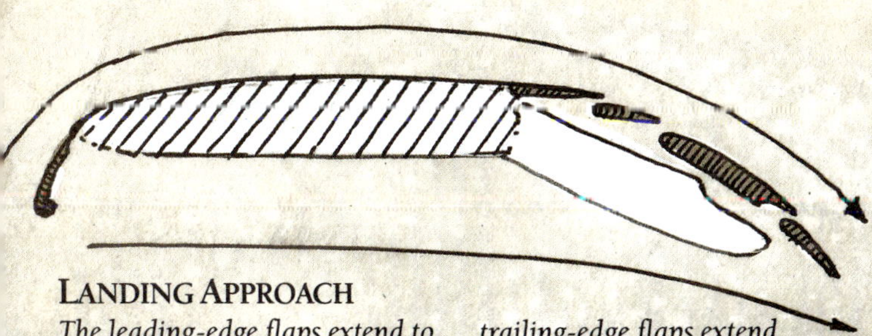

LANDING APPROACH

The leading-edge flaps extend to increase wing area and produce more lift at low speed. The trailing-edge flaps extend and droop to increase drag, slowing the aircraft for landing.

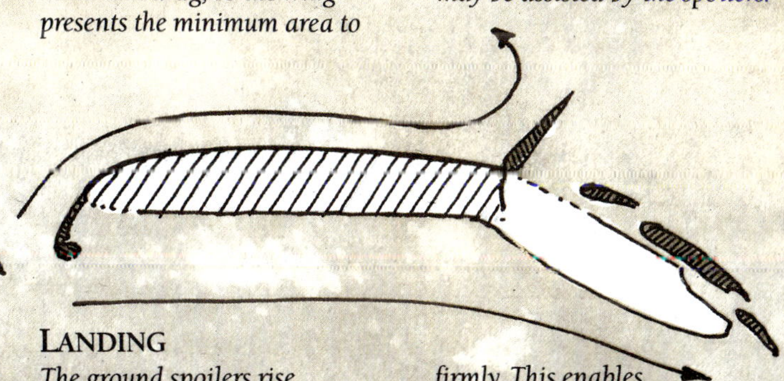

LANDING

The ground spoilers rise immediately on landing to reduce lift and push the aircraft down so that the wheels grip the runway firmly. This enables the brakes to work. The engines may reverse thrust to assist braking.

THE HELICOPTER

With its whirling rotors, a helicopter looks very different to an airplane. Yet, like an airplane, it too uses airfoils for flight. The blades of the helicopter's main rotor have an airfoil shape like the wings of a plane. But whereas a plane has to rush through the air for the wings to develop sufficient lift for flight, the helicopter moves only the rotor blades. As they circle, the blades produce lift to support the helicopter in the air and also to move it in the required direction. The angle at which the blades are set determines how the helicopter flies – hovering, vertical, forward, backward or sideways.

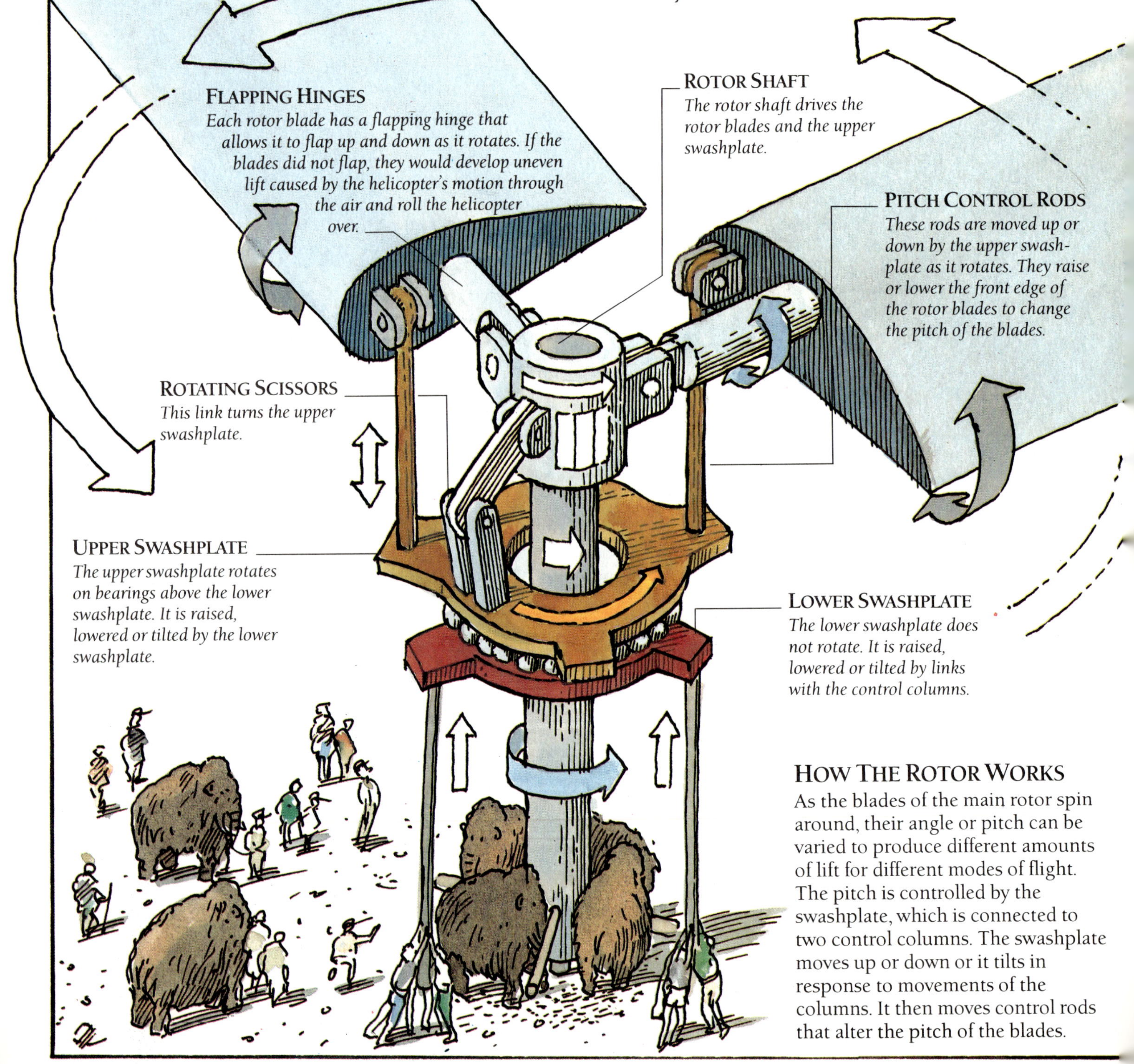

ROTOR BLADES
Most helicopter rotors have from three to six blades. Each is connected to a flapping hinge and a pitch control rod.

FLAPPING HINGES
Each rotor blade has a flapping hinge that allows it to flap up and down as it rotates. If the blades did not flap, they would develop uneven lift caused by the helicopter's motion through the air and roll the helicopter over.

ROTOR SHAFT
The rotor shaft drives the rotor blades and the upper swashplate.

PITCH CONTROL RODS
These rods are moved up or down by the upper swashplate as it rotates. They raise or lower the front edge of the rotor blades to change the pitch of the blades.

ROTATING SCISSORS
This link turns the upper swashplate.

UPPER SWASHPLATE
The upper swashplate rotates on bearings above the lower swashplate. It is raised, lowered or tilted by the lower swashplate.

LOWER SWASHPLATE
The lower swashplate does not rotate. It is raised, lowered or tilted by links with the control columns.

HOW THE ROTOR WORKS
As the blades of the main rotor spin around, their angle or pitch can be varied to produce different amounts of lift for different modes of flight. The pitch is controlled by the swashplate, which is connected to two control columns. The swashplate moves up or down or it tilts in response to movements of the columns. It then moves control rods that alter the pitch of the blades.

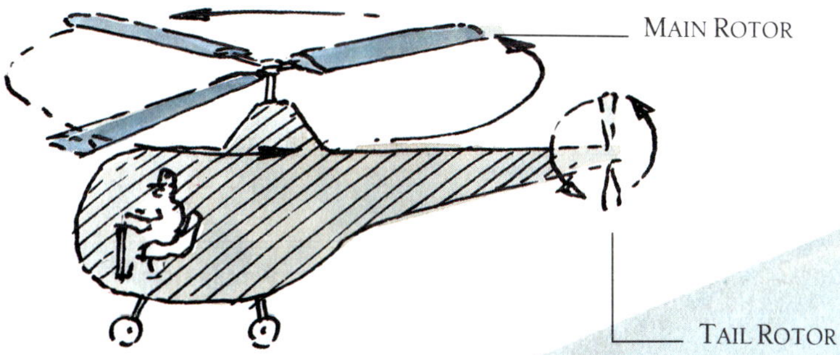

VERTICAL FLIGHT

To ascend, the collective pitch control column raises the swashplate and increases the pitch of all the blades by an equal amount. The rotor lift increases to exceed the helicopter's weight so that the helicopter rises. To descend, the swashplate is lowered. The pitch of all the blades decreases and reduces rotor lift so that the helicopter's weight now exceeds lift and causes it to descend.

HOVERING FLIGHT

The cyclic pitch control column holds the swashplate level, so that each rotor blade has the same pitch and the helicopter does not move forward or backward. The collective pitch control column raises the swashplate so that the pitch of the blades is sufficiently steep for the rotor to produce just enough lift to equal the weight of the helicopter.

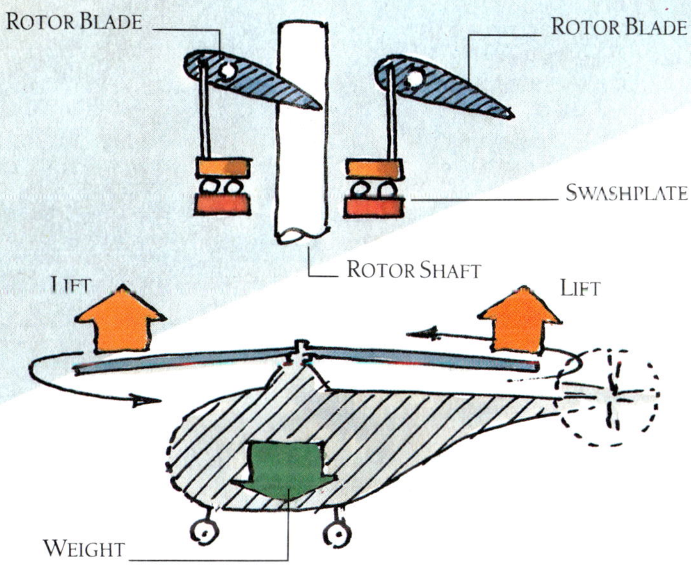

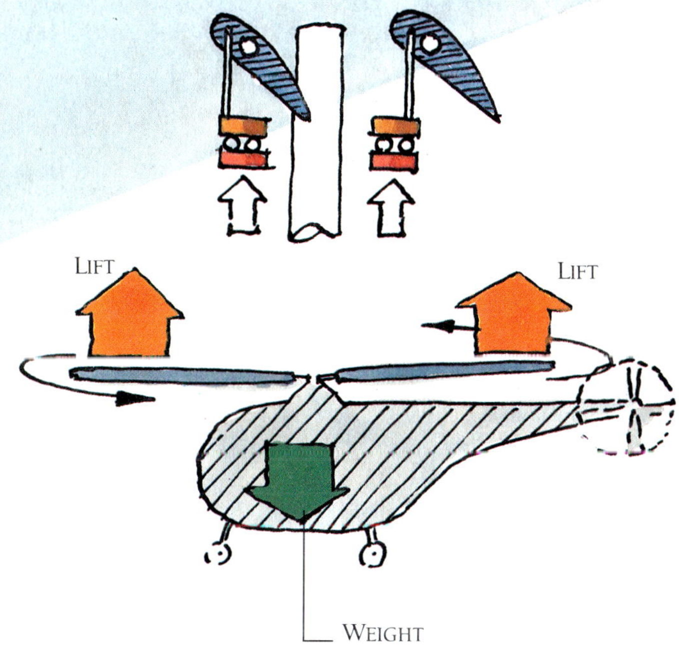

FORWARD FLIGHT

The cyclic pitch control column tilts the swashplate forward. The pitch of each blade increases as it moves behind the rotor shaft then decreases as it moves in front. Lift increases over the back of the rotor, tilting the whole rotor forward. The total rotor lift splits into a raising force that supports the helicopter's weight, and thrust that moves it forward.

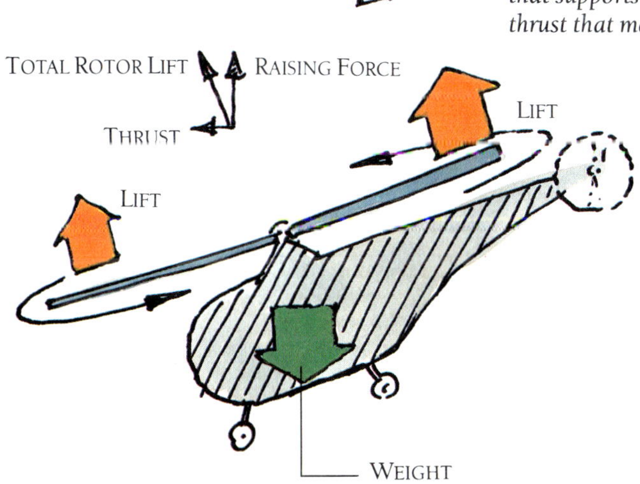

BACKWARD FLIGHT

The cyclic pitch control column tilts the swashplate backward. The pitch of each blade increases as it moves in front of the rotor shaft then decreases as it moves behind. Lift increases over the front of the rotor, producing a backward thrust.

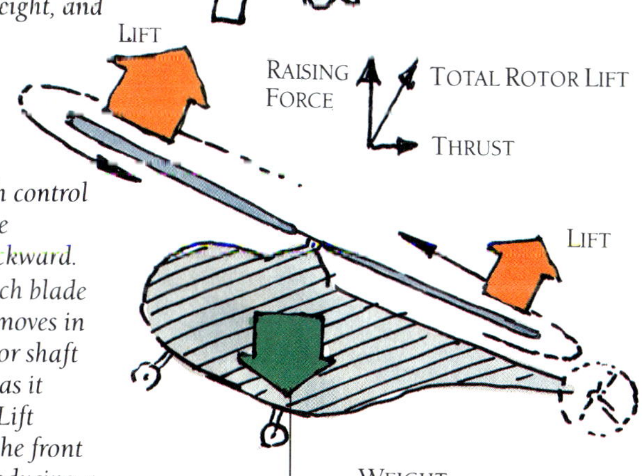

SINGLE-ROTOR HELICOPTER

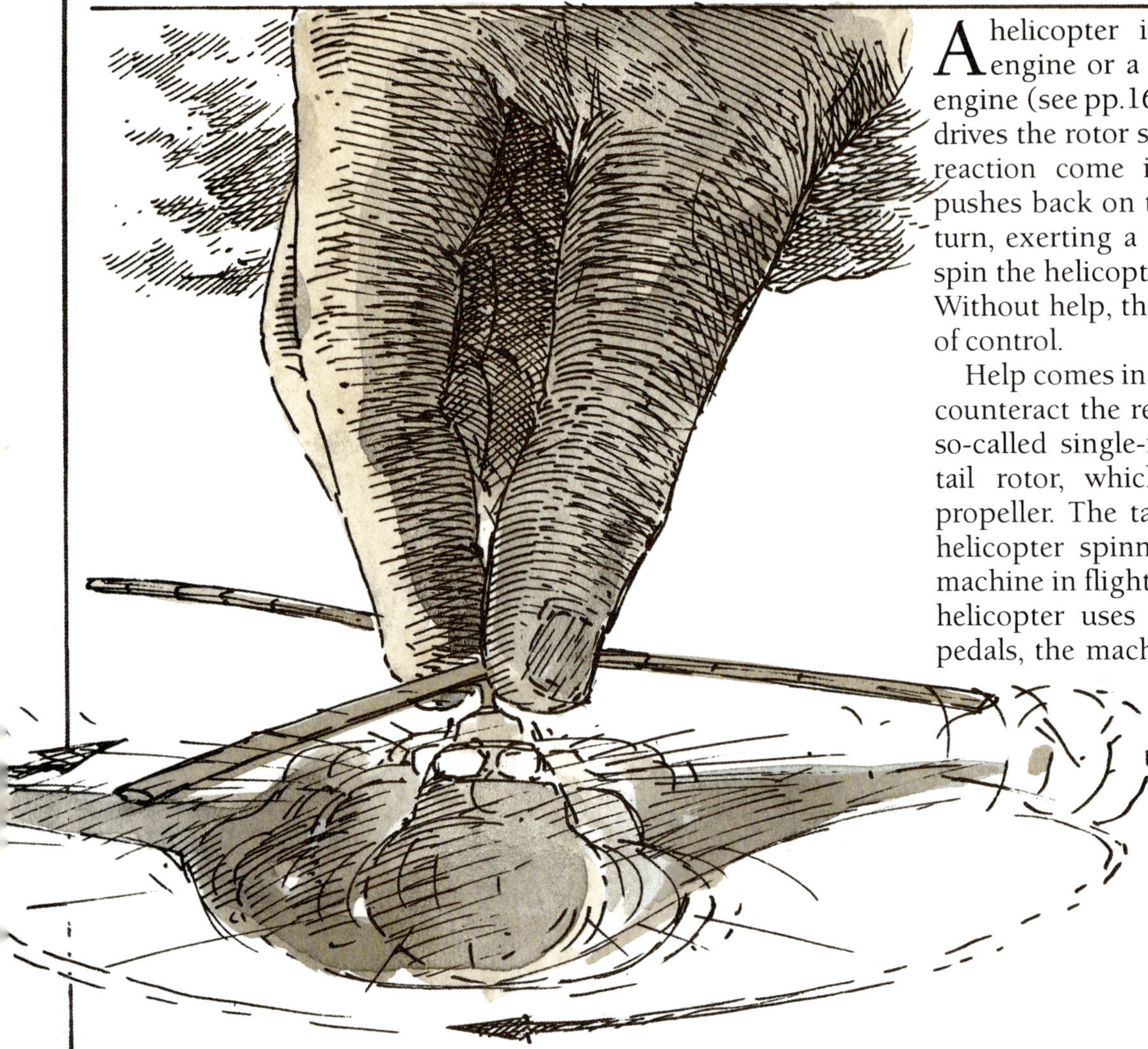

A helicopter is powered by a gasoline engine or a gas turbine similar to a jet engine (see pp.168-9). The engine or turbine drives the rotor shaft, whereupon action and reaction come into play. The rotor shaft pushes back on the helicopter as the blades turn, exerting a powerful force that tries to spin the helicopter in the opposite direction. Without help, the helicopter would spin out of control.

Help comes in the form of another rotor to counteract the reaction of the main rotor. A so-called single-rotor helicopter also has a tail rotor, which produces thrust like a propeller. The tail rotor not only stops the helicopter spinning, but it also steers the machine in flight. Although the pedals that a helicopter uses to steer are called rudder pedals, the machine does not in fact have a rudder: the pedals control the thrust of the tail rotor.

BACKWARD SPIN
If the blades of a helicopter were held still, the reaction of the rotor would make the helicopter spin around in the opposite direction to the blades' normal rotation.

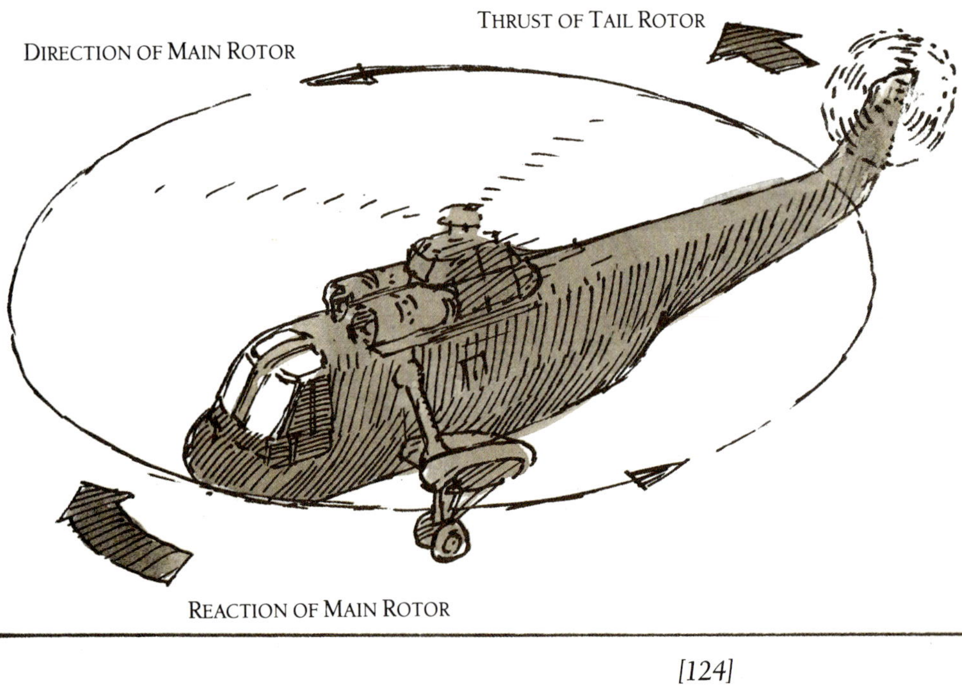

STEERING A SINGLE-ROTOR HELICOPTER
Nomally, the thrust of the tail rotor equals the reaction of the main rotor. The thrust and reaction cancel each out, and no force acts to spin the helicopter. Operating the rudder pedals to increase the thrust makes the extra thrust turn the helicopter in the same direction as the rotor blades. Decreasing the thrust of the tail rotor allows the reaction of the main rotor to turn the helicopter in the opposite direction.

TWIN-ROTOR HELICOPTER

Large helicopters often have two main rotors to give double the lift of a single main rotor and raise a heavy load or more passengers. No tail rotor is needed because the two rotors spin in opposite directions. The reaction of one rotor cancels out the reaction of the other. To turn, the rudder pedals change the speed of the rotors so that one rotor gets more power than the other. The reaction of this rotor increases, and the extra force turns the helicopter.

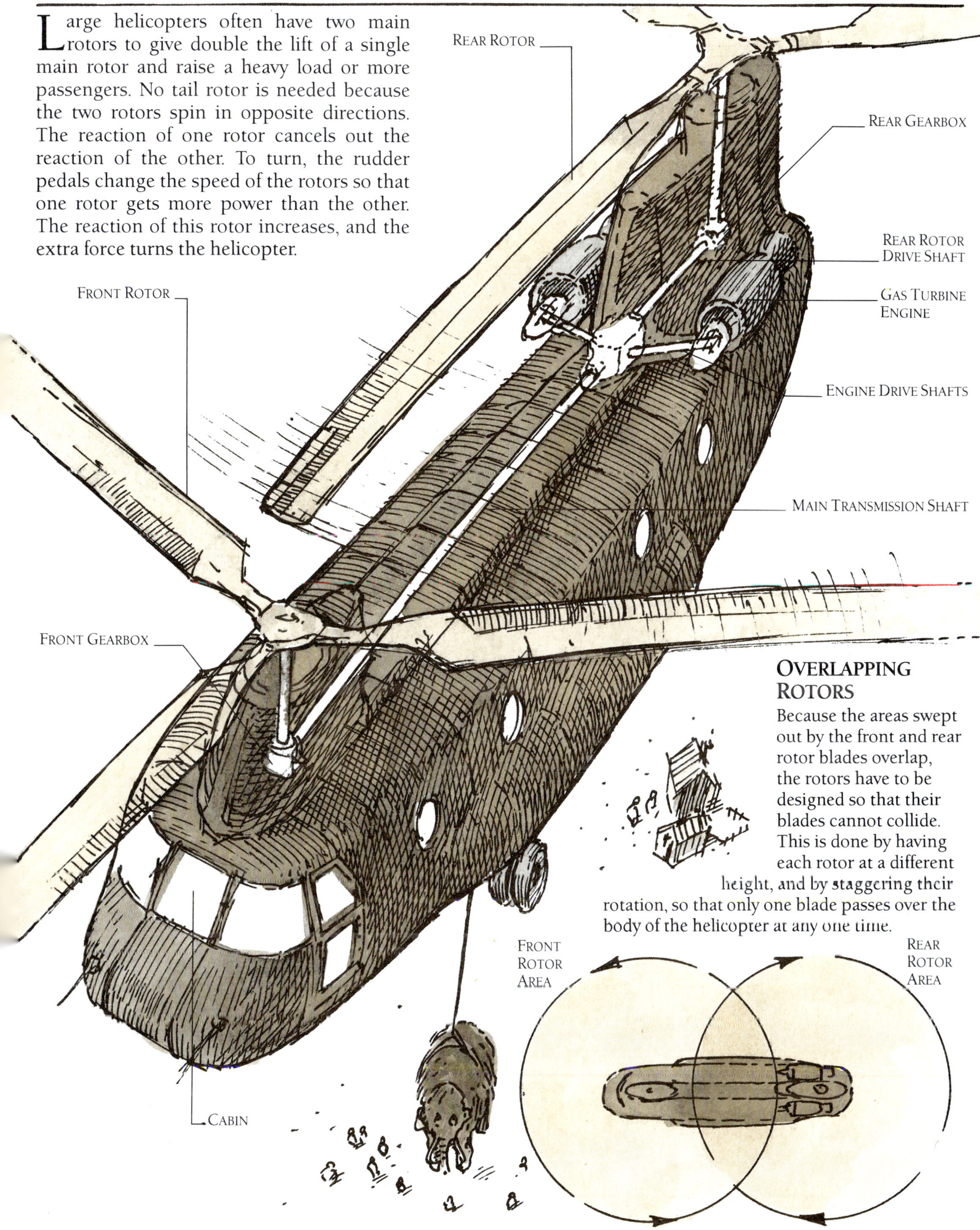

OVERLAPPING ROTORS

Because the areas swept out by the front and rear rotor blades overlap, the rotors have to be designed so that their blades cannot collide. This is done by having each rotor at a different height, and by staggering their rotation, so that only one blade passes over the body of the helicopter at any one time.

THE JUMP JET

The principle of action and reaction (see p.108) is put to use in all powered aircraft, but as a means of propulsion rather than as a direct method of producing lift. Propellers and jet engines move air backward at high speeds, and this pushes back to force the aircraft forward.

By using the downward thrust of its jet engine, the jump jet can dispense with the need for a runway and take off vertically from the ground. When the engine exhausts are swiveled backward, the wings then provide lift in the normal way.

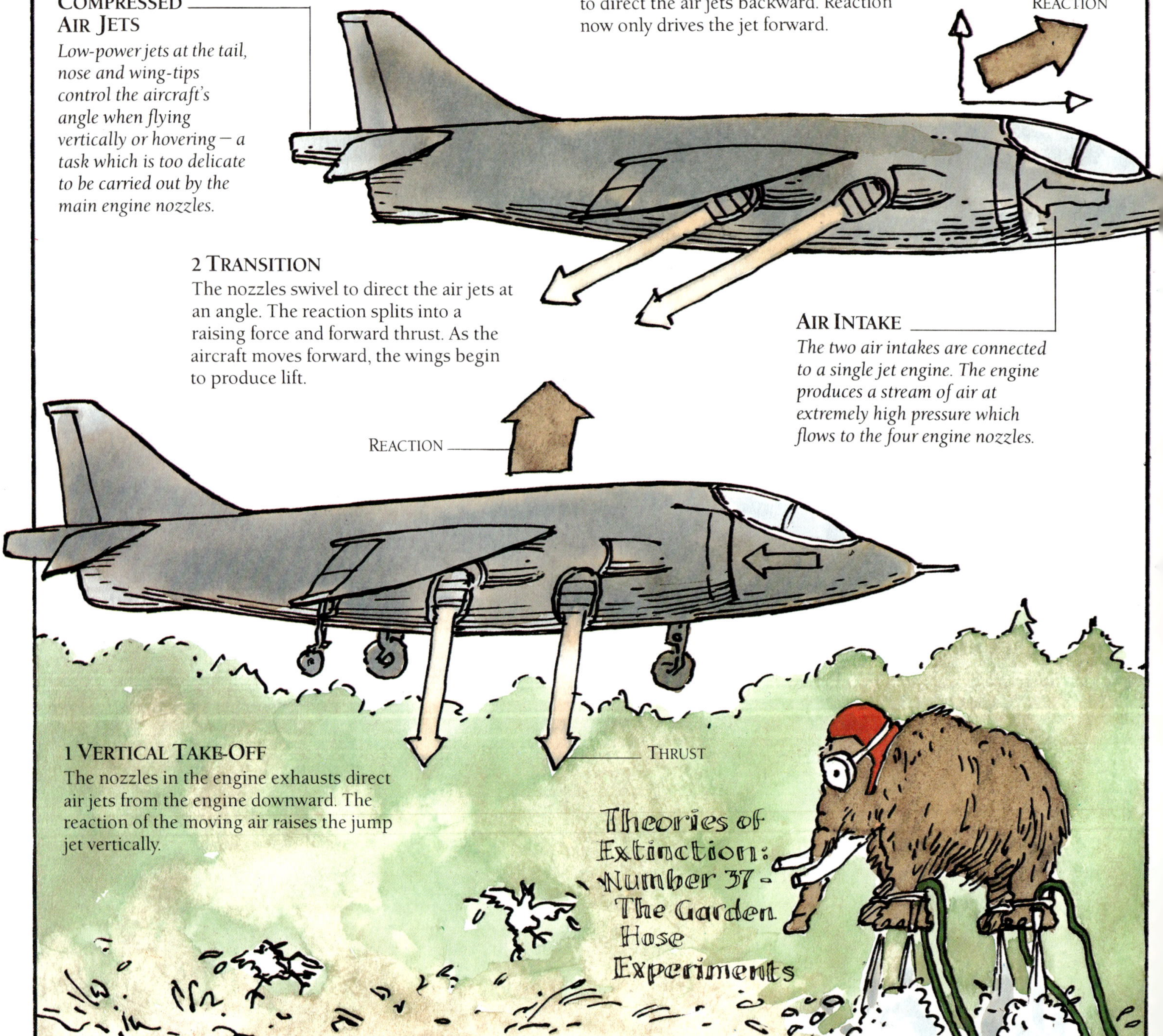

ENGINE NOZZLES

The jump jet has four engine nozzles which can be swiveled to point at any angle from vertical to horizontal. These provide the power for vertical and horizontal flight.

3 FORWARD FLIGHT

As forward speed increases to give sufficient lift for flight, the nozzles swivel to direct the air jets backward. Reaction now only drives the jet forward.

COMPRESSED AIR JETS

Low-power jets at the tail, nose and wing-tips control the aircraft's angle when flying vertically or hovering – a task which is too delicate to be carried out by the main engine nozzles.

2 TRANSITION

The nozzles swivel to direct the air jets at an angle. The reaction splits into a raising force and forward thrust. As the aircraft moves forward, the wings begin to produce lift.

AIR INTAKE

The two air intakes are connected to a single jet engine. The engine produces a stream of air at extremely high pressure which flows to the four engine nozzles.

1 VERTICAL TAKE-OFF

The nozzles in the engine exhausts direct air jets from the engine downward. The reaction of the moving air raises the jump jet vertically.

THE HYDROFOIL

The principles of flight do not only apply to air. An airfoil in fact works better in water, which is denser than air and therefore gives more lift at lower speed. An airfoil used in this way is called a hydrofoil, and this name is also given to a kind of boat that literally flies through the water.

A hydrofoil has a hull like a floating boat, and it does float at rest and low speed. But at high speed, wing-like foils beneath the hull rise in the water and lift the hull above the surface. Freed from friction with the water, a hydrofoil can skim over the waves at two or three times the speed of the fastest floating boats.

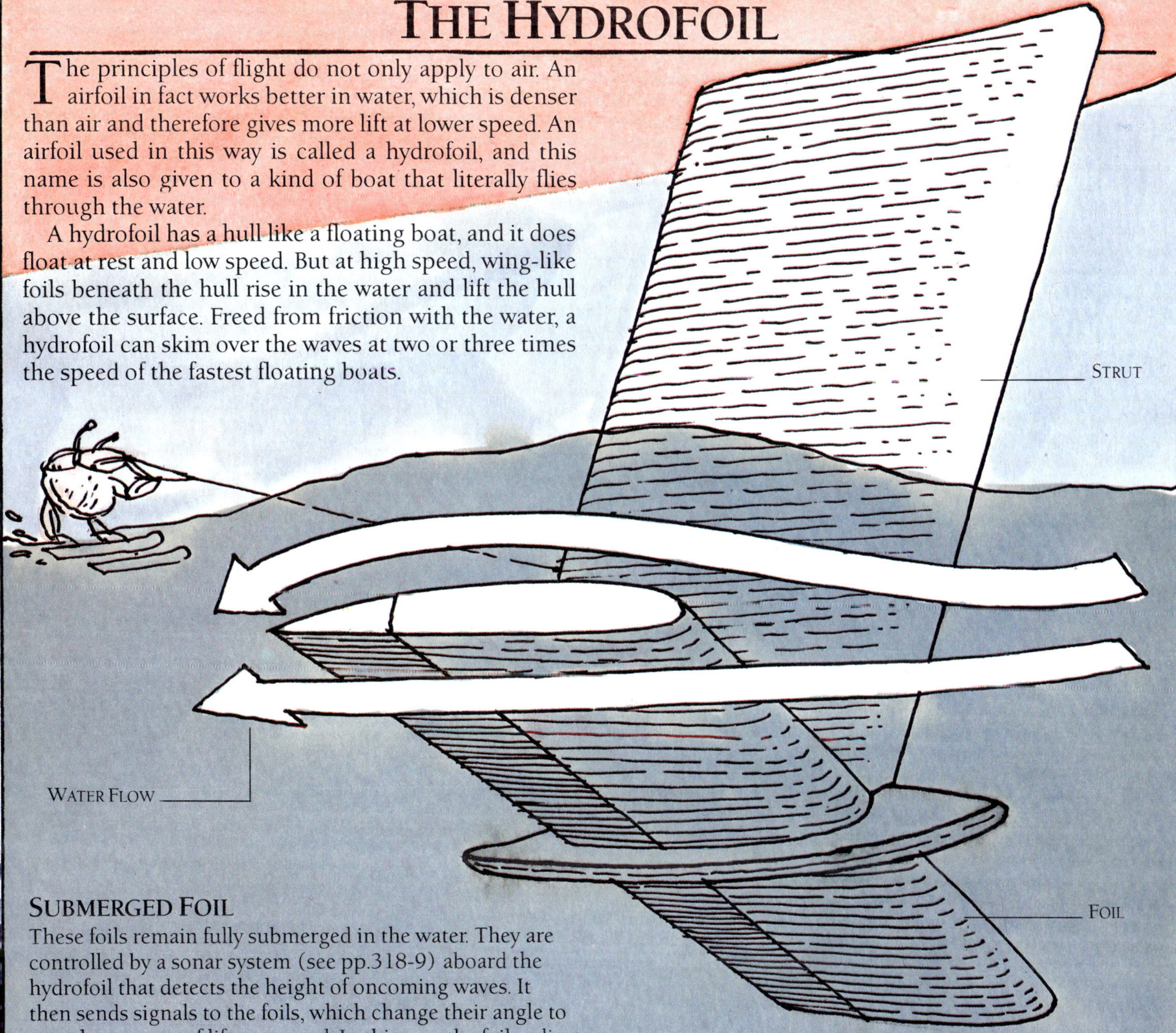

SUBMERGED FOIL

These foils remain fully submerged in the water. They are controlled by a sonar system (see pp.318-9) aboard the hydrofoil that detects the height of oncoming waves. It then sends signals to the foils, which change their angle to vary the amount of lift generated. In this way, the foils adjust lift as the hydrofoil encounters waves, smoothing out the rise and fall and ensuring a steady ride.

SURFACE-PIERCING FOIL

The amount of lift generated by surface-piercing foils depends on the depth of each foil in the water. When the foil is deeper, it generates more lift. This makes the hydrofoil rise as it moves into the crest of a wave. As it enters a trough, more of the foil emerges from the water; lift decreases and the foil sinks. The hydrofoil follows the contours of the waves instead of breaking through them.

PRESSURE POWER

ON FIGHTING FIRES

Through careful study, I have been able to devise a way to improve both the capacity and range of mammoths in fighting fires. First the mammoth is encouraged to drink as much water as it can hold and still get to the scene of the conflagration. Meanwhile a heavy post is set into the ground a short but safe distance from the blaze. The creature is then squeezed against the post in a series of rapid strokes by a large fire-fighter-operated piston.

PUMPS FOR PRESSURE

The events recorded for all time in the parchment above concern the conversion of the mammoth into a primitive but highly effective pump. Pumps are often required to raise the pressure of a fluid (a liquid or a gas), though they may alternatively reduce the pressure. The change in pressure is then put to work, usually to exert a force and make something move or to cause the fluid to flow.

A pump increases pressure by pushing the molecules in the fluid that enters the pump closer together. One way of doing this is to compress the fluid, and this is what is happening to the mammoth. The piston squeezes its stomach, so that the molecules of water inside crowd together. The pressure of the water increases as the molecules exert a greater force on the stomach walls.

If the fluid is able to move, it flows from the pump towards any region that has a lower pressure. The air around the mammoth is at a lower pressure than the water inside. The water pressure therefore forces the water along the trunk, where it emerges in a powerful jet.

SUCTION POWER

A pump may also reduce the pressure of a gas. One way is to increase the volume of the gas so that its molecules become more widely spaced. The mammoth experiences this as the piston is removed, and its empty stomach regains its normal bulk. The pressure of the air inside now becomes less than the pressure of the air outside, and air flows into the mammoth – sucking any nearby object in with it.

PRESSURE AND WEIGHT

Any liquid or gas has a certain pressure by virtue of its weight. When the weight of a liquid or gas presses against a surface within the liquid or gas or against the walls of a container, it creates a pressure on the surface or the walls. Water flows from a tap under pressure because of the weight of the water in the pipe and tank above. Air has a strong pressure because of the great weight of the air in the atmosphere. Suction makes use of this "natural" pressure of the air.

Tests show that my apparatus not only completely empties the mammoth, but also dramatically increases the force with which the water is discharged. The only problem with my design occurs if the piston is released too quickly when the mammoth is empty. Naturally, once the pressure is off, the mammoth expands to its original shape and size, resulting in a deep and powerful inhalation. Anyone or anything standing too close to the animal's trunk during this expansion is likely to be sucked bodily into the animal's interior.

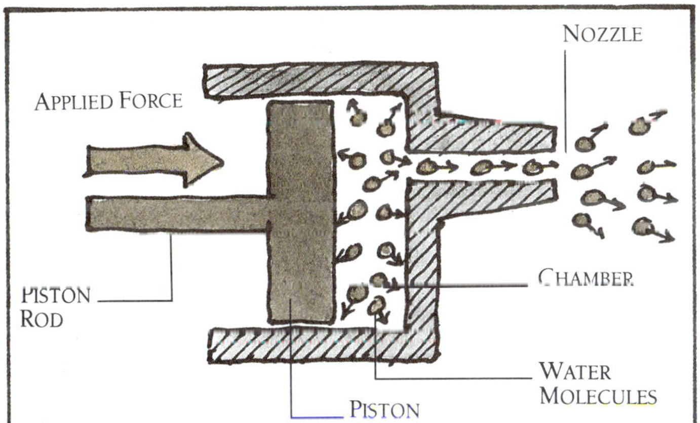

PUMPING OUT

When the piston is pushed in a simple pump, the force creates a high pressure in the water as the water molecules crowd together. The molecules move to any point where the pressure is lower and they are less crowded. This point is the nozzle of the pump, and the water emerges from it in a jet.

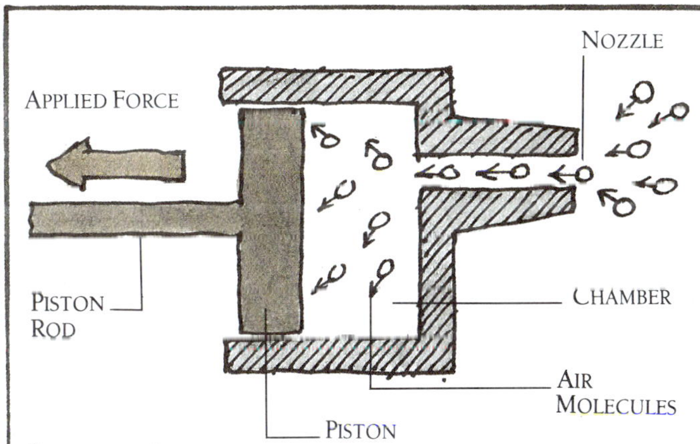

SUCKING IN

As the piston is pulled back, the air pressure in the now empty pump is reduced because the air molecules move apart. The air molecules outside the pump are closer together because the air there is at higher pressure, and so they surge into the pump chamber.

RECIPROCATING PUMPS

PISTON PUMP

In the piston pump, a piston moves up and down inside a cylinder, sucking in water or air at one end and then compressing it to expel it at the other end. A hand-operated water pistol contains the mechanism shown here. A bicycle pump is another simple kind of piston pump.

Pumps increase (or decrease) the pressure of a liquid or gas in two main ways. The piston pump is a reciprocating pump, in which a part such as a piston or diaphragm moves repeatedly to and fro. Rotary pumps compress with a rotating mechanism.

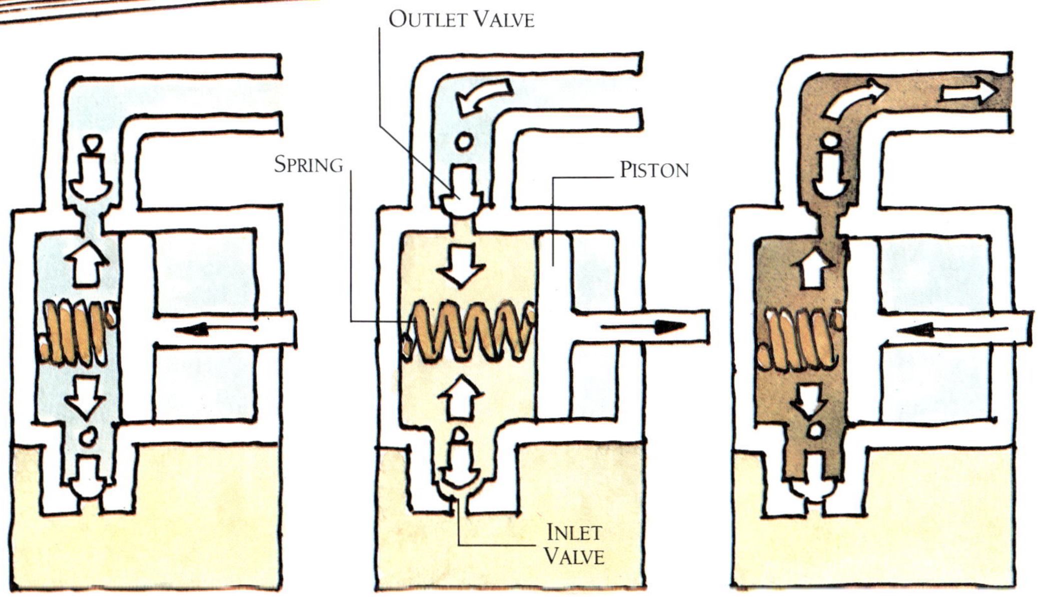

PISTON IN
The piston moves in, increasing the pressure of air in the empty pump. The inlet valve closes, but the outlet valve opens as air escapes.

PISTON OUT
The piston moves back, lowering the air pressure. The outlet valve closes, while the water beneath the pump, which has a higher pressure, flows up into the pump.

PISTON IN
The piston moves in again, increasing the pressure of the water in the pump. The inlet valve closes, but the outlet valve opens to let the water out of the pump.

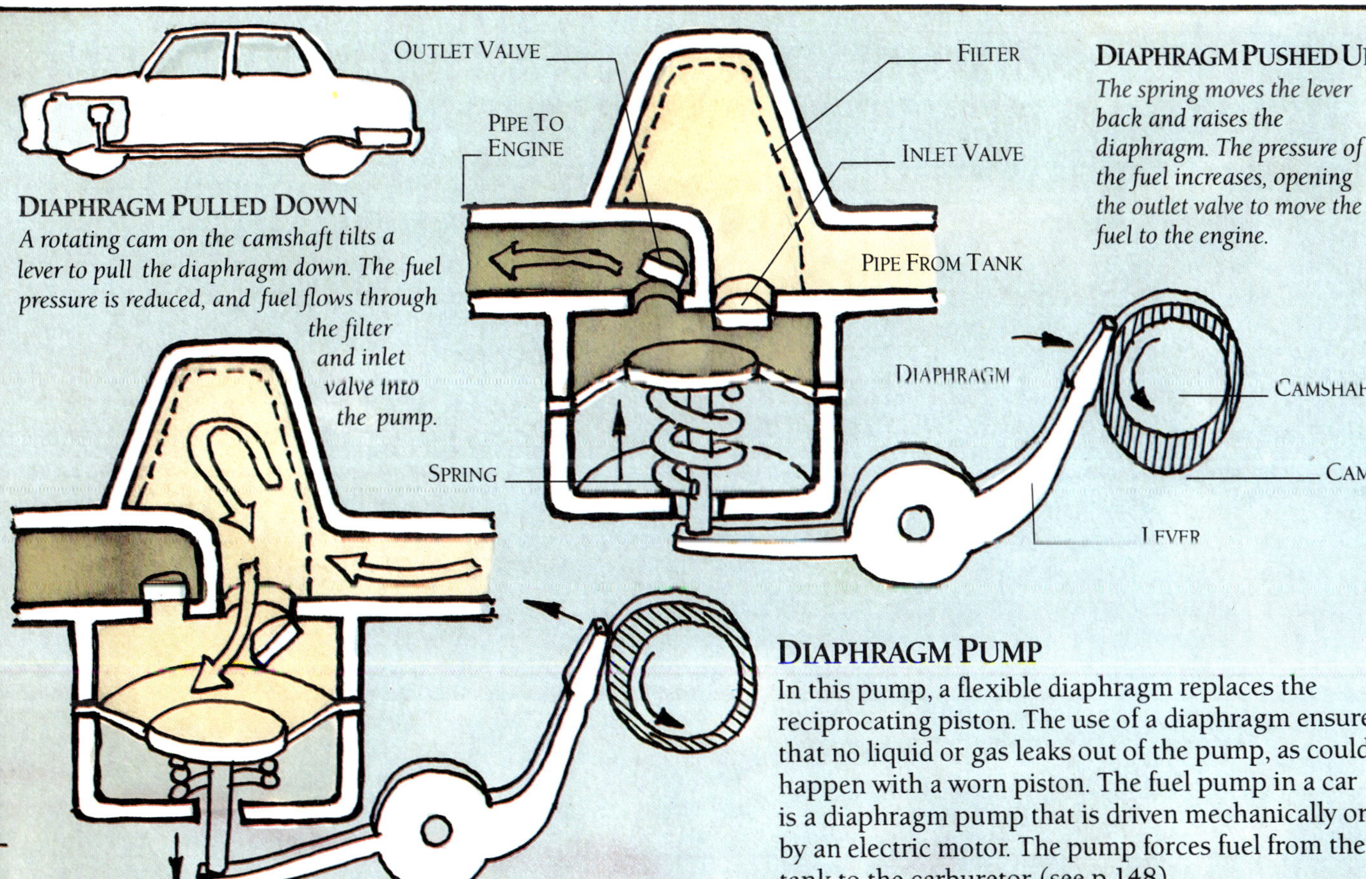

DIAPHRAGM PULLED DOWN
A rotating cam on the camshaft tilts a lever to pull the diaphragm down. The fuel pressure is reduced, and fuel flows through the filter and inlet valve into the pump.

DIAPHRAGM PUSHED UP
The spring moves the lever back and raises the diaphragm. The pressure of the fuel increases, opening the outlet valve to move the fuel to the engine.

DIAPHRAGM PUMP

In this pump, a flexible diaphragm replaces the reciprocating piston. The use of a diaphragm ensures that no liquid or gas leaks out of the pump, as could happen with a worn piston. The fuel pump in a car is a diaphragm pump that is driven mechanically or by an electric motor. The pump forces fuel from the tank to the carburetor (see p.148).

ROTARY PUMPS

GEAR PUMP

The oil that lubricates the engine of a car must be forced at high pressure around channels in the engine (see p.92). A sturdy and durable gear pump is often used to do the job.

The rotating camshaft of the engine (see pp.54-5) normally powers the oil pump, driving a shaft that turns a pair of intermeshing gear wheels inside a close-fitting chamber. The oil enters the pump, where it is trapped in the wheels. The wheels carry the oil around to the outlet, where the teeth come together as they intermesh. This squeezes the oil and raises its pressure as it flows to the outlet. The speed of pumping is directly linked to the speed of the engine.

ROTARY VANE PUMP

This pump contains a chamber with a rotor mounted slightly off-center. The rotor has slots fitted with sliding vanes. As the rotor turns, the vanes are thrown outward against the chamber wall, thereby creating compartments of changing size.

Where the liquid or gas enters the pump, the compartments expand to suck it in. As the fluid is carried around the pump, the compartments get smaller. The fluid is squeezed, and leaves the pump at high pressure. Rotary vane pumps are often used to deliver fuel at gas stations.

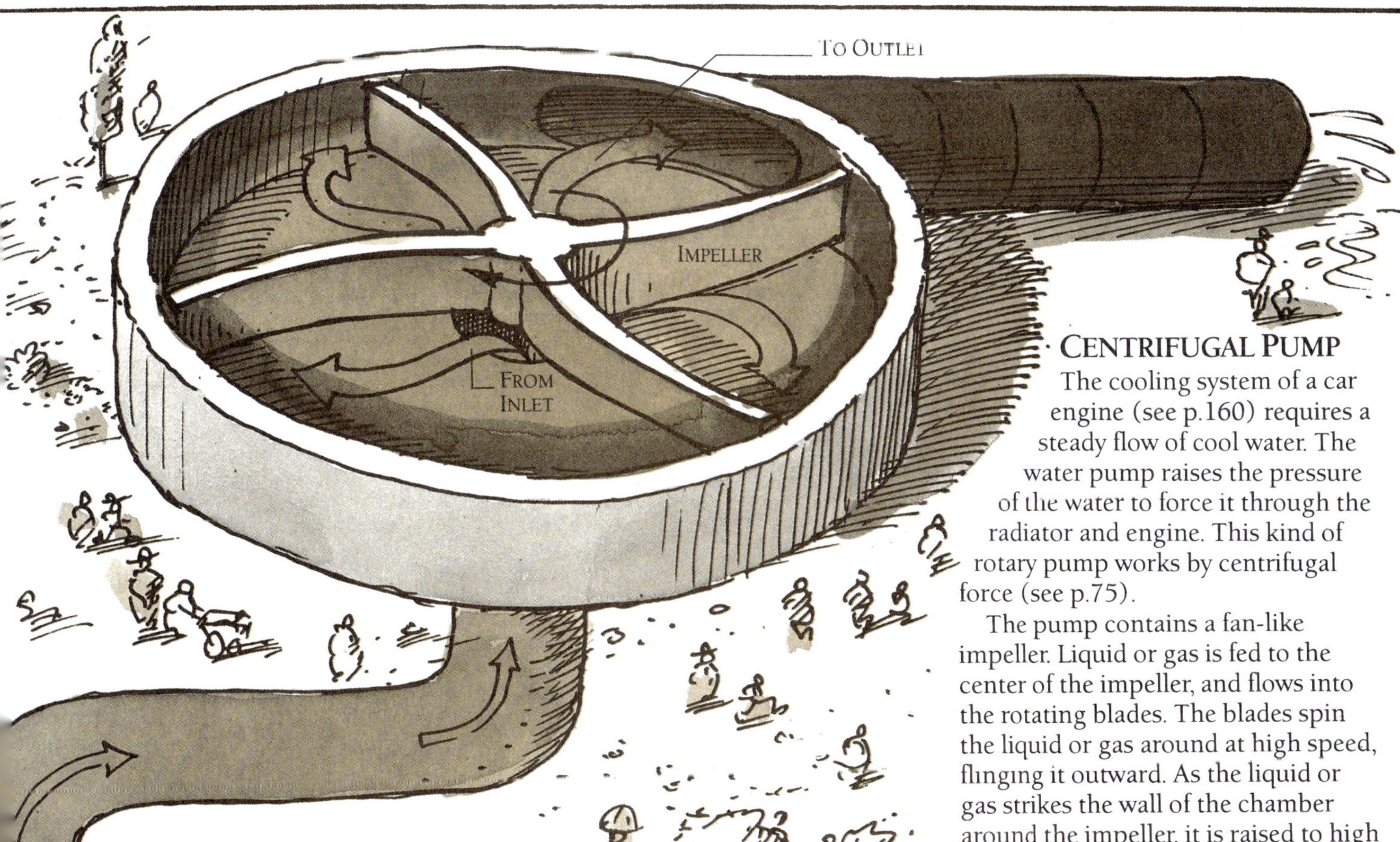

Centrifugal Pump

The cooling system of a car engine (see p.160) requires a steady flow of cool water. The water pump raises the pressure of the water to force it through the radiator and engine. This kind of rotary pump works by centrifugal force (see p.75).

The pump contains a fan-like impeller. Liquid or gas is fed to the center of the impeller, and flows into the rotating blades. The blades spin the liquid or gas around at high speed, flinging it outward. As the liquid or gas strikes the wall of the chamber around the impeller, it is raised to high pressure before it leaves the outlet.

Peristaltic Pump

Most pumps are likely to clog up when used with a liquid such as blood which contains particles. Furthermore, they would damage blood cells. The peristaltic pump, which is used in devices such as heart-lung machines, avoids both these problems.

The pump contains a flexible tube that is repeatedly squeezed by rotating rollers. The rollers push the blood gently along the tube. This pump has the added advantage that the blood does not come into contact with any mechanical parts and so remains clean.

PNEUMATIC MACHINES

SUPPORTING A WEIGHT
The weight that compressed air can support depends on the pressure difference between it and the atmosphere.

DOUBLING THE AREA
The weight that can be supported also depends on the area. Doubling the area doubles the weight that the air can support.

DOUBLING THE PRESSURE
Doubling the pressure on the original area also doubles the weight that the air can support.

Air possesses considerable power when placed under pressure, and when compressed it can be used to drive machines. Pneumatic or air-driven machines all make use of the force exerted by air molecules striking a surface. The compressed air exerts a greater pressure than the air on the other side of the surface, which is at atmospheric pressure. The difference in pressure drives the machine.

HOVERCRAFT
A hovercraft exploits the power of compressed air to lift itself above the surface of the water or ground. Buoyed up by a cushion of air, it can then float and travel rapidly because there is little friction with the water or ground. The hovercraft uses propellers for horizontal movement and rudders for steering. These may operate in the air, as in an aircraft, or underwater like those in a ship. Gas turbines or piston engines drive both the lift fans that compress the air and pump it into the flexible skirt and also the propellers.

THE PNEUMATIC DRILL

The force that lifts a hovercraft above the sea is also put to use on the road, but in the reverse direction. The ear-blasting roar that often accompanies road repairs is produced by the pneumatic drill. This device is fed with compressed air as a source of power, whereas the hovercraft uses its power to produce the compressed air. Although commonly known as a drill, this machine is actually more like an automatic hammer in its method of working, and for this reason it is also known as the air-hammer or jack-hammer.

The high-pressure air that drives the drill is produced by a compressor. This machine uses a pump to deliver the air to the drill through a hose. There it is used to produce a cycle of operations which delivers powerful repeated blows to the tool or blade that hammers into the road surface.

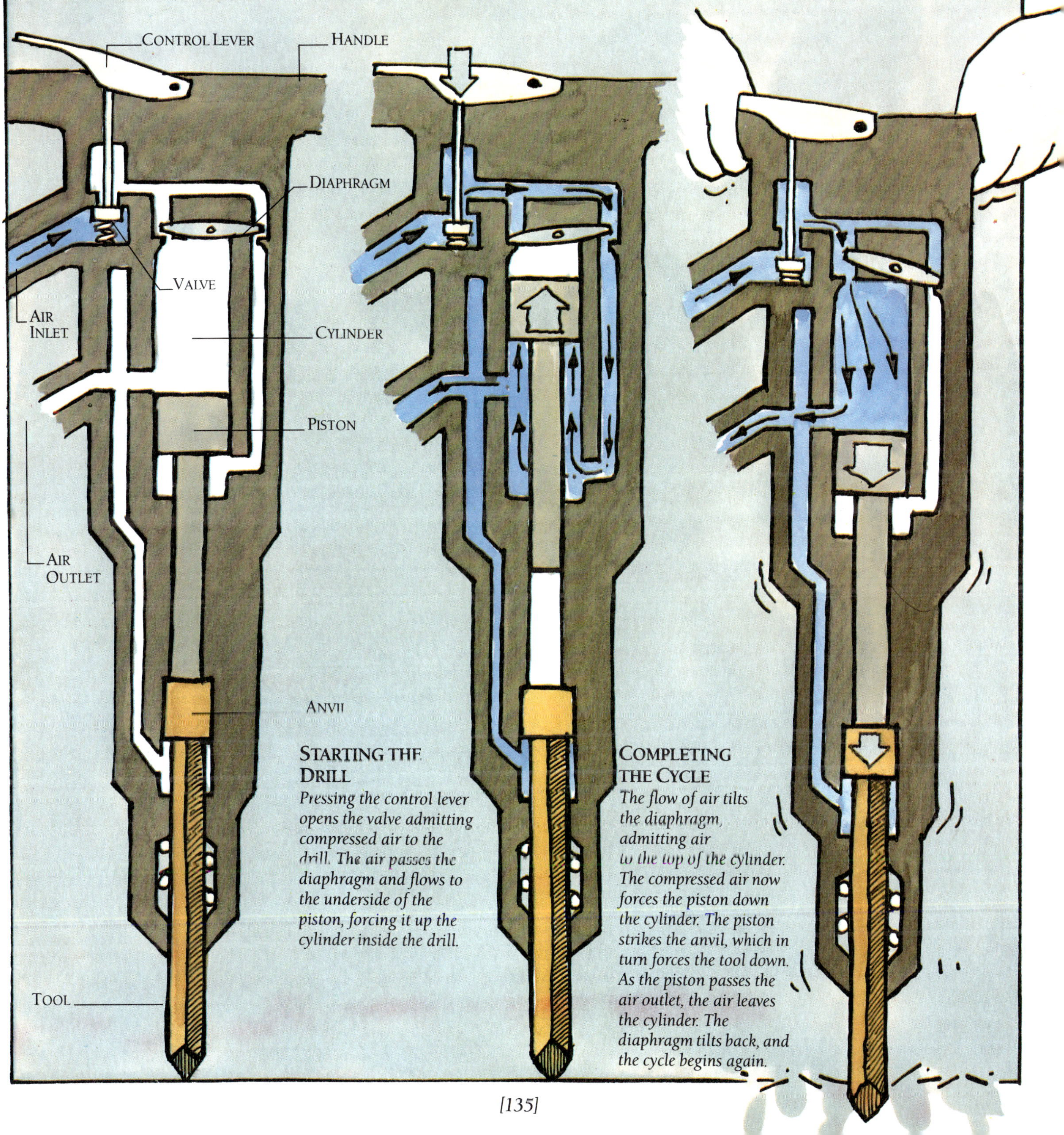

STARTING THE DRILL
Pressing the control lever opens the valve admitting compressed air to the drill. The air passes the diaphragm and flows to the underside of the piston, forcing it up the cylinder inside the drill.

COMPLETING THE CYCLE
The flow of air tilts the diaphragm, admitting air to the top of the cylinder. The compressed air now forces the piston down the cylinder. The piston strikes the anvil, which in turn forces the tool down. As the piston passes the air outlet, the air leaves the cylinder. The diaphragm tilts back, and the cycle begins again.

Hydraulic Machines

A hydraulic machine makes use of pressure in a liquid. It does this with a set of two or more cylinders connected by pipes containing the hydraulic fluid. In each cylinder is a piston. To work the machine, force is applied to one cylinder, which is known as the "master" cylinder. This raises the pressure of the fluid throughout the whole system, and the pistons in the other cylinders – the "slave" cylinders – move out and perform a useful action. The force produced by each slave cylinder depends on its diameter.

Hydraulic machines work on the same principle as levers and gears: the wider the slave cylinder, the greater is the force that it applies, and the shorter is the distance that it moves. As with levers and gears, the converse also applies, so a narrow slave cylinder moves a large distance with reduced force.

Hydraulic Brakes

With the exception of the hand-brake, which is operated by a cable, cars use hydraulic braking systems. The brake pedal moves the piston in the master cylinder, raising the pressure of the brake fluid evenly throughout the system. The high-pressure fluid then makes pistons in the wheel cylinders move out with great force, applying the brake pads or shoes. The friction that this creates (see p.90) then slows the car down.

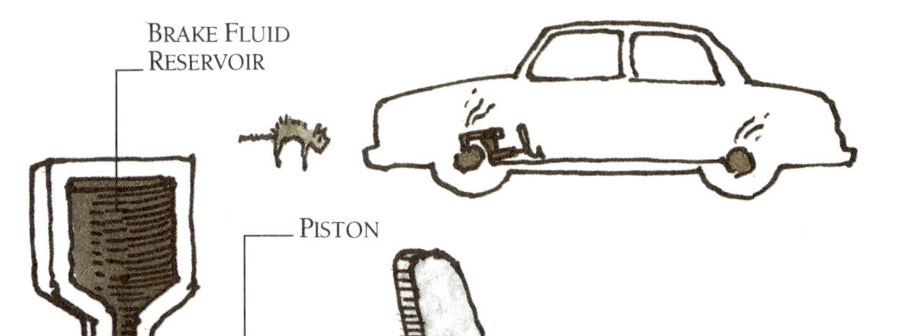

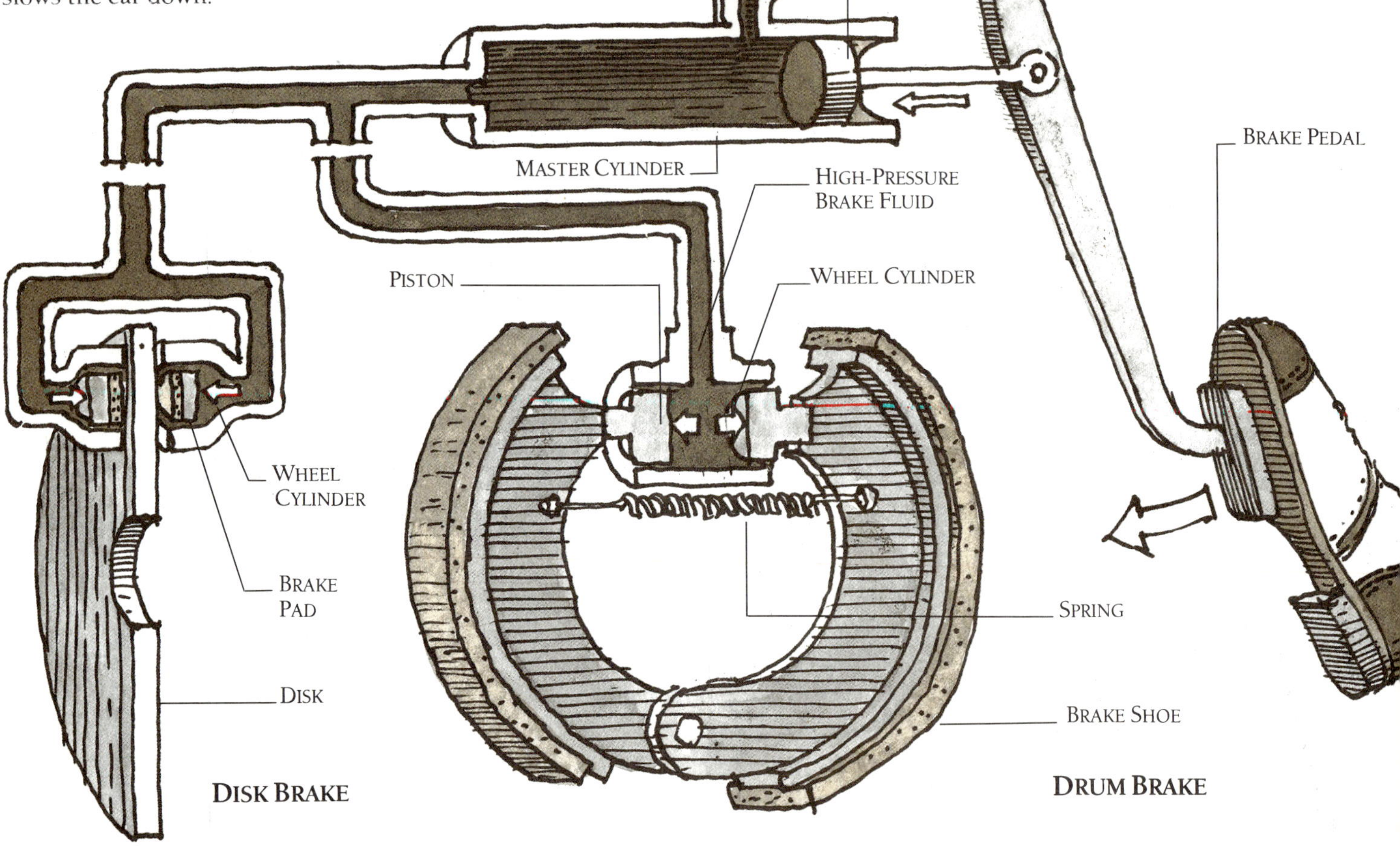

HYDRAULIC RAM

Machines such as excavators work with hydraulic rams. Each ram consists of a piston in a cylinder connected by pipes to a central reservoir of hydraulic fluid. The controls open valves that admit the high-pressure fluid to either side of the piston. In this way, the piston can move in or out with great force and precision.

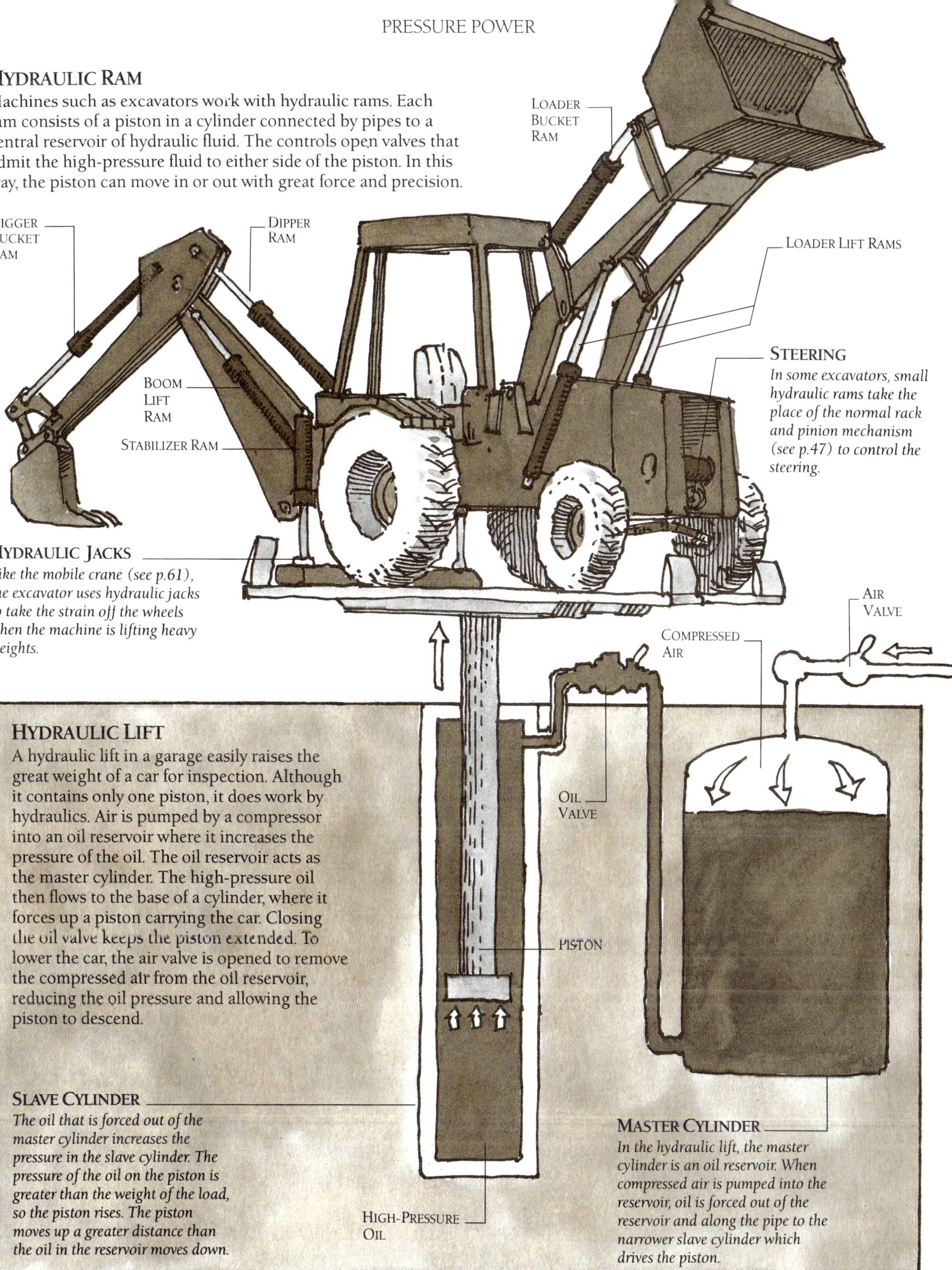

STEERING

In some excavators, small hydraulic rams take the place of the normal rack and pinion mechanism (see p.47) to control the steering.

HYDRAULIC JACKS

Like the mobile crane (see p.61), the excavator uses hydraulic jacks to take the strain off the wheels when the machine is lifting heavy weights.

HYDRAULIC LIFT

A hydraulic lift in a garage easily raises the great weight of a car for inspection. Although it contains only one piston, it does work by hydraulics. Air is pumped by a compressor into an oil reservoir where it increases the pressure of the oil. The oil reservoir acts as the master cylinder. The high-pressure oil then flows to the base of a cylinder, where it forces up a piston carrying the car. Closing the oil valve keeps the piston extended. To lower the car, the air valve is opened to remove the compressed air from the oil reservoir, reducing the oil pressure and allowing the piston to descend.

SLAVE CYLINDER

The oil that is forced out of the master cylinder increases the pressure in the slave cylinder. The pressure of the oil on the piston is greater than the weight of the load, so the piston rises. The piston moves up a greater distance than the oil in the reservoir moves down.

MASTER CYLINDER

In the hydraulic lift, the master cylinder is an oil reservoir. When compressed air is pumped into the reservoir, oil is forced out of the reservoir and along the pipe to the narrower slave cylinder which drives the piston.

SUCTION MACHINES

Reducing the pressure inside a machine causes suction. The pressure of the outside air, which is created by the weight of the atmosphere, is greater than that inside the machine. This difference in pressure can then be put to work. In a vacuum cleaner, the pressure of the outside air forces material into the cleaner. In some kinds of brakes, atmospheric pressure is used to force the brake pad or shoe into position when the brake pedal is pressed.

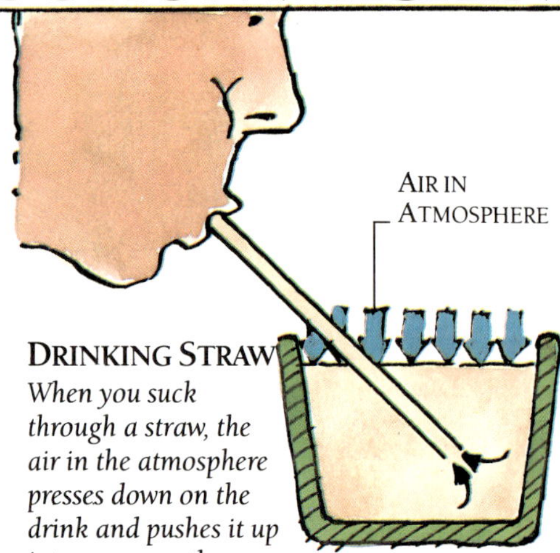

DRINKING STRAW
When you suck through a straw, the air in the atmosphere presses down on the drink and pushes it up into your mouth.

VACUUM CLEANER

Cylinder vacuum cleaners work entirely by suction. An electric motor in the cleaner drives a fan that pumps the air out of the hose. The pressure of the atmosphere pushes air into the cleaning attachment and up the hose, pulling in dust and dirt with it. The dust-laden air then passes through a dust bag, which retains the dust and dirt, before leaving the back of the cleaner. In some cleaners, this air is directed to the base of the machine to form an air cushion like that in a hovercraft (see p.134). The cleaner then hovers above the floor and can be moved easily.

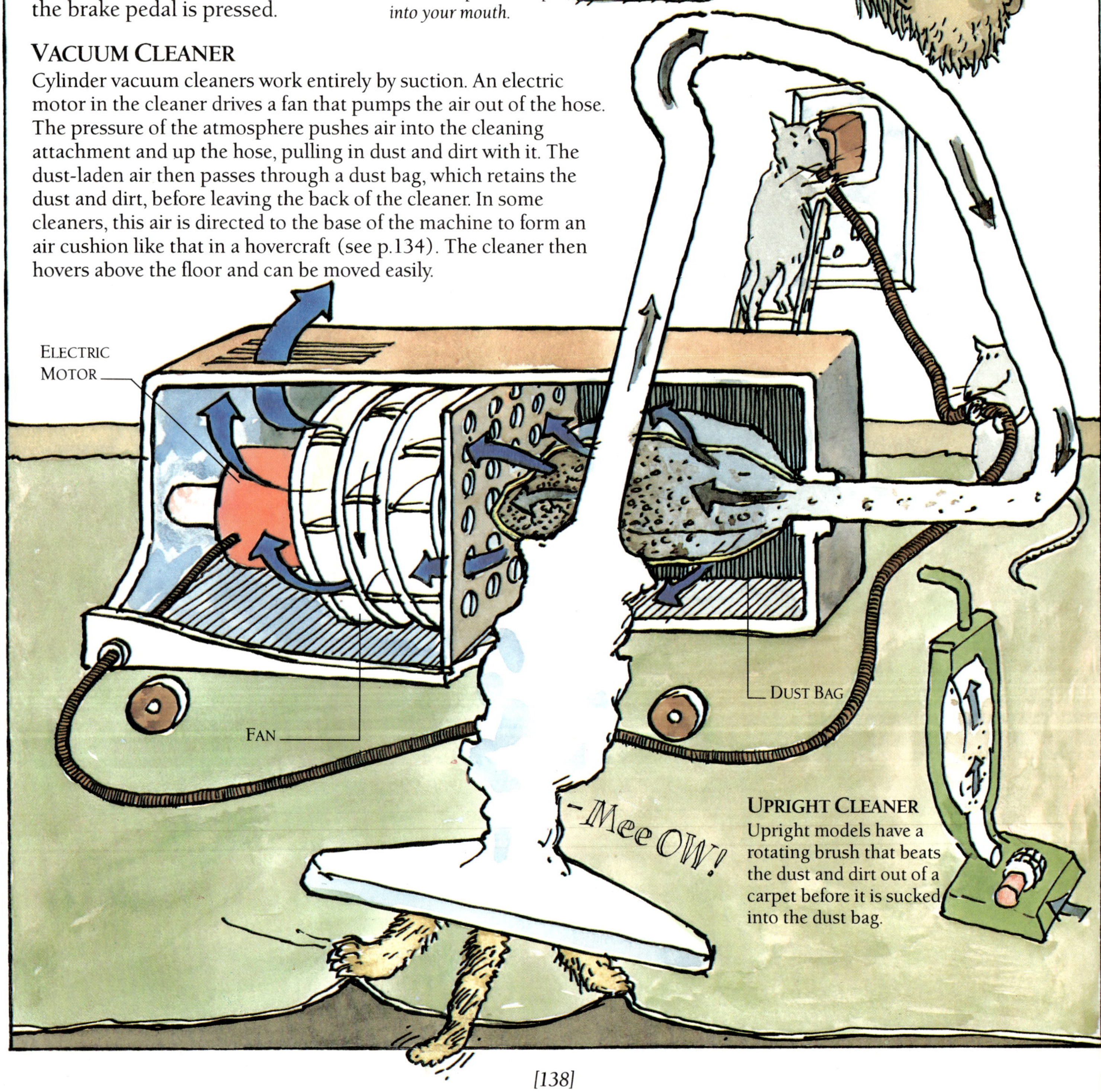

UPRIGHT CLEANER
Upright models have a rotating brush that beats the dust and dirt out of a carpet before it is sucked into the dust bag.

THE AQUALUNG

With the aid of an aqualung or scuba (Self-Contained Underwater Breathing Apparatus), a diver can stay underwater for long periods. This device does away with the need for a diving suit by supplying air at changing pressures during a dive.

The diver's body is under pressure from the surrounding water, which becomes greater the deeper one dives. The air inside the diver's lungs is at about the same pressure as the water. The air in the cylinder is at high pressure. The aqualung's regulator has two stages that reduce the pressure of the air coming from the cylinder to the same pressure as the water so that the diver can breathe in. The first-stage valve, worked by a spring, opens to admit air at a set pressure always greater than water pressure. The second-stage valve, worked by a lever, opens by suction to admit air at water pressure.

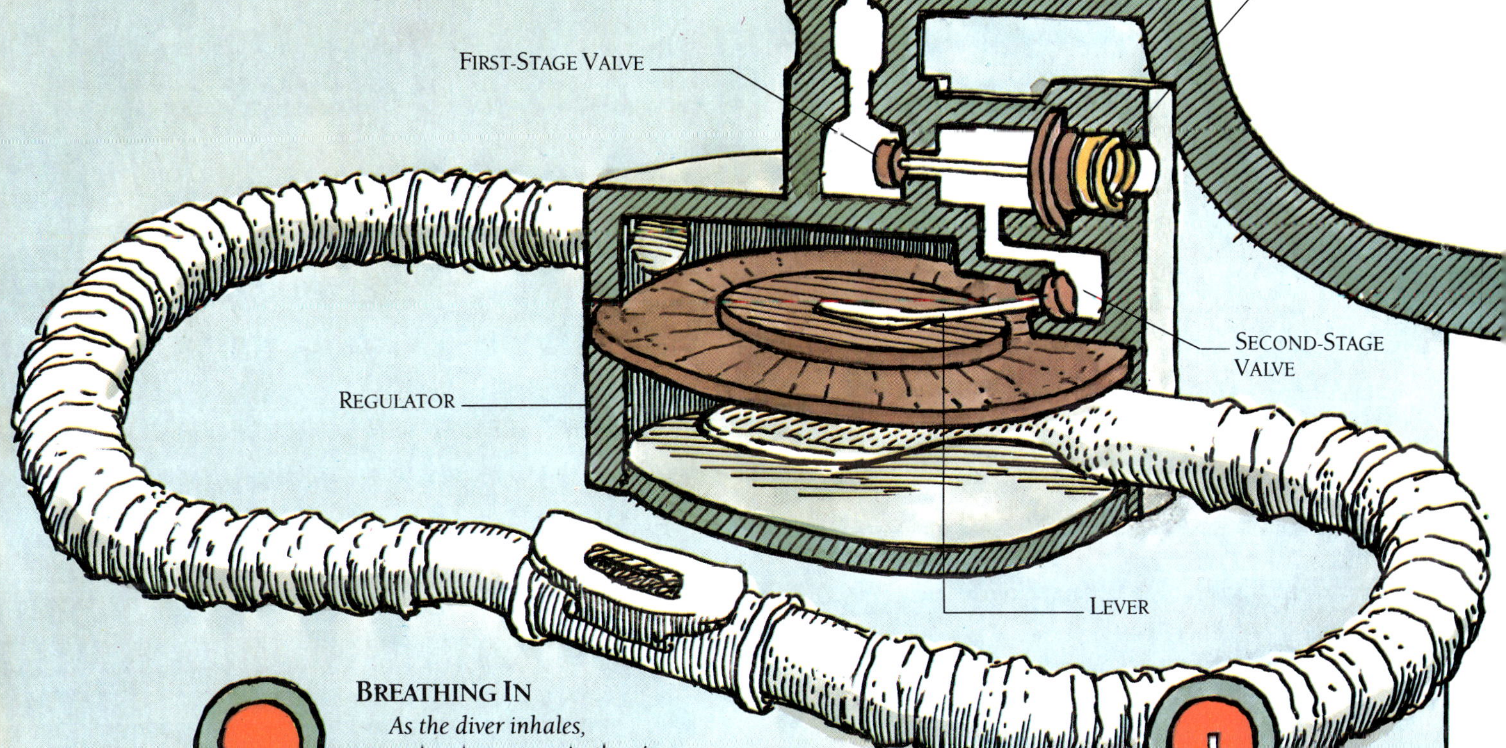

BREATHING IN

As the diver inhales, the air pressure in the air tubes falls. The diaphragm is sucked in, pushed by the greater pressure of water on the outside of the diaphragm. The lever opens the second-stage valve, admitting more air to the diver.

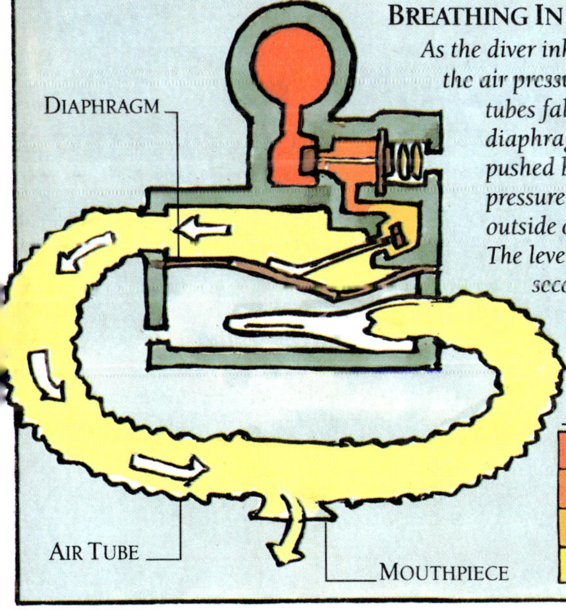

BREATHING OUT

As the diver breathes out, the air pressure in the air tubes rises, pushing the diaphragm down to shut off the incoming air. The one-way valve opens to expel the exhaled air to the sea.

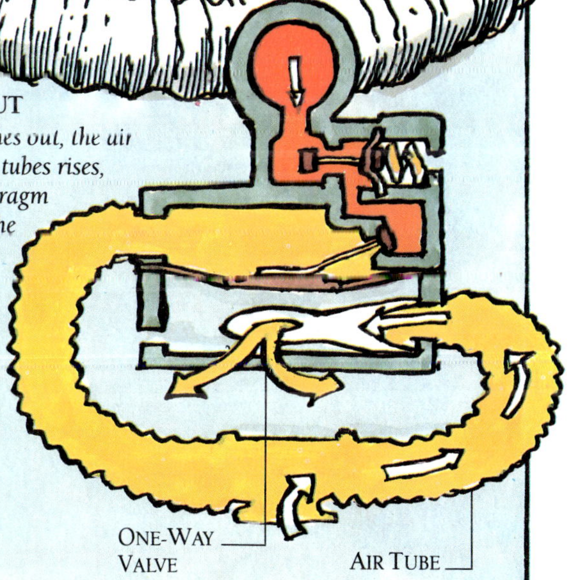

KEY

- AIR FROM CYLINDER
- AIR AT SET PRESSURE
- AIR JUST ABOVE WATER PRESSURE
- AIR JUST BELOW WATER PRESSURE

THE TOILET TANK

Many toilet tanks work with a siphon, which accomplishes the apparently impossible feat of making water (or any other liquid) flow uphill. Provided the open end of the siphon tube is below the level of the surface, the water will flow up the tube, around the bend and then down to the open end. Operating the toilet tank starts the siphon flowing. Once the water begins to double back down the siphon tube, air pressure makes the rest of the water follow it.

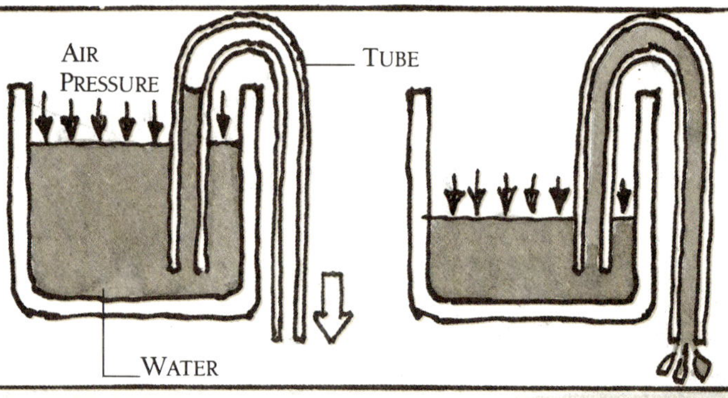

1 THE TANK FLUSHES
After the handle is pressed down, water is lifted up the siphon tube by the disk. The water reaches the bend in the siphon pipe and then travels around it. As it falls, the water in the tank follows it.

2 THE VALVE OPENS
When the water level in the tank falls below the bottom of the bell, air enters the bell and the siphon is broken. By this time, the float has fallen far enough to open the valve, and water under pressure enters to refill the tank and the float begins to rise again.

3 THE VALVE CLOSES
The rising float gradually shuts the valve, cutting off the water supply. Although the tank is full, the water cannot leave through the siphon tube until the handle is pressed down, forming the siphon once again. The float and valve work together to form a self-regulating mechanism.

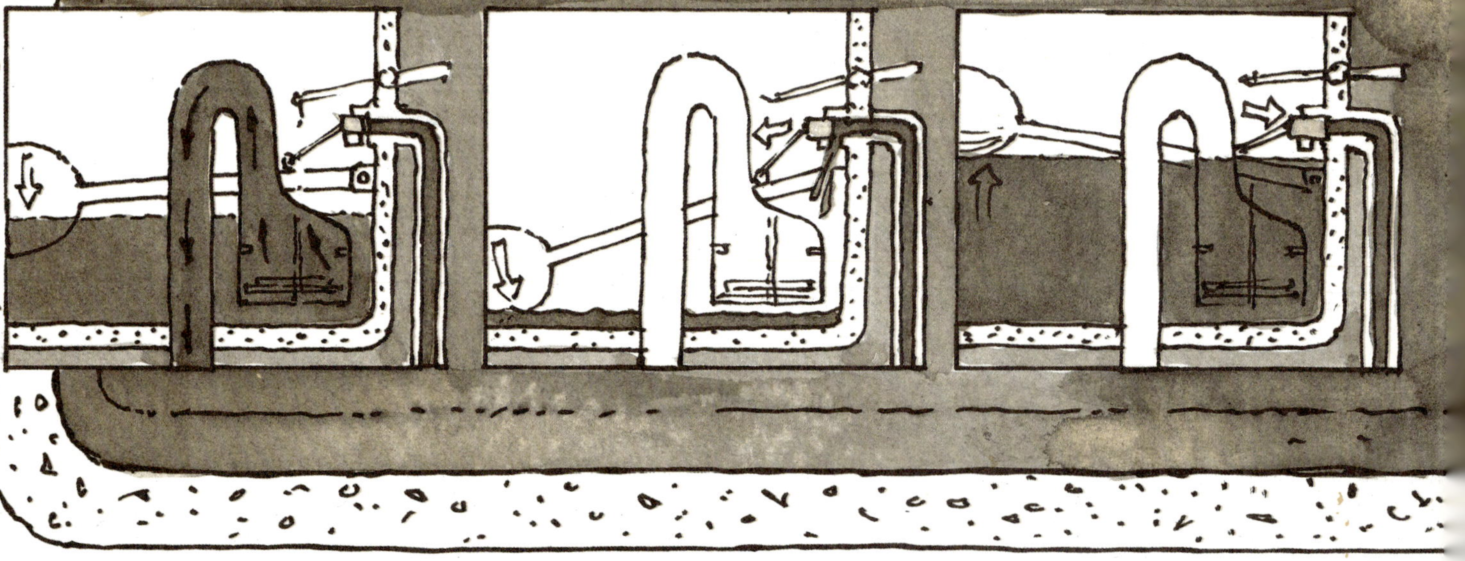

Tank Cover
Siphon Pipe
Handle
Valve
Disk
Bell
Water Pipe

PRESSURE GAUGES

Mechanical pressure gauges respond to the pressure of a fluid, which exerts a force to move a pointer over a dial. One of the simplest is the Bourdon gauge, which is found in the oil-pressure gauge in a car, the pressure gauge on a gas cylinder, and the depth gauge used by a diver. It works like the curled paper tubes you could find yourself blowing into at parties.

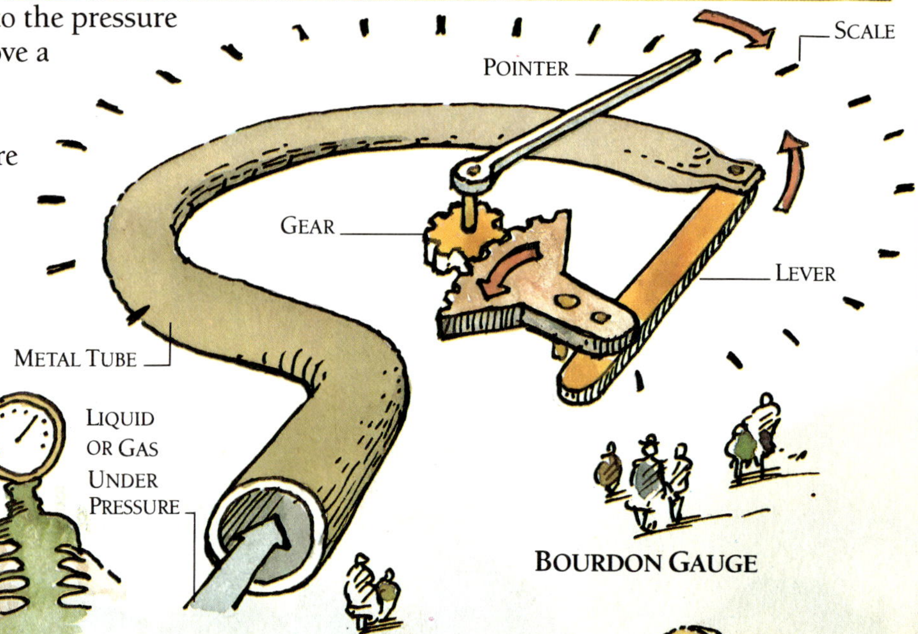

BOURDON GAUGE

ANEROID BAROMETER

A barometer measures changes in the pressure of the air, which is an indicator of the weather ahead. The most common kind is the aneroid barometer.

At the heart of this barometer is a capsule from which air is removed. As the air pressure falls, the spring pulls the side of the capsule outward. The arm rises, causing the rocking bar to slacken the chain. The hairspring unwinds, moving the pointer counter-clockwise until the chain is pulled taut. When the air pressure rises, the capsule contracts and the pointer moves clockwise, winding up the hairspring.

THE WATER METER

Any liquid or gas that is under pressure will flow. By detecting the rate of flow with a meter, the amount of liquid or gas that passes can be measured. A water meter often works rather like a rotary pump in reverse. As the water flows through the meter, it turns the blades of an impeller. The shaft of the impeller turns a worm gear (see p.41) that reduces the speed of the impeller. Sets of gears then turn a pointer and counters that register the total amount of water used.

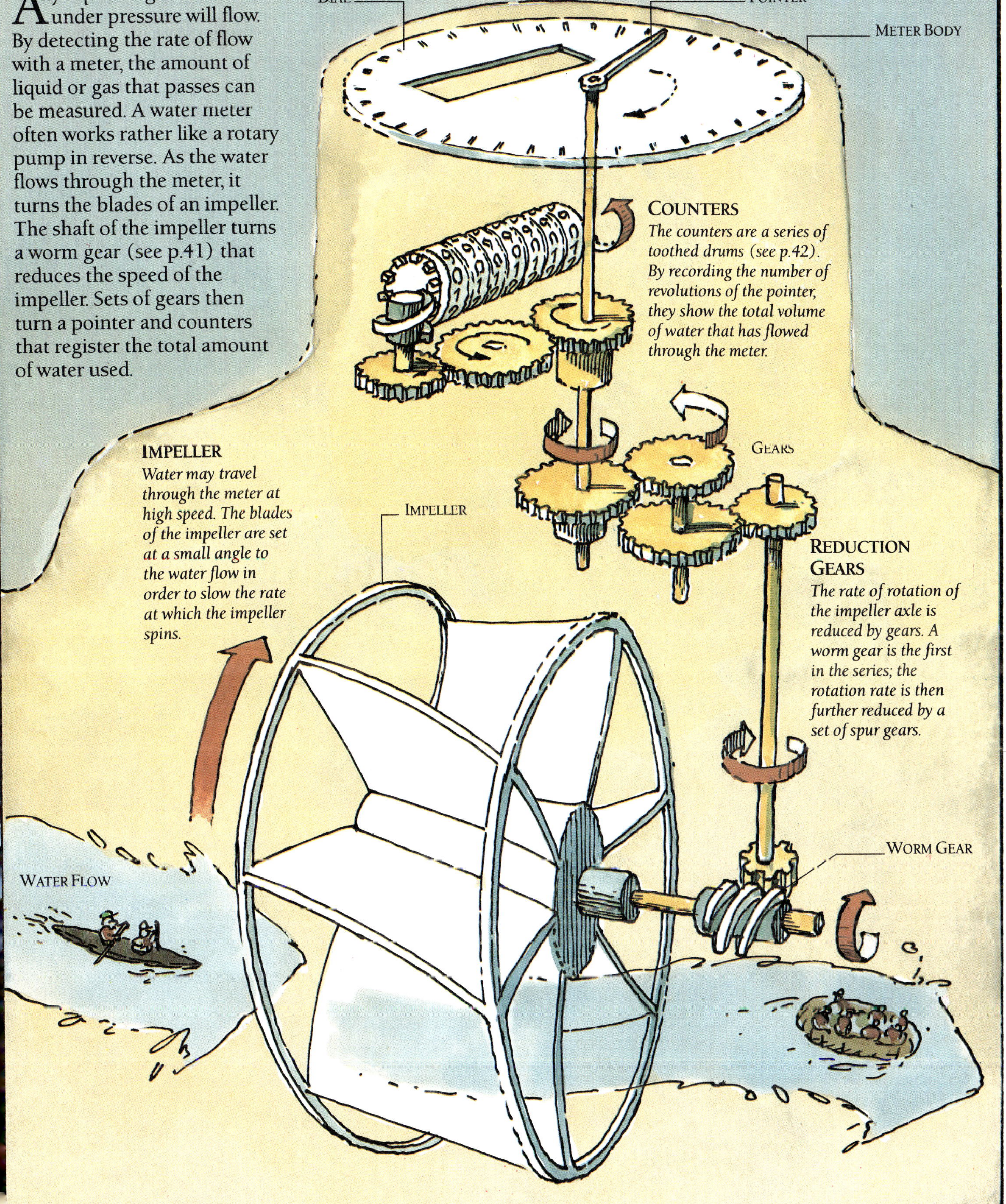

COUNTERS
The counters are a series of toothed drums (see p.42). By recording the number of revolutions of the pointer, they show the total volume of water that has flowed through the meter.

IMPELLER
Water may travel through the meter at high speed. The blades of the impeller are set at a small angle to the water flow in order to slow the rate at which the impeller spins.

REDUCTION GEARS
The rate of rotation of the impeller axle is reduced by gears. A worm gear is the first in the series; the rotation rate is then further reduced by a set of spur gears.

JETS AND SPRAYS

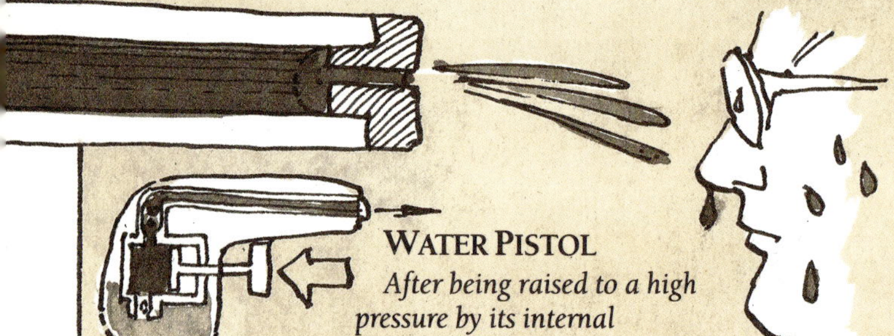

WATER PISTOL
After being raised to a high pressure by its internal piston pump (see pp.130-1), the water is forced out of the nozzle in a powerful jet.

Forcing a liquid through a nozzle requires pressure because the narrow hole restricts the flow. The liquid emerges in a high-pressure jet which may break up into a spray of droplets as it meets the air.

Jets and sprays have many uses, from delivering liquids in a useful form to providing power by action and reaction. In this case gases, rather than liquids, are used. A pump may deliver the fluid to the nozzle, as in a dishwasher, or it may be contained under pressure, as in a spray can.

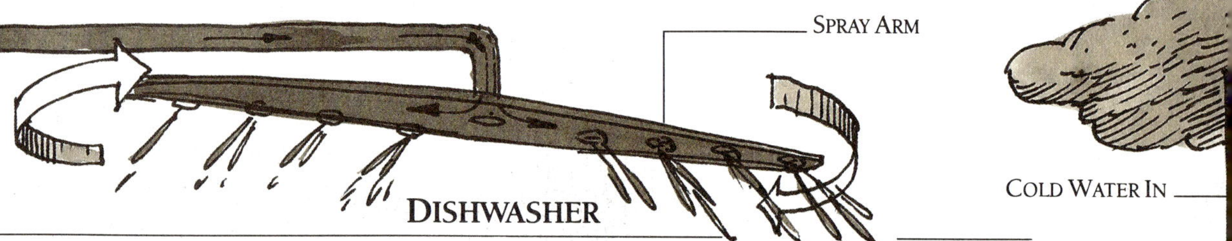

DISHWASHER

A dishwasher uses hot water under pressure both to power its spray arms, and also to do the cleaning itself. To be effective, the water has to be sprayed in powerful jets from all directions so that it reaches all the dishes and utensils. These are then rinsed by jets of clean water before drying.

THE DISHWASHER CYCLE

1 WATER TREATMENT
Cold water enters through a water softener, which treats the water so that the dishes dry without marks.

2 HEATING
The water fills the base of the dishwasher, where it is heated. Detergent is added.

3 WASHING
The hot water is pumped by the wash pump to the rotating spray arms. It sprays the dishes and returns to the base of the dishwasher, where it is recycled after being filtered.

4 RINSING AND DRYING
After washing, the dirty water is pumped out of the dishwasher and goes to the drain. The dishes are then rinsed and dried.

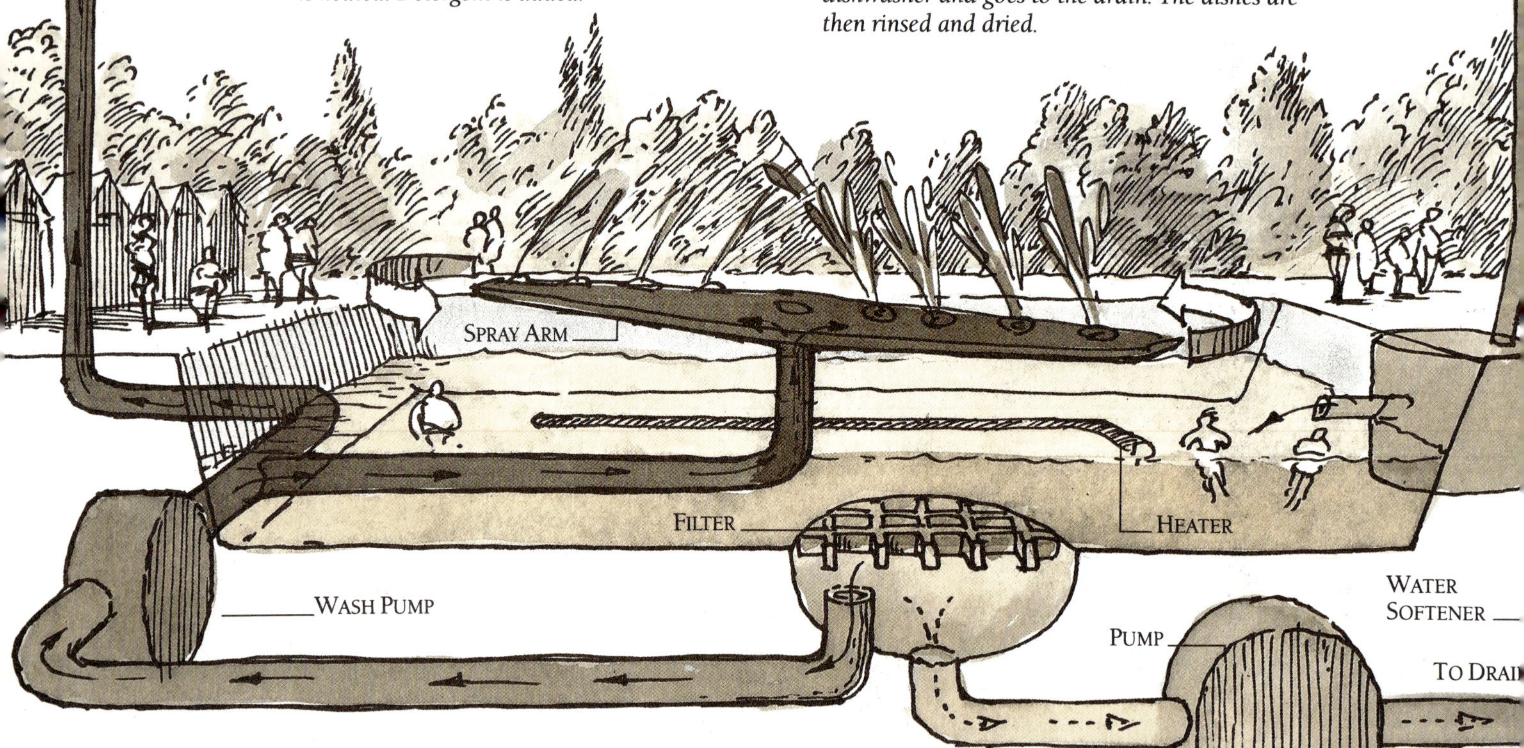

MANNED MANEUVERING UNIT

Whenever a jet or spray is produced, a force is generated that acts in the reverse direction to the flow of the fluid. This is an example of action and reaction (see p.108). It causes the spray arms of a dishwasher to rotate, and is also made use of in the manned maneuvering unit (MMU). This vehicle enables astronauts to fly around in space. It is propelled by jets of nitrogen gas which spurt from small nozzles. These jets make the nozzles move backward, thereby pushing or turning the MMU in the desired direction.

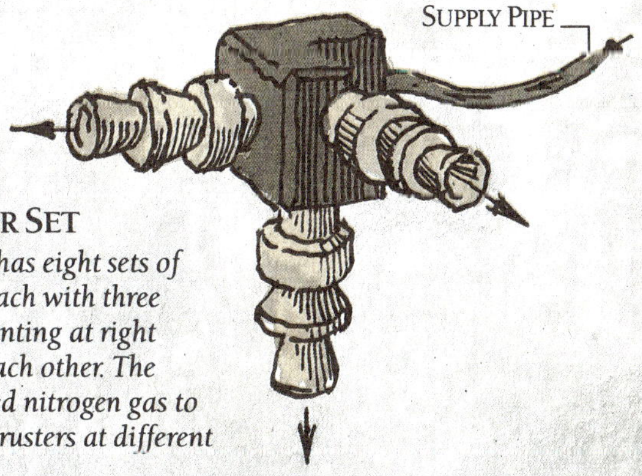

THRUSTER SET
The MMU has eight sets of thrusters, each with three nozzles pointing at right angles to each other. The controls feed nitrogen gas to different thrusters at different pressures.

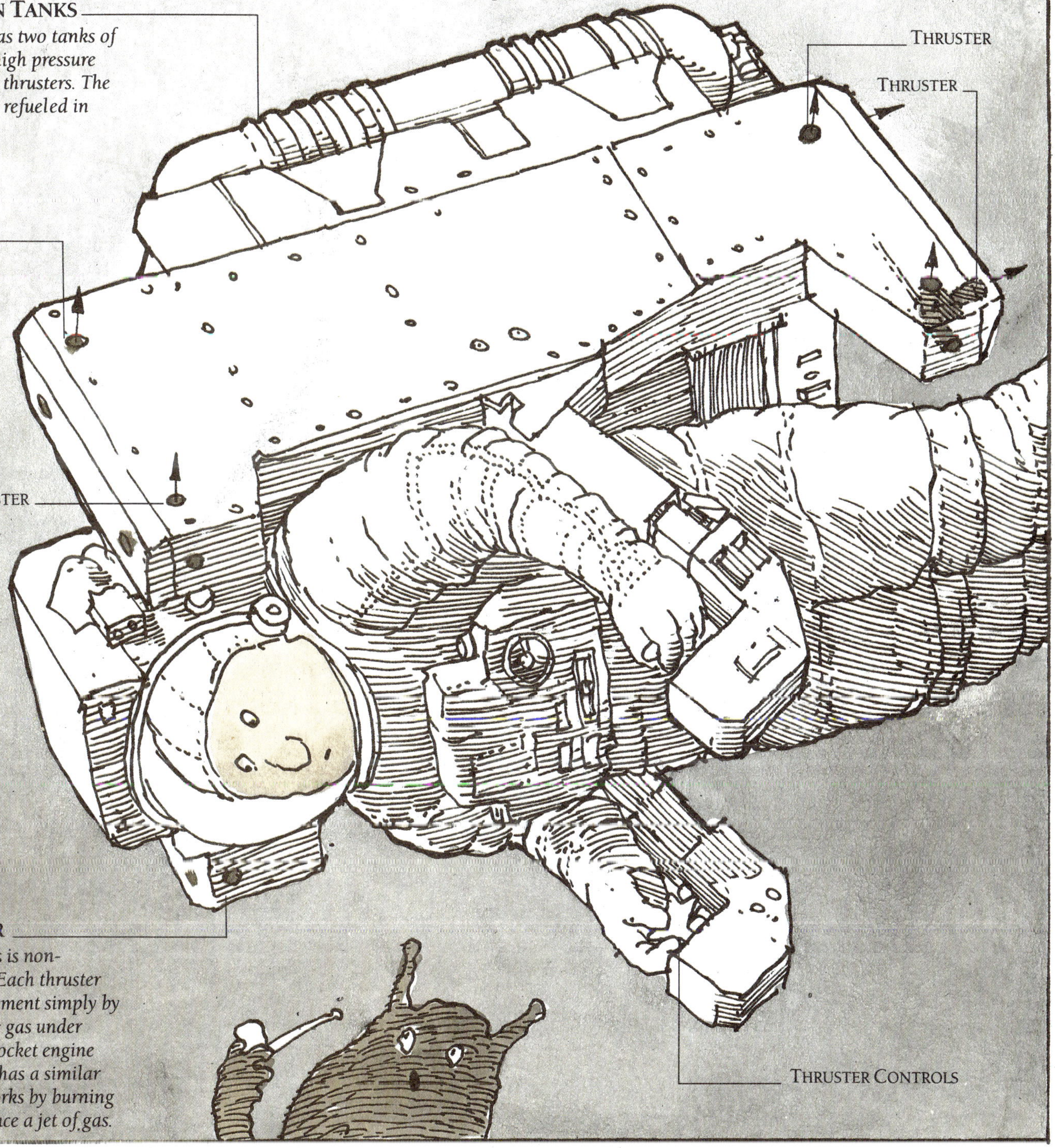

NITROGEN TANKS
The MMU has two tanks of nitrogen at high pressure that feed the thrusters. The tanks can be refueled in space.

THRUSTER
Nitrogen gas is non-flammable. Each thruster creates movement simply by releasing the gas under pressure. A rocket engine (see p.170) has a similar effect but works by burning fuel to produce a jet of gas.

THE SPRA…

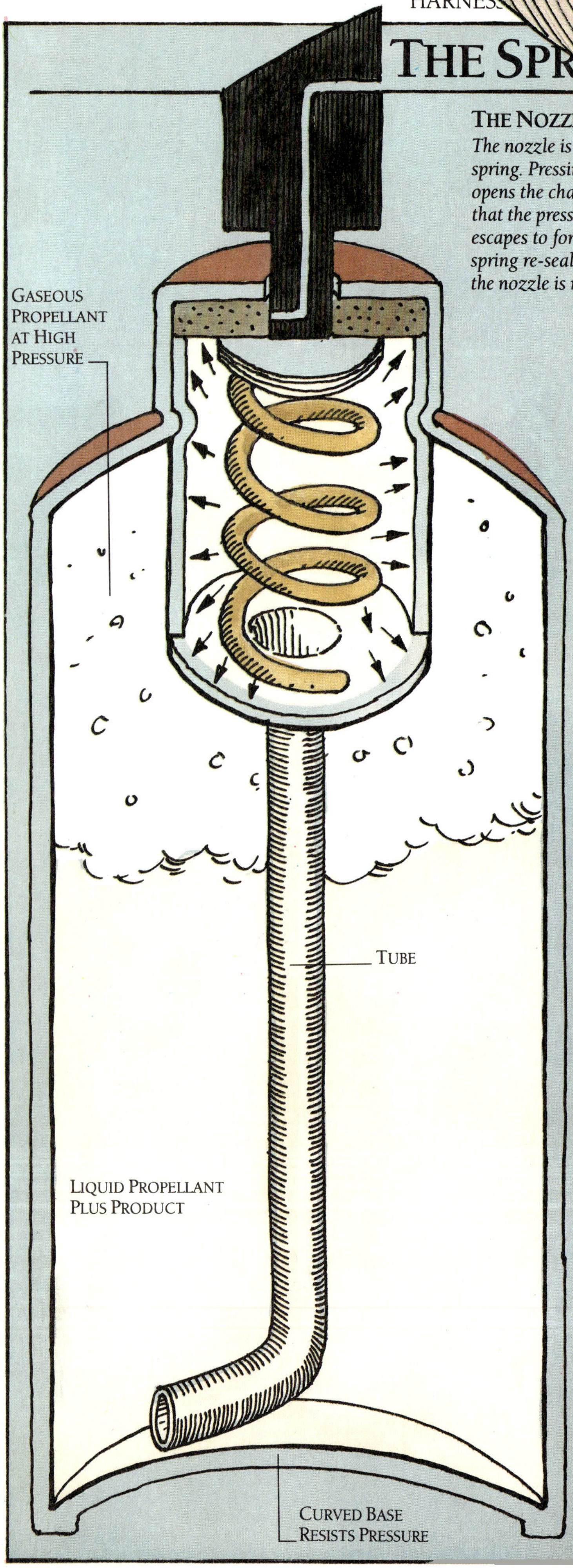

THE NOZZLE

The nozzle is held shut by a spring. Pressing it down opens the channel inside so that the pressurized liquid escapes to form a spray. The spring re-seals the can when the nozzle is released.

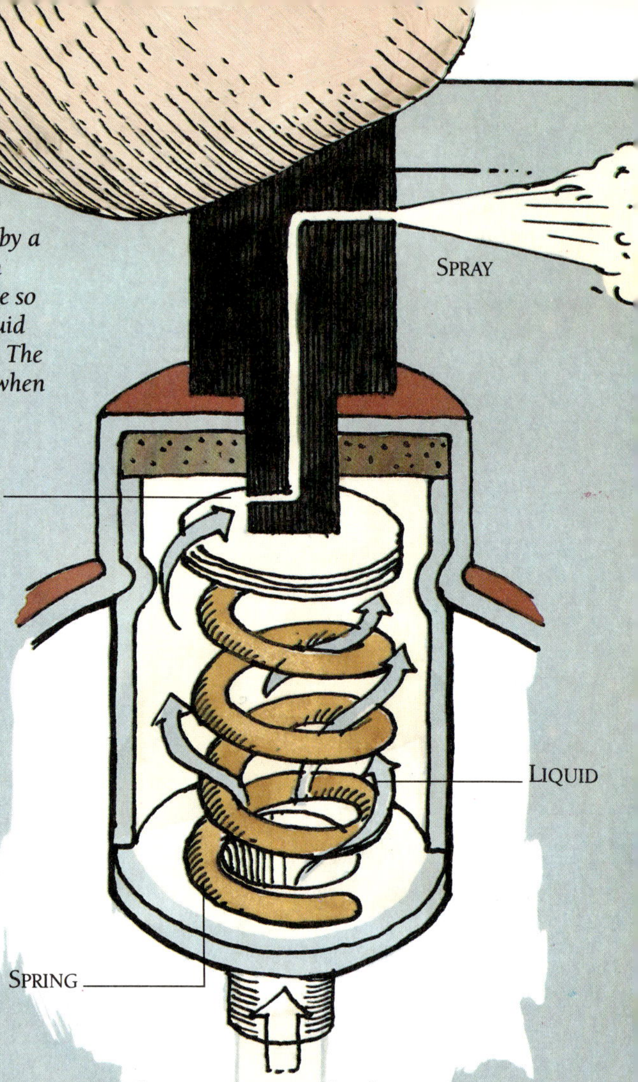

Spray cans produce an aerosol, the technical term for a very fine spray. They do this by means of a pressurized propellant, which is a liquid that boils at everyday temperatures. Inside the can, a layer of gaseous propellant forms over the liquid as it boils. The gas pressure increases, and eventually it becomes so high that boiling stops. When the nozzle is pressed, the gas pressure forces the product up the tube in the can and out of the nozzle in a spray or foam. The propellant may emerge as well but, now under less pressure, it immediately evaporates.

THE FIRE EXTINGUISHER

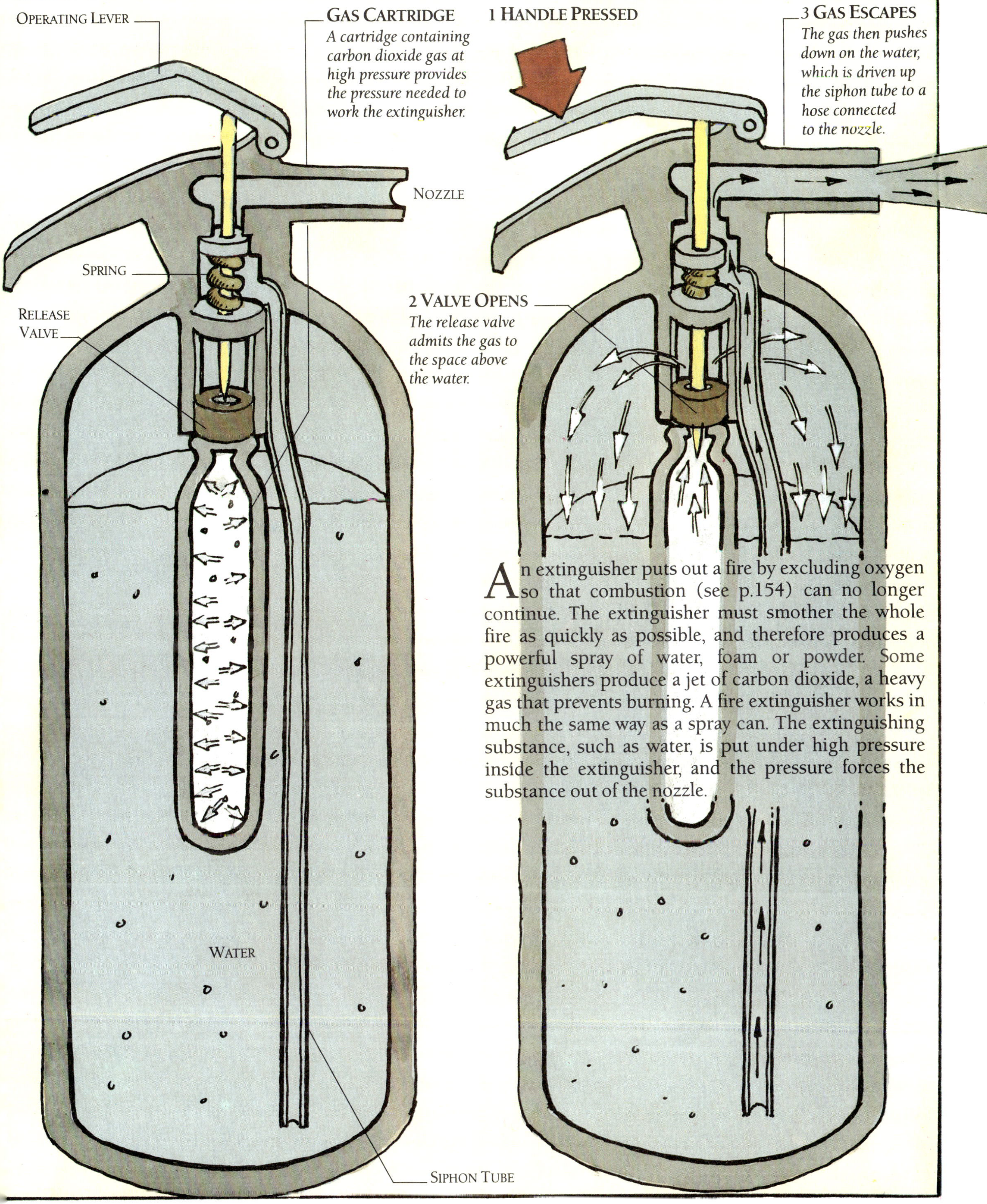

An extinguisher puts out a fire by excluding oxygen so that combustion (see p.154) can no longer continue. The extinguisher must smother the whole fire as quickly as possible, and therefore produces a powerful spray of water, foam or powder. Some extinguishers produce a jet of carbon dioxide, a heavy gas that prevents burning. A fire extinguisher works in much the same way as a spray can. The extinguishing substance, such as water, is put under high pressure inside the extinguisher, and the pressure forces the substance out of the nozzle.

THE CARBURETOR

The carburetor feeds gasoline and air to a car's engine in a precise but variable mixture. It works by suction. The pistons suck air in through a narrow section called a venturi. The air speeds up, so its pressure falls – just as happens with an aircraft wing (see p.115). The low-pressure air sucks gasoline out of a nozzle to form a spray, and this mixture of gasoline and air goes to the cylinders in the engine.

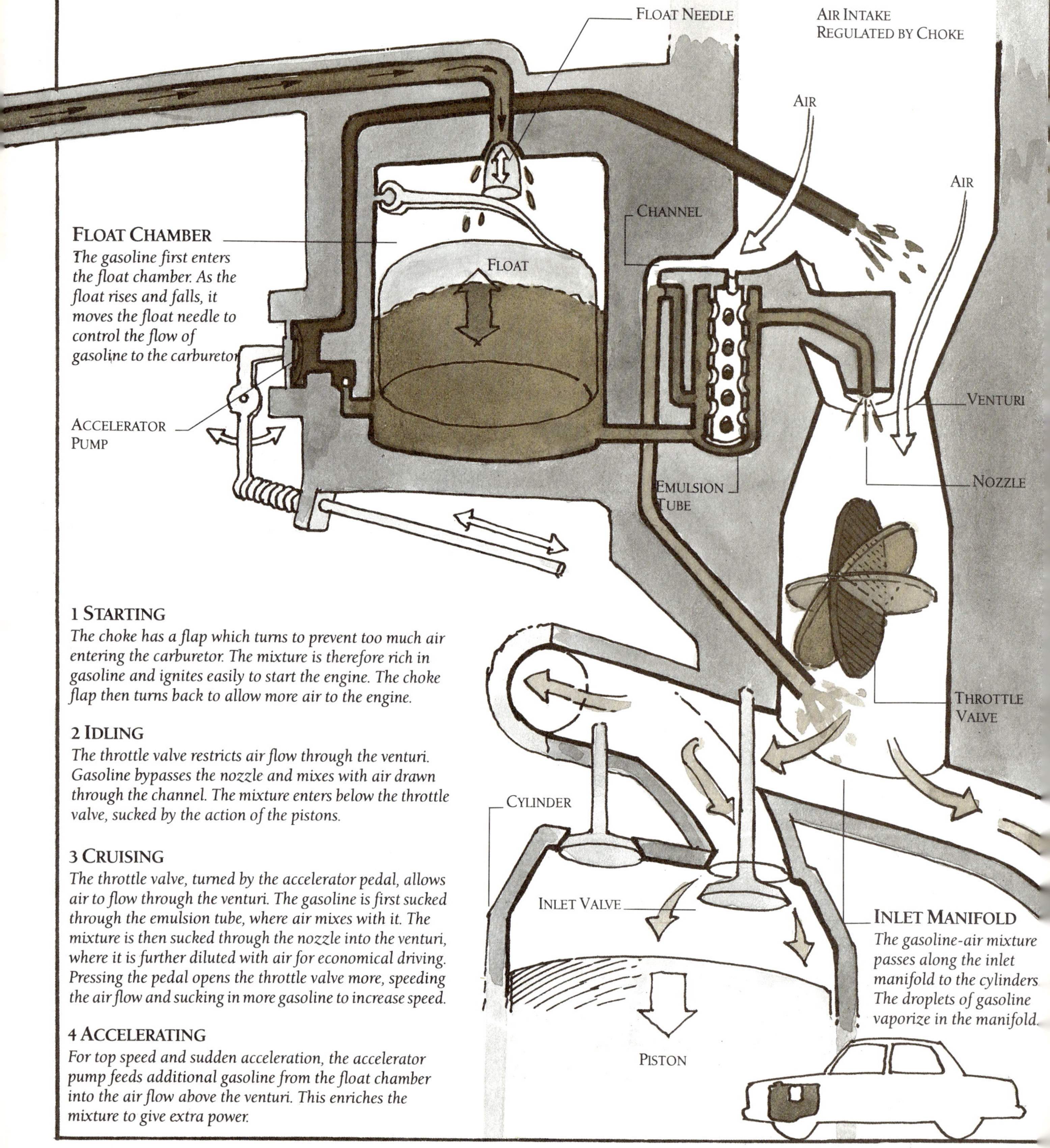

1 STARTING
The choke has a flap which turns to prevent too much air entering the carburetor. The mixture is therefore rich in gasoline and ignites easily to start the engine. The choke flap then turns back to allow more air to the engine.

2 IDLING
The throttle valve restricts air flow through the venturi. Gasoline bypasses the nozzle and mixes with air drawn through the channel. The mixture enters below the throttle valve, sucked by the action of the pistons.

3 CRUISING
The throttle valve, turned by the accelerator pedal, allows air to flow through the venturi. The gasoline is first sucked through the emulsion tube, where air mixes with it. The mixture is then sucked through the nozzle into the venturi, where it is further diluted with air for economical driving. Pressing the pedal opens the throttle valve more, speeding the air flow and sucking in more gasoline to increase speed.

4 ACCELERATING
For top speed and sudden acceleration, the accelerator pump feeds additional gasoline from the float chamber into the air flow above the venturi. This enriches the mixture to give extra power.

PENS

Many pens work by capillary action, which occurs in very narrow tubes or channels. A liquid flows up such a tube because the pressure inside is lowered. Air pressure then forces the liquid up the tube. The low pressure is caused by forces between the molecules at the liquid's surface in the tube.

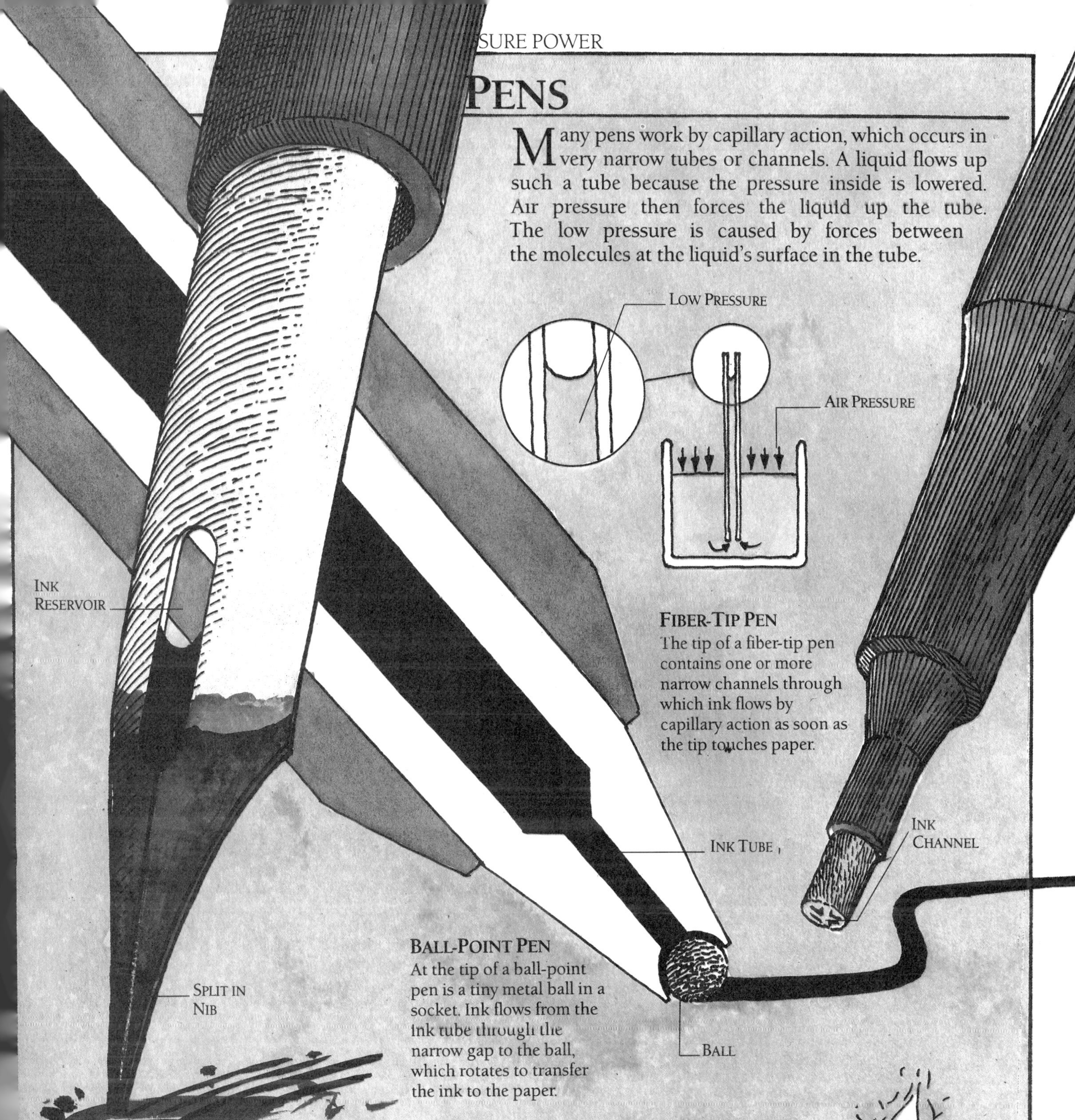

FIBER-TIP PEN
The tip of a fiber-tip pen contains one or more narrow channels through which ink flows by capillary action as soon as the tip touches paper.

BALL-POINT PEN
At the tip of a ball-point pen is a tiny metal ball in a socket. Ink flows from the ink tube through the narrow gap to the ball, which rotates to transfer the ink to the paper.

DIP PEN
A dip pen's nib is split into two halves. The halves meet at the point of the nib, but a little way above it they separate to form an ink reservoir that fills as the pen is dipped in the ink. Capillary action together with gravity conducts ink from the reservoir down the narrow split in the nib to the paper.

CAPILLARY ACTION IN A CANDLE
A candle relies on capillary action to keep the flame supplied with molten wax. The wick of a candle is made up of many fine fibers wound closely together. The gaps between these fibers act just like a collection of narrow tubes to draw molten wax upward.

EXPLOITING HEAT

fig. 1

fig. 2

ON THE USES OF MAMMOTH HEAT

There are two things that mammoths enjoy above all else (with the possible exception of swamp grass). They are working at some useful task and sleeping.

During my travels, I have come across a number of situations in which the two have been successfully combined to the benefit of both man and beast.

In figure 1, heat absorbed during a long sleep in the Sun or created by chewing swamp grass is used to warm water stored in the animal's trunk. When the trunk is secured vertically, the warmest water rises to the top, making it readily available.

In figure 2, the animal is shown performing its bed-warming function. Heat absorbed or created during the day is transferred from the mammoth to the bed in anticipation of its human occupant. To rouse the beast either a mouse is slipped under the covers, or the bed's would-be occupant makes squeaking noises. In either case the terrified beast is quickly displaced.

THE NATURE OF HEAT

The mammoth receives heat from the Sun in the form of invisible heat rays and makes heat inside its vast bulk by the consumption of swamp grass and other elephantine foods. The heat travels through its body and warms its skin. In the trunk, the heated water rises of its own accord.

Heat is a form of energy that results in the motion of molecules. Molecules are constantly on the move in everything and the faster they move, the hotter is their possessor. So when anything receives heat energy, its molecules speed up; removing heat energy slows them down. Heat travels in three ways – by radiation, conduction and convection.

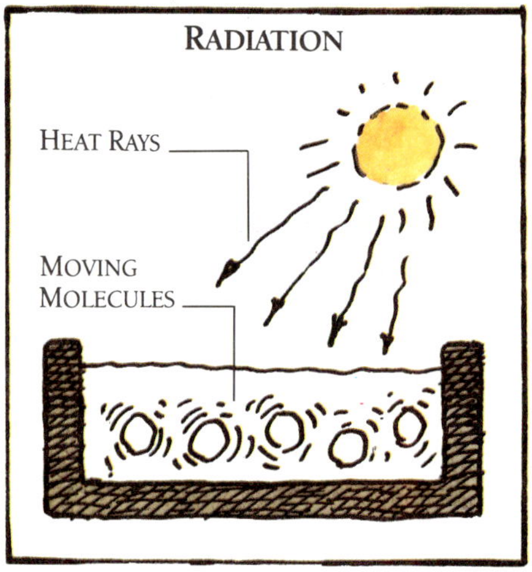

RADIATION

Hot things radiate heat rays, which are received by cooler objects. This form of heat transfer is called thermal radiation. The heat rays make the molecules speed up so that the object gets hotter. As the molecules move about with greater energy, they strike other molecules and make these move faster too.

fig. 3

In figure 3, a hot sleepy mammoth is employed as a clothes press. To operate the mammoth, one worker tickles the beast behind the ear with a feather. As the mammoth rolls over onto its back in anticipation of having its stomach scratched, a second worker places the garments to be pressed onto the warm spot. When the tickling stops, the mammoth resumes its original position. (I have observed that if the tickling stops before the switching of garments has been completed, the result can be disastrous.)

Figure 4 shows a further development on the principle of the clothes press. In this case, the weight and heat of one or more mammoths is employed to make and cook "Big Mamms". These wafer-thin burgers have become particularly popular with the young and are available with a variety of toppings.

fig. 4

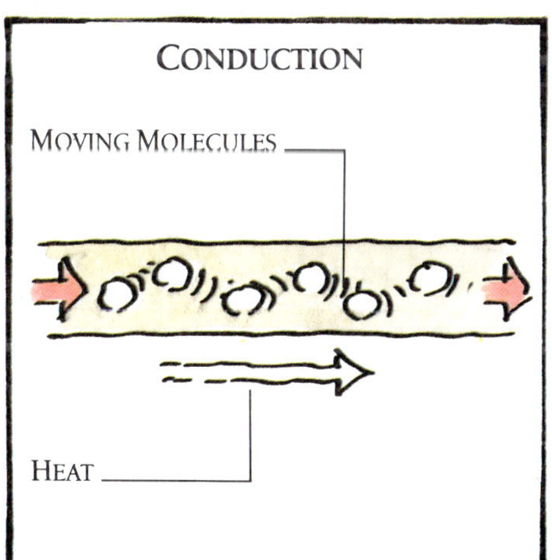

CONDUCTION

The molecules in solids move by vibrating to and fro within fixed limits. In conduction, heat spreads through solids by increasing the speed of this vibration.

CONVECTION

In liquids and gases the molecules move about. When heated, they also move further apart. A heated liquid or gas expands and rises, while a cooled liquid or gas contracts and sinks. This movement, which is known as convection, spreads the heat.

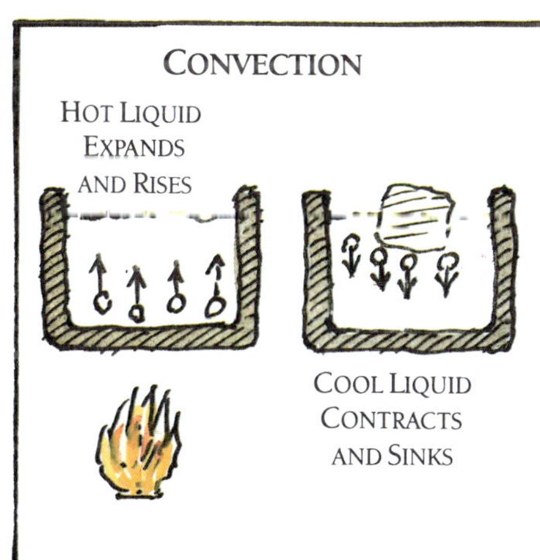

HEAT WAVES

The Sun bombards us with a whole range of energy-carrying rays, particularly light rays and infra-red rays. These rays, and also microwaves, have similar characteristics: they pass straight through some substances, they are reflected by others and they are absorbed by the remainder. Objects that absorb rays become hot, and this is made use of in a variety of devices including the solar heater and the microwave oven.

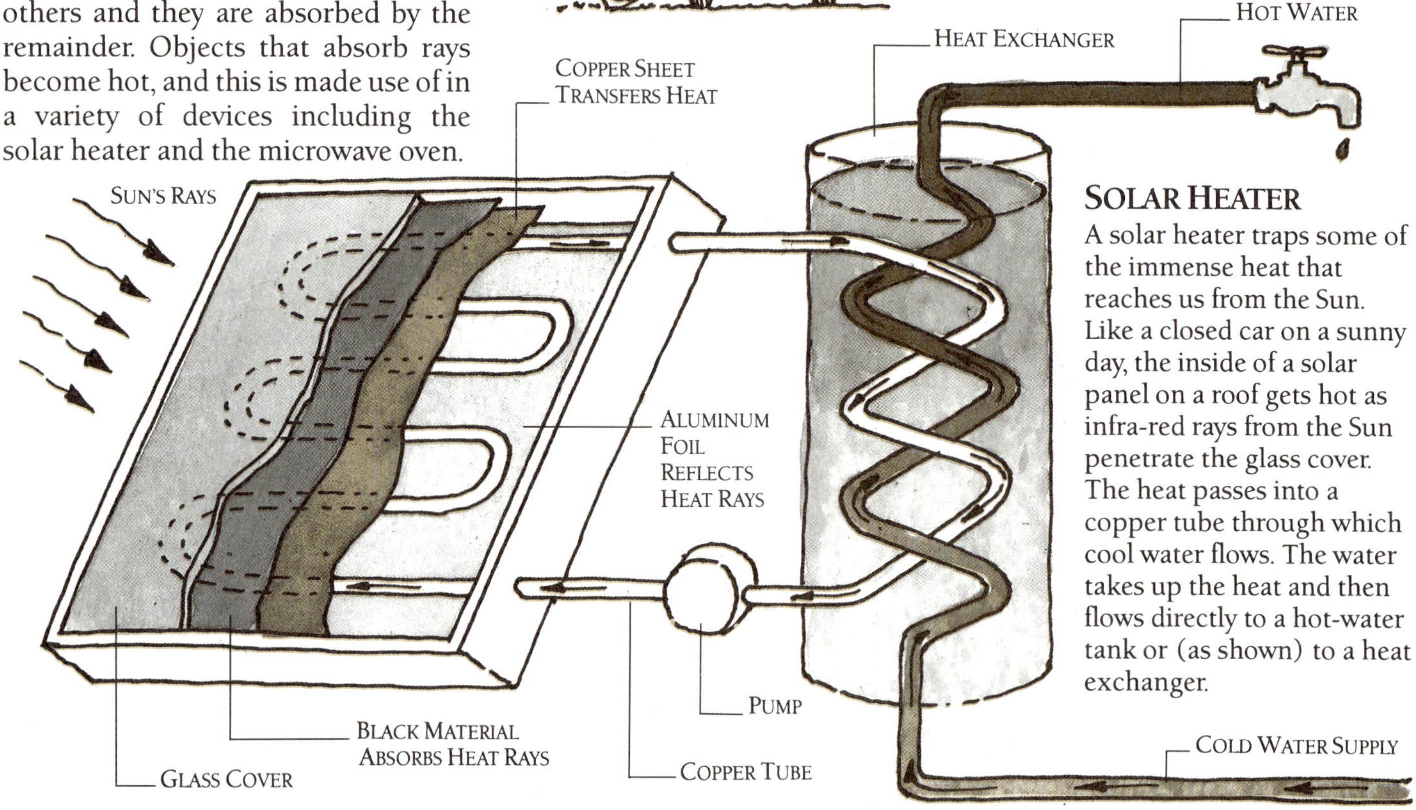

SOLAR HEATER

A solar heater traps some of the immense heat that reaches us from the Sun. Like a closed car on a sunny day, the inside of a solar panel on a roof gets hot as infra-red rays from the Sun penetrate the glass cover. The heat passes into a copper tube through which cool water flows. The water takes up the heat and then flows directly to a hot-water tank or (as shown) to a heat exchanger.

MICROWAVE OVEN

A magnetron produces a beam of microwaves, which have high heating power. The beam strikes a spinning fan, which reflects the waves onto the food from all directions. They pass through the container and enter the food, heating it throughout and cooking the food evenly and quickly.

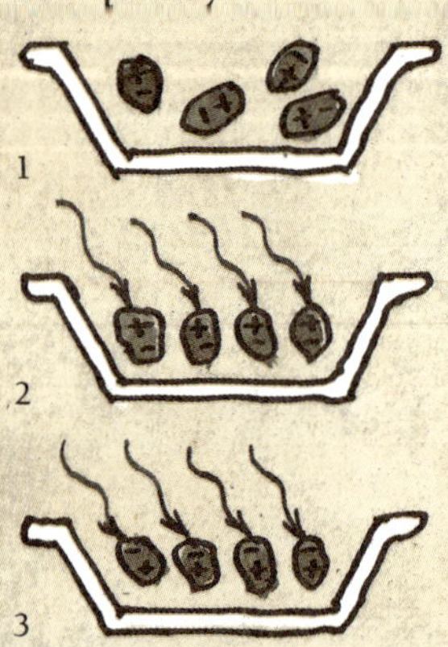

MICROWAVE HEATING

The microwaves strike molecules of water in the food (1). Each wave of energy causes the water molecules to align (2) and then reverse alignment (3). The extremely rapid and repeated twisting produces heat.

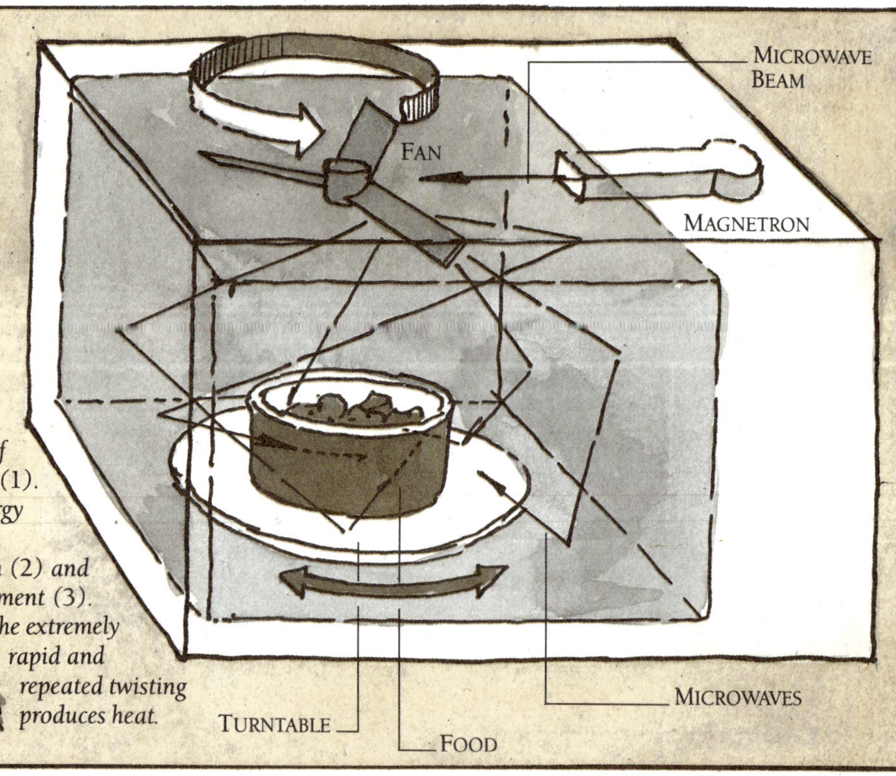

THE VACUUM FLASK

A vacuum flask can keep drinks piping hot – or icy cold – for hours on end. It does this by preventing as much movement of heat as possible, either out of or into the flask.

Inside the flask is a double-walled container of glass or steel. The walls are silvered on the inside to reflect heat rays (which behave like light rays) so that rays cannot leave or enter the flask. Between the container walls is a vacuum, which prevents heat conduction through the walls. The container support and stopper are made of an insulating material, such as cork, that reduces conduction.

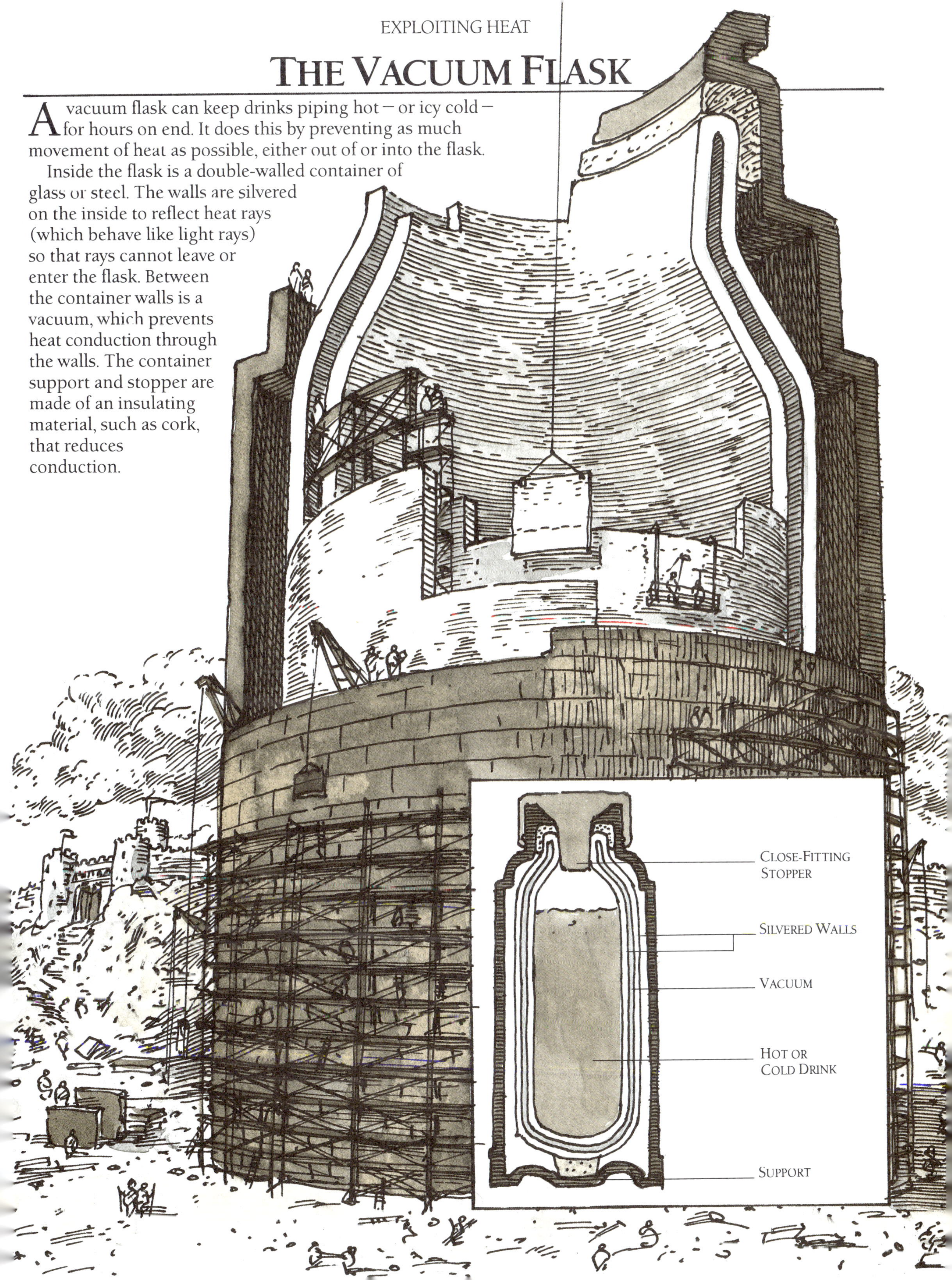

COMBUSTION MACHINES

FIRE AND WATER

Hydrogen burns in oxygen to create great heat, producing water. Molecules of hydrogen and oxygen both contain pairs of atoms. At a sufficiently high temperature each oxygen molecule collides violently with two hydrogen molecules. The collision breaks the molecules apart, and the atoms reform as two fast-moving water molecules.

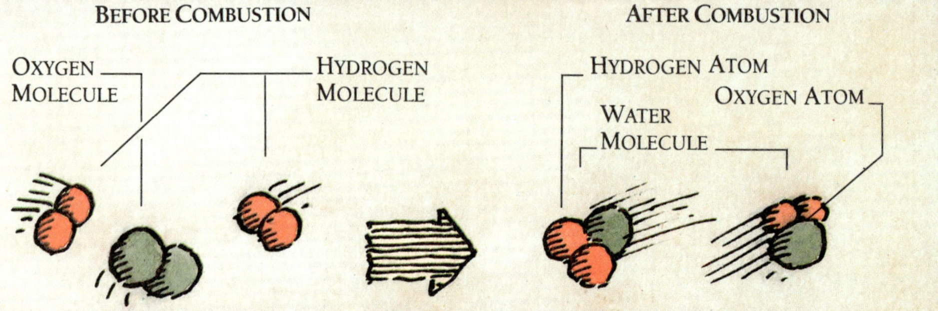

Combustion, or burning, is one common source of heat. Combustion machines allow two substances, often a fuel and oxygen in the air, to react so that the heat they produce is harnessed. The production of heat may be the sole purpose of the machine, as in a gas fire, or the heat produced may then be put to use, as in a welding torch or an engine.

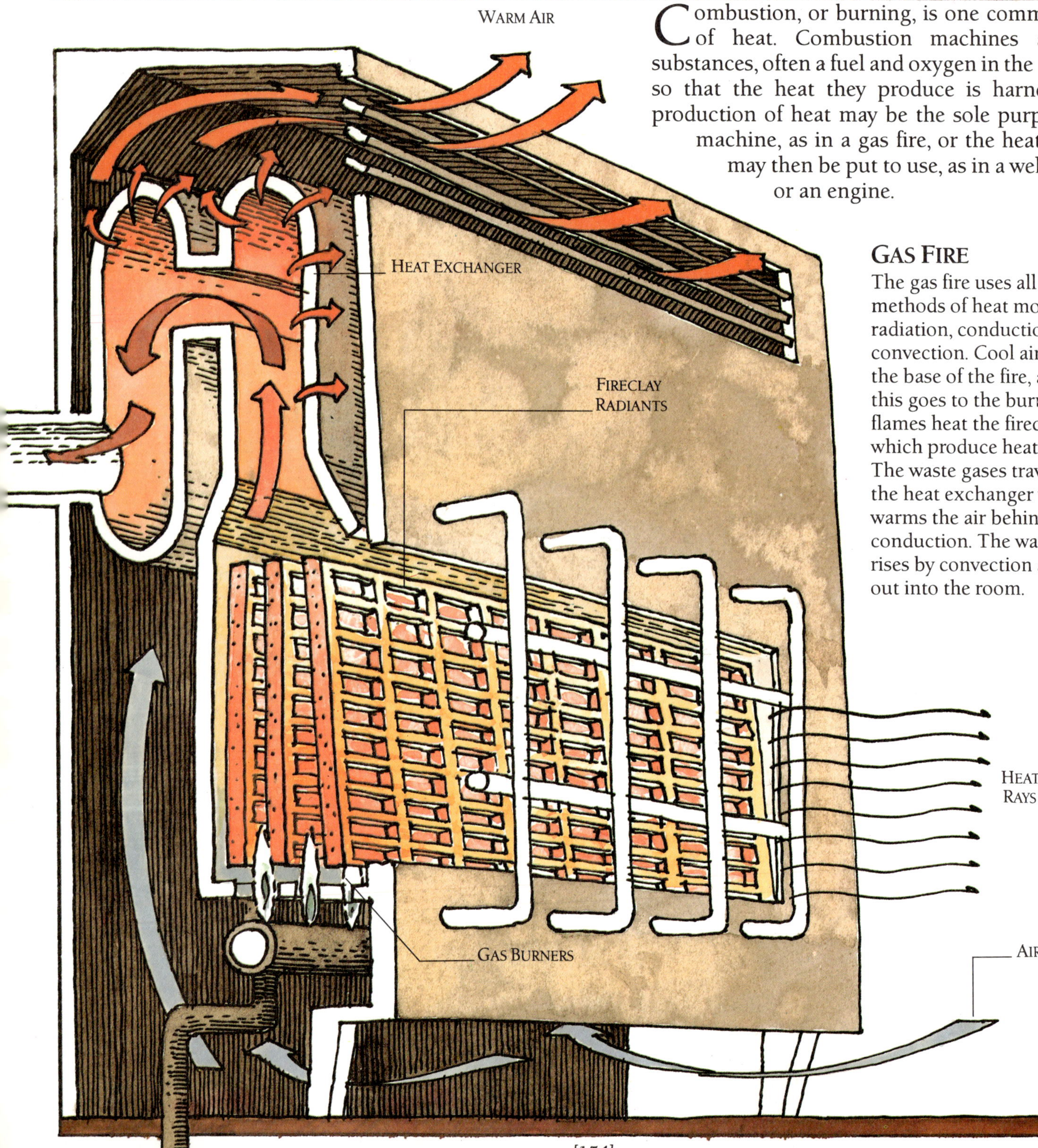

GAS FIRE

The gas fire uses all three methods of heat movement – radiation, conduction and convection. Cool air enters at the base of the fire, and some of this goes to the burners. The flames heat the fireclay radiants, which produce heat by radiation. The waste gases travel through the heat exchanger which warms the air behind the fire by conduction. The warmed air rises by convection and flows out into the room.

Welding Torch

Metal parts can be joined with great strength by melting or welding them together. A welding torch may work by combustion. Cylinders of oxygen and hydrogen, acetylene or another gaseous fuel are connected to the torch. The oxygen and fuel burn to give a very hot flame.

In electric welding, a strong electric current or a hot electric spark heats the two parts at the join.

Filler Rod

The welder uses a filler rod to add metal to the join.

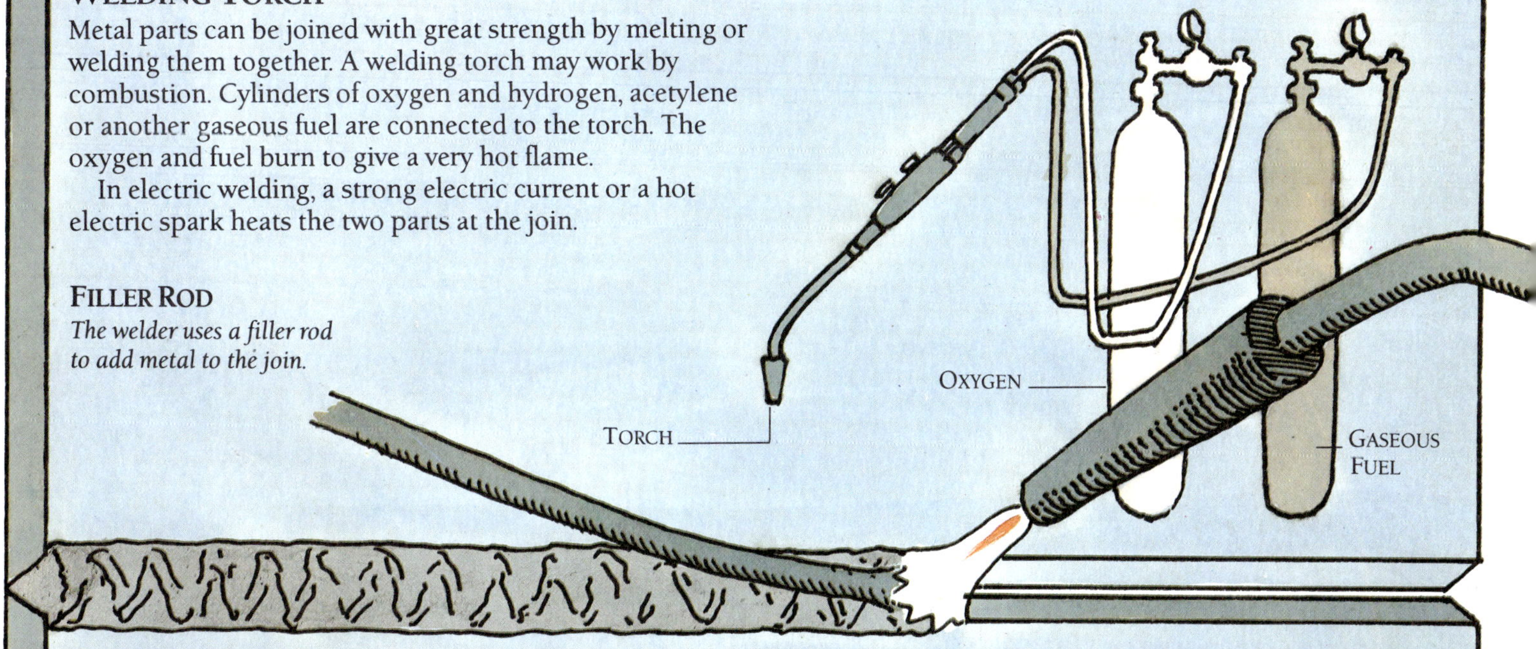

Gun Cartridge

A cartridge contains two explosives to fire a bullet from the barrel of a gun at very high speed. Explosives are materials that burn very quickly; they produce lots of gas that rapidly expands with great power.

The first explosive in a cartridge (the detonator) is a sensitive explosive that is set off by the firing mechanism of the gun. It ignites the second explosive (the propellant). The resulting gas can only expand in one direction, and it drives the bullet out of the cartridge and along the barrel.

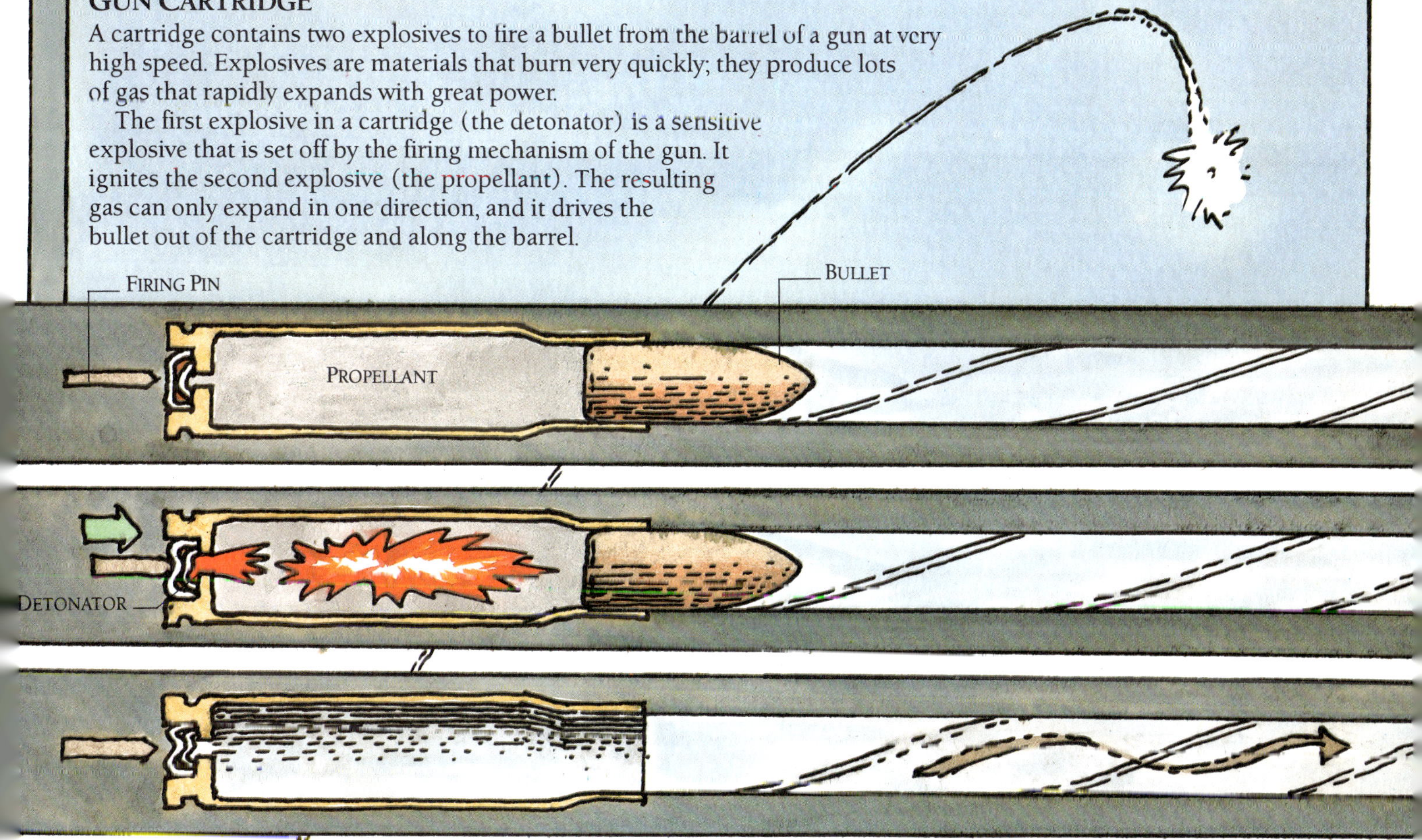

Flares

Combustion gives light as well as heat but is nowadays mostly used for emergency lighting only. Flares contain materials that burn brightly to light up the night. When used as distress signals, flares also contain chemicals that produce intense colors.

BLAST FURNACE AND STEEL CONVERTER

Steel depends on combustion at several points in its manufacture. It is basically iron mixed with a precise but small quantity of carbon, and it is made from iron ore and carbon in the form of coke. Iron ore is a compound of iron and oxygen. To remove the oxygen and free the iron, the ore is heated with coke in a blast furnace. The oxygen in the iron is released and taken up by the coke during combustion.

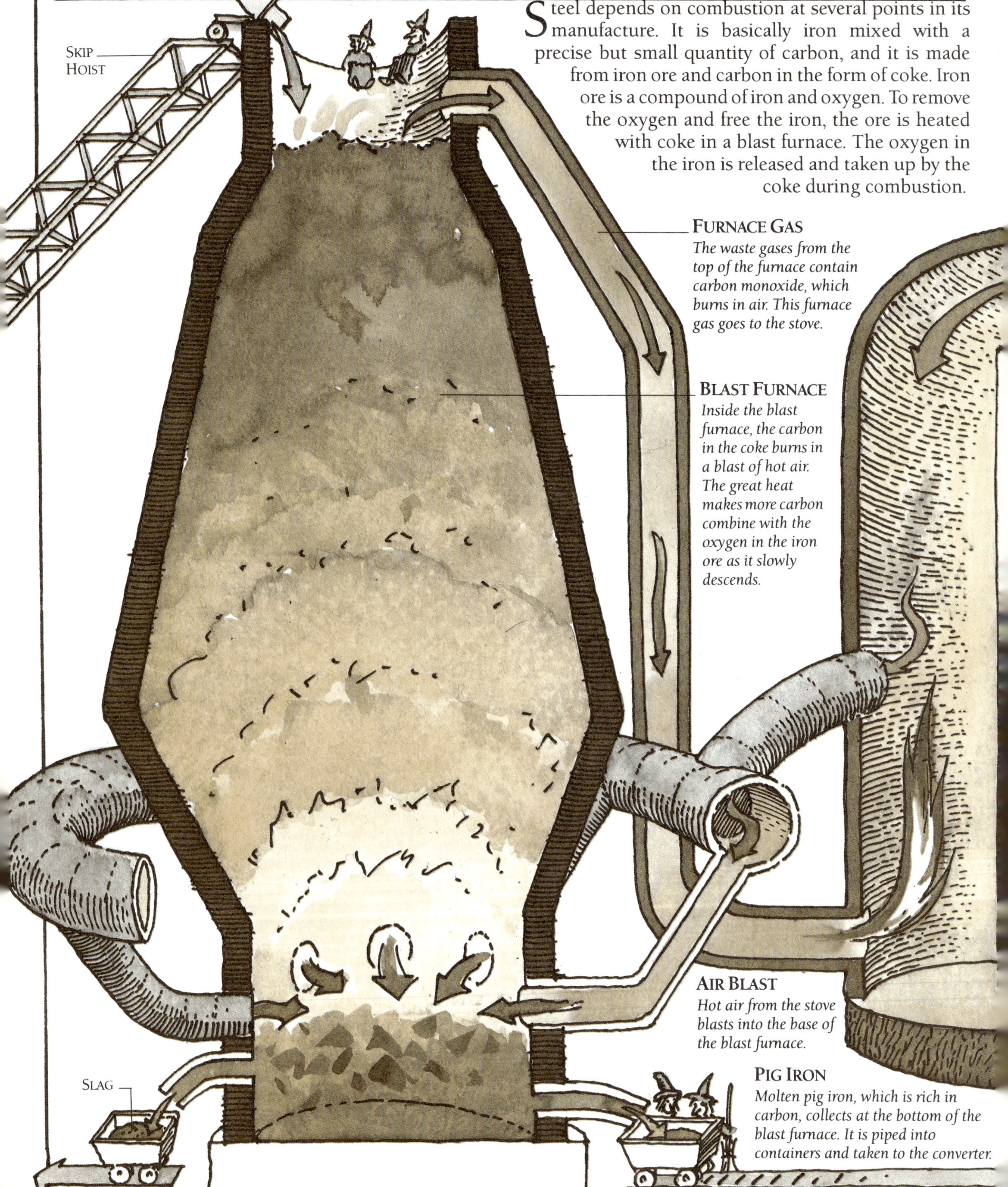

In the blast furnace, the iron mixes with too much carbon to make good steel. A steel converter removes this extra carbon. The most common kind of converter blows oxygen gas onto the molten iron. The oxygen burns away the extra carbon to give steel. Scrap steel may be added to the converter for recycling.

Other kinds of steel converters include the open hearth furnace, in which flames of burning fuel play on a charge of iron to burn away the excess carbon, and electric furnaces powered by a strong electric current.

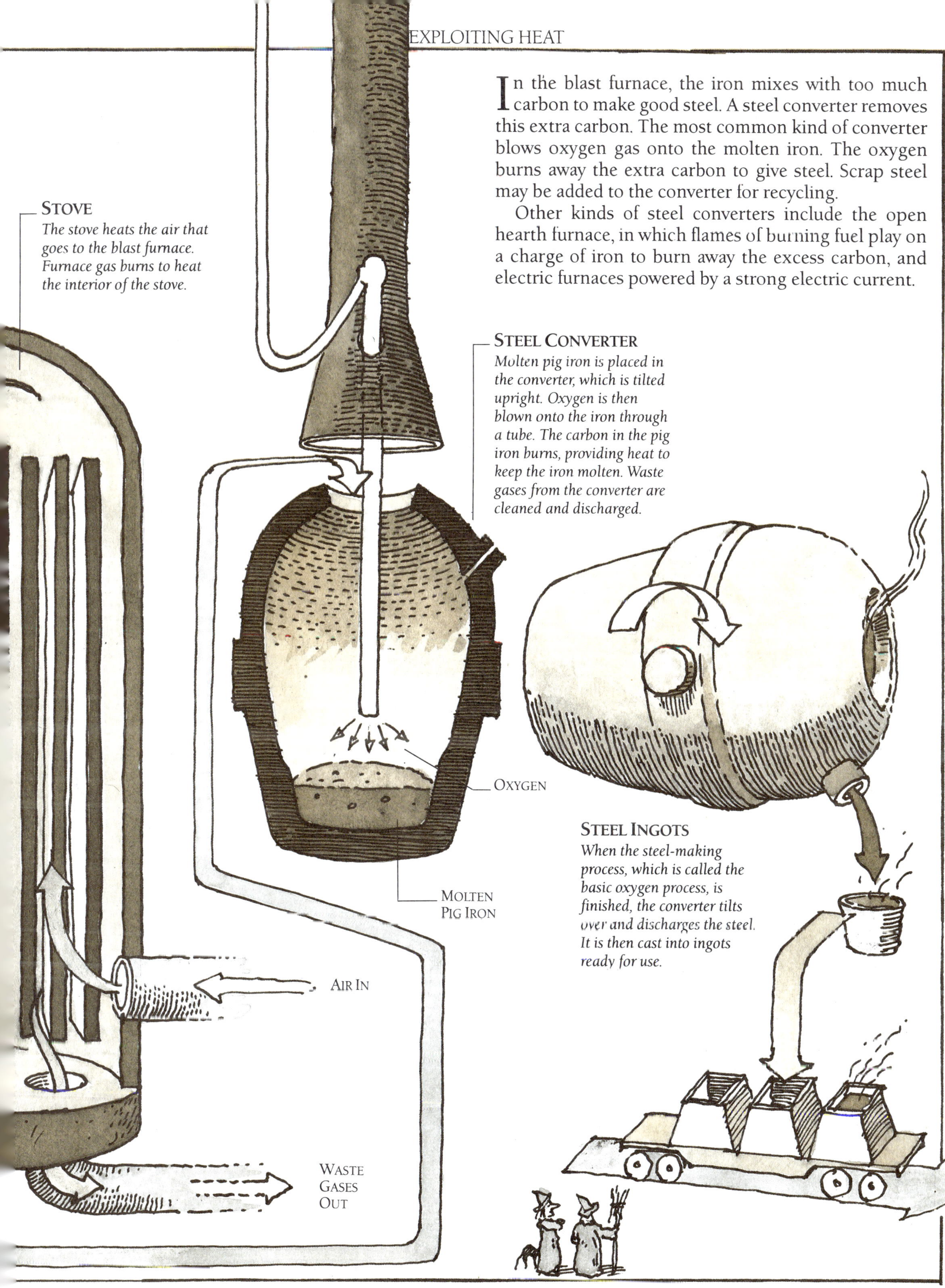

STOVE
The stove heats the air that goes to the blast furnace. Furnace gas burns to heat the interior of the stove.

STEEL CONVERTER
Molten pig iron is placed in the converter, which is tilted upright. Oxygen is then blown onto the iron through a tube. The carbon in the pig iron burns, providing heat to keep the iron molten. Waste gases from the converter are cleaned and discharged.

STEEL INGOTS
When the steel-making process, which is called the basic oxygen process, is finished, the converter tilts over and discharges the steel. It is then cast into ingots ready for use.

ELECTRIC HEAT

No form of heating is as convenient as electric heating. It is available at the click of a switch and is totally clean to use, although its generation may produce polluting waste through combustion and nuclear fission (see p.174).

Like all other sources of heat, electricity hastens the motion of molecules, giving them extra energy that appears as heat. When an electric current flows along a wire, billions of tiny particles called electrons move from atom to atom along the wire. The electrons are smaller than the atoms, and jostle the atoms as they pass. The vibration of the metal atoms increases, and the wire gets hotter.

Many machines contain electric heating elements that work in this way. Heat may radiate from the element, as in an electric fire, or the element may be enclosed in an electrically insulated container that heats water, for example, by conduction and convection.

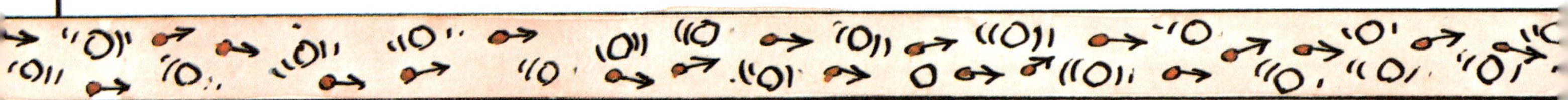

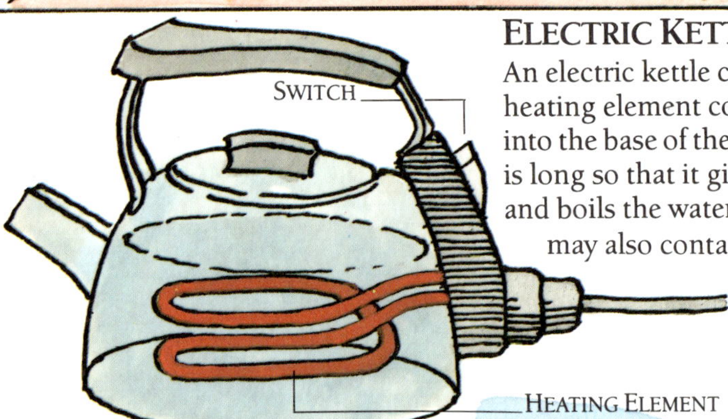

ELECTRIC KETTLE

An electric kettle contains a long heating element coiled so that it fits into the base of the kettle. The element is long so that it gives plenty of heat and boils the water quickly. The kettle may also contain a thermostat (see p.162). This stops the supply of current to the element when the water boils so that the kettle will not boil dry if unattended. The thermostat may also cut off the power if the kettle is switched on without any water so that the element does not overheat.

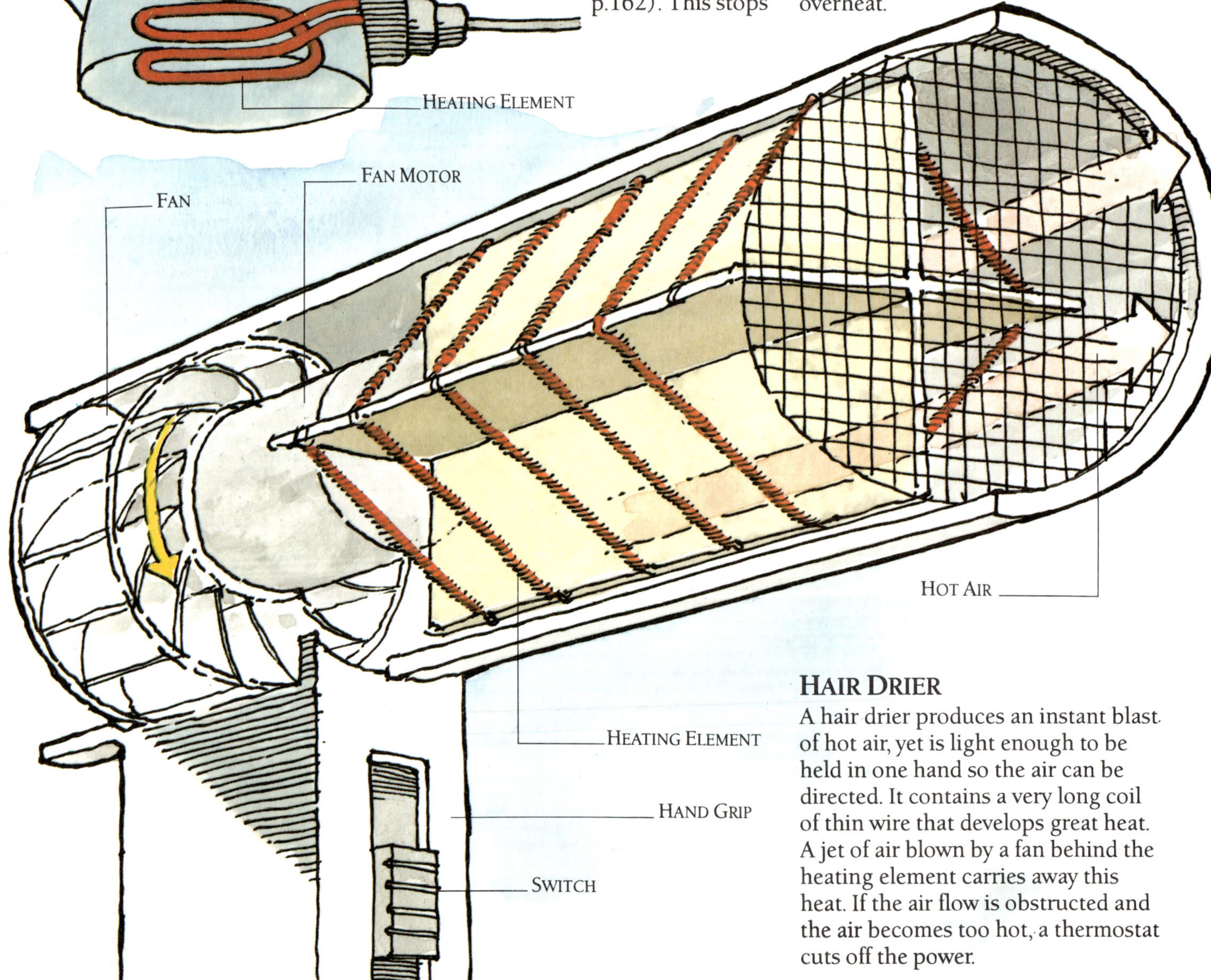

HAIR DRIER

A hair drier produces an instant blast of hot air, yet is light enough to be held in one hand so the air can be directed. It contains a very long coil of thin wire that develops great heat. A jet of air blown by a fan behind the heating element carries away this heat. If the air flow is obstructed and the air becomes too hot, a thermostat cuts off the power.

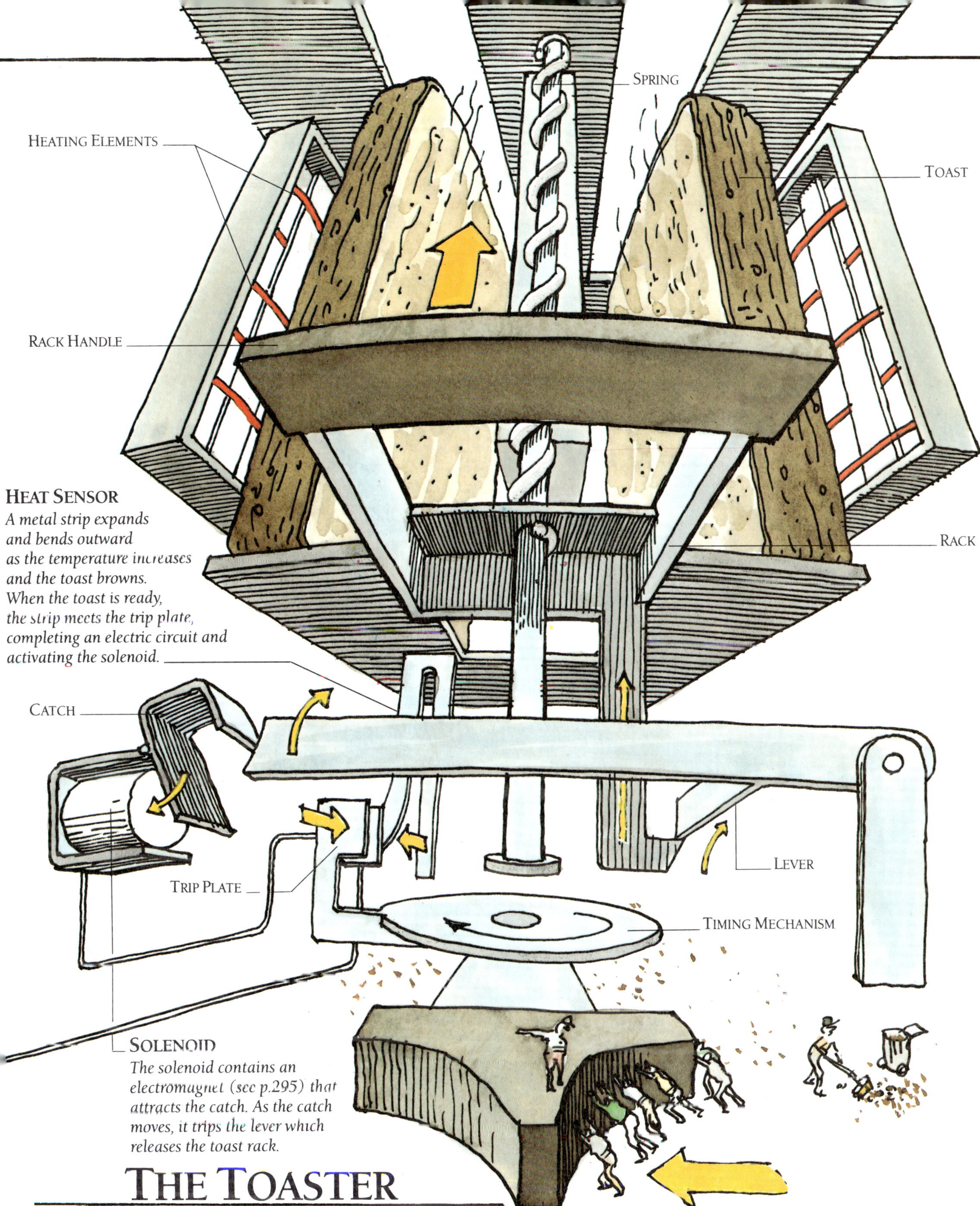

THE TOASTER

An electric toaster is designed to pop up toast browned to perfection. The slices of bread descend into the toaster on a spring-loaded rack. This switches on heating elements that brown all sides. A timing mechanism then switches off the elements when the toast is ready. The rack, released by an electromagnetic catch, springs back up.

REFRIGERATOR

The refrigerator is a machine that makes heat move. It takes heat out of the inside and moves it to the outside. The heat flows into the air and the inside, having lost heat, becomes cold. Refrigerators work by evaporation. When a liquid turns to vapor, it loses heat and gets colder. This is because the molecules of vapor need energy to move and leave the liquid. This energy comes from the liquid; the molecules left behind have less energy and so the liquid becomes colder.

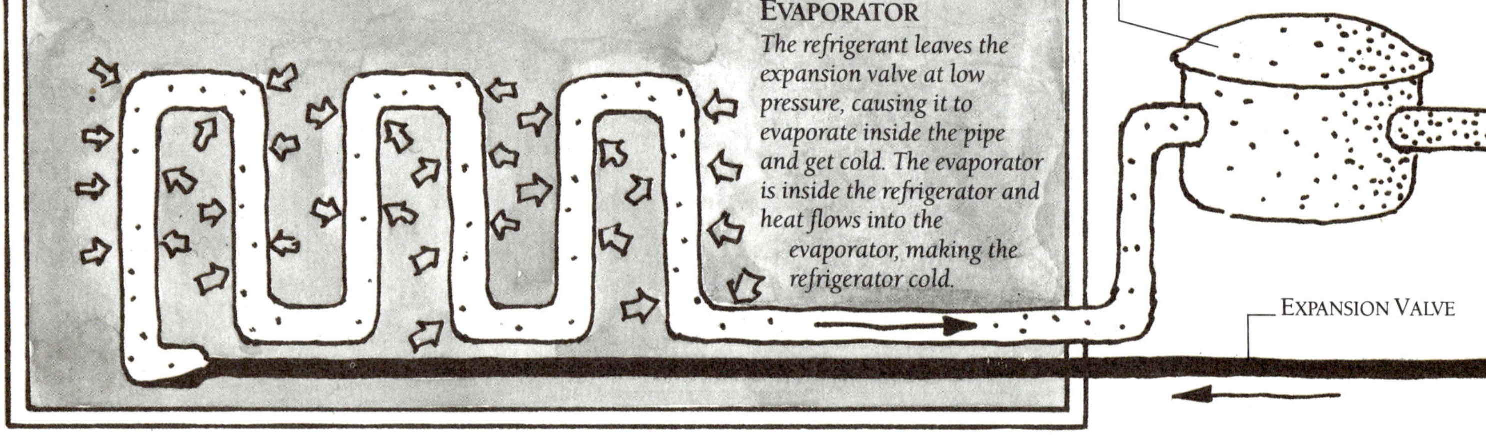

CAR COOLING SYSTEM

Most cars have water-cooled engines. A pump (see p.133) drives the water around channels inside the engine. The hot water then passes through the thermostat to the radiator, where it loses its heat to the air before returning to the pump. In some cars, the hot water also flows through the car's heater.

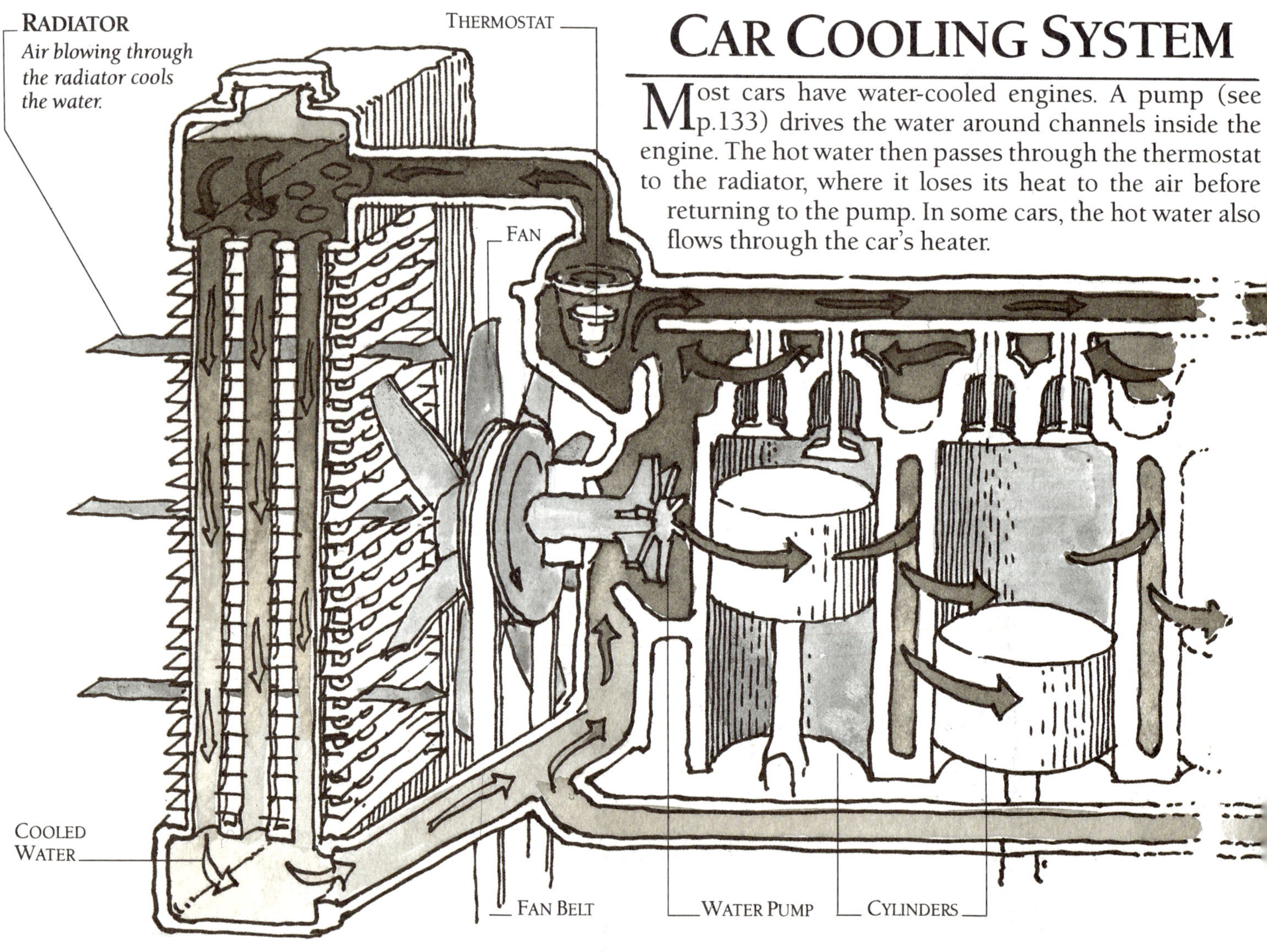

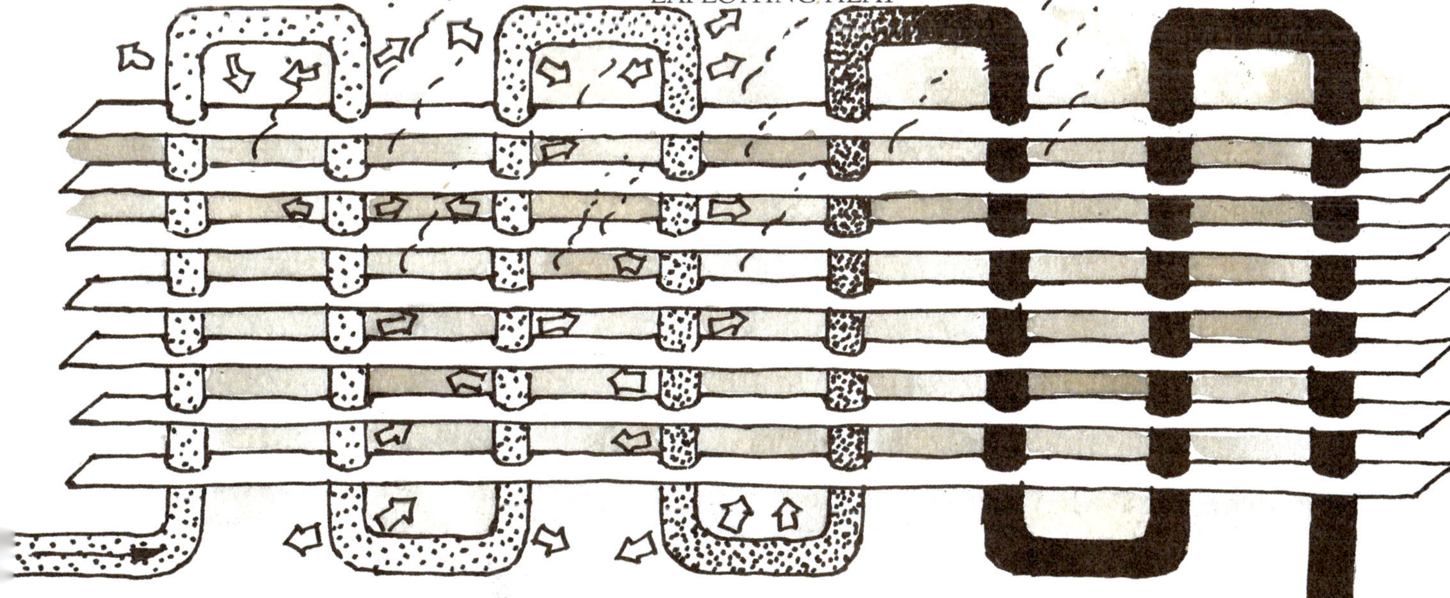

CONDENSER

The refrigerant vapor leaves the compressor at high pressure. As it flows through the condenser, the high pressure causes the vapor to condense back to liquid refrigerant. As this happens the vapor gives out heat, making the condenser warm. The condenser is at the back of the refrigerator, and the heat flows into the air around the refrigerator.

HEATER

The car heater may be part of the cooling system. It contains a heat exchanger, in which the hot water from the engine heats air driven by a fan in the passenger compartment. The warm air then returns to various parts of the passenger compartment.

AIR CONDITIONER

This machine works in the same basic way as a refrigerator. A compressor circulates a refrigerant from an evaporator through a condenser and expansion valve and back to the evaporator. The evaporator is placed over a fan that extracts hot and humid air from the room. It takes heat from the air, making its moisture condense into water droplets. The cool dry air then returns to the room. A fan removes the heat from the condenser outside the room.

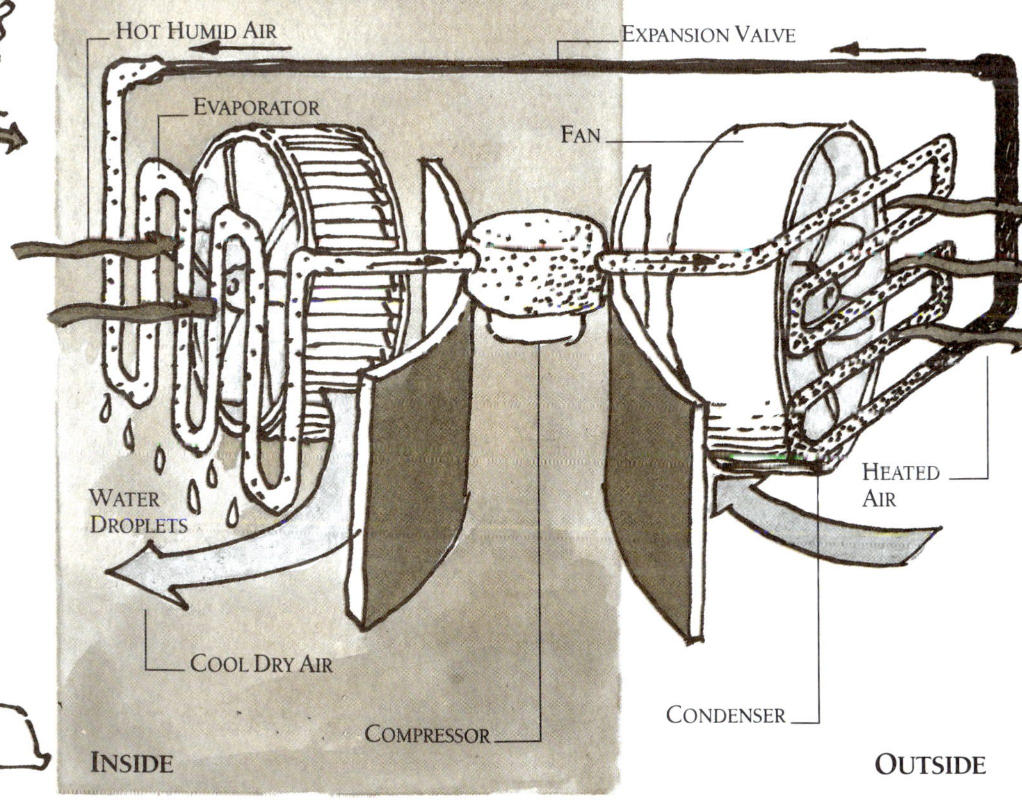

THERMOSTATS

Thermostats are devices that regulate heaters and cooling machines, repeatedly turning them on and off so that they maintain the required temperature. They work by expansion and contraction. As something heats up, its molecules move further apart. The object expands in size. When the object cools, the force pulling the molecules together reasserts itself; the molecules close ranks and the object contracts.

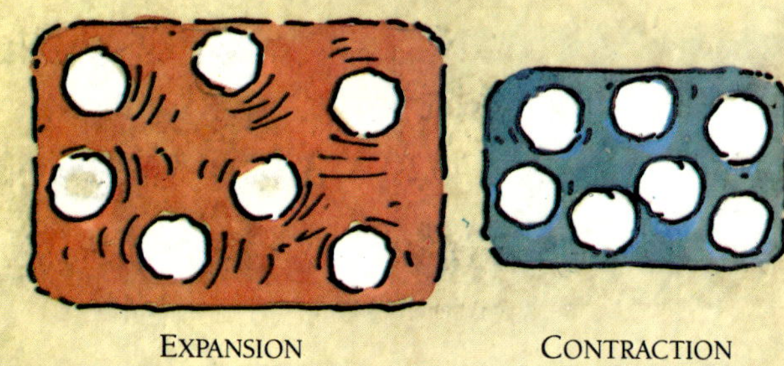

BIMETAL THERMOSTAT

This common thermostat contains a strip of two different metals, often brass and iron. The metals expand and contract by different amounts. The bending produced by heating or cooling the thermostat can be used to activate a heater switch.

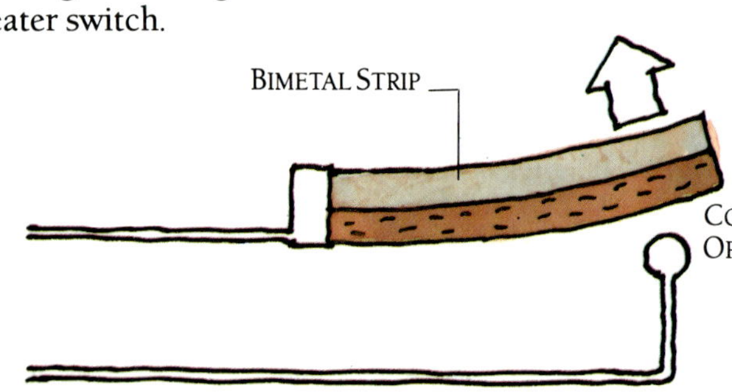

SWITCH OPEN

The strip bends as it gets hotter, opening the contact. The current stops flowing and the heater switches itself off.

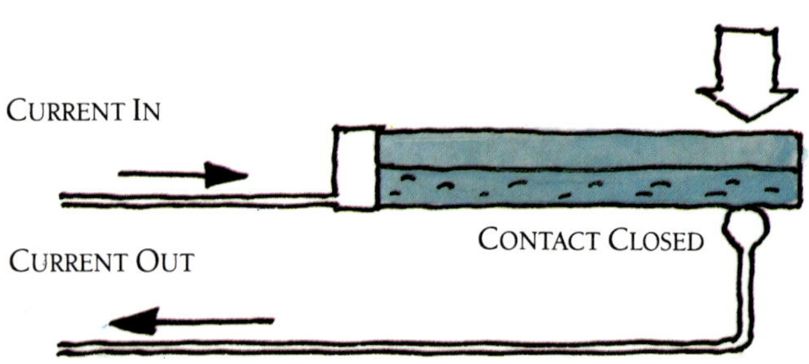

SWITCH CLOSED

The strip bends back as it cools and makes contact. The current passes and the heater switches itself back on.

ROD THERMOSTAT

Gas ovens and heaters often contain rod thermostats. The control is connected to a steel rod housed in a brass tube. The tube expands or contracts more than the rod, which closes and opens a valve in the flow of gas.

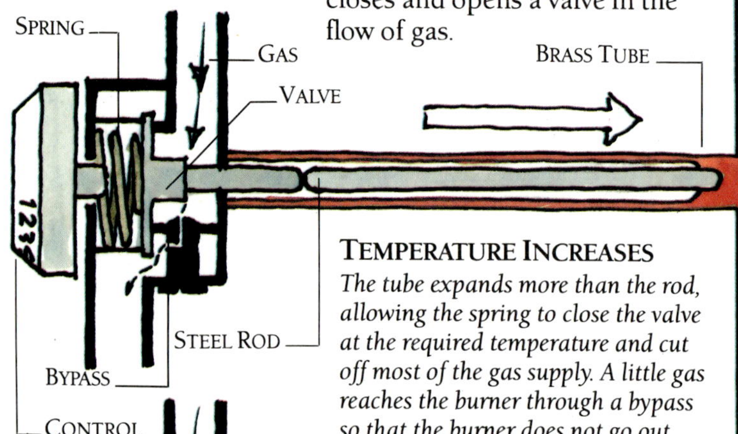

TEMPERATURE INCREASES

The tube expands more than the rod, allowing the spring to close the valve at the required temperature and cut off most of the gas supply. A little gas reaches the burner through a bypass so that the burner does not go out, which would be dangerous.

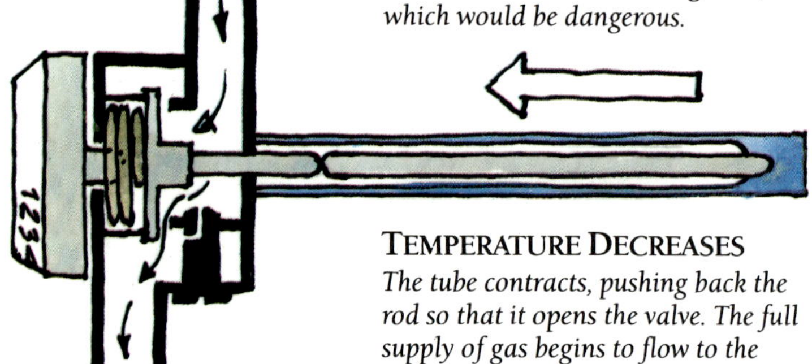

TEMPERATURE DECREASES

The tube contracts, pushing back the rod so that it opens the valve. The full supply of gas begins to flow to the burner.

CAR THERMOSTAT

The thermostat in a car cooling system (see p.160) controls the flow of cooling water to the radiator. Most car thermostats contain wax, which melts when the water gets hot. The wax expands, opening a valve in the water flow. A spring closes the valve when the water cools and the wax solidifies.

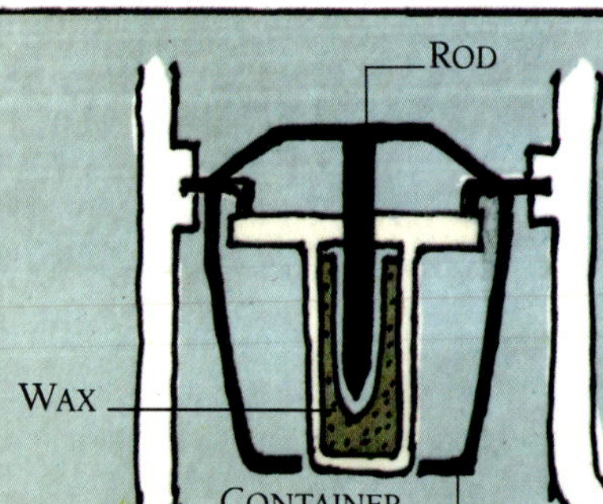

VALVE CLOSED

When the engine is cool, the rod is seated in the wax inside the brass container.

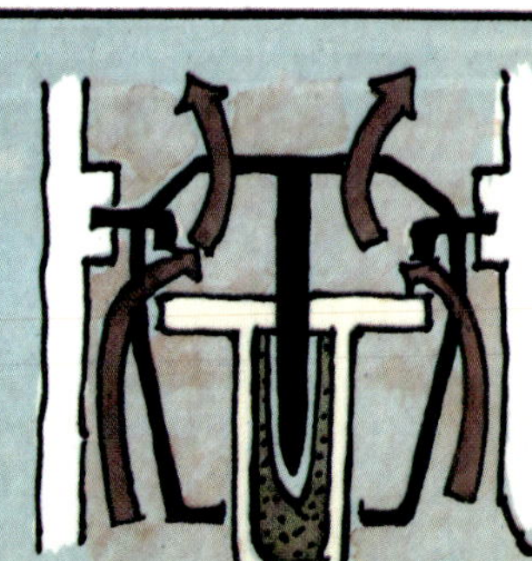

VALVE OPEN

The wax melts and expands, pushing against the rod and forcing the container down.

THERM

As things expand or contract, they change size by an amount that depends on the temperature. A rise of twenty degrees, for example, gives twice the expansion produced by ten degrees. Expansion and contraction can therefore be used to measure temperature.

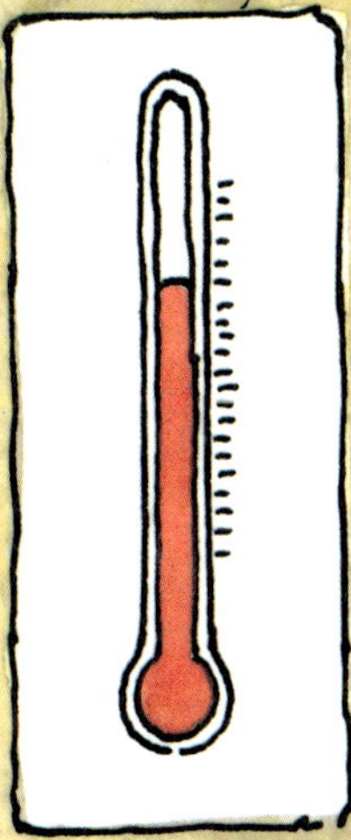

In a common thermometer (*left*), colored alcohol or mercury rises in a narrow tube as the liquid gets hotter and expands. The level falls as it gets colder and contracts. The maximum-minimum thermometer (*right*) makes use of both to record extremes of temperature.

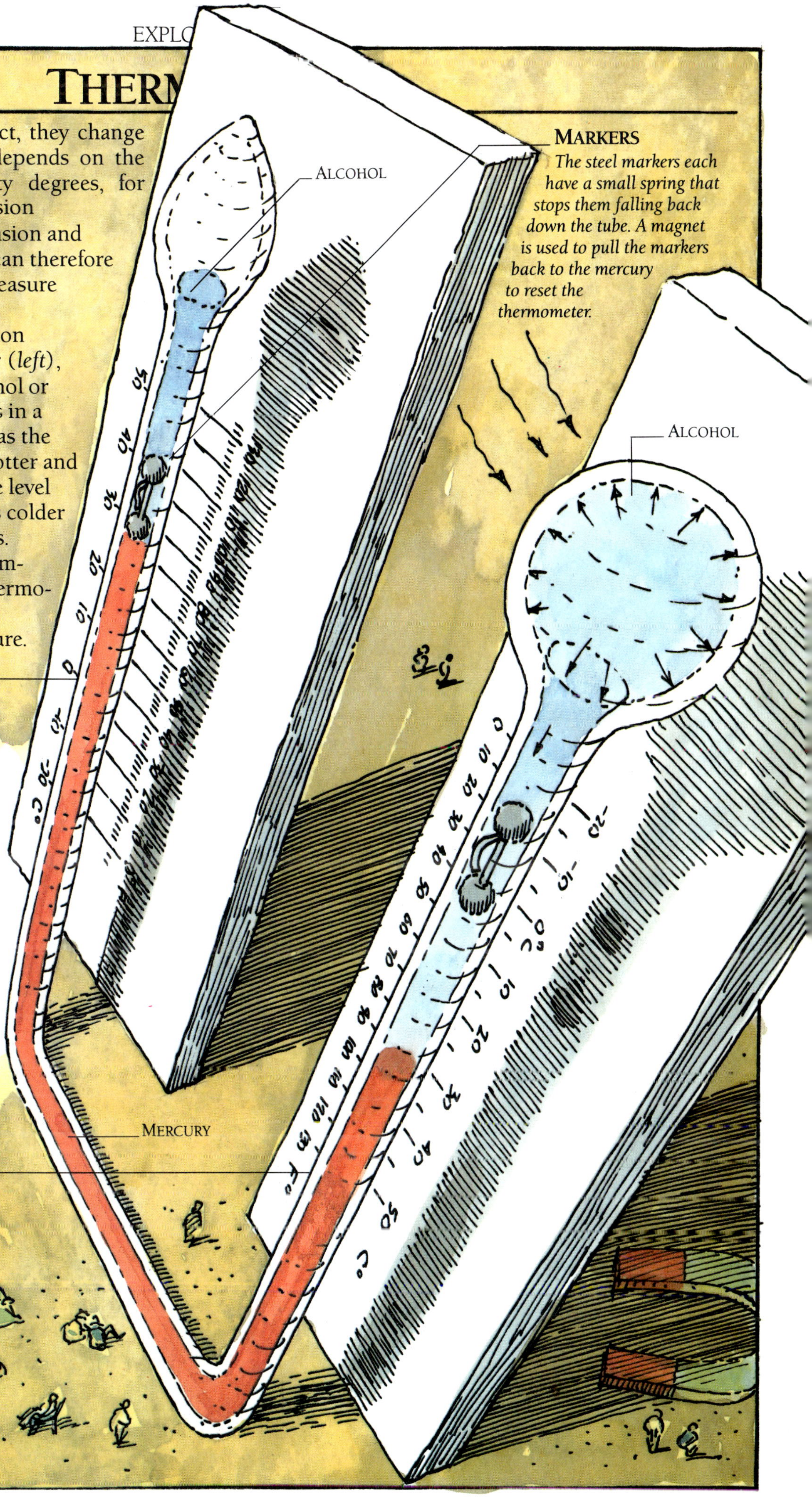

MAXIMUM TEMPERATURE

The U-shaped tube contains alcohol with mercury in the center. At high temperatures, the alcohol in the bulb above the minimum scale expands, pushing the mercury up the maximum scale. A metal marker remains at the highest point reached.

MINIMUM TEMPERATURE

The alcohol in the bulb above the minimum scale contracts. The air in the other bulb pushes the mercury up the minimum scale, moving the marker up the scale.

THE GASOLINE ENGINE

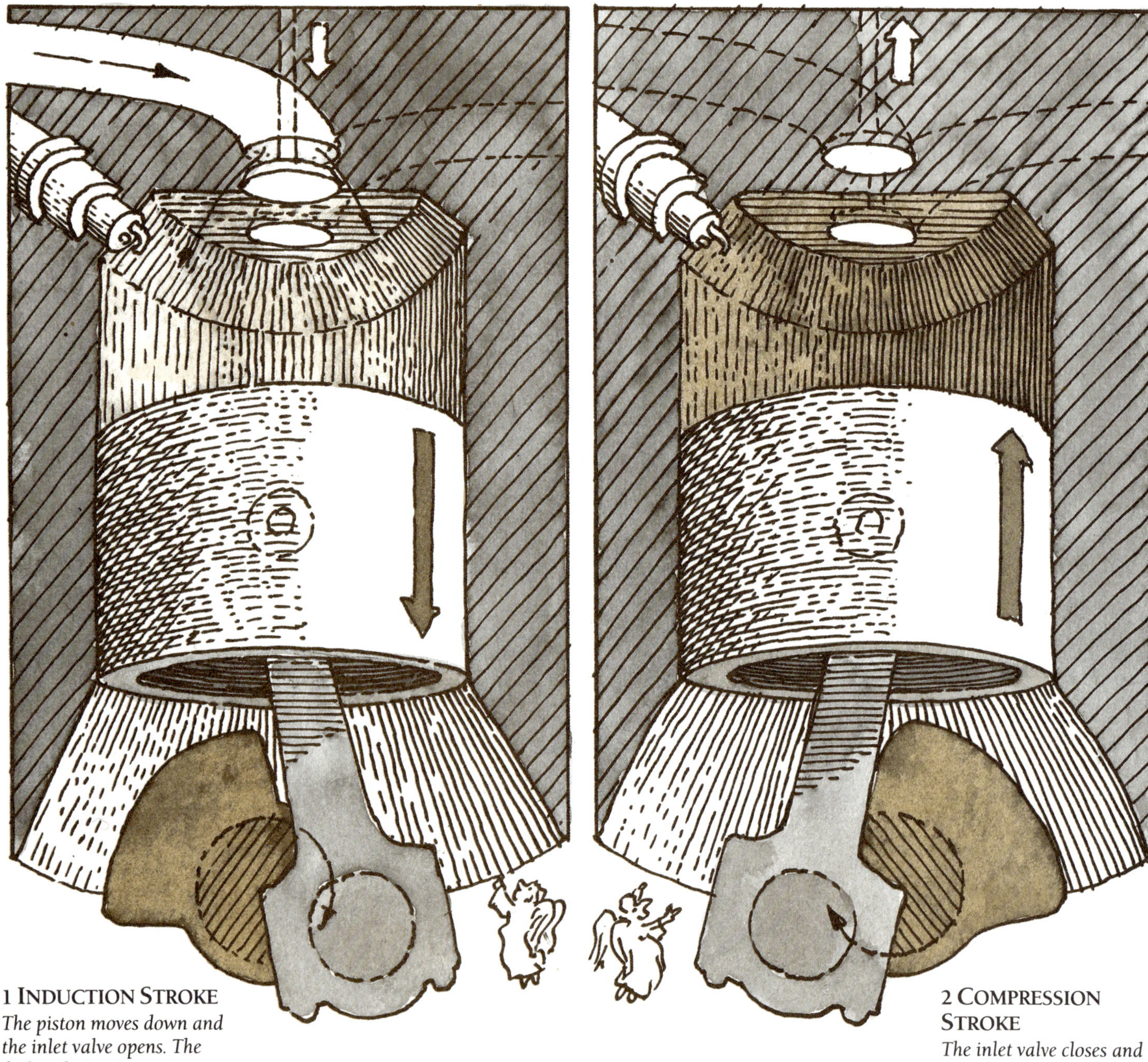

1 INDUCTION STROKE
The piston moves down and the inlet valve opens. The fuel and air mixture is sucked into the cylinder.

2 COMPRESSION STROKE
The inlet valve closes and the piston moves up. The mixture is compressed.

In the gasoline engine, we put heat to use by converting it into motive power. A gasoline engine is often called an internal combustion engine, but this means only that the fuel burns inside the engine. The jet engine and rocket engine are also internal combustion engines.

A gasoline engine works by burning a mixture of gasoline and air in a cylinder containing a piston. The heat produced causes the air to expand and force down the piston, which turns a crankshaft linked to the wheels.

Most cars have a four-stroke engine. A stroke is one movement of the piston, either up or down. In a four-stroke engine, the engine repeats a cycle of actions (shown above) in which the piston moves four times. Many light vehicles, such as motorcycles, have two-stroke engines. This kind of engine is simpler in construction than a four-stroke engine, but not as powerful. A two-stroke engine has no valves. Instead there are three ports in the side of the cylinder that the piston opens and closes as it moves up and down.

The diesel engine is similar to the gasoline engine, but runs on a heavier grade of fuel. The inlet valve admits only air, and the fuel is sprayed into the cylinder at the end of the compression stroke. The cylinder has no spark plug, but high compression makes the air in the cylinder so hot that the fuel ignites of its own accord as soon as it is sprayed in.

3 POWER STROKE
The electric spark plug produces a spark and the fuel ignites, forcing the piston back down the cylinder.

4 EXHAUST STROKE
The outlet valve opens and the piston rises, pushing the exhaust gases out of the cylinder.

THE EXHAUST AND SILENCER

The exhaust gases leave the engine at high pressure, and would produce intolerable noise if allowed to escape directly. The exhaust manifold therefore conducts the gases to the silencer, where they pass through a series of holes in metal plates or tubes. This reduces the pressure of the gases so that they leave the silencer quietly.

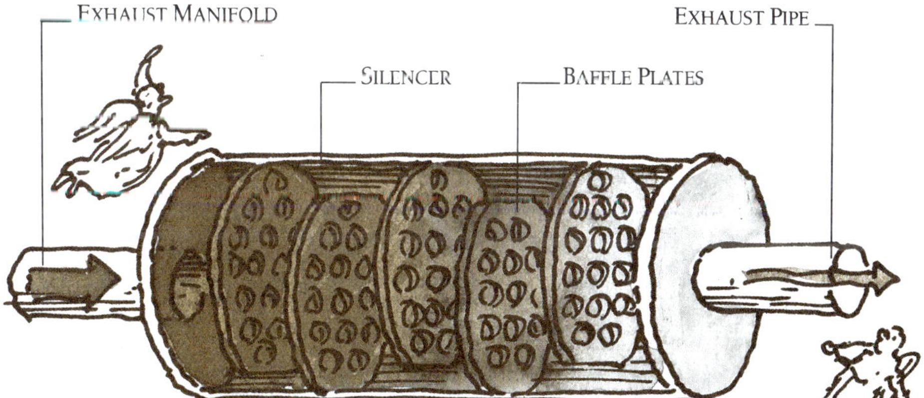

STEAM POWER

The first engine to make use of heat to drive a machine was the steam engine. It employed steam raised in a boiler to drive a piston up and down a cylinder. This engine was vital in the development of the Industrial Revolution, but is now obsolete.

However, the age of steam is by no means over because steam power provides us with the bulk of our electricity. Thermal power stations, which burn fuels such as coal (shown here) and oil, contain steam turbines to drive the electricity generators – as do nuclear power stations (see pp.178-9). All power stations are designed to pass as much energy as possible from the fuel to the turbines.

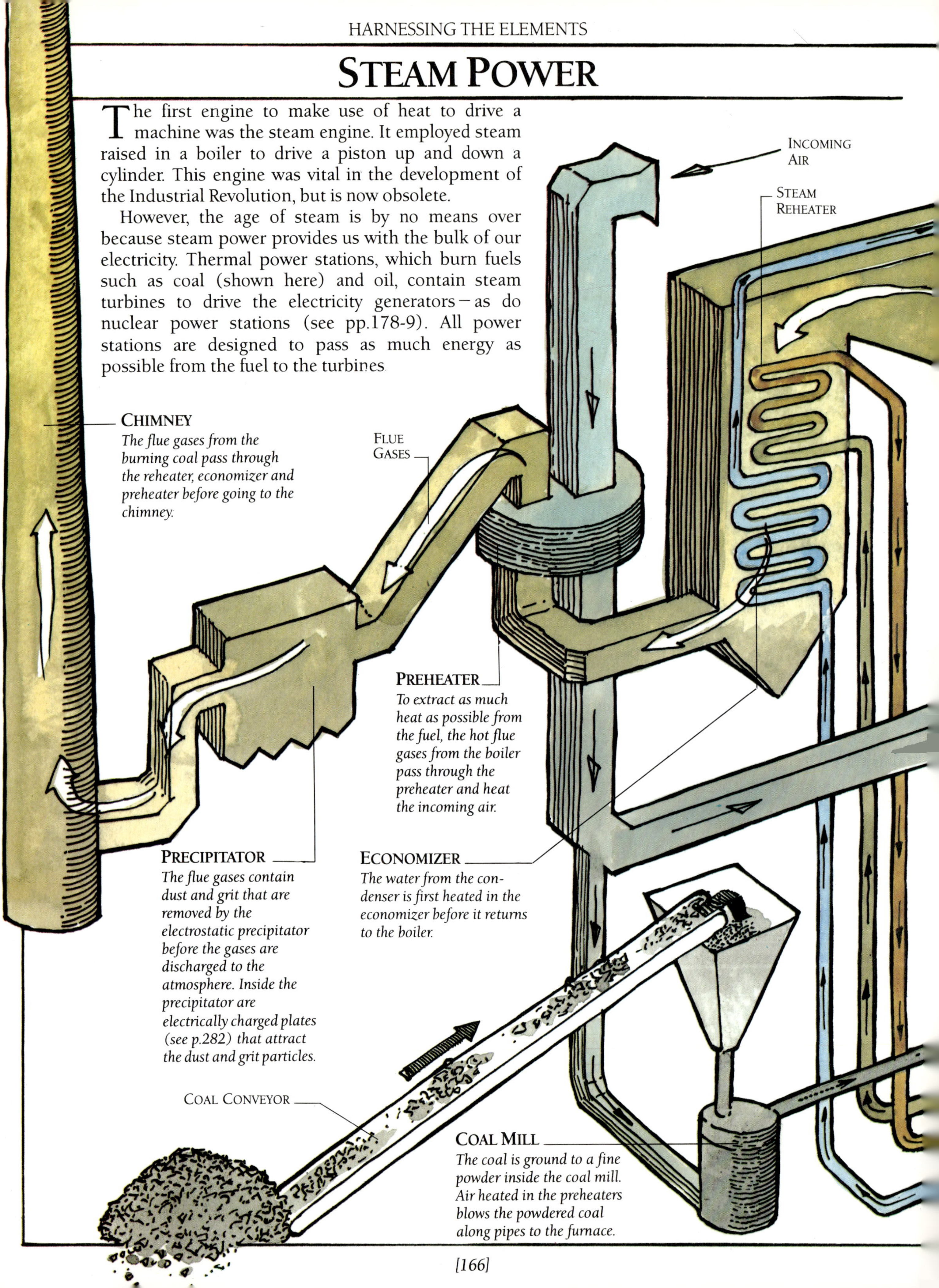

CHIMNEY
The flue gases from the burning coal pass through the reheater, economizer and preheater before going to the chimney.

PREHEATER
To extract as much heat as possible from the fuel, the hot flue gases from the boiler pass through the preheater and heat the incoming air.

PRECIPITATOR
The flue gases contain dust and grit that are removed by the electrostatic precipitator before the gases are discharged to the atmosphere. Inside the precipitator are electrically charged plates (see p.282) that attract the dust and grit particles.

ECONOMIZER
The water from the condenser is first heated in the economizer before it returns to the boiler.

COAL MILL
The coal is ground to a fine powder inside the coal mill. Air heated in the preheaters blows the powdered coal along pipes to the furnace.

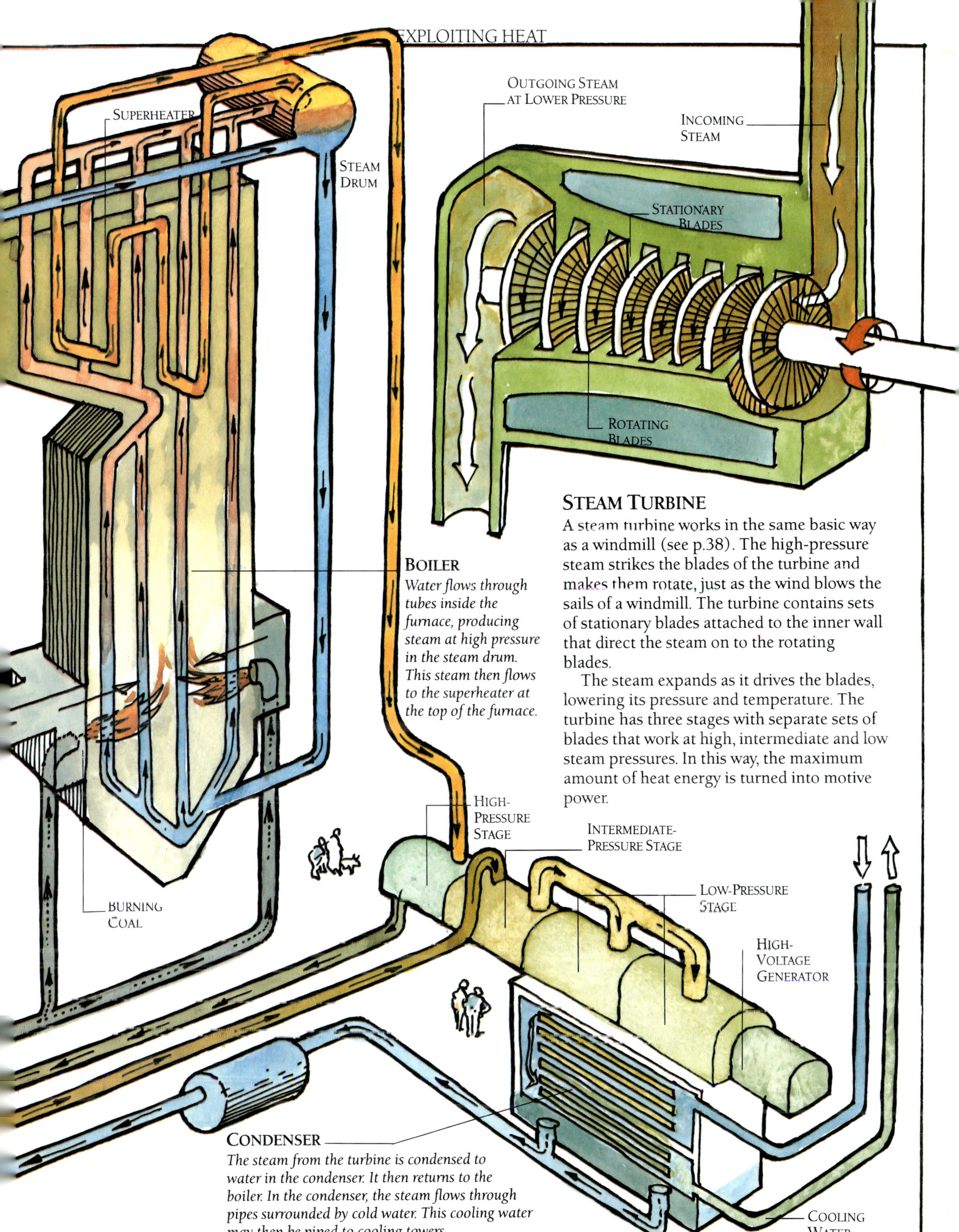

BOILER
Water flows through tubes inside the furnace, producing steam at high pressure in the steam drum. This steam then flows to the superheater at the top of the furnace.

STEAM TURBINE

A steam turbine works in the same basic way as a windmill (see p.38). The high-pressure steam strikes the blades of the turbine and makes them rotate, just as the wind blows the sails of a windmill. The turbine contains sets of stationary blades attached to the inner wall that direct the steam on to the rotating blades.

The steam expands as it drives the blades, lowering its pressure and temperature. The turbine has three stages with separate sets of blades that work at high, intermediate and low steam pressures. In this way, the maximum amount of heat energy is turned into motive power.

CONDENSER
The steam from the turbine is condensed to water in the condenser. It then returns to the boiler. In the condenser, the steam flows through pipes surrounded by cold water. This cooling water may then be piped to cooling towers.

THE JET ENGINE

Without the jet engine, many of us would have little experience of flight. Superior both in power and economy to the propeller engine, it has made mass worldwide air travel possible.

A jet engine sucks air in at the front and ejects it at high speed from the back. The principle of action and reaction (see p.108) forces the engine forward as the air streams backward. The engine is powered by heat produced by burning kerosene or paraffin.

THE TURBOFAN

The engine that drives big airliners is a turbofan engine. At the front of the engine, a large fan rotates to draw air in. Some of this air then enters the compressors, which contain both rotating and stationary blades. The compressors raise the pressure of the air, which then flows to the combustors or combustion chambers. There, flames of burning kerosene heat the air, which expands. The hot, high-pressure air rushes toward the exhaust, but first passes through turbines which drive the compressors and the fan.

The rest of the air sucked in by the fan passes around the compressors, combustors and turbines. It helps to cool and quieten the engine, and then joins the heated air. A large amount of air speeds from the engine, driving the aircraft forward with tremendous force.

Compressor Shafts
Heated Air
Exhaust
Combustor
Compressor Turbines
Compressors
Fan Turbine
Outgoing Bypass Air
Bypass Duct

ROCKET ENGINES

The rocket is the simplest and most powerful kind of heat engine. It burns fuel in a combustion chamber with an open end. The hot gases produced expand greatly and rush from the open end or exhaust at high speed. The rocket moves forward by action and reaction (see p.108) as the gases exert a powerful force on the chamber walls.

Rockets can work in space because, unlike other heat engines, they do not require air for combustion. Their fuel burns without the need for an external oxygen supply.

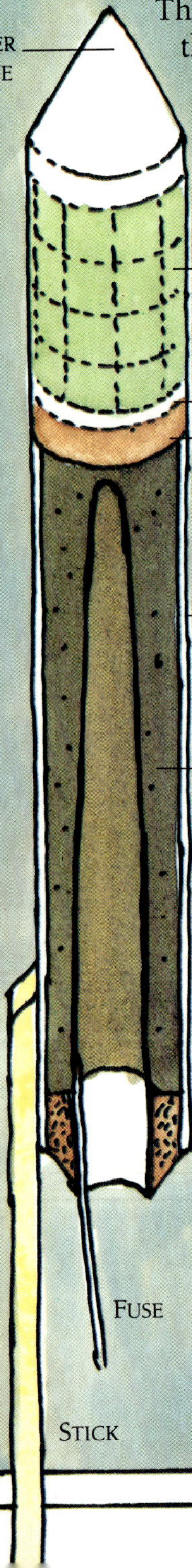

FIREWORK ROCKET

Firework rockets are the simplest form of heat engine. They are packed with a propellant, a powder that burns fiercely. The smoke and hot gases stream from the base and drive the rocket upward, while the long stick keeps the rocket's flight straight. The propellant is slowly consumed by combustion, and finally the burning powder ignites an explosive charge which expels the glowing stars.

SOLID-FUEL ROCKET

Many spacecraft are launched by solid-fuel boosters, which are rocket engines that, like firework rockets, contain a solid propellant. A circular or star-shaped channel runs down the center of the propellant. The propellant burns at the surface of this channel, so the channel is the combustion chamber. A solid-fuel booster develops more power if the channel is star-shaped. This is because the channel's area is larger, and a greater volume of hot gases is produced. Solid-fuel rockets can produce great power but, once ignited, they cannot be shut down; they fly until all the propellant has burned.

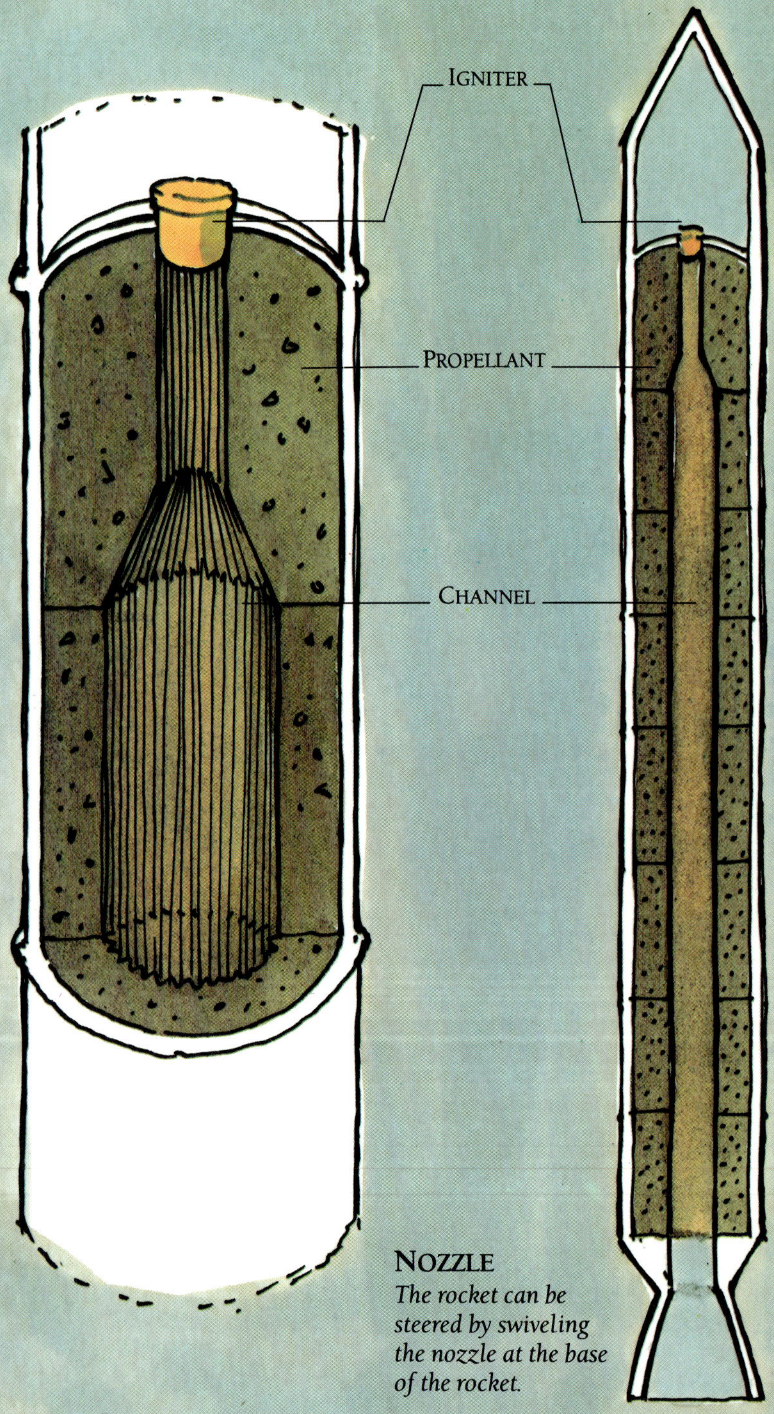

NOZZLE

The rocket can be steered by swiveling the nozzle at the base of the rocket.

LIQUID-FUEL ROCKET

Spacecraft that require repeated firings of their engines, often for maneuvering in space, have liquid-fuel rocket engines. Unlike solid propellants, liquid propellants are fed to the combustion chamber and are burned for as long as necessary. The propellants consist of two liquids, usually called the fuel and the oxidizer. Liquid hydrogen and liquid oxygen are often used.

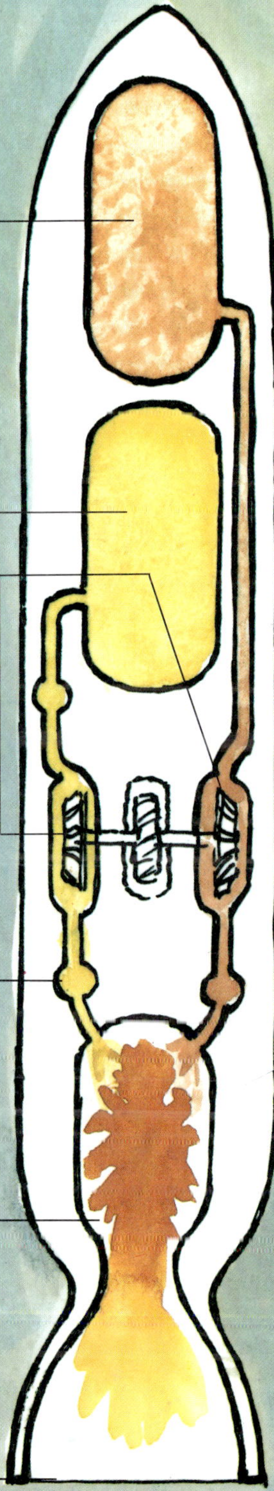

PUMPS
The rocket may contain pumps that feed the propellants to the combustion chamber. The pumps are driven by a turbine often powered by gas produced by the propellants. In some rockets, the propellants are delivered under gas pressure, so pumps are not needed.

VALVES
These control the flow of the propellants, enabling the engine to work at different degrees of power.

COMBUSTION CHAMBER
The propellants generally have to be ignited in the combustion chamber, but some rockets use propellants which ignite on contact.

SPACE SHUTTLE

The space shuttle has five principal rocket engines. These are two huge solid-fuel boosters fixed alongside the orbiter, and three liquid-fuel main engines at the rear of the orbiter. The external tank contains liquid hydrogen and liquid oxygen for the main engines. These five engines only take the shuttle into space. Other smaller liquid-fuel engines are used to attain and leave orbit and to maneuver the shuttle in space.

NUCLEAR POWER

ON THE GIFT THAT KEPT ON GIVING

During my travels, I once became snowbound in a town that had completely exhausted its fuel supply. On one bitterly cold morning, I awoke to learn that an enormous concrete mammoth had somehow appeared outside the gate. An excited crowd quickly surrounded it, and my professional opinion was sought.

Attached to the mammoth's long flexible trunk I found a note which stated that this gigantic machine was a gift from a friend. If treated properly, the note continued, the mammoth would give all the heat, in the form of steam, that the town would ever need. In return, all the machine required was plenty of water and an occasional pellet from a bag provided. Scribbled at the bottom of the note was a reminder to bury in heavy containers all the waste material that discharged periodically from the rear of the contraption. The note was not signed.

NUCLEAR REACTIONS

The mechanical mammoth is able to supply such prodigious amounts of energy from so little fuel because it is a nuclear machine. Inside is a nuclear reactor that converts the fuel into heat, but it does not burn the nuclear fuel.

Burning or combustion is a chemical reaction. The elements present in the fuel and the oxygen in the air merely rearrange themselves as burning progresses. The new arrangement of elements, which yields ash, smoke and waste gases, has less energy then the original fuel and oxygen. The leftover energy appears as heat.

A nuclear reaction exploits the elements to produce heat in a totally different manner.

When a nuclear reaction occurs, the elements in the fuel do not remain the same. Instead, the nuclear reactor changes the elements in the fuel into other elements. The waste products of the nuclear reaction have less mass than the fuel, and the lost mass turns into heat energy. A nuclear reaction in fact creates energy; it does not convert one form of energy into another. A little mass gives a lot of energy, which is why nuclear power is so abundant.

We followed the instructions faithfully. A large amount of water was pumped into the concrete creature after which a few pellets were tossed in. It seemed no time at all before, to loud cheers from the assembled populace, clouds of steam began puffing from the mammoth's trunk. A piping system was promptly attached from the trunk to every building in the town. Thereafter, even on the coldest days, everyone, myself included, was warm and cosy. When the waste issued forth, we took it in turns to seal and bury them as the note had instructed.

As the winter wore on, however, the disposal teams grew less and less inclined to turn out to bury the waste which, after all, seemed harmless enough. Fortunately, I was able to devise a most ingenious way to solve the problem of unsightly waste. A large hole was cut in the wall through which the front end of the mammoth was pushed. Since the rear end of the mammoth stood outside the wall, the waste products could be ignored.

There were those in the town, who remained suspicious of the concrete mammoth. They wondered not only how it worked but also where it had come from; could it really be as beneficial as it seemed? But by announcing that I would open the machine up and lead conducted tours of the mechanism, I was able to allay these fears.

However, spring arrived, the snow cleared and I was on my way again before the promise could be kept. As I left, I noticed that the trees near the waste had yet to burst into leaf.

CONTROLLING NUCLEAR POWER

The people are right to be wary of the nuclear mammoth. A nuclear reaction can, if uncontrolled, produce an enormous explosion as the immense release of energy takes place almost instantaneously. This occurs in nuclear weapons. Nuclear reactors achieve the controlled release of nuclear power. However, the reaction itself and also its waste products give out harmful rays known as radiation, and the reactor has to be encased in concrete and the waste stored well away from people for safety. The production of radiation is also a nuclear reaction, for it changes the elements in the waste. However, it takes place at a much slower rate than in a nuclear reactor, and the radioactive waste can continue to emit harmful radiation for many years.

It is ironic then that we all depend ultimately on nuclear power, because the heat and light that support life on Earth come from a gigantic nuclear reaction that is taking place in the Sun. Indeed, the production of energy on such a vast scale is possible only by nuclear reactions. Every second, the Sun losses four million tons of its mass to sustain its huge output of energy.

NUCLEAR FISSION

Nuclear power gets its name because the process of power production takes place inside the nucleus. Each atom of fuel contains a central particle called the nucleus, which is itself made up of even smaller particles called protons and neutrons.

The kind of nuclear reaction that happens inside a nuclear reactor is called nuclear fission. The fuel is uranium or plutonium, two very heavy elements which have many protons and neutrons in their nuclei. Fission starts when a fast-moving neutron strikes a nucleus. The nucleus cannot take in the extra neutron, and the whole nucleus breaks apart into two smaller nuclei. Several neutrons are also released and these go on to break more nuclei, which produce more neutrons and so on. Because the first neutron sets off a chain of fissions, the nuclear reaction is called a chain reaction. Without control, it can multiply rapidly and produce enormous heat in a fraction of a second.

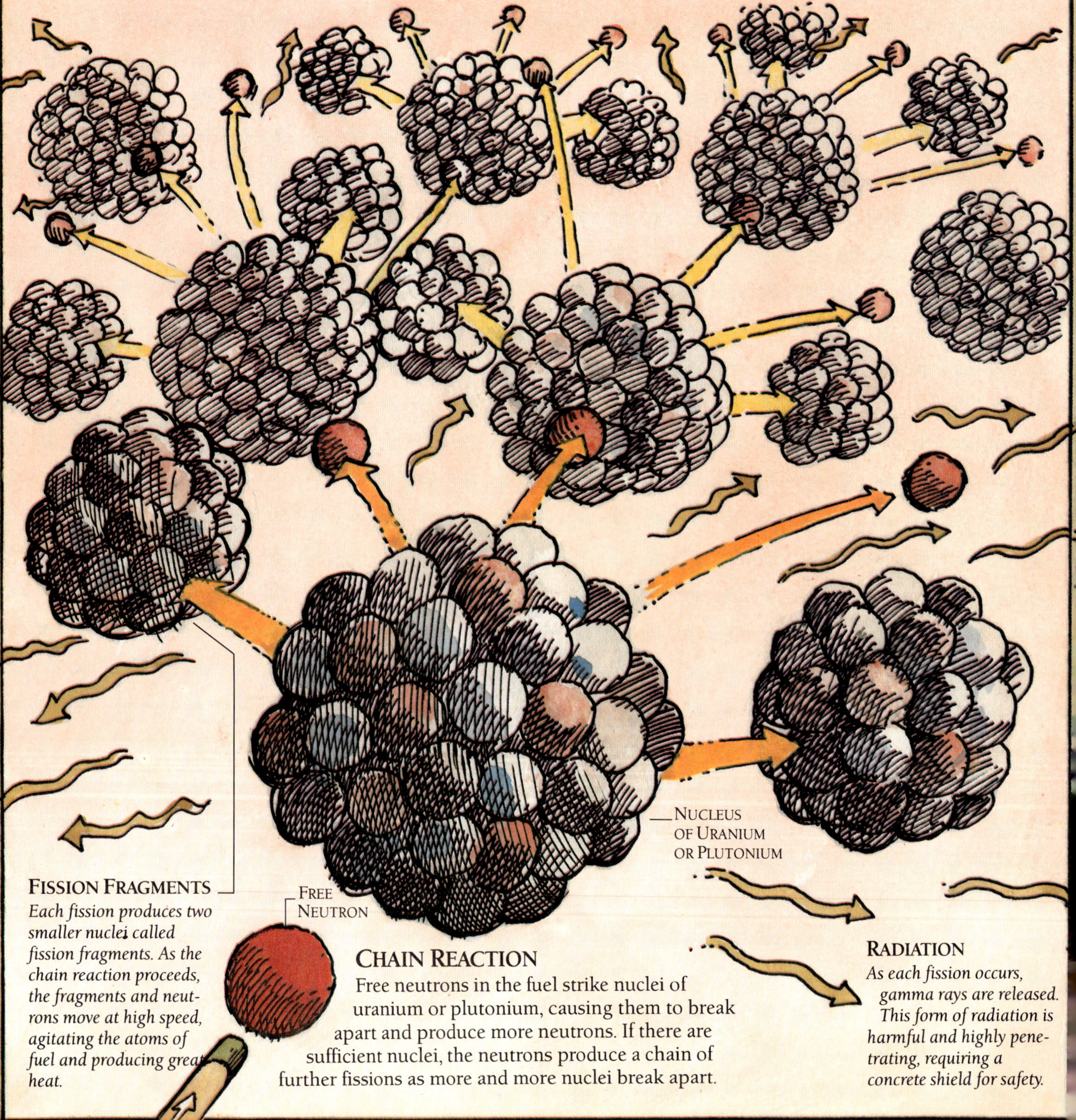

FISSION FRAGMENTS
Each fission produces two smaller nuclei called fission fragments. As the chain reaction proceeds, the fragments and neutrons move at high speed, agitating the atoms of fuel and producing great heat.

CHAIN REACTION
Free neutrons in the fuel strike nuclei of uranium or plutonium, causing them to break apart and produce more neutrons. If there are sufficient nuclei, the neutrons produce a chain of further fissions as more and more nuclei break apart.

RADIATION
As each fission occurs, gamma rays are released. This form of radiation is harmful and highly penetrating, requiring a concrete shield for safety.

NUCLEAR FUSION

Nuclear power can be produced by a process called fusion as well as by fission. In this kind of reaction, the nuclei of the fuel come together and do not break apart. Unlike fission, nuclear fusion occurs only with small atoms whose nuclei contain very few protons and neutrons. The gaseous fuel consists of two different forms of hydrogen, which is the lightest of the elements. To produce nuclear fusion, pairs of nuclei meet so that their protons and neutrons fuse together and become a single nucleus. A spare neutron is left over. The fused nuclei and neutrons move off at high speed, producing great heat. Radiation is not emitted, but the neutrons are harmful.

To get the nuclei to meet and fuse, the atoms must be banged together with tremendous force. This can only be done by heating the fuel to temperatures of millions of degrees. Nuclear fusion powers the Sun, and it also occurs in thermonuclear weapons.

NUCLEAR WEAPONS

ATOM BOMB

An atom bomb is better known as a fission bomb because it works by nuclear fission. The bomb contains a hollow sphere of uranium or plutonium. The sphere is too big to initiate a chain reaction because neutrons that occur naturally escape from the surface of the sphere without causing fission.

To detonate the bomb, a source of neutrons is shot by the detonator into the center of the sphere. Explosives then crush the sphere around the neutron source. The neutrons cannot now escape. A chain reaction occurs and fission flashes through the uranium or plutonium in a fraction of a second. The bomb explodes with a power equal to thousands of tons (kilotons) of TNT. Intense radiation is also produced.

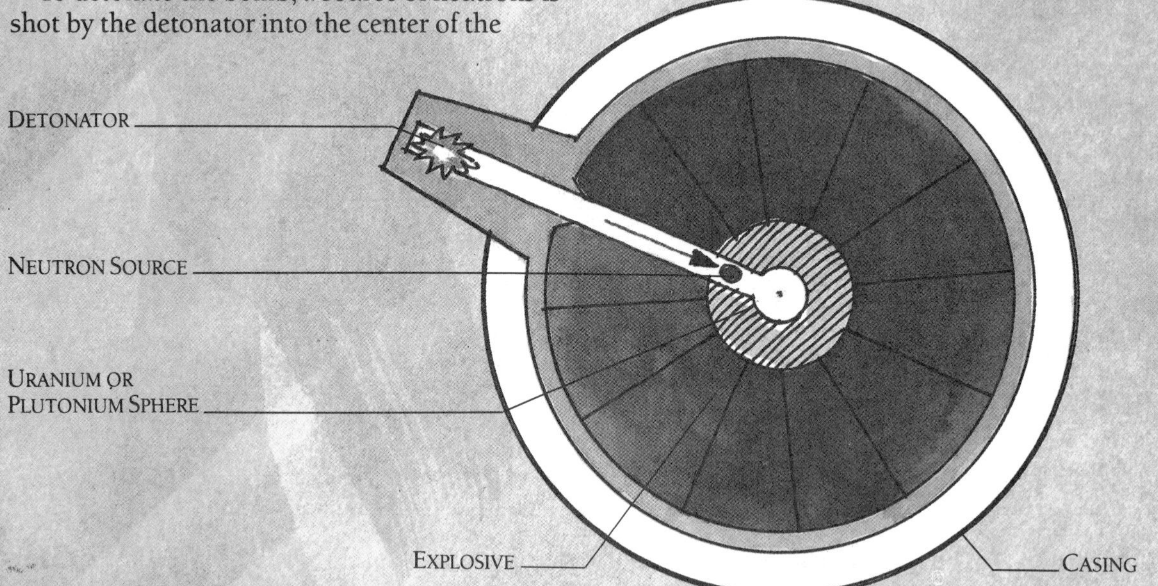

HYDROGEN BOMB

The hydrogen bomb or H-bomb is a thermonuclear weapon which works partly by nuclear fusion. Two forms of hydrogen – deuterium and tritium – are compressed at very high temperature to produce instant fusion. These conditions of ultra-high temperature and pressure can only be created by a fission bomb, which is used to trigger fusion in a thermonuclear weapon. Explosives crush all the nuclear materials around a neutron source to detonate the bomb.

Some thermonuclear weapons also contain a jacket of uranium, which produces a blast equal to millions of tons (megatons) of TNT. The neutron bomb, on the other hand, is a fusion weapon of relatively low power that produces penetrating neutrons. The neutrons released by the bomb would kill people while most buildings would survive the weak blast.

NUCLEAR TESTING

The fallout or debris produced by the explosion of a nuclear weapon is so radioactive that the weapons must be tested in chambers dug deep underground in remote areas. In this way, the atmosphere is not contaminated by the fallout.

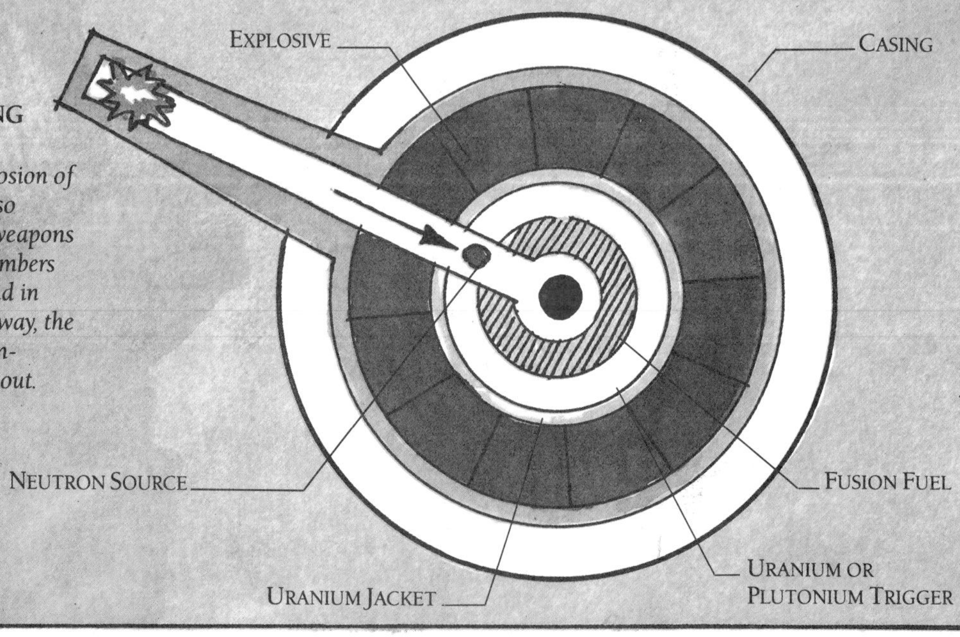

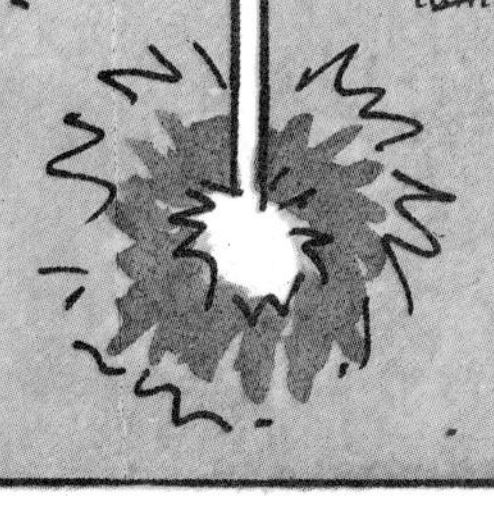

FALLOUT

A future nuclear war would not only reduce cities and towns to ruins. Fallout from the nuclear explosions would spread through the atmosphere, bombarding the land with lethal amounts of radiation. The only means of escape would be to live in deep underground shelters away from the fallout. This imprisonment would have to last until the radiation decreased to an acceptable level, which could take many years. Even then, climatic changes, shortage of food and the threat of disease would make life above ground a grim business.

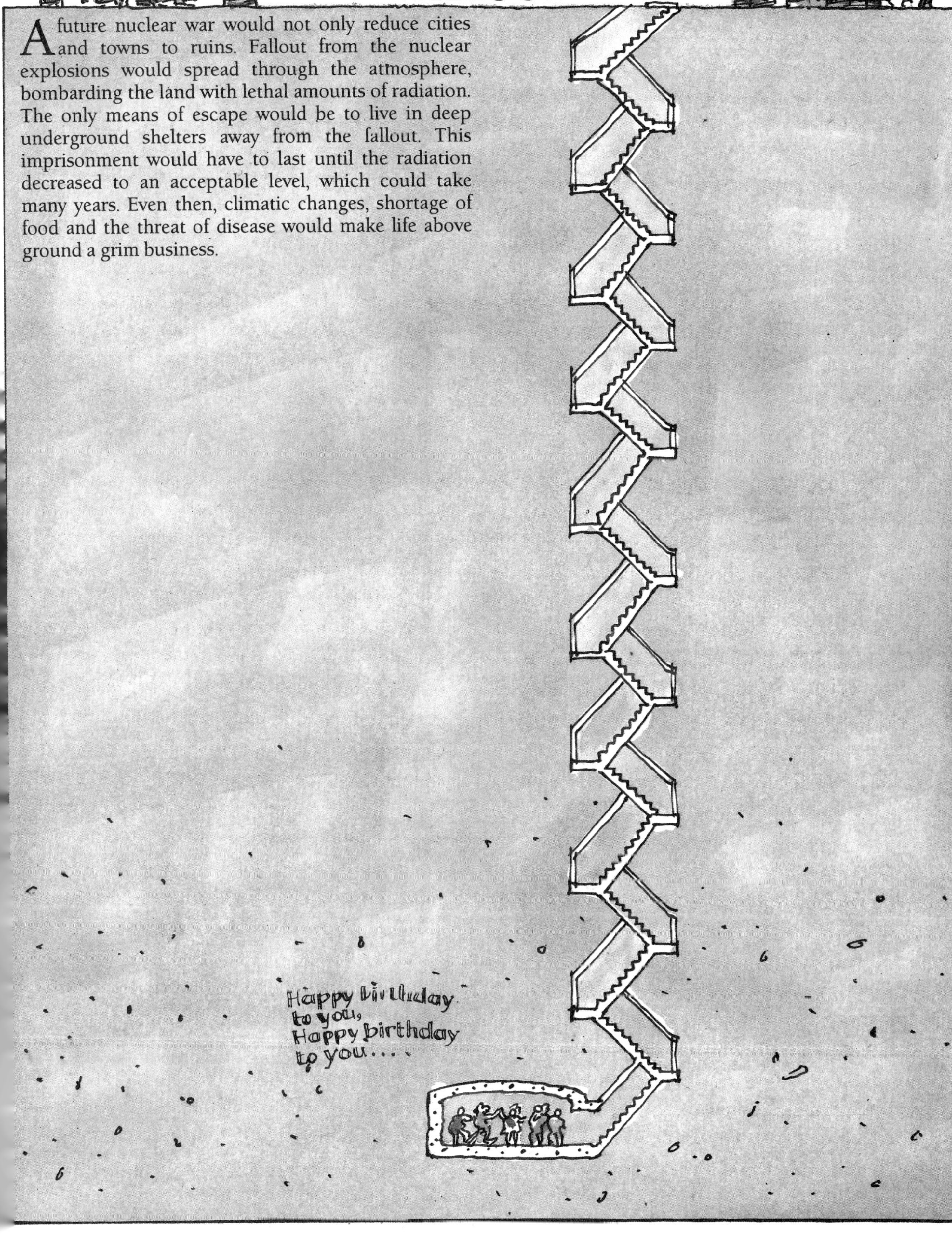

Nuclear Reactor

The heart of a nuclear power station is its nuclear reactor. Here, immense heat is generated by the fission of uranium fuel. The heat is transferred from the reactor to a steam generator, where it boils water to steam. The rest of the nuclear power station works in the same way as one powered by coal (see pp.166-7). In all nuclear reactors, a liquid or gas flows through the core of the reactor and heats up. Its purpose is to take away the heat generated by fission in the reactor core, so it is called a coolant. The main kind of nuclear reactor used in nuclear power stations, the pressurized water reactor (PWR), uses water as the coolant.

Fuel Rods

The fuel consists of pellets of uranium dioxide loaded into long metal tubes. Clusters of these fuel rods are then inserted into the reactor core.

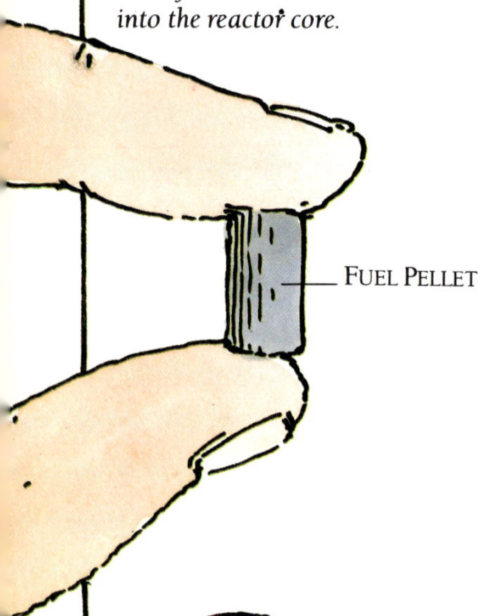

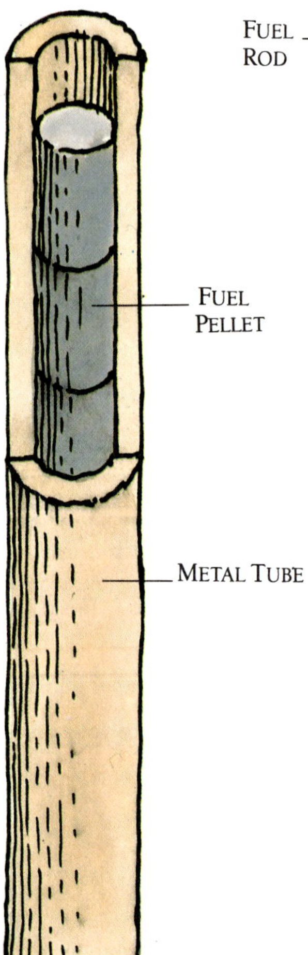

Control Rods

Among the fuel rods are control rods, which contain a substance that absorbs neutrons. Moving the rods in or out of the core controls the flow of neutrons so that fission progresses steadily and provides a constant supply of heat. The reactor is shut down by fully inserting the control rods.

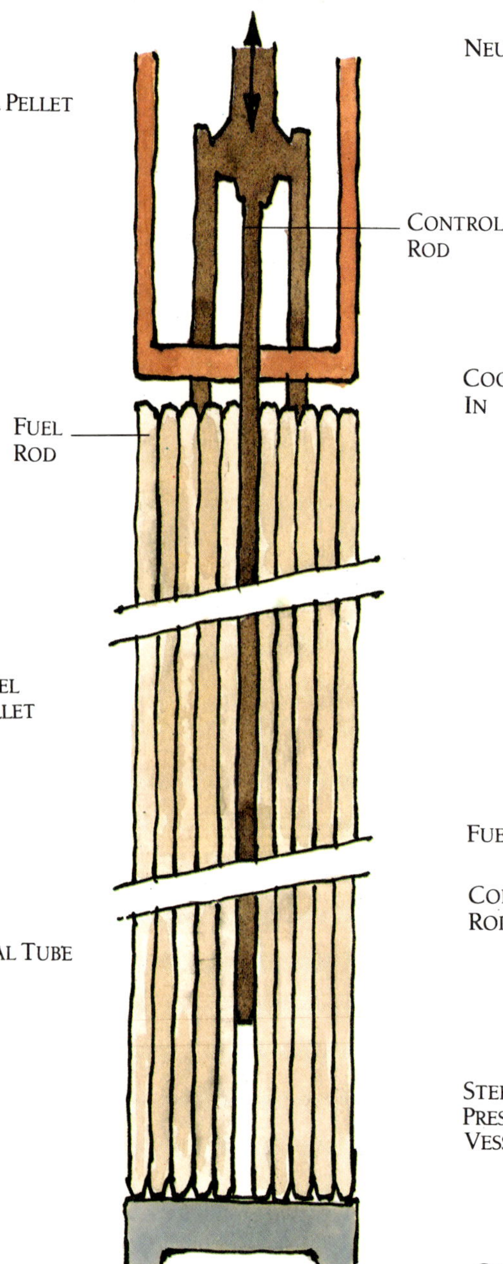

Reactor Core

A steel pressure vessel surrounds the core of the pressurized water reactor, which contains the fuel rods and control rods. Neutrons, which occur naturally, start a chain reaction in the fuel and fast neutrons are produced by fission. The coolant (pressurized water) flowing through the core slows the neutrons down. The slow neutrons cause further fission and keep the chain reaction going. Heat produced by fission is passed to the coolant.

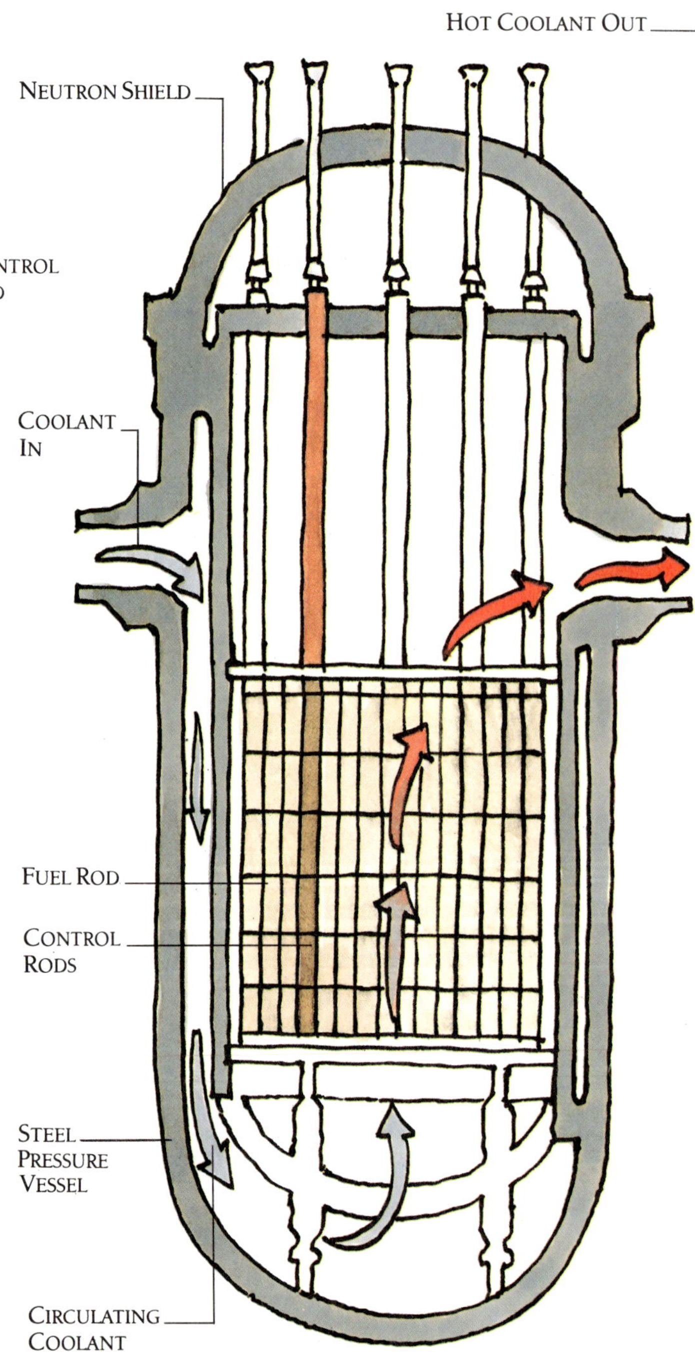

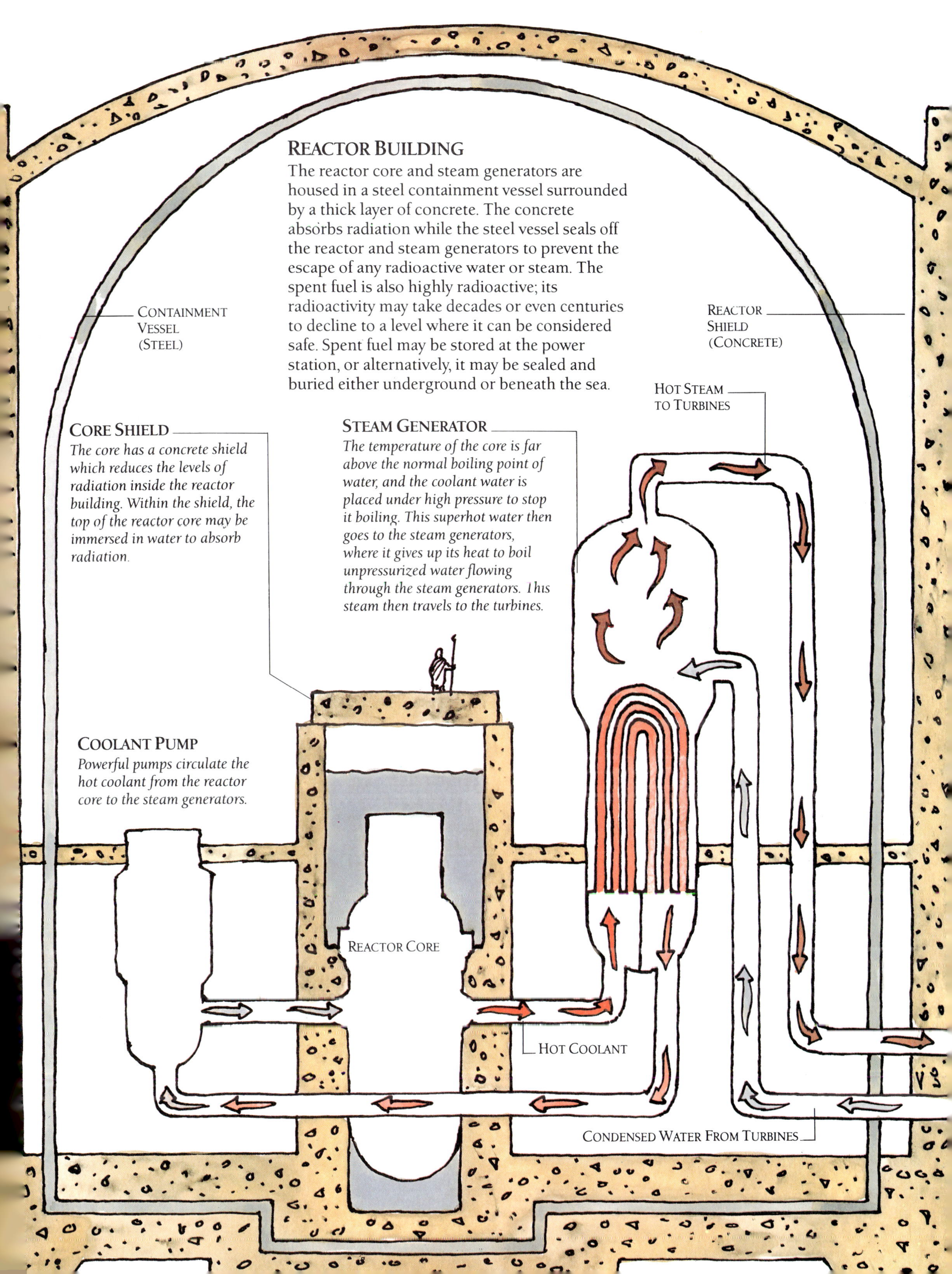

REACTOR BUILDING

The reactor core and steam generators are housed in a steel containment vessel surrounded by a thick layer of concrete. The concrete absorbs radiation while the steel vessel seals off the reactor and steam generators to prevent the escape of any radioactive water or steam. The spent fuel is also highly radioactive; its radioactivity may take decades or even centuries to decline to a level where it can be considered safe. Spent fuel may be stored at the power station, or alternatively, it may be sealed and buried either underground or beneath the sea.

CORE SHIELD

The core has a concrete shield which reduces the levels of radiation inside the reactor building. Within the shield, the top of the reactor core may be immersed in water to absorb radiation.

STEAM GENERATOR

The temperature of the core is far above the normal boiling point of water, and the coolant water is placed under high pressure to stop it boiling. This superhot water then goes to the steam generators, where it gives up its heat to boil unpressurized water flowing through the steam generators. This steam then travels to the turbines.

COOLANT PUMP

Powerful pumps circulate the hot coolant from the reactor core to the steam generators.

FUSION POWER

Nuclear fusion could provide us with almost unlimited power. The fuels for fusion come from materials that are common. Deuterium is made from water and tritium is produced from lithium, which is a metal that occurs widely in minerals. All that is needed is a machine to make them fuse under controlled conditions.

In practice, these conditions are extremely difficult to achieve. The two gases must be heated to a temperature of hundreds of millions of degrees, and kept together for a few seconds. No ordinary container can hold them, and several different systems based on magnetic fields or lasers are being tried.

However, progress is being made; fusion has been achieved on a limited scale but the amount of energy produced is much less than the energy fed into the fusion machine to create the conditions. Scientists hope that fusion power will advance to become reality early in the next century. If so, we shall possess a source of energy that not only has tremendous power but uses fuels that are abundant. Although a fusion reactor would not be likely to explode and release radioactivity, it would produce radioactive waste in the form of discarded reactor components.

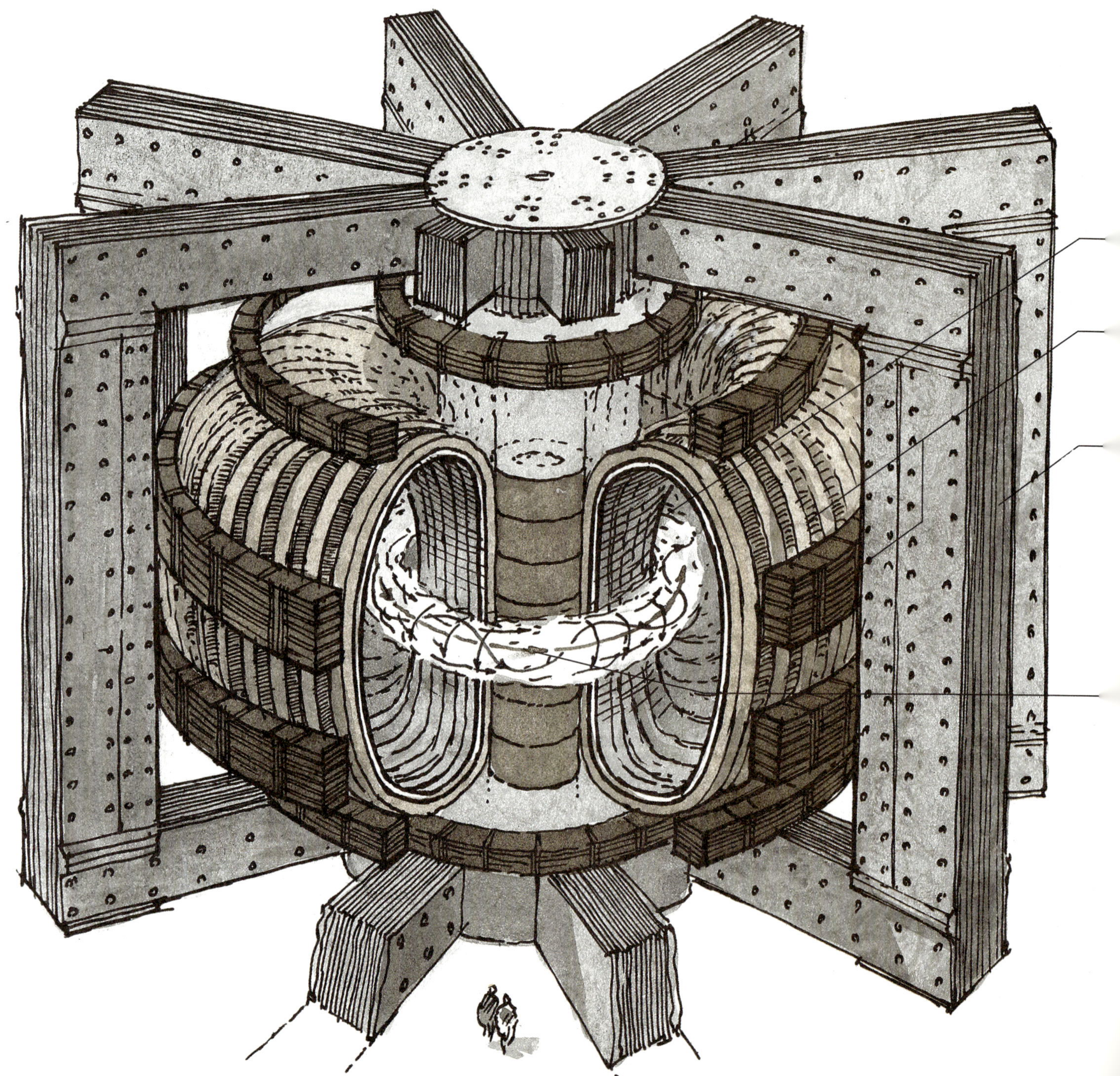

THE TOKAMAK

Most fusion research uses a machine called a tokamak, which was originally developed in Russia. At its heart is a torus – a doughnut-shape tube that contains the gases to be fused. A huge electrical transformer and coils of wire surround the tube. The transformer produces an electric current in the gases, which heats them up to produce an electrically charged mixture, or plasma. At the same time, strong magnetic fields produced by the current and the coils act on the hot gases.

The magnetic fields (see pp.294-5) confine the gases to the center of the torus so that they do not touch the walls. They can then become very hot indeed and begin to fuse. Extra heating can be achieved by bombarding the gases with powerful radio waves, and by injecting beams of particles into the torus.

TORUS

The torus contains a vacuum into which the fuel gases are injected.

MAGNETIC FIELD COILS

These coils are wound around the torus and are supplied with a powerful electric current. A magnetic field is created in the torus.

TRANSFORMER

Electric current supplied to the transformer coils at the center of the machine is stepped up by the transformer coils to create a powerful current in the plasma. This current heats the plasma and produces a second magnetic field around the plasma. The two magnetic fields combine to give a field that confines the plasma to the center of the torus.

PLASMA

The gases fed into the torus are heated to such high temperatures that they become a plasma, a form of superhot gas that is affected by magnetism. The magnetic field squeezes the plasma into a narrow ring at the centre of the torus. The high temperature and pressure cause fusion to occur.

FUSION REACTOR

This is how a fusion reactor of the future could work. Deuterium and tritium are fed into the torus, where they fuse together. Fusion produces non-radioactive helium, which leaves the torus, and high-energy neutrons. Around the torus is a blanket of lithium metal. The neutrons enter the blanket and convert some of the lithium into tritium, which is extracted and goes to the torus. The neutrons also heat up the blanket. This heat is removed by a heat exchanger and goes to a boiler to raise steam for electricity generation. The reactor shield absorbs the low-energy neutrons leaving the blanket.

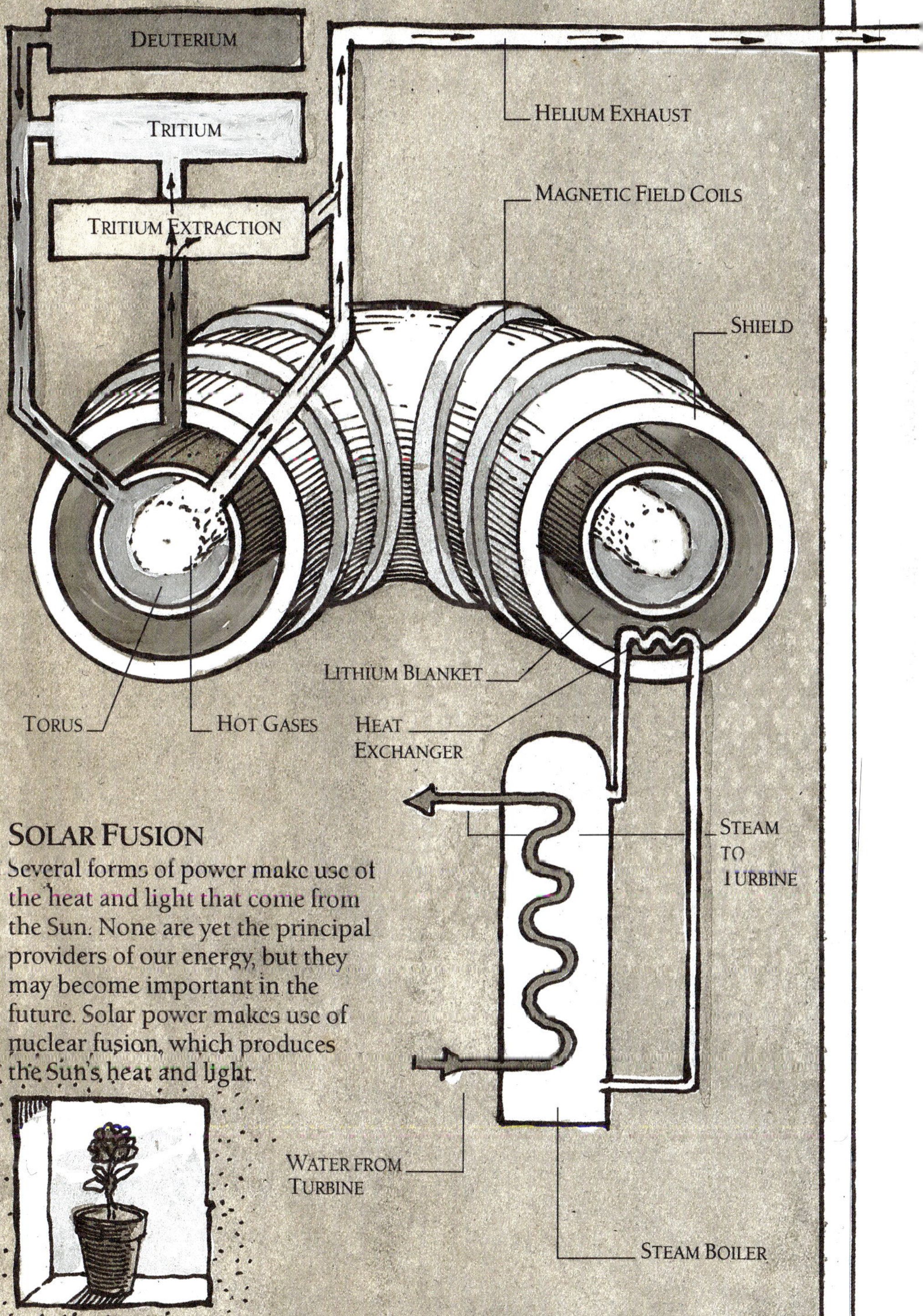

SOLAR FUSION

Several forms of power make use of the heat and light that come from the Sun. None are yet the principal providers of our energy, but they may become important in the future. Solar power makes use of nuclear fusion, which produces the Sun's heat and light.

THE·FIRST SON·ET·LUMIÈRE —ATLANTIS

PART 3

WORKING WITH WAVES

CONTENTS

THE THIRD PART
OF
THE GREAT WORK
Herein Follows In Which
MYSTERIES
OF LIGHT & SOUND
including acts of MAMMOTH MAGNIFICATION,
TUSK TAPPING & BOULDER HURLING
are relentlessly explored with the help
OF THE
GREAT WOOLY MAMMOTH
being necessarily unfettered by
COMMON SENSE
As observed and recorded during my experiments
for the illumination of future generations
and the general advantage of all
MANKIND

INTRODUCTION

At every moment of our lives, we are all bombarded with waves of energy. Painful though this may sound, it is actually nothing to get alarmed about, because most of this energy passes by us, or in some cases, right through us, without having any harmful effect.

Many of these waves collide with us without our having any idea of their existence. However, not all escape our notice. Through our senses, we can detect a small but important part of this ceaseless barrage. We can feel heat energy through our skin, we can see light energy with our eyes, and we can detect sound energy with our ears. But with the help of the machines described in this part of *The Way Things Work* we can do far more than this: we can communicate over unimaginable distances, bring hidden worlds – both microscopic and astronomic – into view, and reconstruct sights and sounds that would otherwise be locked away in the past.

Stretching Our Senses

Machines that work with waves use wave energy to amplify and extend our eyes and ears. Telescopes and microscopes upgrade the lenses in our eyes to reveal the extraordinary amount of fine detail that is actually present in light rays, but which unaided our eyes cannot see. Printing and photography put words and pictures on paper in full color, while in holograms, lasers exploit the clash of light waves to produce astonishing images that are so real you think you

can put your hands around them. Methods of recording sound and moving images recreate waves of sound and light to produce a potent means of illusion.

What most of these machines do is quite easy to describe, because many of them, for example the camera, record player, tape recorder, video recorder and telephone, are familiar objects found in almost every home. More difficult is understanding how wave energy allows them to do it.

ENERGY ON THE MOVE

When a sewing machine or a gasoline engine is used, it is easy to see where the energy comes from and where it goes to. Machines that work with waves are different. You cannot hold waves of energy in order to examine them, and to make things more tricky, energy waves behave according to a separate set of principles from those that govern physical matter.

The important feature of energy waves is that, although sometimes they may be conducted through matter, it is only the energy itself that moves. When a stone is dropped into a pond, for example, the ripples spread out from the point where the stone hits the water. But these miniature waves are not made up of water traveling outward. Instead, the water at the surface of the pond just rises and falls, and only the energy moves outward. The waves used by machines work in just the same way.

Every passing wave consists of a regular rise and fall of energy. The distance between successive energy rises is the wavelength, and the rate at which they pass is the wave's frequency. Both are very important in our perception of waves.

WAVES THROUGH MATTER

The machines in the following pages use two very different types of waves. Of the two, sound waves are easier to understand because they consist of vibrations in matter. They can only travel through matter – air, water, glass, steel, bricks and mortar; if it can be made to vibrate, sound will travel through it.

An individual sound wave is a chain of vibrating molecules – the tiny particles in the air, water or solid materials. When a loudspeaker vibrates, the molecules in the air around it also vibrate. But like the water in the pond, the molecules do not themselves move with the sound. Instead, they just pass on the energy. Regions of high and low pressure move through the air and spread out from the source.

Sound is simply our perception of this vibration. If something vibrates faster than about 20 times a second, we can hear it – this is the deepest note that human ears can detect. As the vibration speeds up, the pitch gets higher. At 20,000 vibrations a second, the pitch becomes too high for us to hear, but not too high for machines such as the ultrasound scanner, which uses high-pitched sound in the same way as a flying bat to create an image built up of echoes.

WAVES THROUGH SPACE

The second category of waves includes light and radio waves, which although they may seem very different, are actually variations on a common theme. Light and radio waves are members of a family known as electromagnetic waves. These mobile forms of energy are often called rays instead of waves – heat rays are also family members. The only way these waves differ is in their frequency.

Rather than vibrating molecules, electromagnetic waves – light, heat rays and radio waves – consist of vibrating electric and magnetic fields. Because these fields can exist in empty space, electromagnetic waves can travel through nothingness itself.

Like sound waves, each wave has a particular frequency. In light, we see different frequencies as different colors just as higher and lower sound frequencies give treble and bass notes. However, there the similarity ends. Electromagnetic waves travel, quite literally, at the speed of light, while sound waves crawl along at a millionth of that speed.

COMMUNICATING WITH WAVES

In traveling to us and through us, waves and rays may bring not just energy but also communicate meaning. Waves that are constant, for example as in the beam of a flashlight, cannot convey any information. But if that beam is interrupted, or if its brightness can be made to change, then it can carry a message. This is how all wave-borne communications work. Patterns of energy arrive from energy sources that are high or low, loud or soft, light or dark, one color or another. In this way, sound and light rays bring us music, voices, words on a page and expressions on faces. By converting one kind of wave into another, and also by storing their energy, waves can be made to carry sounds and images around the world and far beyond it – and to transport them through time. The machines in the following pages show something of the vast range of wave communications – from a telephone conversation with a next-door neighbor to the feeble signals from a space probe hurtling towards the Solar System's distant edge.

LIGHT AND IMAGES

ON SEEING THINGS

My life as an inventor has not been without its setbacks. Perhaps the most distressing was the failure of my athletic trophy business. Having perfected the folding rubber javelin and the stunning crystal discus, I entrusted their production to an apprentice. His initial enthusiasm however soon gave way to strange delusions of giant mammoths.

LIGHT RAYS

All sources of light produce rays that stream out in all directions. When these rays strike objects, they usually bounce off them. If lights rays enter our eyes, we either see the source of the light or the object that reflected the rays toward us. The angle of the rays gives the object its apparent size.

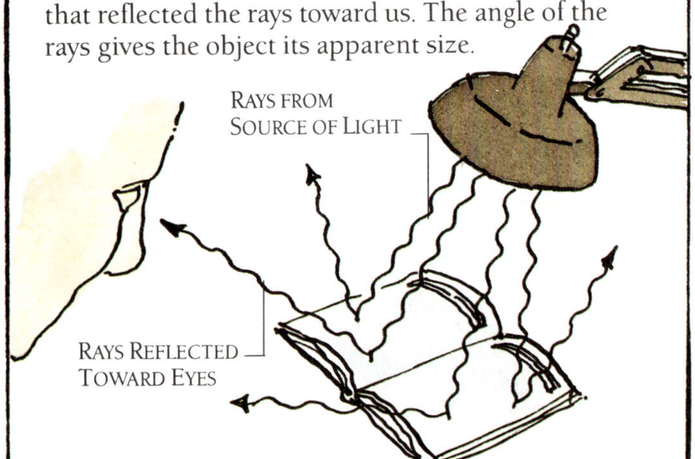

EYESIGHT

The lens of the eye bends the light rays that come from an object. It forms an image of the object on the light-sensitive retina of the eye, and this image is then changed to nerve impulses which travel to the brain. The image is in fact upside down on the retina, but the brain interprets it as upright.

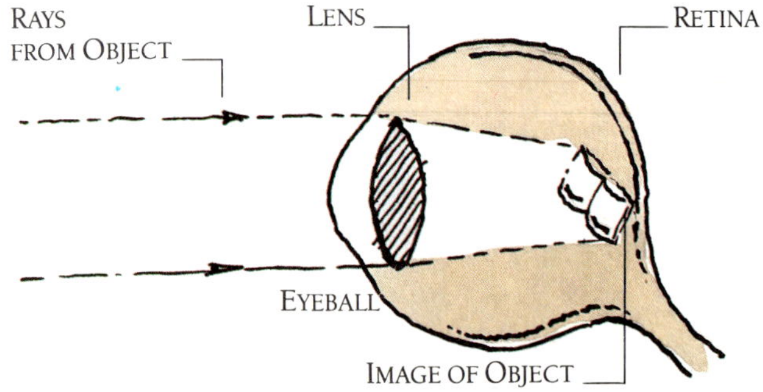

Assuming that he was simply overworked, I reduced his hours and improved ventilation in the workshop. But his condition deteriorated and one day he confronted me in my laboratory, claiming that miniature mammoths had invaded the premises. He insisted that a procession of these creatures was making its way across the wall, accompanied by a trail of smoke. Within the hour, word reached us that the workshop and all its contents had mysteriously burned to the ground. I realized that the frightened youth must have knocked over a candle as he fled, and although very disappointed at the loss, I decided to humor him and attribute the disaster to the spirits.

Forming Images

As light rays enter and leave transparent materials such as glass, they bend or refract. Seen through a lens, a nearby object appears to be much bigger because the rays enter the eye in a wider angle than they would without it. This is why the mammoth's eye is magnified by the discus.

Lenses can also throw images onto a surface. Cones of rays from every point on the object are bent by the lens to meet at the surface. The cones cross, inverting the mammoths, while the sun's rays meet to form a hot spot on the wall.

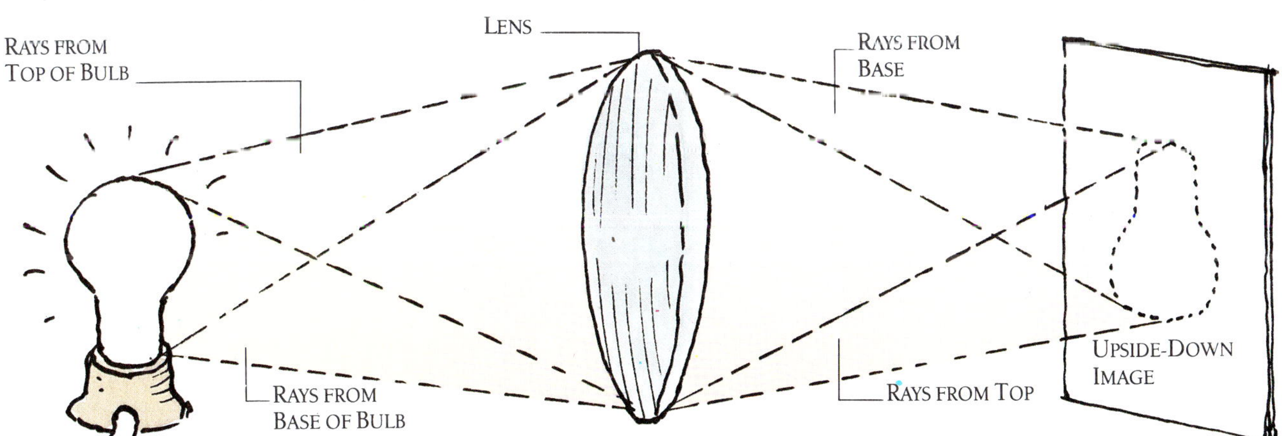

LIGHTING

There are two basic methods of producing artificial light. The first is to heat something so hot that it glows. The flame of a candle or oil lamp contains particles of carbon that have been made white-hot by the combustion of the wax or oil. In a light bulb, the filament is heated so much that it glows. The second method is to pass an electric current through a gas or vapor so that the gas or vapor lights up. Both methods cause electrons, the tiny charged particles inside atoms, to emit energy in the form of light rays.

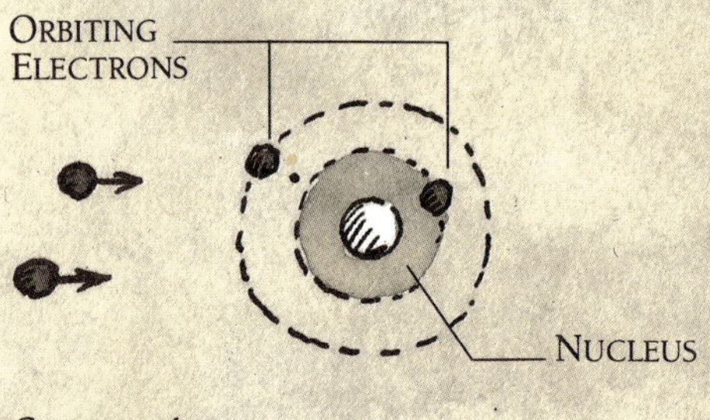

STABLE ATOM
Inside an atom, electrons move in a number of concentric orbits around the nucleus.

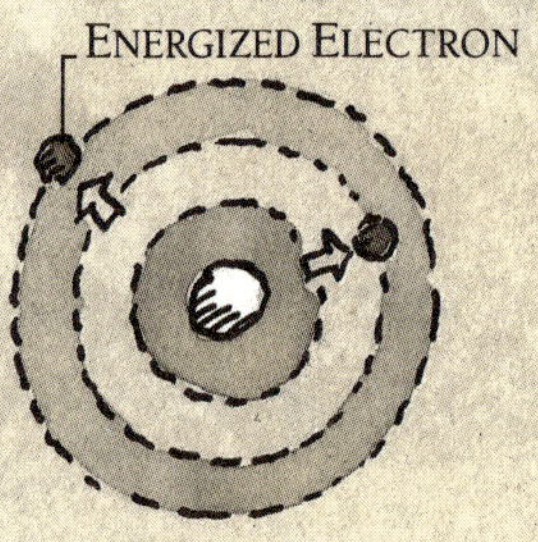

ELECTRONS MOVE OUT
Heat or electricity provides enough energy to make the electrons "jump" to higher orbits.

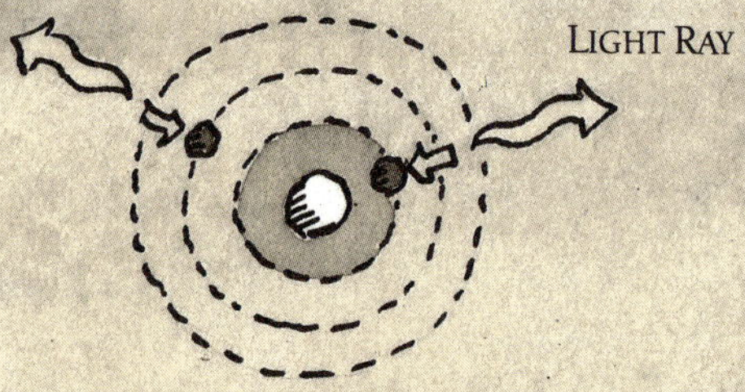

ELECTRONS FALL BACK
When the electrons fall back, their extra energy is emitted as a ray of light.

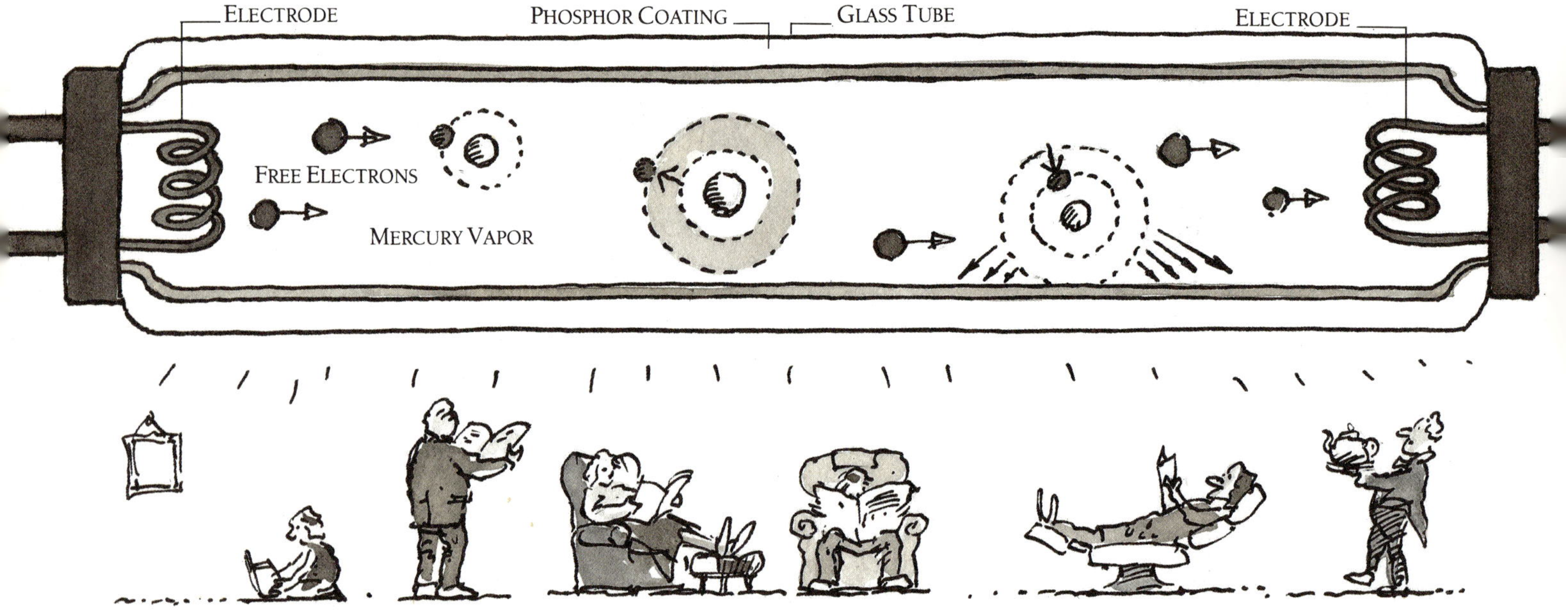

FLUORESCENT LAMP

A fluorescent lamp contains a glass tube that glows with white light when an electric current is passed through it. At each end of the tube are electrodes that are heated by the current and emit free electrons. The electrons strike atoms of mercury vapor in the tube, and cause the atoms to emit rays of ultraviolet light. The ultraviolet rays, which are invisible, strike a phosphor coating on the inside of the tube. The rays energize the electrons in the phosphor atoms, and the atoms emit white light. The conversion of one kind of light into another is known as fluorescence.

STREET LIGHT

The color of fluorescent street lights depends on the substance inside the tube. Sodium lights contain sodium vapor which glows a bright yellow-orange when electricity is passed through it. Neon signs work with a number of gases: neon itself glows red.

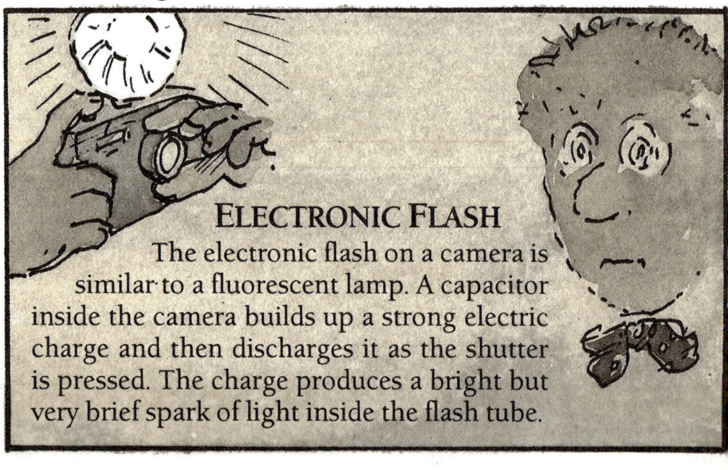

ELECTRONIC FLASH

The electronic flash on a camera is similar to a fluorescent lamp. A capacitor inside the camera builds up a strong electric charge and then discharges it as the shutter is pressed. The charge produces a bright but very brief spark of light inside the flash tube.

LIGHT BULB

An electric light bulb consists of a filament of tungsten wire wound in a tight coil. The passage of electricity through the filament heats the coil so that it becomes white hot. The filament reaches a temperature of about 4,500°F (2,500°C). Tungsten is chosen because it has a very high melting point and will not melt as it heats up. The bulb contains an inert gas such as argon to prevent the metal combining with oxygen in the air, which would cause the filament to burn out. The gas is usually under reduced pressure.

In modern light bulbs each coil of the filament is often made up of even tinier coils. The filament is therefore very long but very thin. This arrangement increases its light output.

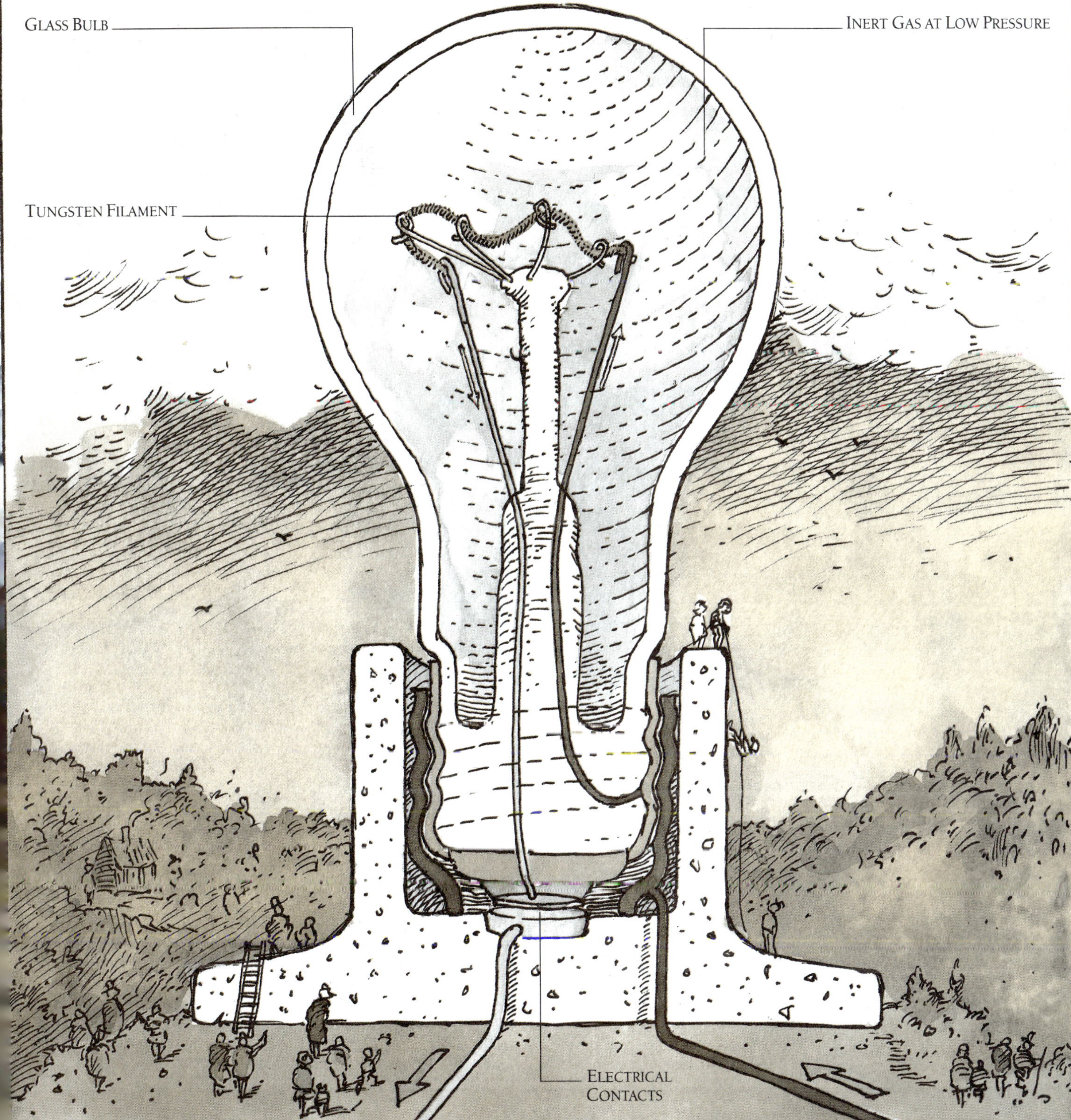

ADDING COLORS

Many of the color images we see are not quite what they seem. Instead of being composed of all the colors that we perceive, they are actually made of three primary colors mixed together. Images that are sources of light, such as color television pictures (see p.262), combine colors by "additive" mixing. Stage lights produce a range of colors by additive mixing of three primary colors at various brightnesses.

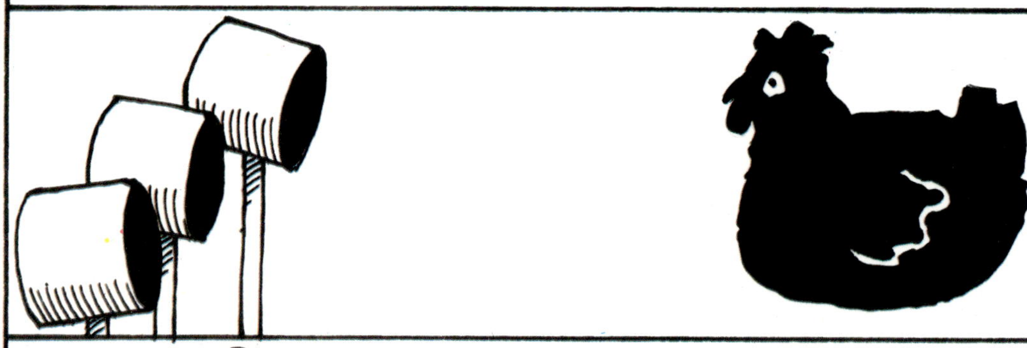

BLACK

The three primary colors in additive mixing are red, green and blue. When no light is produced, there are no colors to mix together and the result is an absence of light – or black.

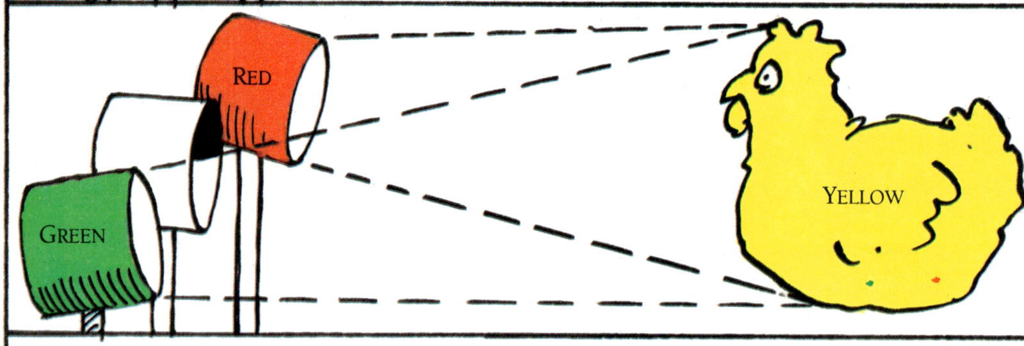

YELLOW

When green and red lights illuminate a white object, they mix together to color the object yellow. In a television picture, green and red dots or stripes light up and the eye fuses them to see yellow.

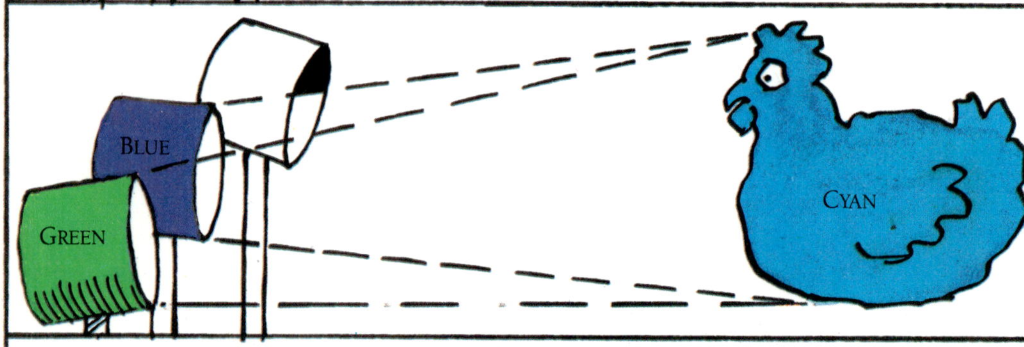

CYAN

An equal mixture of two primary colors is called a secondary color. Yellow is a secondary color and so is cyan, which is produced by mixing blue and green lights.

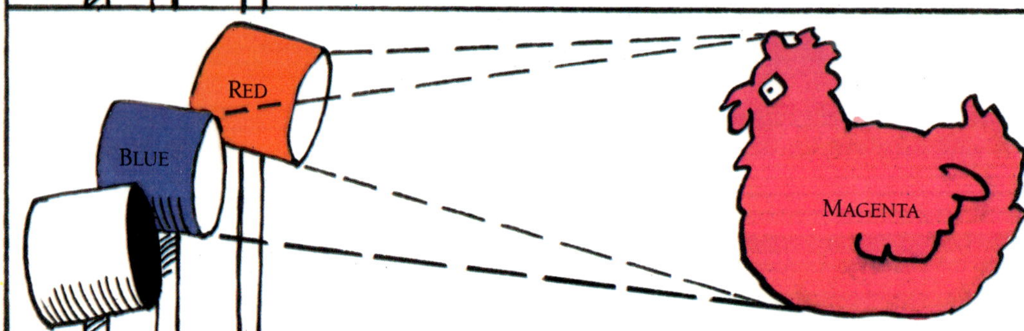

MAGENTA

Magenta is a third secondary color, produced by mixing red and blue. Other colors are formed by mixing the primary colors in different proportions.

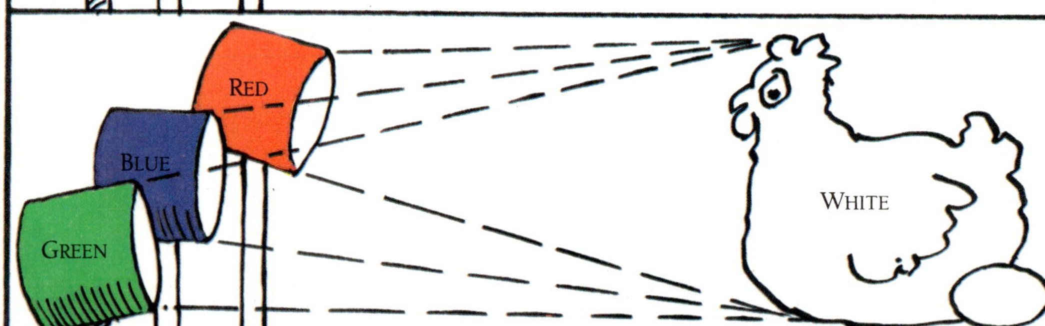

WHITE

White is produced by mixing all three primary colors together. White light is given by an equal mixture of red, green and blue lights.

SUBTRACTING COLORS

Images produced by mixing printing inks (see pp.226-7) and paints form colors by "subtractive" mixing. This gives different colors to additive mixing because the pictures themselves are not sources of light. The pictures reflect some of the primary colors in the white light that illuminates them, and absorb or subtract the other primary colors. We see the reflected primary colors added together.

WHITE

A white surface reflects all the light falling on it and absorbs none. No subtraction takes place and all three colors are reflected, mixing together to give white.

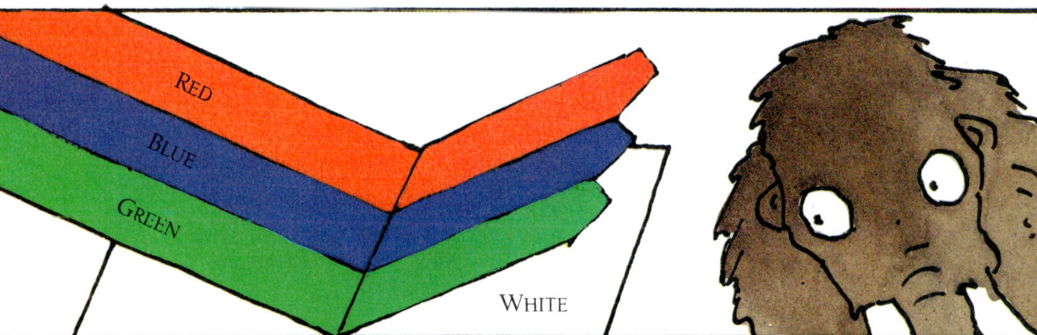

YELLOW

A yellow surface absorbs the blue light in the white light striking it. Blue is subtracted and red and green are reflected, which combine to give yellow.

CYAN

A cyan surface subtracts red from the light that illuminates it. Blue and green are reflected and combine together to give cyan. Mixing yellow and cyan subtracts blue and red respectively, leaving green.

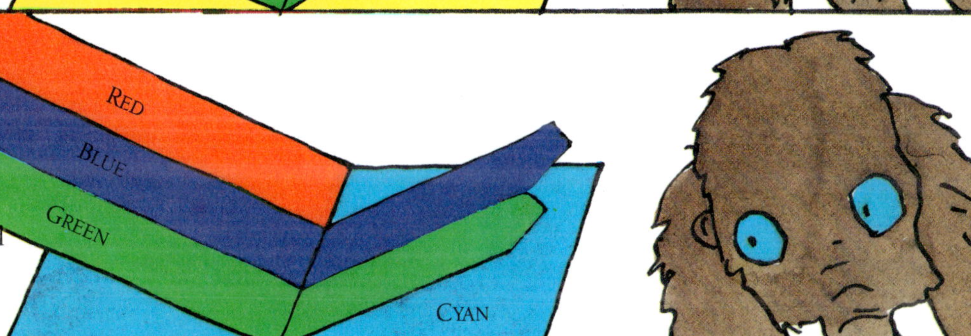

MAGENTA

A magenta surface absorbs green from the white light that strikes it. Red and blue are reflected and mix together to give magenta. Mixing magenta with yellow subtracts green and blue, leaving red.

BLACK

A black "color" is given by a pigment that absorbs all the colors falling on it. All three primary colors are subtracted and none reflected, causing the surface to appear black.

MIRRORS

A flat mirror reflects the light rays which strike it so that the rays leave the surface of the mirror at exactly the same angle that they meet it. The light rays enter the eye as if they had come directly from an object behind the mirror, and we therefore see an image of the object in the mirror. This image is a "virtual" image: it cannot be projected on a screen. It is also reversed. Images formed by two mirrors, as in the periscope, are not reversed because the second mirror corrects the image.

PERISCOPE

The periscope makes it possible to see around corners. It has one mirror to capture light rays from an object and sends them to another mirror which directs the rays into the eye.

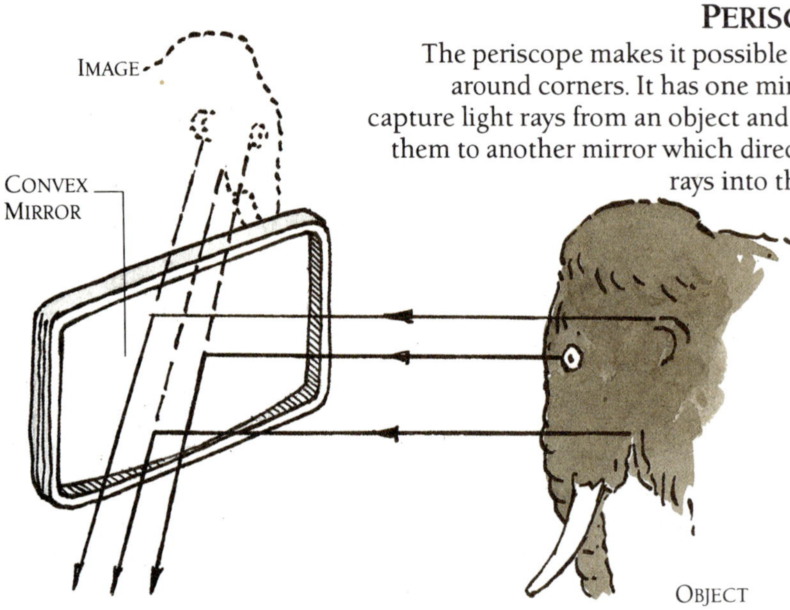

DRIVING MIRROR

A driving mirror is a convex mirror, which curves toward the viewer. It reflects light rays from an image so that they diverge. The eye sees an image which is reduced in size, giving the mirror a wide field of view.

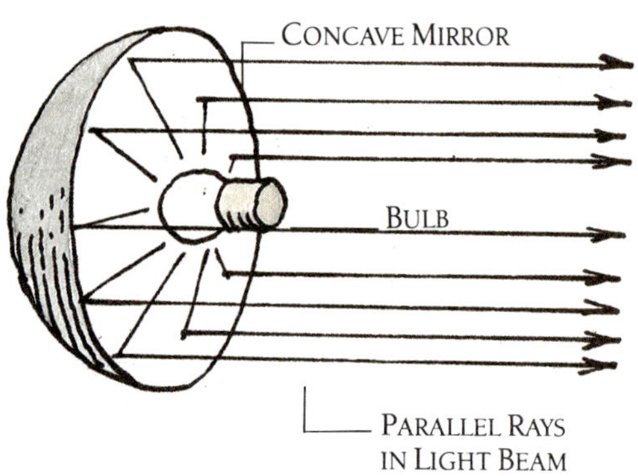

HEADLIGHT MIRROR

In headlights and flashlights, a concave mirror is placed behind the bulb. The light rays are reflected by the curved surface so that they are parallel and form a narrow and bright beam of light.

ENDOSCOPE

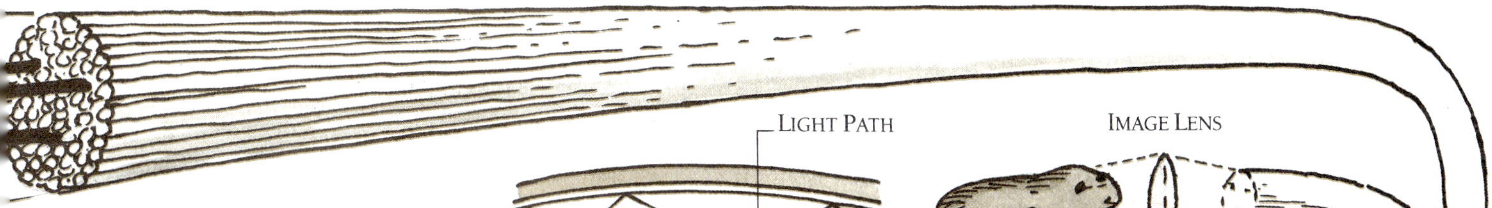

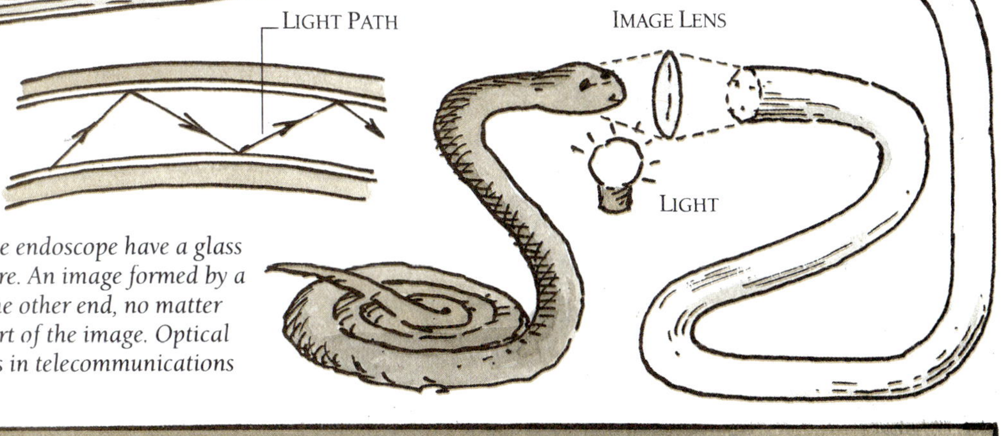

INTERNAL REFLECTION

Fiber optic devices depend on internal reflection, which allows light to pass along a narrow filament of very pure glass. The light-conducting fibers used by instruments such as the endoscope have a glass coating that reflects light rays along the fiber core. An image formed by a lens on one end of a cable of fibers appears at the other end, no matter how much the cable twists. Each fiber carries part of the image. Optical fibers also carry light signals over long distances in telecommunications *(see pp.250-1).*

By using an endoscope, a doctor can easily see what is going on inside a body without cutting it open. A narrow tube containing fiber optic cables or guides is inserted into a channel in the body, such as the throat. Light guides transmit light along the fibers to light up the interior. The image guide sends a picture of the interior back along the tube, where it is viewed in an eyepiece. The tube also contains air and water pipes as well as a channel for small surgical instruments. Wires control the bending of the tube.

CONNECTOR

Light sources and air, water and suction pipes are attached to the connector, which links them to the endoscope tube.

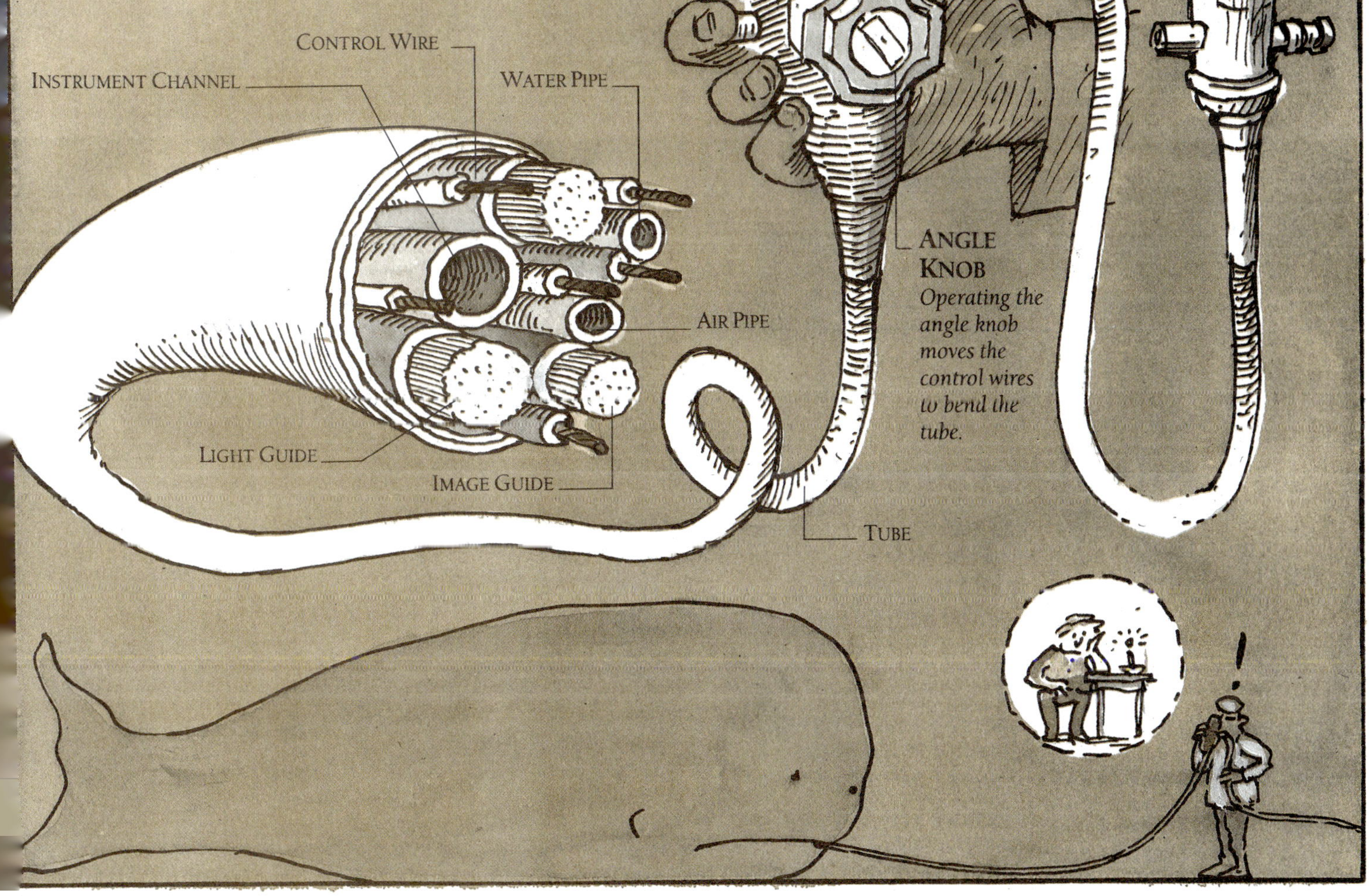

ANGLE KNOB

Operating the angle knob moves the control wires to bend the tube.

LENSES

Lenses are of great importance in devices that use light. Optical instruments such as cameras, projectors, microscopes and telescopes all produce images with lenses, while many of us see the world through lenses that correct poor sight. Lenses work by refraction, which is the bending of light rays that occurs as rays leave one transparent material and enter another. In the case of lenses, the two materials involved are glass and air. Lenses in glasses and contact lenses are used to supplement the lens in the eye (see p.190) when it cannot otherwise bend the rays by the angle required to form a sharp image.

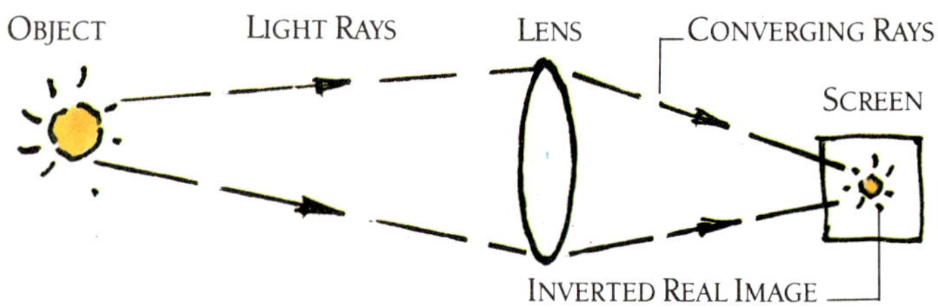

CONVEX LENS

A convex lens is thicker at the center than the edges. Light rays from an object pass through it and converge to form a "real" image – one that can be seen on a screen.

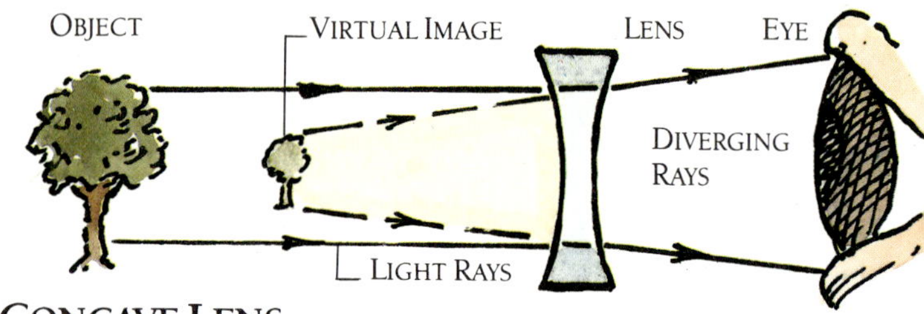

CONCAVE LENS

A concave lens is thicker at the edges than the center. It makes light rays diverge. The eye receives these rays and sees a smaller "virtual" image (see p.196) of the object.

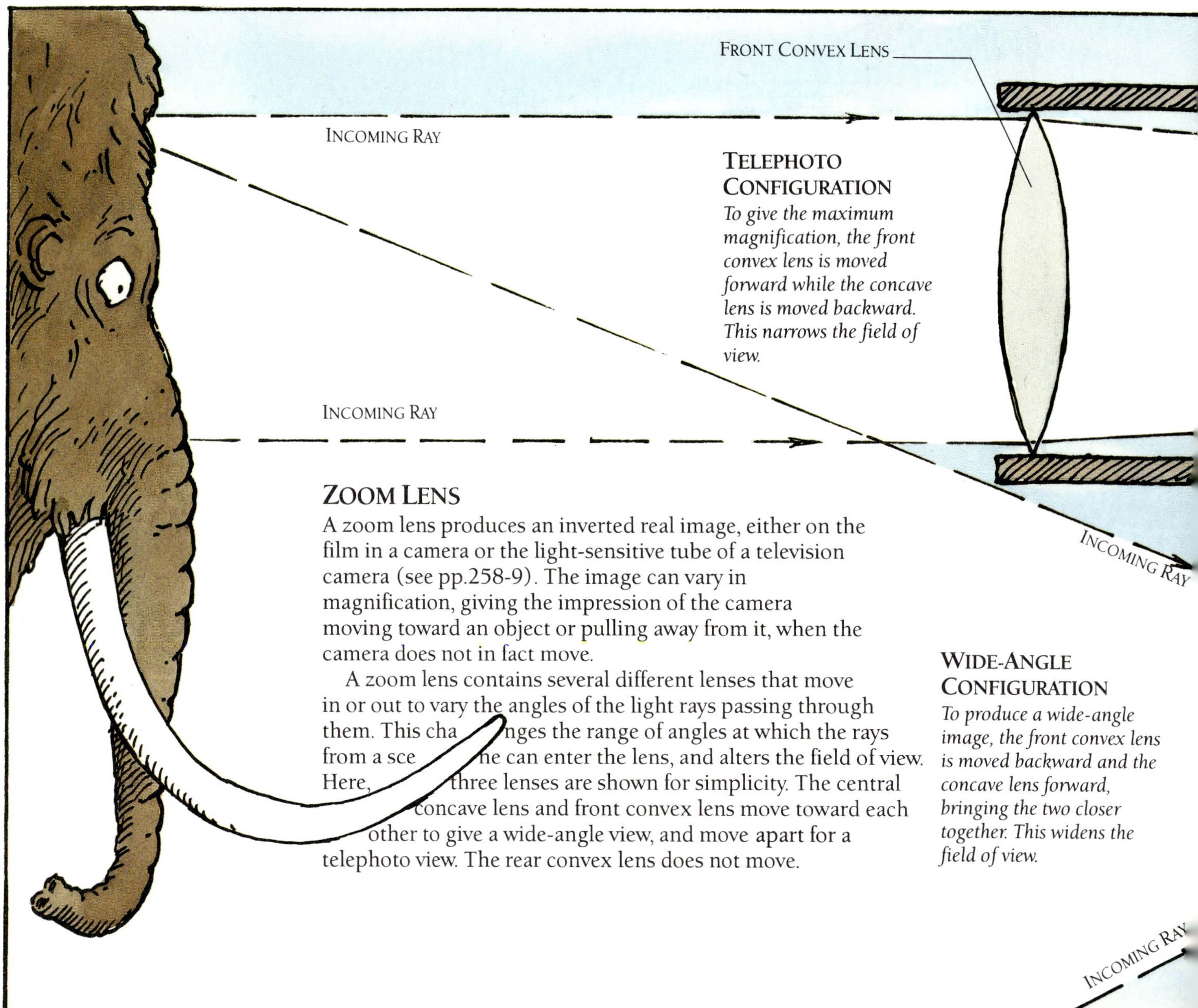

TELEPHOTO CONFIGURATION

To give the maximum magnification, the front convex lens is moved forward while the concave lens is moved backward. This narrows the field of view.

ZOOM LENS

A zoom lens produces an inverted real image, either on the film in a camera or the light-sensitive tube of a television camera (see pp.258-9). The image can vary in magnification, giving the impression of the camera moving toward an object or pulling away from it, when the camera does not in fact move.

A zoom lens contains several different lenses that move in or out to vary the angles of the light rays passing through them. This changes the range of angles at which the rays from a scene can enter the lens, and alters the field of view. Here, three lenses are shown for simplicity. The central concave lens and front convex lens move toward each other to give a wide-angle view, and move apart for a telephoto view. The rear convex lens does not move.

WIDE-ANGLE CONFIGURATION

To produce a wide-angle image, the front convex lens is moved backward and the concave lens forward, bringing the two closer together. This widens the field of view.

MAGNIFYING GLASS

A magnifying glass is a large convex lens. When held near a small object, a magnified virtual image can be seen in the lens. The lens makes the rays from the object converge as they enter the eye. The part of the brain that deals with vision always assumes that light rays arrive at the eye in straight lines. For this reason, it perceives the object as being larger than it really is.

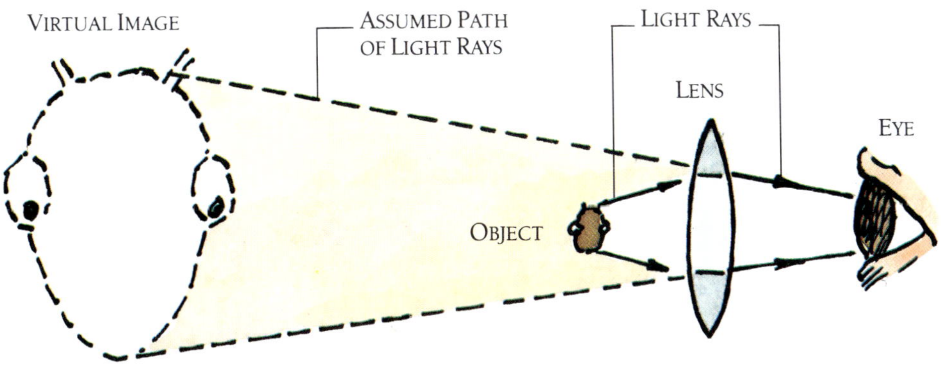

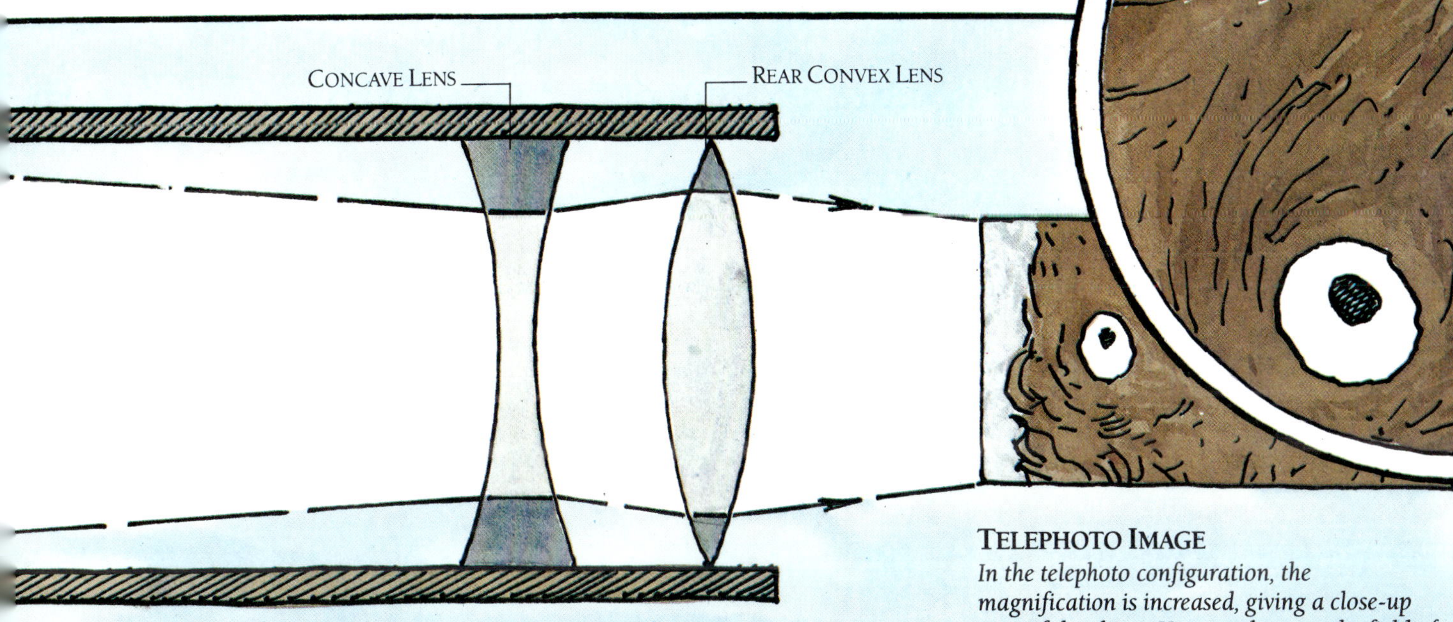

TELEPHOTO IMAGE

In the telephoto configuration, the magnification is increased, giving a close-up view of the object. However, because the field of view is decreased, only a small part of it can be seen.

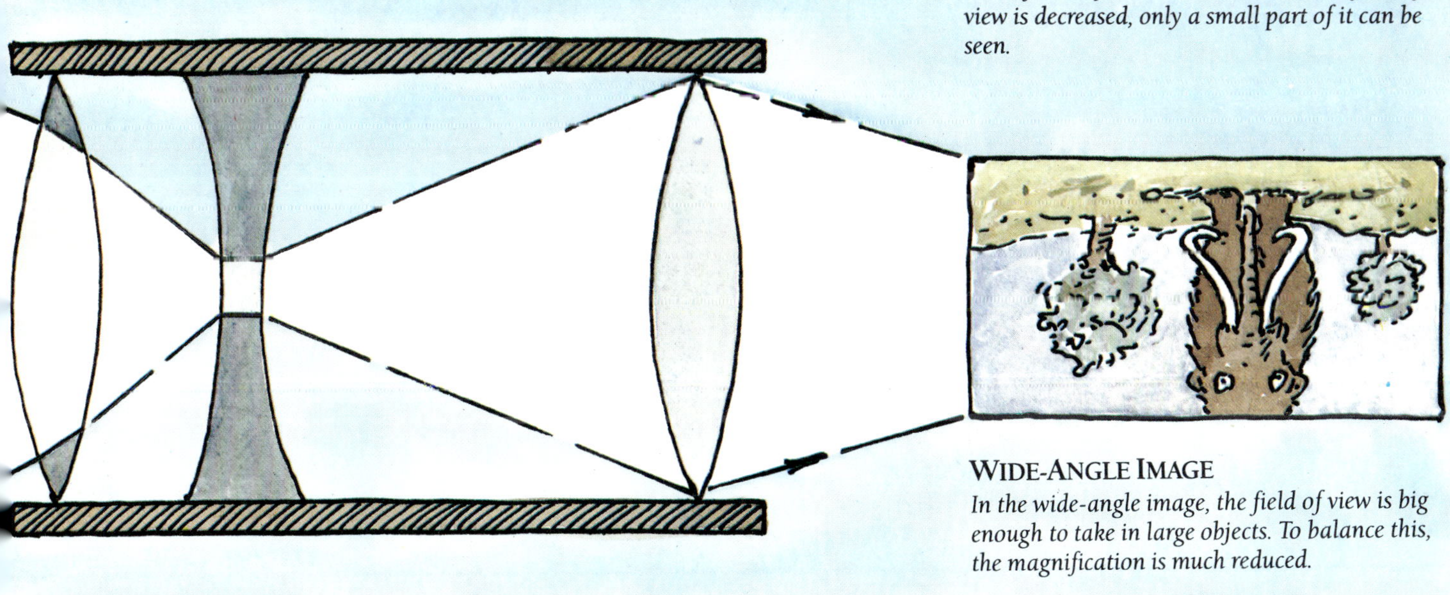

WIDE-ANGLE IMAGE

In the wide-angle image, the field of view is big enough to take in large objects. To balance this, the magnification is much reduced.

TELESCOPES

A telescope gives a close-up view of a distant object, which, in the case of an astronomical telescope viewing a far-off planet or galaxy, is very distant indeed. Most telescopes work in the same basic way, which is to produce a real image of the object inside the telescope tube. The eyepiece lens then views this image in the same way as a magnifying glass (see p.199). The viewer looks at a very close real image, which therefore appears large. The degree of magnification depends mainly on the power of the eyepiece lens.

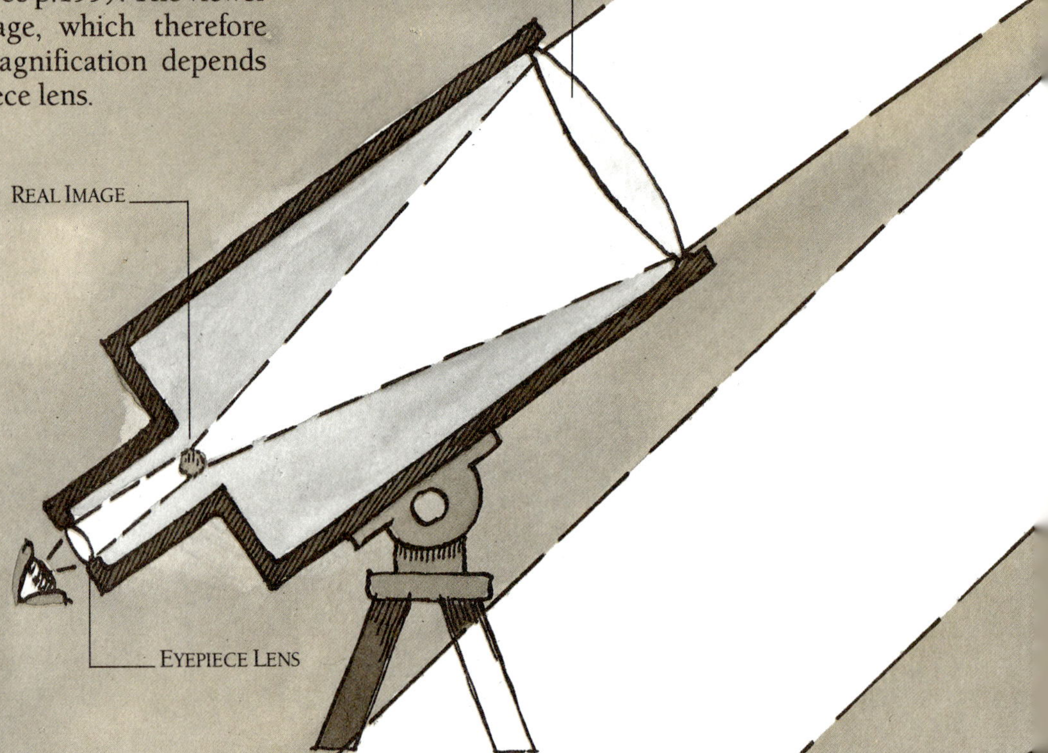

REFRACTING TELESCOPE

In a refracting telescope, an objective lens forms the real image that is viewed by the eyepiece lens. The image is upside down but this is not important in astronomy. A terrestrial telescope gives an upright view. It contains an extra convex lens that forms an upright real image and the eyepiece lens views this image.

REFLECTING TELESCOPE

In a reflecting telescope, a large concave primary mirror forms the real image that is then viewed by an eyepiece lens. Usually, a secondary mirror reflects the rays from the primary mirror so that the real image forms beneath the mirror or to the side. This is more convenient for viewing.

Reflecting telescopes are important in astronomy because the primary mirror can be very wide. This enables it to collect a lot of light, making faint objects visible. Collecting light from an object is often more important than magnifying it because distant stars do not appear bigger even when magnified.

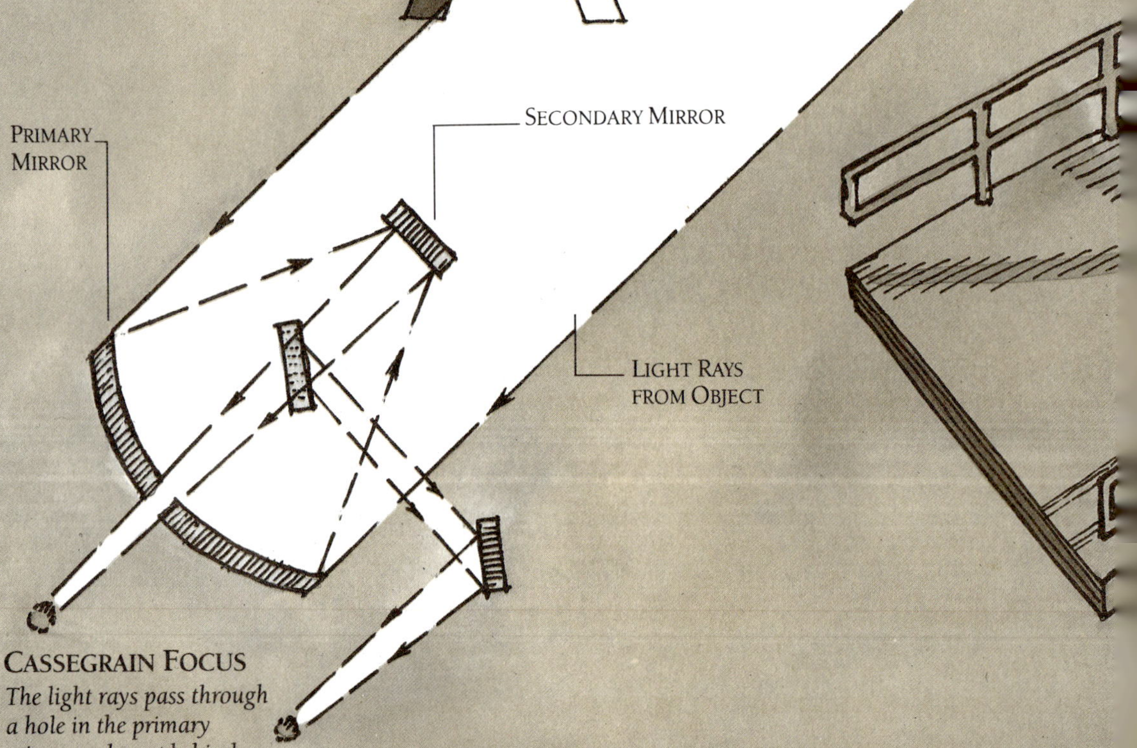

CASSEGRAIN FOCUS
The light rays pass through a hole in the primary mirror and meet behind it to form the real image. This is then viewed with an eyepiece lens or photographed with a camera.

COUDÉ FOCUS
Two extra mirrors are inserted to form the real image at the side of the telescope, where it can be easily viewed or photographed.

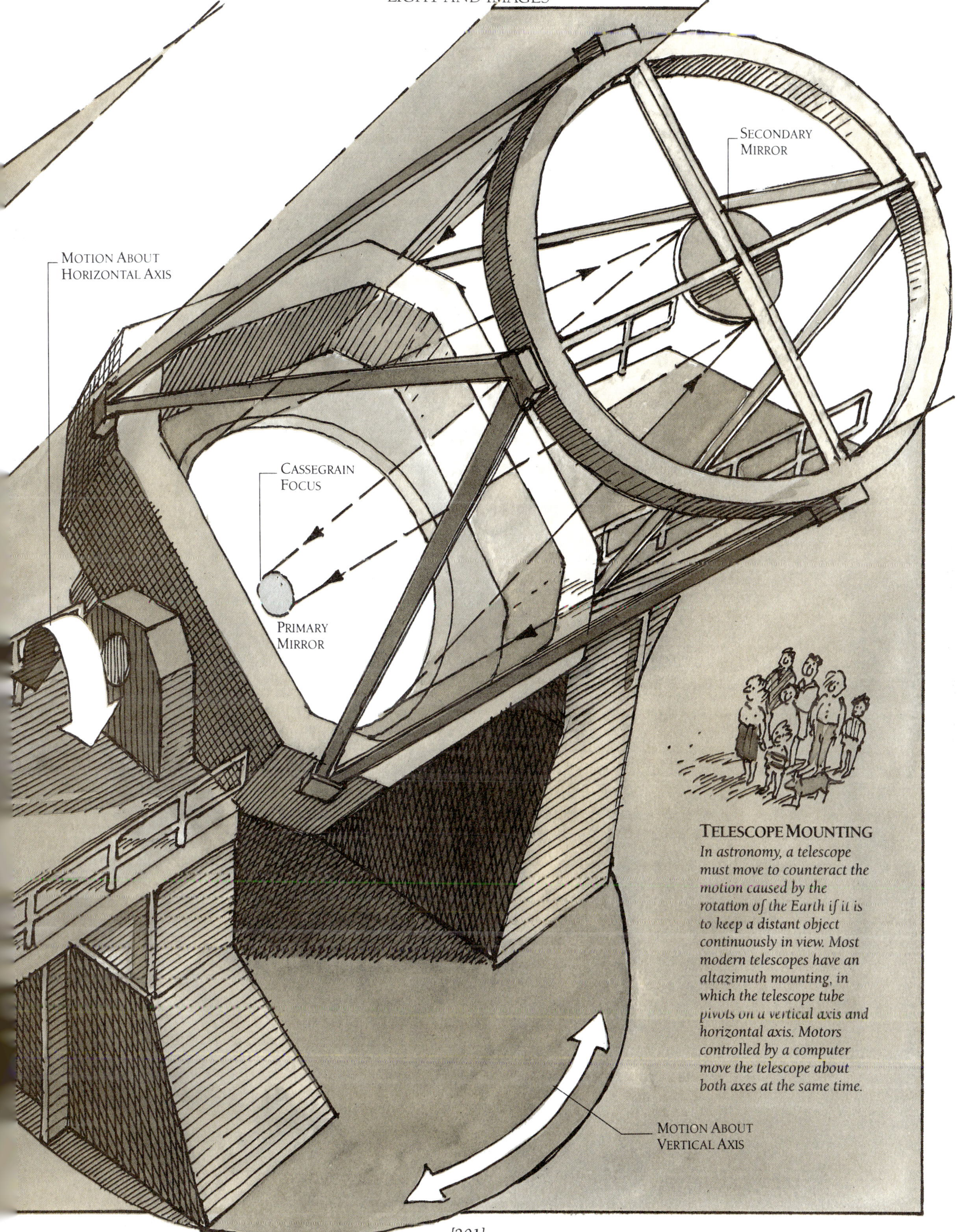

TELESCOPE MOUNTING
In astronomy, a telescope must move to counteract the motion caused by the rotation of the Earth if it is to keep a distant object continuously in view. Most modern telescopes have an altazimuth mounting, in which the telescope tube pivots on a vertical axis and horizontal axis. Motors controlled by a computer move the telescope about both axes at the same time.

BINOCULARS

A pair of binoculars is basically two small refracting telescopes that together produce a stereoscopic or three-dimensional view. Each eye sees a separate close-up view, but the brain combines them to perceive an image that has depth.

Binoculars are different from telescopes in one respect. They contain a pair of prisms between the objective and eyepiece lenses. The faces of the prisms reflect the light rays internally so that an upright non-reversed image is seen. The prisms also lengthen the light path between the lenses, which narrows the field of view and increases magnification in a short tube. In addition, the two objective lenses are further apart than the eyes, which enhances stereoscopic vision.

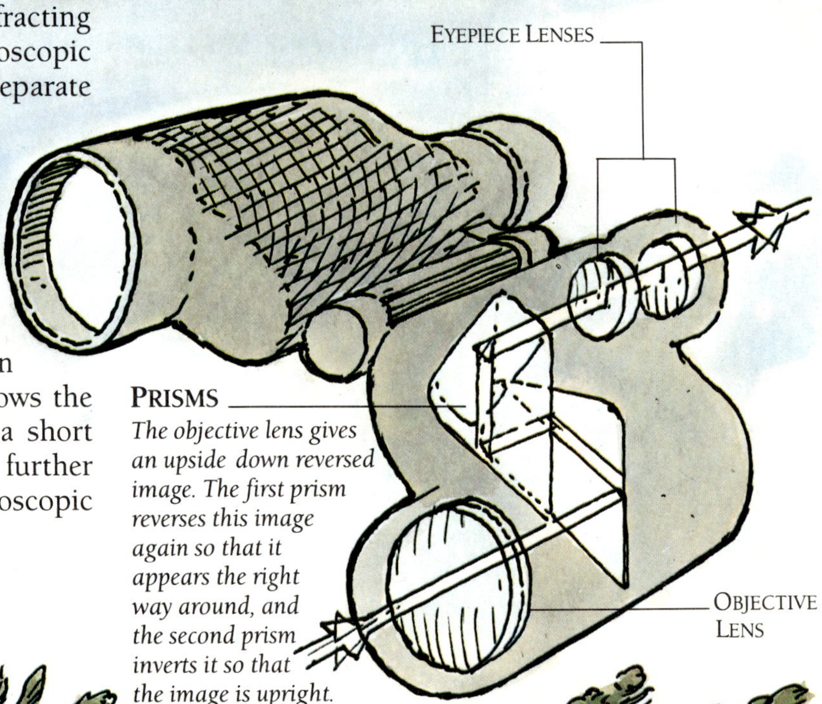

PRISMS
The objective lens gives an upside down reversed image. The first prism reverses this image again so that it appears the right way around, and the second prism inverts it so that the image is upright.

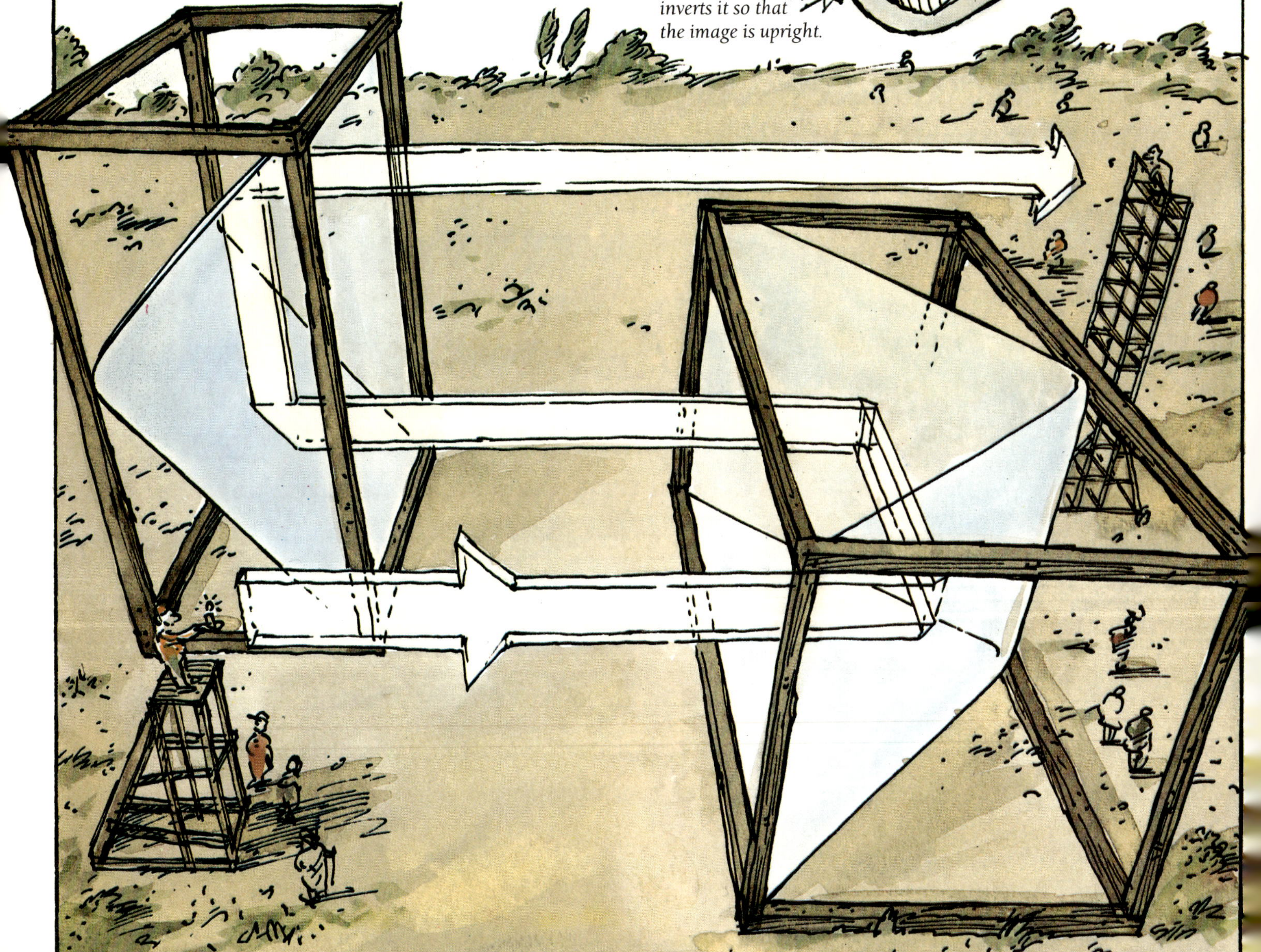

MICROSCOPES

An optical microscope (*left*) gives a highly enlarged view of an object that is invisible to the unaided eye. The microscope works in the same way as a refracting telescope, but the object or specimen is very close to the objective lenses instead of being distant. The objective lenses form an enlarged real image of the specimen near the eyepiece lenses, and this image is viewed through the eyepiece lenses which further enlarge it. The specimen is illuminated by a beam of light reflected from a mirror and concentrated by condenser lenses.

MAGNETIC CONDENSER
The condenser concentrates the electrons into a beam that strikes the specimen.

MAGNETIC OBJECTIVE
The objective deflects the electrons that pass through the specimen. Denser or thicker parts of the specimen allow fewer electrons through.

MAGNETIC PROJECTOR
The projector further deflects the electrons to form an electron image on the fluorescent screen.

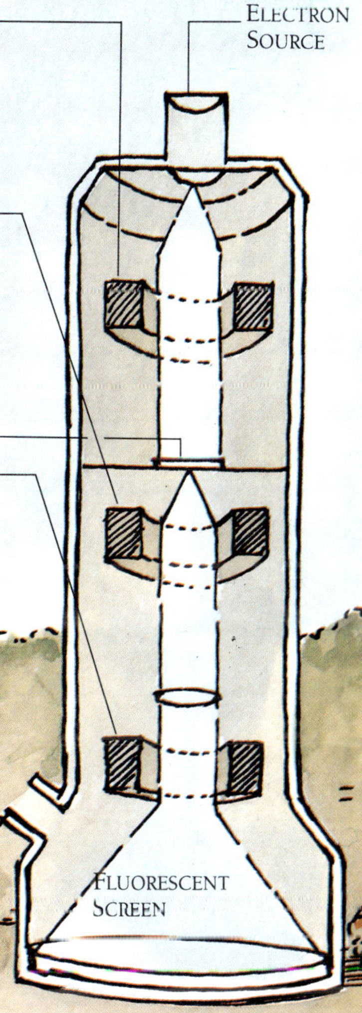

ELECTRON MICROSCOPE

An optical microscope magnifies as much as 2,000 times, but an electron microscope (*above*) can make things look a million times bigger. Instead of using light, it uses a beam of moving electrons (see p.192). It has magnetic lenses, which are electric coils that produce magnetic fields to deflect the electrons in the same way that glass lenses bend light rays. In the transmission electron microscope (shown here), the beam passes through the specimen. In the scanning electron microscope, the beam is reflected from the specimen.

POLARIZED LIGHT

Light rays are electromagnetic waves: their energy consists of vibrating electric and magnetic fields (see p.255). In normal light rays, these fields vibrate in planes at random angles. In polarized light, all the rays vibrate in the same plane. The direction of this plane is the plane in which the electric field vibrates. Polarizing filters are found in, among other things, anti-glare sunglasses and liquid crystal displays.

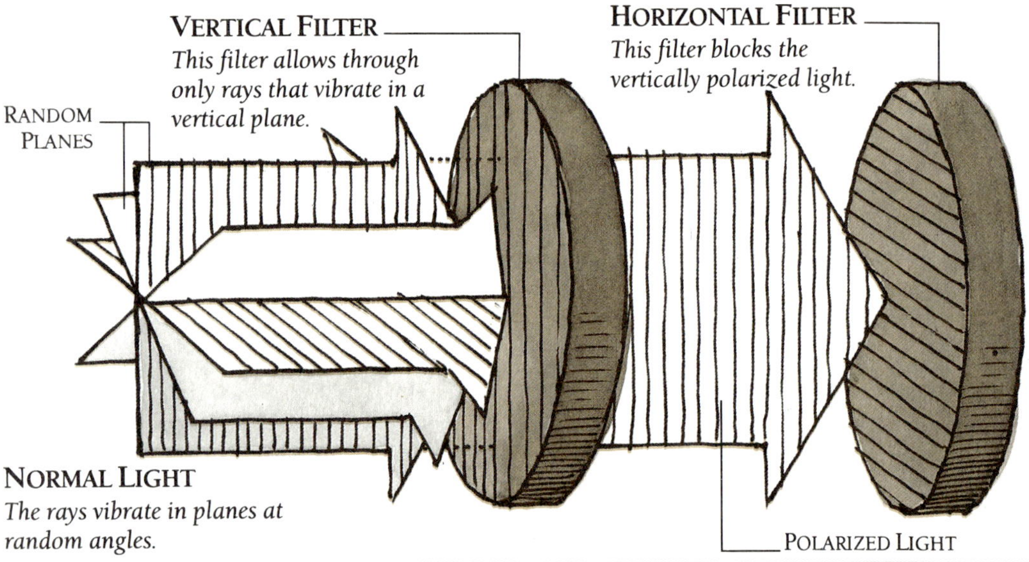

POLARIZING FILTERS

A polarizing filter blocks all rays except those vibrating in a certain plane. If polarized light strikes a filter whose plane is at right angles to the plane of the rays, then no light passes.

Polarizing sunglasses work in this way. Light reflected from shiny surfaces is partly polarized, and the sunglasses are polarizing filters. They block the polarized light and reduce glare.

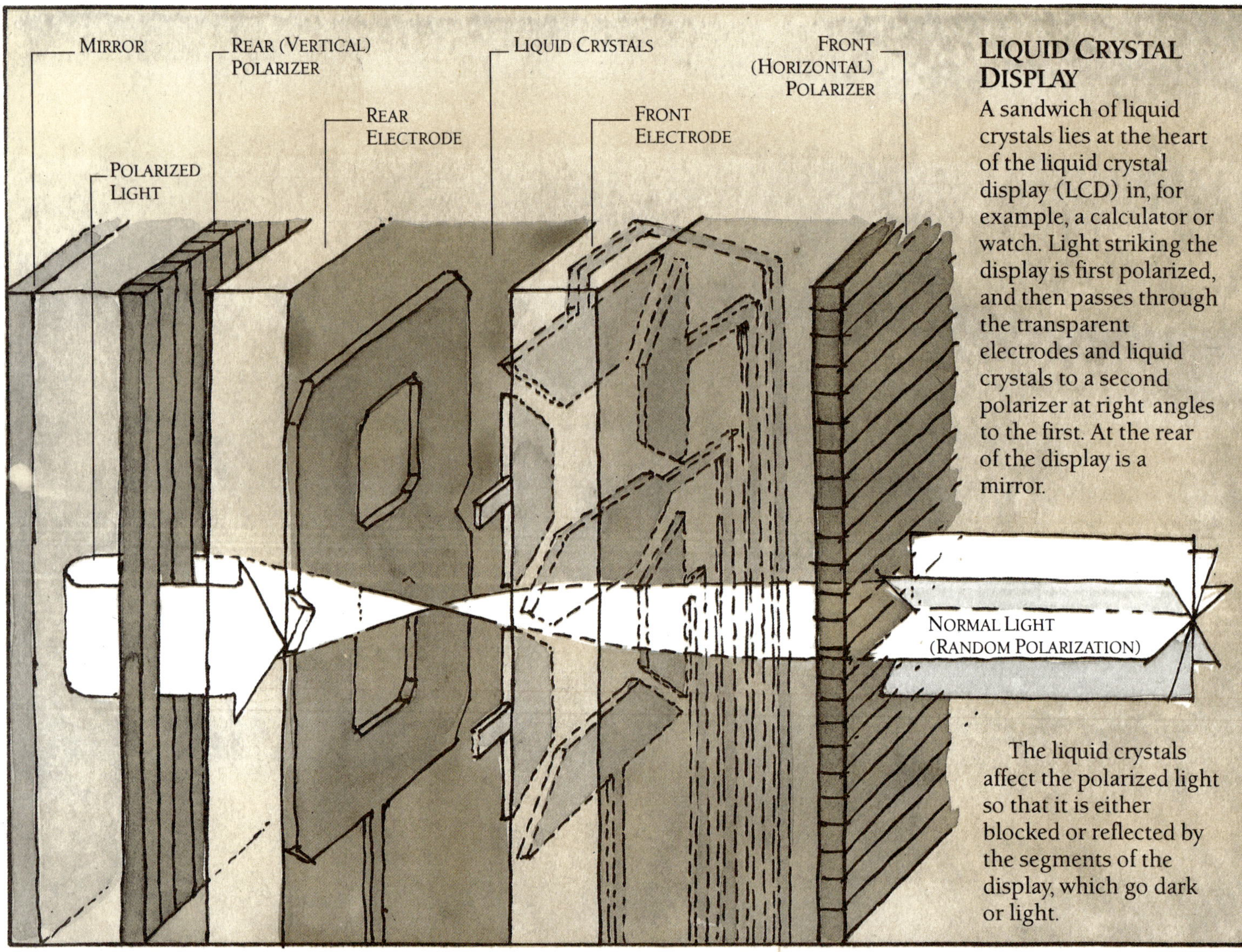

LIQUID CRYSTAL DISPLAY

A sandwich of liquid crystals lies at the heart of the liquid crystal display (LCD) in, for example, a calculator or watch. Light striking the display is first polarized, and then passes through the transparent electrodes and liquid crystals to a second polarizer at right angles to the first. At the rear of the display is a mirror.

The liquid crystals affect the polarized light so that it is either blocked or reflected by the segments of the display, which go dark or light.

LIQUID CRYSTALS

Liquid crystals are liquid materials with molecules arranged in patterns similar to those of crystals. The molecules are normally twisted and when polarized light passes through liquid crystals, its plane of vibration twists through a right angle.

A weak electric current changes the pattern of molecules in liquid crystals. It causes the molecules to line up so that polarized light is no longer affected. The liquid crystals are sandwiched between two transparent electrodes, which pass light rays and deliver the electric current.

By arranging liquid crystals in separate segments, numbers and letters can be produced in a liquid crystal display. The display is controlled by microchips (see pp.336-7).

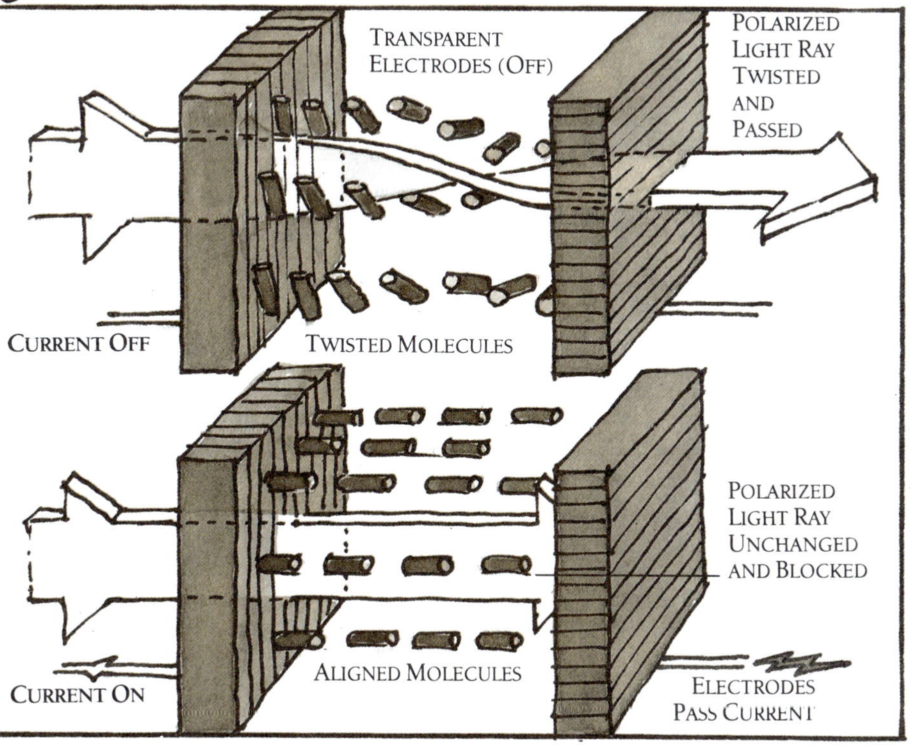

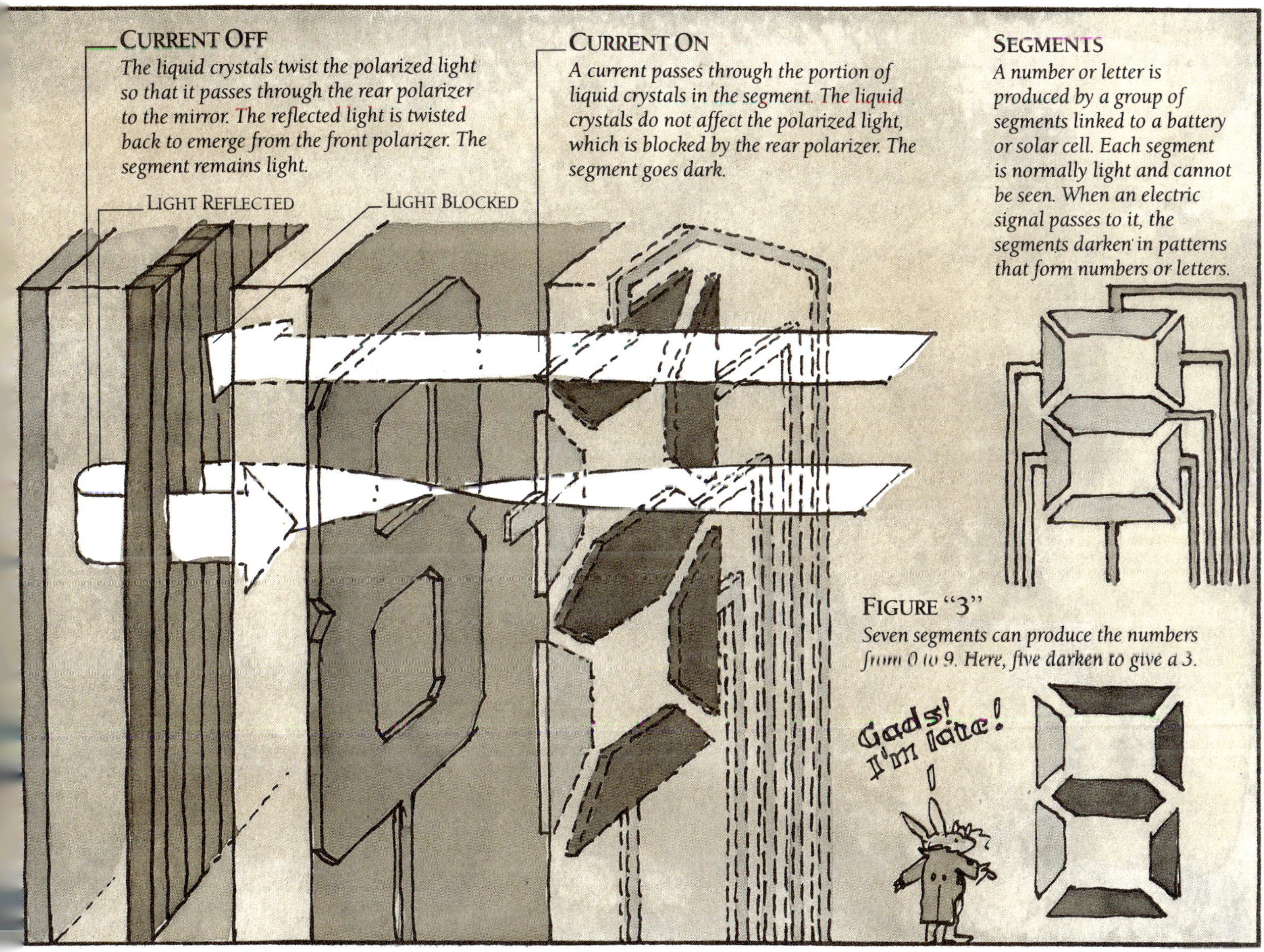

LASER

A laser produces a narrow beam of very bright light, either firing brief pulses of light or forming a continuous beam. Laser stands for Light Amplification by Stimulated Emission of Radiation. Unlike ordinary light, laser light is "coherent", meaning that all the rays have exactly the same wavelength and are all in phase, vibrating together to produce a beam of great intensity.

A laser beam may either be of visible light, or of invisible infra-red rays. Visible light lasers are used in digital recording and fiber-optic communications as well as in surveying and distance measurement, and give results of very high quality and accuracy. The intense heat of a powerful infra-red laser beam is sufficient to cut metal.

1 Exciting the Atoms

In a laser, energy is first stored in a lasing medium, which may be a solid, liquid or gas. The energy excites atoms in the medium, raising them to a high-energy state. One excited atom then spontaneously releases a light ray. In a gas laser, shown here, electrons in an electric current excite the gas atoms.

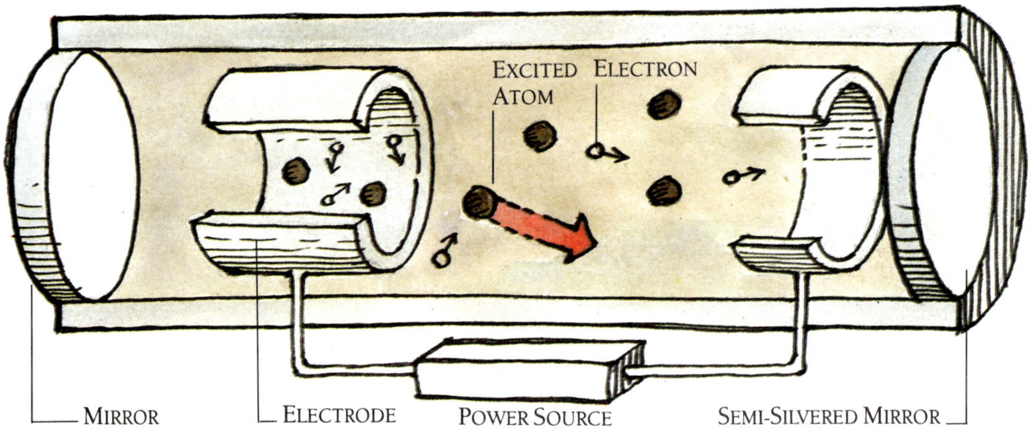

2 Light Builds Up

The ray of light from the excited atom strikes another excited atom, causing it also to emit a light ray. These rays then strike more excited atoms, and the process of light production grows. The mirrors at the ends of the tube reflect the light rays so that more and more excited atoms release light.

3 The Laser Fires

As each excited atom emits a light ray, the new ray vibrates in step with the ray that strikes the atom. All the rays are in step, and the beam becomes bright enough to pass through the semi-silvered mirror and leave the laser. The energy is released as laser light.

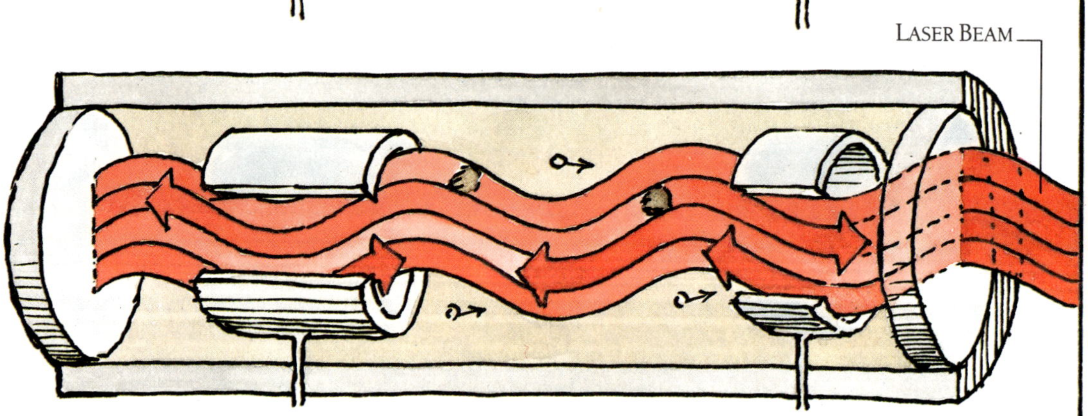

Gas Laser

A gas laser produces a continuous beam of laser light as the gas atoms absorb energy from the electrons moving through the gas and then release this energy as light.

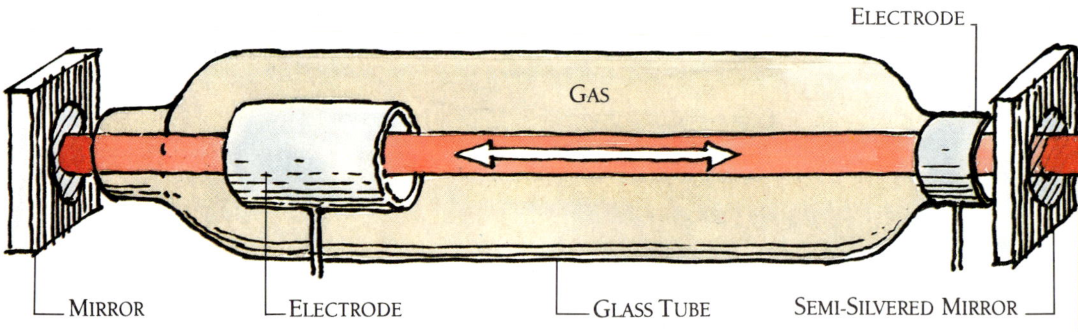

HOLOGRAPHY

One very important application of laser is holography, the production of images that are three-dimensional and that appear to have depth just like a real object. Holography requires light of a single exact wavelength, which can only be produced by a laser.

In holography, the light beam from a laser is split into two beams. One beam, the object beam, lights up the object. The second beam, the reference beam, goes to a photographic plate or film placed near the object. When developed, the plate or film becomes a hologram, in which a three-dimensional image of the object can be seen (see pp.208-9).

MAKING A HOLOGRAM

The photographic plate or film receives laser light from the object and from the reference beam. The arrangement here produces a reflection hologram, which gives an image in ordinary light. For a transmission hologram, which is viewed with a laser, the two beams strike the same side of the plate or film.

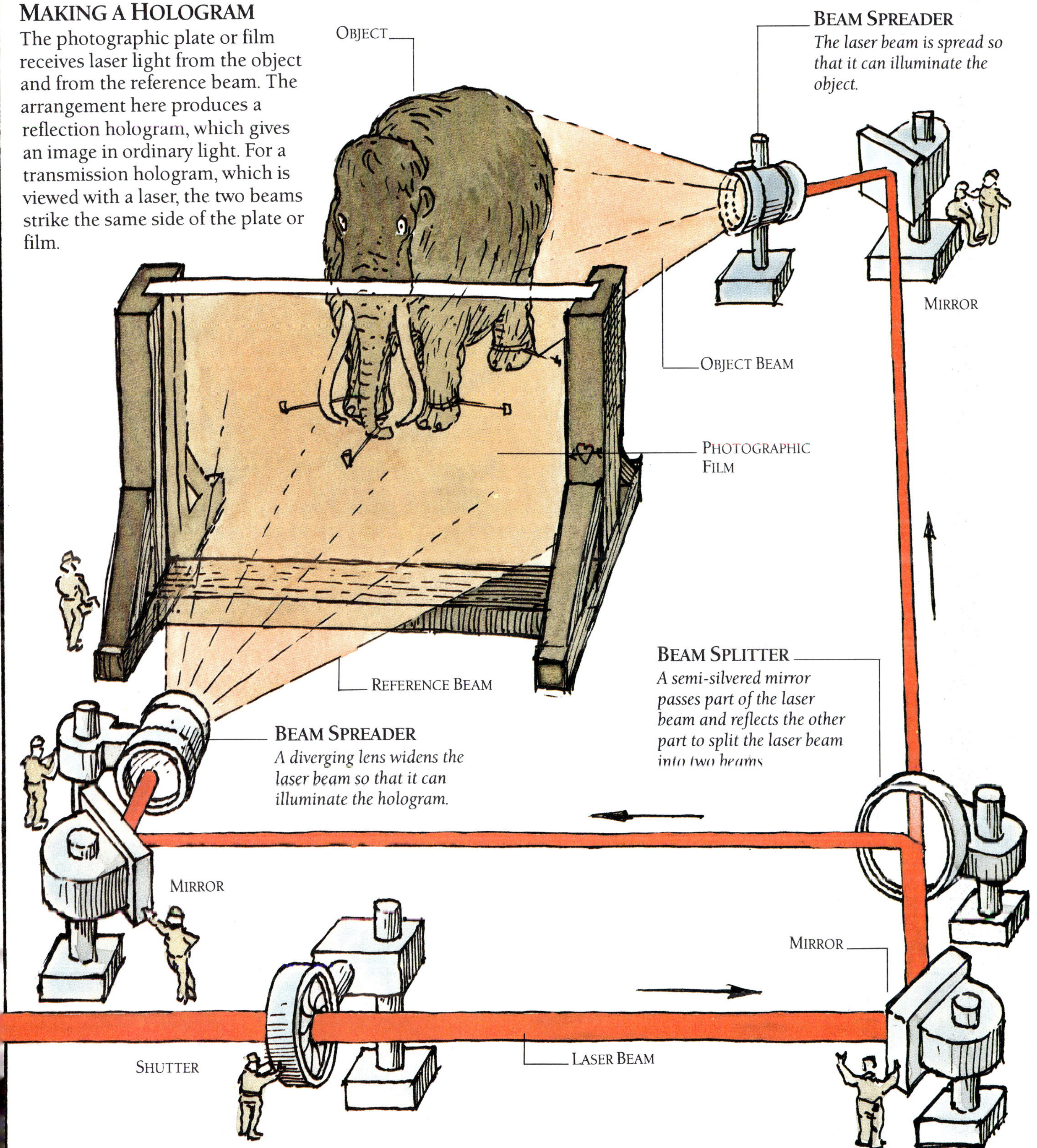

HOLOGRAM

A reflection hologram is made with a photographic plate or film and laser light (see p.207). In the plate or film, light first reflected by the object meets light coming directly from the laser. Each pair of rays – one from every point on the surface of the object and one in the reference beam – interferes. The two rays give light if the interference is "constructive" or they cancel each other out to give dark if the interference is "destructive". Over the whole hologram, an interference pattern forms as all the pairs of rays meet. This pattern depends on the energy levels of the rays coming from the object, which vary with the brightness of its surface.

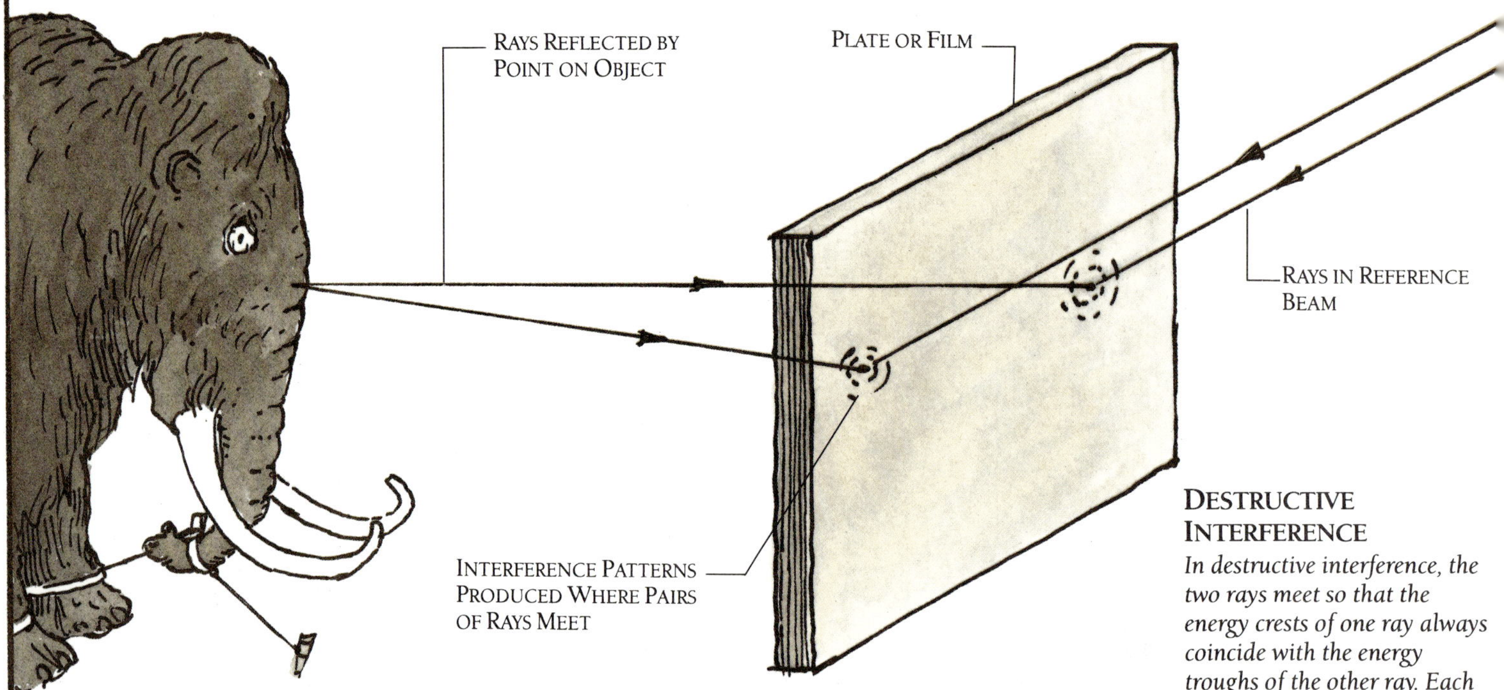

DESTRUCTIVE INTERFERENCE

In destructive interference, the two rays meet so that the energy crests of one ray always coincide with the energy troughs of the other ray. Each crest cancels out a trough, and there is no light at the interference point.

CONSTRUCTIVE INTERFERENCE

When two light rays meet, they interfere. The energy level of each ray rises and falls like a wave. In constructive interference, the energy crests or energy troughs always coincide, producing bright light at the interference point.

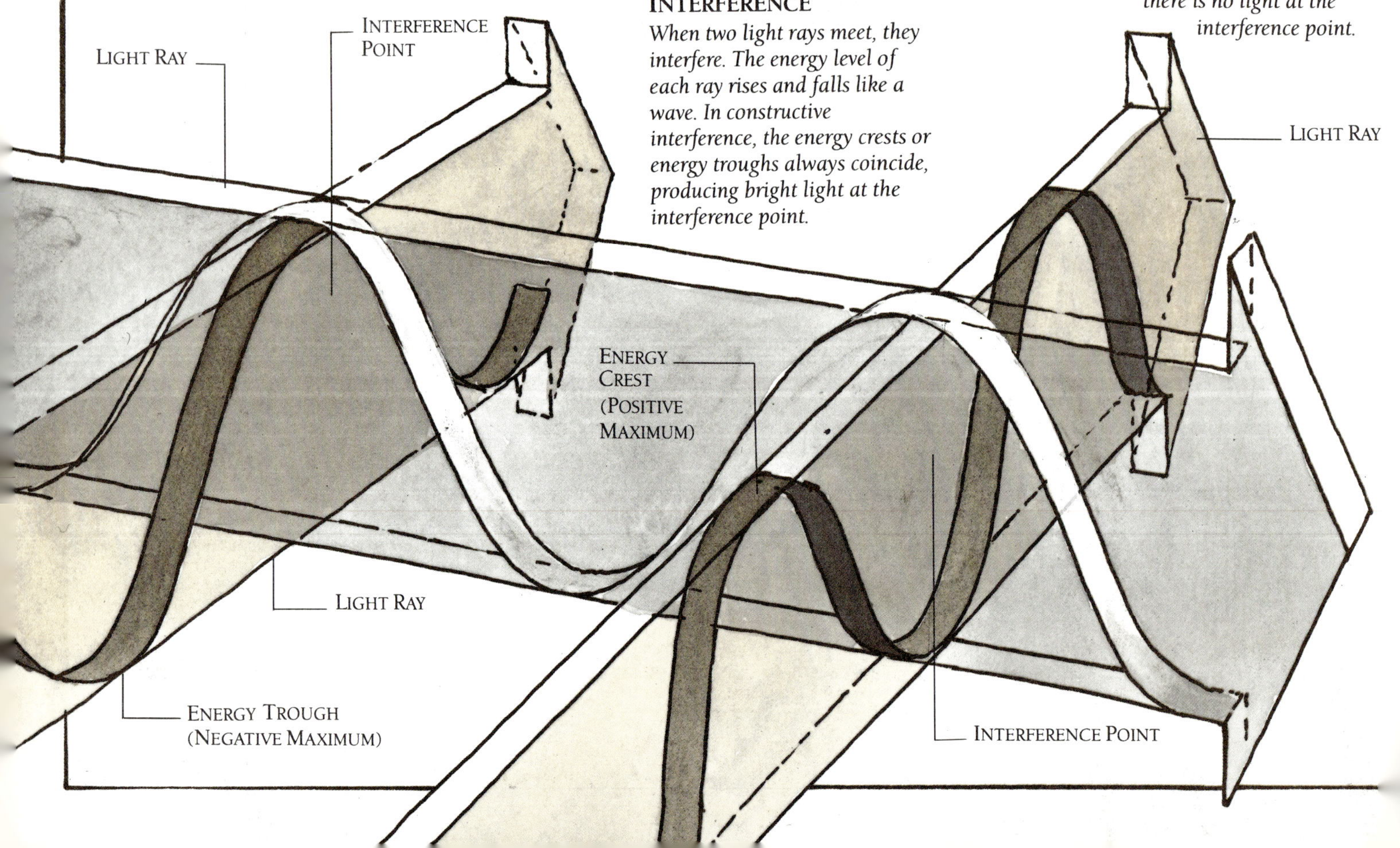

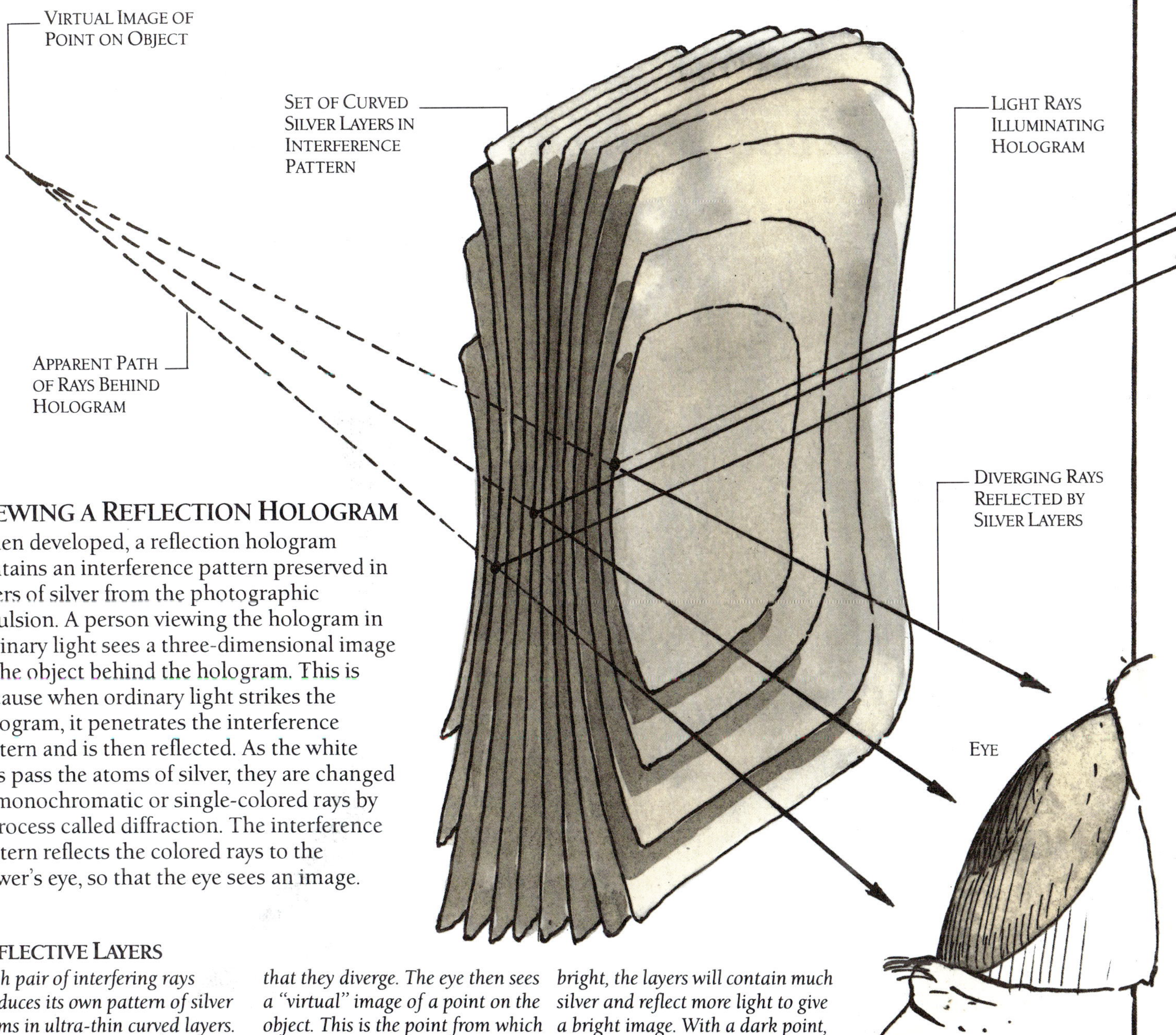

VIEWING A REFLECTION HOLOGRAM

When developed, a reflection hologram contains an interference pattern preserved in layers of silver from the photographic emulsion. A person viewing the hologram in ordinary light sees a three-dimensional image of the object behind the hologram. This is because when ordinary light strikes the hologram, it penetrates the interference pattern and is then reflected. As the white rays pass the atoms of silver, they are changed to monochromatic or single-colored rays by a process called diffraction. The interference pattern reflects the colored rays to the viewer's eye, so that the eye sees an image.

REFLECTIVE LAYERS

Each pair of interfering rays produces its own pattern of silver atoms in ultra-thin curved layers. When light rays illuminate the hologram, the layers reflect them so that they diverge. The eye then sees a "virtual" image of a point on the object. This is the point from which the light ray came that formed the set of layers. If that point was bright, the layers will contain much silver and reflect more light to give a bright image. With a dark point, the layer contains less silver and reflects less light.

SEEING IN DEPTH

In a hologram, each eye sees many points formed by different sets of layers in the interference pattern. This gives an image of the object. The two eyes look at different parts of the hologram and so see separate images of the object. The brain combines them to give a three-dimensional image.

The image in each eye is produced by different parts of the hologram formed by rays that left the original object at different angles. Each side of the hologram is formed by rays coming from that side of the object. Moving your head therefore brings another side of the object into view and your view of the image changes.

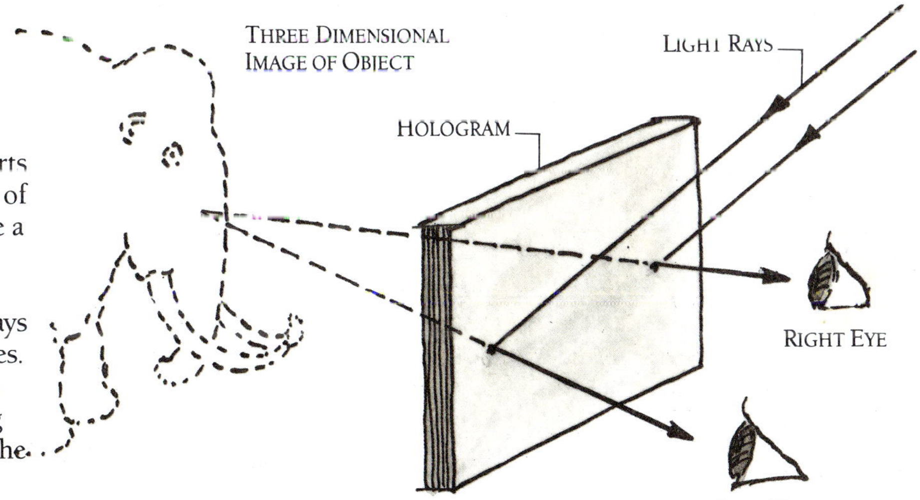

PHOTOGRAPHY

ON MAMMOTH PICTURES

While playing golf one day, I noticed that the grass in the specified caddy waiting areas was considerably lower and less green than the grass in the sunlight. I played on, but my mind was no longer on the game. If the image of a mammoth could be made on the grass accidentally, I reasoned, then perhaps images of other things could be made intentionally?

Returning to my workshop, I begged the assistance of the family next door for my first experiment. I asked them to sleep in a line on the grass outside. They were reluctant. I offered to pay them and they were soon snoring away.

PRESERVED IN SILVER

Rather than grass, black-and-white photography uses silver to preserve images permanently. Tiny crystals of light-sensitive silver compounds are suspended in an emulsion of gelatin, which coats a transparent plastic film. In a camera, the lens forms an image of a scene on the film. This exposure to light, even for a fraction of a second, is enough to start a chemical change in the crystals, and they begin to break down into black specks of metallic silver. Developing the film then completes the formation of silver to produce a negative of the scene.

EXPOSURE
When the light from the bright parts of the image is thrown onto the film, it makes the crystals in the emulsion start breaking down into silver.

DEVELOPMENT
The developer changes the crystals exposed to light into silver, while the fixer dissolves the unexposed crystals.

NEGATIVE
Black silver is left where the scene was bright, while clear film is left where it was dark.

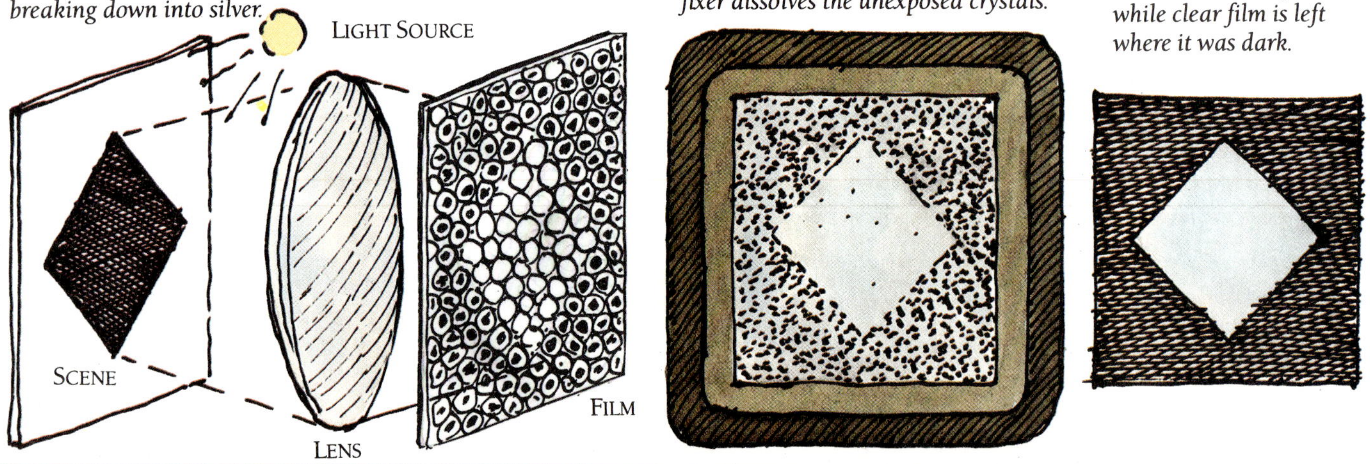

fig 1

fig 2

fig 3

I had them return for the next five days and lie in exactly the same spot. By the end of the week I had a perfect image of my neighbors. The procedure soon caught on, and eventually even school groups could be seen lying motionless on the workshop's lawns.

However, several drawbacks arose which I had not forseen. The images required continuous trimming once the subjects left the picture. They were also difficult to display as well as being astronomically expensive to frame. Had I been able to shrink people before they were recorded, I am convinced that my discovery would have had a bright future.

PRINTING THE PICTURE

A print is made on photographic printing paper, which bears a light-sensitive emulsion like that on a film. The negative is used to expose the paper, often with an enlarger that projects a magnified image of the negative. The paper is then developed to create a picture made up of black silver. Light areas in the negative now appear dark in the print and vice-versa – the same as in the original scene. Using a different process, a positive picture can be made directly on photographic film or paper without producing a negative.

EXPOSURE

An image of the negative is formed on printing paper by placing the negative in an enlarger. Moving the lens enables the image size to be enlarged.

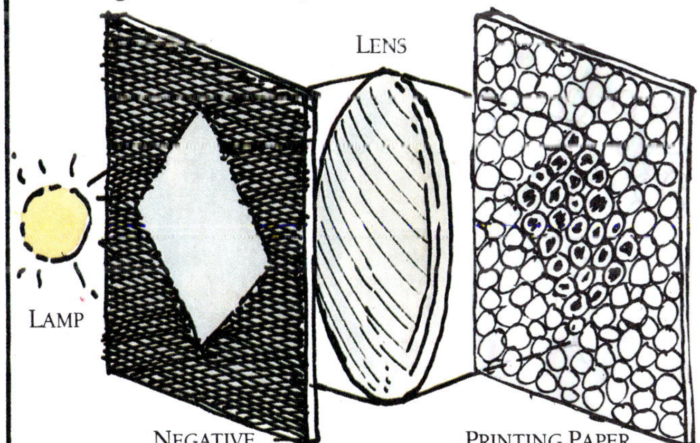

DEVELOPMENT

The paper is developed and fixed in red light, to which the emulsion is not sensitive.

POSITIVE

Light areas on the negative create black silver on the print, while dark areas create white paper.

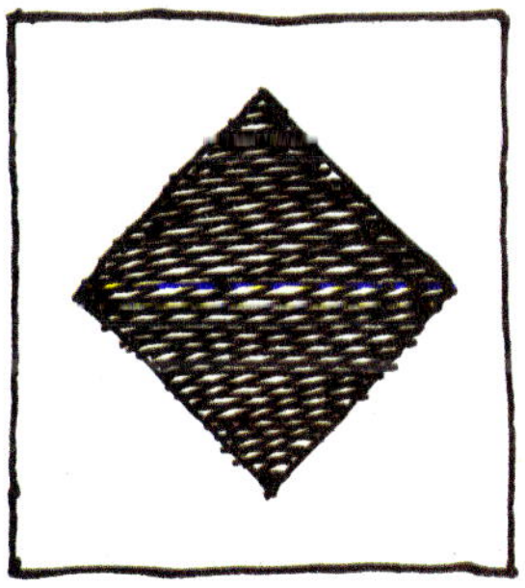

THE SINGLE-LENS REFLEX CAMERA

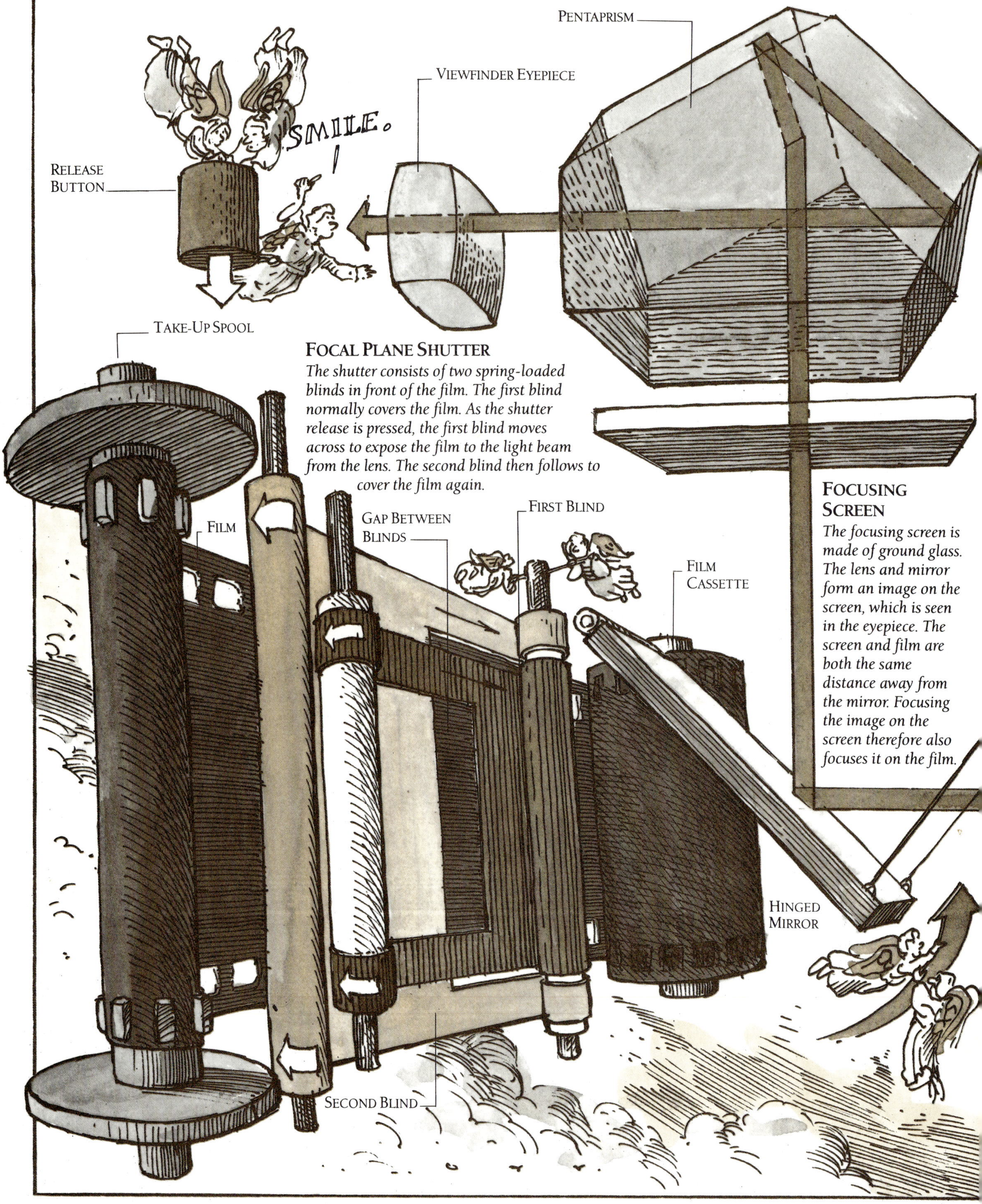

FOCAL PLANE SHUTTER

The shutter consists of two spring-loaded blinds in front of the film. The first blind normally covers the film. As the shutter release is pressed, the first blind moves across to expose the film to the light beam from the lens. The second blind then follows to cover the film again.

FOCUSING SCREEN

The focusing screen is made of ground glass. The lens and mirror form an image on the screen, which is seen in the eyepiece. The screen and film are both the same distance away from the mirror. Focusing the image on the screen therefore also focuses it on the film.

In some cameras, two different sets of lenses are used one to view the image, and one to throw it onto the film. With this kind of camera, focus and aperture have to be set by measurement rather than by eye. The single-lens reflex (SLR) camera is so named because it uses a single collection of lenses both for viewing and for taking the picture. The SLR camera has a hinged mirror that rests at an angle of 45 degrees in front of the film, and reflects the light beam from the lens onto a focusing screen above the mirror. The image forms on this screen, and light from it is reflected by the faces of the pentaprism into the viewfinder eyepiece. The various reflections turn the viewfinder image upright and the right way round. When the shutter release is pressed, the mirror rises and the light beam strikes the film.

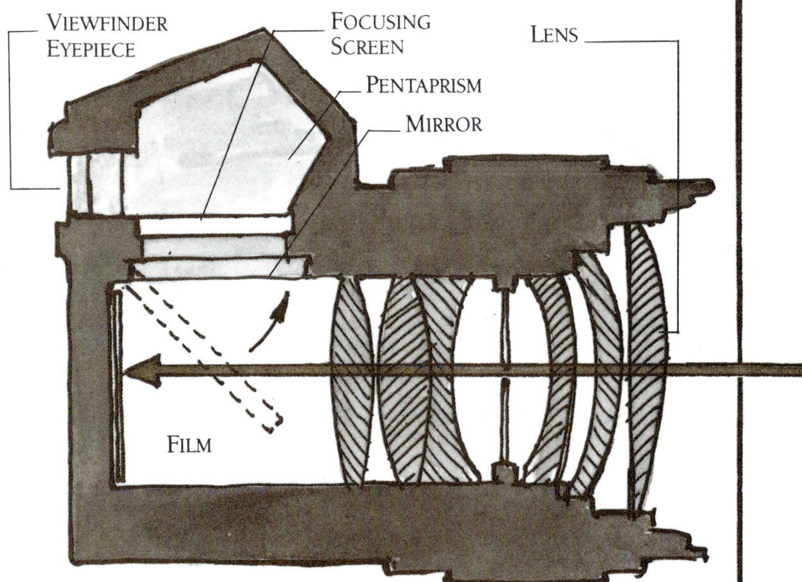

IRIS DIAPHRAGM
This controls the aperture of the lens, regulating the amount of light that enters the camera. The diaphragm is at the centre of the lens, and contains a set of hinged blades that move to open or close the central hole. Opening the diaphragm, which gives a lower f number, lets more light in.

LENS
A high-quality lens is made up of several lens elements that work together to form a sharp image on the film. The field of view depends on the focal length of the lens, which is the distance from the lens to the film when an object at infinity is in focus. The lens surfaces are often coated to reduce reflections.

COLOR PHOTOGRAPH

No matter how multicolored prints or slides may appear, they are made of only the three secondary colors (see p.195) arranged in layers. When you look at a photograph, light passes through the layers and combines to give full color. Developing a print film produces a color negative, while in a slide, a process called color reversal (*below*) forms a positive color image on the film.

1 UNEXPOSED SLIDE FILM
The film contains three color-sensitive layers. The one in the middle is sensitive to green.

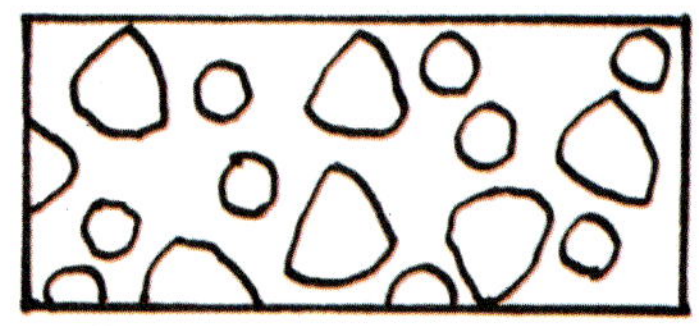

2 FIRST DEVELOPER
Magenta light contains no green. It does not expose the layer, so silver develops.

3 COLOR DEVELOPER
This attaches a magenta dye to the silver particles.

4 SILVER DISSOLVED
Dissolving the silver leaves the layer colored magenta.

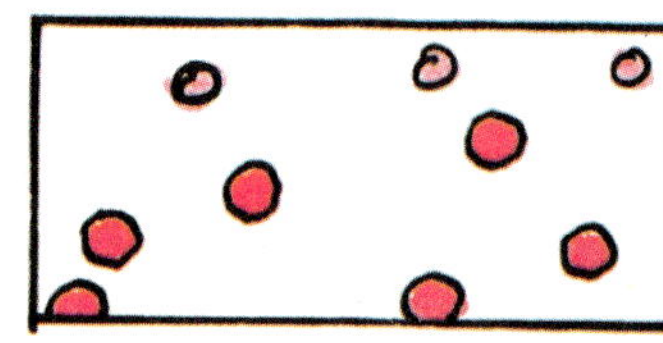

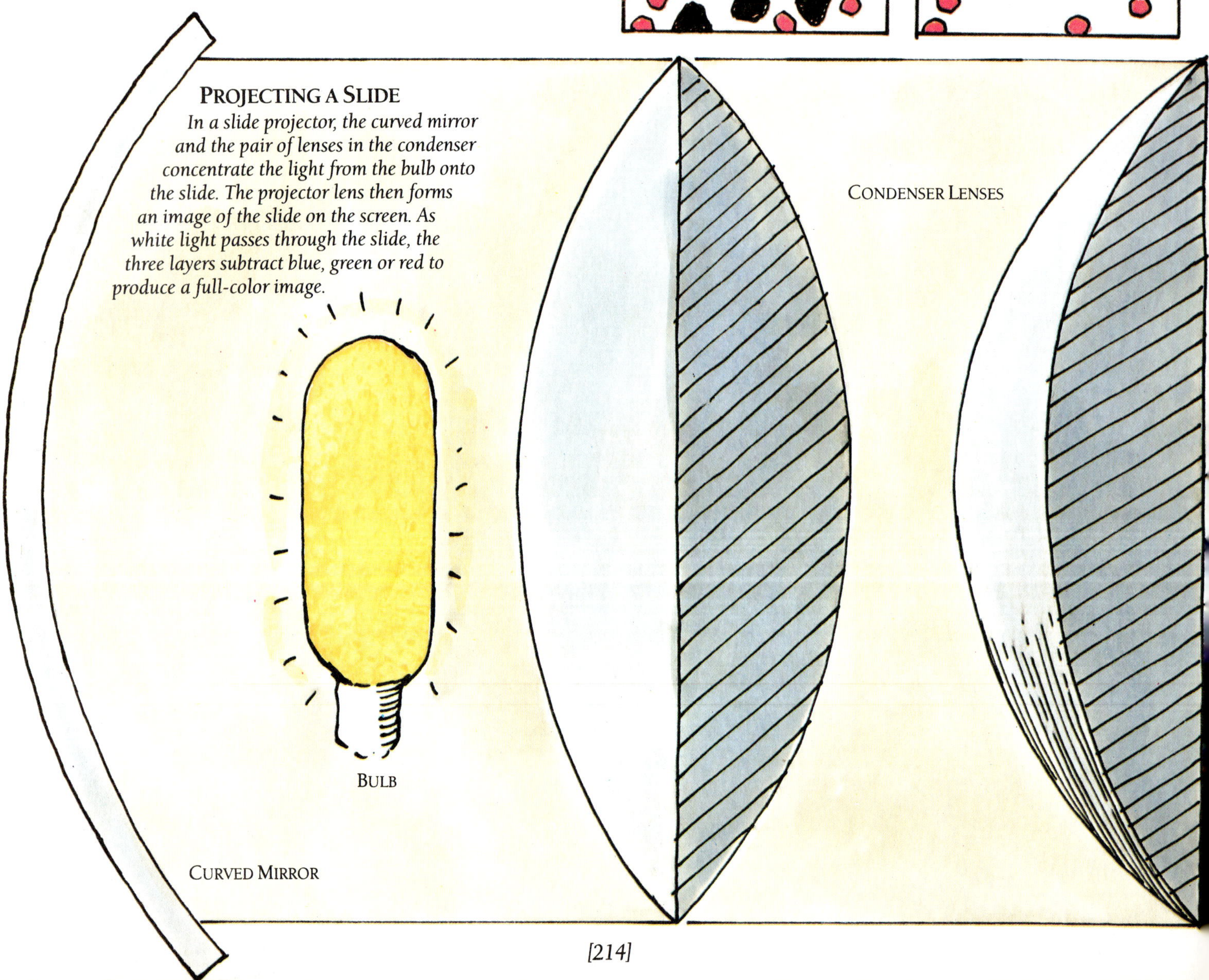

PROJECTING A SLIDE
In a slide projector, the curved mirror and the pair of lenses in the condenser concentrate the light from the bulb onto the slide. The projector lens then forms an image of the slide on the screen. As white light passes through the slide, the three layers subtract blue, green or red to produce a full-color image.

THE THREE-LAYER SYSTEM

Each of the color-sensitive layers in a color film is similar to a black-and-white film, except that the top layer is sensitive only to blue light, the middle layer to green and the bottom layer to red. The three layers detect the amounts of these colors in the image formed on the color film by the camera lens.

Developers for color films contain dye couplers, which attach dyes to the silver that forms in the emulsion during development. The silver is then dissolved, leaving a layer of dye. The top layer becomes yellow, the middle layer magenta and the bottom layer cyan.

In color slide film, color reversal, shown on the opposite page, changes the *unexposed* layers into layers of dye. Green, for example, exposes only the middle layer, so the first and third layers become yellow and cyan. These two layers mix to give green. Each piece of film becomes a color slide.

In color print film, the *exposed* layers are changed into layers of dye. Yellow, for example, appears as a mixture of magenta and cyan which gives blue. The negative is then printed on color paper, which contains the same three layers as a film, and which is developed in the same way.

ALMOST INSTANT PICTURES

Immediately on taking a photograph with an instant camera, a plastic sheet emerges. After a minute, the picture begins to appear and a little later a full-color photograph is in your hand. Instant photography uses basically the same process as a color slide (see p.214). The film contains layers of silver compounds sensitive to blue, green and red light that develop to give layers of yellow, magenta and cyan dyes. It also contains the developing chemicals. As an instant photograph develops, the three dyes leave their layers and move up through the film. They collect in a layer beneath the surface, where the dyes mix to form the picture.

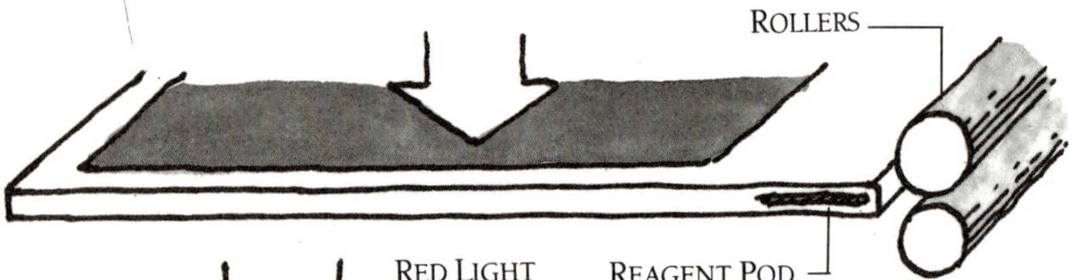

1 FILM EXPOSED

The film contains nine separate chemical layers. When it is exposed, light penetrates to the layers of silver compounds sensitive to blue, green and red light. Here, red light is shown exposing the red layer.

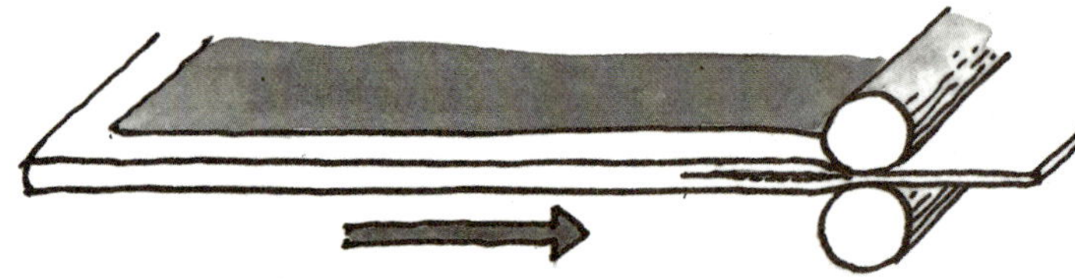

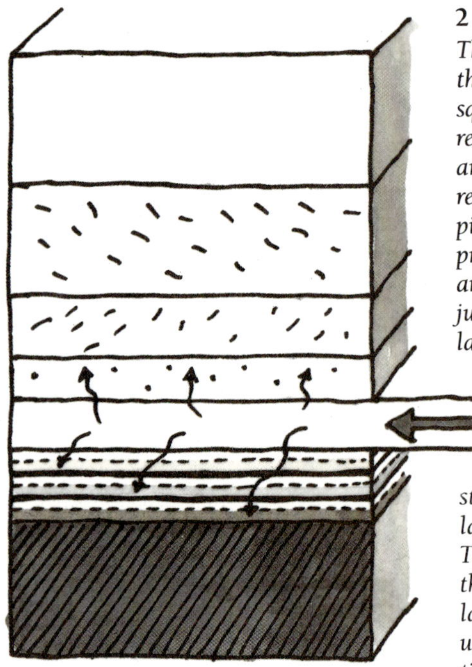

2 FILM LEAVES CAMERA

The sheet of film passes through the rollers, which squeeze a thin pod of reagent in the borders around the picture area. The reagent contains white pigment, opacifiers that prevent light passing, water and alkali. It enters the film just above the light-sensitive layers, and the opacifiers stop light getting to these layers as the film emerges. The reagent spreads down through the light-sensitive layers, but its progress upward is slowed by the timing layer.

3 IMAGE BEGINS TO FORM

The reagent activates the dye developers in the light-sensitive layers. Where the silver compounds were not exposed, the dyes dissolve and begin to diffuse through the other layers to the image layer. But in the exposed layers, the developer produces silver, which prevents the dye moving. Here, the cyan dye in the exposed red-sensitive layer cannot move. Yellow and magenta dyes formed in the other two unexposed layers diffuse upward.

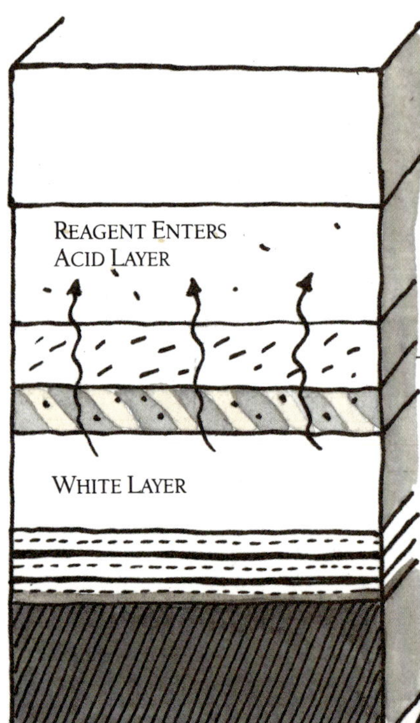

4 IMAGE CLEARS

The reagent reaches the acid layer, which neutralizes the alkali and makes the opacifiers clear. The dyes in the image layer are seen against the white pigment, giving a clear picture.

OTO BOOTH

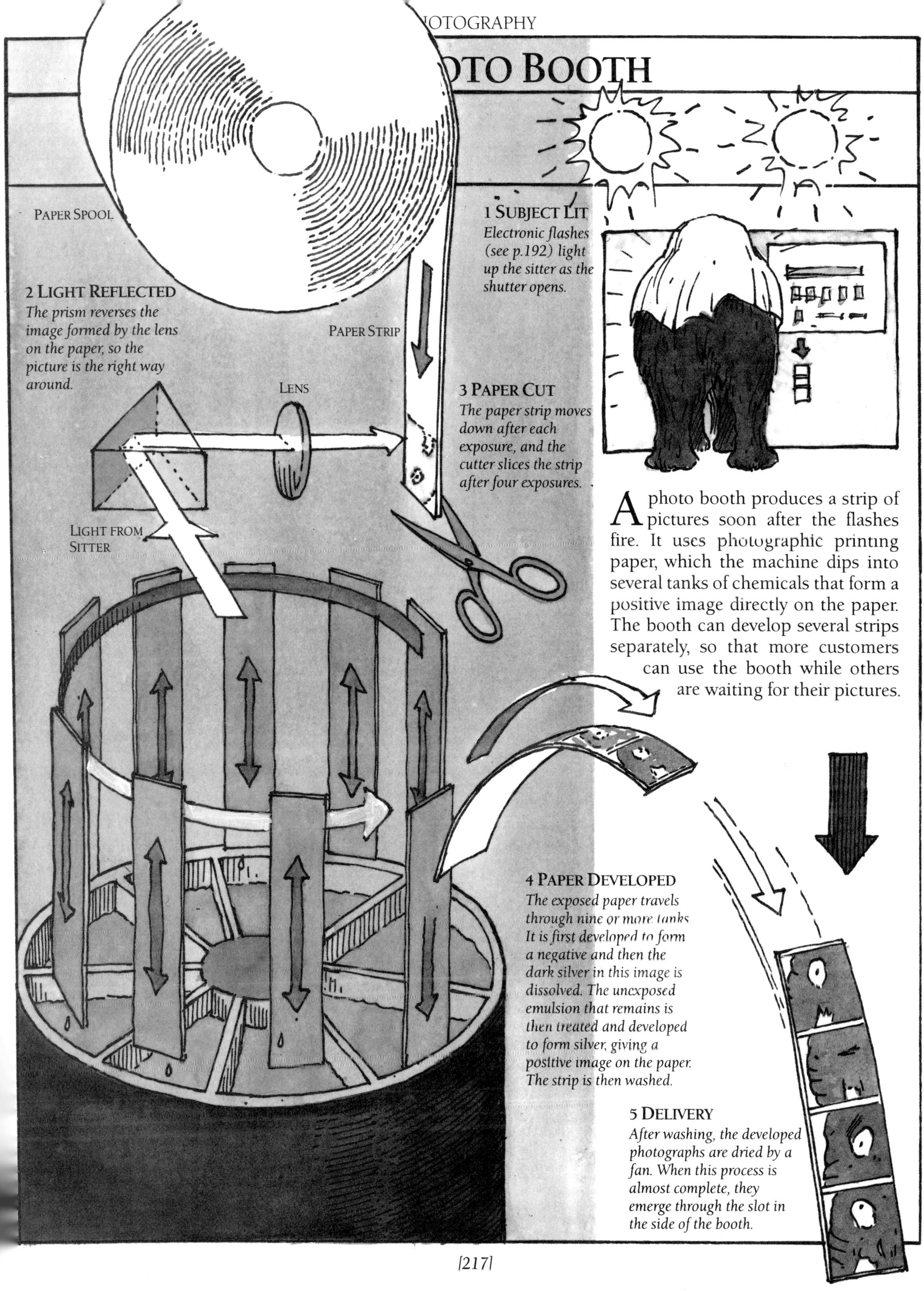

A photo booth produces a strip of pictures soon after the flashes fire. It uses photographic printing paper, which the machine dips into several tanks of chemicals that form a positive image directly on the paper. The booth can develop several strips separately, so that more customers can use the booth while others are waiting for their pictures.

MOVIE CAMERA

Moving pictures depend on illusion, not only in the acting but in our eyes. A movie film consists of a series of still pictures seen in rapid succession. After it leaves the screen, each image persists in the eye until the next image arrives. Our eyes merge the separate images together to give an illusion of motion.

A movie camera photographs a sequence of still images on a strip of film. It normally takes 24 pictures or frames every second. The film is still when each picture is taken, and moves on between exposures.

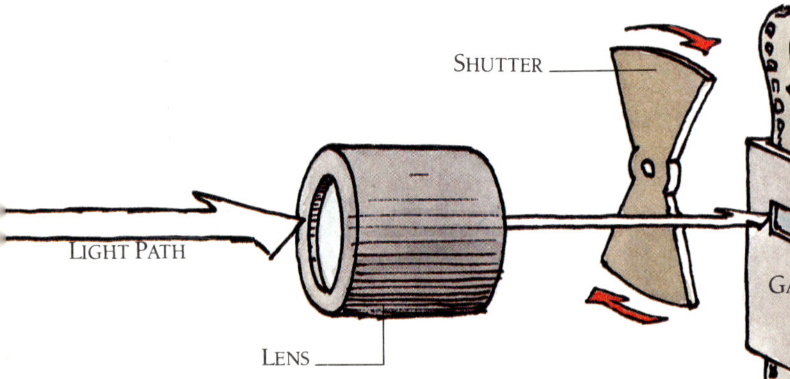

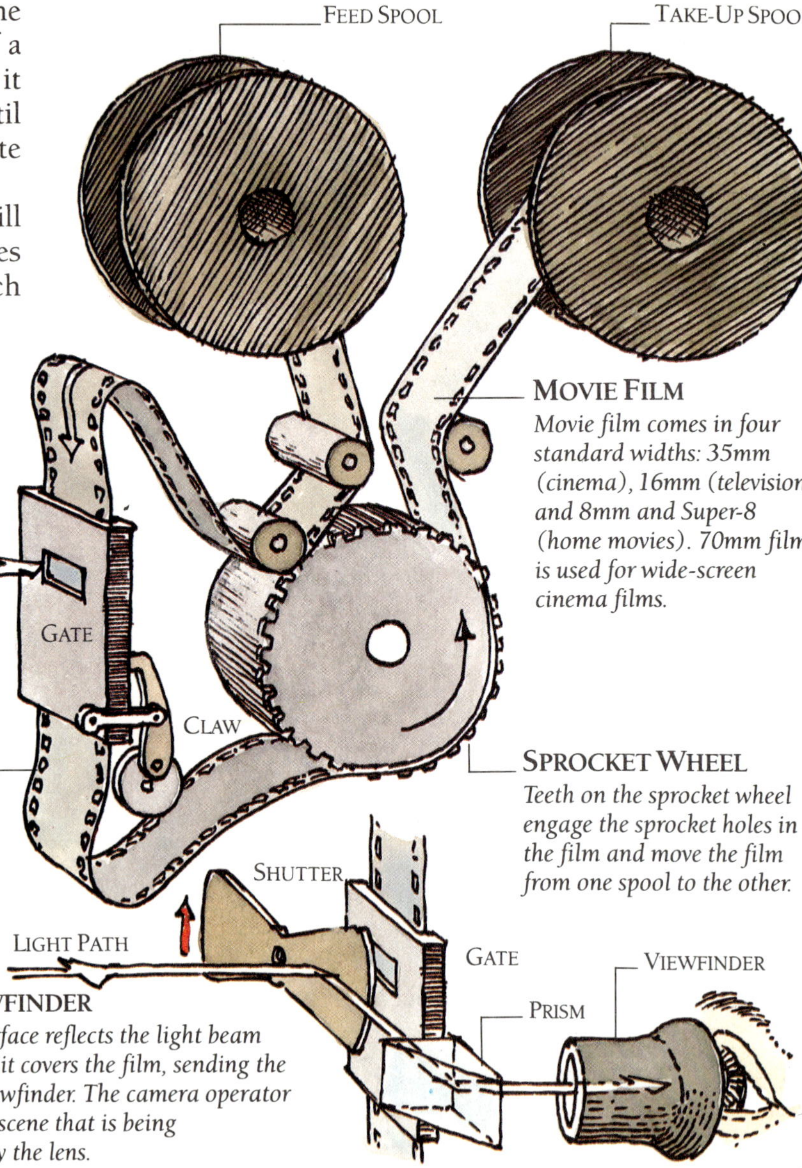

MOVIE FILM
Movie film comes in four standard widths: 35mm (cinema), 16mm (television) and 8mm and Super-8 (home movies). 70mm film is used for wide-screen cinema films.

FILM LOOP
The film makes a loop before and after the gate. The loops enable the film to be moved intermittently through the gate by the claw.

SPROCKET WHEEL
Teeth on the sprocket wheel engage the sprocket holes in the film and move the film from one spool to the other.

REFLEX VIEWFINDER
The shutter's surface reflects the light beam from the lens as it covers the film, sending the light into the viewfinder. The camera operator sees the moving scene that is being photographed by the lens.

MAKING MOVIES

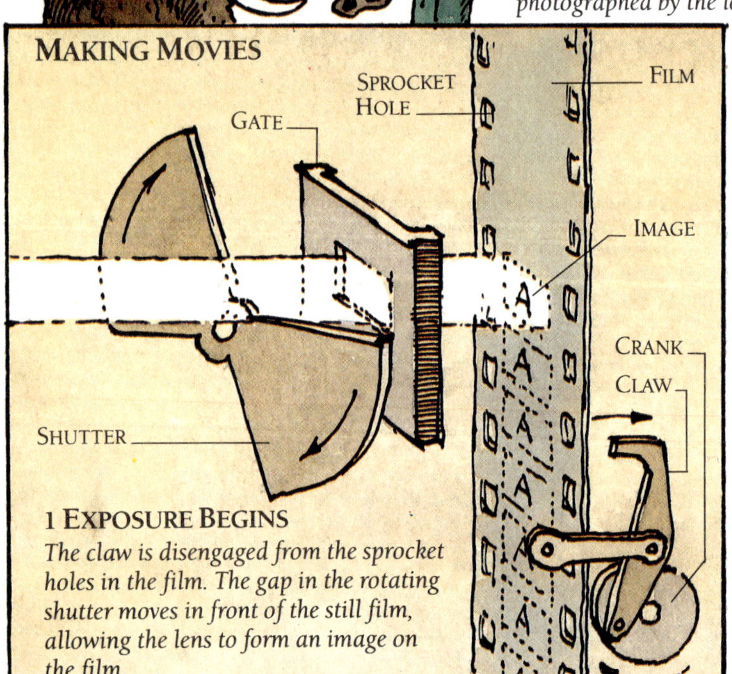

1 EXPOSURE BEGINS
The claw is disengaged from the sprocket holes in the film. The gap in the rotating shutter moves in front of the still film, allowing the lens to form an image on the film.

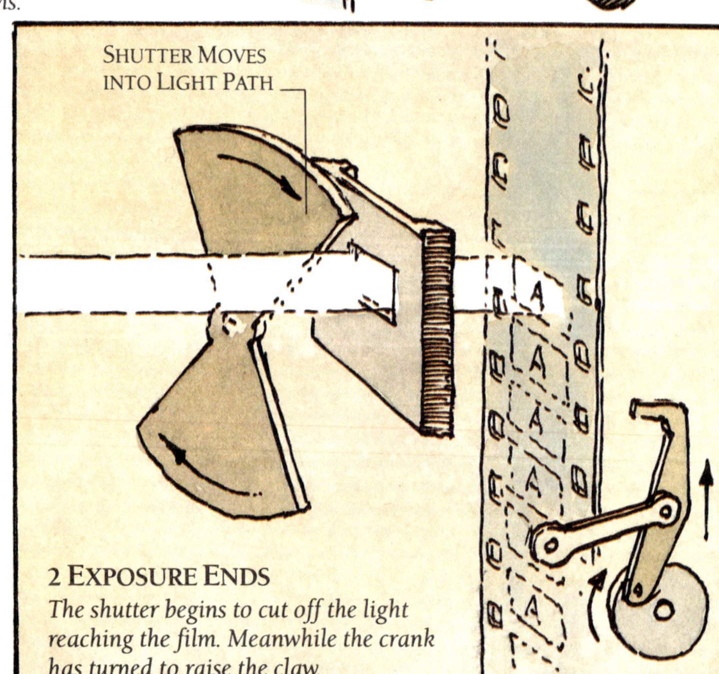

2 EXPOSURE ENDS
The shutter begins to cut off the light reaching the film. Meanwhile the crank has turned to raise the claw.

MOVIE PR

A movie projector works like a movie camera in reverse. A claw moves the film intermittently through the projector gate, allowing the lens to project a still image on the screen for a fraction of a second. A rotating shutter allows light to reach the film, and then cuts off the light source so that the image disappears as the film moves. As in the movie camera, the projector shows 24 frames in every second. But it projects each frame twice so that the eye sees 48 separate pictures, which reduces flicker.

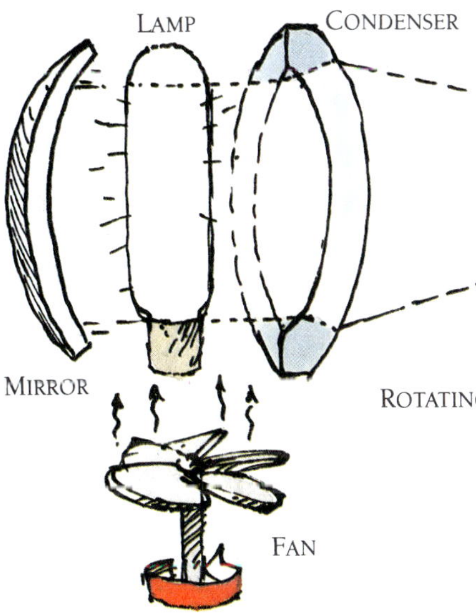

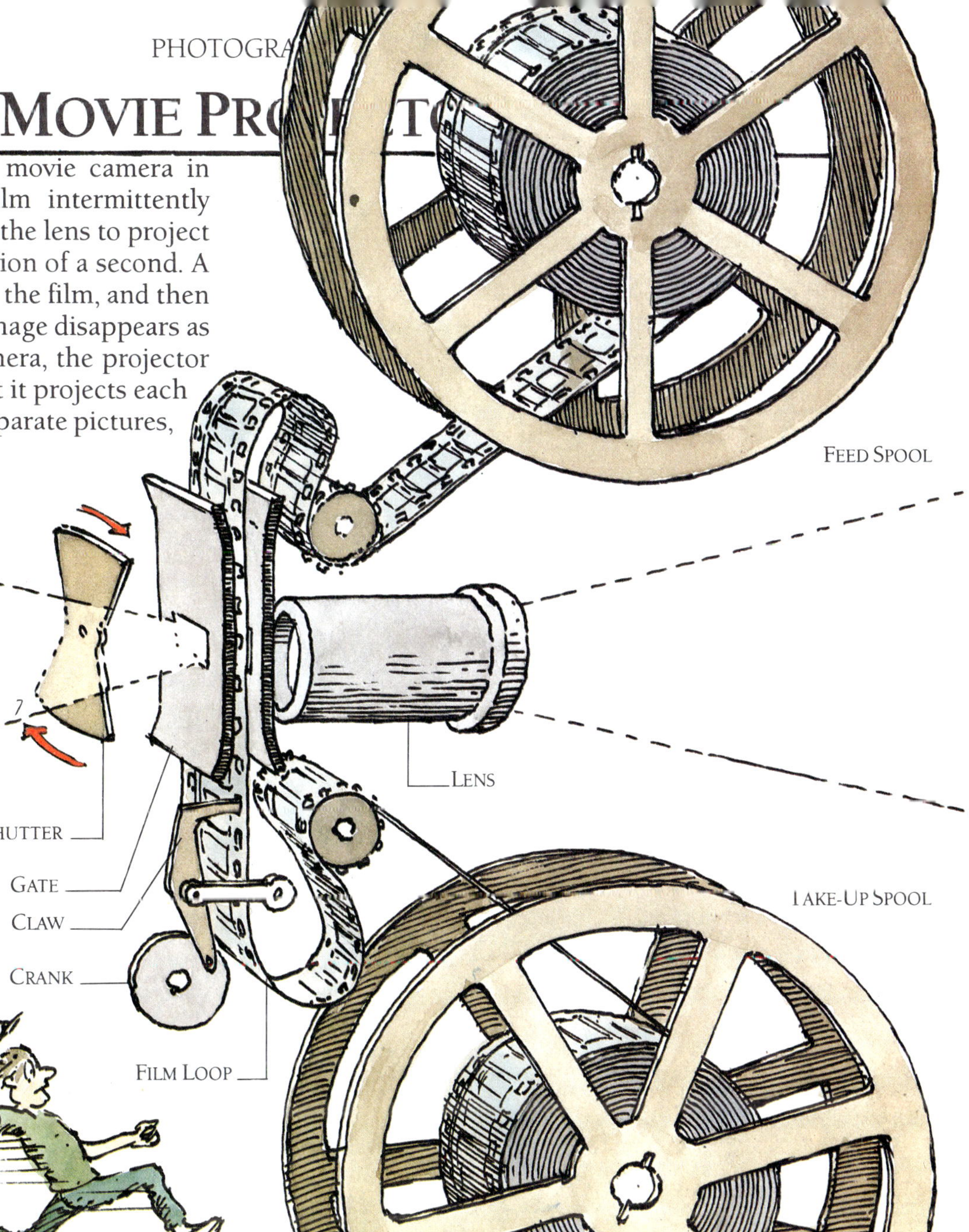

LIGHT SOURCE

The light from a bright lamp is reflected by a curved mirror and concentrated by a condenser to illuminate the film. A fan cools the lamp.

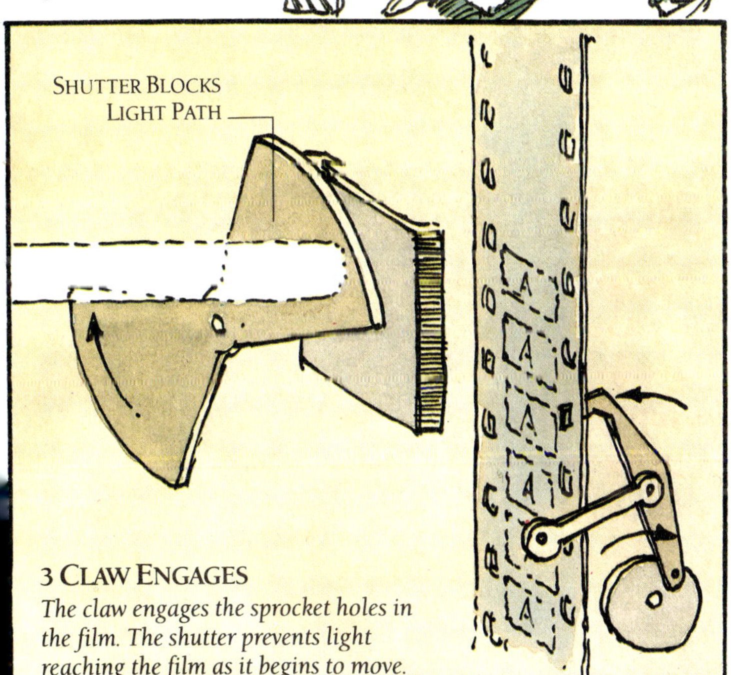

3 CLAW ENGAGES

The claw engages the sprocket holes in the film. The shutter prevents light reaching the film as it begins to move.

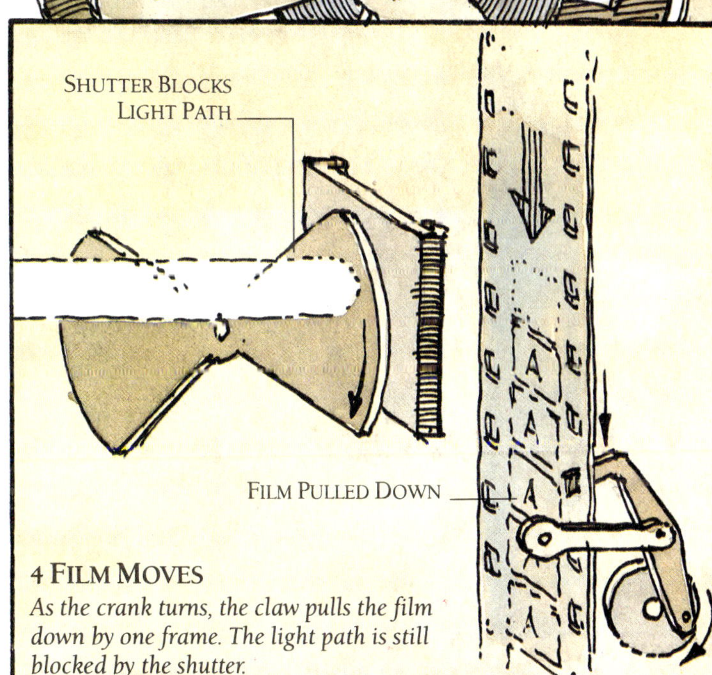

4 FILM MOVES

As the crank turns, the claw pulls the film down by one frame. The light path is still blocked by the shutter.

PRINTING

MODERN METHODS OF PRINTING

The mammoth mint works by a printing process known as letterpress, the oldest of the three main methods now in use. The other two are gravure and lithography. Here we show flat printing plates for simplicity. Printing presses often have curved plates that rotate to print multiple copies on sheets or strips of paper, but the principles involved are the same.

LETTERPRESS

1 The plate has raised letters.

2 Ink sticks to the letters.

3 Paper is pressed against the plate.

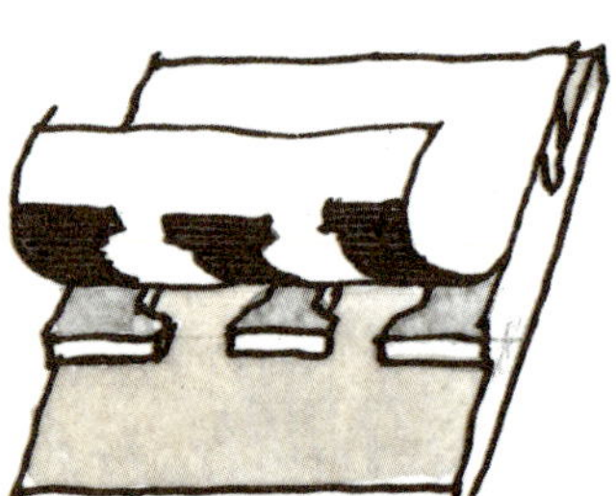

4 The ink transfers to the paper.

GRAVURE

1 The plate has recessed letters.

2 Ink fills the letter recesses.

3 Paper is pressed against the plate.

4 The ink transfers to the paper.

ON A MAMMOTH MINT

Following a rash of particularly skillful boulder forging, I was asked to suggest a more secure and if possible more portable medium of exchange. The result was my mammoth mint.

High quality leaves of predetermined size were carefully centered on a mat, one at a time, by trunk suction. A large pad containing a herbal dye of my own concoction was kept at the required level of moisture by the chief squirter. After pressing a patterned stamp down on the pad, the master minter then brought the same stamp down onto the pre-centered leaf transferring the impression. Each leaf was then thoroughly dried, checked and counted before shipment to one of several mammoth banks.

Although technically flawless, the mint suffered insurmountable staffing difficulties. Losses which I had initially attributed to pilfering were later explained by the toothsome character of the new currency to mammoths.

The processes here show printing in a single color. In full-color printing, a number of inks are applied one after the other by different rollers. Three colored inks, together with black, can produce a complete range of hues by the process of color subtraction.

LITHOGRAPHY

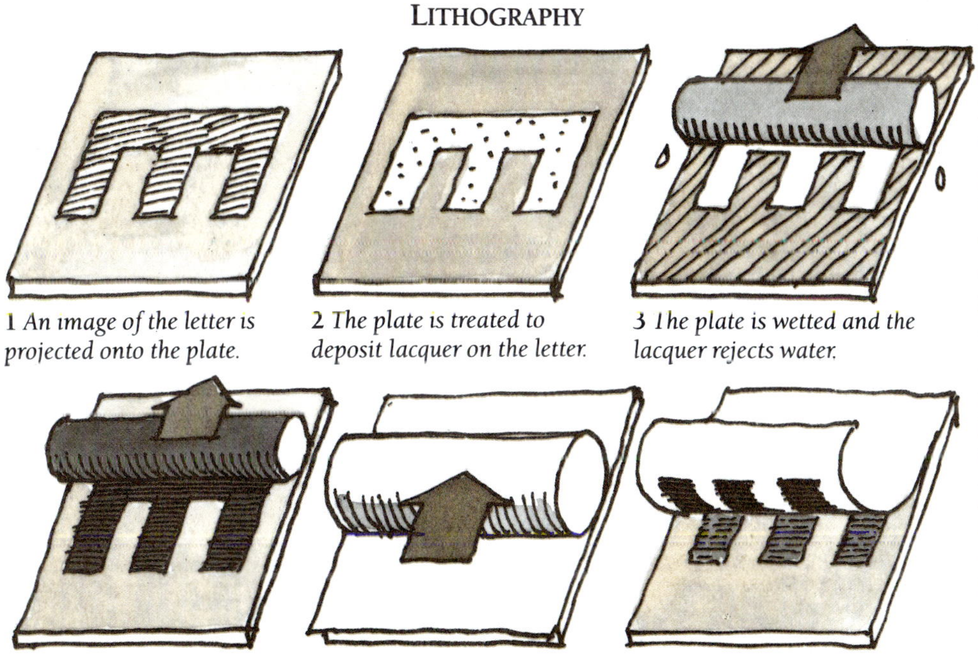

1 *An image of the letter is projected onto the plate.*

2 *The plate is treated to deposit lacquer on the letter.*

3 *The plate is wetted and the lacquer rejects water.*

4 *The plate is inked. The lacquer accepts ink, but the wet surface rejects ink.*

5 *Paper is pressed against the printing plate.*

6 *The ink transfers from the lacquer to the paper.*

PRINTING ON STONE

Lithography first used stone printing plates, hence its name which means "writing on stone". An artist can draw on the stone with a greasy substance that attracts ink, and the ink is transferred to paper in a press. Modern litho printing uses light-sensitive plates on which text and images can be deposited by photography. In offset litho printing, the ink is first transferred to a cylinder that then prints the paper.

PAPERMAKING

Printing is of little use without paper. A sheet of paper is a flattened mesh of interlocking plant fibers, mainly of wood and cotton. Making paper involves reducing a plant to its fibers, and then aligning them and coating the fibers with materials such as glues, pigments and mineral fillers.

1 FELLING
Trees are felled and then transported to paper mills as logs.

2 DEBARKING
The bark has first to be stripped off the logs without damaging the wood.

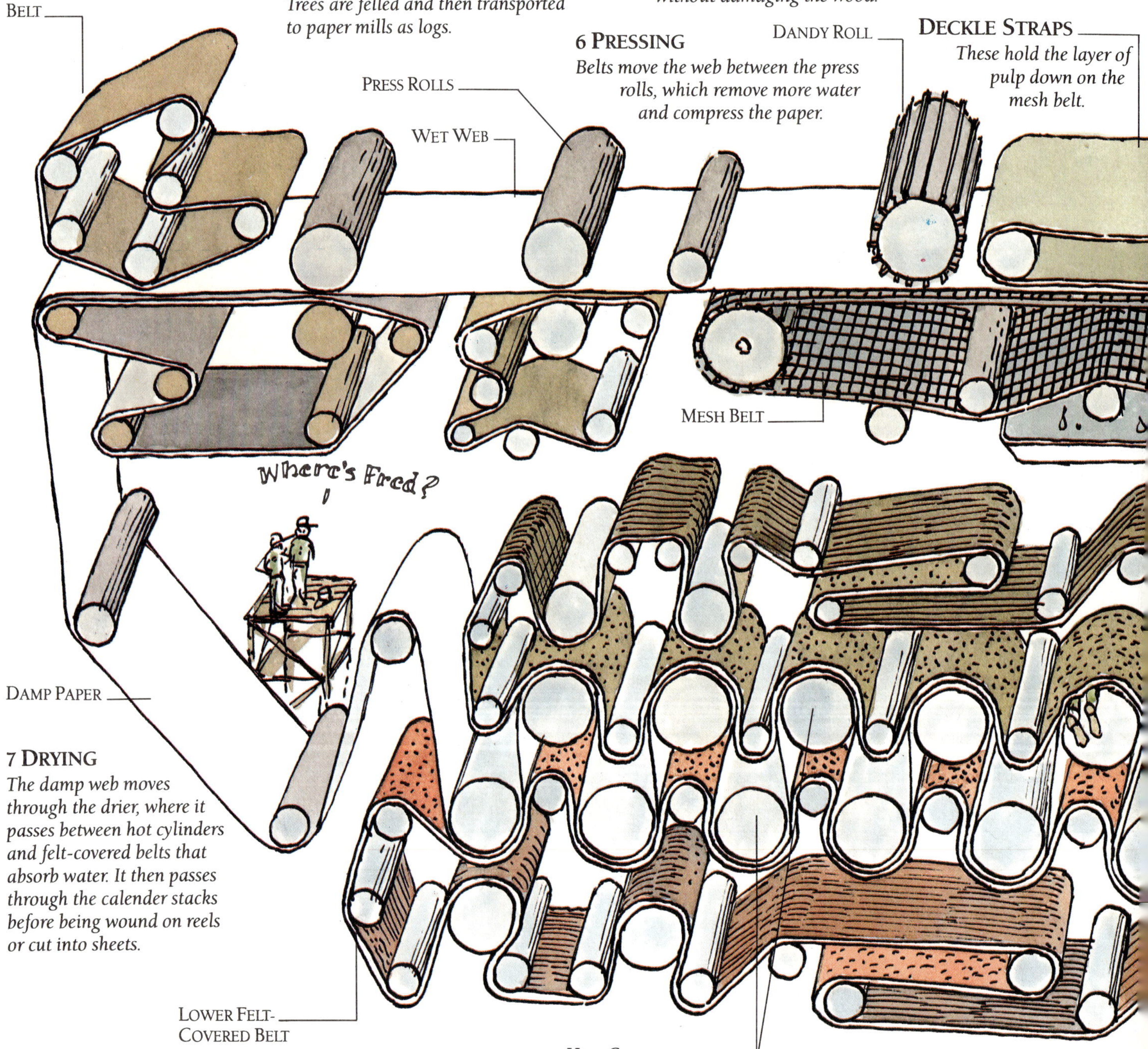

6 PRESSING
Belts move the web between the press rolls, which remove more water and compress the paper.

7 DRYING
The damp web moves through the drier, where it passes between hot cylinders and felt-covered belts that absorb water. It then passes through the calender stacks before being wound on reels or cut into sheets.

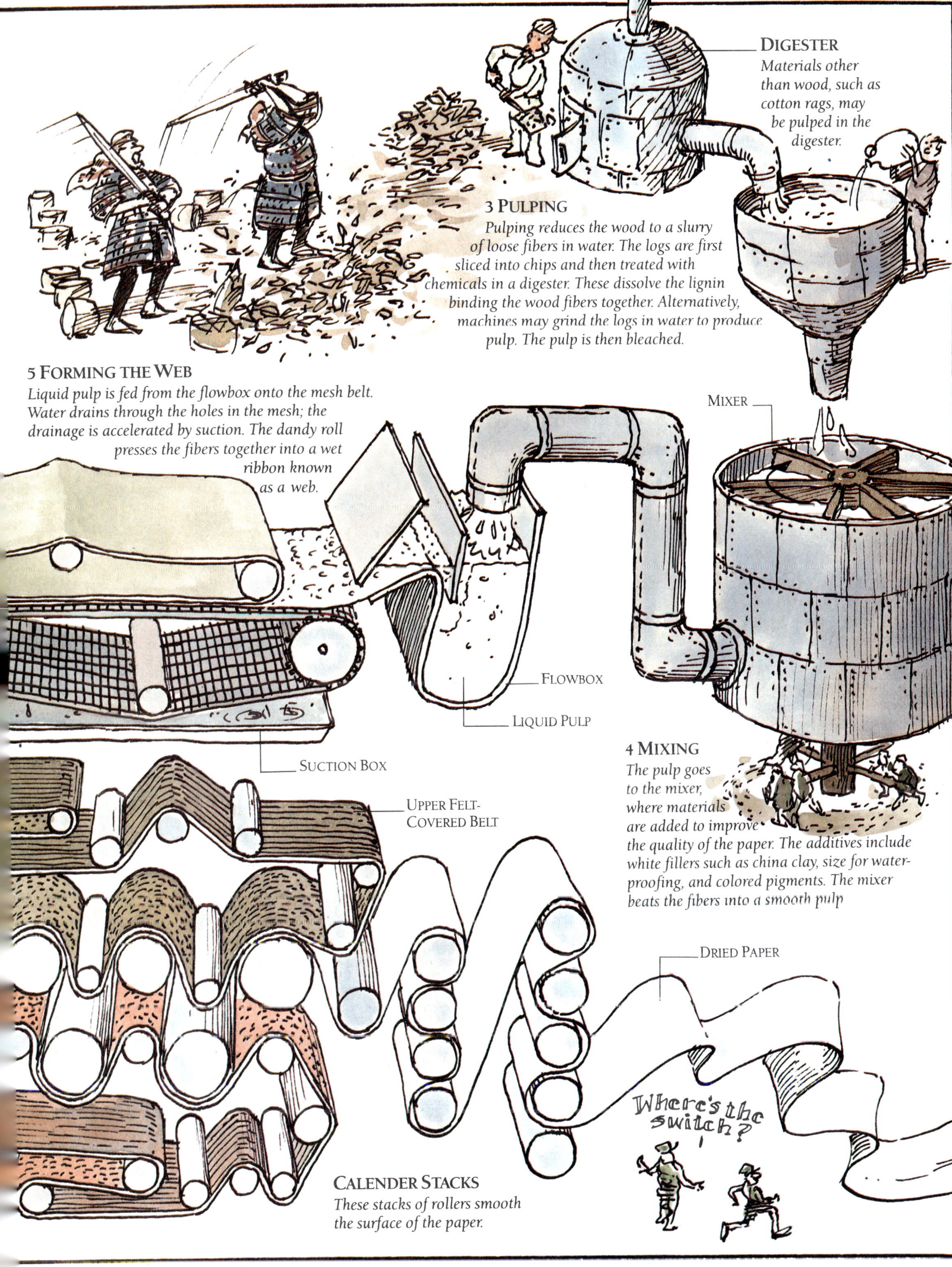

DIGESTER
Materials other than wood, such as cotton rags, may be pulped in the digester.

3 PULPING
Pulping reduces the wood to a slurry of loose fibers in water. The logs are first sliced into chips and then treated with chemicals in a digester. These dissolve the lignin binding the wood fibers together. Alternatively, machines may grind the logs in water to produce pulp. The pulp is then bleached.

5 FORMING THE WEB
Liquid pulp is fed from the flowbox onto the mesh belt. Water drains through the holes in the mesh; the drainage is accelerated by suction. The dandy roll presses the fibers together into a wet ribbon known as a web.

4 MIXING
The pulp goes to the mixer, where materials are added to improve the quality of the paper. The additives include white fillers such as china clay, size for waterproofing, and colored pigments. The mixer beats the fibers into a smooth pulp

CALENDER STACKS
These stacks of rollers smooth the surface of the paper.

PRINTING PLATE

When a book like *The Way Things Work* is printed, all the colors in it are produced by just four inks. These are the three secondary colors – yellow, magenta and cyan (see p.195) – and black. Each ink is printed by a separate printing plate. Producing these plates involves two processes, color scanning and typesetting. Color scanning breaks the illustrations down into the three secondary colors and black, while typesetting converts the text into print.

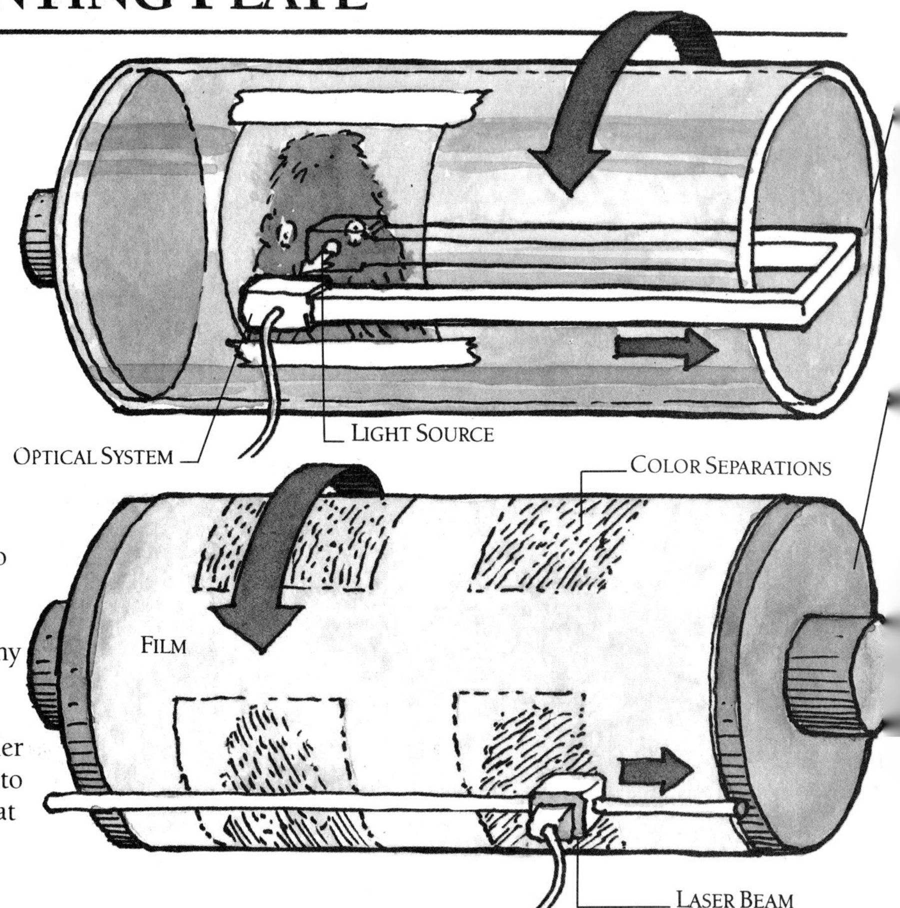

COLOR SCANNING

A color scanner breaks down each illustration into the four separate images or separations which together will re-create the full-color picture. In each separation, the picture is broken up into many separate dots. The computer in the scanner controls the size of the dots, producing large dots where the ink must be dense and small dots in other parts of the separation. Breaking the picture up into dots enables the printing plate to print any color at any intensity.

TYPESETTING

The text of this book has been set into print using a laser typesetter. The operator keys in the text, together with codes that specify the typeface and type size. The operator's keyboard is linked to a computer whose memory contains all the type characters in a variety of different typefaces and sizes. The computer then sends signals to the typesetter, in which a laser beam forms lines of text on film or paper.

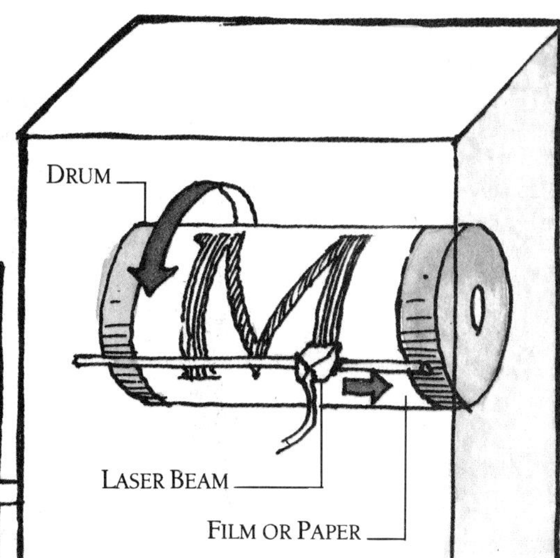

KEYBOARD

The text appears on a screen as the operator keys it into the typesetter. Alternatively, a disk from the author's word processor may by inserted directly into the typesetter.

COMPUTER

The computer converts the signals produced by the keyboard into signals that operate the laser.

LASER TYPESETTER

Photographic film or paper rotates on a drum as a laser beam moves across the film or paper. The computer switches the beam on and off so that each character is formed by a set of close-spaced vertical lines. The film or paper is then developed.

SCANNING DRUM
The picture – either a color transparency or piece of artwork – rotates on the scanning drum. The optical system moves across the picture, detecting its color and brightness and breaking them down into a large number of dots. Signals from the optical system go to the scanner's computer where they are stored.

RECORDING DRUM
Signals from the computer operate the recording drum. Film is placed around the drum, which rotates. The computer-controlled laser beam moves across the film to create four separations made of lines of dots.

COMPLETED SEPARATIONS
The yellow, magenta and cyan separations are black-and-white images formed by the amounts of each color in the original picture. The black separation is an ordinary black-and-white image of the picture. The lines of dots are scanned at different angles to prevent patterns becoming visible in the printed picture.

PLATE PRODUCTION

The printing plates are made from negative or positive films containing the text and color separations. The black film (shown here) may contain both the black separation and the text, or each may have its own film. The plate is coated with a light-sensitive substance. Light is shone through the film to expose the plate, which is then developed so that the text and picture form in the coating. The plate is then treated with chemicals, which penetrate parts of the coating and create the text characters and picture dots on the plate. The kind of chemical treatment depends on whether the plate is for letterpress, gravure or litho printing. Three more printing plates are then made from the other color separations in the same way. Each plate may contain a number of pages.

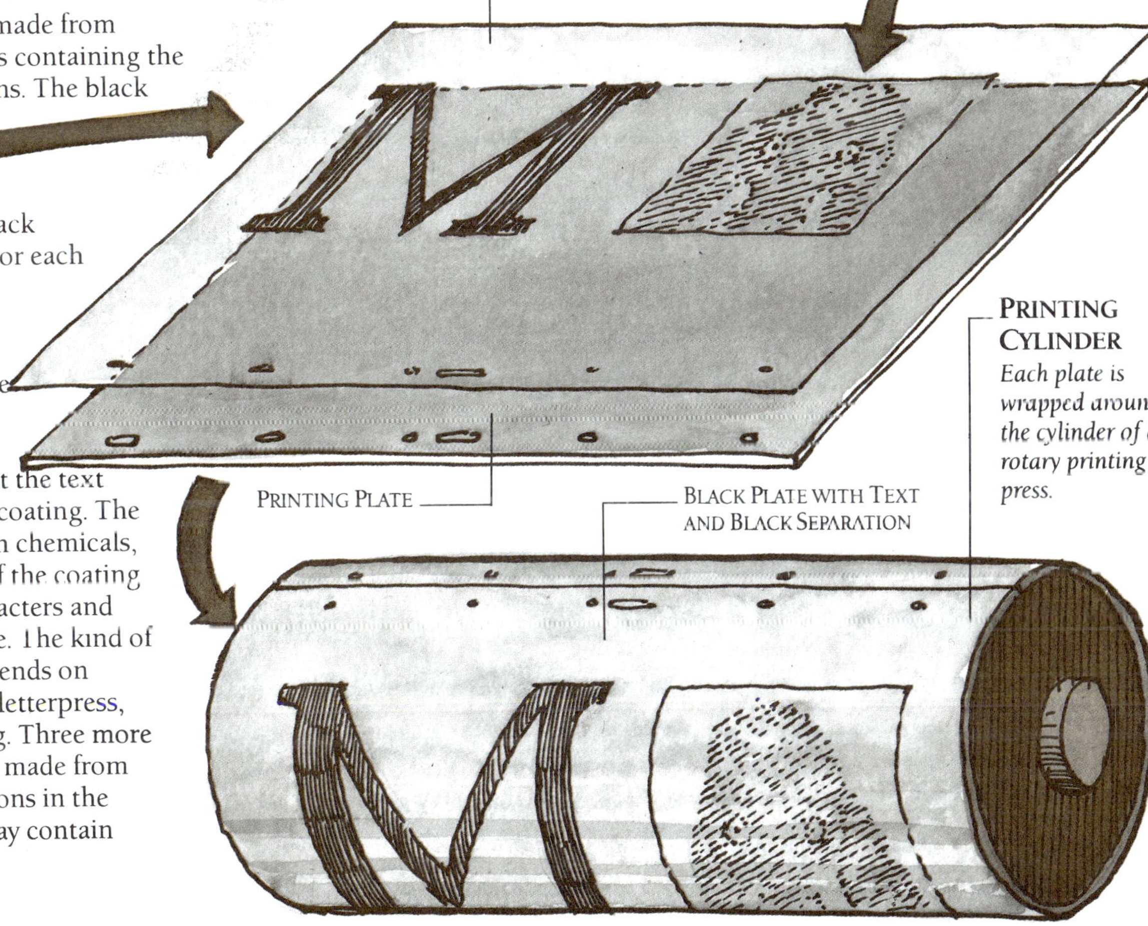

PRINTING CYLINDER
Each plate is wrapped around the cylinder of a rotary printing press.

PRINTING PRESS

The printing press, as its name implies, prints by pressing paper against an inked plate. Large printing presses are rotary machines in which the printing plate is fitted around a cylinder. As the cylinder rotates, cut sheets or a web (continuous strip) of paper pass rapidly through the press and are printed while on the move. Presses which print in color have four or more printing cylinders so that the color separations are printed immediately one after the other. Quick-drying inks prevent smudging.

SHEET-FED OFFSET PRESS

This book, like many books and magazines, has been printed by offset lithography, a process which combines speed with quality printing. Sheet-fed presses are mainly used for printing books because print quality is very high. Sheets of paper are fed into the press and pass through four printing units that print in cyan, magenta, yellow and black. The three colors form color pictures, while the black plate adds contrast to the pictures and prints black text. The sheets are printed first on one side, and are then fed back into the machine for printing on the reverse side.

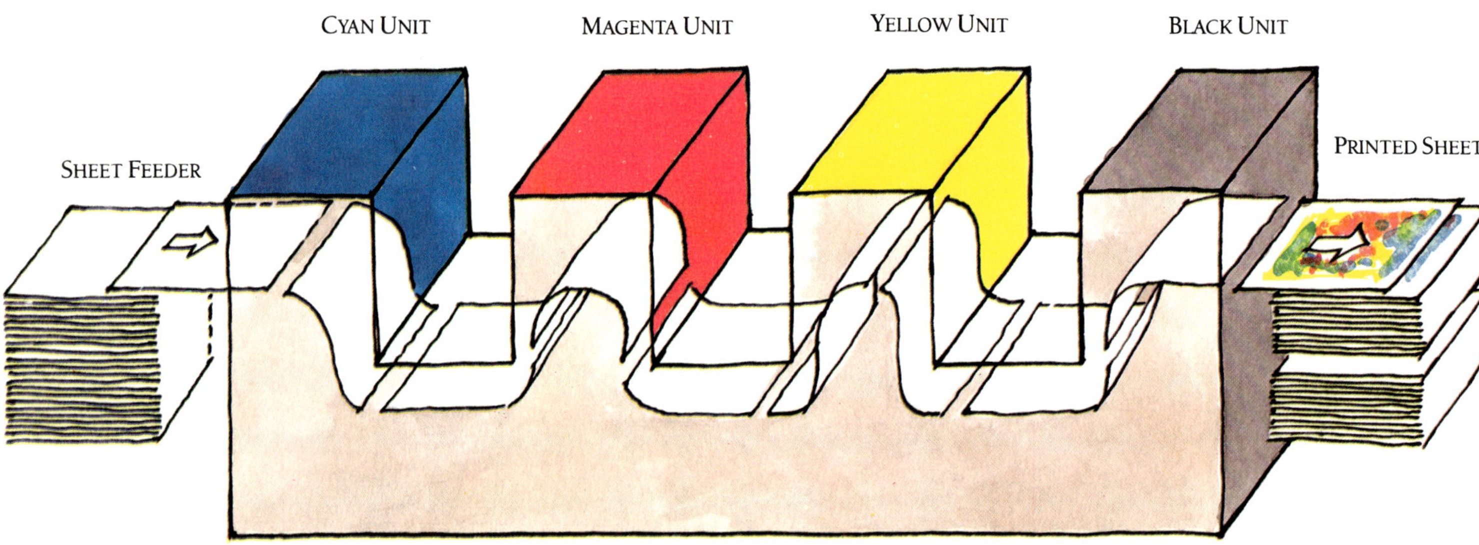

WEB OFFSET PRESS

Web offset presses achieve very high speed as well as good quality, and are often used to print magazines and newspapers. Large reels feed the web into the press, which is then printed with four or more colors. Each printing unit usually contains two sets of printing cylinders so that both sides of the paper are printed at the same time. After leaving the press, the web continues on to folding and cutting machines (see p.228)

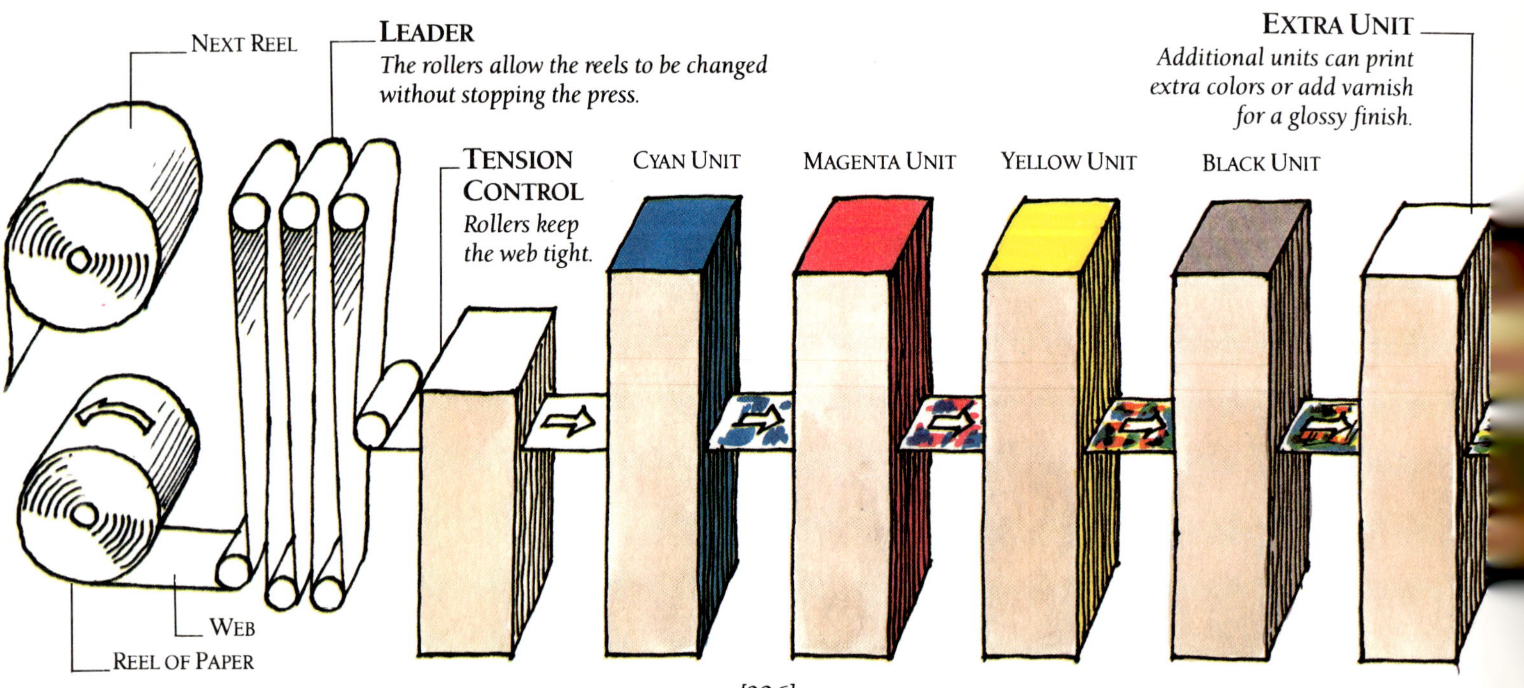

Offset Litho Printing

The dampening roller first wets the printing plate on the plate cylinder, which is then inked by the inking rollers. The ink then transfers to the rubber blanket cylinder, and the impression cylinder presses the paper against the blanket cylinder. The blanket cylinder prints the paper, its flexible rubber surface overcoming any irregularities in the paper.

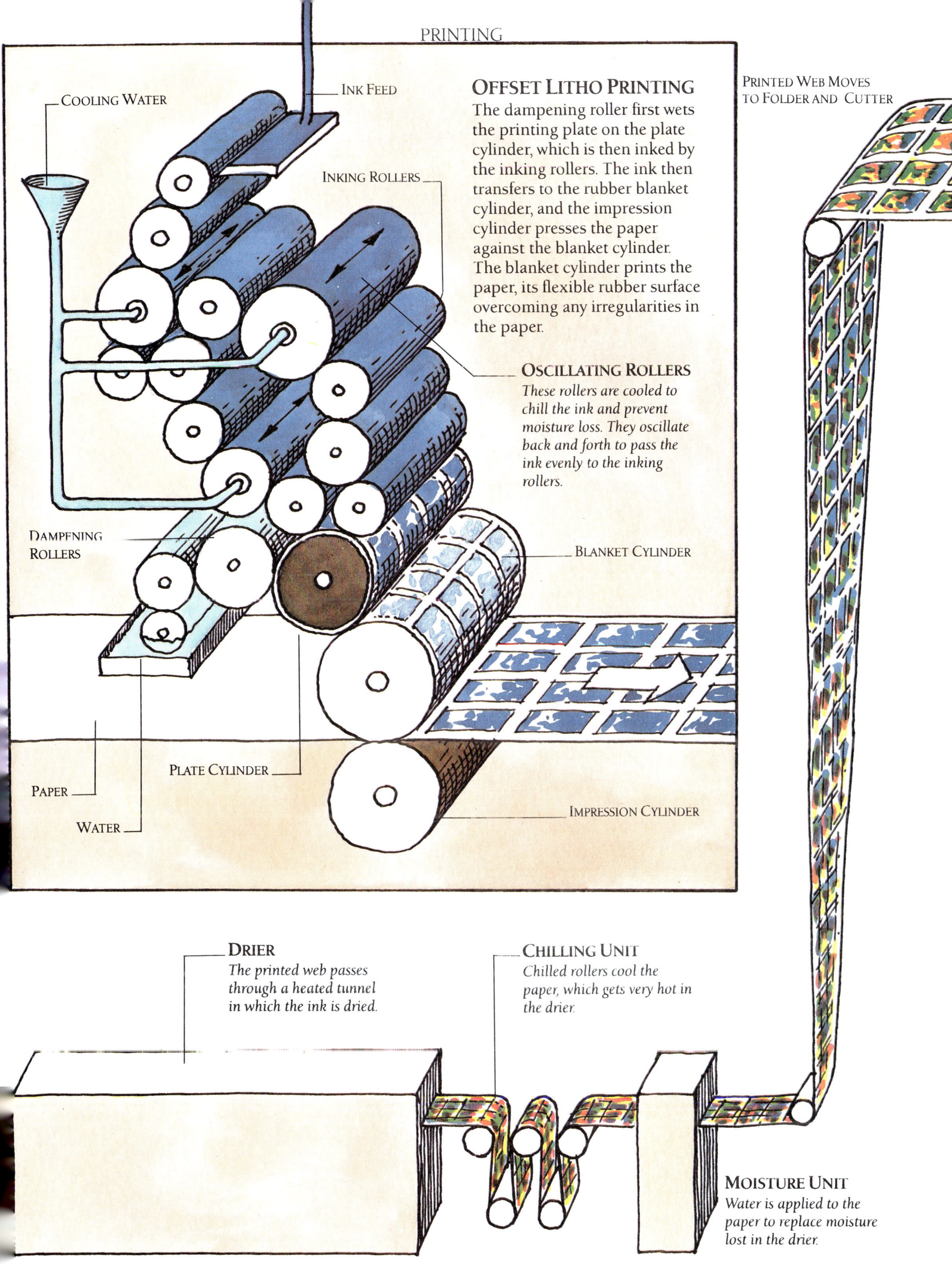

Oscillating Rollers
These rollers are cooled to chill the ink and prevent moisture loss. They oscillate back and forth to pass the ink evenly to the inking rollers.

Drier
The printed web passes through a heated tunnel in which the ink is dried.

Chilling Unit
Chilled rollers cool the paper, which gets very hot in the drier.

Moisture Unit
Water is applied to the paper to replace moisture lost in the drier.

BOOKBINDING

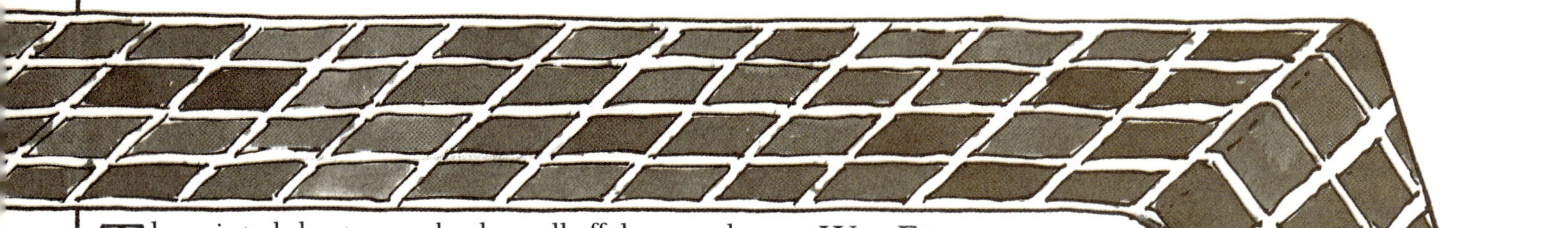

The printed sheets or webs that roll off the press have to be folded and, if necessary, cut to produce sections of the book called signatures. Then all the signatures in the book must be collated, or assembled in the correct order. Next, the signatures are bound together and their edges trimmed. Finally, the cover – which is printed separately – is attached and the book is ready to use.

SHEET FOLDER

A sheet from a sheet-fed press usually contains one signature and is folded several times.

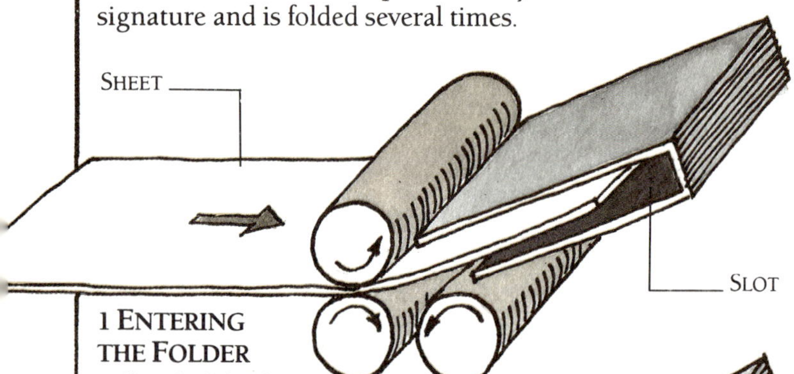

1 ENTERING THE FOLDER
Rollers feed the sheet into the slot of the folder, which stops it moving.

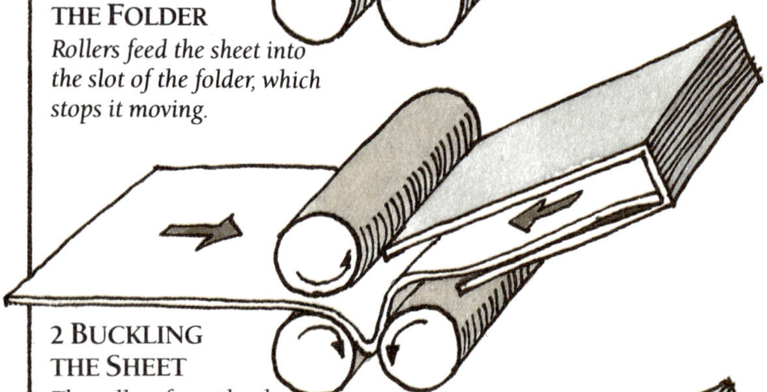

2 BUCKLING THE SHEET
The rollers force the sheet forward so that it begins to buckle in the center.

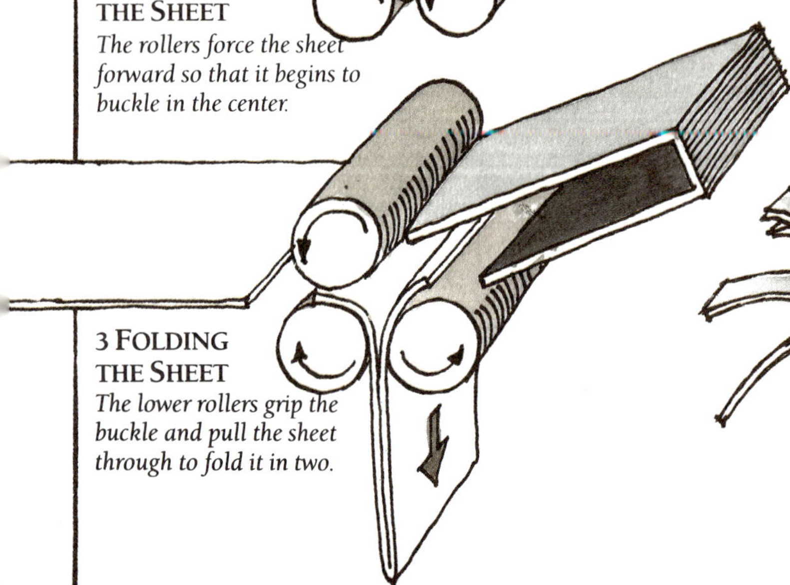

3 FOLDING THE SHEET
The lower rollers grip the buckle and pull the sheet through to fold it in two.

WEB FOLDER

Web signatures are printed one after another, and the folder separates each signature as well as folding it.

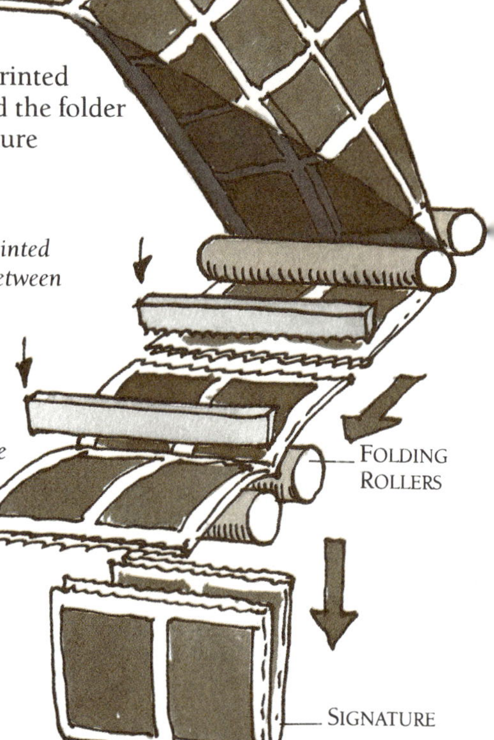

1 FIRST FOLD
The web passes over a pointed metal "nose" and then between rollers that fold the web along the center.

2 SEPARATION
A serrated blade pierces the folded web so that the signature is torn loose.

3 SECOND FOLD
A folder blade pushes the center of the signature between a pair of folding rollers.

4 THIRD FOLD
The signature is folded again and the pages are now in the correct order.

FAN WHEEL

Signatures are fed into the fan wheel, and the wheel delivers them to a conveyor belt which takes them to be bound into books.

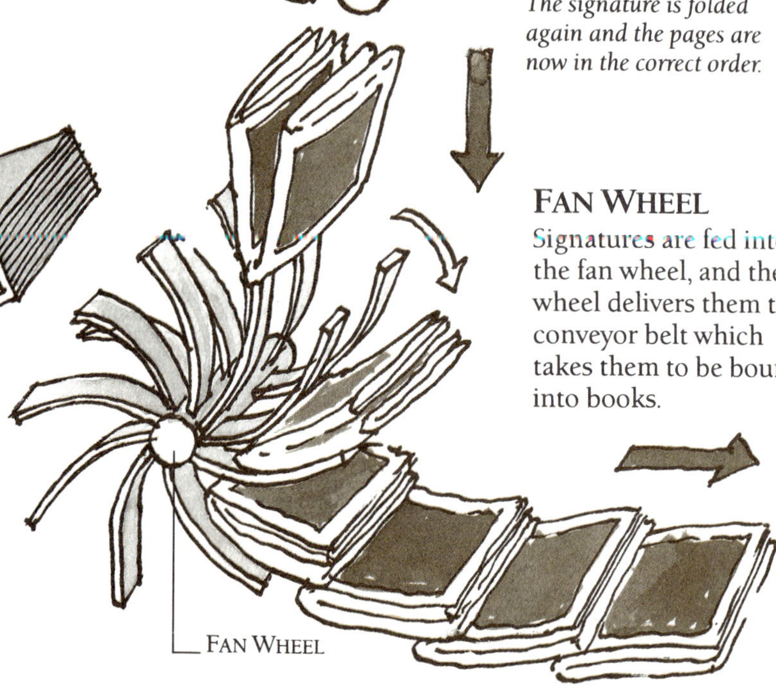

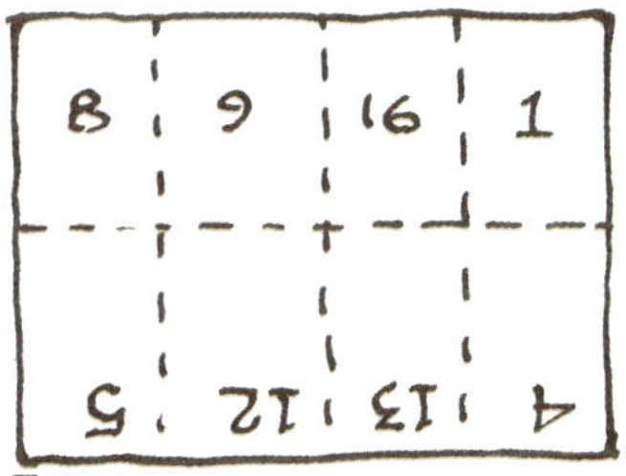

FRONT

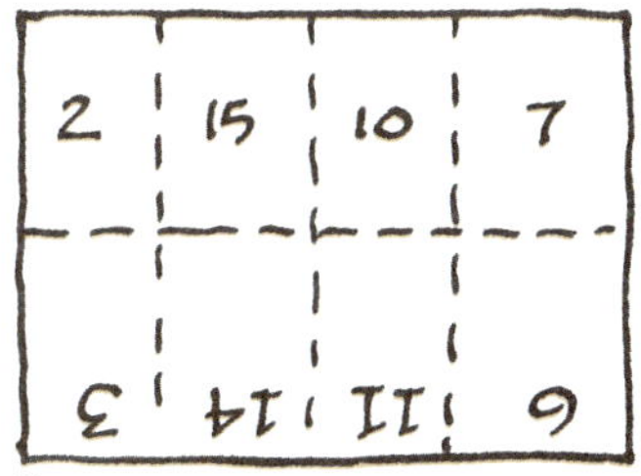

BACK

SIGNATURES

The pages in the signature are printed on the sheet or web in a particular order. When folded the right way the pages in each signature will be in the correct sequence. Signatures may contain various numbers of pages: most books have signatures of 16, 24 or 32 pages.

FIRST FOLD

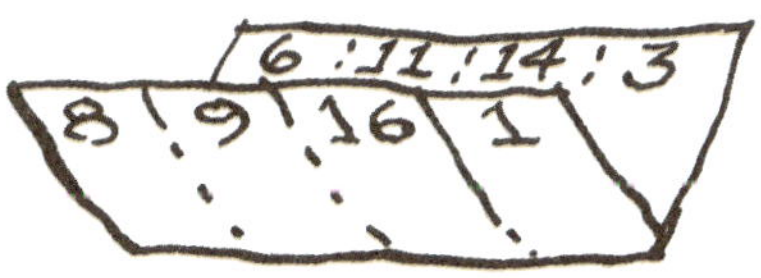

SECOND FOLD

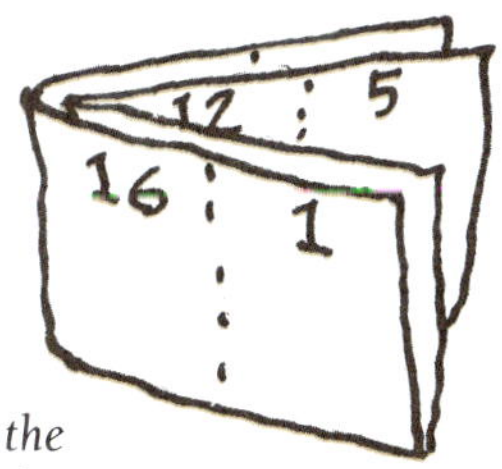

THIRD FOLD

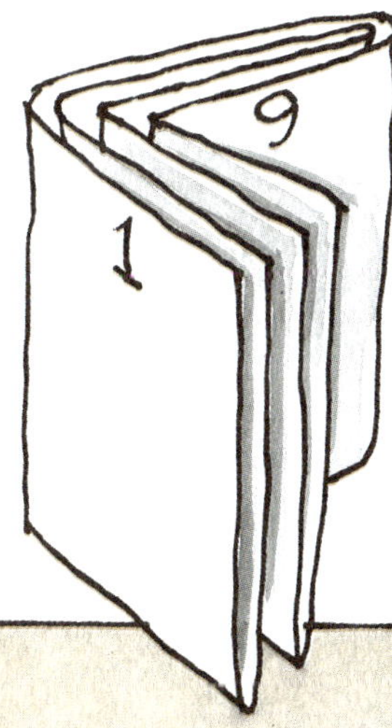

16-PAGE SIGNATURE

The sheet or web is four pages wide and the signature two pages deep. It is folded in the center three times.

HAND BINDING

1 *The set of signatures is aligned in the correct order.*

2 *The backs of the signatures are sewn together.*

3 *Glue is applied to hold the signatures together. The pages are then trimmed.*

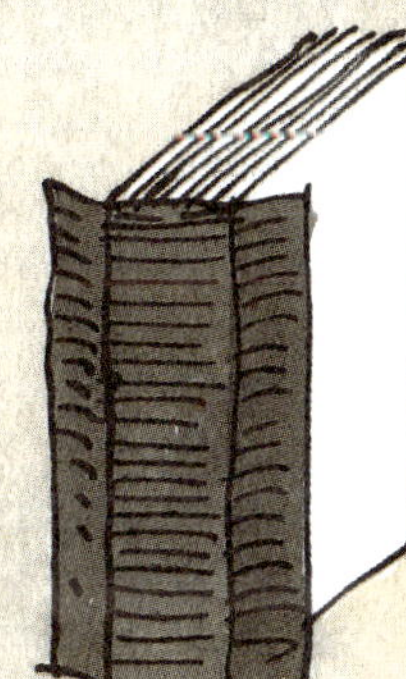

4 *A lining is glued to the spine (back) of the book.*

5 *The case (cover) is glued to the lining.*

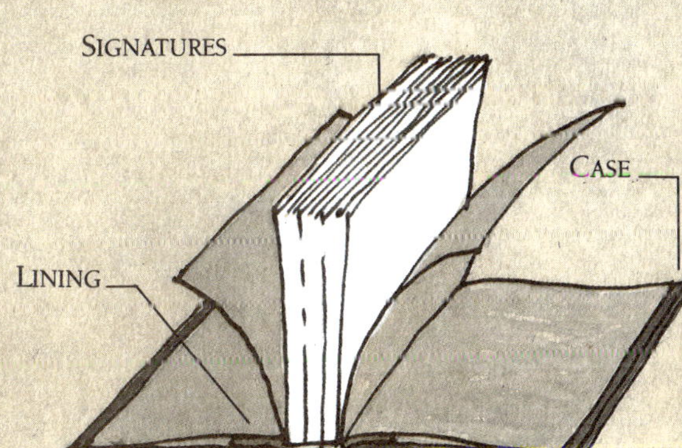

THE FINISHED BOOK

Machine binding follows much the same sequence of operations as hand binding, although sometimes gluing is used without sewing.

SOUND AND MUSIC

ON PLAYING THE MAMMOTH

While I do not profess to understand the "modern" music, I have long been involved in the development of the mammoth as an instrument. In my earliest experiments, a trio of courageous musicians produced the most remarkable assortment of sounds from a single properly tuned and securely tethered beast. The tusks, when struck by wooden mallets, gave a rich melodic chime. The great belly, played with leather-covered mallets, offered a sonorous thud. The tail, secured by a flexible tree trunk, produced a soothing twang when plucked. By moving the tree trunk to either stretch or relax the tail, the plucker could achieve many different notes. But perhaps the most extraordinary sound was that produced voluntarily by the animal itself. As the mammoth slipped into the spirit of the music, it issued periodic trumpet blasts from its great trunk. The trio became a quartet in which man and nature achieved an unforgettable harmony.

MAKING SOUND

All sound producers emit sound by making something vibrate. As a vibrating object moves to and fro, it sets up sound waves in the air. The waves consist of alternate regions of high and low pressure, which are known as compressions and rarefactions. As the object's surface moves forward into the air, it produces a compression. The surface then moves back, producing a rarefaction. Together each compression and rarefaction makes up a sound wave, and the waves move out in all directions at high speed. The stronger the vibrations, the greater the pressure difference between each compression and rarefaction and the louder the sound.

The vibrations that set up sounds can be produced in a number of different ways. The simplest is hitting an object: the energy from the blow vibrates the object and these vibrations are transmitted to the air. Plucking a taut string (or tail) makes it vibrate, while releasing air under pressure into a hollow tube (such as a trunk) can also set up vibrations in the air.

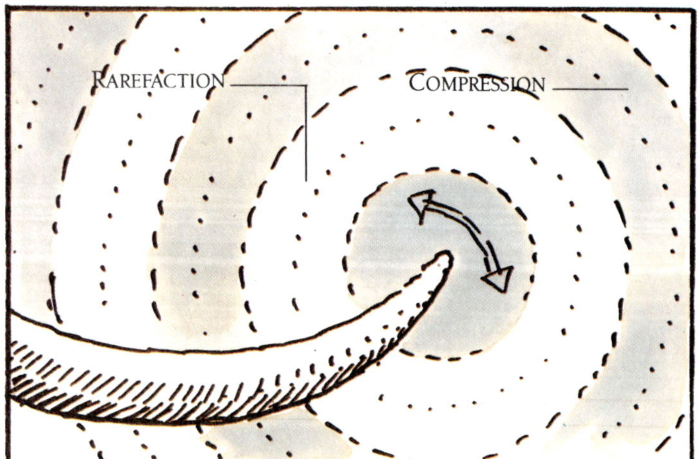

SETTING UP SOUND WAVES
Hitting an object like a tusk makes it vibrate, and this vibration is then transmitted to the air around the object. The vibration needed to create an audible sound wave has to have a rate of more than 20 compressions and 20 rarefactions per second.

More recent experiments have focused on the mammoth as an ensemble instrument. Perhaps the best known of these undertakings was my arrangement for four mammoths, tethered in order of size. Although the instruments often grew restless during rehearsals, the twelve musicians, comprising four tusk-tappers, four stomach-thumpers and four tail-twangers, became highly proficient at playing them. The performance was a feast not only for the ears but also for the eyes.

The popularity of massed mammoth music reached its peak with the creation of the Mammoth Tabernacle Choir. While I personally never saw or heard it, I am assured that the effect, especially at close range, was nothing short of stunning.

SOUND REFLECTIONS

As well as traveling directly to listeners, sound waves may also bounce off nearby surfaces. The ear receives a mixture of the direct sound and echoes. If the reflecting surfaces are fairly distant, the reflected sound will take much longer to reach the ear and separate echoes will be heard.

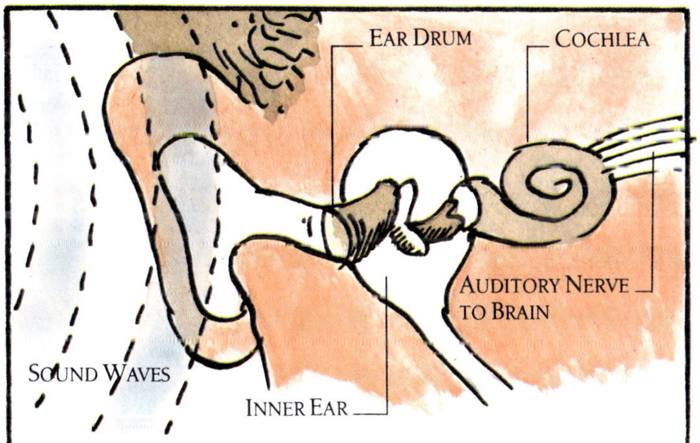

HEARING

As sound waves enter the ear, the pressure differences between successive compressions and rarefactions set the ear drum vibrating. These vibrations pass to the cochlea in the inner ear, where they are converted into electric signals. The signals travel along the auditory nerve to the brain, and the sound is heard.

WOODWIND INSTRUMENTS

Woodwind instruments are not necessarily made of wood, many of them, like the saxophone, being metal, but they do require wind to make a sound. They consist basically of a tube, usually with a series of holes. Air is blown into the top of the tube, either across a hole or past a flexible reed. This makes the air inside the tube vibrate and give out a note. The pitch of the note depends on the length of the tube, a shorter tube giving a higher note, and also on which holes are covered. Blowing harder makes the sound louder.

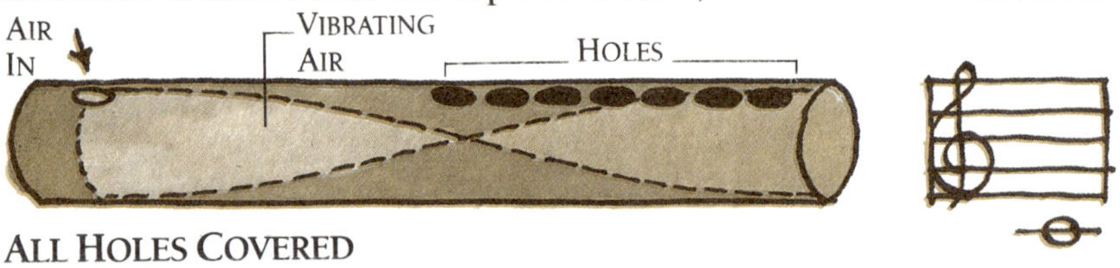

ALL HOLES COVERED

Covering all seven holes in a simple pipe makes the air in the whole tube vibrate, giving the note middle C.

FIRST THREE HOLES COVERED

This shortens the vibrating air column to two-thirds of the tube, giving the higher note G.

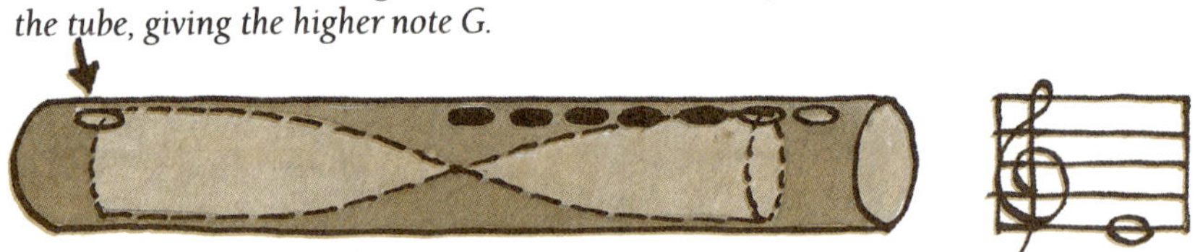

FIRST FIVE HOLES COVERED

This extends the vibrating air column to four-fifths of the total length of the tube, giving an E.

KEYS AND CURVES

To produce deep notes, woodwind instruments have to be quite long. The tube is therefore curved so that the player can hold the instrument, as in this alto saxophone. Keys allow the fingers to open and close holes all along the instrument.

EDGE-BLOWN WOODWINDS

In the flute and recorder, the player blows air over an edge in the mouthpiece. This sets the air column inside the instrument vibrating.

SINGLE-REED WOODWINDS

In the clarinet and saxophone, the mouthpiece contains a single reed that vibrates to set the air column inside the instrument vibrating.

DOUBLE-REED WOODWINDS

The oboe, cor anglais and bassoon have a mouthpiece made of a double reed that vibrates to set the air column inside the instrument vibrating.

FINGERHOLES

In a short and simple woodwind instrument, such as the recorder, the fingers can cover all the holes directly.

PADS

Several woodwinds have holes that are larger than the fingers, requiring the fingers to press pads to cover the holes.

KEYS

Holes that are out of reach of the fingers are covered by pressing sprung keys attached to pads.

BRASS INSTRUMENTS

Brass instruments are in fact mostly made of brass, and consist of a long pipe that is usually coiled and has no holes. The player blows into a mouthpiece at one end of the pipe, the vibration of the lips setting the air column vibrating throughout the tube. The force of the lips varies to make the vibrating column divide into two halves, three thirds, and so on. This gives an ascending series of notes called harmonics. Opening extra lengths of tubing then gives other notes that are not in this harmonic series.

LOW PRESSURE

With low lip pressure, the air column vibrates in two halves and each half gives the note middle C. The length of the tube is therefore twice as long as a woodwind instrument sounding the same note.

INCREASED PRESSURE

Raising the lip pressure makes the air column vibrate in three thirds. Each vibrating section is two-thirds the length of the previous section, raising the pitch of the note to G.

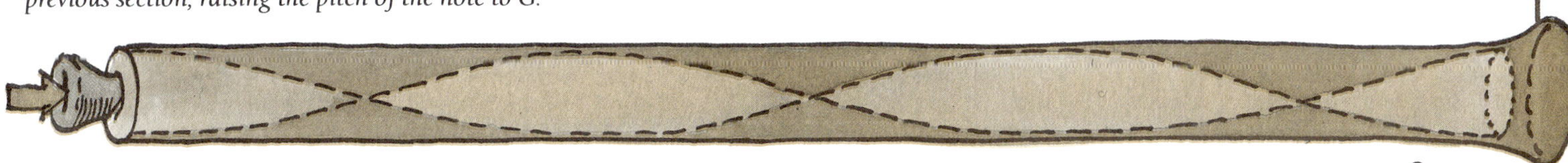

INCREASED LENGTH

To play an E, which is not in the harmonic series, the player keeps the air vibrating in three thirds and increases the total length of the tube. Each vibrating section becomes four-fifths the length for middle C.

THE TROMBONE

The trombone has a section of tubing called a slide that can be moved in and out. The player pushes out the slide to lengthen the vibrating air column and produce notes that are not in the harmonic series.

THE TRUMPET

The trumpet has three pistons that are pushed down to open extra sections of tubing and play notes that are not in the harmonic series. Up to six different notes are obtained by using different combinations of the three pistons.

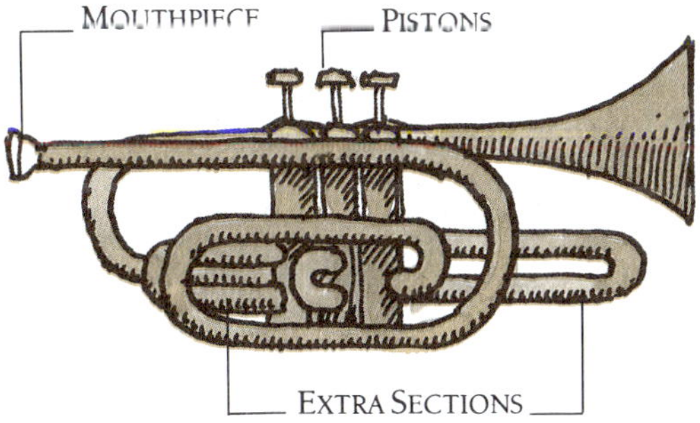

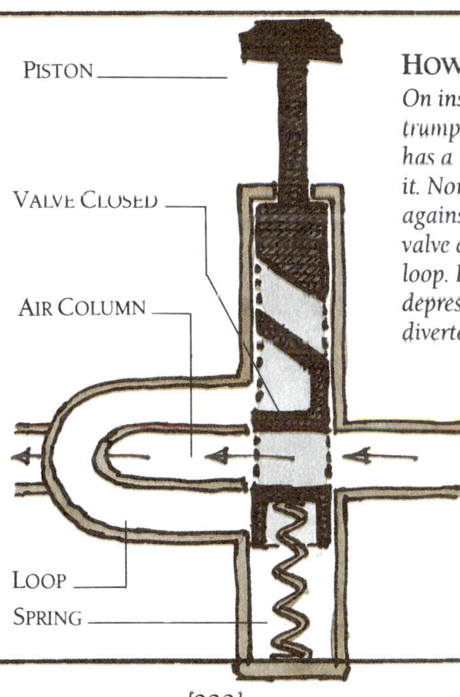

HOW VALVES WORK

On instruments such as the trumpet and tuba, each valve has a loop of tubing attached to it. Normally, the spring pushes against the piston, keeping the valve closed and shutting off the loop. But when the piston is depressed, the air column is diverted through the loop.

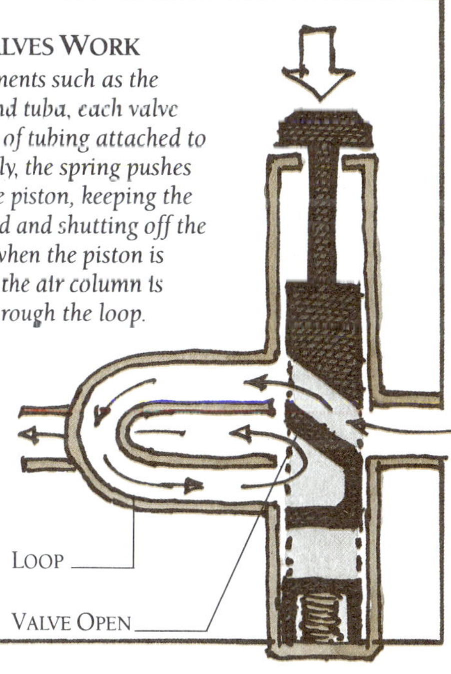

STRING AND PERCUSSION INSTRUMENTS

String instruments form a large group of musical instruments which includes the violin family and guitar, and also harps, zithers and the piano. All these instruments make a sound by causing a taut string to vibrate. The string may be bowed, as with the violin family, plucked as in guitars, harps and zithers, or struck by a hammer as in the piano (see pp.30-1). The pitch of the note produced depends on three factors – the length, weight and tension of the string. A shorter, lighter or tighter string gives a higher note.

In many string instruments, the strings themselves do not make much sound. Their vibration is passed to the body of the instrument, which resonates to increase the level of sound that is heard.

Percussion instruments are struck, usually with sticks or mallets, to make a sound. Often the whole instrument vibrates and makes a crack or crash, as in castanets and cymbals. Their sound does not vary in pitch and can only be made louder or softer. Drums contain stretched skins, which may vibrate to give a pitched note. As with strings, tightening the skin makes the note higher in pitch and smaller drums give higher notes.

Tuned percussion instruments, such as the xylophone, have sets of bars that each give a definite note. The pitch of the note depends on the size of the bar, a smaller bar giving a higher pitch.

THE KETTLE-DRUM

Kettledrums or timpani make sounds with a definite pitch, which can be varied. Turning screws at the side of the head pulls the hoop downward to tighten the skin and raise the pitch, or slackens the skin to lower the pitch.

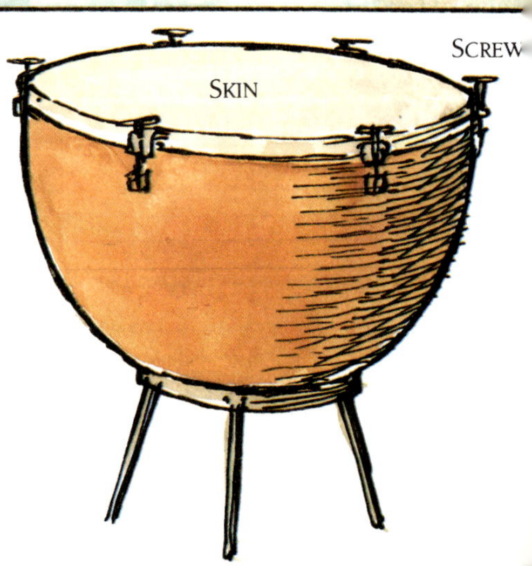

THE VIOLIN

The violin and its relatives are the most expressive of string instruments. The violin has four strings of different weights. These are wound around tuning pegs to produce the correct amount of tension, and they sound four "open" notes when they are plucked or bowed. The performer stops the strings to obtain other notes, pressing one or more strings against the fingerboard to shorten the section that vibrates, thus raising the pitch of the string.

The front and back of the violin are connected by a short sound post, which transmits vibrations to the back. The whole body vibrates and the sound emerges through the f-shaped sound holes on the front of the instrument.

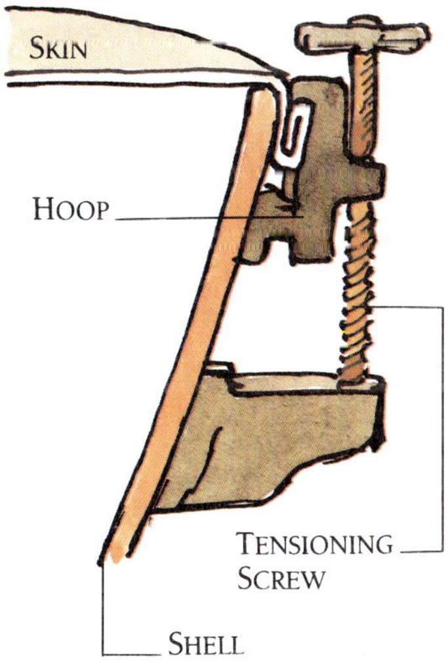

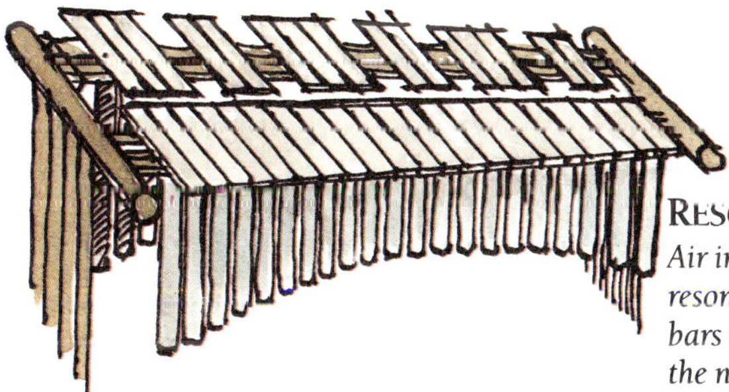

RESONATORS
Air inside tubes called resonators under the bars vibrates to make the notes louder.

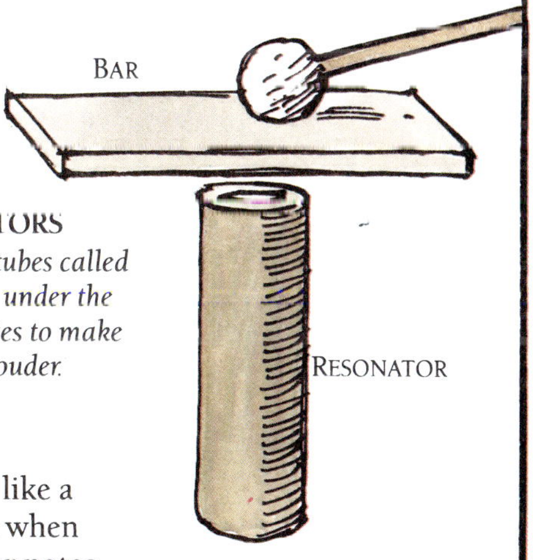

THE XYLOPHONE

The xylophone and similar instruments such as the vibraphone and marimba have sets of bars arranged like a piano keyboard. Each bar gives out a particular note when struck with a mallet, the longer bars sounding deeper notes.

MICROP

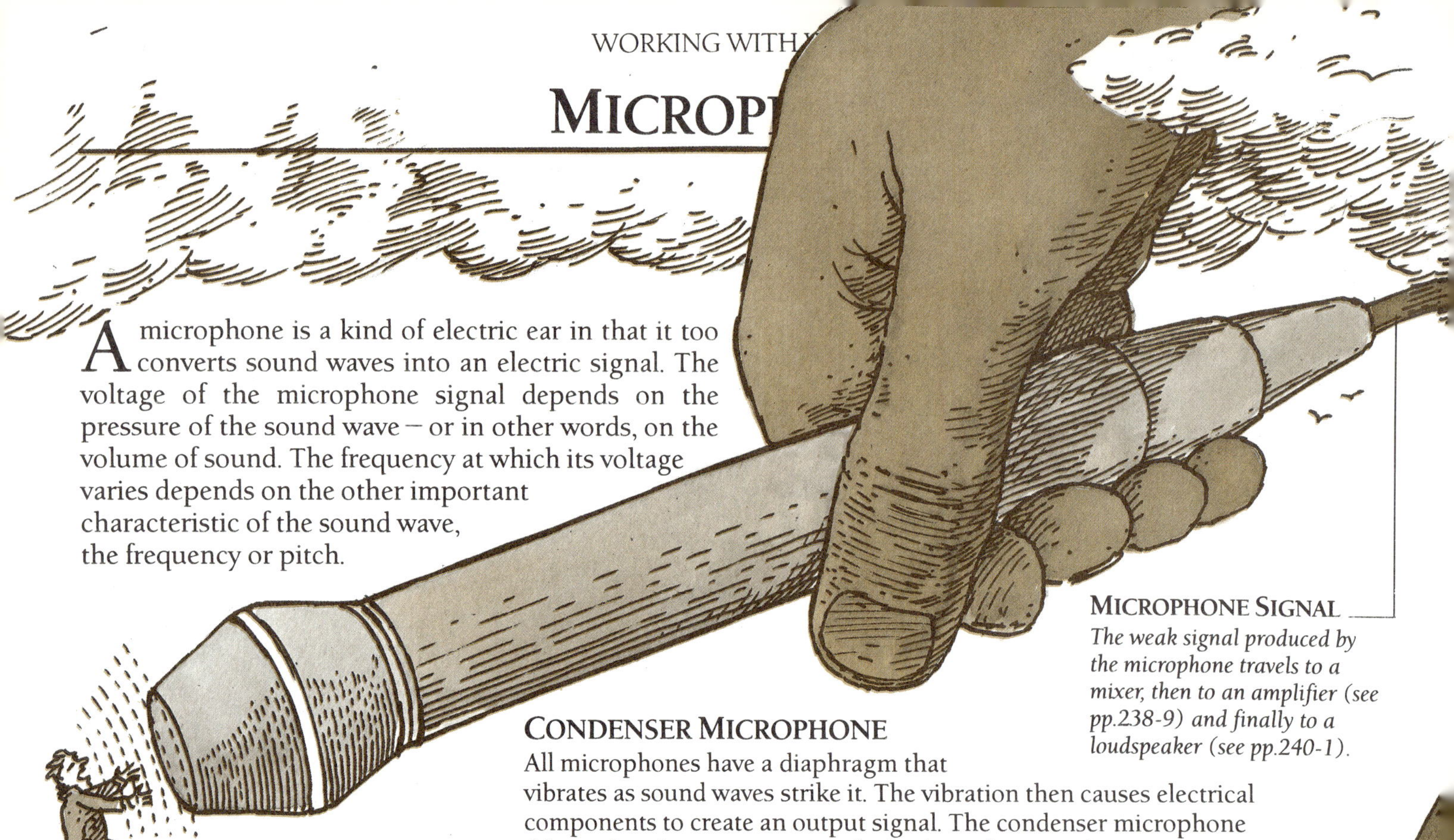

A microphone is a kind of electric ear in that it too converts sound waves into an electric signal. The voltage of the microphone signal depends on the pressure of the sound wave – or in other words, on the volume of sound. The frequency at which its voltage varies depends on the other important characteristic of the sound wave, the frequency or pitch.

MICROPHONE SIGNAL
The weak signal produced by the microphone travels to a mixer, then to an amplifier (see pp.238-9) and finally to a loudspeaker (see pp.240-1).

CONDENSER MICROPHONE
All microphones have a diaphragm that vibrates as sound waves strike it. The vibration then causes electrical components to create an output signal. The condenser microphone (shown here) uses a capacitor for high-quality sound.

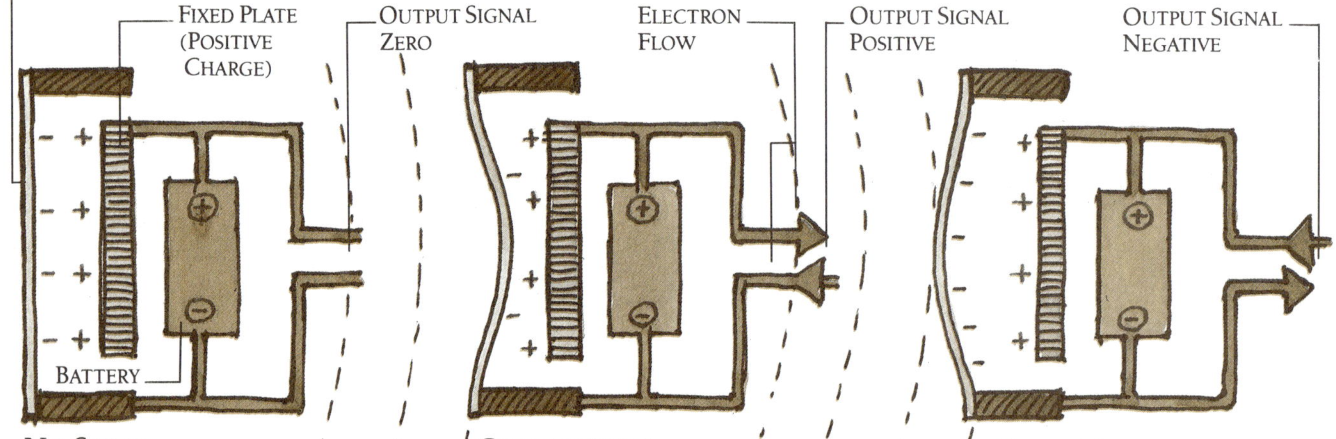

NO SOUND
The battery produces equal charges on the diaphragm and fixed plate. Together, they form a capacitor. No further current flows.

COMPRESSION
As the diaphragm moves in, the plate attracts electrons from the diaphragm. Electrons in the output signal flow to the diaphragm.

RAREFACTION
As the diaphragm moves out, electrons in the diaphragm repel each other and flow away from it. The output signal reverses.

SYNTHESIZER

Electronic music makes great use of the synthesizer, which is an instrument that produces an electric sound signal similar to that of a microphone. Inside the synthesizer are electronic components that create the signal. The keyboard controls the voltage rate or frequency of the signal to determine the pitch of the sound, which emerges from a loudspeaker connected to the synthesizer.

ECTRIC GUITAR

An electric guitar makes little sound of its own. Playing the metal strings makes pick-ups beneath the strings generate electric sound signals. The signals go to volume and tone controls, which determine the loudness and the kind of sound, and then to an amplifier and loudspeaker.

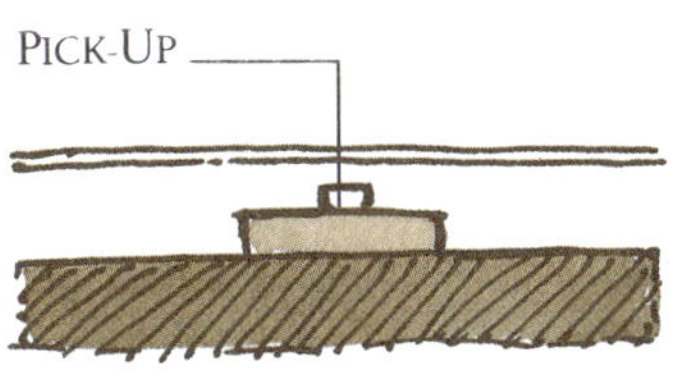

STRING AT REST
The pick-up produces no signal.

MOVING STRING
When the string moves out (above) or in (below) it produces a signal in the pick-up.

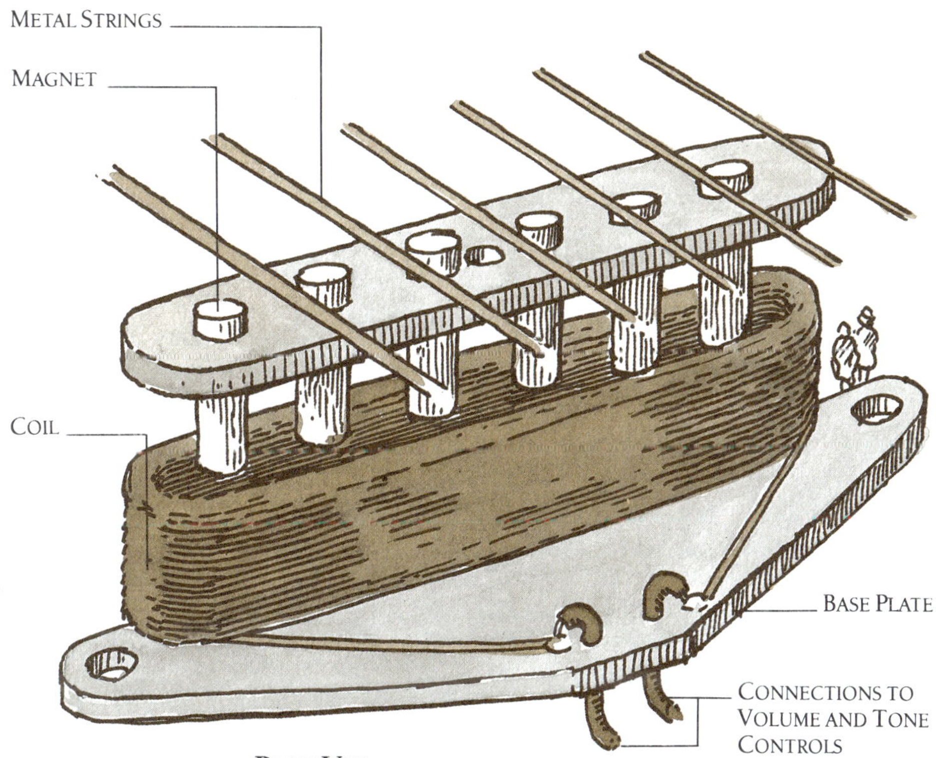

PICK-UPS
Magnets in the pick-up produce magnetic fields around the metal strings and the coil of wire. As the metal strings vibrate, they cause the fields to vary in strength. The changing field in turn creates a varying electric current in the coil (see pp.304-5), and this is the sound signal that goes to the guitar controls.

MIXER

A mixer takes sound signals from several different sources and mixes them together. The tone and volume of each signal is controlled so that a good sound balance results. One combined signal (or two for stereo sound) then goes to the amplifier and loudspeakers.

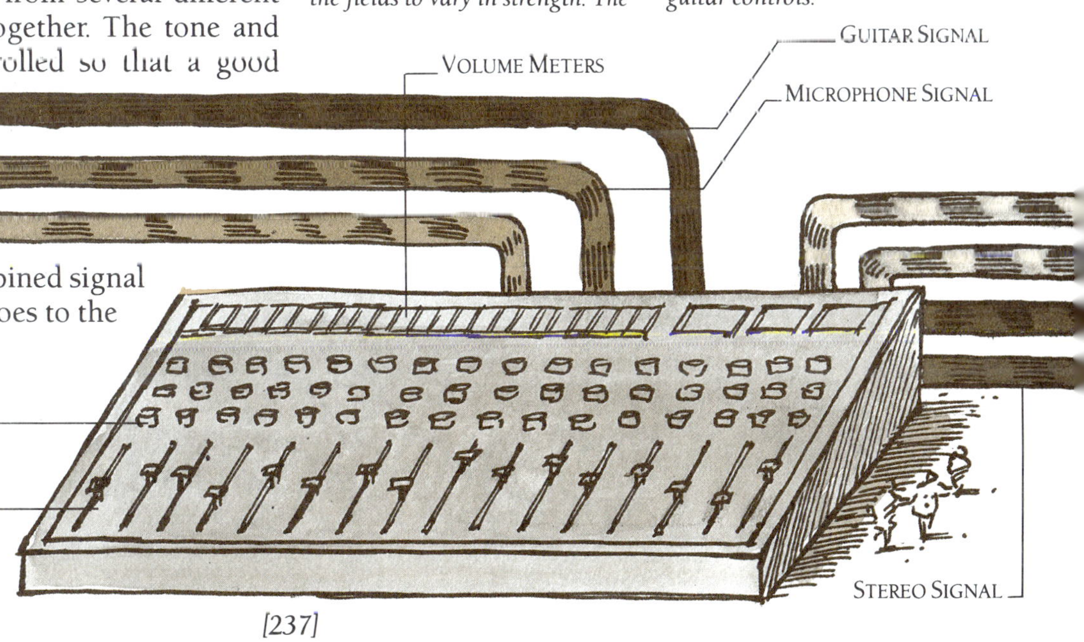

AMPLIFIER

An amplifier increases the voltage of a weak signal from a microphone, mixer, electric instrument, radio tuner or tape replay head, making it powerful enough to drive a loudspeaker or earphone. It works by using the weak signal to regulate the flow of a much stronger current, which normally comes from a battery or the electricity supplied. The key components that regulate the flow of the strong current are usually transistors. These two pages show the principles of amplification with a basic single-transistor amplifier.

RAREFACTION IN SOUND WAVE

A transistor is a small sandwich of two types of semiconductor, so-called because their conductivity changes as the transistor works. The two *n*-type (negative) pieces have some free electrons, while the *p*-type (positive) piece has "holes" into which the electrons can fit. The three pieces are known as the emitter, base and collector. When the microphone diaphragm moves out, electrons from the weak sound signal fill holes in the *p*-type semiconductor. This blocks the electrons from the power supply.

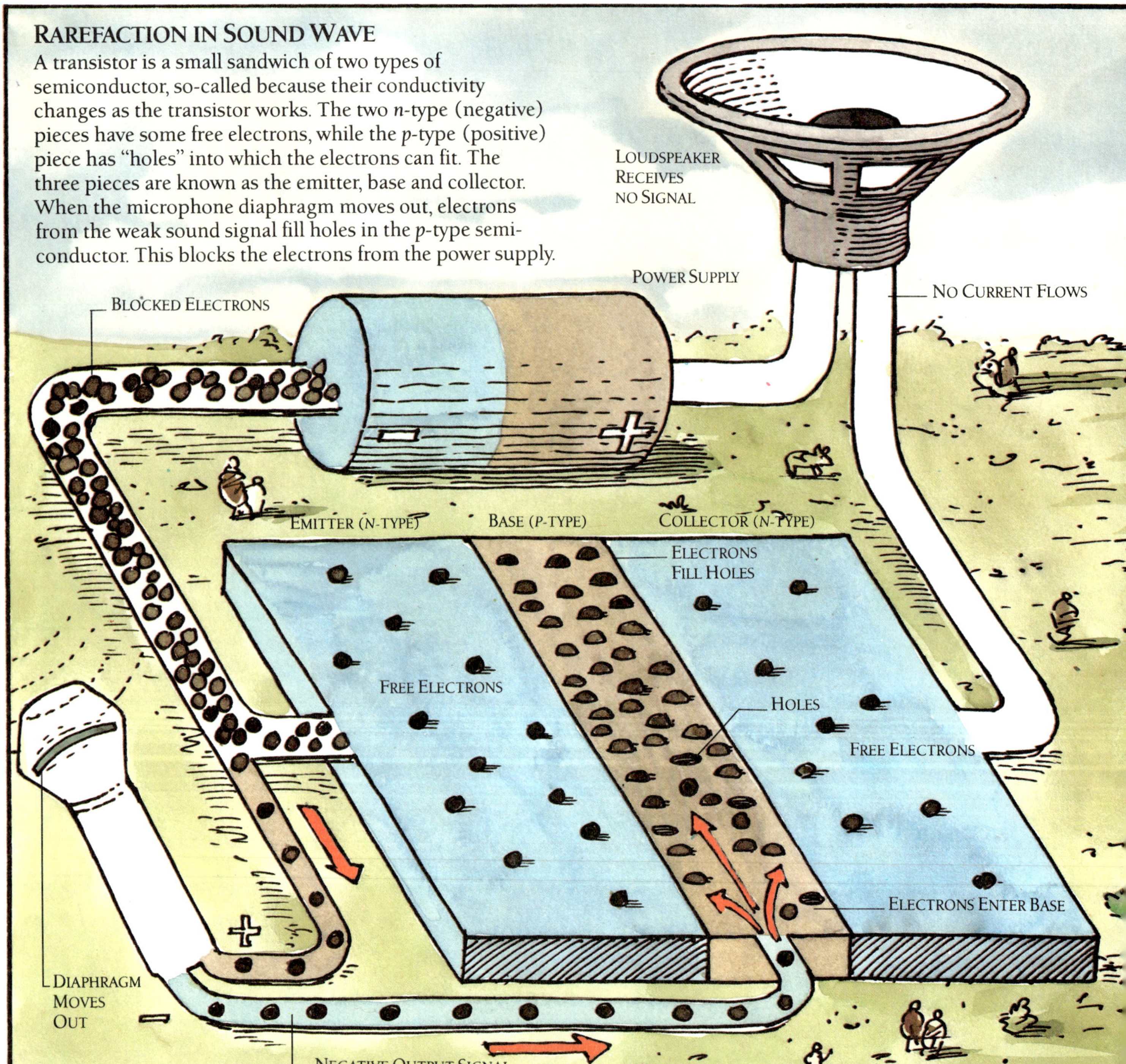

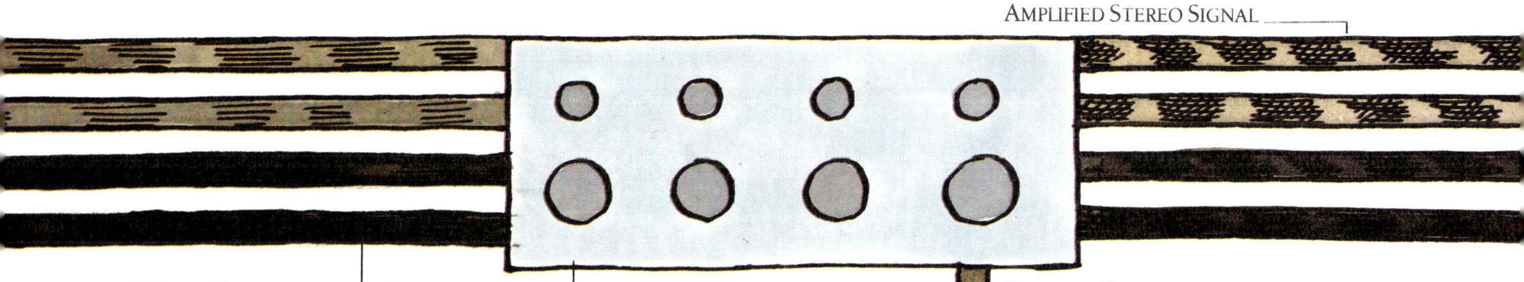

NCOMING WEAK SIGNAL
n stereo sound, four wires conduct he weak incoming signal to the mplifier – a pair of wires for each hannel.

AMPLIFIER
An amplifier usually contains many transistors and other components that enable the amount of amplification and also the tone of the sound to be varied.

POWER SUPPLY
This provides the energy that is needed to amplify the signal.

COMPRESSION IN SOUND WAVE

Vhen the microphone diaphragm is pushed in by a ompression in the sound wave, it reverses the flow of lectrons in the weak signal. Electrons leave the base emiconductor in the center of the "sandwich" and create oles. Forced by the power supply, many electrons enter hese holes from the emitter and then move on into the ollector. The result is a flow of electrons much larger than hat in the weak signal, but exactly in step with it: the weak ignal has been amplified.

LOUDSPEAKERS

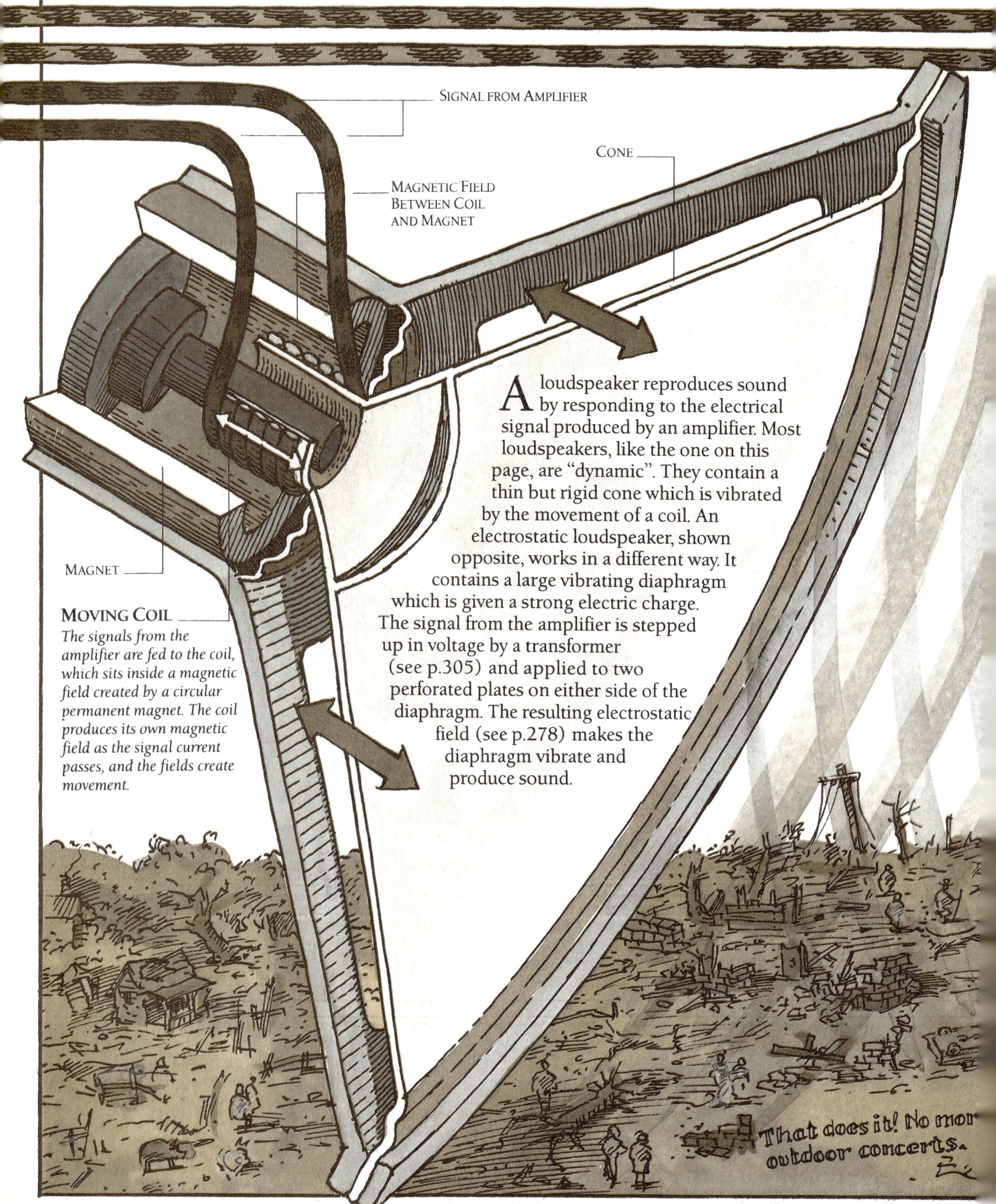

A loudspeaker reproduces sound by responding to the electrical signal produced by an amplifier. Most loudspeakers, like the one on this page, are "dynamic". They contain a thin but rigid cone which is vibrated by the movement of a coil. An electrostatic loudspeaker, shown opposite, works in a different way. It contains a large vibrating diaphragm which is given a strong electric charge. The signal from the amplifier is stepped up in voltage by a transformer (see p.305) and applied to two perforated plates on either side of the diaphragm. The resulting electrostatic field (see p.278) makes the diaphragm vibrate and produce sound.

MOVING COIL
The signals from the amplifier are fed to the coil, which sits inside a magnetic field created by a circular permanent magnet. The coil produces its own magnetic field as the signal current passes, and the fields create movement.

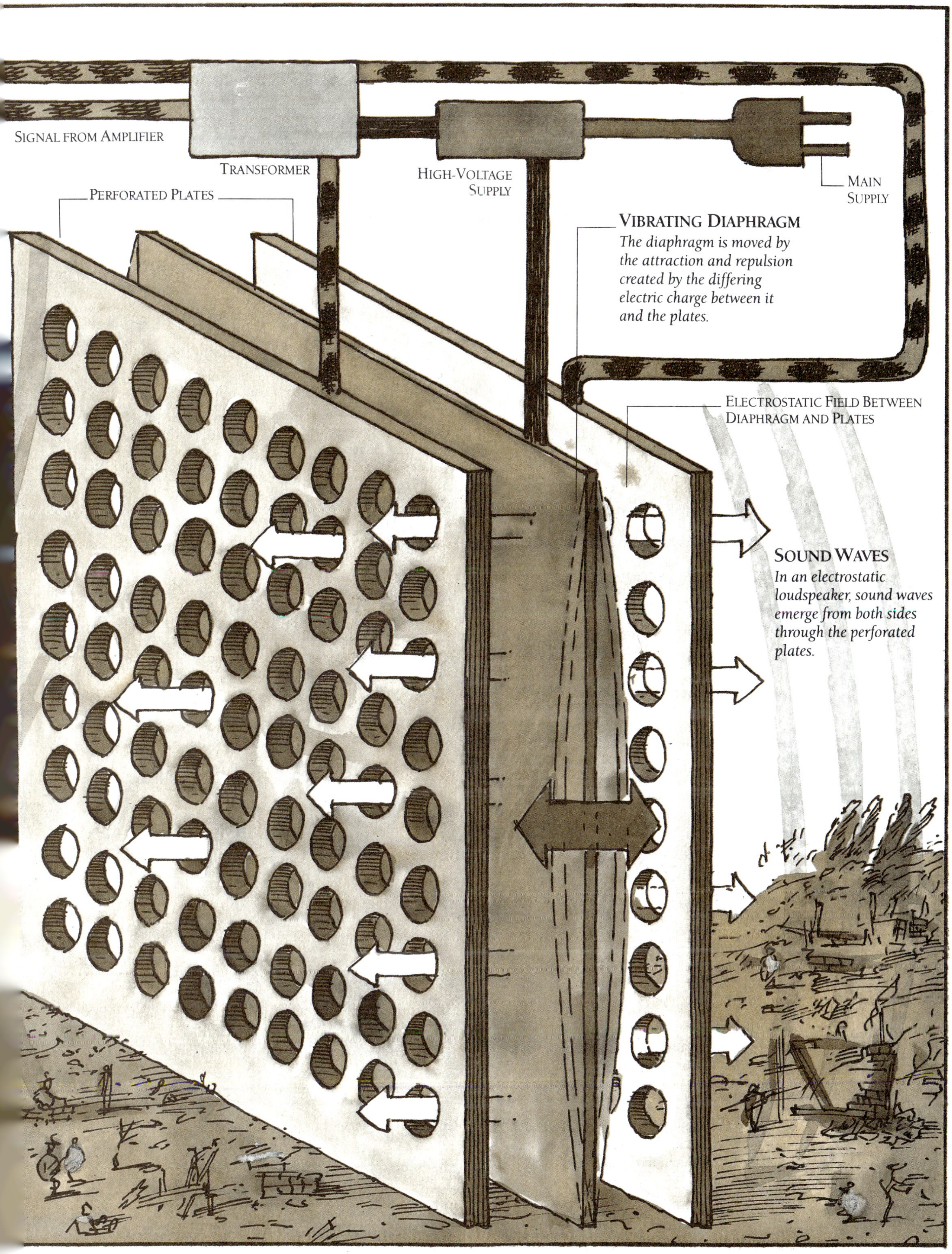
SIGNAL FROM AMPLIFIER
TRANSFORMER
HIGH-VOLTAGE SUPPLY
MAIN SUPPLY
PERFORATED PLATES
VIBRATING DIAPHRAGM
The diaphragm is moved by the attraction and repulsion created by the differing electric charge between it and the plates.
ELECTROSTATIC FIELD BETWEEN DIAPHRAGM AND PLATES
SOUND WAVES
In an electrostatic loudspeaker, sound waves emerge from both sides through the perforated plates.

SOUND RECORDING

There are two basic methods of recording voices and music – analog and digital. In analog recording, the recording medium varies continuously in a way that is similar to or analogous to the incoming signal. In digital recording, the signal is sampled electronically and recorded as a rapid sequence of separate coded measurements. Both analog and digital recording preserve the varying voltage of the sound signal produced by a microphone, but of the two, digital recording is the more accurate. In addition, a certain amount of electrical noise or hiss always enters the recording process. Digital recording is insensitive to this noise, whereas analog recording requires noise reduction systems.

SOUND SIGNAL

The curve represents the varying voltage of the electrical sound signal produced when a sound wave strikes a microphone. The varying levels of the voltage are produced by the varying pressures of the sound wave, so the curve also represents the changing energy of the sound wave. The voltage varies within a limited range, from silence to maximum volume.

STEREO

In stereophonic sound, two separate tracks or channels of sound are recorded – one to the left and one to the right. When the two channels are reproduced through loudspeakers the sounds seem to have locations in space.

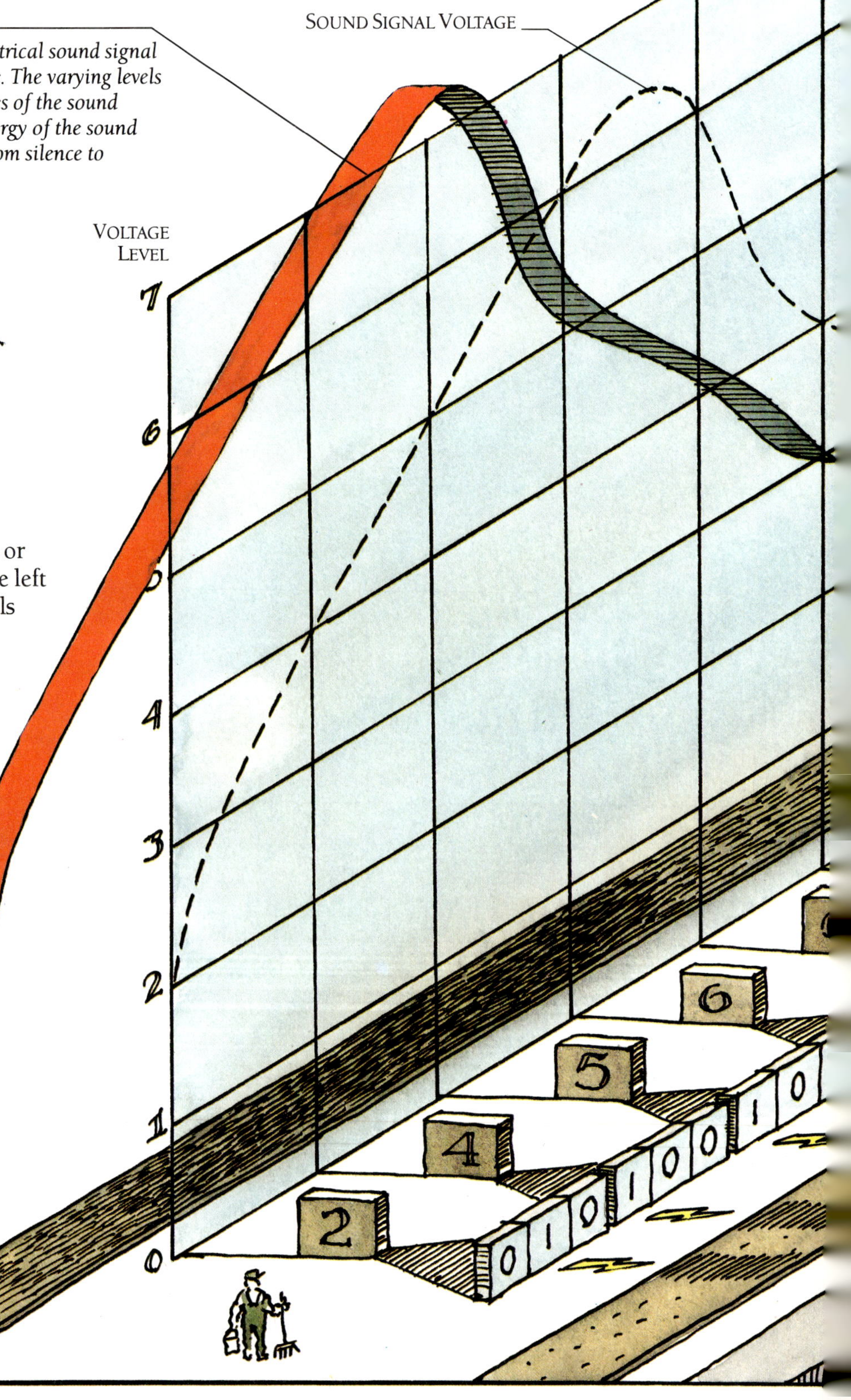

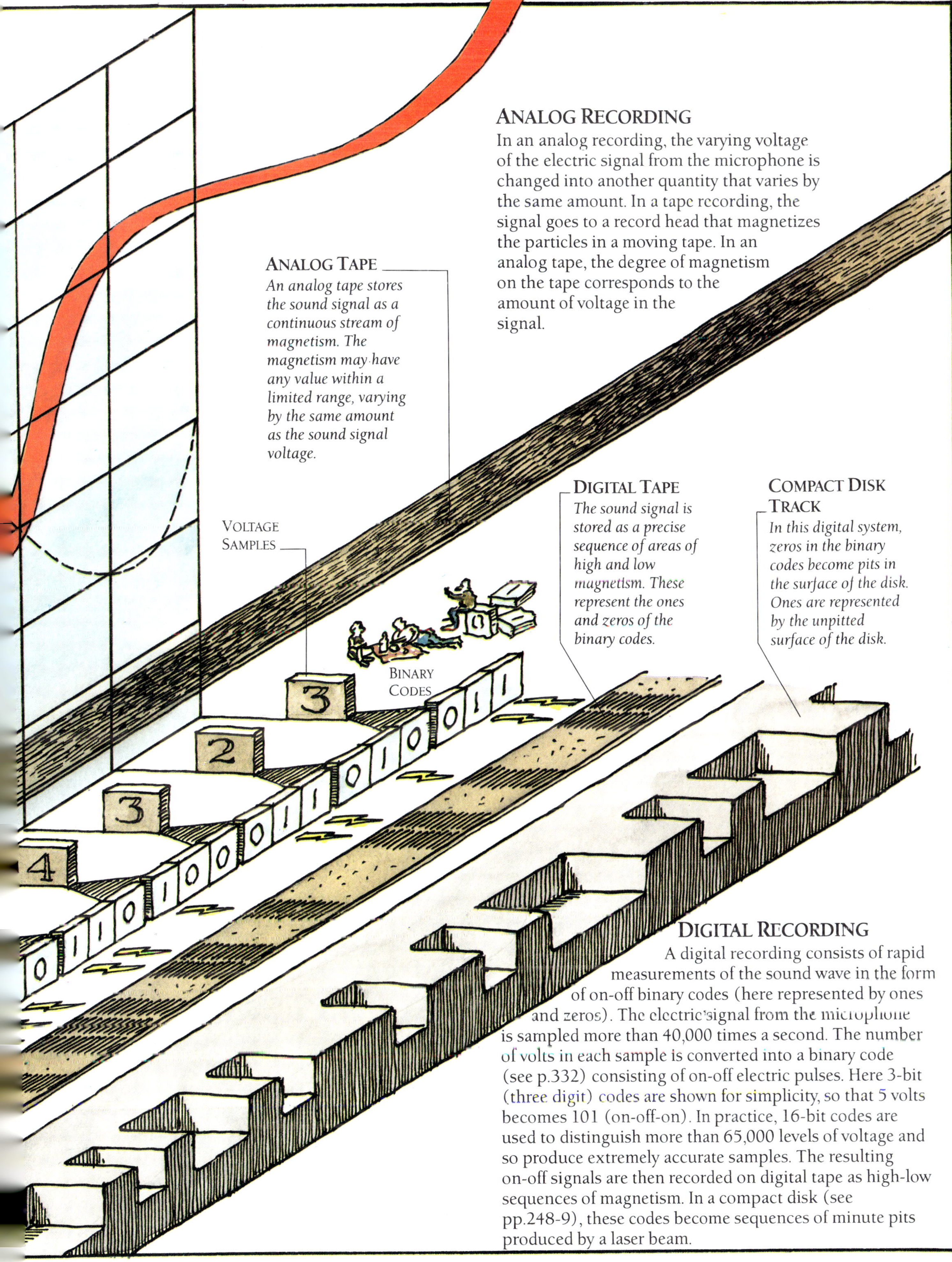

ANALOG RECORDING

In an analog recording, the varying voltage of the electric signal from the microphone is changed into another quantity that varies by the same amount. In a tape recording, the signal goes to a record head that magnetizes the particles in a moving tape. In an analog tape, the degree of magnetism on the tape corresponds to the amount of voltage in the signal.

DIGITAL RECORDING

A digital recording consists of rapid measurements of the sound wave in the form of on-off binary codes (here represented by ones and zeros). The electric signal from the microphone is sampled more than 40,000 times a second. The number of volts in each sample is converted into a binary code (see p.332) consisting of on-off electric pulses. Here 3-bit (three digit) codes are shown for simplicity, so that 5 volts becomes 101 (on-off-on). In practice, 16-bit codes are used to distinguish more than 65,000 levels of voltage and so produce extremely accurate samples. The resulting on-off signals are then recorded on digital tape as high-low sequences of magnetism. In a compact disk (see pp.248-9), these codes become sequences of minute pits produced by a laser beam.

PRESSURE PAD
The pressure pad is mounted in the cassette, and presses the tape against the playback head when the cassette is inserted into the player.
ERASE HEAD
A high-frequency electric signal is fed to the erase head when recording. It produces a magnetic field (see p.295) that alternates rapidly, disorienting the magnetic particles on the tape and erasing any previous recording.
TAPE SPOOL
UNMAGNETIZED TAPE
MAGNETIZED TAPE
GUIDE PILLAR

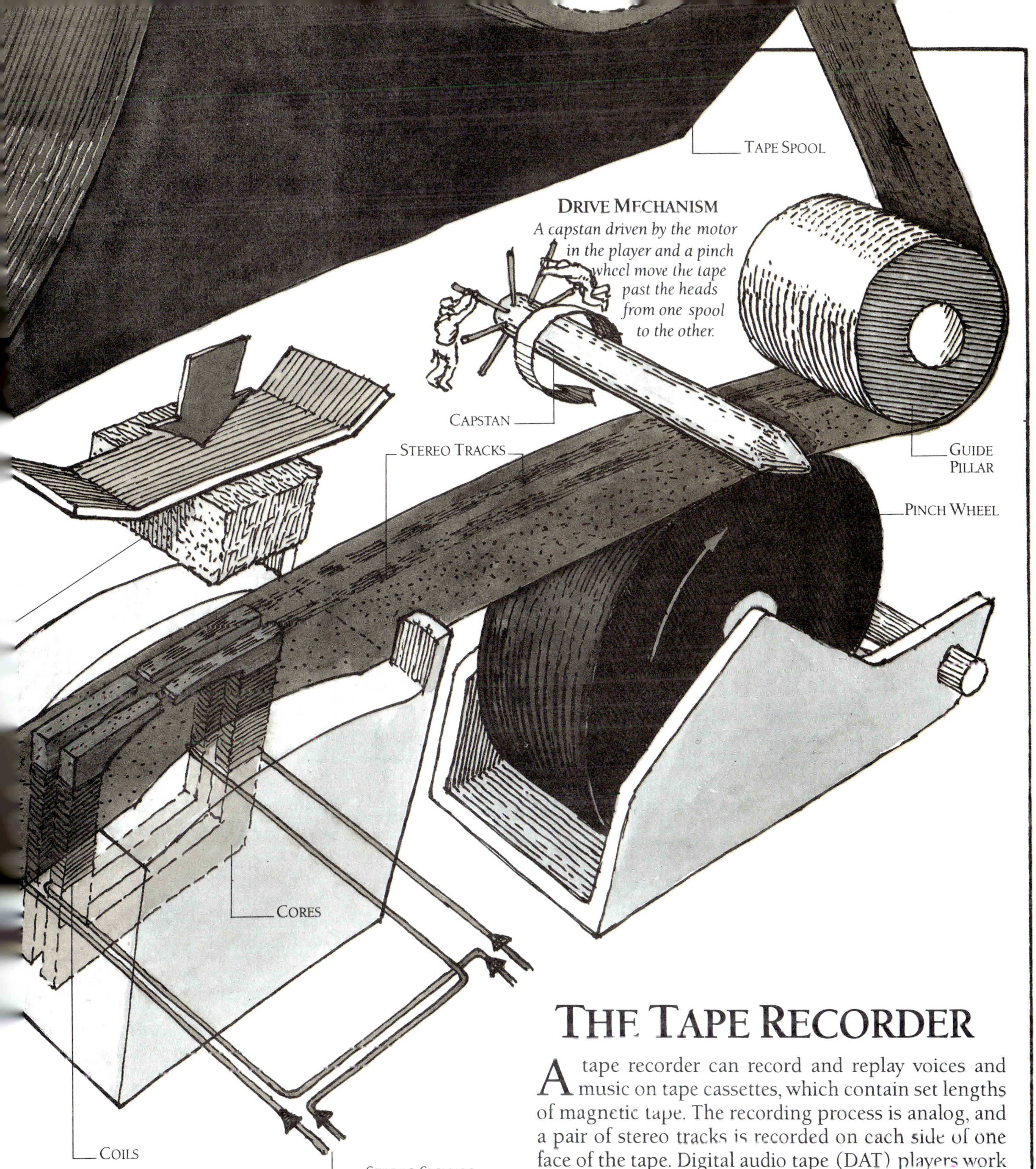

THE TAPE RECORDER

A tape recorder can record and replay voices and music on tape cassettes, which contain set lengths of magnetic tape. The recording process is analog, and a pair of stereo tracks is recorded on each side of one face of the tape. Digital audio tape (DAT) players work in the same way as video recorders (see pp.260-1) and have rotating heads.

On inserting the cassette into the machine, the centers of the tape spools fit over the spindles in the tape recorder. Pressing the PLAY button brings the playback head and drive mechanism into contact with the tape. The tape moves and the head records or plays the sounds. When recording, the erase head wipes off any previous recording on the tape.

PLAYBACK HEAD

A coil of wire is wound around each core, causing it to act as an electromagnet. When recording, two stereo electric signals are amplified and go to the pair of coils in the head. They produce magnetic fields that magnetize the particles in the tape. On playback, the magnetic particles in each track produce a pair of stereo electric signals in the coils, and these go to an amplifier and pair of loudspeakers or earphones to reproduce the sound.

THE RECORD PLAYER

A record player takes disks that revolve at 33⅓ or 45 revolutions per minute, each side containing one spiral groove. The recording system is analog, the number and depth of the contours in the groove wall corresponding to the frequency and loudness of the recorded sound.

The record rests on a rotating turntable, and the pick-up arm in the player has a cartridge with a stylus that rests in the groove and vibrates as the record revolves. The vibrations of the stylus make the cartridge produce a stereo electric signal. This signal then goes to an amplifier and pair of loudspeakers to reproduce the recorded sound.

BELT DRIVE
Many turntables are driven by a belt that runs around a drive spindle turned by the motor. This system prevents motor vibration reaching the record.

MAKING A RECORD

A metal stereo master disk is made from a stereo master tape. The master disk is made by a cutter that produces a spiral groove on its surface. Plastic copies of the master disk are then pressed.

CUTTING HEAD

The cutting head has two blades that vibrate at right angles in response to the stereo signals on the master tape. The head moves across the disk, and the blades cut into the walls of the groove so that the right-hand signal is recorded in one wall and the left-hand signal in the other.

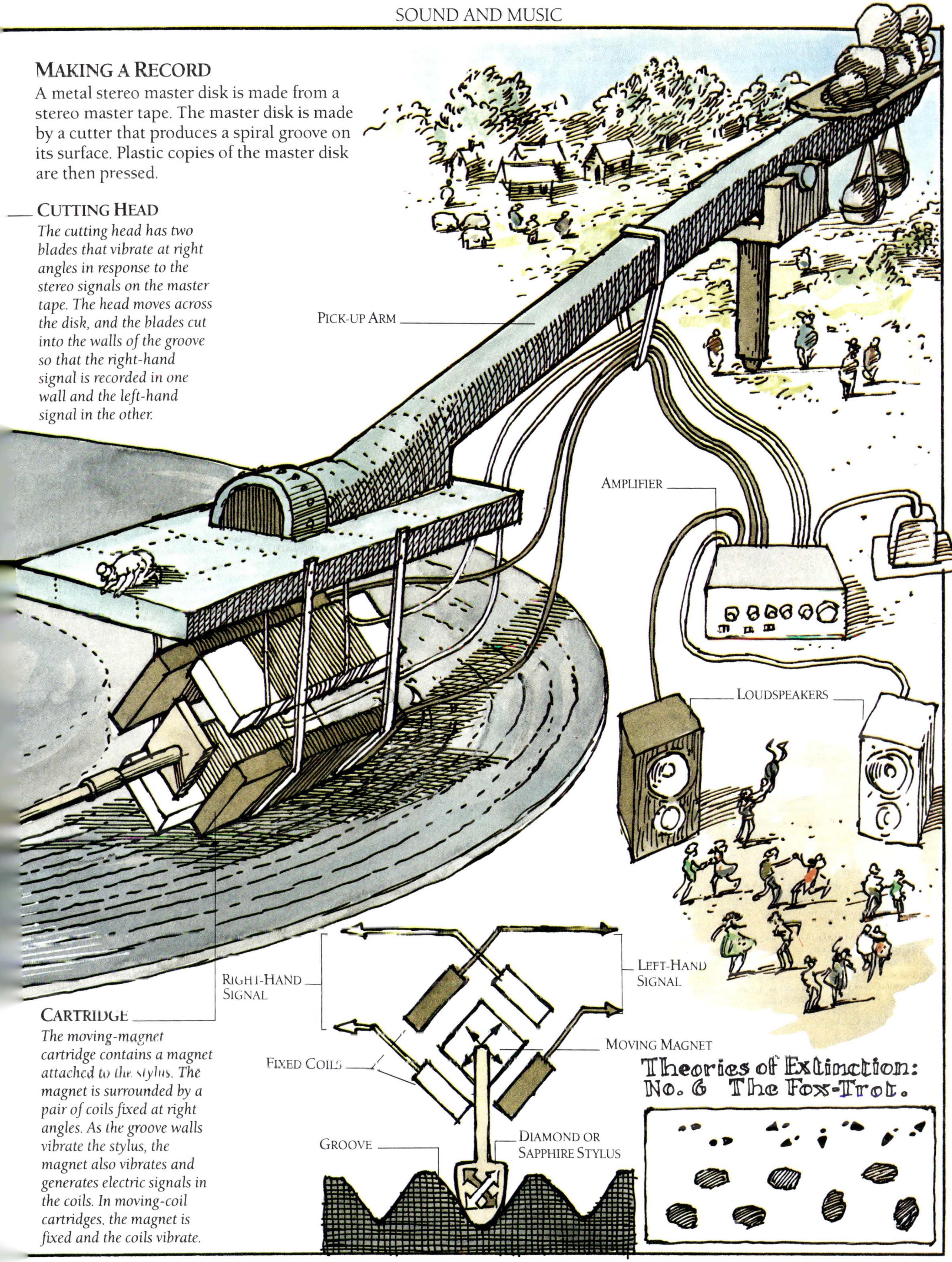

CARTRIDGE

The moving-magnet cartridge contains a magnet attached to the stylus. The magnet is surrounded by a pair of coils fixed at right angles. As the groove walls vibrate the stylus, the magnet also vibrates and generates electric signals in the coils. In moving-coil cartridges, the magnet is fixed and the coils vibrate.

THE COMPACT DISK PLAYER

Unlike a record player, a compact disk player is a digital device. A compact disk contains a spiral track of binary codes in the form of sequences of minute pits (see pp.242-3). The disk is under 5 inches (12.5cm) across, but the track is thinner than a hair and has a length of several miles. The disk rotates at a playing speed which varies from 500 revolutions per minute at the center, where the track starts, to 200 revolutions per minute at the edge. The linear speed of the disk is constant as it passes over the optical read-out system that decodes the tracks.

OPTICAL READ-OUT SYSTEM
A system of mirrors and lenses fires a beam of laser light at the spiral track on the underside of the disk. As the disk rotates, the beam moves across the disk from the center to the edge. The light-sensitive photodiode (see p.292) produces on-off code signals, which are converted into a stereo electric signal.

TRACK CODES
The sequence of pits in the track on a compact disk contain several kinds of codes. Left-hand and right-hand stereo sound codes are recorded alternately, and the sequences also contain codes that control the motor speed and give timings.

READING ZERO (OFF)
The laser beam enters a pit in the track and is not reflected, so that the photodiode produces no signal.

READING ONE (ON)
The track surface reflects the light beam to the photodiode, which produces an electric signal.

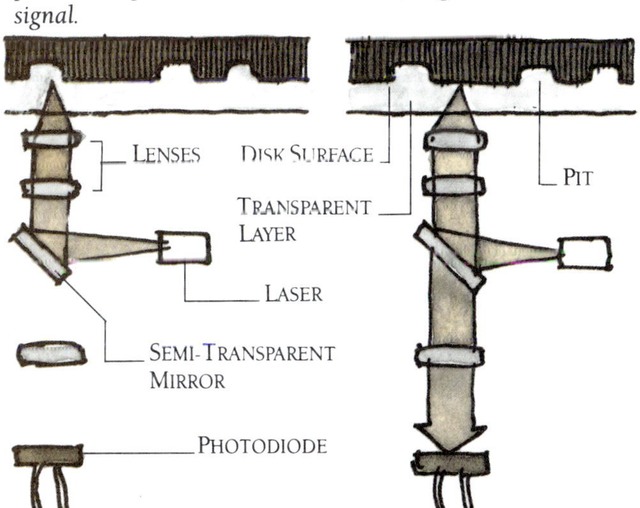

FILM SOUND

Film sound is often recorded by an analog system which, like the compact disk, uses light. The electric signal from the microphone goes to a recorder in which a light beam forms a track at the side of the film. The beam's intensity or width varies with signal voltage.

On projection, a lamp shines a beam of light through the sound track onto a photodiode (see p.292), which produces a varying electric signal.

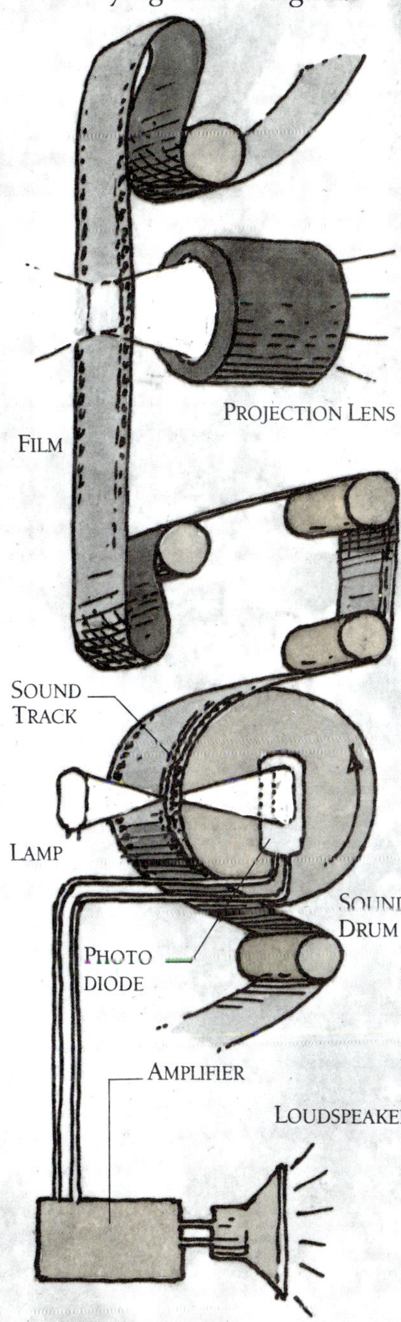

TELECOMMUNICATIONS

ON THE CONVEYING OF MESSAGES

While on a mammoth watch in the mountainous southern area, I was asked for some advice in the matter of communication between remote villages. It appeared that the age-old system of conveying messages – which relied on catapulting couriers from one place to another – was critically threatened by a shortage of both volunteers and also paper. After inspecting the catapults and calculating certain distances and elevations, I devised a completely new system. Instead of relying on dwindling manpower, I suggested that the messages could be carried through the air in the form of stones.

INSTANT SOUNDS AND IMAGES

Telecommunications are communications at a distance beyond the range of unaided hearing or eyesight. In order to send messages without delay over long distances, a fast-moving signal carrier is required. The method of telecommunication recorded above uses catapulted rocks as the signal carriers. The rocks are hurled aloft in a sequence that encodes a message, and when they land, the sequence is decoded and the message read.

Modern telecommunications use electricity, light and radio waves as signal carriers. These have the great advantage that they travel almost instantly. The simplest way of making them carry a message is to interrupt them; this is the basis of Morse code, which uses combinations of short and long signals to code letters and numbers. In more advanced methods of telecommunication, the signal carrier is "modulated" by combining it with a signal produced by a sound or image. The modulated signal is then sent to a receiver. A detector in the receiver extracts the signal from the signal carrier and reproduces the sound or image.

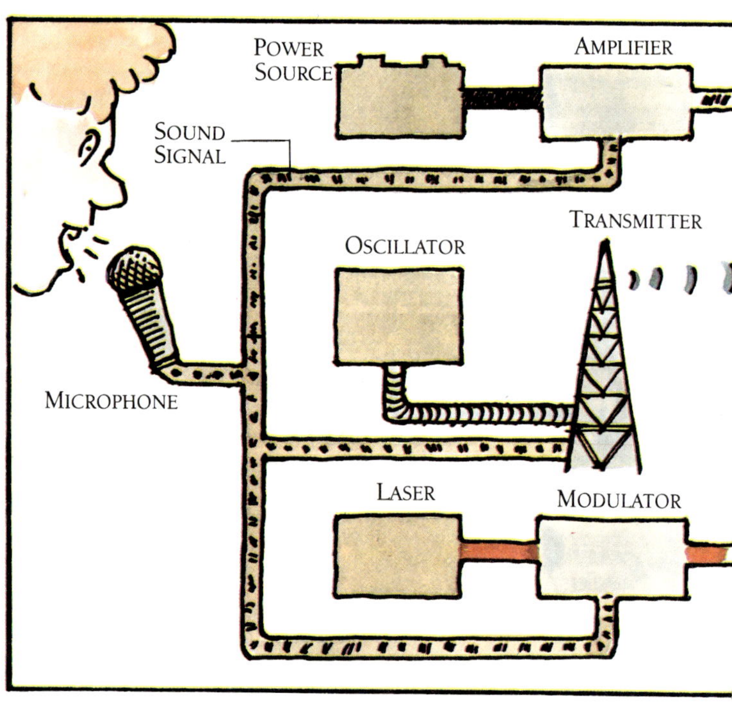

My system worked as follows. Stones of predetermined size were launched in particular combinations – each combination representing a letter of the alphabet. The various combinations were observed as they arrived and then translated back into words by a trained translator. Safety was assured by installing a large metal funnel in the center of each village to catch the incoming messages.

The technical aspects of the system worked perfectly. However, I had completely overlooked the villagers' atrocious spelling. So frequent were unintentional insults that all forms of communication eventually ceased.

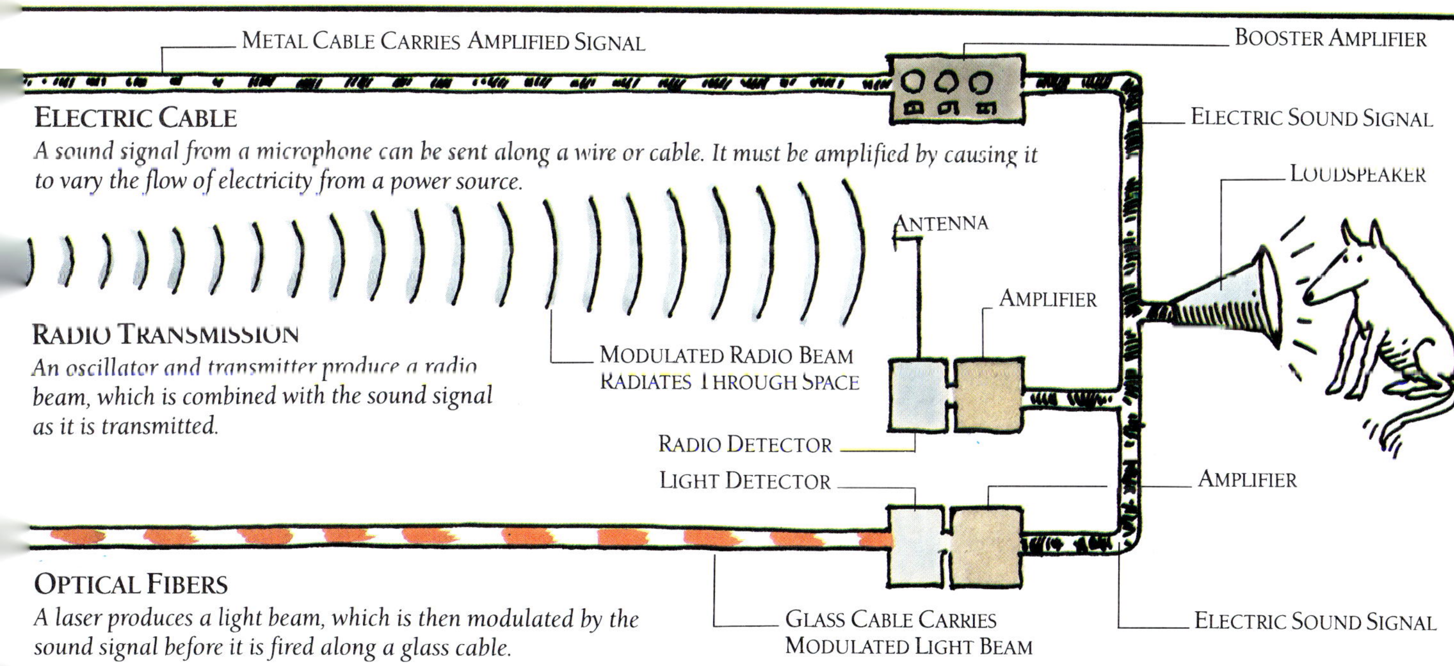

ELECTRIC CABLE

A sound signal from a microphone can be sent along a wire or cable. It must be amplified by causing it to vary the flow of electricity from a power source.

RADIO TRANSMISSION

An oscillator and transmitter produce a radio beam, which is combined with the sound signal as it is transmitted.

OPTICAL FIBERS

A laser produces a light beam, which is then modulated by the sound signal before it is fired along a glass cable.

THE TELEPHONE

A telephone handset contains a microphone in the mouthpiece and a small loudspeaker in the earpiece. A vibrating diaphragm alters the resistance (see p.286) of carbon granules in the microphone, causing the current flowing through the microphone to vary. Dialing a number connects two handsets through a network of exchanges. At the handset, sound signals are in the form of electrical signals carried along metal wires. In the telephone network, they may be converted to light signals carried along fiber-optic cables, or radio signals transmitted over radio or microwave links. Modern telephone systems convert speech into digital signals. These are processed by "multiplexing" so that more than one conversation can be carried on an individual line.

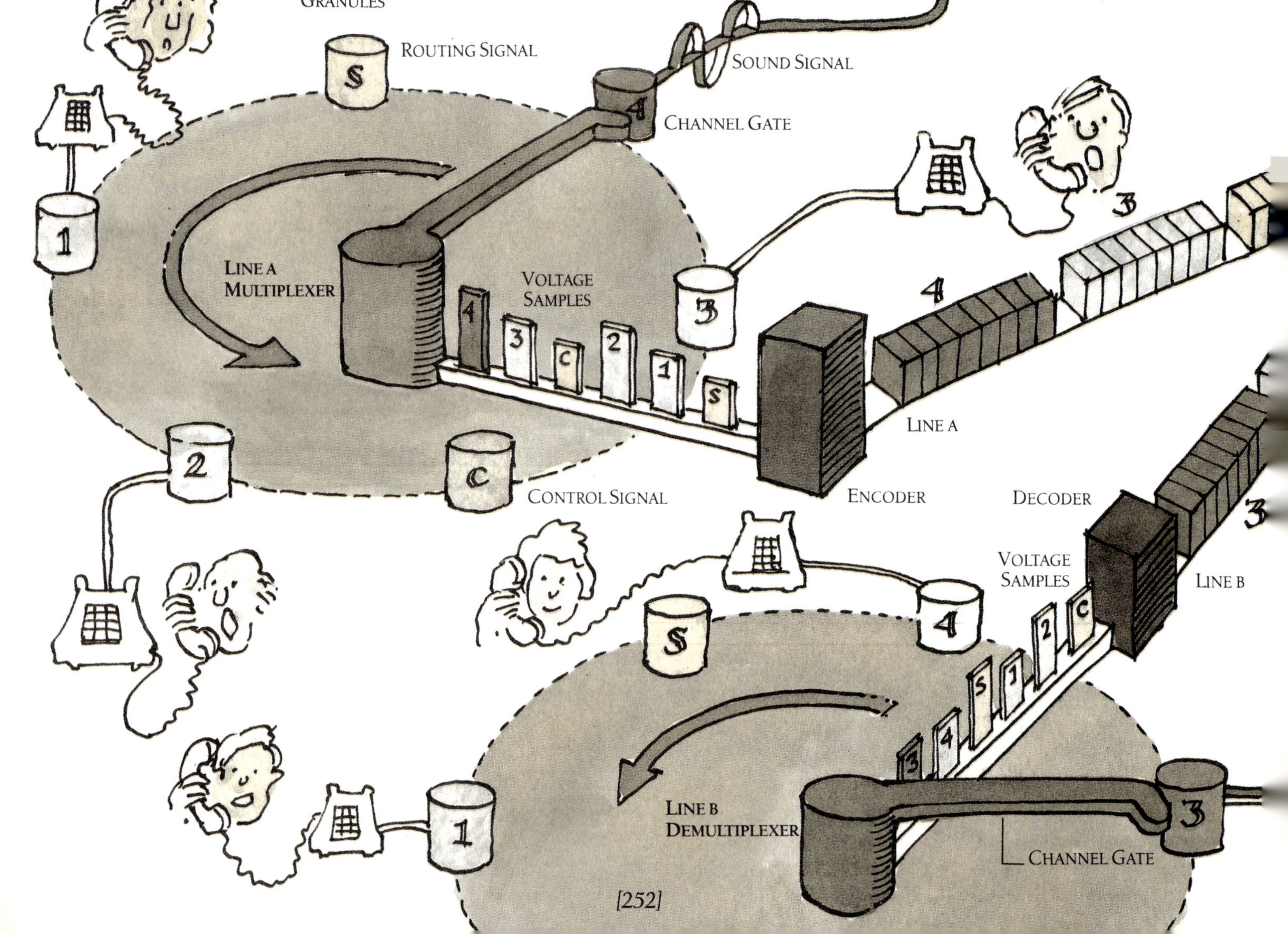

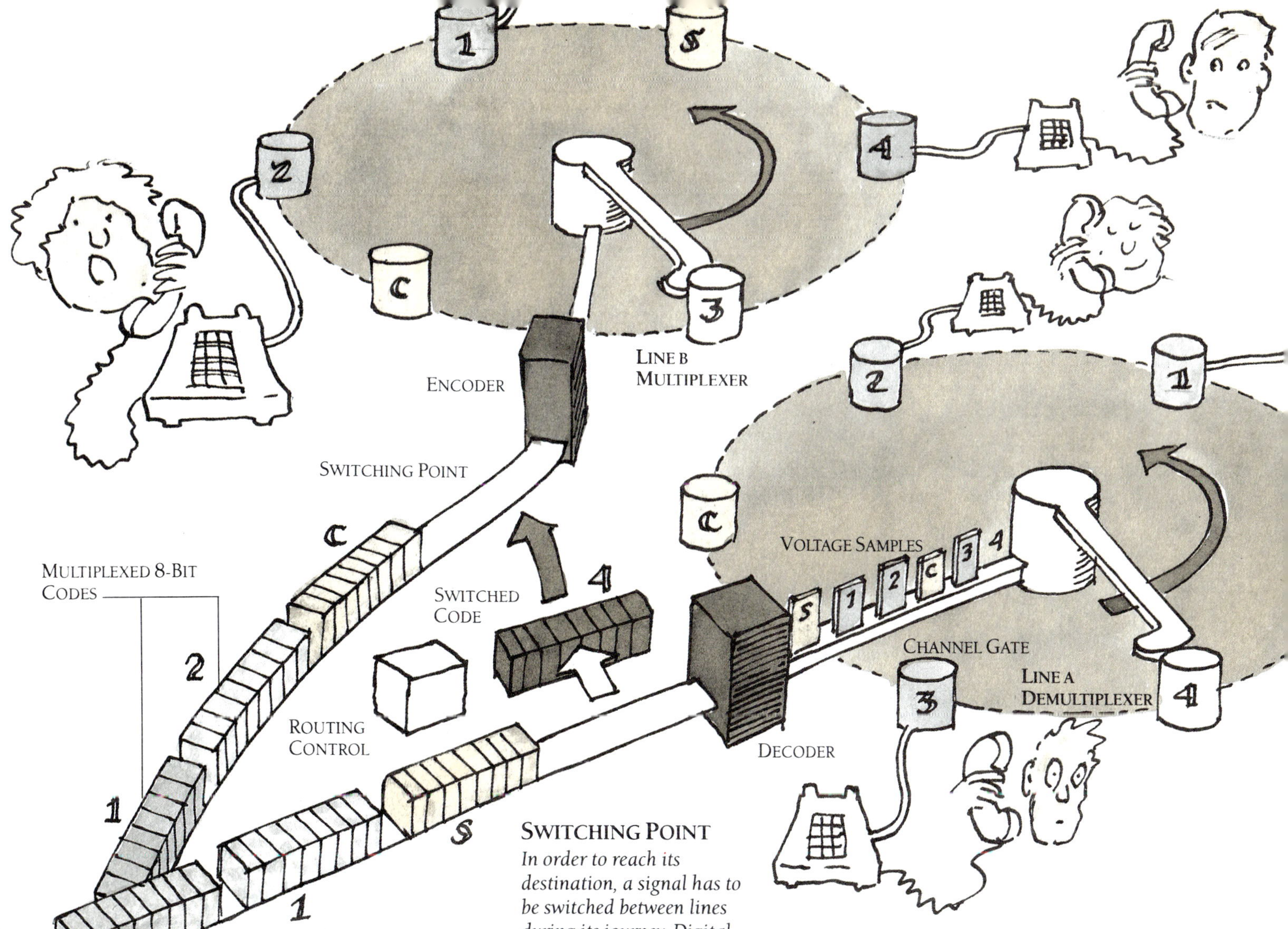

Switching Point

In order to reach its destination, a signal has to be switched between lines during its journey. Digital switching makes use of routing signals to send the 8-bit codes of a single call from one line to another.

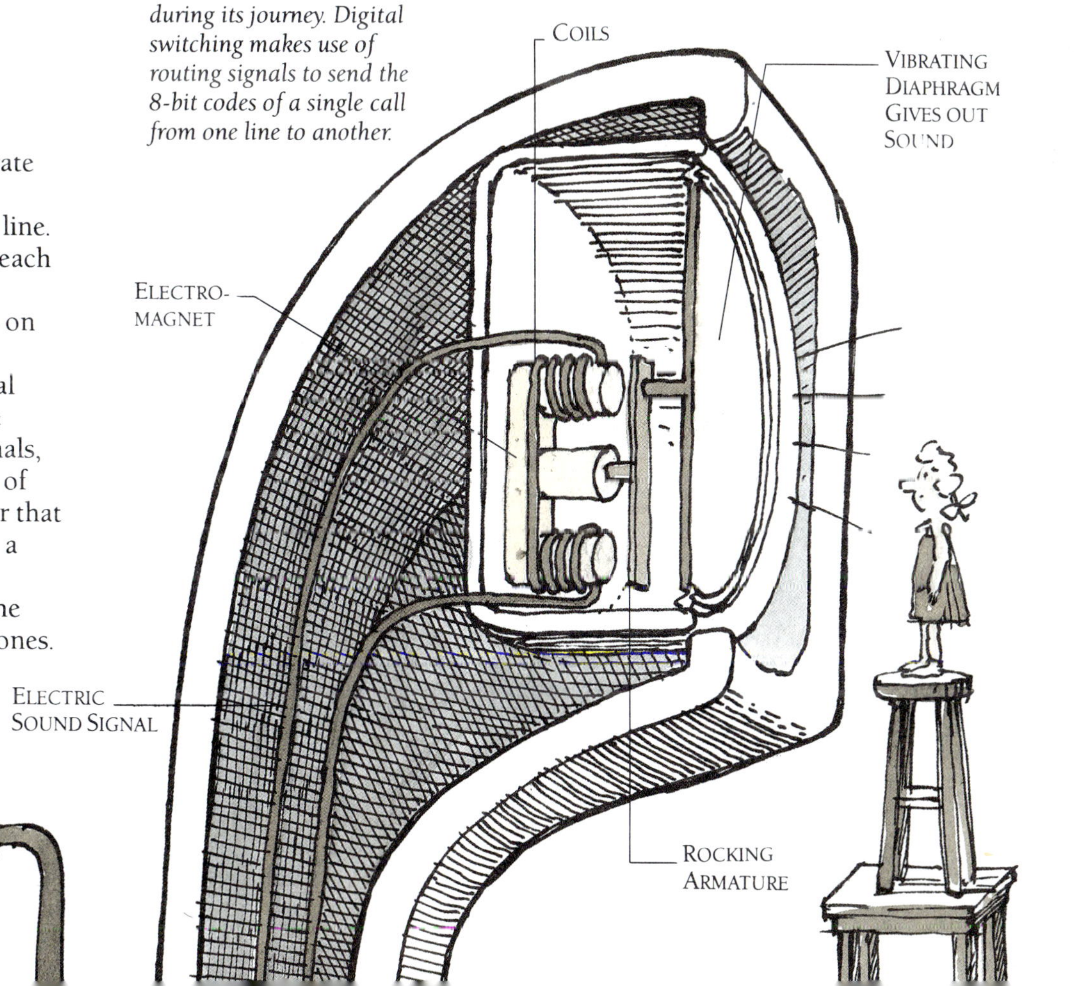

Multiplexing

Multiplexing enables separate telephone calls to be sent simultaneously along a single line. Here, two lines – A and B – are each carrying four calls.

The signals from the telephones on each line are sampled in turn by the multiplexer in the same way as a digital recording (see pp.242-3). The voltage samples, plus control and routing signals, are then encoded in binary. Sequences of codes travel along the lines in the order that they were sampled. At the destination, a decoder converts the codes back into samples, and a demultiplexer routes the samples through to the correct telephones.

RADIO TRANSMITTER

Radio waves are produced by feeding an electric signal to the mast or antenna of a transmitter. The signal makes the electrons in the metal atoms of the mast or antenna change energy levels and emit radio waves. They do this in the same way that electrons in atoms emit light rays (see p.192). Radio transmitters broadcast radio waves that are modulated. This means that the original sound signal is superimposed on the radio wave so that the radio wave "carries" the sound.

Like all waves, radio waves have a particular frequency or wavelength. Frequency is a measure of the number of waves that are transmitted per second, and it is measured in hertz. Wavelength is a measure of the length of each complete wave, and is expressed in metres. Frequency and wavelength are directly linked: a radio wave with a high frequency has a short wavelength, and one with a low frequency has a long wavelength.

SOUND SIGNAL

A microphone responds to sound waves by producing an electric sound signal that changes in voltage at the same rate or frequency. The curved line represents the variation of voltage in the signal.

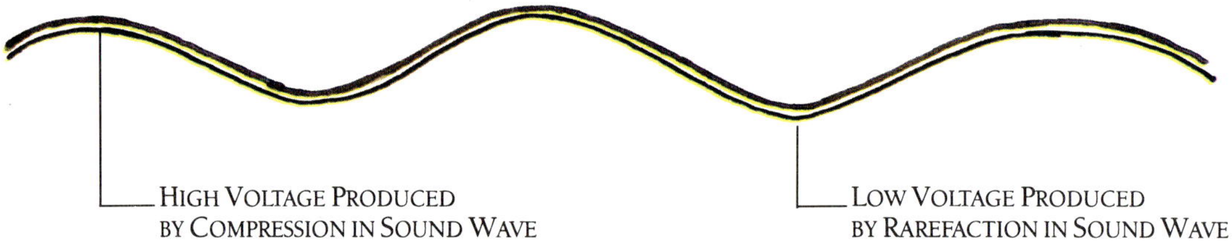

CARRIER SIGNAL

The radio wave that carries the sound signal is called the carrier wave. It is produced by a radio-frequency (rf) carrier signal, which is an electric signal generated by a component called an oscillator. The frequency of the rf signal is constant and very much greater than the frequency range of the sounds being carried.

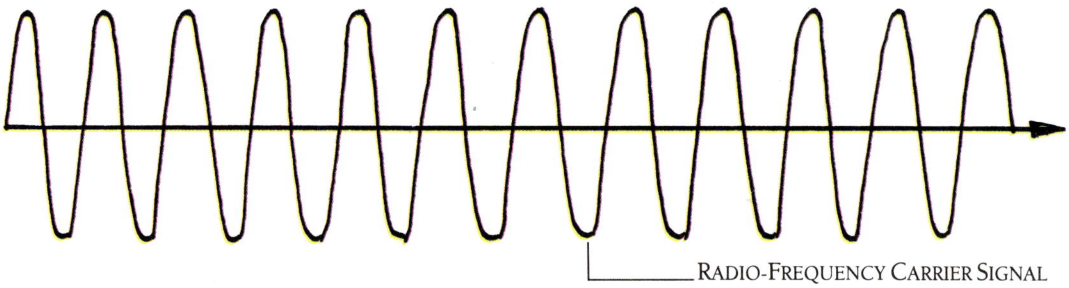

MODULATED SIGNAL

The sound signal from the microphone and the rf carrier signal from the oscillator are amplified and then combined in the modulator of the transmitter. This is done by amplitude modulation (AM) or by frequency modulation (FM). In AM radio, the waves are modulated so that the amplitude (energy level) of the carrier wave varies at the same frequency as the changing voltage in the sound signal. In FM radio, the waves are modulated so that that frequency of the carrier wave varies with the voltage level of the sound signal.

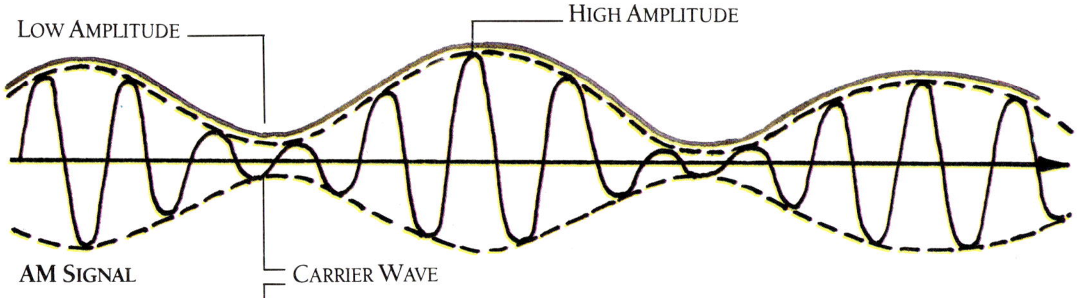

AM SIGNAL

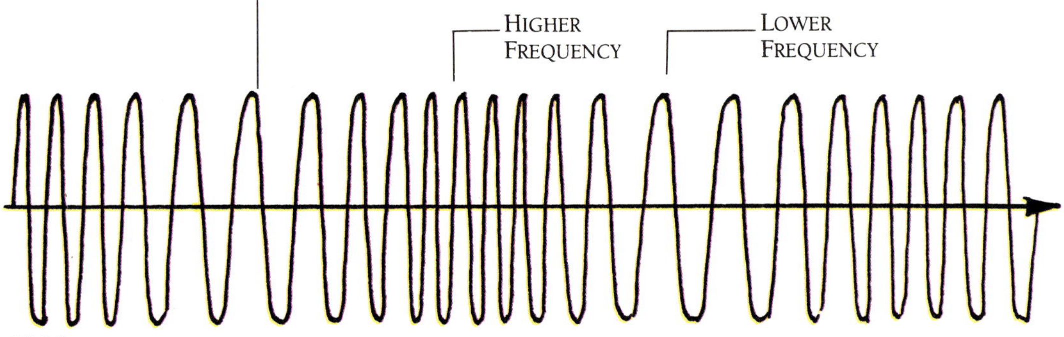

FM SIGNAL

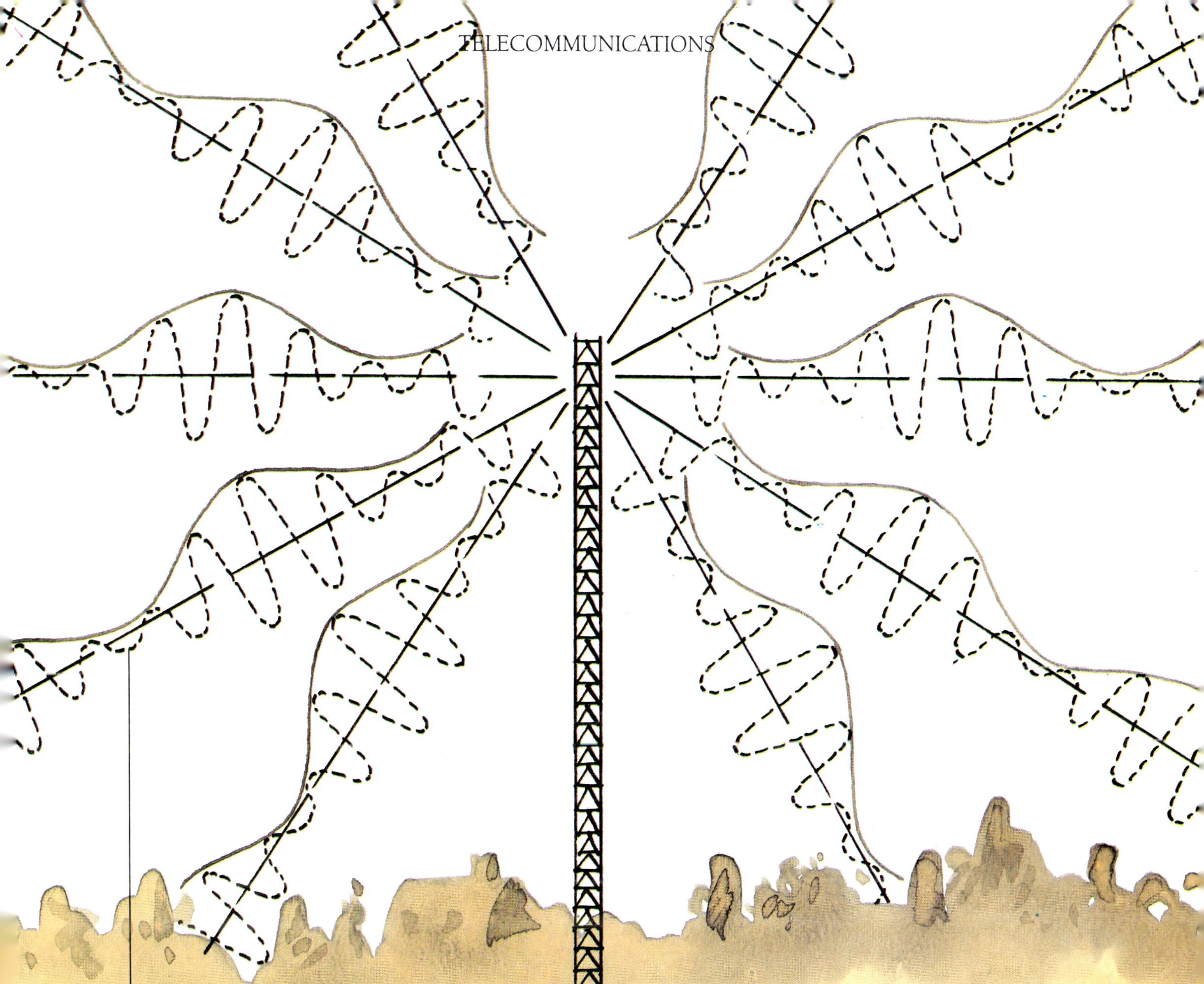

ANALOG AND DIGITAL RADIO

The AM waves shown here carry an analog sound signal, but AM and FM radio can also carry digital signals (see pp.212-3). Microwave beams, radio waves of extremely high frequency, carry telephone calls as digital signals (see p.252).

ELECTROMAGNETIC WAVES

Radio waves are part of a large family of rays and waves known as electromagnetic waves. They consist of electric and magnetic fields that vibrate at right angles to each other. Both vibrate at the same frequency.

Light rays are also electromagnetic, and so too are radar, microwaves, infra-red rays, ultraviolet rays and X-rays. All electromagnetic waves move at the speed of light, which is 186,000 miles per second (300,000 kilometers per second). They travel through air and space.

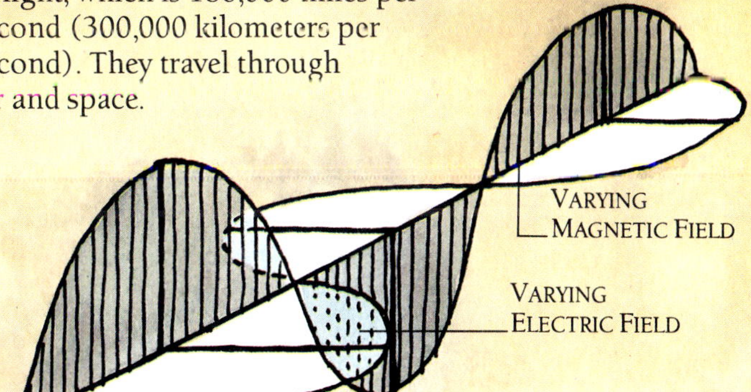

AMPLIFICATION AND TRANSMISSION

The modulated signal next goes to a powerful amplifier, which sends it to the mast or antenna of the transmitter. Radio carrier waves, which are modulated in exactly the same way as the modulated signal, radiate from the transmitter. A radio mast broadcasts several carrier waves at different frequencies, each carrying a different sound signal. Every radio station or channel broadcasting from the transmitter has a different frequency.

RADIO RECEIVER

A radio receiver is essentially a transmitter in reverse. Radio waves strike the antenna connected to the receiver. They affect the metal atoms, producing weak electric carrier signals in the antenna. The receiver then selects the carrier signal of the required station or channel. It extracts the sound signal from the carrier signal, and this signal goes to an amplifier and loudspeaker to reproduce the sound.

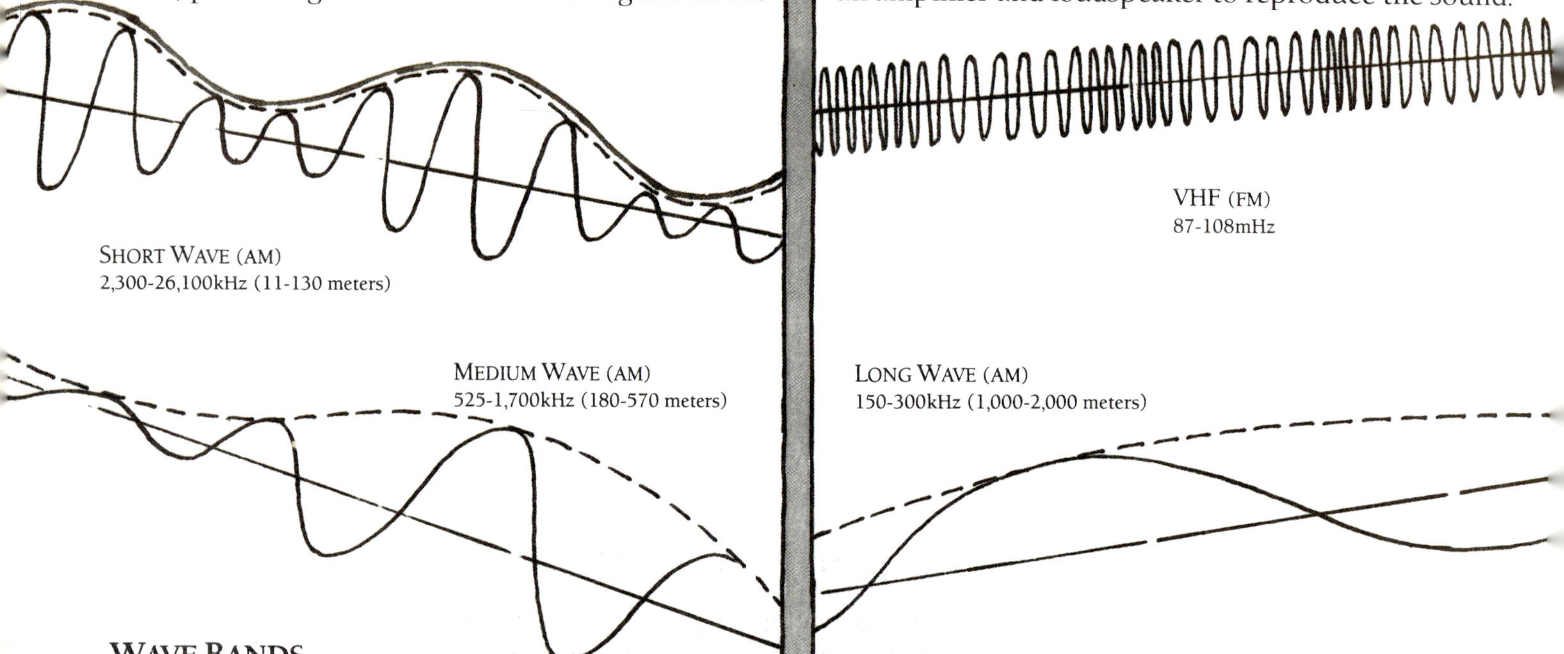

WAVE BANDS

Radio waves are broadcast in several wave bands, often called long wave, medium wave, short wave and VHF (Very High Frequency) bands. Each band contains a range of radio frequencies or wavelengths, and each station or channel has its own particular frequency or wavelength within a band.

Stations are shown on a radio dial either in meters, which indicates the wavelength of the carrier wave, or in hertz, which gives its frequency. Radio frequencies are so high that they are shown in kilohertz (kHz) or megahertz (mHz), which are thousands and millions of hertz. The long, medium and short wave bands broadcast in AM; VHF radio is FM.

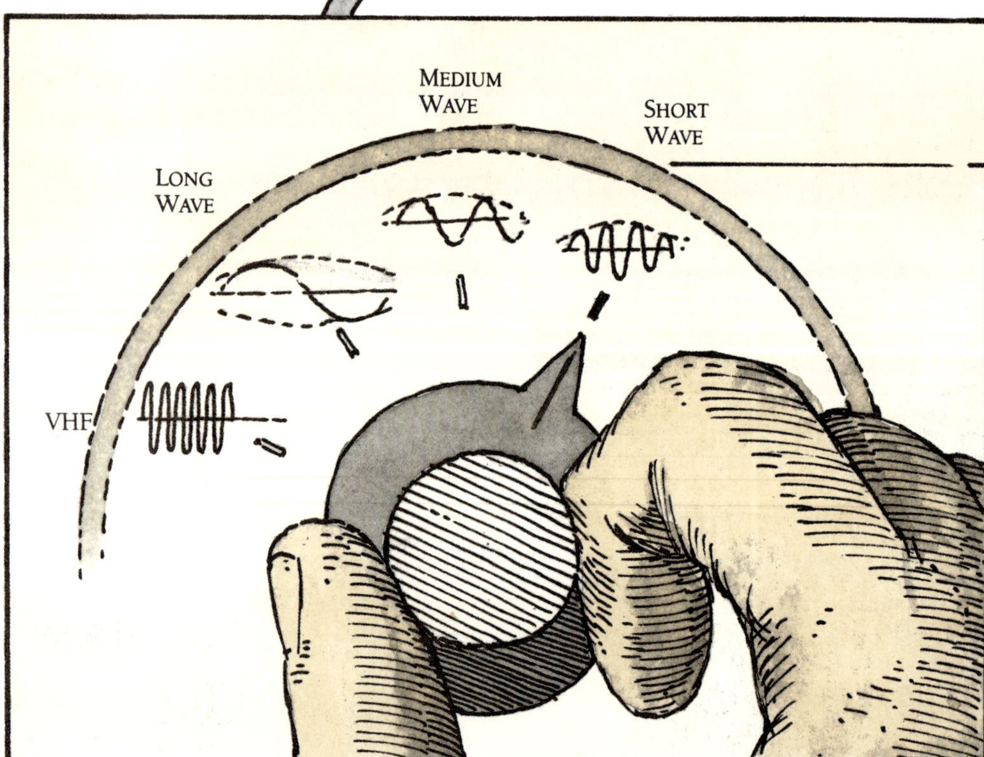

REPRODUCING THE SOUND

The carrier signal selected by the tuner is modulated with the original sound signal. The detector in the receiver removes the carrier frequency to produce the sound signal. This signal is then amplified and goes to the loudspeaker in the receiver. In stereo radio, left and right signals are combined and broadcast on one carrier. A stereo receiver has a decoder that separates them into two sound signals, while an ordinary receiver reproduces the combined signal.

TUNER

The tuner selects a particular station or channel by removing all other frequencies. The required carrier signal passes through the tuner and then goes to the detector.

Radio Signals

VHF and Medium Waves

VHF waves (*below*) travel a short distance, bouncing off the ground or large objects.

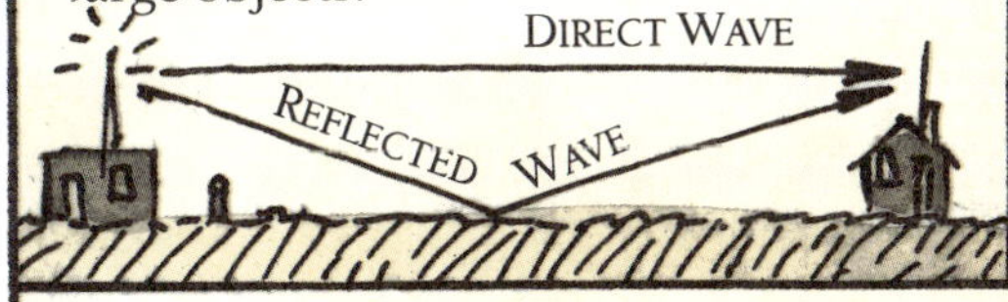

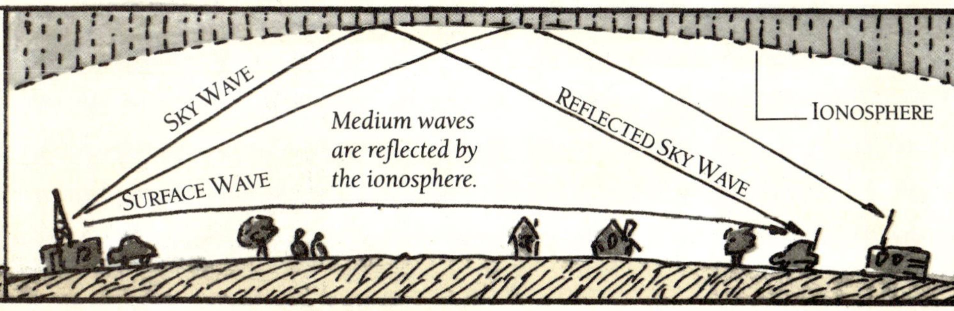

Long Waves

A surface wave curves around the Earth's surface, giving a range of thousands of miles or kilometers.

Short Waves

Multiple reflections of a sky wave between the ionosphere and the Earth's surface give worldwide communications by short waves.

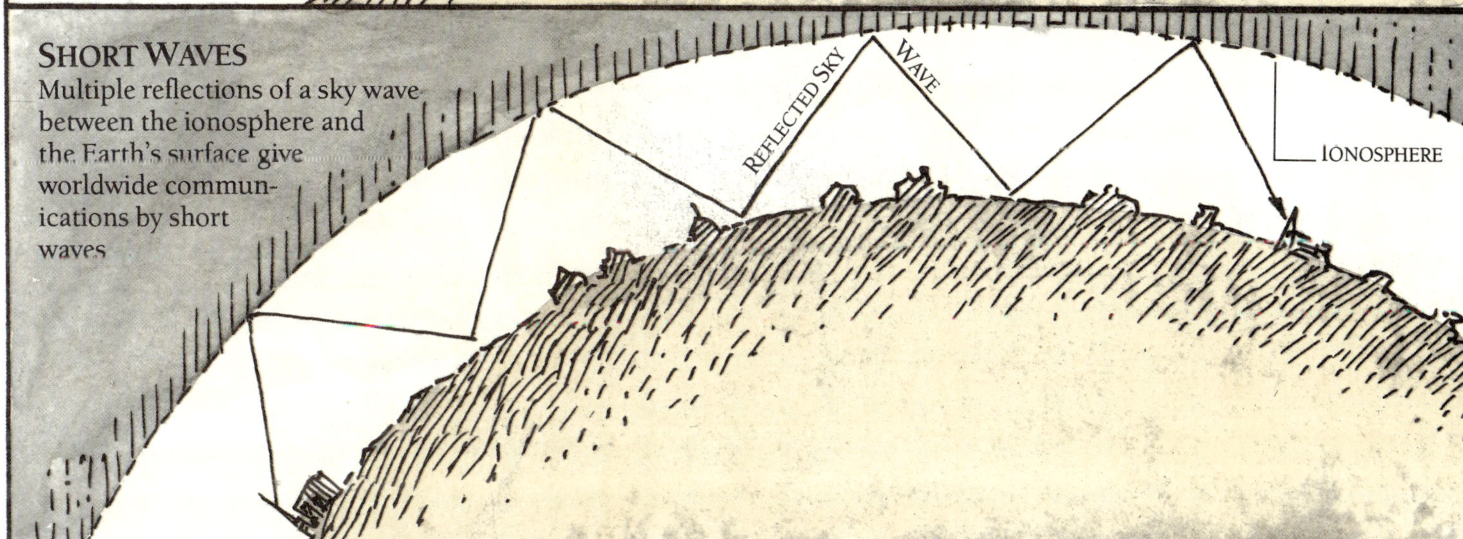

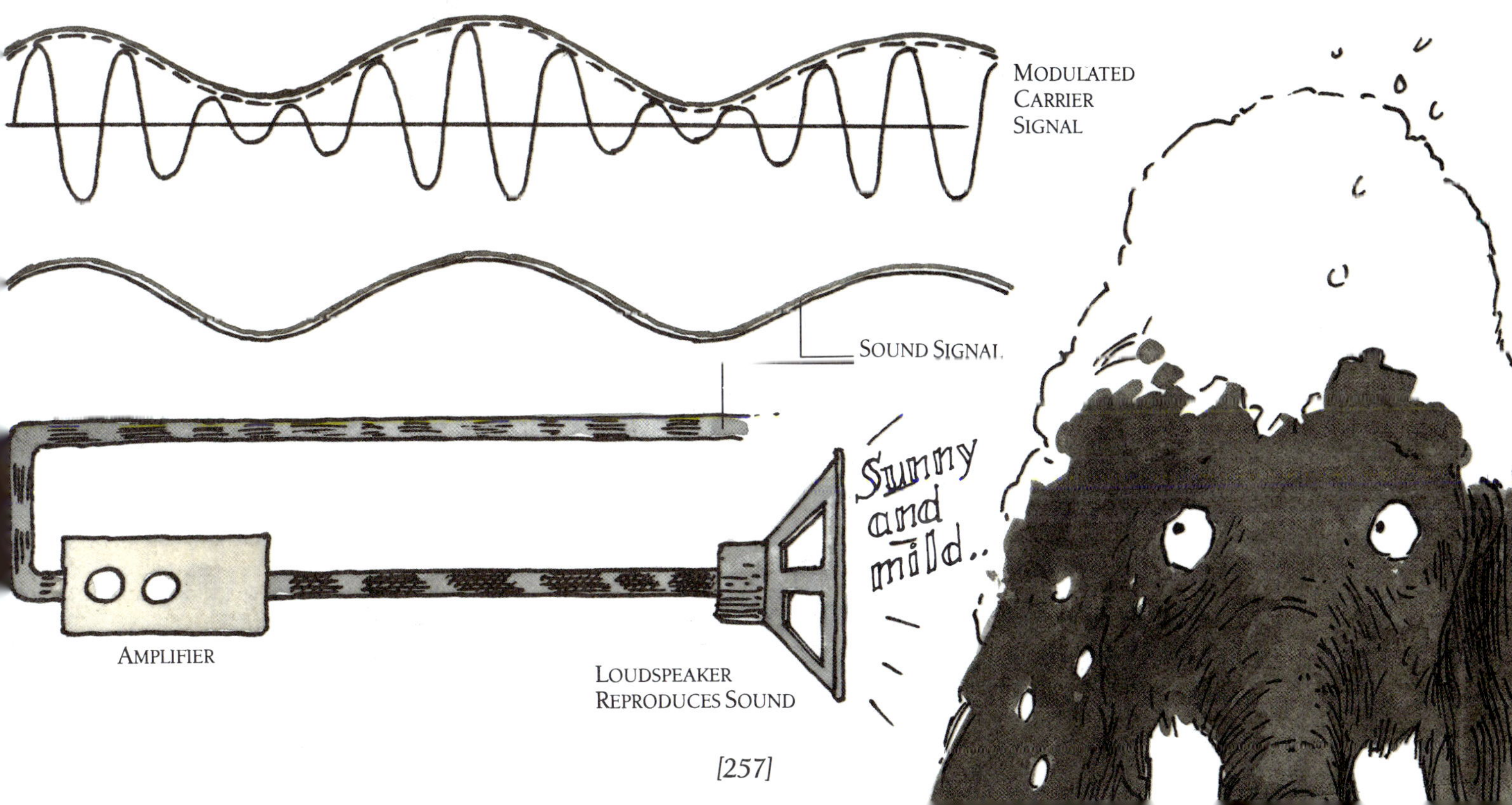

TELEVISION CAMERA

Television transmits a sequence of 30 still images per second, and the eyes merge these into a moving picture in the same way as they do with a movie film (see p.218). A color television camera produces three images of each still picture in the three primary colors — red, green and blue. Each image forms on a tube that converts the light in the image into an electric signal. The tube scans the image, splitting it into 525 horizontal lines. It puts out an electric signal whose voltage varies with the brightness of the image along each line. The signals from the three tubes are then combined into one electric video signal, which is transmitted to the home by radio or cable.

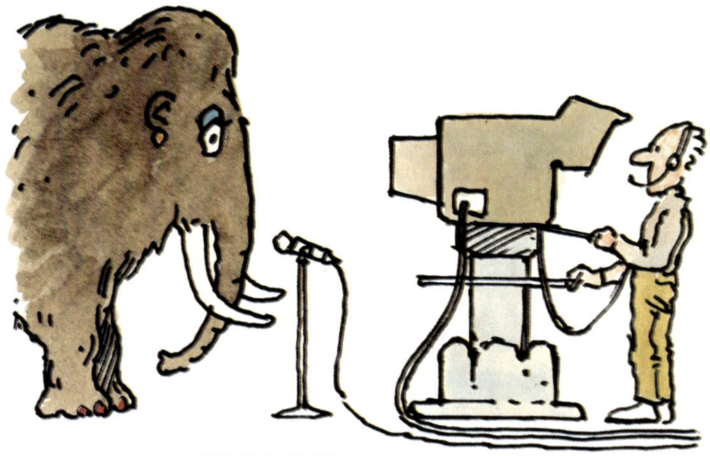

STUDIO CAMERA

Mirrors separate the image formed by the lens into three images in red, green and blue. Video cameras may contain a single color tube that separates the picture without using mirrors.

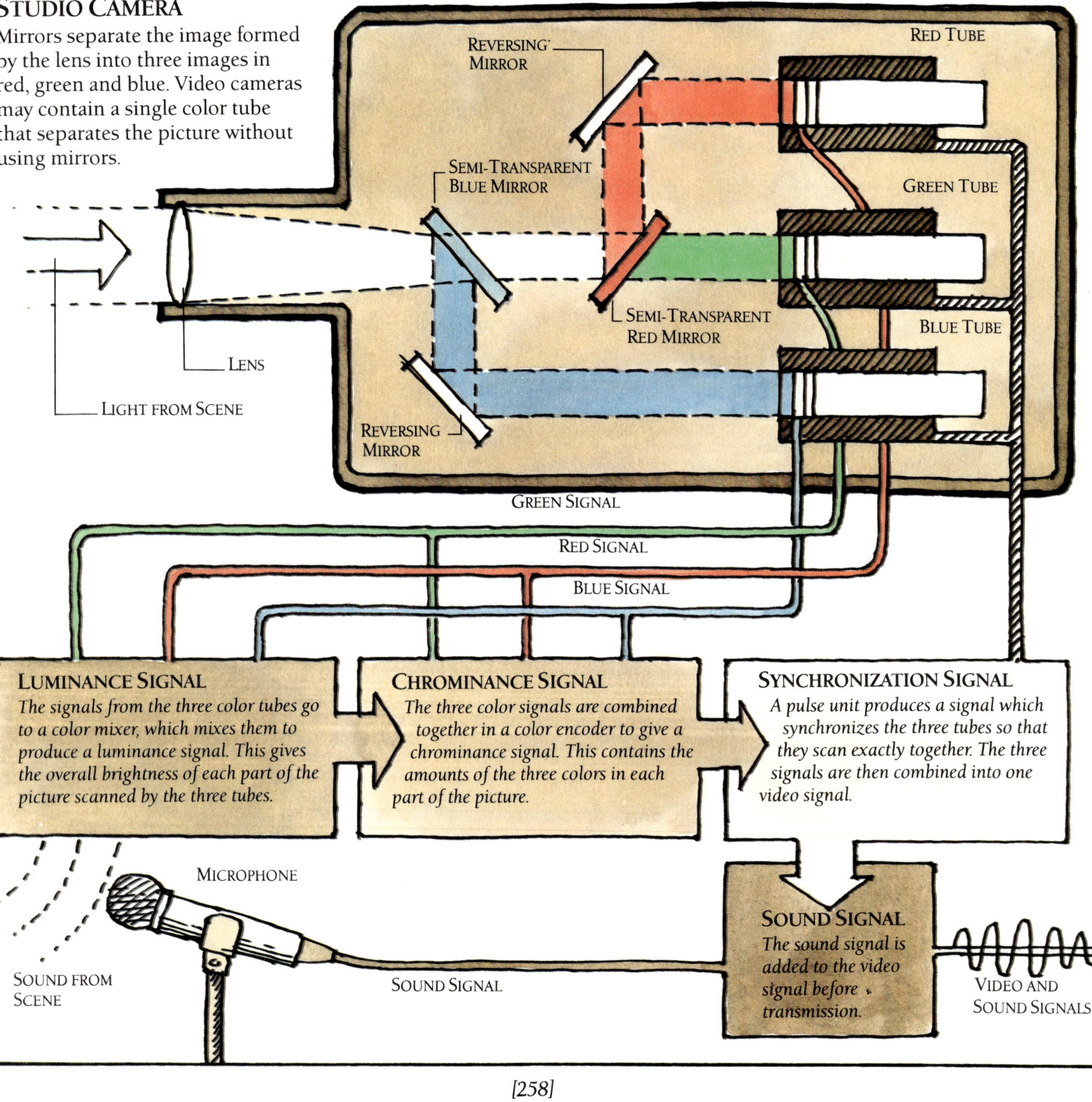

Vidicon Tube

In a vidicon picture tube, the image formed by a lens strikes the target plate. An electron beam scans the back of the target plate, causing a variation in the voltage supplied to the front of the plate through the transparent conductive layer. The voltage depends on the amount of light at the point where the beam strikes the target plate.

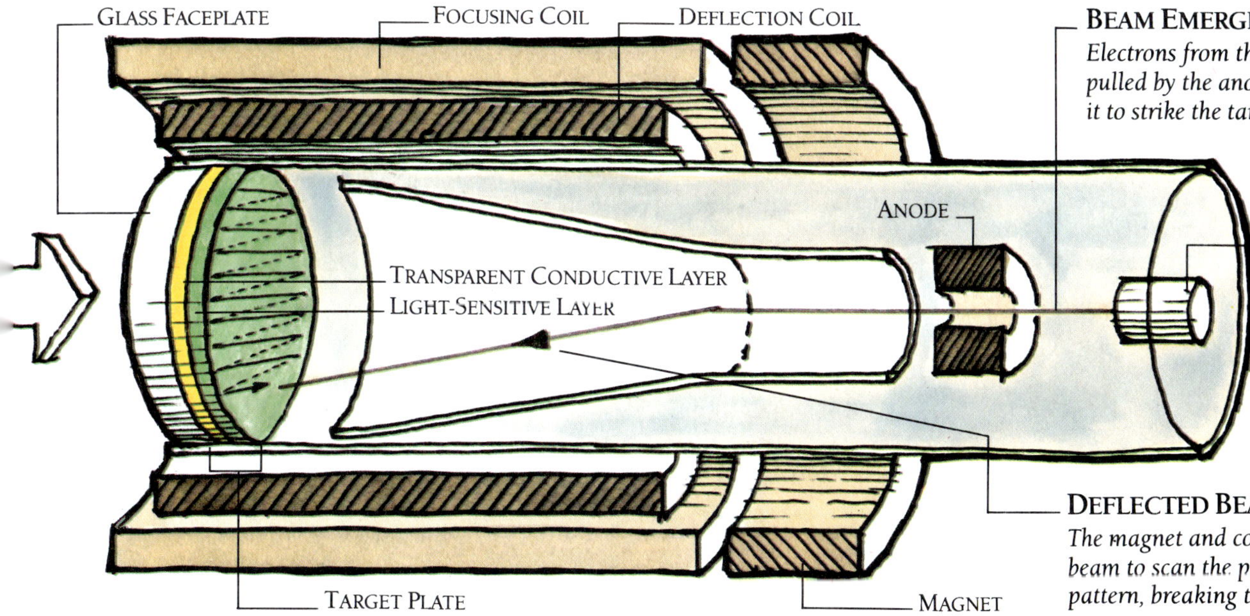

Beam Emerging from Gun
Electrons from the electron gun are pulled by the anode, passing through it to strike the target plate.

Deflected Beam
The magnet and coils deflect the beam to scan the plate in a zigzag pattern, breaking the picture into lines.

1 Light-Sensitive Layer Charges
The back and front of the light-sensitive layer are first charged through the conductive layer.

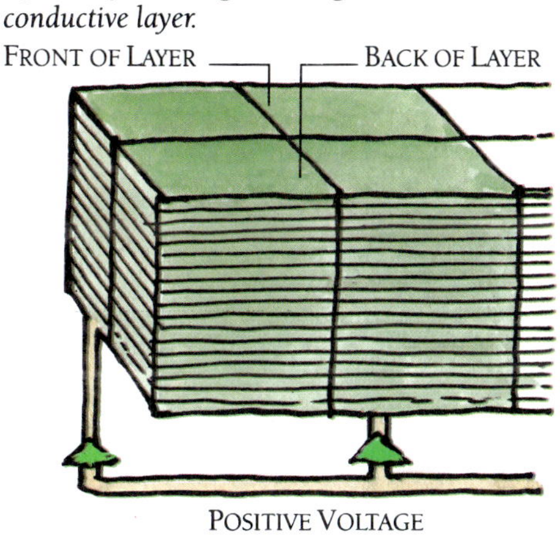

2 Electron Beam Scans
Electrons strike the back of the light-sensitive layer, neutralizing the positive charge there.

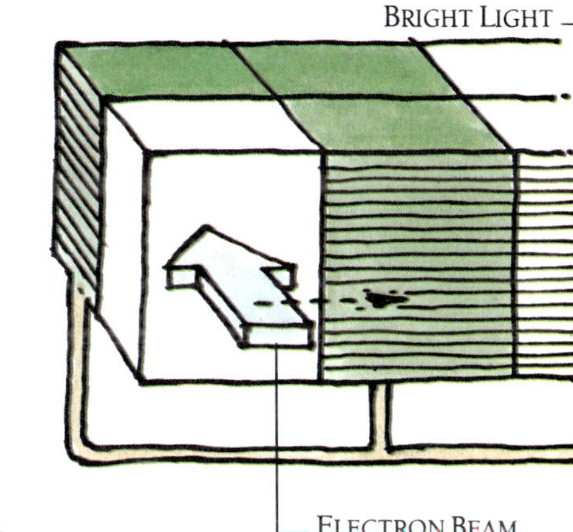

3 Light Alters the Charge
Bright light striking an area on the front of the layer reduces its resistance so much that all the charge flows to the back.

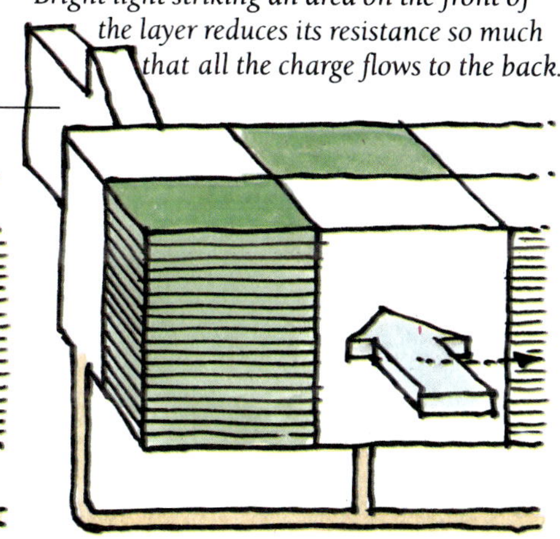

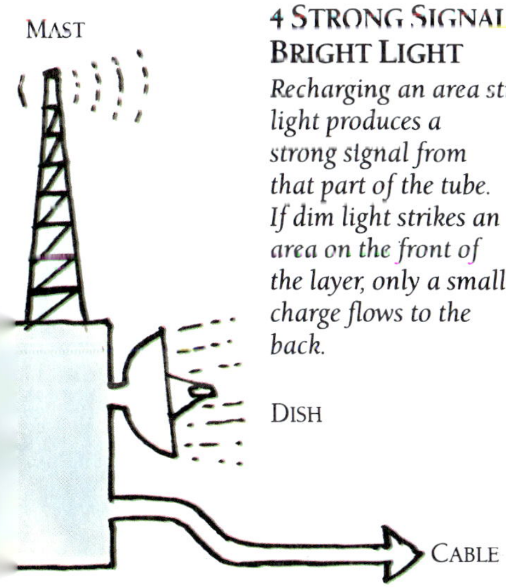

4 Strong Signal from Bright Light
Recharging an area struck by bright light produces a strong signal from that part of the tube. If dim light strikes an area on the front of the layer, only a small charge flows to the back.

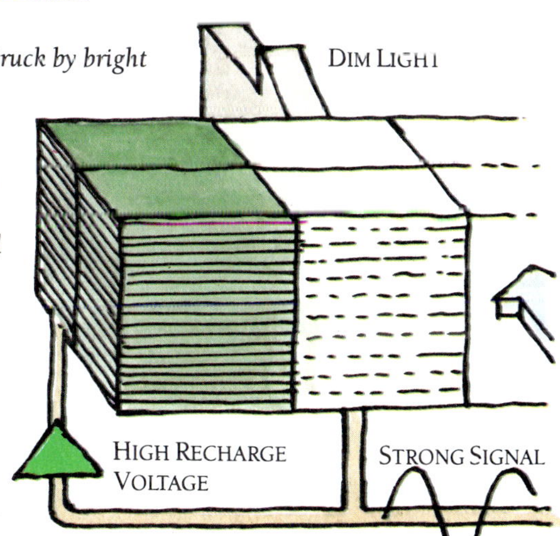

5 Weak Signal from Dim Light
Recharging an area struck by dim light produces a weak signal from that part of the tube.

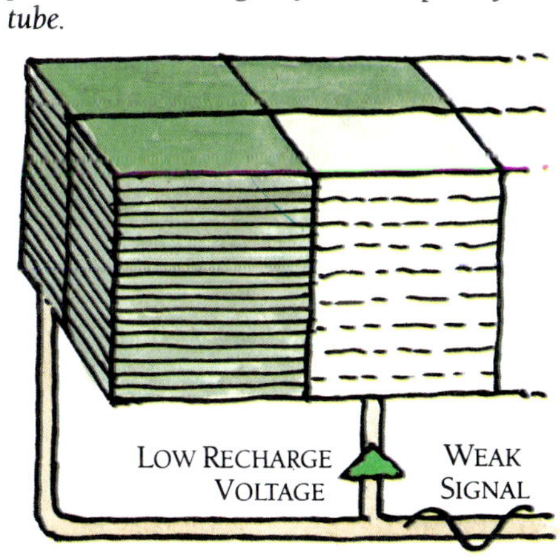

VIDEO RECORDER

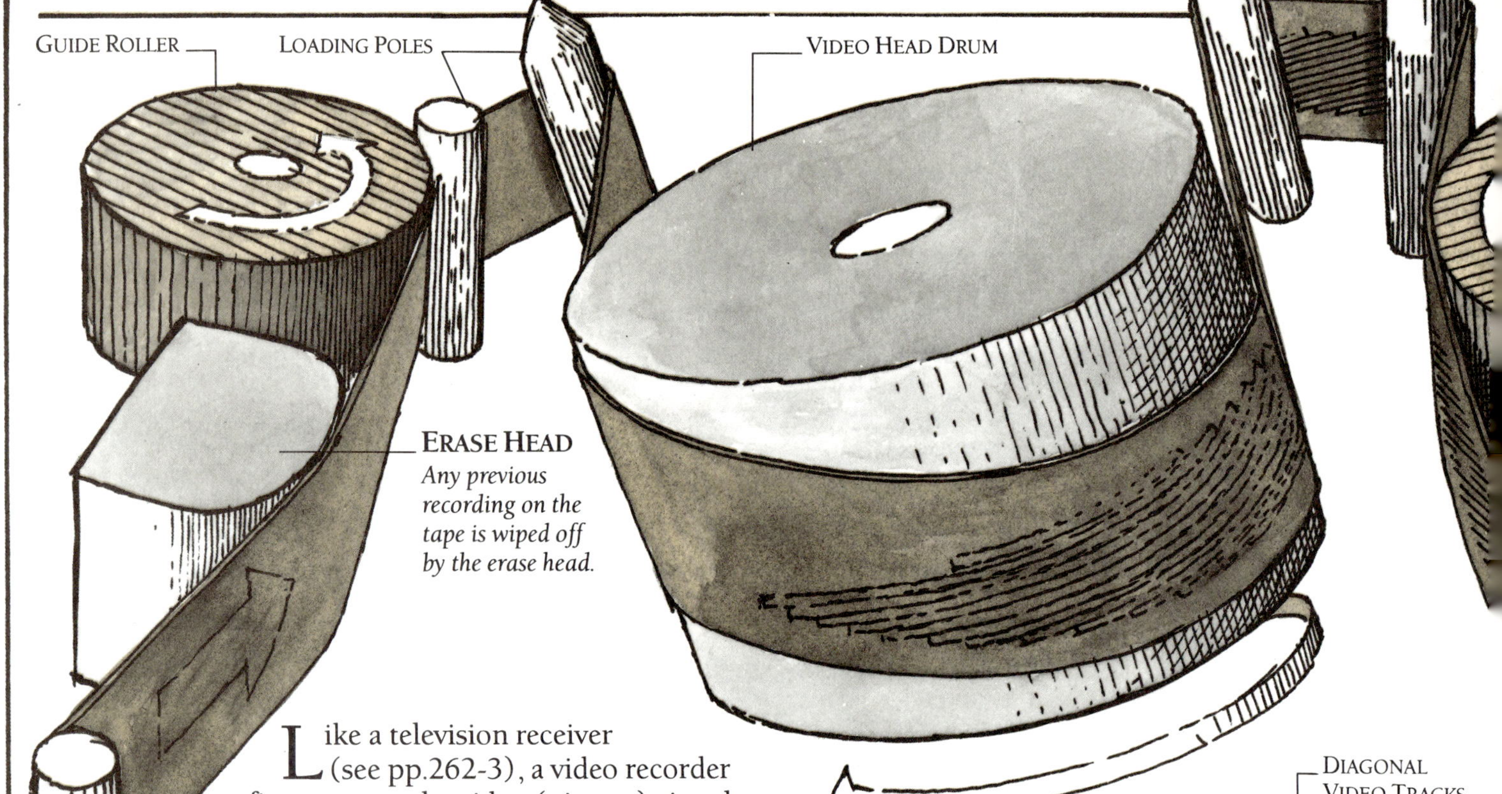

Like a television receiver (see pp.262-3), a video recorder first extracts the video (picture) signal from the television carrier signal that is either broadcast or cabled to the home. But instead of sending the video signal to the television tube to be seen, the video recorder preserves the signal on tape. It does this with a tape cassette in much the same way that a tape recorder records a sound signal (see pp.244-5).

The video signal goes to a record-replay head that records it on the magnetic tape, and replays the tape to send the recorded signal to a television set. However, the tape must pass the head at a very high speed to record a video signal, because so much information has to be stored. The head therefore rotates rapidly as the tape passes, recording video signals in diagonal tracks across the tape.

The VHS (Video Home System) arrangement shown here has become the main system used in home video cassette recorders. Other video systems use the same basic method of a rotating head to record the signals on tape.

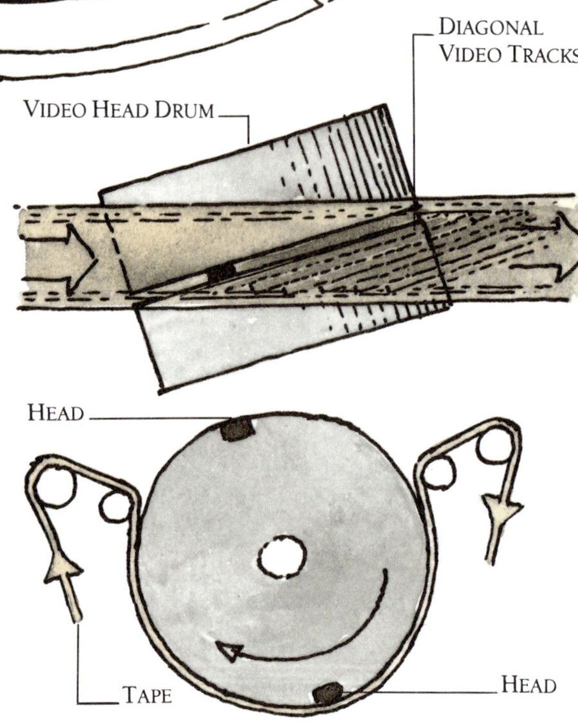

ROTATING VIDEO HEADS
The drum contains two video record/replay heads on opposite sides of the drum, and it is tilted relative to the tape. The pair of heads records one complete picture in each revolution in the form of a pair of diagonal tracks across the tape.

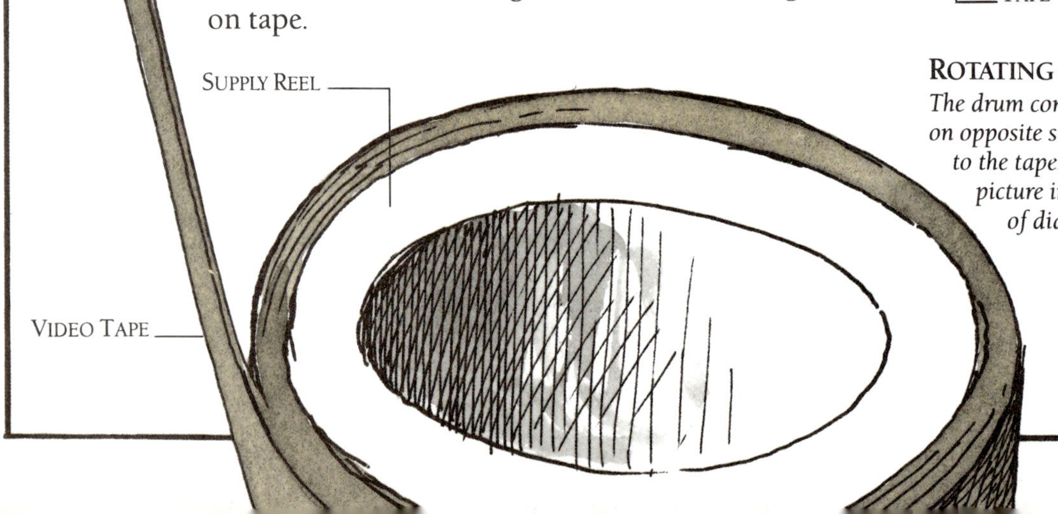

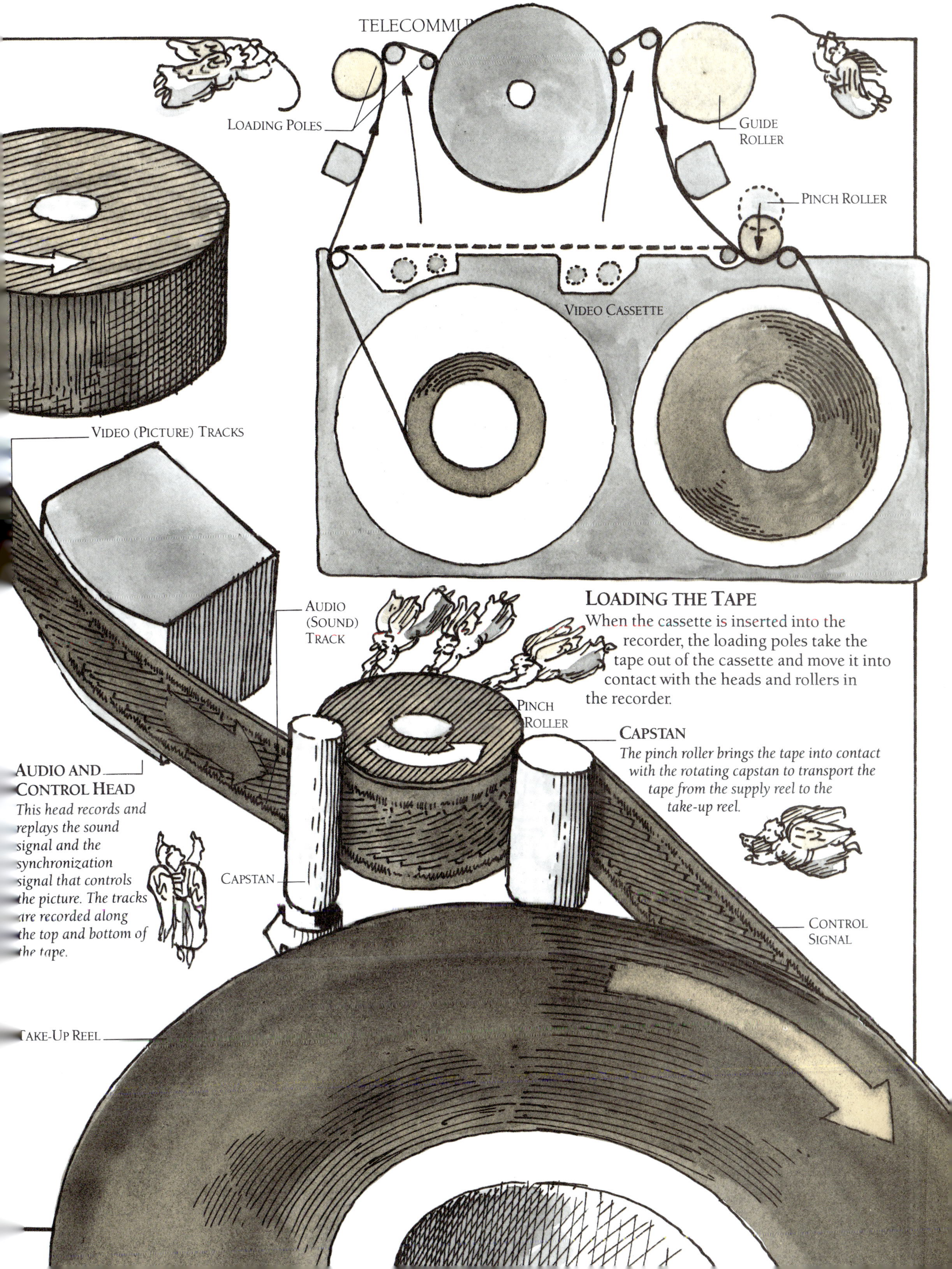

LOADING THE TAPE

When the cassette is inserted into the recorder, the loading poles take the tape out of the cassette and move it into contact with the heads and rollers in the recorder.

CAPSTAN

The pinch roller brings the tape into contact with the rotating capstan to transport the tape from the supply reel to the take-up reel.

AUDIO AND CONTROL HEAD

This head records and replays the sound signal and the synchronization signal that controls the picture. The tracks are recorded along the top and bottom of the tape.

TELEVISION SET

A television set receives a video signal from a television station or video recorder. It works like a television camera in reverse to form a succession of still pictures on the screen. It does this by scanning in the same way as a camera tube to build up the picture in horizontal lines on the screen. In a color picture, each line contains a series of red, green and blue stripes. At viewing distance, the lines and stripes cannot be made out. The eye merges them together, and we see a sharp picture in full color (see p.194).

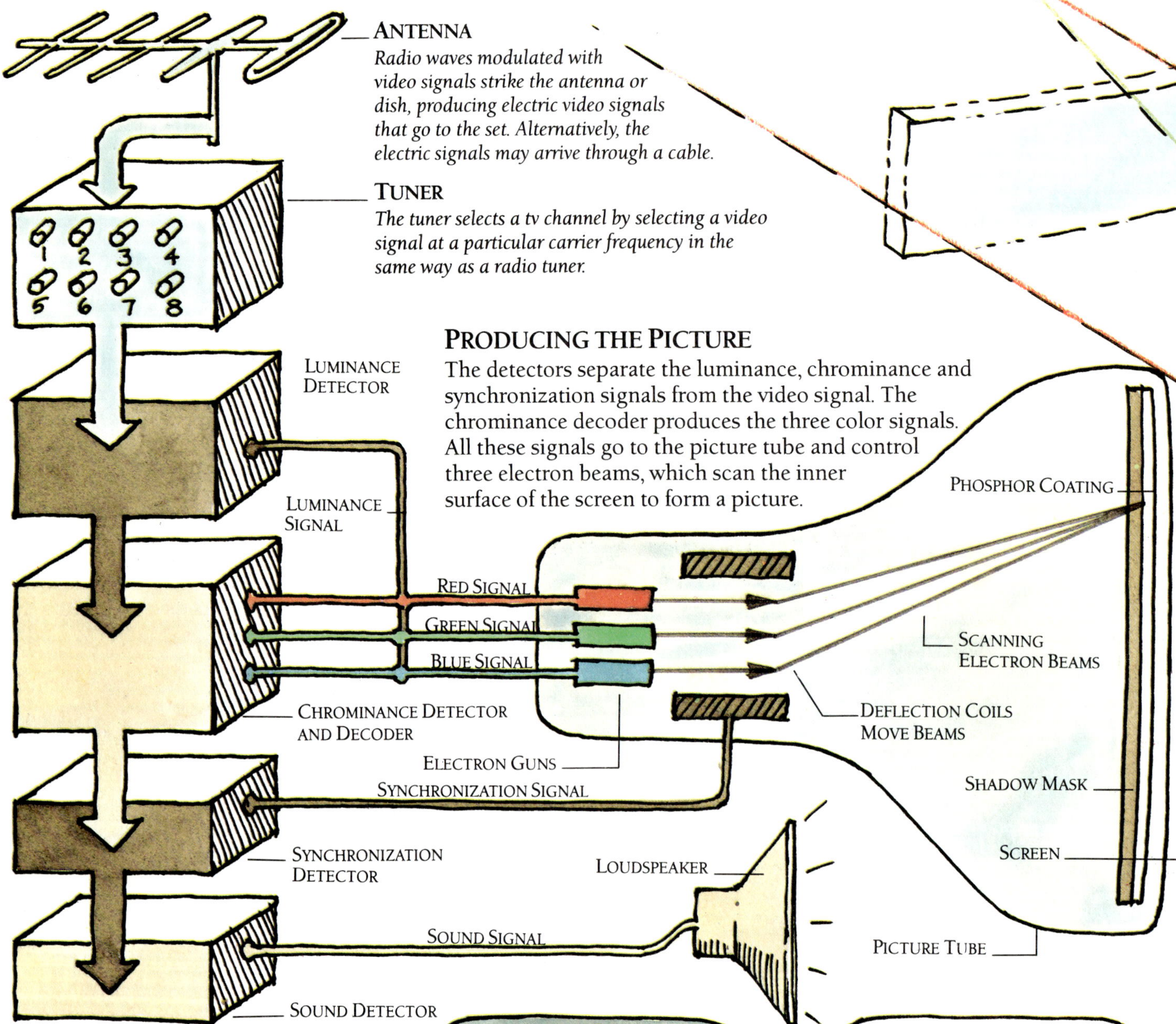

PRODUCING THE PICTURE

The detectors separate the luminance, chrominance and synchronization signals from the video signal. The chrominance decoder produces the three color signals. All these signals go to the picture tube and control three electron beams, which scan the inner surface of the screen to form a picture.

INTERLACING

Each still picture is made up of two scans consisting of alternate lines. The camera and picture tubes first scan the odd-numbered lines and then scan the picture again to form the even-numbered lines. We see 60 scans a second, which reduces flicker in the moving picture.

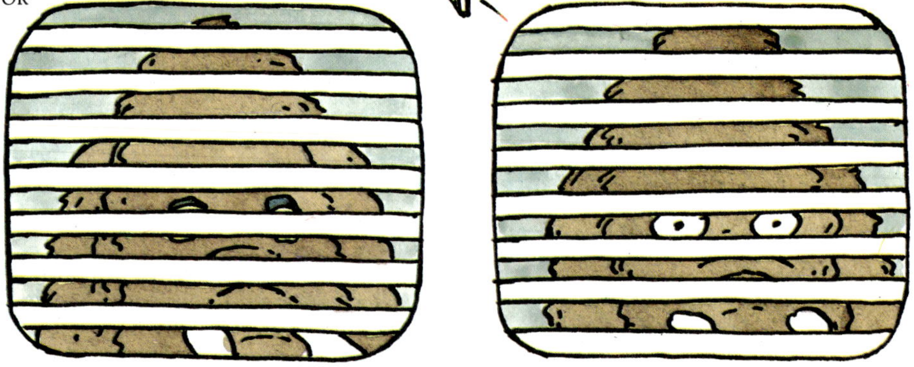

FIRST SCAN   SECOND SCAN

COLOR SCREEN

The screen contains tiny stripes of phosphors that light up in red, green or blue. The three electron beams, one for each color, scan across the shadow mask behind the screen. The mask contains holes that allow each beam to strike only stripes of the correct color. These light up as each beam passes, and vary in brightness with the strength of the beam. A black-and-white set contains only one electron beam, which is controlled by the luminance and synchronization signals, and no shadow mask.

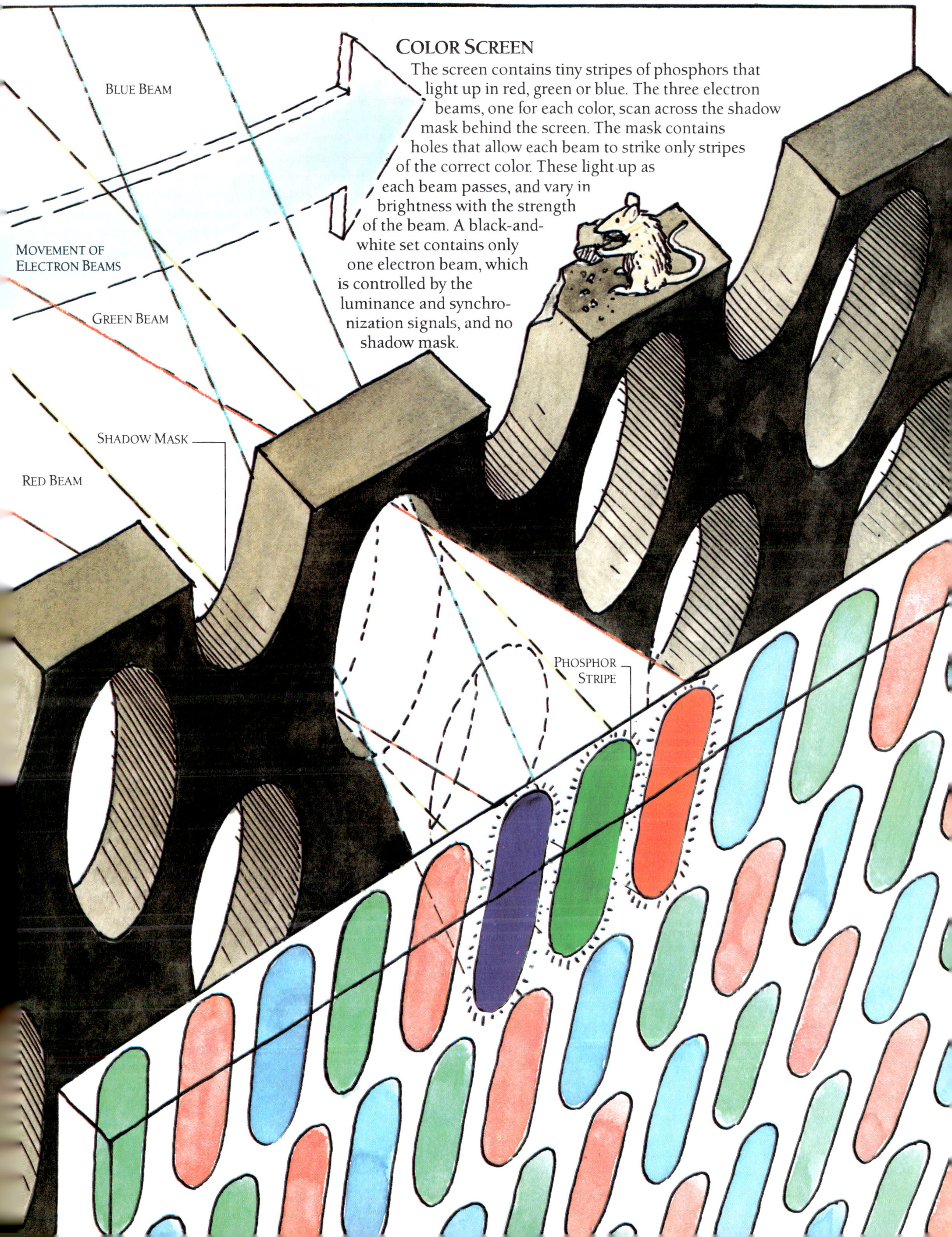

SATELLITES

Artificial satellites orbit the Earth, communicating with us from a unique vantage point high above the atmosphere. Weather and spy satellites look down and astronomy satellites peer outward while communications satellites link distant parts of our planet and beam television channels to our homes. Some satellites have orbits that criss-cross the Earth, while others are "parked" in geostationary orbits, so that they always remain above a particular point on the equator. All satellite communications, whether voices, pictures or information, reach us from space by radio.

RADIO DISHES

Many satellites and ground stations have radio dishes that transmit and receive signals. The curved dish reflects outgoing signals from the central horn to form a narrow beam, and reflects signals in an incoming beam to meet at the central horn.

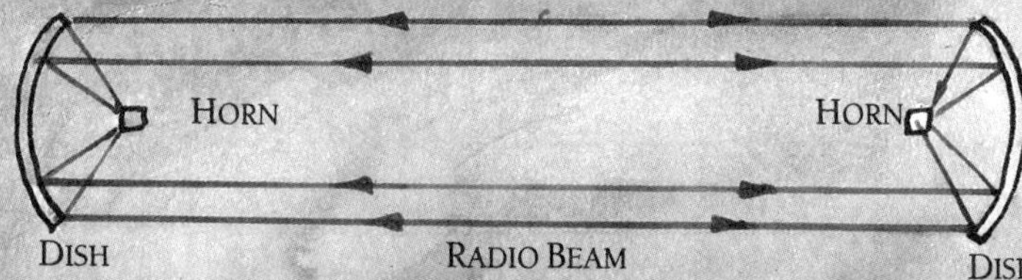

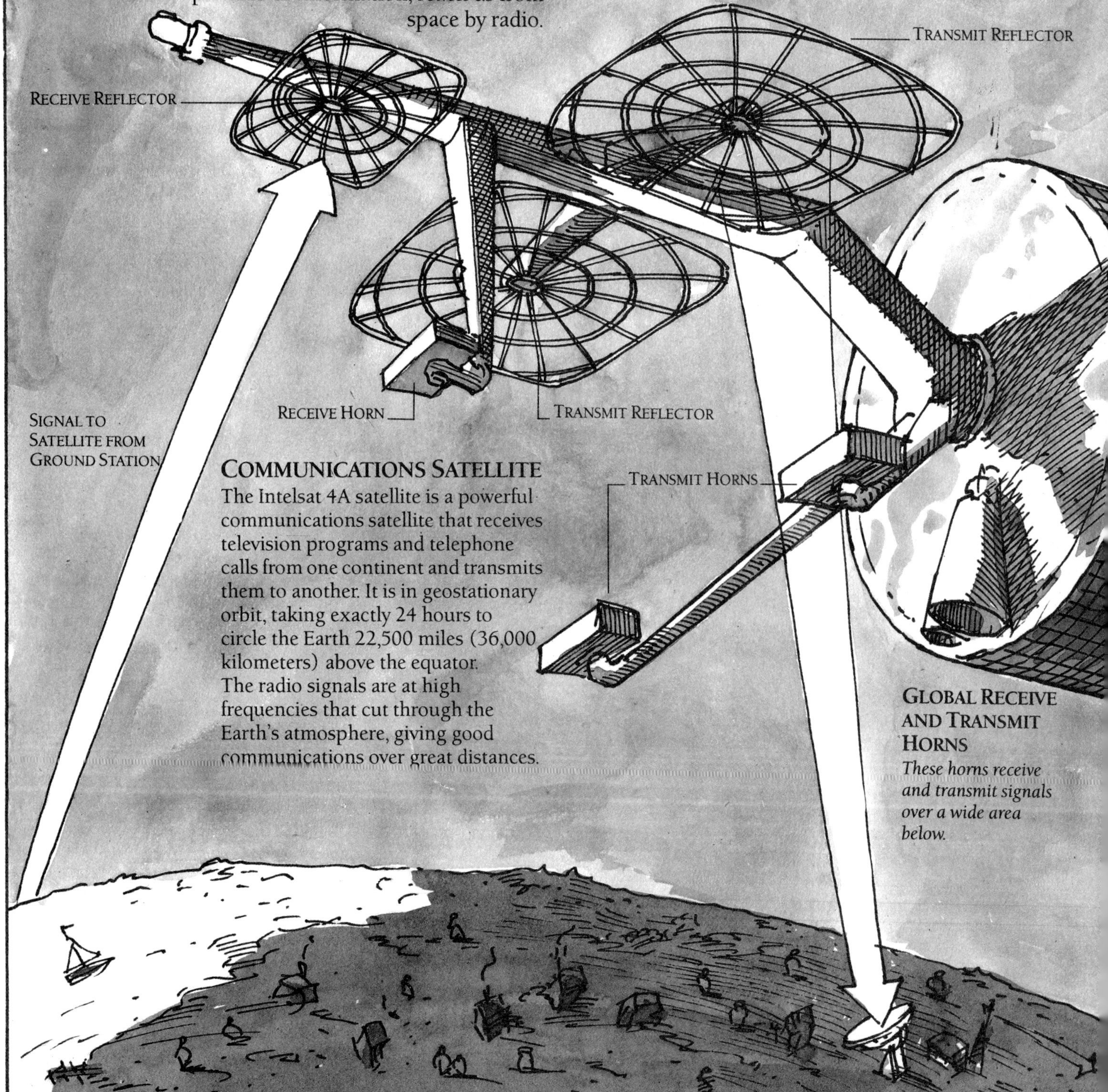

COMMUNICATIONS SATELLITE

The Intelsat 4A satellite is a powerful communications satellite that receives television programs and telephone calls from one continent and transmits them to another. It is in geostationary orbit, taking exactly 24 hours to circle the Earth 22,500 miles (36,000 kilometers) above the equator. The radio signals are at high frequencies that cut through the Earth's atmosphere, giving good communications over great distances.

GLOBAL RECEIVE AND TRANSMIT HORNS

These horns receive and transmit signals over a wide area below.

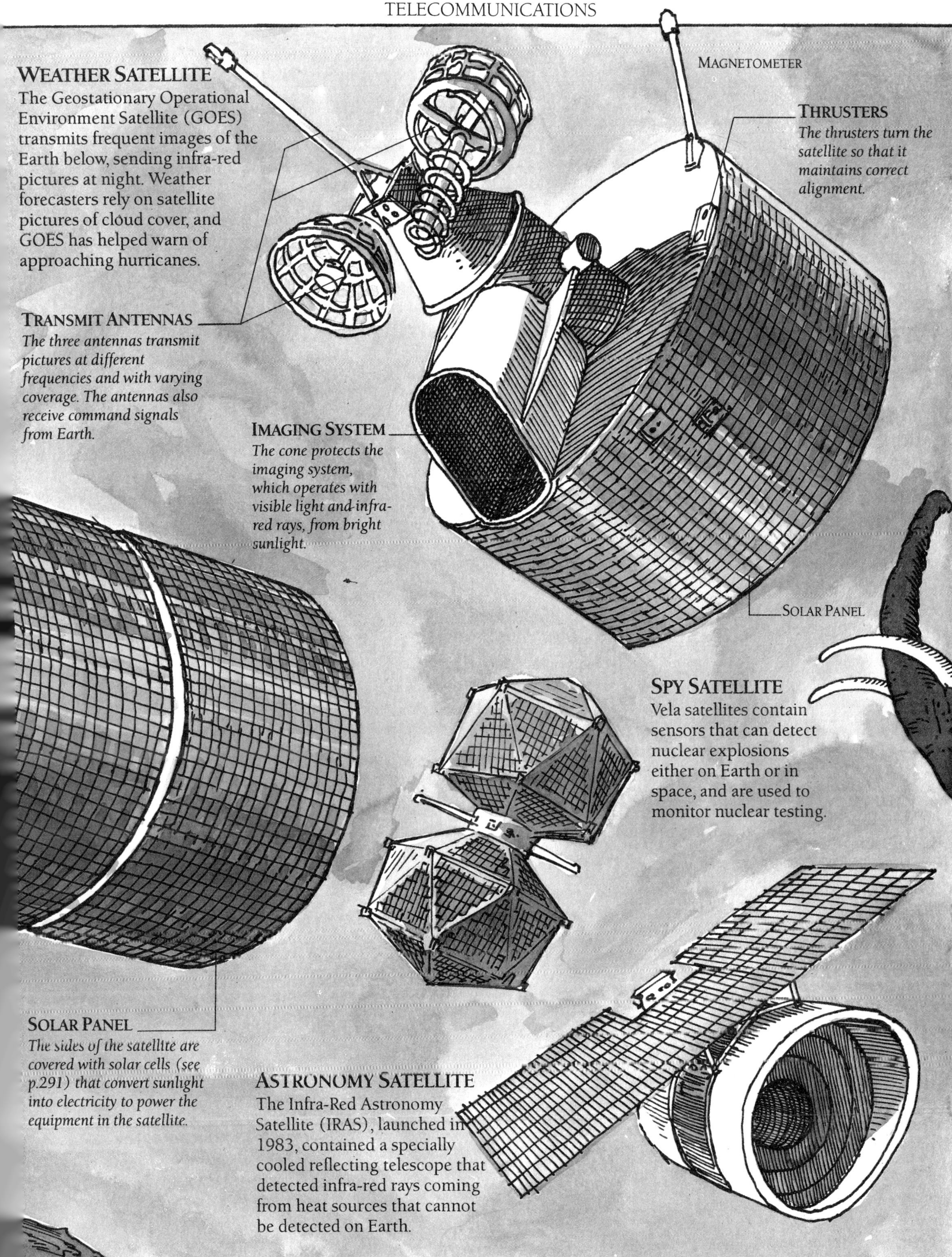

WEATHER SATELLITE
The Geostationary Operational Environment Satellite (GOES) transmits frequent images of the Earth below, sending infra-red pictures at night. Weather forecasters rely on satellite pictures of cloud cover, and GOES has helped warn of approaching hurricanes.

THRUSTERS
The thrusters turn the satellite so that it maintains correct alignment.

TRANSMIT ANTENNAS
The three antennas transmit pictures at different frequencies and with varying coverage. The antennas also receive command signals from Earth.

IMAGING SYSTEM
The cone protects the imaging system, which operates with visible light and infra-red rays, from bright sunlight.

SPY SATELLITE
Vela satellites contain sensors that can detect nuclear explosions either on Earth or in space, and are used to monitor nuclear testing.

SOLAR PANEL
The sides of the satellite are covered with solar cells (see p.291) that convert sunlight into electricity to power the equipment in the satellite.

ASTRONOMY SATELLITE
The Infra-Red Astronomy Satellite (IRAS), launched in 1983, contained a specially cooled reflecting telescope that detected infra-red rays coming from heat sources that cannot be detected on Earth.

RADIO TELESCOPE

Many objects in the Universe send out radio waves, and a radio telescope can be used to detect them. A large curved metal dish collects the radio waves and reflects them to a focus point above the center of the dish, rather as the curved mirror of a reflecting telescope gathers light waves from space (see p.200). At this point, an antenna intercepts the radio waves and turns them into a weak electric signal. The signal goes to a computer. Radio telescopes detect very weak waves, and can also communicate with spacecraft.

By detecting radio waves coming from galaxies and other objects in space, radio telescopes have discovered the existence of many previously unknown bodies. It is possible to make visible images of radio sources by scanning the telescope or a group of telescopes across the source. This yields a sequence of signals from different parts of the source, which the computer can process to form an image. Differences in frequency of the signals give information about the composition and motion of the radio source.

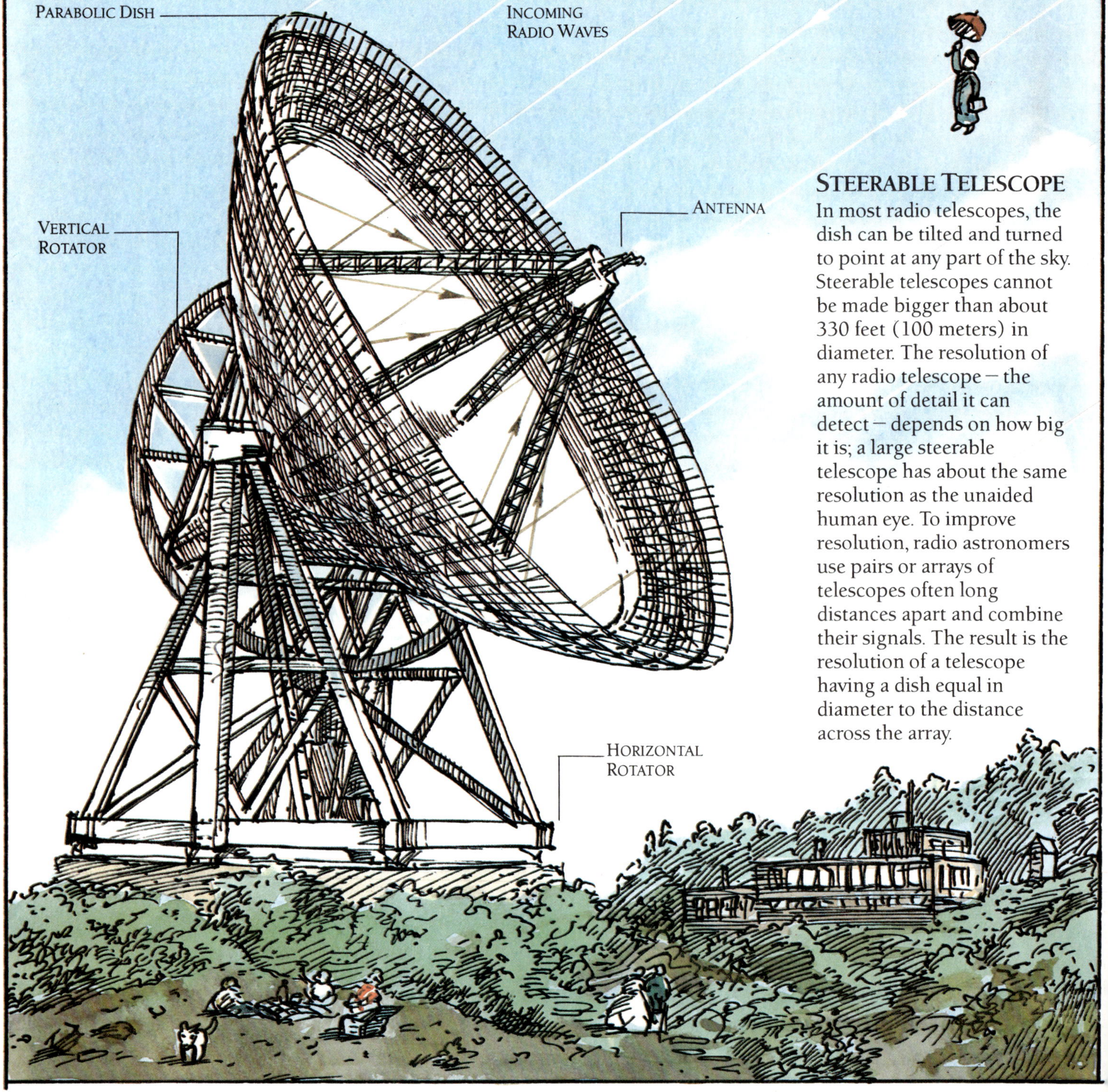

STEERABLE TELESCOPE

In most radio telescopes, the dish can be tilted and turned to point at any part of the sky. Steerable telescopes cannot be made bigger than about 330 feet (100 meters) in diameter. The resolution of any radio telescope – the amount of detail it can detect – depends on how big it is; a large steerable telescope has about the same resolution as the unaided human eye. To improve resolution, radio astronomers use pairs or arrays of telescopes often long distances apart and combine their signals. The result is the resolution of a telescope having a dish equal in diameter to the distance across the array.

SPACE TELESCOPE

The Hubble space telescope is part optical telescope and part satellite. It is due to be launched by the space shuttle, and promises to revolutionize astronomy because it will operate outside the atmosphere, which hampers any observations made from the ground. The space telescope will orbit the Earth, observing distant stars and galaxies in the total clarity of space. It will peer seven times further into the Universe than we can see now, and will also detect very faint objects. The telescope will be able to "see" far back in time by observing ancient light waves from the most distant galaxies. Among these may be light waves produced just after the big bang that blew the Universe into existence some 15 billion years ago.

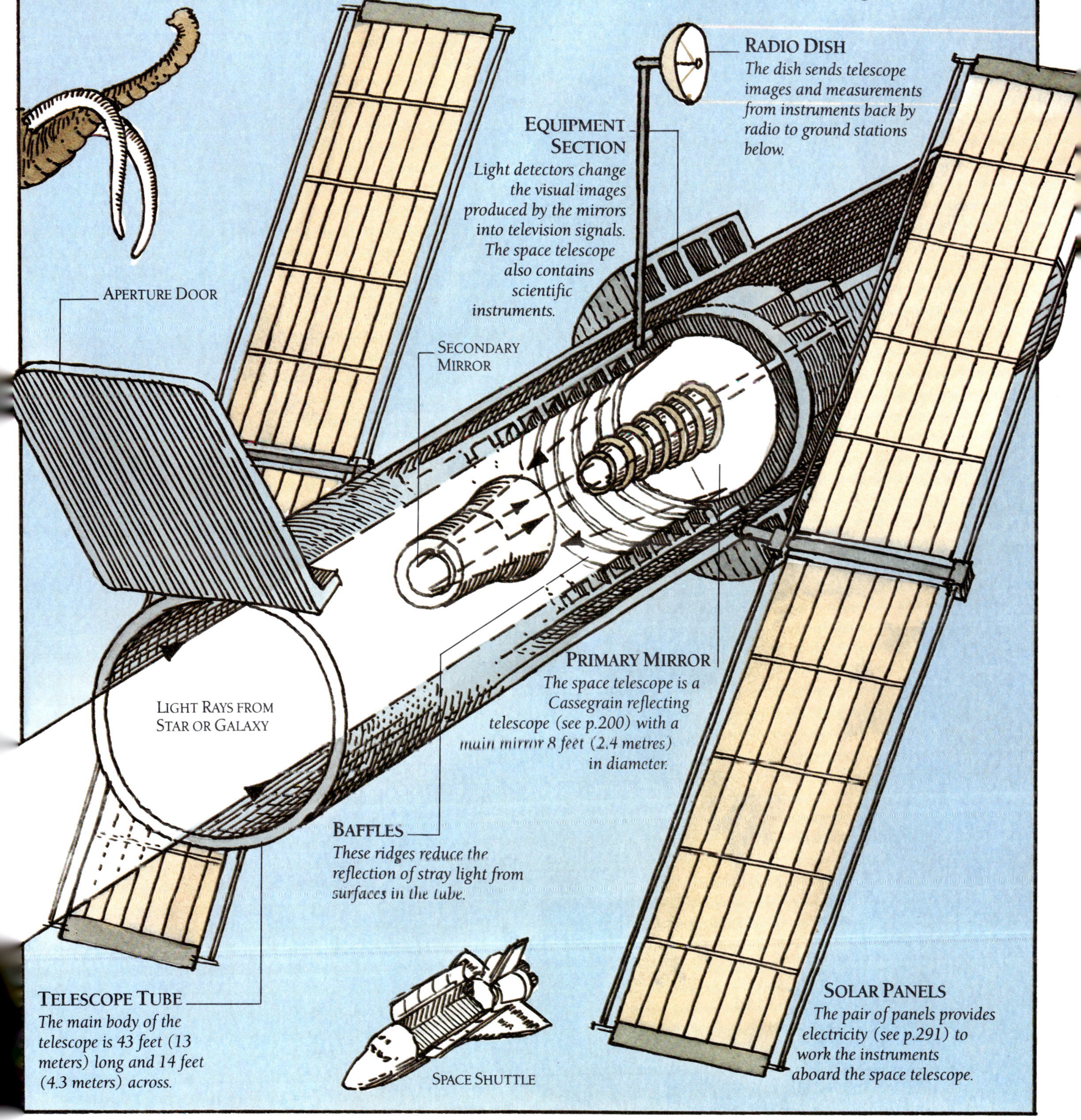

Space Probes

The ultimate limits of communications are reached with space probes, which have sent us detailed pictures and information from almost every planet in the Solar System. These probes have also surveyed many of the moons that orbit distant worlds, and one has flown to the very heart of Halley's comet.

Radio waves, traveling at the speed of light, have brought their discoveries to Earth. Such is their speed that signals can reach us from the furthest worlds in just a few hours. These signals include geological and atmospheric data, and video signals which can be decoded on Earth to produce television pictures. However, the signals are extremely weak, and several ground stations must train large dishes on a probe to pick them up. The stations also beam powerful command signals back to the space probe. Computers can correct errors and remove noise in the signals to enhance the images, providing us with detailed pictures of some of the most amazing sights in the Universe.

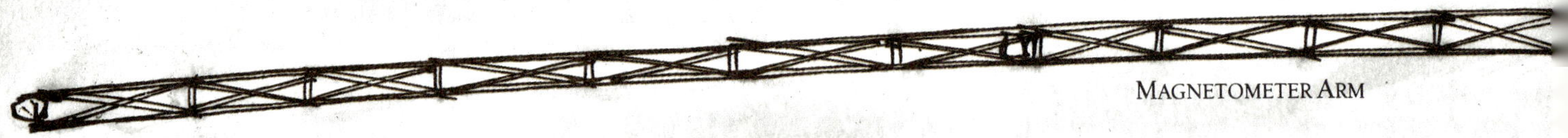

Viking

The two Viking probes are the most successful of the space probes that have landed on planets. Launched from Earth in 1975, they first orbited Mars to choose landing sites and then landed on the "red planet" almost a year after launch. Both probes found a rust-colored desert littered with rocks. They both extended a scoop to collect Martian soil, returning it to the probe for analysis. The tests examined the soil for signs of living things, but found no evidence that Mars has life.

Other instruments tested the air and measured conditions on Mars. All the information and pictures were transmitted back to Earth by radio dishes. Some signals returned directly, while others were relayed by sections of the probes that had remained in orbit around Mars.

VOYAGER

The most traveled of all space probes are the two Voyager probes, which left Earth in 1977. Both visited Jupiter and Saturn, while Voyager 2 went on to Uranus in 1986 and is due to arrive at Neptune in 1989. Both probes worked perfectly, sending back superb pictures of these giant planets and examining their strange rings and moons at close quarters. Their instruments have also studied the atmosphere of the planets, measured the magnetic fields and radio waves that they produce, and detected particles in space.

The Voyagers are "sling-shot" probes: they do not land on a planet but use its gravity to pull them into a path toward their next objective and to speed them on their way. In the late 1990s, both probes will leave the Solar System and begin a long journey out toward the stars. Should any alien being encounter them, the probes carry news of the human beings that sent them and the sights and sounds of their home planet.

THE DISCOVERY OF MAGNETIC NORTH

PART 4

ELECTRICITY & AUTOMATION

Contents

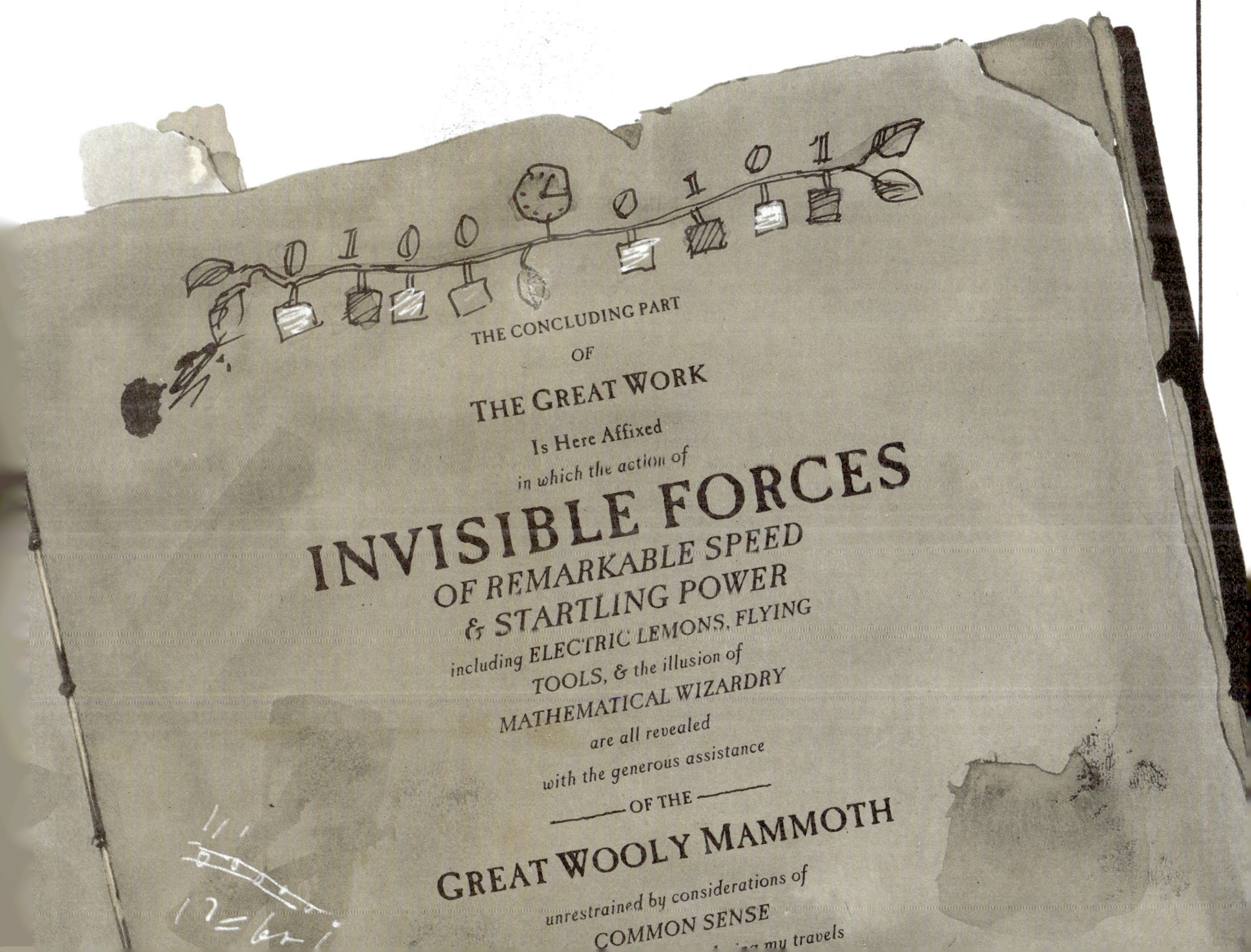
THE CONCLUDING PART
OF
THE GREAT WORK
Is Here Affixed
in which the action of
INVISIBLE FORCES
OF REMARKABLE SPEED
& STARTLING POWER
including ELECTRIC LEMONS, FLYING
TOOLS, & the illusion of
MATHEMATICAL WIZARDRY
are all revealed
with the generous assistance
OF THE
GREAT WOOLY MAMMOTH
unrestrained by considerations of
COMMON SENSE

INTRODUCTION

THE POWER BEHIND ELECTRICITY comes from the smallest things known to science. These are electrons, tiny particles within atoms, that each bear a minute electric charge. If a million million of them were lined up, they would scarcely reach across the head of a pin. When an electric current flows through a wire, these tiny particles surge through the metal in unimaginable numbers. In a current of 1 ampere, sufficient to light a flashlight bulb for example, 6 million million million electrons pass any point in just one second. Each electron moves relatively slowly, but the movement itself is passed from electron to electron at the speed of light.

If the nineteenth century was the heyday of mechanical machines, then the twentieth century belongs to machines powered by electricity. This does not mean that the age of mechanical machines is behind us. Machines that move will always be needed for doing work but machines that think, operated by electricity, are increasingly what controls them.

EXPLOITING ELECTRONS

The machines in this part of *The Way Things Work* use electricity in a number of different ways. In many electrical machines, moving electrons and magnetic fields are intimately linked. As soon as electrons start to move along a wire, they create a magnetic field around them. Quite why they do this cannot easily be explained, but the fact that they do is borne out in all machines that use electric motors, because all these make use of magnetism. So too do electric

generators, which produce our supply of electricity in the first place.

Electrons also produce electric fields, which have the same ability to attract and repel as magnetic ones. Some machines in the following pages, such as the photocopier and the air ionizer, work by shifting electrons about so that electrical attraction and repulsion comes into play.

Yet more machines, like the calculator and the computer, use electrons as a means to carry information. But despite these differences, the principles that govern the flow of electricity in all machines are just the same. The electrons always need energy to make them move. They always travel in a set direction (perversely from negative to positive) and at a set speed. Furthermore, they will always produce particular effects while they are on the move. One of these, as we have already seen in Part 2, is heat; magnetism is another.

ELECTRICITY AND MOVEMENT

As a source of power, electricity has no rival. It is clean, silent, can be turned on and off instantly, and can be fed easily to where it is needed.

Electric machines that produce movement are extraordinarily diverse. At first sight, there is little similarity between, for example, a quartz wristwatch and an electric locomotive. However, both use the motive force produced by the magnetic effects of an electric current – although the current used by a train is hundreds of thousands of times greater than that which flows in a watch.

Like all electrical machines, those that use electricity to produce movement take only as much power as they need. An electric motor will only take a set amount of current.

This means that one source can power many machines with each one taking only the current it needs and no more.

ELECTRICITY FOR SIGNALS

In the same way that energy in waves can be made to carry information, so too can energy in the form of electricity. Like light and radio waves, electricity travels almost instantaneously, so the message arrives at its destination with little or no delay.

Machines like calculators and computers that make use of electricity to carry information are known as electronic machines. This means that they work by controlling the movements of electrons. Because electrons are so small, it is possible to make the components that control them very small indeed. The components of a computer can therefore be highly miniaturized and assembled in complex circuits to give the computer its extraordinary range of abilities. A single microchip can store the street map of any capital city – something big enough for you the get lost in. Yet it is so small that you could very easily lose it!

MACHINES THAT CONTROL THEMSELVES

The paramount importance of electric signals is to control machines – not just to switch them on and off but to provide information and instructions that govern the way in which they work.

Sensors and detectors are often the source of these signals. They can detect the presence of physical objects, like metal or smoke, and then can also measure quantities, such as speed. Linked with powerful computers, electrical machines

can be used to control mechanical machines and process information faster, more reliably and more accurately than ever humans can.

An air journey, for example, shows how many tasks they have taken away from human hands. Even before you board a plane, a computer will have booked your place and probably printed your ticket, while scanners have checked your baggage. Although the aircraft itself is not basically an electrical machine – a paper dart can fly, for example – an airliner could not even leave the ground without electricity to start its engines and power its on-board systems.

Once in the air, you travel in safety because electrically operated radar enables air traffic controllers to guide the pilot, and you get to your destination mainly because the autopilot has kept the plane on course. The human crew will have trained on a computer-controlled flight simulator, learning how to cope with everyday matters and emergencies thrown at them by machine. Even mundane creature comforts – fresh air, light, music and movies – all require electricity.

More and more aspects of our everyday lives now depend in some way on the flow of billions of electrons flowing through countless switches in machines that can almost think. Just as in the last century when the development of mechanical machines seemed to be unbounded, so today the growth of electronic machines seems unlimited. In the computer and robot, which are explored in the closing pages of this book, mankind is advancing towards the ultimate machine – one that will be able to perform any task required of it.

ELECTRICITY

ON MAMMOTH ATTRACTION

One day, I happened upon a mammoth whose hair had been lovingly combed. The hairdresser, in fact, was just about to return her creation to its owner. No sooner had the perfectly coiffed animal stepped into the street, however, than a combination of litter, loose laundry and stray cats flew into the air and secured themselves to the startled beast's freshly combed coat. It is common knowledge that a well groomed individual is more attractive, but never before had I seen this so forcefully illustrated.

STATIC ELECTRICITY

All things are made up of atoms, and within atoms are even smaller particles called electrons. Electrons each have an electric charge, and this charge, which is considered to be negative, is the fundamental cause of electricity.

Static electricity is so-called because it involves electrons that are moved from one place to another rather than ones that flow in a current. In an object with no static electric charge, all the atoms have their normal number of electrons. If some of the electrons are then transferred to another object by, for example, vigorous rubbing or brushing, the other object becomes negatively charged while the object that loses electrons becomes positively charged. An electric field is set up around each object.

Unlike charges always attract each other and like charges always repel each other. This is the reason why the mammoth finds itself festooned with trash after its brushing, and why a comb rubbed with a cloth will attract pieces of paper. Rubbing or brushing creates a charge and therefore an electric field. The field affects objects nearby, producing an unlike charge in them, and the unlike charges are drawn together.

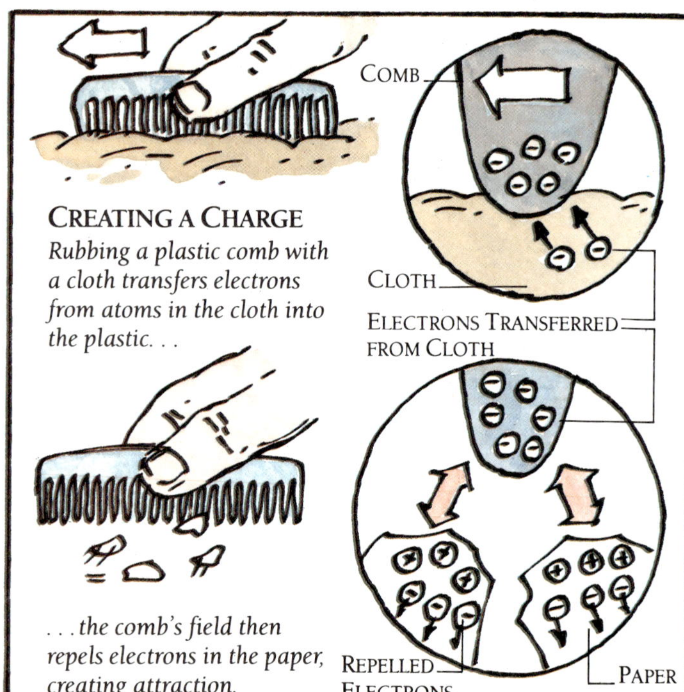

CREATING A CHARGE

Rubbing a plastic comb with a cloth transfers electrons from atoms in the cloth into the plastic. . .

. . . the comb's field then repels electrons in the paper, creating attraction.

ON MAMMOTH LEMONS

At harvest time, I once watched with great admiration as lemons were gathered with mammoth assistance. Large specimens were harpooned, the mammoths being equipped with copper lances, and their riders with zinc ones – a lightweight improvement of my own devising. During my visit, the riders did complain of suffering powerful shudderings which they somehow attributed to their new equipment, but I was able to assure them that of course there could be no connection.

As each team boldly rode into action, the air was almost electric.

CURRENT ELECTRICITY

Current electricity is produced by electrons on the move. Unlike static electricity, current electricity can only exist in a conductor – that is, a material such as a metal that allows electrons to pass freely through it.

In order to make electrons move, a source of energy is needed. This energy can be in the form of light, heat, or pressure, or it can be the energy produced by a chemical reaction. Chemical energy is the source of power in a battery-powered circuit. The mammoth and its rider suffer a surge of electric current because they inadvertently form this type of circuit. Lemons contain acid, which reacts with the zinc and copper in the lances. Atoms in the acid take electrons from the copper atoms and transfer them to the zinc atoms. The electrons then flow through the materials connected to the two metal lances. The zinc lance, which releases the negatively charged electrons, is the negative terminal of the lemon battery. The copper lance, which receives the electrons, is the positive terminal. Whereas an ordinary lemon would not produce sufficient electrons to give a big current, the giant lemon yields enough to produce a violent shock.

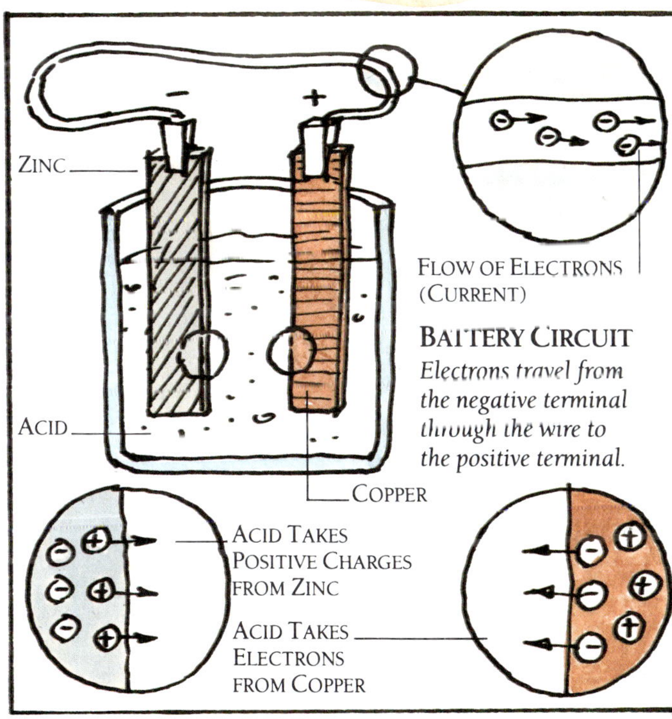

BATTERY CIRCUIT
Electrons travel from the negative terminal through the wire to the positive terminal.

DOCUMENT
TONER BRUSHES APPLY TONER TO DRUM
MIRRORS
LENS
STRIP OF DOCUMENT
DRUM
DRUM CHARGER
CLEANER
FIRST ERASE LAMP
This lamp removes the charge on the drum.
TRANSFER CHARGER
The transfer charger applies negative charge to the piece of paper so that it attracts the toner particles.
SECOND ERASE LAMP
This removes the charge on the drum after the toner has been deposited on the paper.

THE PHOTOCOPIER

Static electricity enables a photocopier to produce almost instant copies of documents. At the heart of the machine is a metal drum which is given a negative charge at the beginning of the copying cycle. The optical system then projects an image of the document on the drum. The electric charge disappears where light strikes the metal surface, so only dark parts of the image remain charged. Positively charged particles of toner powder are then applied to the drum. The charged parts of the drum attract the dark powder, which is then transferred to a piece of paper. A heater seals the powder to the paper, and a warm copy of the document emerges from the photocopier. A color copier works in the same basic way, but scans the document with blue, green and red filters. It then transfers toner to the paper in three layers colored yellow, magenta and cyan. The three colors overlap to give a full color picture (see p.195).

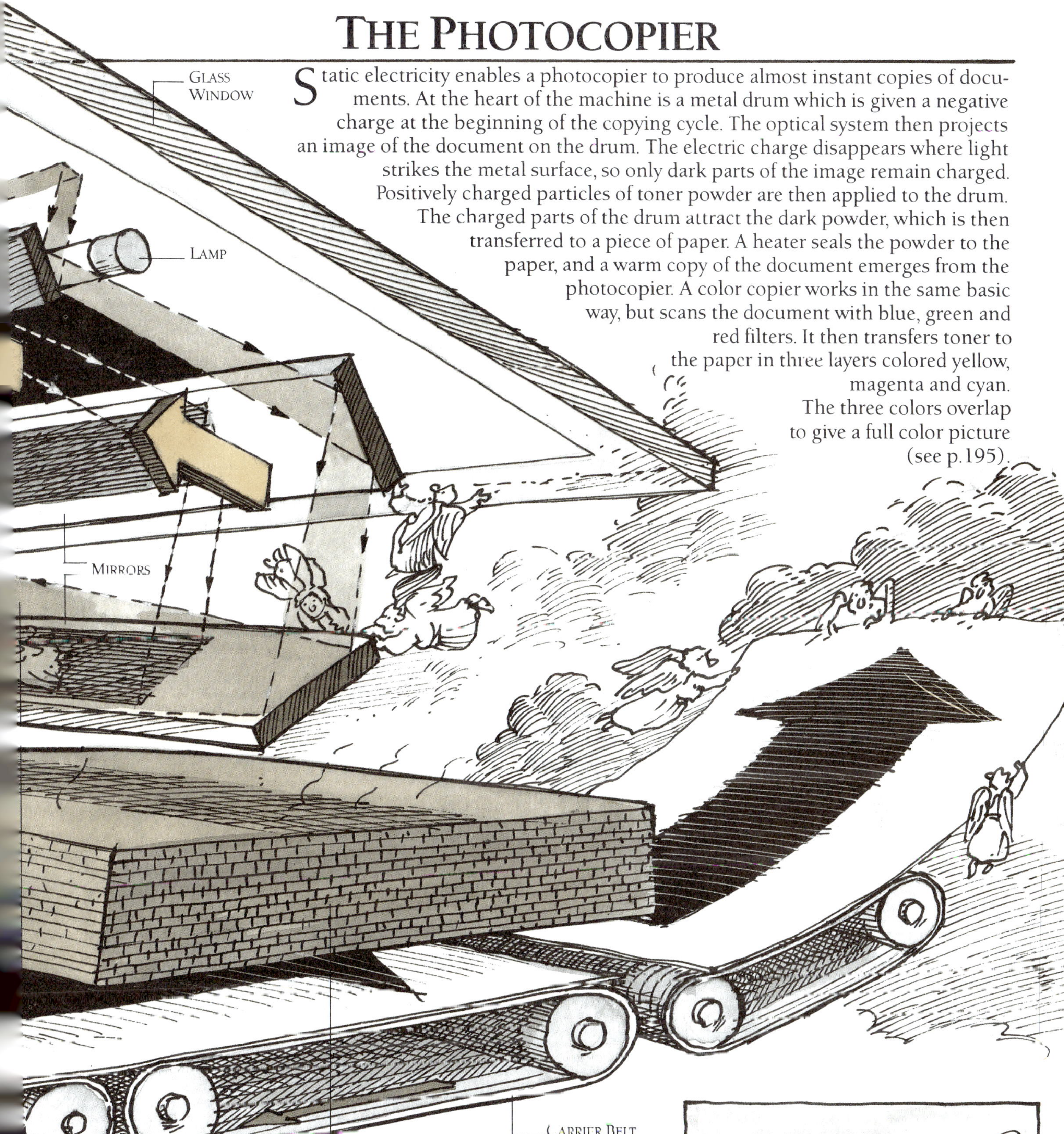

HEATER
The heater warms the paper so that the toner particles soften and are pressed into the surface of the paper.

OPTICAL SYSTEM
Beneath the glass window, a lamp, set of mirrors and a lens scan the document, moving across it to project a strip onto the rotating drum. The optical system may enlarge or reduce the size of the image on the drum.

AIR CLEANER

The most effective kind of air cleaner uses an electrostatic precipitator to remove very fine particles, such as cigarette smoke and pollen, from the air in a room. The precipitator works by giving a positive charge to particles in the air and then trapping them with a negatively charged grid. The cleaner may also contain filters to remove dust and odors, and finally an ionizer to add negative ions to the clean air.

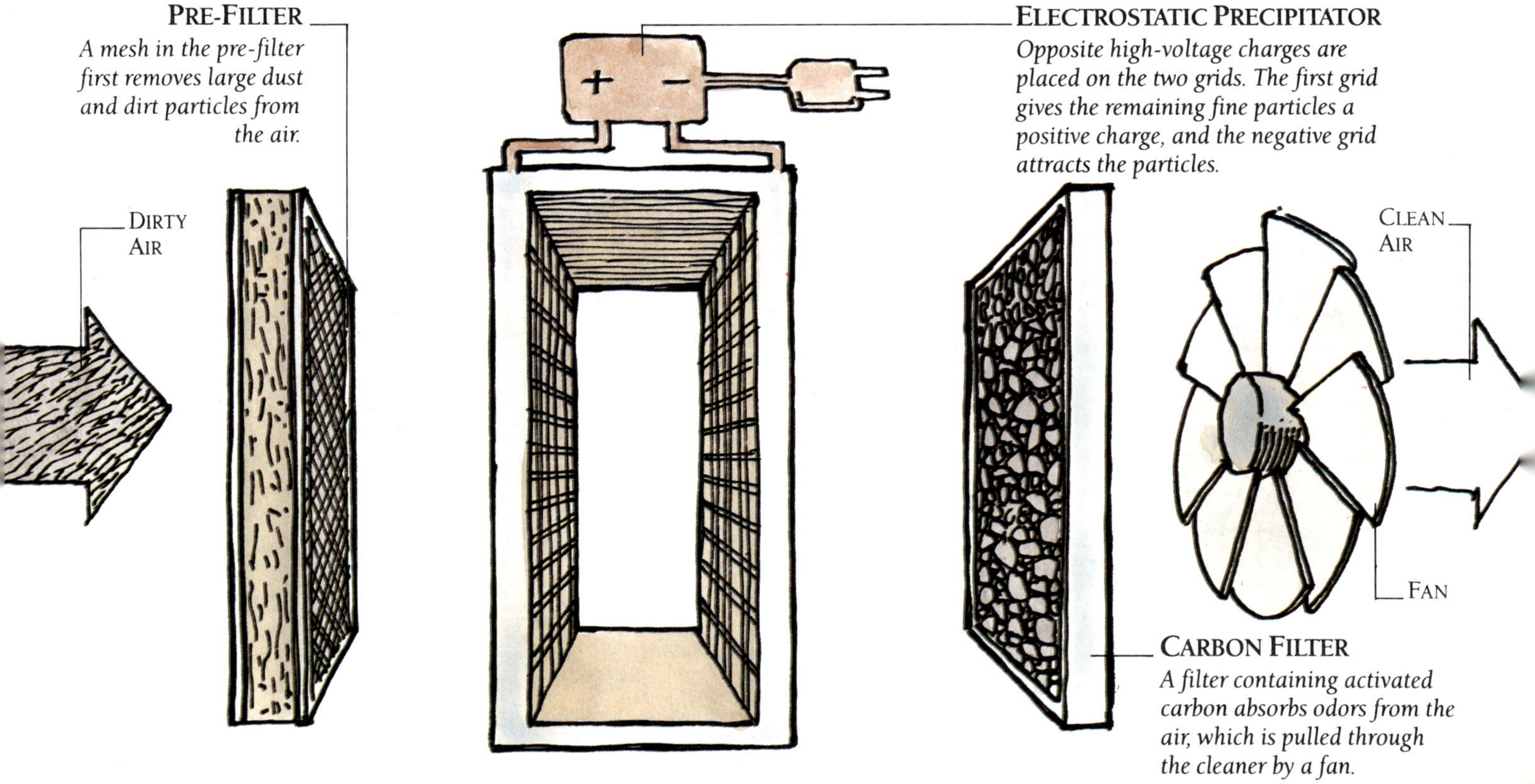

LIGHTNING CONDUCTOR

CHARGE BUILD-UP

A thunderstorm creates regions of strong negative electric charge at the base of clouds. These charges cause strong positive charges to form in the ground.

LIGHTNING DISCHARGE

The very strong electric fields produce ions and free electrons in the air. The air can then conduct electricity and a flash of lightning surges through it.

IONIZER

Atoms that have an electrical charge are called ions. Ions occur naturally; they make up many solid substances and they are also found in the atmosphere. Air that contains a high concentration of negative ions is reputed to be beneficial; ionizers are designed to produce them. An ionizer supplies a strong negative charge to one or more needles. An intense electric field is developed at the point of a needle, and it creates ions in the atoms in the air. Positive ions are attracted to the needle, while negative ions flow outward.

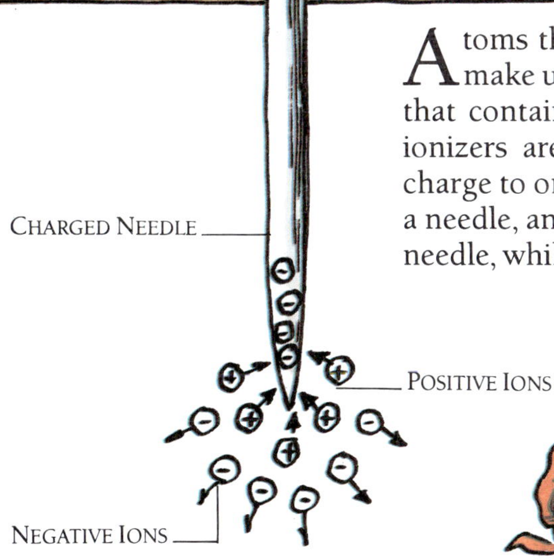

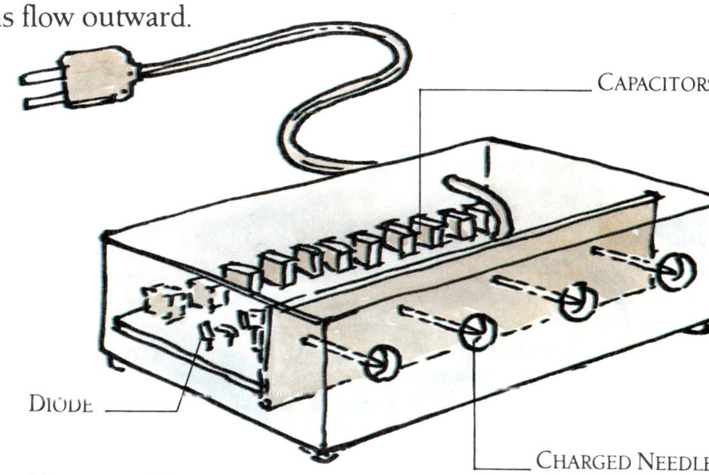

VOLTAGE MULTIPLIER

This converts the alternating current of the electricity supply to a high-voltage direct current that charges the ionizer needles. The diodes change the alternating current to direct current (see p.287) which charges the capacitors. The capacitors store increasing amounts of charge to raise the voltage.

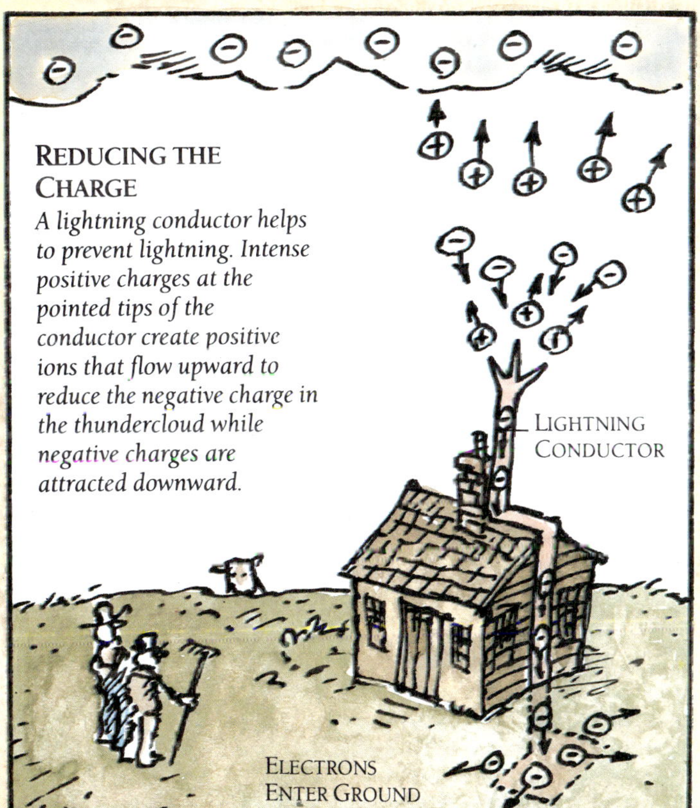

REDUCING THE CHARGE

A lightning conductor helps to prevent lightning. Intense positive charges at the pointed tips of the conductor create positive ions that flow upward to reduce the negative charge in the thundercloud while negative charges are attracted downward.

CONDUCTING THE CHARGE TO EARTH

If lightning does strike, it tends to follow the ion path and hits the lightning conductor. The powerful current flows down the cable and enters the ground without causing any damage.

ANTI-STATIC GUN

PIEZOELECTRICITY

Exerting pressure on certain crystals and ceramics can cause them to produce an electric charge. This effect is called piezoelectricity, from the Greek word meaning to press, and it is put to use in several electrical devices.

In many substances, the atoms are in the form of ions (see p.283) which are held together very tightly by their electric charges. Quartz, for example, has positive silicon ions and negative oxygen ions. Pressing it displaces the ions so that negative ions move toward one side of the crystal and positive ions toward the other. The opposite faces develop negative and positive charges, which can be very powerful. The reverse happens too: applying an electric signal to a crystal makes it vibrate at a precise natural frequency, as in a quartz oscillator.

NORMAL QUARTZ CRYSTAL

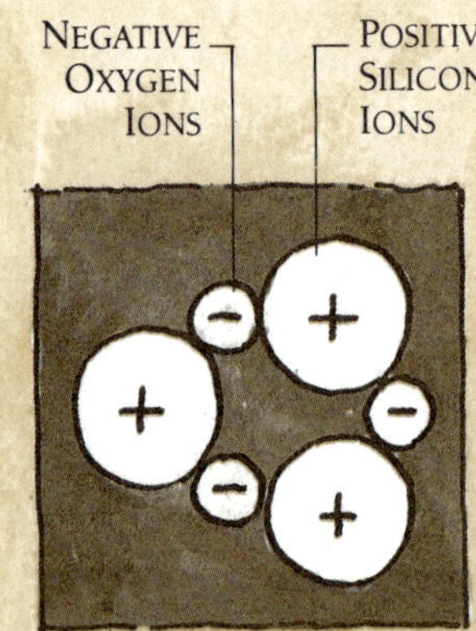

CRYSTAL UNDER PRESSURE

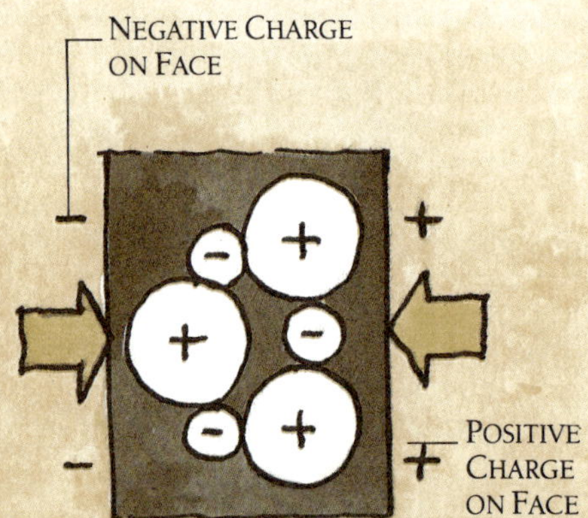

QUARTZ OSCILLATOR

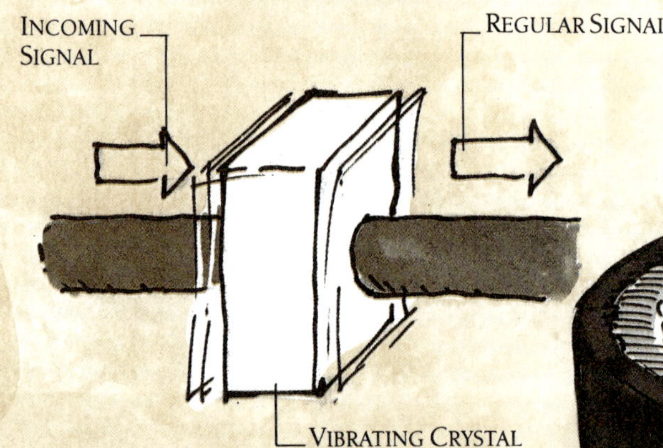

Handling a record causes a negative charge of static electricity to build up on the record. This charge attracts dust particles, which settle in the grooves of the record and make it deteriorate.

An anti-static gun helps to keep a record free from dust by neutralizing this charge. On pressing the trigger, the needle sprays out a stream of positive ions. These are formed by a high-voltage positive charge fed to the needle by piezoelectric ceramic elements connected to the trigger.

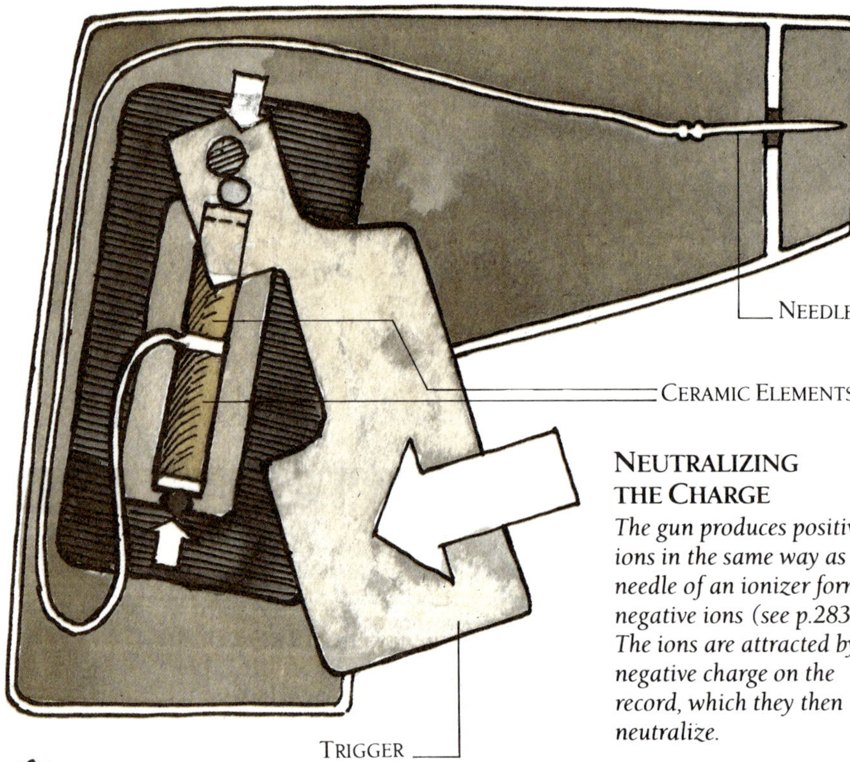

NEUTRALIZING THE CHARGE

The gun produces positive ions in the same way as the needle of an ionizer forms negative ions (see p.283). The ions are attracted by the negative charge on the record, which they then neutralize.

Quartz Clock

Piezoelectricity provides a simple method of accurate time-keeping. Many clocks and watches contain a quartz crystal oscillator which controls the hands or display. Power from a small battery makes the crystal vibrate and it gives out pulses of current at a very precise rate or frequency. A microchip reduces this rate to one pulse per second, and this signal controls the motor that turns the hands or activates the display.

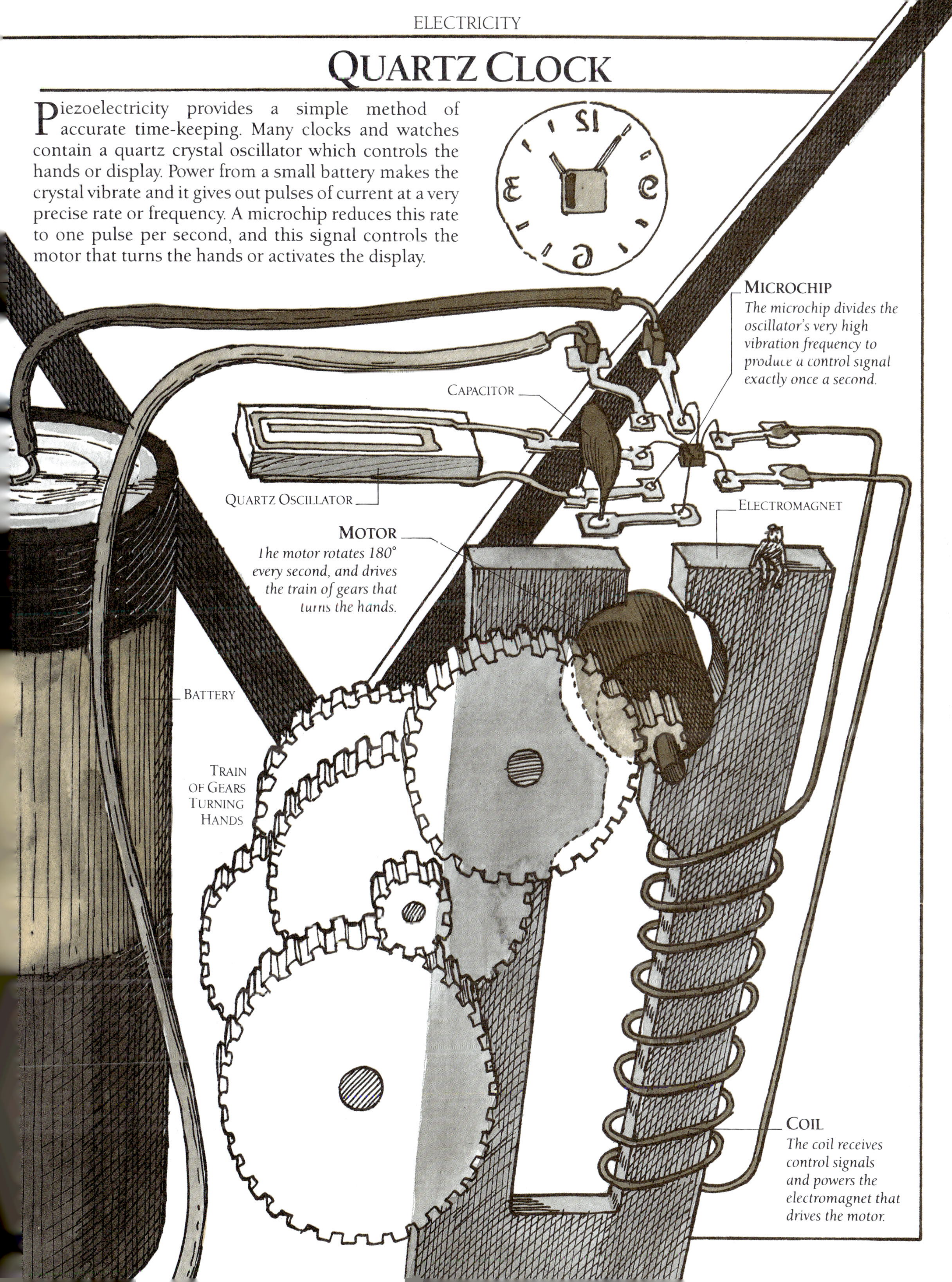

THE CURRENT CART

Because electricity cannot be seen as it flows around a circuit, it is easier to understand by comparing it with something else. The machine on this page is a water-powered equivalent of an electric circuit. Water, rather than electrons, circulates and provides power. Each part of the cart has a counterpart in the simple circuit on the opposite page.

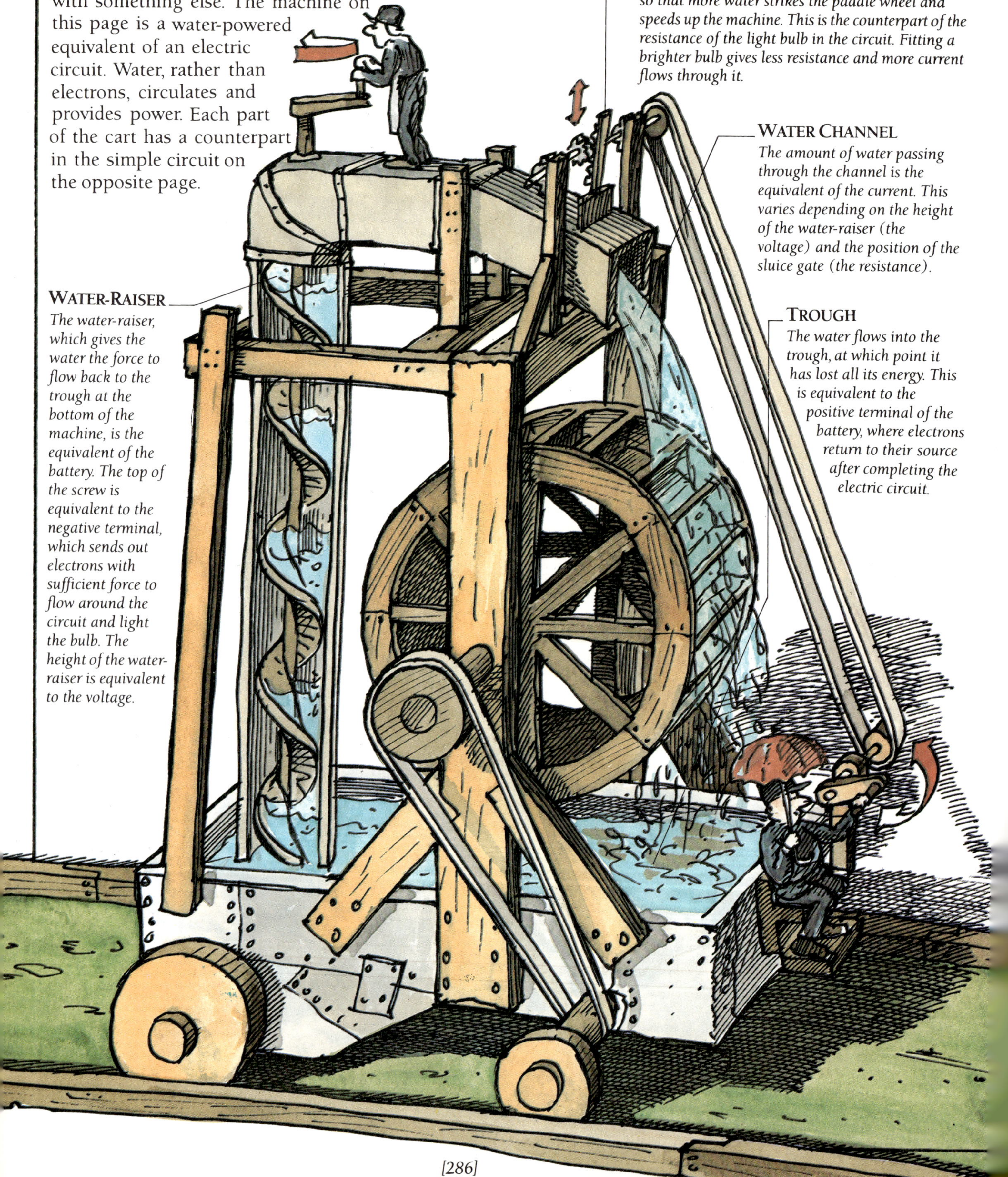

SLUICE GATE

Opening the sluice gate increases the flow of water so that more water strikes the paddle wheel and speeds up the machine. This is the counterpart of the resistance of the light bulb in the circuit. Fitting a brighter bulb gives less resistance and more current flows through it.

WATER CHANNEL

The amount of water passing through the channel is the equivalent of the current. This varies depending on the height of the water-raiser (the voltage) and the position of the sluice gate (the resistance).

WATER-RAISER

The water-raiser, which gives the water the force to flow back to the trough at the bottom of the machine, is the equivalent of the battery. The top of the screw is equivalent to the negative terminal, which sends out electrons with sufficient force to flow around the circuit and light the bulb. The height of the water-raiser is equivalent to the voltage.

TROUGH

The water flows into the trough, at which point it has lost all its energy. This is equivalent to the positive terminal of the battery, where electrons return to their source after completing the electric circuit.

ELECTRIC CIRCUIT

All devices and machines powered by current electricity contain an electric circuit. A source of electricity, usually a battery or generator, drives electrons through a wire to the part of the machine that provides power or releases energy. The electrons then return along a wire to the source and complete the circuit. The source produces a certain number of volts, which is a measure of the electrical force that sends the electrons around the circuit. The current, which is the amount of electricity that flows, is measured in amps or amperes. The working part of the circuit has a resistance measured in ohms.

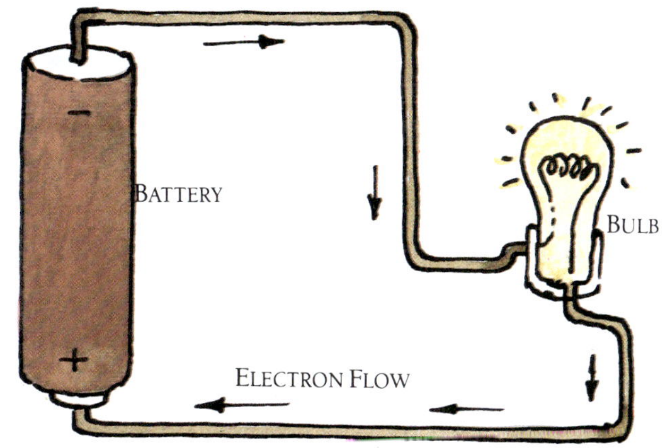

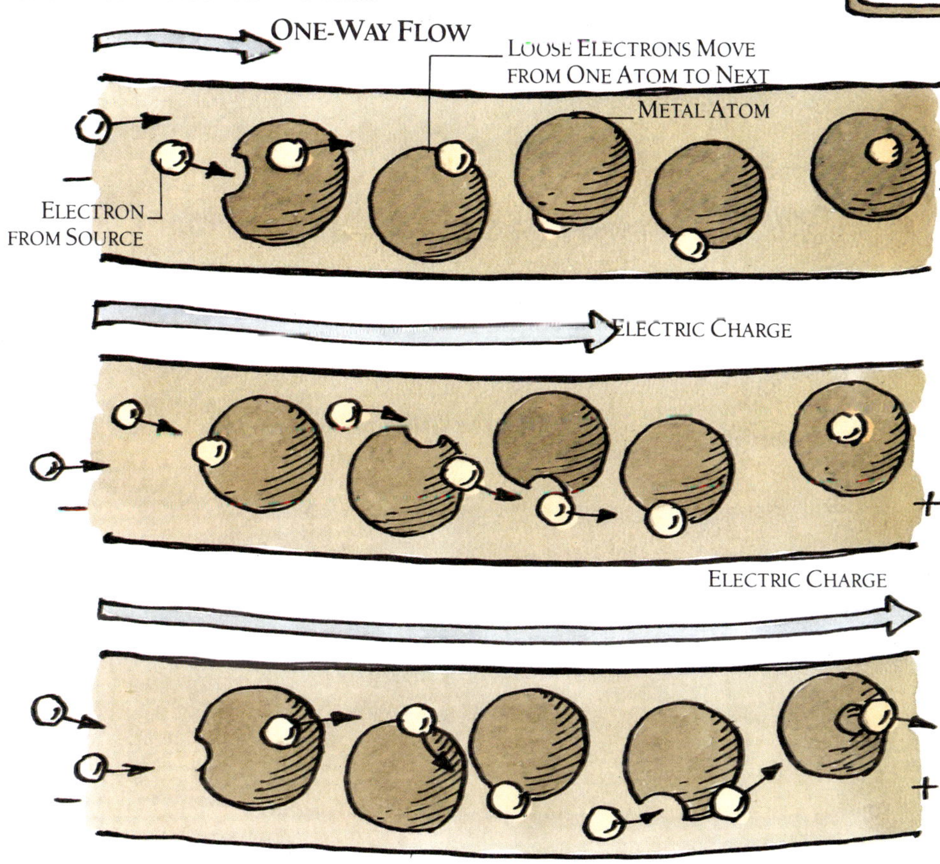

DIRECT CURRENT (DC)

The electric current produced by a battery and solar cell is direct current. The electrons flow in one direction from the negative terminal of the source to the positive terminal. Although individual electrons move very slowly, the electric charge travels very much faster. This is because the arriving electrons collide with loose electrons in the metal atoms, making them leave one atom and collide with the next. Like shunting railroad cars, the shift in electrons progresses very rapidly along the wire, making the electric charge move very quickly.

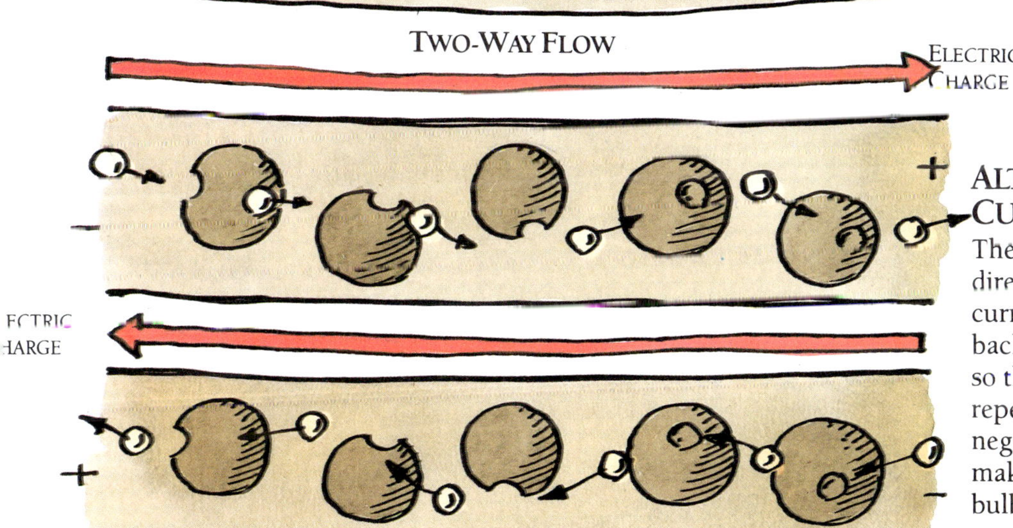

ALTERNATING CURRENT (AC)

The mains supply is usually not direct current but alternating current. Here, the electrons move back and forth 60 times a second, so that the terminals of the supply repeatedly change from positive to negative and vice-versa. This makes no difference to a light bulb, which lights up when the current flows in either direction.

BATTERIES

A battery produces an electric current when its terminals are connected to each other to form a circuit. All batteries contain two electrodes and an electrolyte, which produces the chemical reaction with the electrodes resulting in a current. In "dry" batteries, the electrolyte is a paste of powdered chemicals. "Wet" batteries, like those in cars, contain a liquid electrolyte.

A battery's voltage depends on the metals that are used in its electrodes.

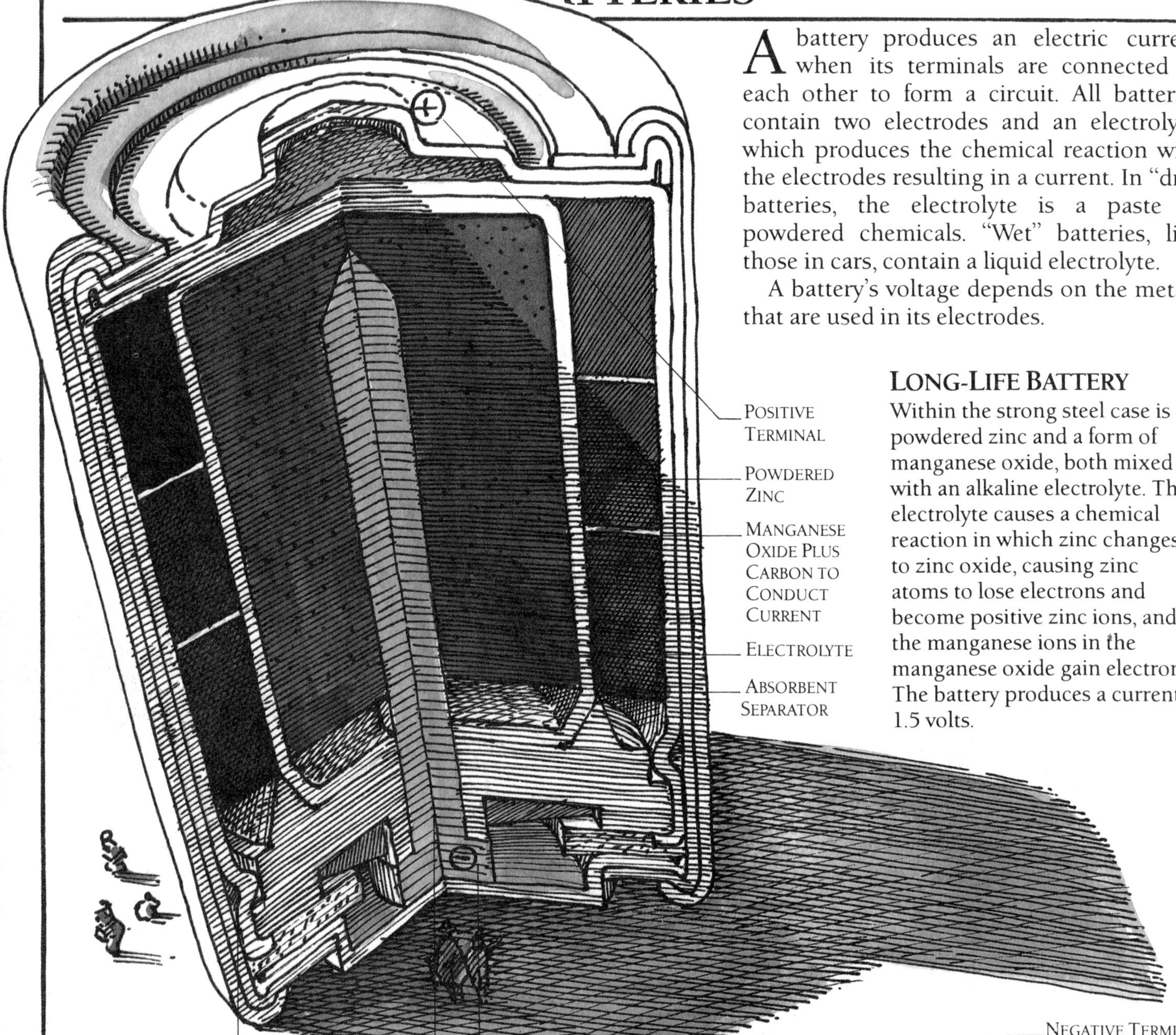

LONG-LIFE BATTERY

Within the strong steel case is powdered zinc and a form of manganese oxide, both mixed with an alkaline electrolyte. The electrolyte causes a chemical reaction in which zinc changes to zinc oxide, causing zinc atoms to lose electrons and become positive zinc ions, and the manganese ions in the manganese oxide gain electrons. The battery produces a current of 1.5 volts.

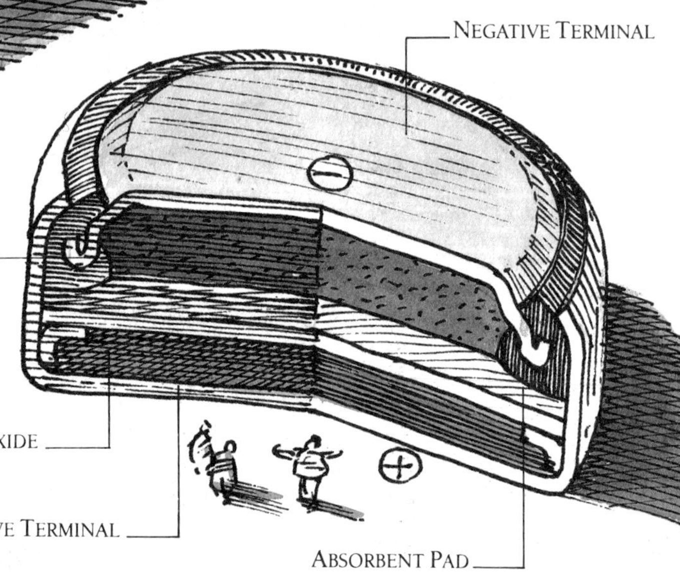

BUTTON BATTERY

The battery contains powdered zinc and mercury oxide with an alkaline electrolyte. The zinc loses electrons as it becomes zinc oxide, while the mercury atoms gain electrons as the mercury oxide changes to mercury. The battery produces a current of 1.35 volts.

CAR BATTERY

The battery in a car is designed to produce the strong current needed to turn the starter motor (see p.77). It does this by using a number of cells linked together. When running, the engine turns a generator which feeds current back into the battery to recharge it.

A car battery contains plates of lead oxide and lead metal, immersed in a sulfuric acid electrolyte. As the battery produces current, both kinds of plate change to lead sulfate. Feeding a current into the battery reverses the chemical reaction.

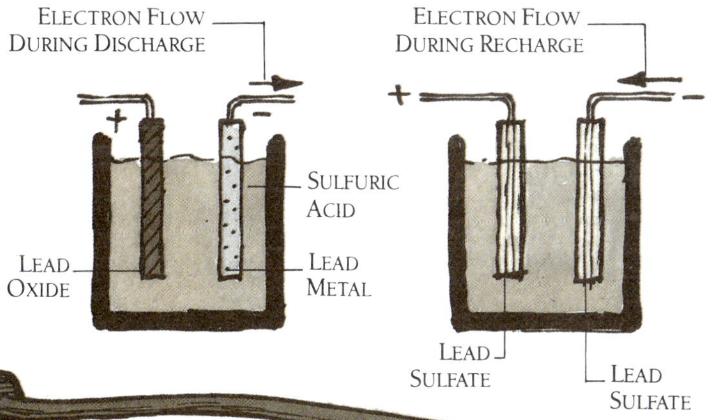

CAR TEMPERATURE GAU

Electrical temperature gauges and thermometers depend on the changing resistance of a heat-sensitive element. The resistance varies with temperature, so that the amount of current flowing depends on how hot the element gets.

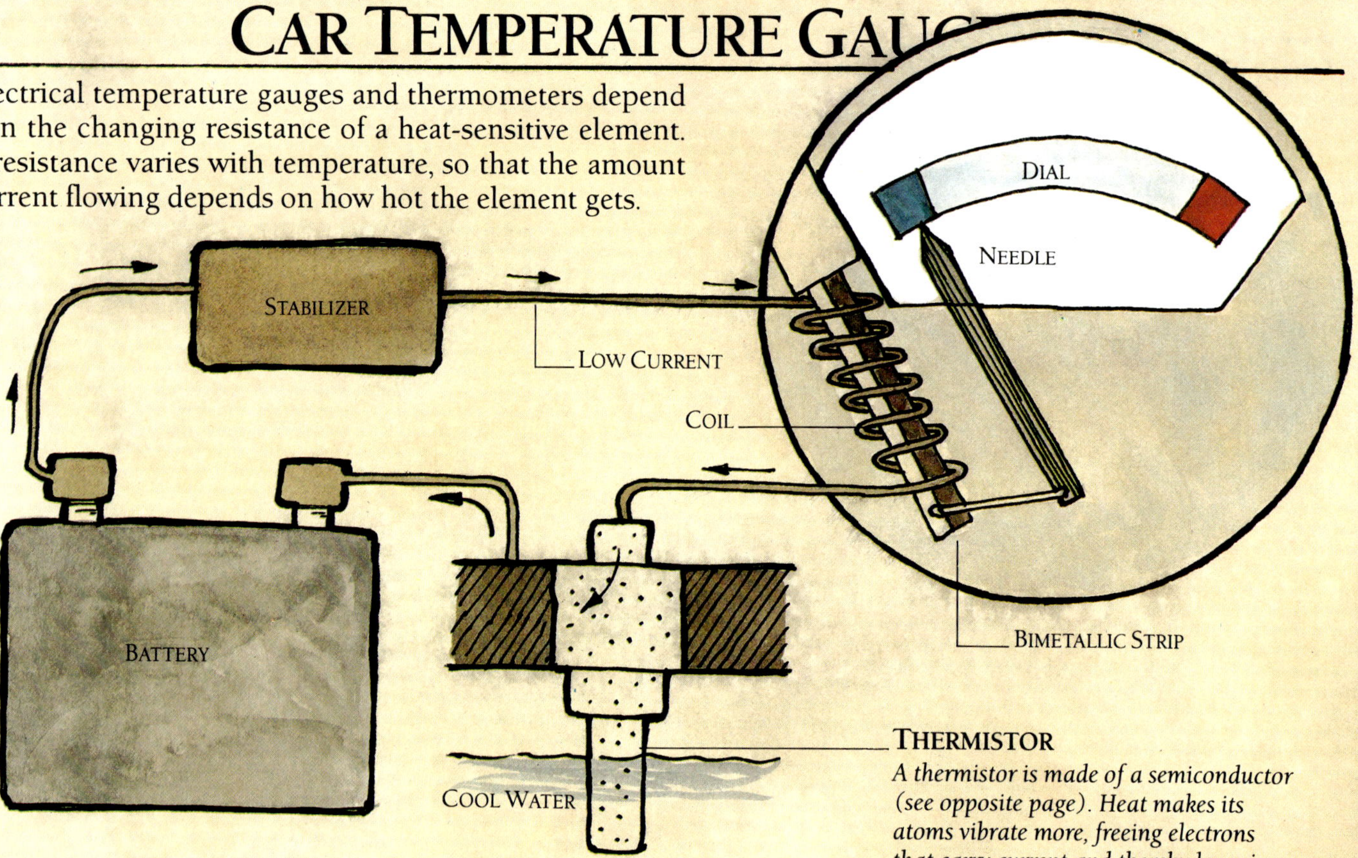

THERMISTOR

A thermistor is made of a semiconductor (see opposite page). Heat makes its atoms vibrate more, freeing electrons that carry current and thereby lowering its resistance. The stabilizer ensures that a constant voltage is fed to the thermistor.

ENGINE COOL

Before the engine has warmed up (*above*), only a small current flows through the gauge. From the battery, it passes through the stabilizer, coil and the thermistor in the water jacket of the car's engine. The thermistor's high resistance restricts the current and the needle indicates that the engine is cool.

ENGINE HOT

As the water in the engine heats up (*below*), the resistance of the thermistor decreases. This enables a larger current to flow through it, and the current heats the coil in the gauge. The heat bends the bimetallic strip (see p.162) which is linked to the needle.

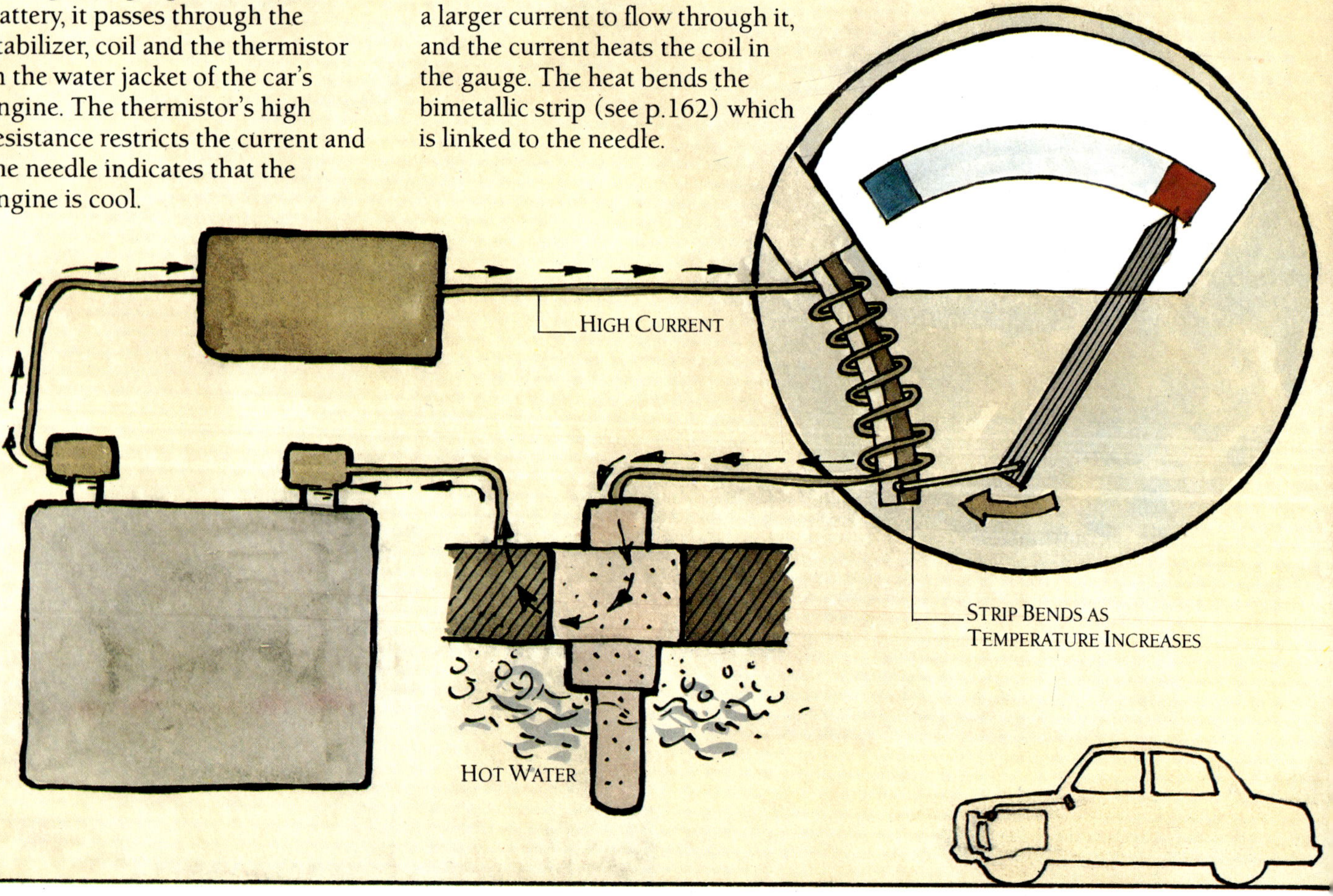

SOLAR CELL

A solar cell turns light into electricity. Large panels of cells power satellites while strips of a few cells provide the much smaller current needed to power calculators. Like many electronic devices, solar cells depend on semiconductors. These are materials in which the flow of electrons can be controlled – in this case, to generate a low current. Each cell contains two layers of different types of silicon. The silicon atoms are arranged in a lattice in which other atoms containing extra or fewer atoms are inserted.

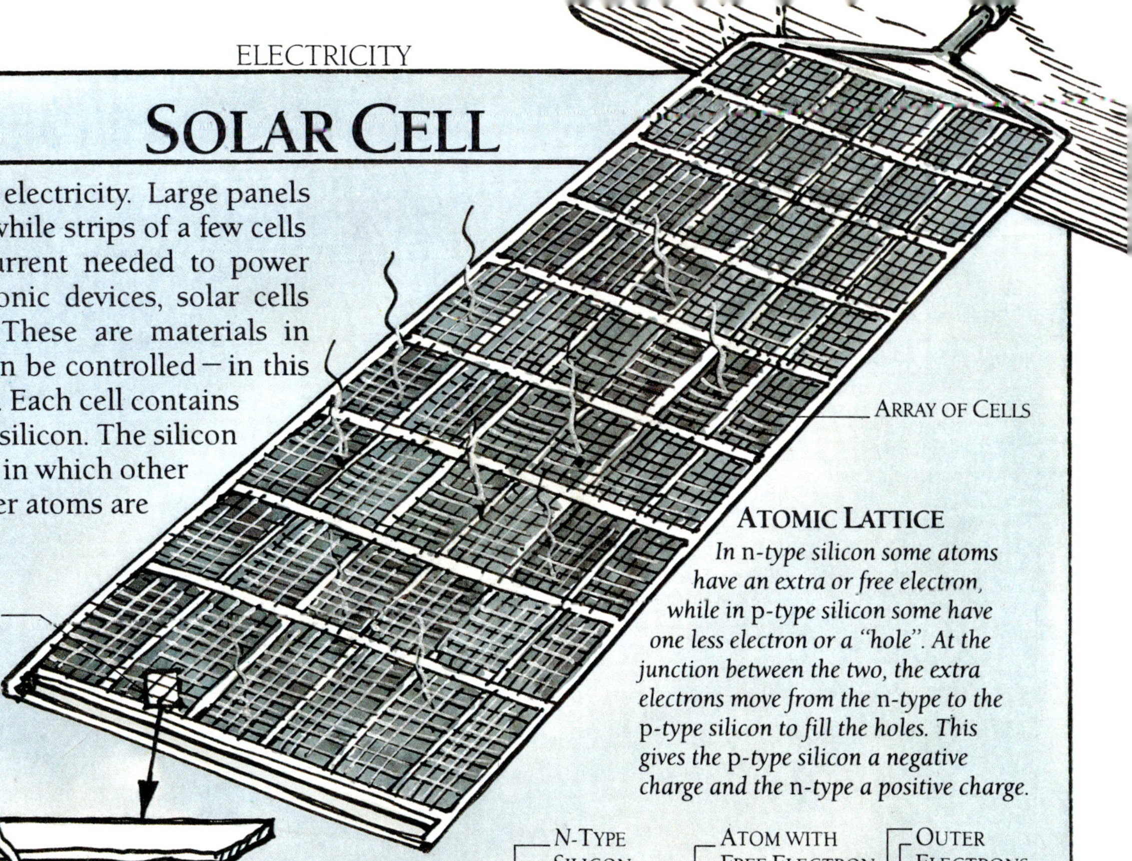

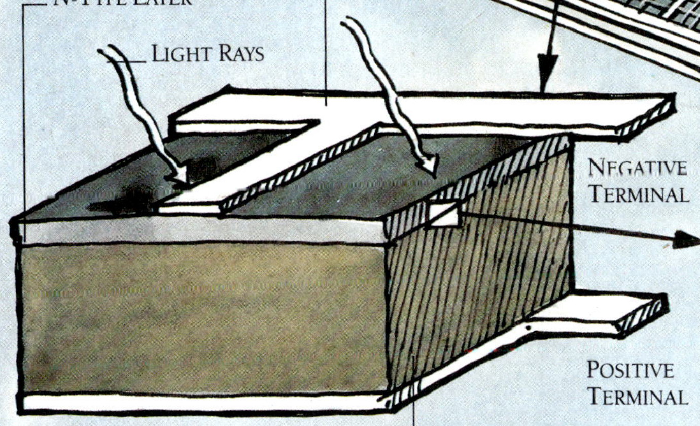

ATOMIC LATTICE

In n-type silicon some atoms have an extra or free electron, while in p-type silicon some have one less electron or a "hole". At the junction between the two, the extra electrons move from the n-type to the p-type silicon to fill the holes. This gives the p-type silicon a negative charge and the n-type a positive charge.

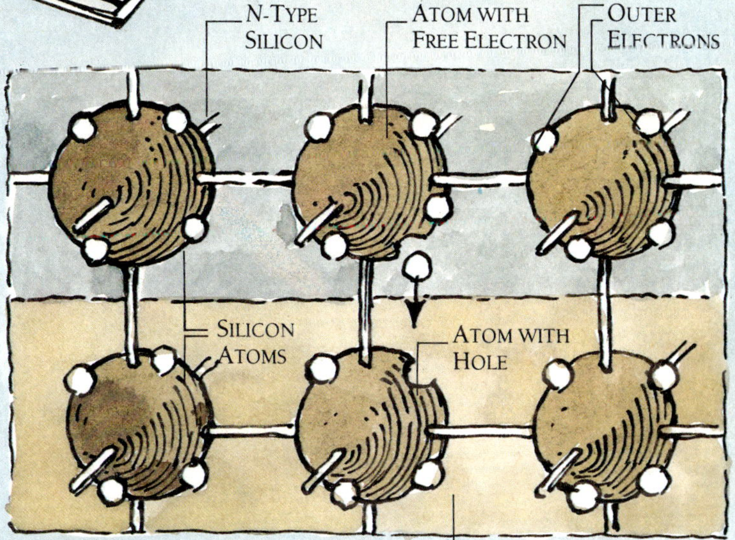

INSIDE A CELL

An individual solar cell (above) is made of two kinds of silicon – an upper n-type layer and a lower p-type layer. When light strikes the cell (below), the rays penetrate the silicon and free electrons from the atoms. The charges on the two layers make the electrons move. The electrons are collected by the contact and the cell generates a current as the electrons flow.

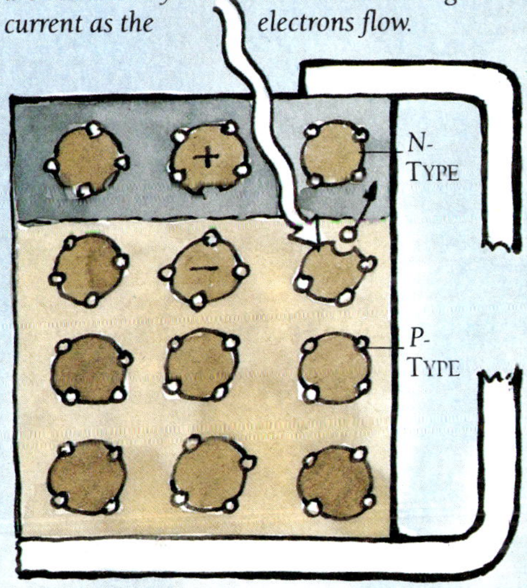

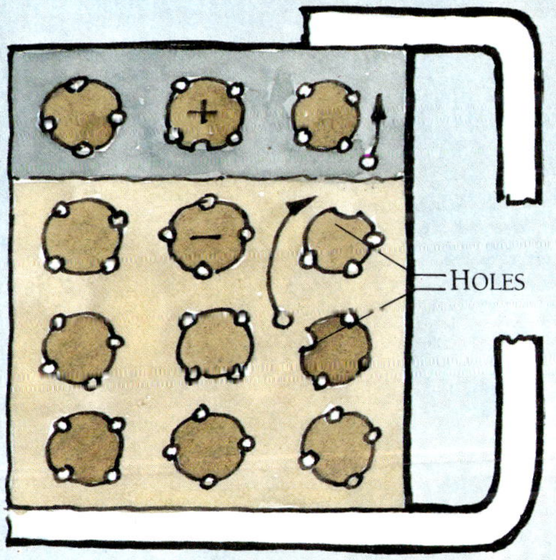

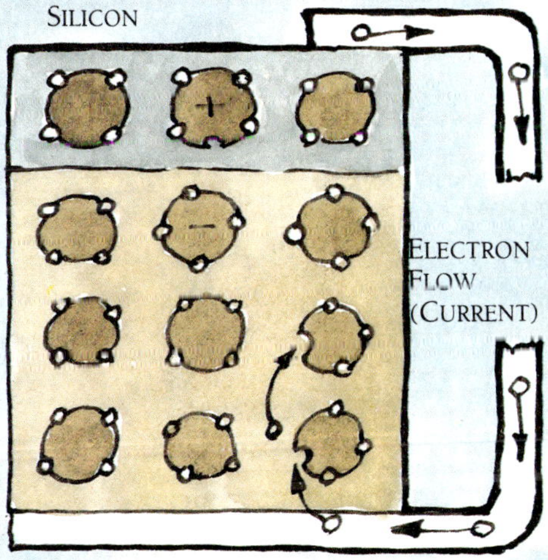

LIGHT STRIKES THE CELL

The light ray frees an electron which is pulled into the n-type layer by the positive charge there.

FILLING THE HOLE

An electron from an adjoining atom moves upward to fill the hole left by the freed electron.

CURRENT FLOWS

Electrons produce a current as light frees them. Returning electrons fill the holes that they have left.

REMOTE CONTROL UNIT

DIODE

A diode allows current to flow in one direction but not in the other. It consists of a *p-n* semiconductor junction (see p.291). When a positive terminal is connected to the *p*-type layer (*far right*), the positive charge of the terminal attracts electrons and a full current flows. On reversing the connections (*right*), the negative charge of the *p*-type layer opposes electron flow. A low current flows as a few electrons freed by atomic vibrations cross the juction.

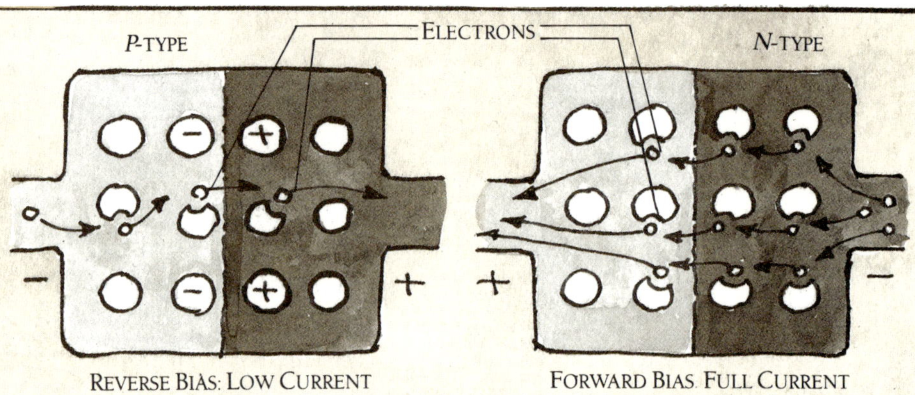

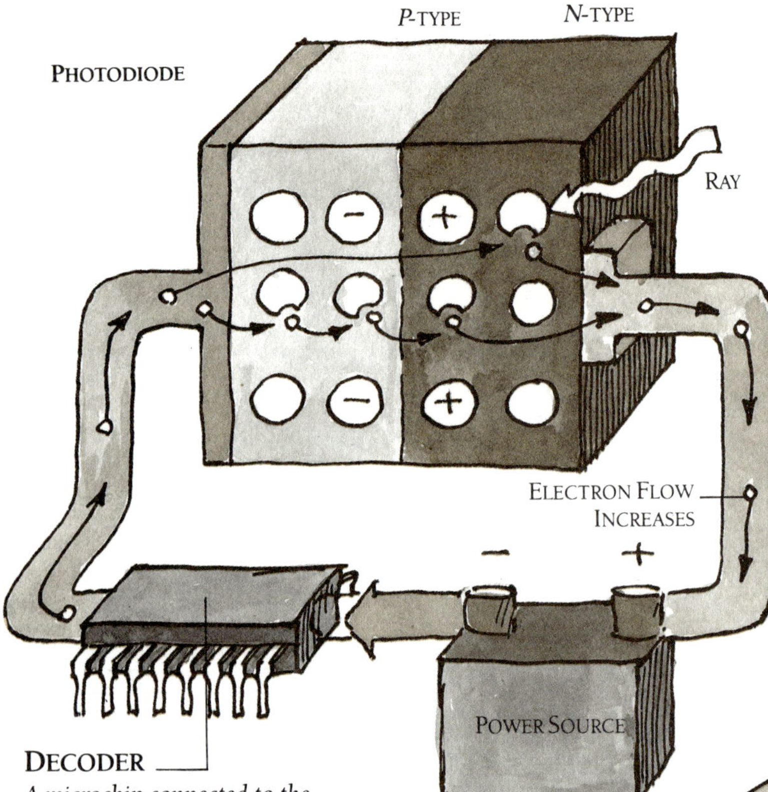

DECODER
A microchip connected to the photodiode receives a series of electrical pulses in binary code as the beam flashes on and off. The chip decodes this signal. A compact disk player (see pp.248-9) works in a similar way.

Pressing a button on the remote control unit for a television or video recorder transmits a beam of invisible infra-red rays to the set. The beam contains a signal, for example to change channel, in a series of electrical pulses in binary code (see pp.332-3). The receiver unit in the set detects the signal and decodes it. Both the transmitter and receiver units work with components known as diodes, but in each case the diodes work in opposite ways.

RECEIVER UNIT

The receiver unit contains a photodiode, which is a diode sensitive to light or infra-red rays. It is connected in reverse bias so that normally only a low current flows through it. When rays strike the diode, they free some electrons, increasing the current to produce a signal which goes to the decoder.

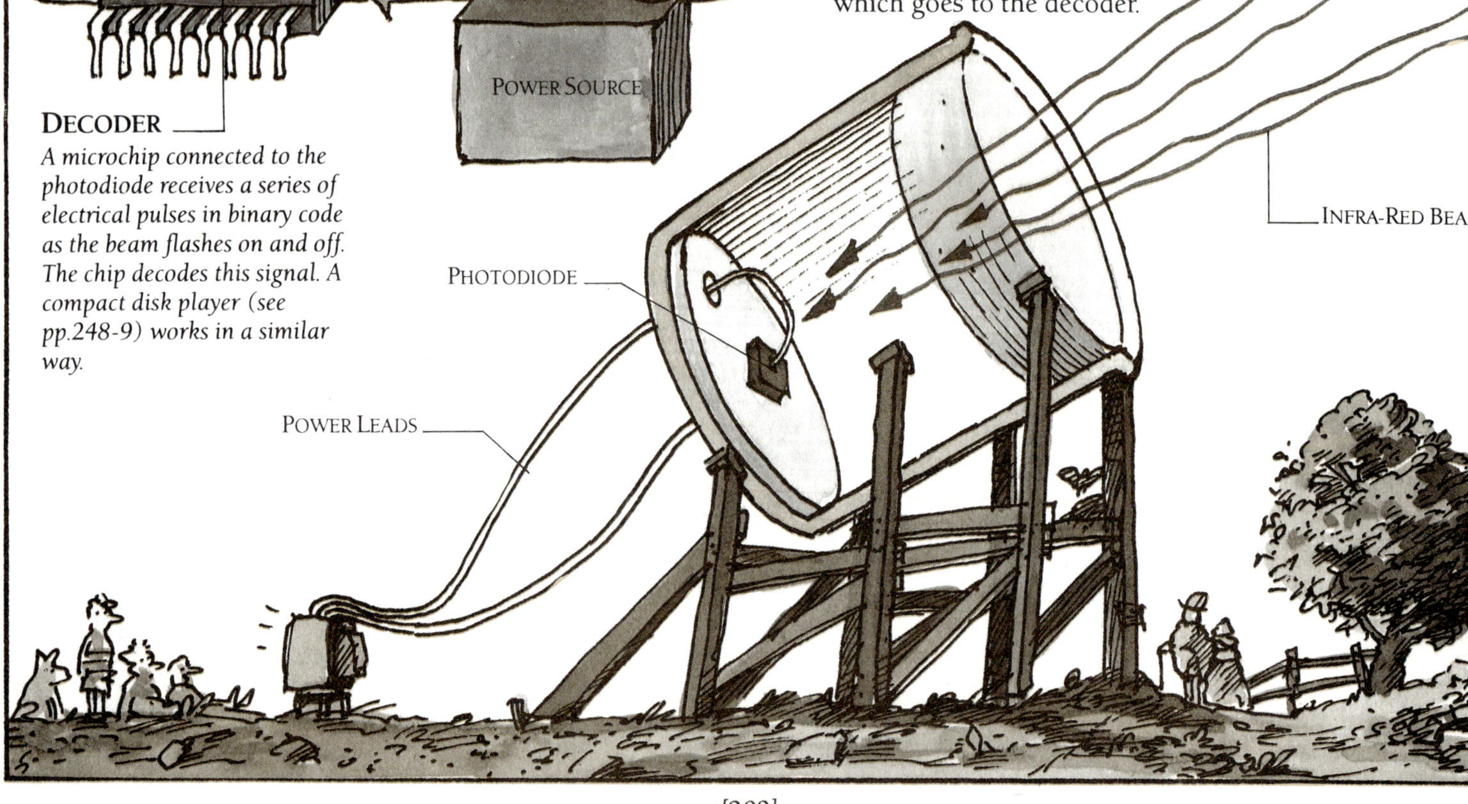

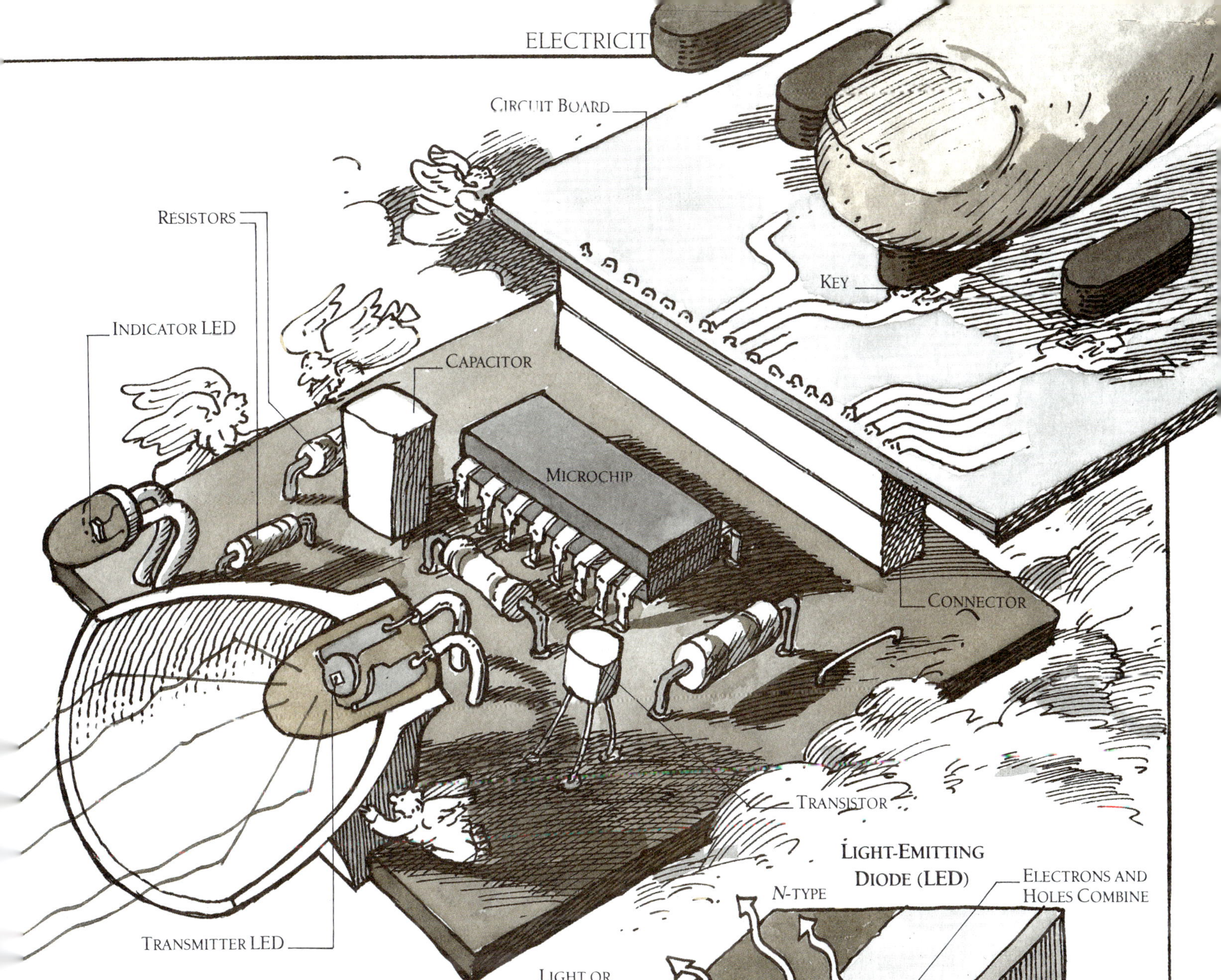

TRANSMITTER UNIT

The hand-held transmitter unit contains keys and electronic components similar to those in a calculator (see pp.336-7). Pressing a key routes a signal to the encoder chip, which sends a series of electrical pulses to the LED (light-emitting diode). The pulses form a signal in binary code, and the LED flashes on and off to send the signal to the receiver. An indicator LED lights up as the key is pressed.

A light-emitting diode is connected to a power source in forward bias. Electrons leaving the semiconductor atoms create holes that are then filled by arriving electrons. As the electrons and atoms combine, they produce light or infra-red rays.

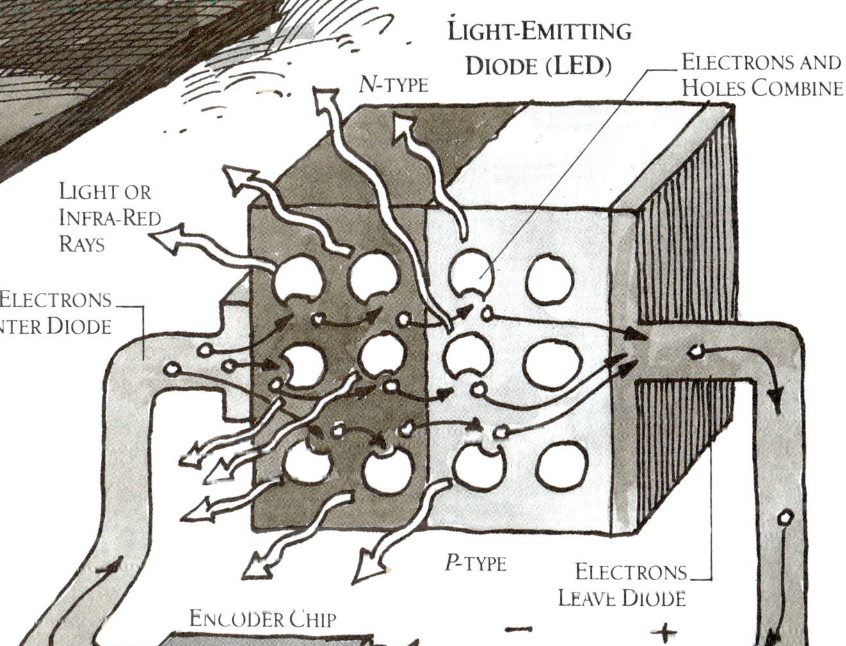

MAGNETISM

ON SHOEING A MAMMOTH

Working mammoths wear out their shoes with great rapidity, so it was with extreme interest that I watched a blacksmith fitting new improved shoes to a volunteer beast. The test had mixed results. Shoe wear was reduced to zero, but only because a strange and powerful attraction between opposite shoes prevented all movement on the part of the wearer.

WHERE NORTH MEETS SOUTH

A magnet is a seemingly ordinary piece of metal or ceramic that is surrounded by an invisible field of force which affects any magnetic material within it. All magnets have two poles. When magnets are brought together, a north pole always attracts a south pole, while pairs of like poles repel each other. Bar magnets are the simplest permanent magnets. Horseshoe magnets, which have such an unfortunate effect when used as mammoth footwear, are bar magnets bent so that their poles are brought close together.

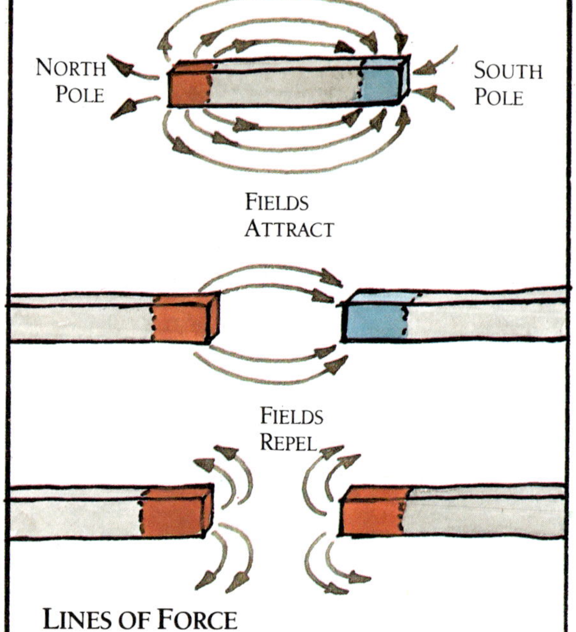

LINES OF FORCE

Because a magnetic field cannot be seen, lines are used to show the direction of the field.

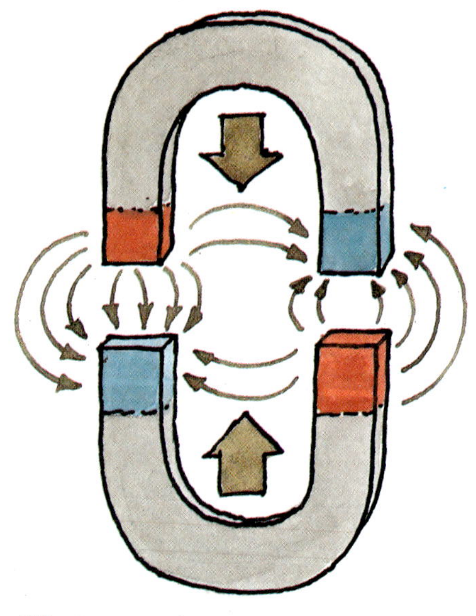

MAGNETIC ATTRACTION

The lines of force extend from the north pole of one magnet to the south pole of the other, pulling the magnets together.

ON A MAMMOTH CLOTHES-DRIER

The problem of how to dry out weatherproof clothing worn by working mammoths in damp climates has long taxed my ingenuity. On one occasion, I designed a hollow drier modeled on the form of a standing mammoth, which was intended to prevent shrinkage of the garments. I accordingly had a blacksmith put my plans into effect, and in no time he was happily coiling some sturdy wire around an iron bar supported on wooden legs. What happened next was both startling and inexplicable. A sudden thunderstorm swept overhead and a bolt of lightning hit one end of the coil, and at that very instant all the blacksmith's tools flew through the air and attached themselves to the work in progress. The project was promptly abandoned.

ELECTRICAL MAGNETS

When an electric current flows through a wire, a magnetic field is produced around it. The field produced by a single wire is not very strong, so to increase it, the wire is wound into a coil. This concentrates the magnetic field, especially if an iron bar is placed in the center of the field. Electromagnets can be very powerful – as the blacksmith finds out. A sudden burst of current momentarily transforms his clothes-drier into a powerful electromagnet which attracts all nearby iron objects to its poles.

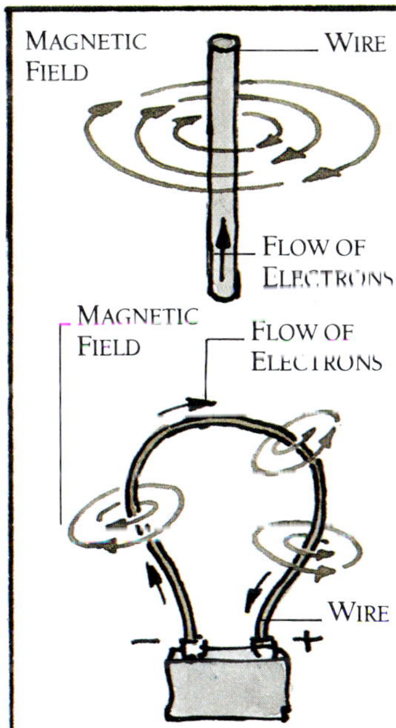

SINGLE WIRE
The lines of force form circles around the wire.

COIL OF WIRE
The lines of force of all the loops in a coil combine to produce a field that is similar to the field around a bar magnet. The poles of the electromagnet are at either end of the coil.

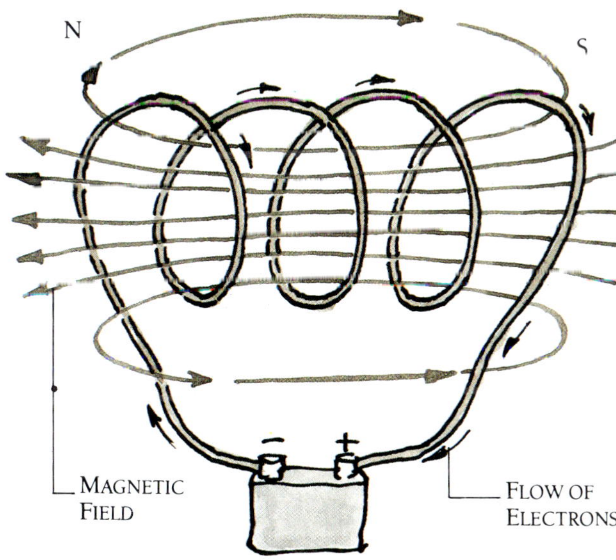

MAGNETS AT WORK

MAGNETIC COMPASS

The Earth has its own magnetic field. A compass needle will align itself so that it point toward the north and south magnetic poles, along lines of force which run in the direction of the field. The magnetic poles are situated away from the geographical poles.

MAGNETIC INDUCTION

A magnet is able to pick up a piece of steel or iron because its magnetic field flows into the metal. This turns the metal into a temporary magnet, and the two magnets then attract each other.

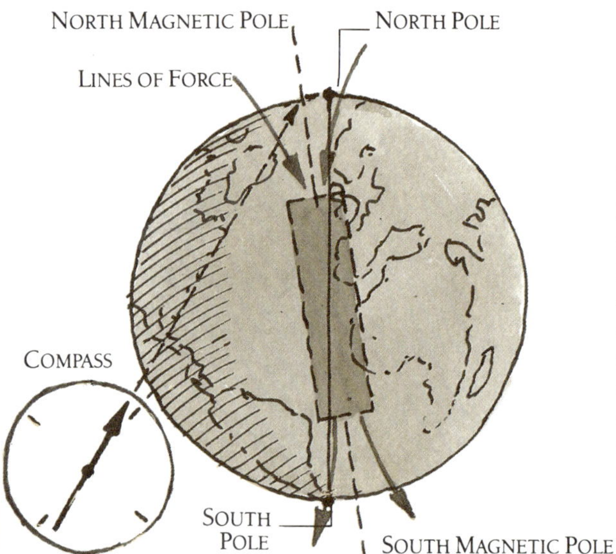

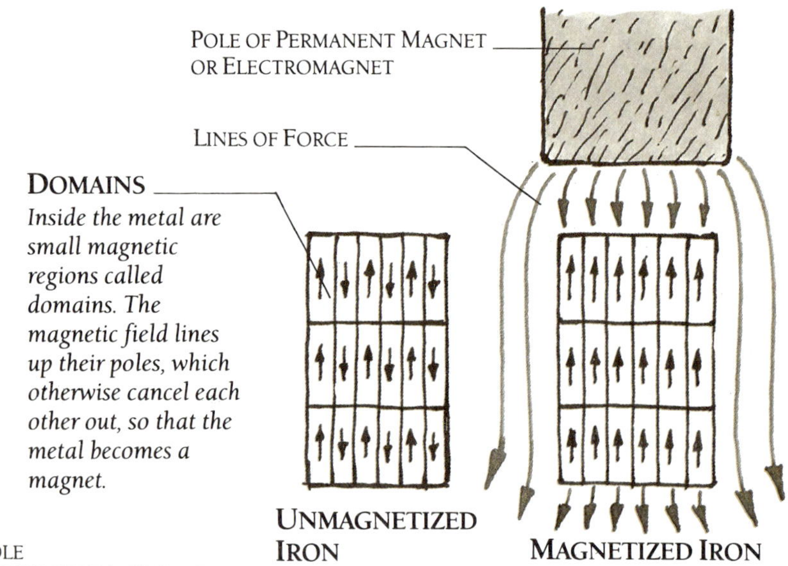

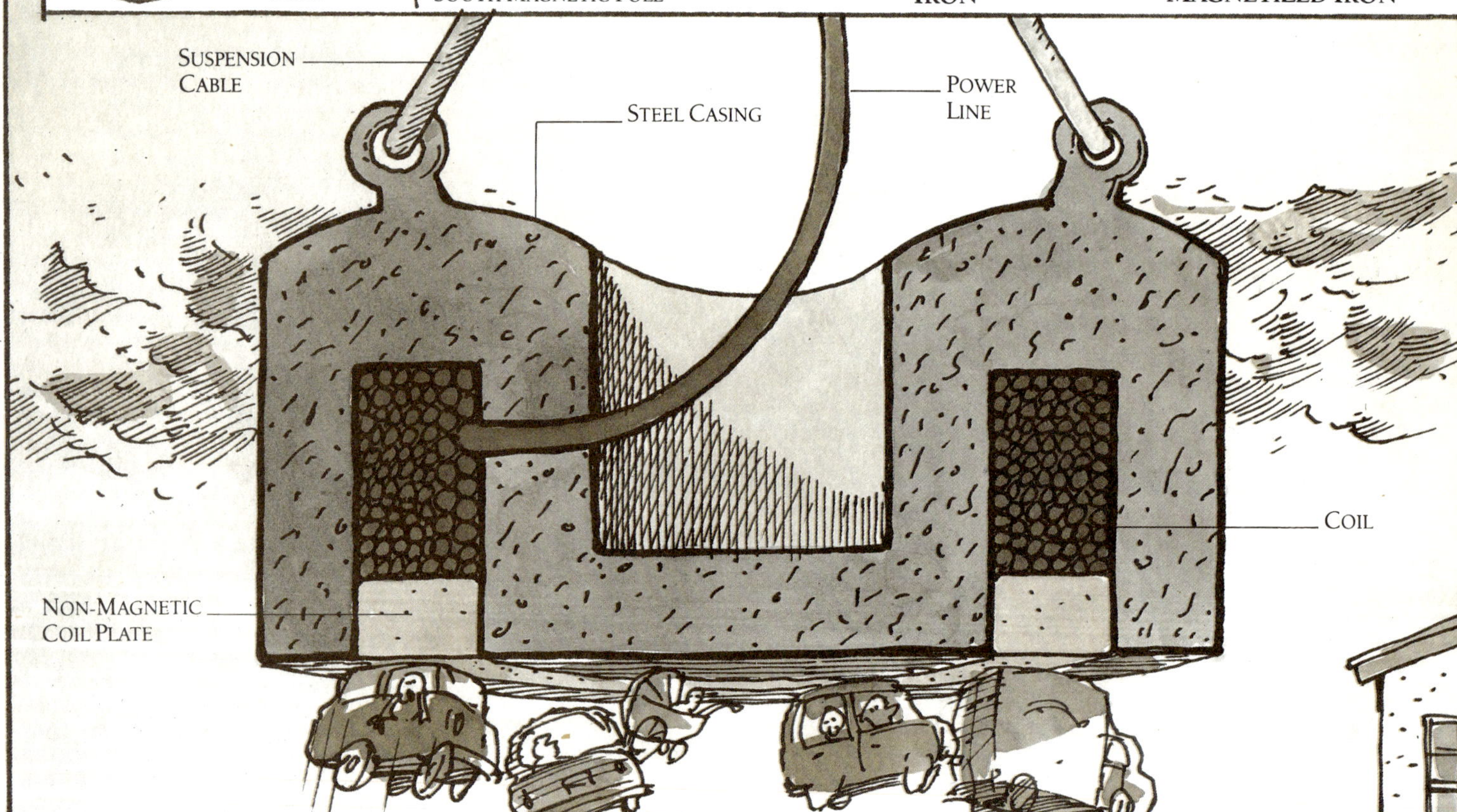

ELECTROMAGNET

An electromagnet is a coil of wire wound around an iron core. When current flows through the coil, it creates a magnetic field. The strength of the field depends on the current. Large electromagnets are strong enough to lift scrapped cars; much smaller electromagnets are used medically for tasks such as extracting metal splinters.

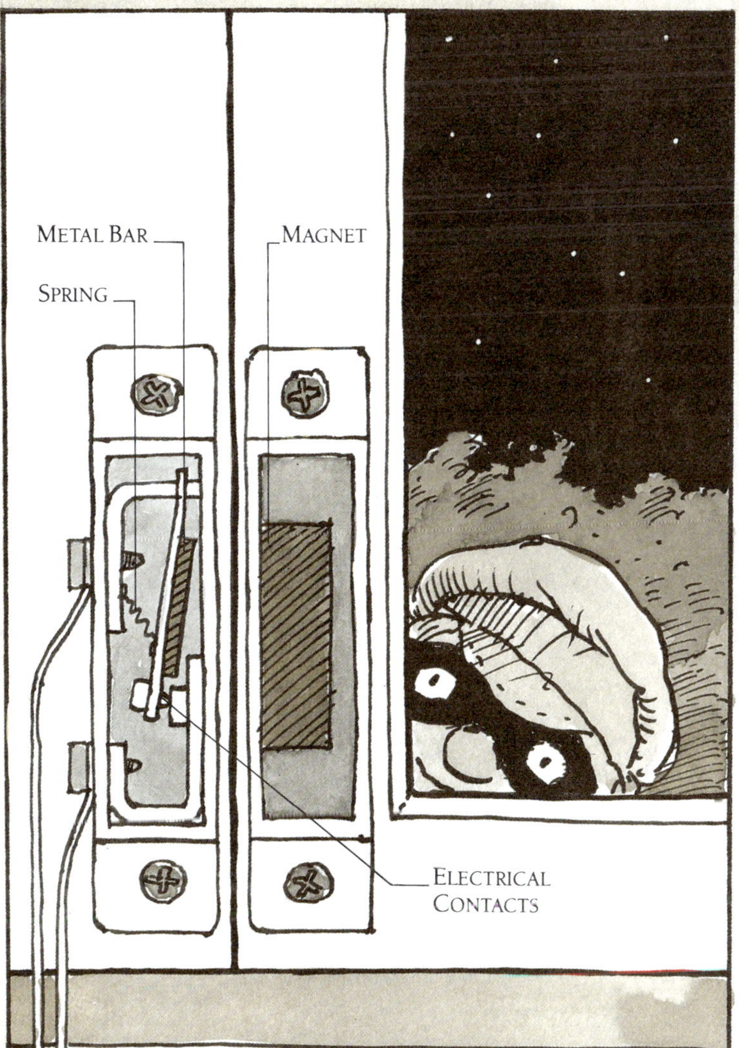

BURGLAR ALARM

A magnetic sensor can detect the opening of a door or window. A permanent magnet is mounted on the window or door and a special switch on the frame. When the window or door is closed, the magnetic field attracts a metal bar, keeping the switch on.

ALARM SOUNDS

If the window or door is opened, the magnet moves and no longer attracts the metal bar. The spring pulls the bar back, opening the contacts. This cuts the circuit, which activates a mechanism that rings the alarm.

MAGNETIC MACHINES

A great number of machines contain electromagnets. Many use them in their electric motors (see pp.300-1) to provide power. Electromagnets are also used to store signals in tape recorders, video recorders and computer disk drives, to produce sound in bells, buzzers, loudspeakers and telephones, and to deflect electron beams in television cameras and receivers.

THE ELECTRIC BELL

One of the many everyday uses of electromagnetism is the electric bell. The button at the door is an electric switch that sends current from a power source such as a battery to the striking mechanism. This makes a hammer move back and forth several times a second, sounding a metal bell An electromagnet and a spring alternately pull the hammer.

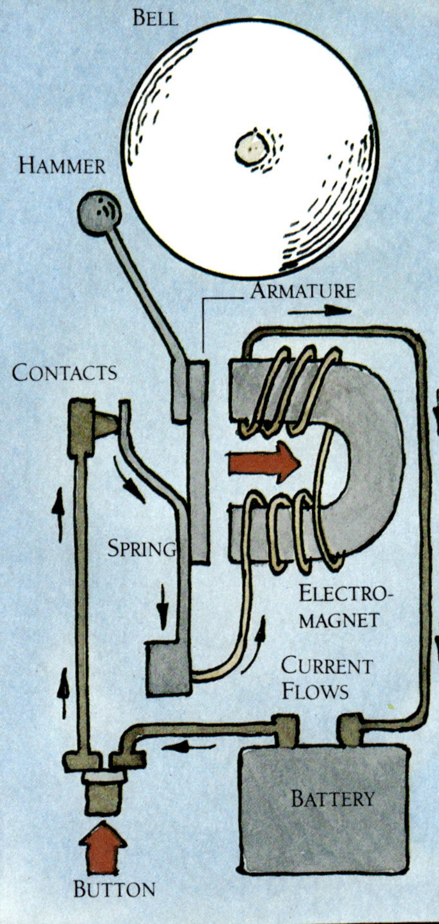

PRESSING THE BUTTON

When the button is pressed, the contacts are first closed. Current flows through the contacts and the spring to the electromagnet, which produces a magnetic field. This field attracts the iron armature, which moves toward the electromagnet against the spring and makes the hammer strike the bell.

THE BELL SOUNDS

As the hammer strikes the bell, the movement of the armature opens the contacts. The current stops flowing to the electromagnet, which loses its magnetism. The spring pulls the armature back, and the hammer moves away from the bell. The contacts then close again, and the cycle repeats itself for as long as the button is pressed.

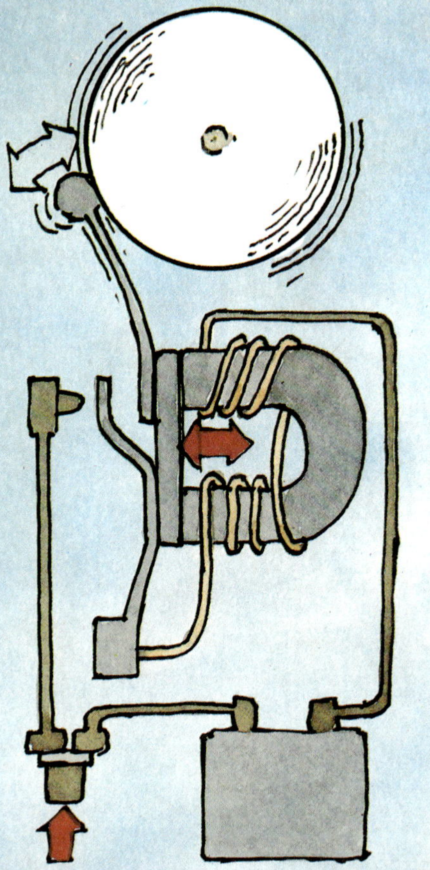

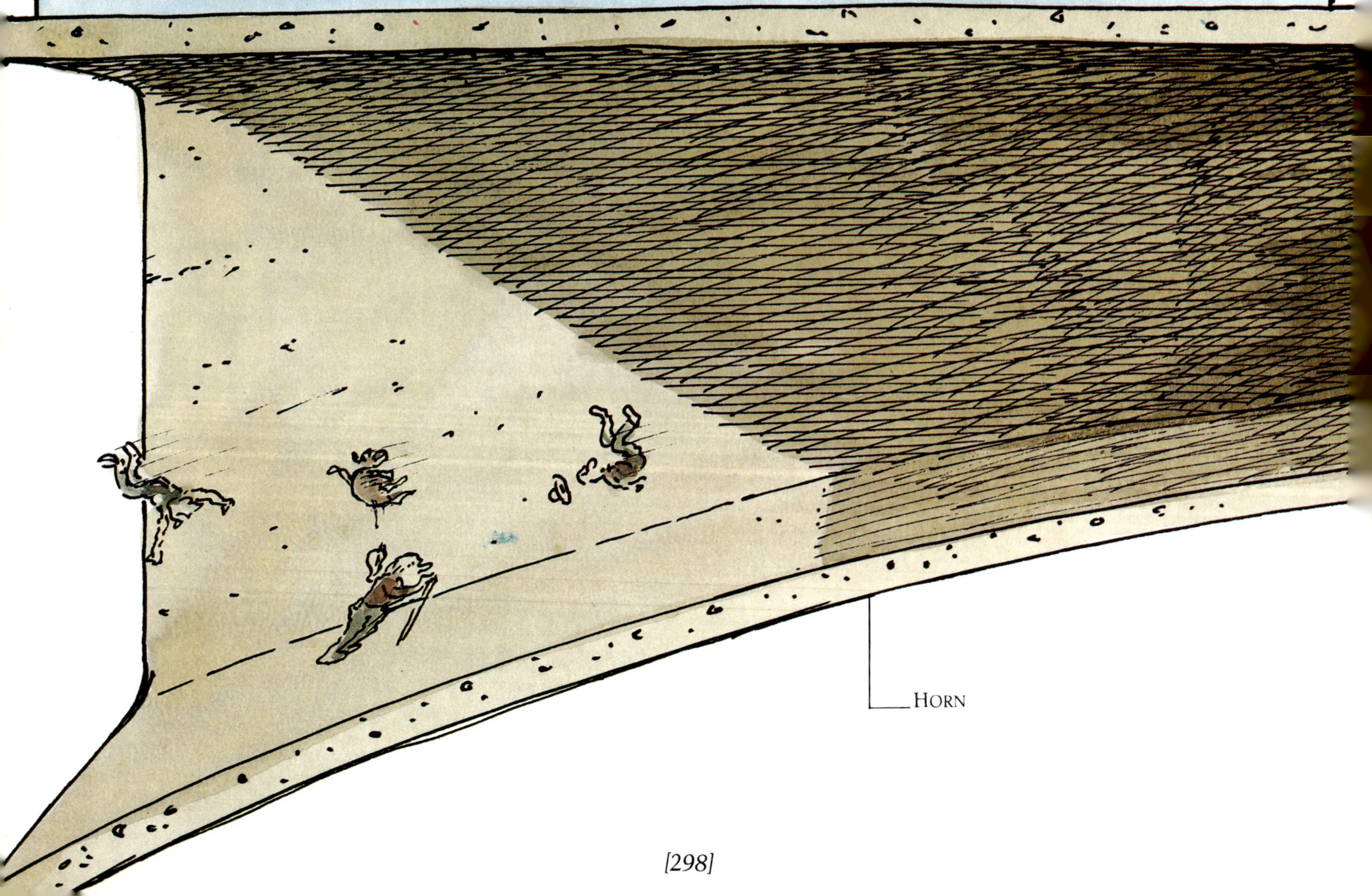

THE ELECTRIC HORN

The horn of an automobile is another example of the use of magnetism to produce sound by a simple vibration. The mechanism of a horn is rather similar to that of an electric bell, with a set of contacts repeatedly closing and opening to interrupt the flow of current to an electromagnet. Here, an iron bar moves up and down inside the coil of the electromagnet as the magnetic field switches on and off. The bar is attached to a diaphragm, which vibrates rapidly and gives out a loud sound.

The horn, as here, may have an actual bell-shaped horn attached to the diaphragm. This resonates to give a penetrating note and projects the sound forward.

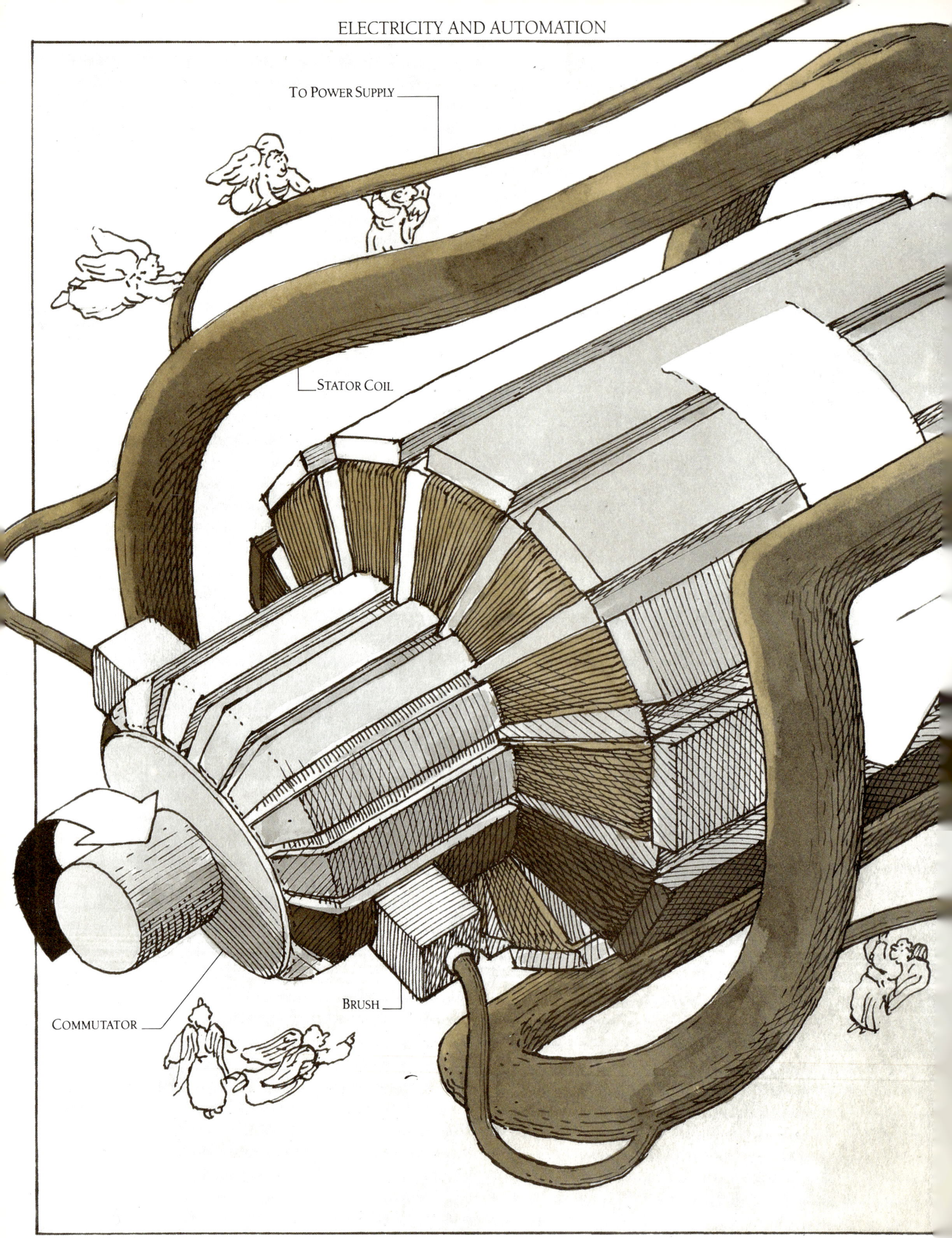
TO POWER SUPPLY
STATOR COIL
COMMUTATOR
BRUSH

ELECTRIC MOTOR

The electric motor is the most convenient of all sources of motive power. It is clean and silent, starts instantly, and can be built large enough to drive the world's fastest trains or small enough to work a watch. Its source of energy can be delivered along wires from an external power source or contained in small batteries.

There are several different kinds of electric motors. Many household machines contain the universal motor shown here. This motor gets its name because it can run on both the direct current provided by a battery and the alternating current from the electricity supply. It is like a direct current motor but has an electromagnet called a stator instead of a permanent magnet, and a rotating set of coils called a rotor. When direct current is used, the rotor field reverses every half turn and the stator field does not. With alternating current, the opposite happens. Either way, the rotor keeps on turning.

ROTOR
The central rotor contains several coils. As it rotates, each coil is in turn supplied with current by the brushes on the commutator.

STATOR
The stator contains coils that are fed with the electric current supplied to the rotor. This produces the magnetic field that interacts with the field of the electrified rotor coil.

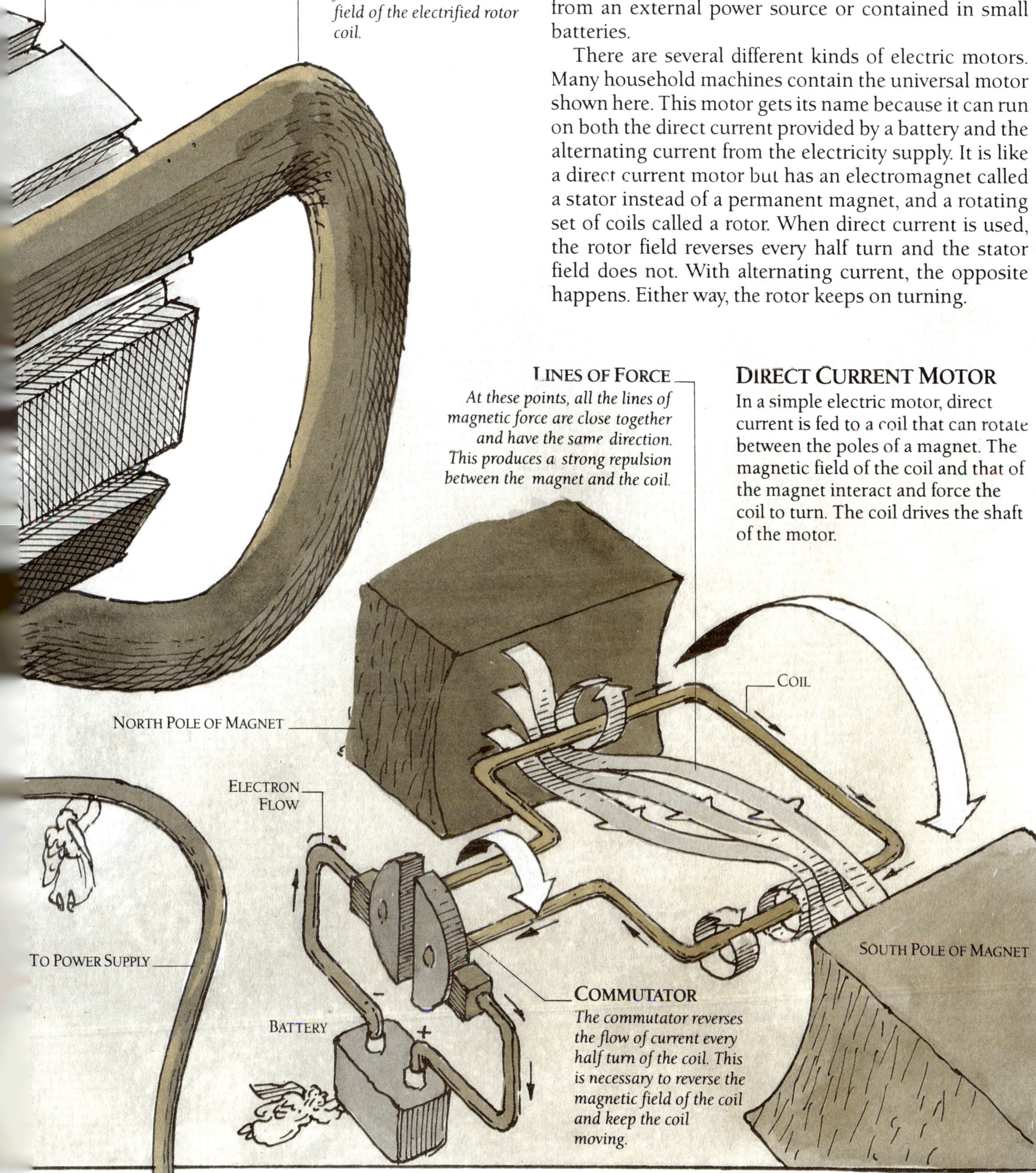

DIRECT CURRENT MOTOR

In a simple electric motor, direct current is fed to a coil that can rotate between the poles of a magnet. The magnetic field of the coil and that of the magnet interact and force the coil to turn. The coil drives the shaft of the motor.

LINES OF FORCE
At these points, all the lines of magnetic force are close together and have the same direction. This produces a strong repulsion between the magnet and the coil.

COMMUTATOR
The commutator reverses the flow of current every half turn of the coil. This is necessary to reverse the magnetic field of the coil and keep the coil moving.

MAGLEV TRAIN

A maglev has no wheels, instead using magnetic fields to levitate itself above a track. Thus freed from friction with the rails, the train can float along the track. The train shown here uses the attractive system of levitation, in which electromagnets attached to the train run below the suspension rail and rise toward it to lift the train.

LINEAR INDUCTION MOTOR

A form of electric motor called an induction motor drives the maglev train. Coils on the train generate a magnetic field in which the poles shift along the train. The field induces electric currents in the reaction rail, which in turn generates its own magnetic field. The two fields interact so that the shifting field pulls the floating train along the track.

The disk drive of a computer uses electromagnetism to "write" or store programs and data. The read-write head converts electric code signals from the computer into magnetic codes recorded on the surface of the disk; the drive then reverses this process to "read" the disk (see p.340).

A disk drive contains two electric motors – a disk motor to rotate the disk at high speed and a head motor to move the head across the disk. The drive

FLOPPY DISK

A floppy disk is inserted into the drive by hand. The disk is protected by a sleeve in which a window is cut to expose the surface of the disk. Inside the drive, the head travels along the window as the disk rotates inside the sleeve.

TIMING HOLE

Some disks contain a hole through which a light shines onto a detector so that the drive can find the required sectors.

STEPPER MOTOR

The stepper motor in a diskdrive contains a rotor that is a permanent cylindrical magnet with many poles around its circumference. It rotates inside two sets of stator coils, each of which has a row of metal teeth. Sending an electric current to a coil (*right*) magnetizes its teeth with alternate north and south poles. Reversing the current (*far right*) reverses the sequence of the poles. The two rows of teeth on the upper and lower stators are placed out of alignment, and the rotor moves to position each pole with a pair of overlapping teeth having the opposite pole. Signals from the drive controller to the stator coils change the teeth poles so that the rotor turns to follow them.

DISK DRIVE

must work with very great precision, because a tiny error in the position of the head could corrupt the program or data and stop the computer. The head is therefore moved by a stepper motor which, instead of turning constantly, obeys a control signal to rotate by an exact amount.

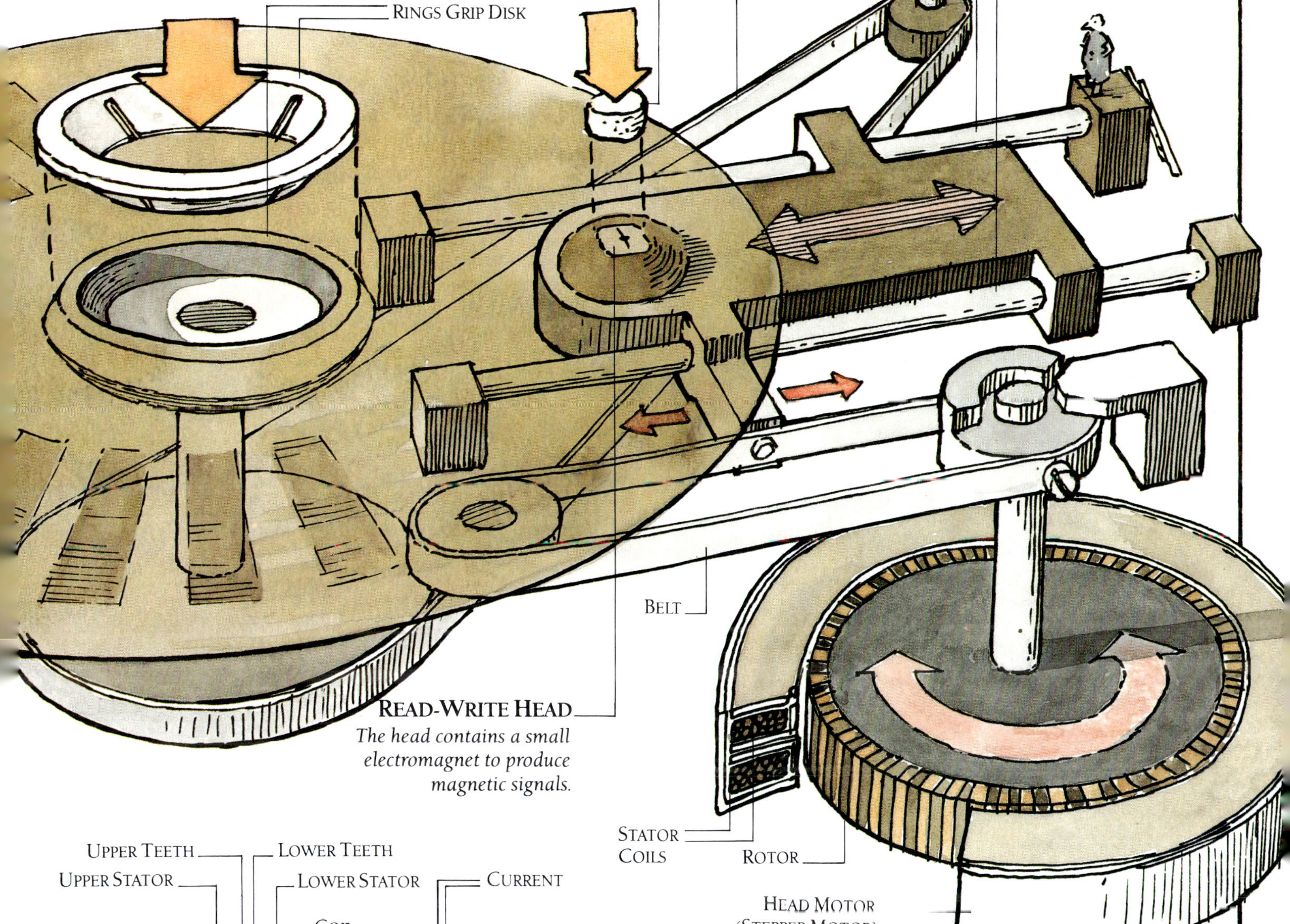

READ-WRITE HEAD
The head contains a small electromagnet to produce magnetic signals.

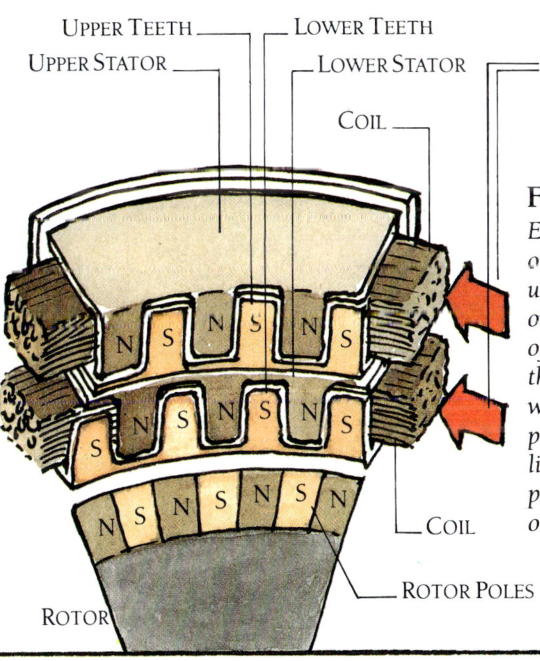

FIRST SIGNAL
Each north pole on the rotor lines up with an overlapping pair of south poles on the stator teeth, while each south pole on the rotor lines up with a pair of north poles on the teeth.

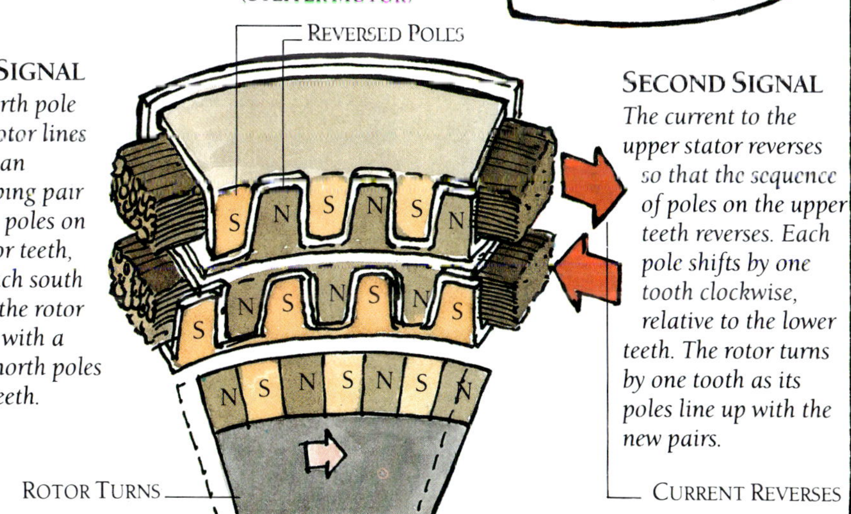

SECOND SIGNAL
The current to the upper stator reverses so that the sequence of poles on the upper teeth reverses. Each pole shifts by one tooth clockwise, relative to the lower teeth. The rotor turns by one tooth as its poles line up with the new pairs.

ELECTRIC GENERATOR

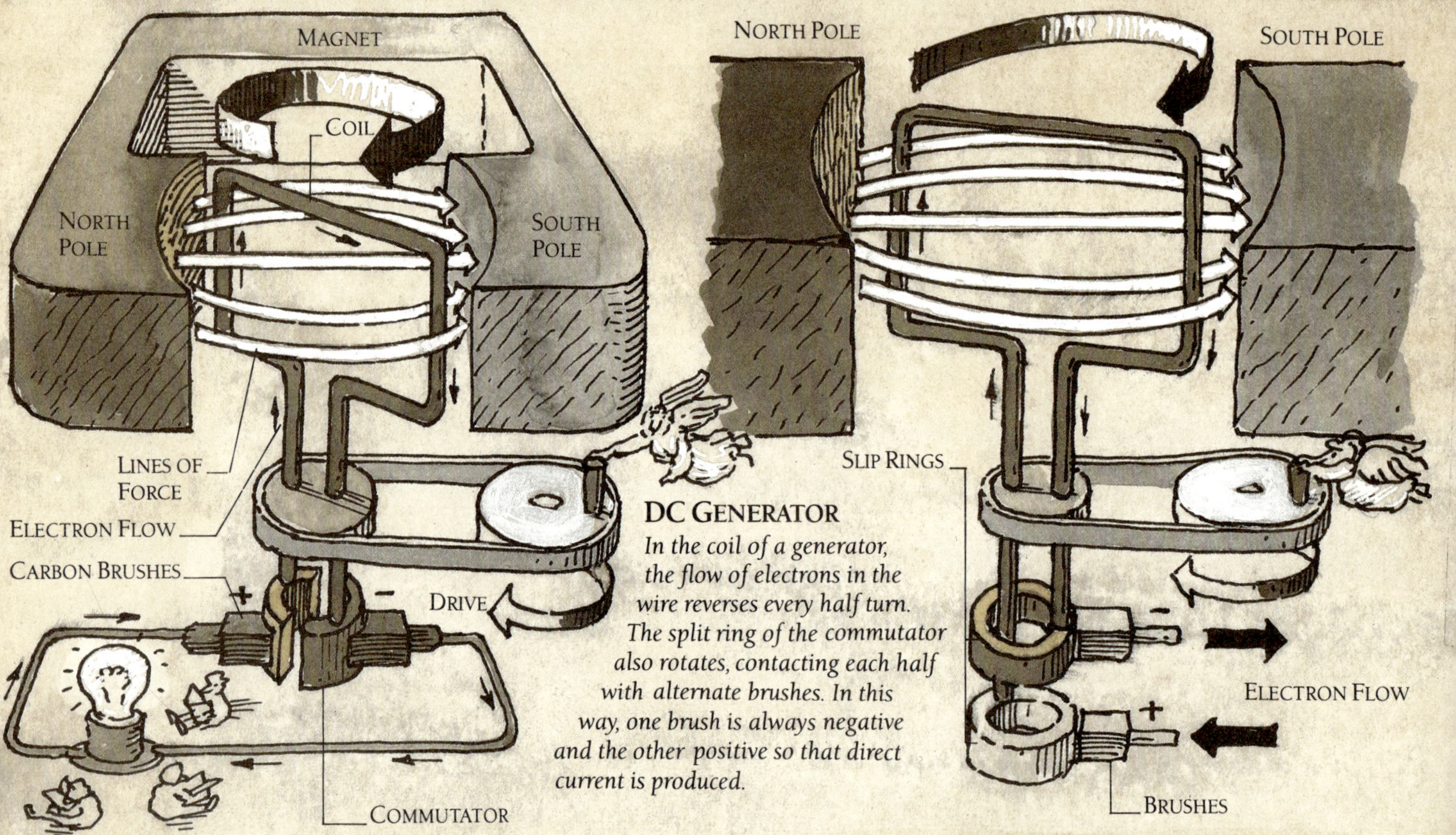

DC GENERATOR
In the coil of a generator, the flow of electrons in the wire reverses every half turn. The split ring of the commutator also rotates, contacting each half with alternate brushes. In this way, one brush is always negative and the other positive so that direct current is produced.

An electric generator works by electromagnetic induction – it uses magnetism to make electricity. The power source spins a coil between the poles of a magnet or electromagnet. As it cuts through the lines of force, an electric current flows through the coil.

AC GENERATOR – FIRST HALF TURN
An alternating current (AC) generator contains two slip rings connected to the end of the coil. As the current reverses in the coil, an alternating current emerges from the brushes. When part of the coil cuts the lines of force near the magnet's north pole, the electrons move up the wire, producing a positive charge at the lower slip ring.

POWER SUPPLY

The large generators in power stations are powered by steam turbines (see p.167) or water turbines (see p.37). The electricity reaches our homes through a network of power lines. Initially the current travels at a very high voltage because much less energy is lost in transmission than if the voltage were low. Transformers then step down the voltage to different levels for use in industry and in the home.

POWER LINE
At high voltage, the current is capable of sparking considerable distances through air. For safety, the lines are suspended from high pylons by long insulators.

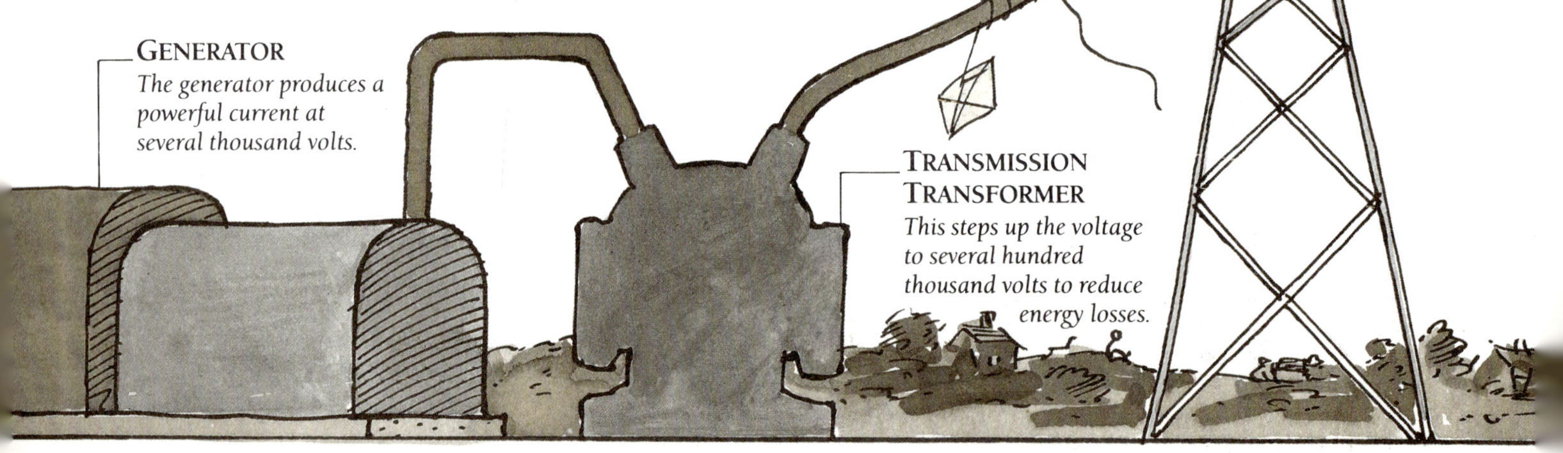

GENERATOR
The generator produces a powerful current at several thousand volts.

TRANSMISSION TRANSFORMER
This steps up the voltage to several hundred thousand volts to reduce energy losses.

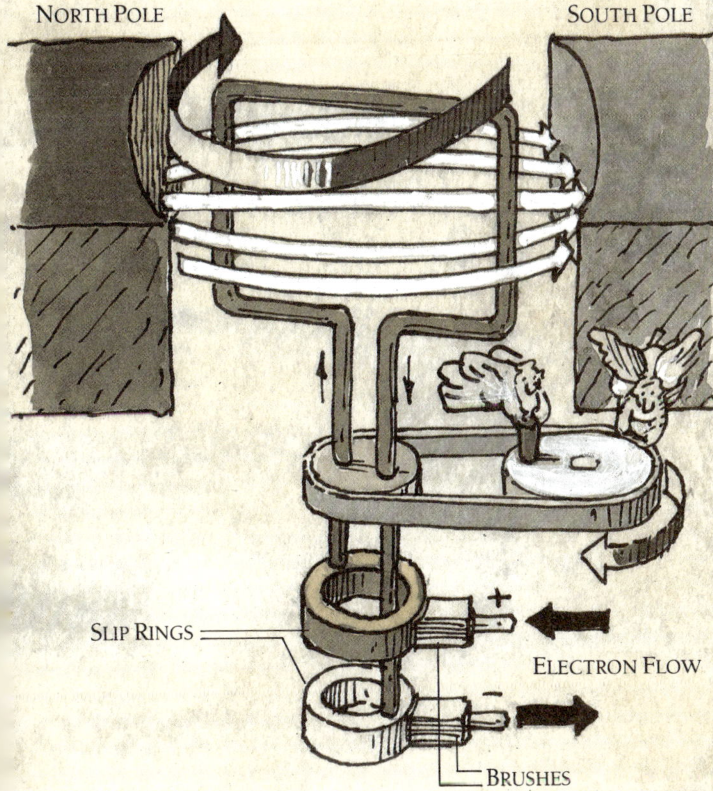

AC GENERATOR – SECOND HALF TURN

The same part of the coil has now turned to cut the lines of force near the magnet's south pole. Electrons now flow down the wire to produce a negative charge at the lower slip ring, reversing the current flow. The frequency of the current reversal produced by an AC generator depends on the speed at which the coil rotates.

TRANSFORMER

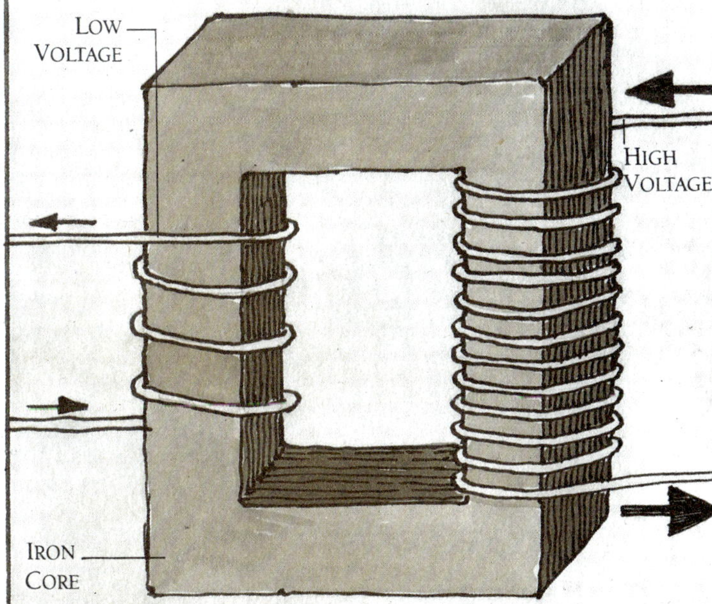

A transformer changes the voltage of an alternating current. The input current goes to a primary coil wound around an iron core. The output current emerges from a secondary coil also wound around the core. The alternating input current produces a magnetic field that continually switches on and off. The core transfers this field to the secondary coil, where it induces an output current. The degree of change in voltage depends on the ratio of turns in the coils; the tranformer shown here steps up or steps down the voltage three times.

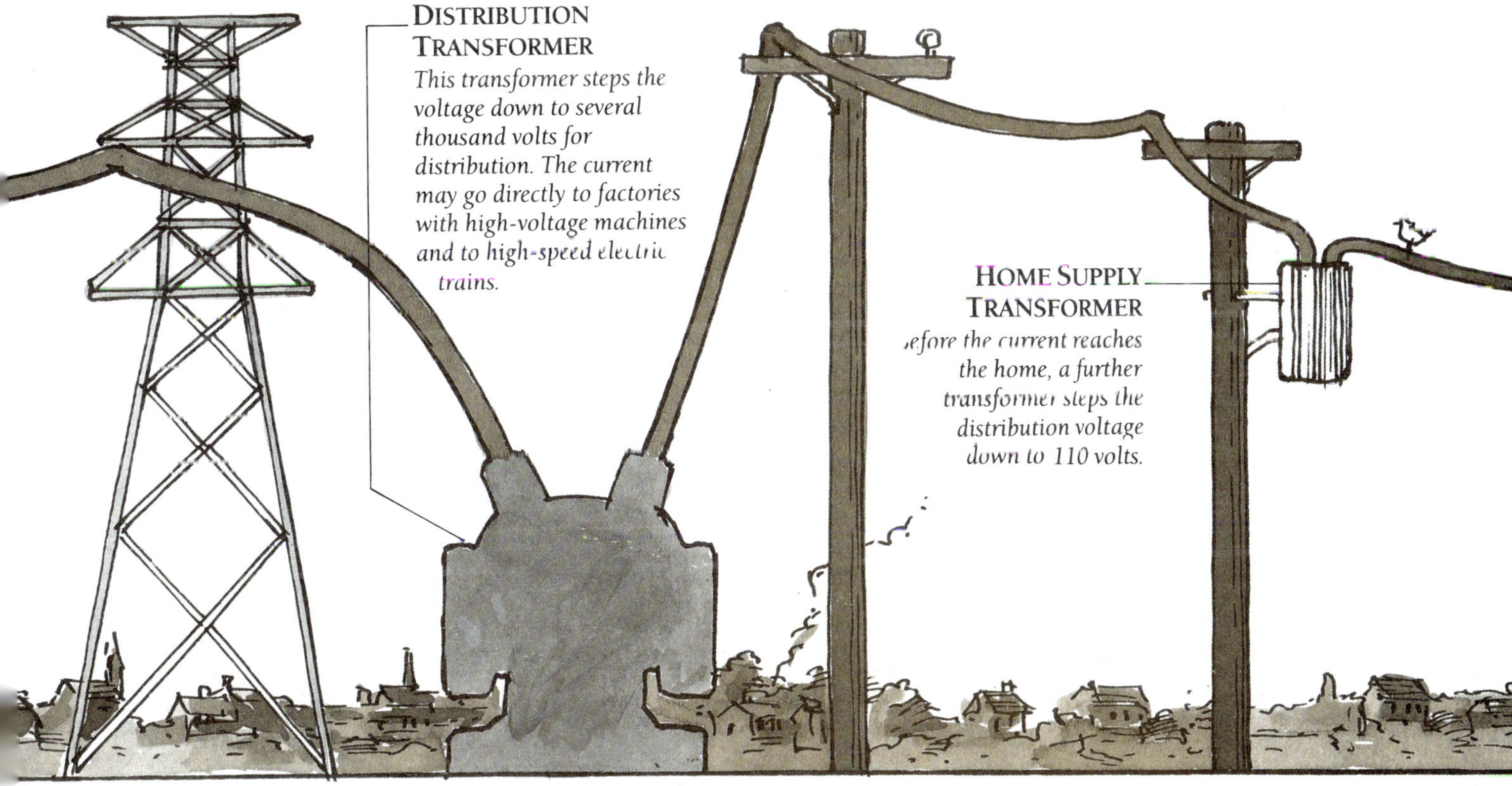

DISTRIBUTION TRANSFORMER

This transformer steps the voltage down to several thousand volts for distribution. The current may go directly to factories with high-voltage machines and to high-speed electric trains.

HOME SUPPLY TRANSFORMER

Before the current reaches the home, a further transformer steps the distribution voltage down to 110 volts.

THE TWO-WAY SW

Once a supply of electricity has entered the home, it is metered and then distributed to outlets and light switches. The two-way switch is a common part of a domestic electric circuit. In it, a pair of switches is connected so that pressing either turns a light on or off. Each switch has two sets of contacts linked by a pair of wires. Moving the switch up or down closes one set of contacts and opens the other set. To turn the light on, the sets of contacts at the ends of either of the two wires must close.

Many appliances have a third wire connected to their metal casing which links through the wiring circuit to the ground. If the appliance is faulty and a live wire touches the case, the strong current is immediately conducted to the ground.

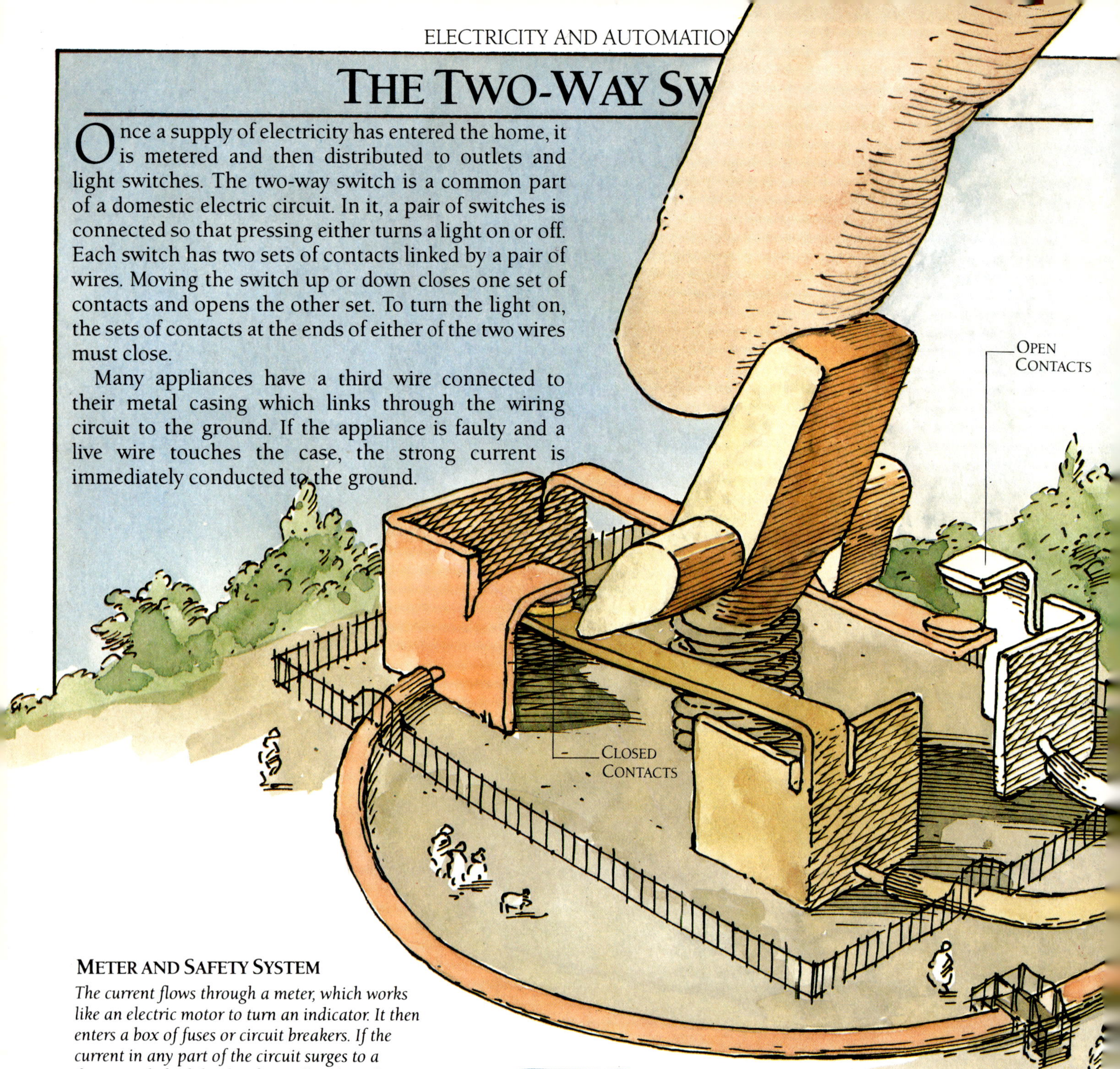

METER AND SAFETY SYSTEM

The current flows through a meter, which works like an electric motor to turn an indicator. It then enters a box of fuses or circuit breakers. If the current in any part of the circuit surges to a dangerously high level, a fuse will melt and break the circuit. A circuit breaker cuts off the supply by using an electromagnetic switch activated by the high current.

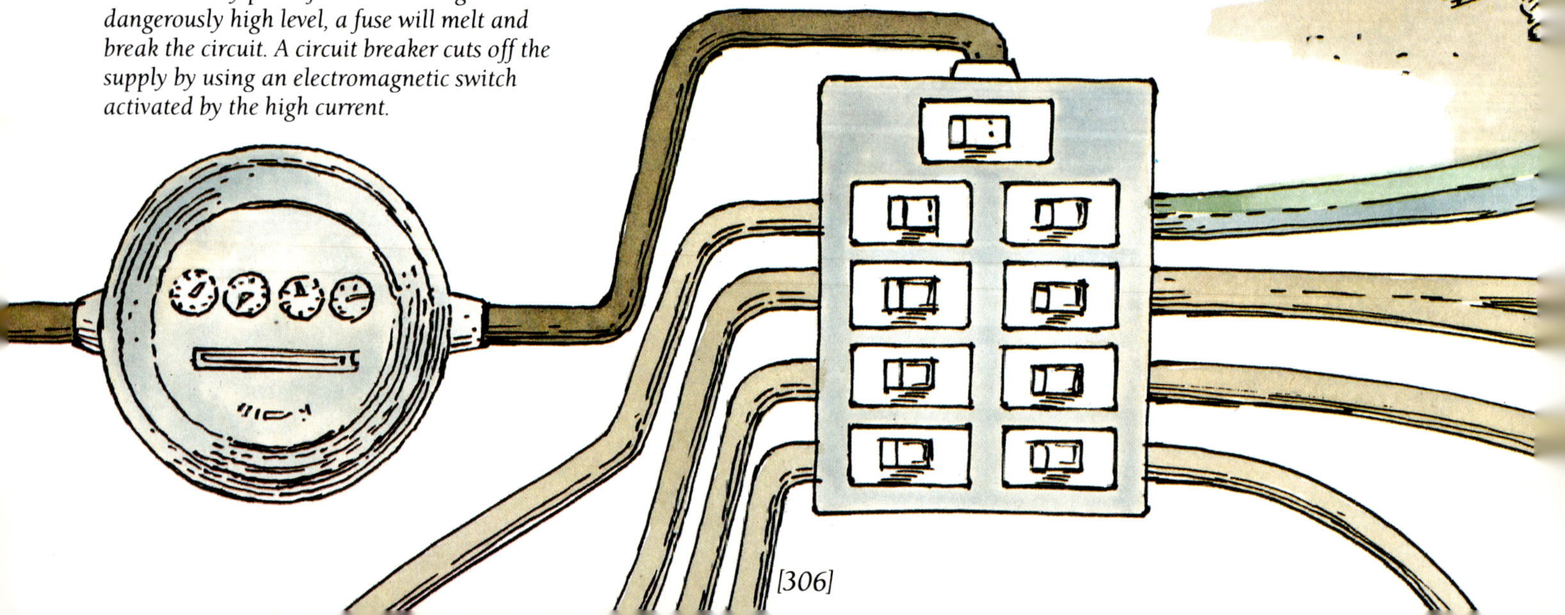

OPEN CONTACTS
CLOSED CONTACTS
POWER SOCKETS
EARTH SOCKET
EARTH WIRE

CAR IGNITION SYSTEM

Electromagnetism enables a car to start and also keeps it running by producing the sparks that ignite the fuel. At a twist of the ignition key, the starter motor draws direct current from the battery to start the engine. Producing the powerful magnetic field needed in the starter motor requires a hefty current, one which is too strong to pass through the ignition switch. So a solenoid, activated by a low current passing through the ignition switch, passes a high current to the starter motor.

In electromechanical ignition systems, like the one shown here, the contact breaker in the distributor opens and interrupts the supply of low-voltage current to the induction coil. The magnetic field around the primary winding collapses, inducing a high voltage in the secondary winding. The distributor then passes the current to the spark plugs. In electronic ignition, the contact breaker is replaced by an electronic switch.

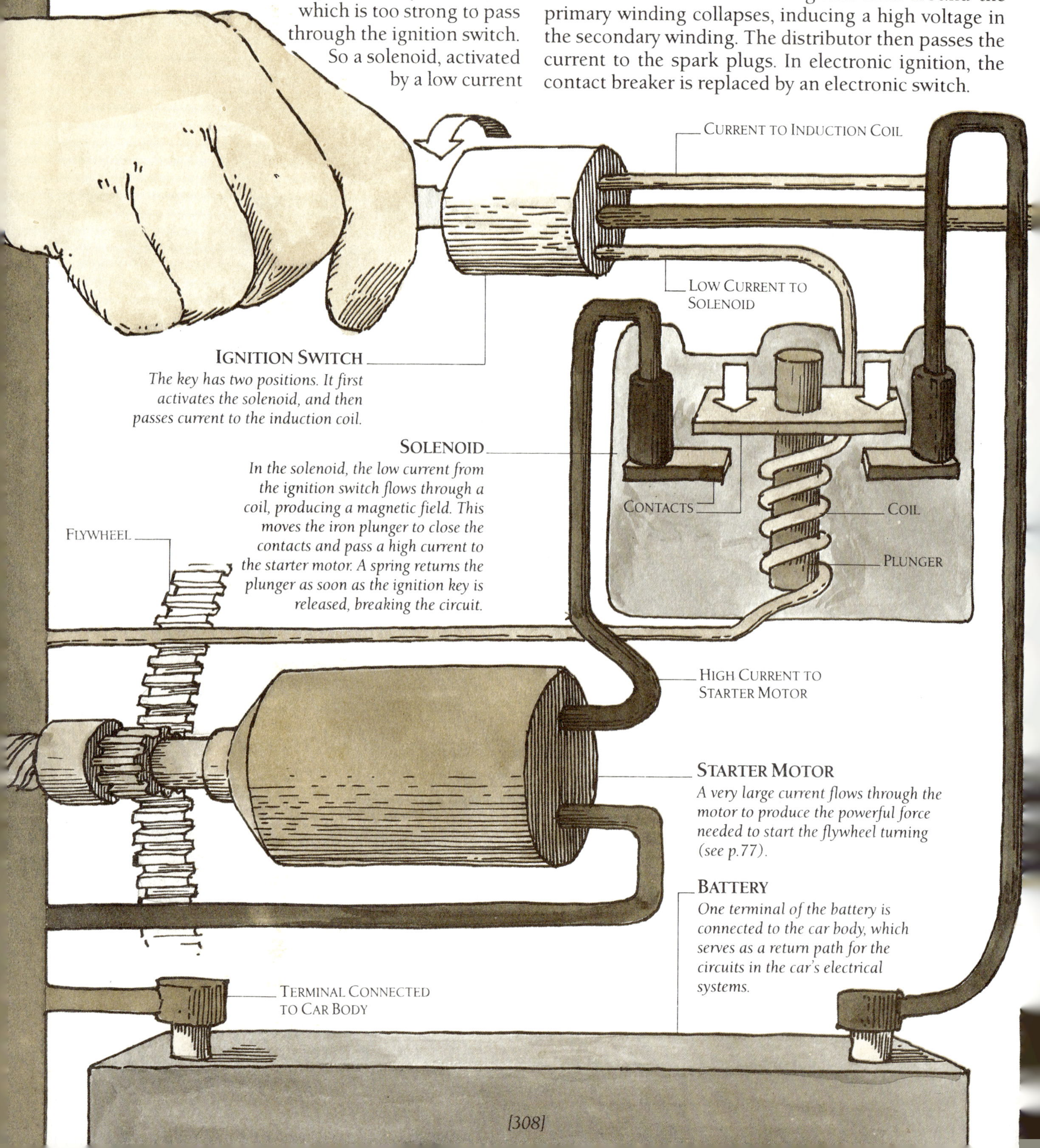

IGNITION SWITCH
The key has two positions. It first activates the solenoid, and then passes current to the induction coil.

SOLENOID
In the solenoid, the low current from the ignition switch flows through a coil, producing a magnetic field. This moves the iron plunger to close the contacts and pass a high current to the starter motor. A spring returns the plunger as soon as the ignition key is released, breaking the circuit.

STARTER MOTOR
A very large current flows through the motor to produce the powerful force needed to start the flywheel turning (see p.77).

BATTERY
One terminal of the battery is connected to the car body, which serves as a return path for the circuits in the car's electrical systems.

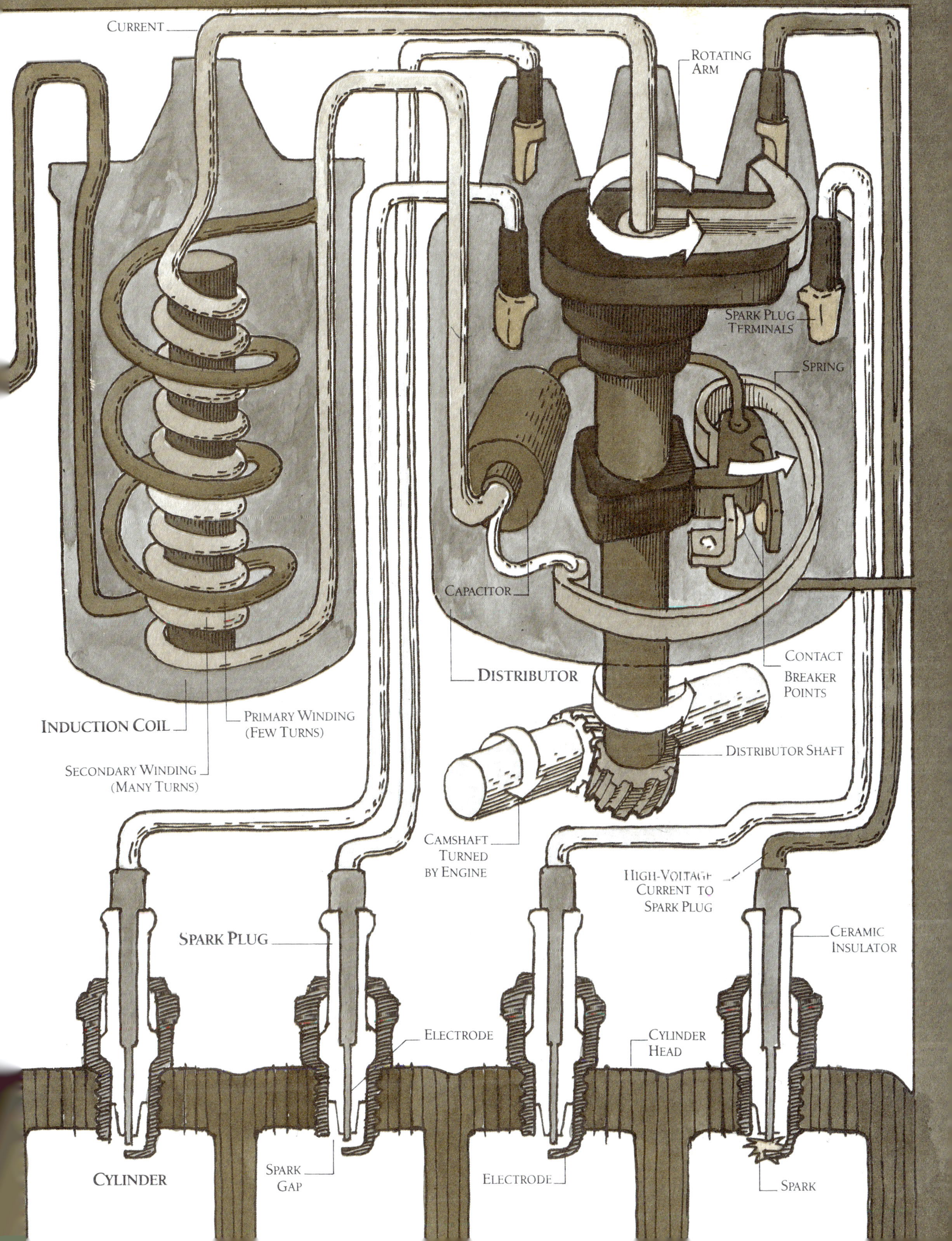

CURRENT
ROTATING ARM
SPARK PLUG TERMINALS
SPRING
CAPACITOR
DISTRIBUTOR
CONTACT BREAKER POINTS
INDUCTION COIL
PRIMARY WINDING (FEW TURNS)
SECONDARY WINDING (MANY TURNS)
DISTRIBUTOR SHAFT
CAMSHAFT TURNED BY ENGINE
HIGH-VOLTAGE CURRENT TO SPARK PLUG
SPARK PLUG
CERAMIC INSULATOR
ELECTRODE
CYLINDER HEAD
CYLINDER
SPARK GAP
ELECTRODE
SPARK

SENSORS AND DETECTORS

ON MAMMOTH SENSITIVITY

Emotionally and physically, mammoths are highly sensitive creatures. Their physical sensitivity can be exploited in numerous ways, assuming always that their emotional sensitivity can be controlled. A selection of such applications is here depicted. In figure 1, the trunk of a sleeping mammoth is used as a pressure-operated alarm to frighten away burglars.

In figure 2, the trunk of a sleeping mammoth is secured to the ceiling to act as a smoke detector. Plants obscure the creature's bulk and also provide it with occasional snacks.

In figures 3, 4 and 5, a highly trained mammoth is used as a metal detector. Once a piece of luggage has been tested, there is no question about the location of bulky items. Chances are that at least some of them are metal.

DISCOVERY AND MEASUREMENT

Sensors and detectors are devices that are used to detect the presence of something and often to measure it. Alarm systems sense the direct evidence of unwanted visitations, such as the tell-tale tread of a burglar or the airborne particles of smoke from a fire. Other sensors and detectors employ penetrating rays or magnetic fields to locate and reveal objects that cannot be seen. Measuring instruments, from seismographs to radar speed traps, are sensors and detectors that react to something specific and then register its quantity.

Sensors and detectors are also very important as essential components of automatic machines. Many machines, for example the autopilot in an aircraft, use feedback. This means that their sensors measure the machine's performance and then feed the results back to control the power output. This in turn affects the performance, which is measured by the sensors... and so on in an endless loop. By sensing their own performance, automatic machines keep within set limits. The mammoth-powered weight-sensing ski lift is a simple automatic machine.

SEISMOGRAPH

Seismographs locate the origin of earthquakes and measure their force. Earthquakes generate waves that travel through the Earth and vibrate the ground as they arrive at its surface. A seismograph is very sensitive to these vibrations, and it uses a heavy mass with a high inertia (see p.74) to detect their movement. The mass itself does not move, but detectors move in relation to it. The simple seismograph shown here operates mechanically; more advanced seismographs have detectors that work electromagnetically.

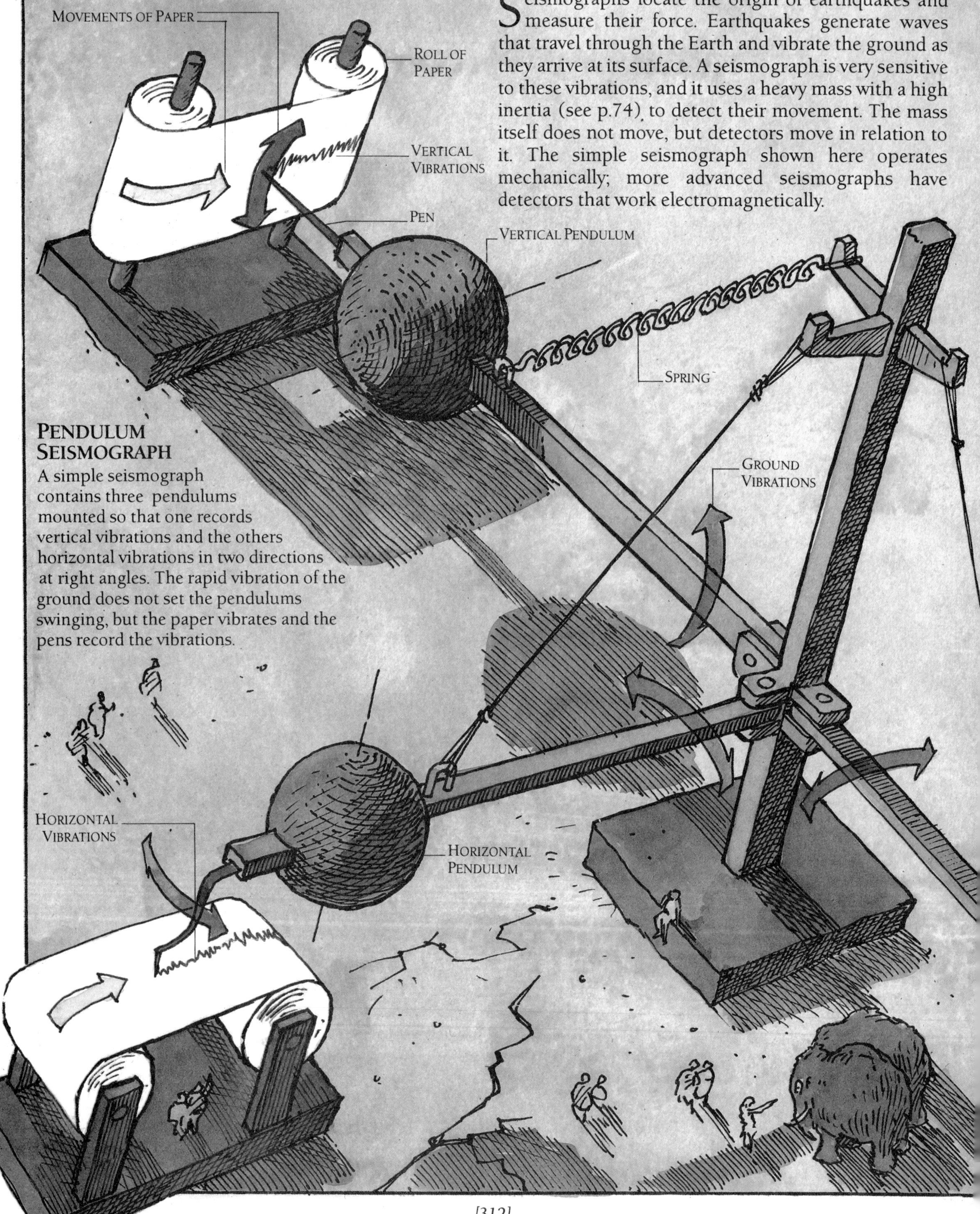

PENDULUM SEISMOGRAPH

A simple seismograph contains three pendulums mounted so that one records vertical vibrations and the others horizontal vibrations in two directions at right angles. The rapid vibration of the ground does not set the pendulums swinging, but the paper vibrates and the pens record the vibrations.

SEISMIC WAVES

Several kinds of vibration travel through the Earth. They move along the crust or through deeper regions, which affect their progress. Comparing arrival times at different locations allows their origin to be found.

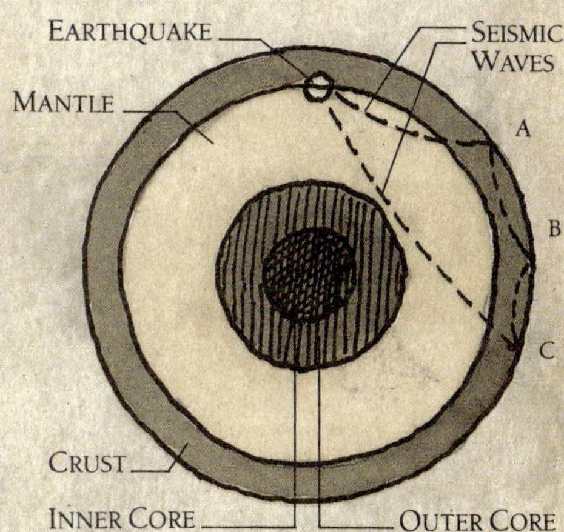

LOCATION A *Primary waves are followed by secondary and long waves.*

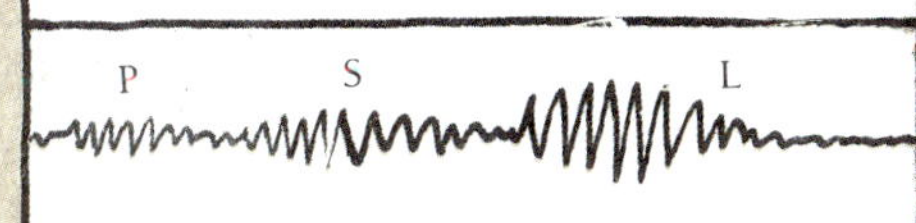

LOCATION B *The vibrations are more spread out if the waves have come further.*

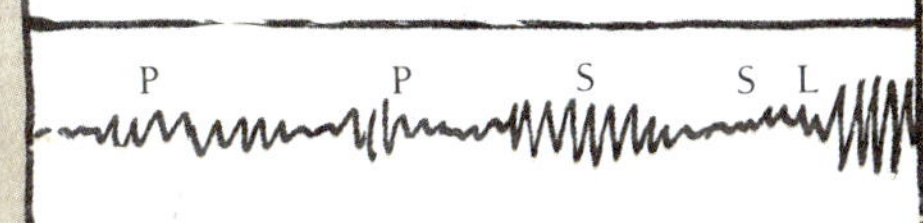

LOCATION C *Two sets of waves arrive, one after the other.*

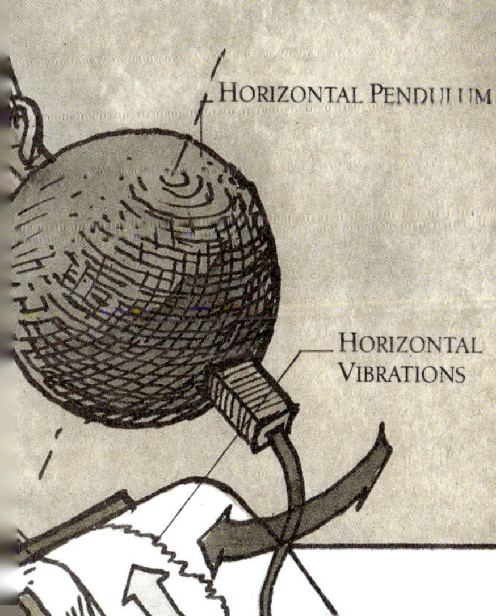

AUTOPILOT

The autopilot of an aircraft keeps it on the same course by correcting for any drifting that occurs during flight. Accelerometers mounted on a level platform stabilized by gyroscopes (see pp.80-1) sense any force that acts to change the plane's direction or height. They work by inertia. The spring-mounted armature in each accelerometer tends to remain still as coils beneath it move, and the relative movement induces an electric signal in the outer coils. The combined signals from the accelerometers then go to the controls of the aircraft, which respond by bringing the aircraft back on course.

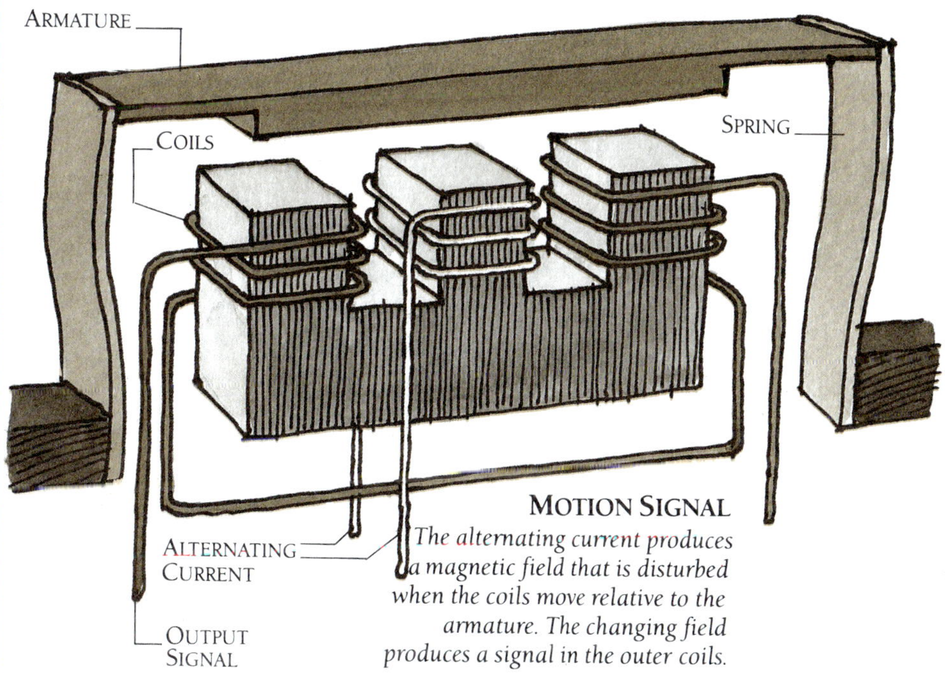

MOTION SIGNAL
The alternating current produces a magnetic field that is disturbed when the coils move relative to the armature. The changing field produces a signal in the outer coils.

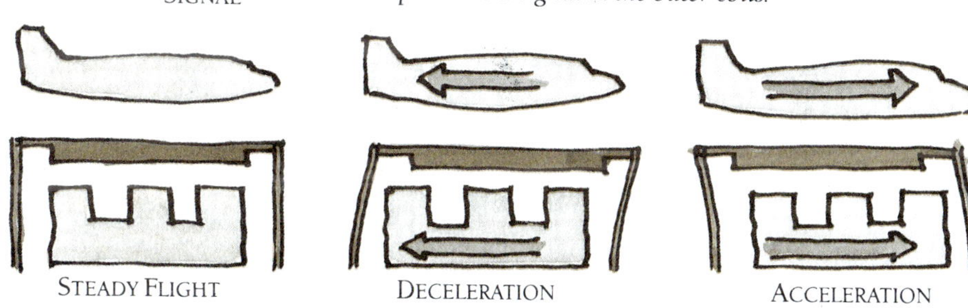

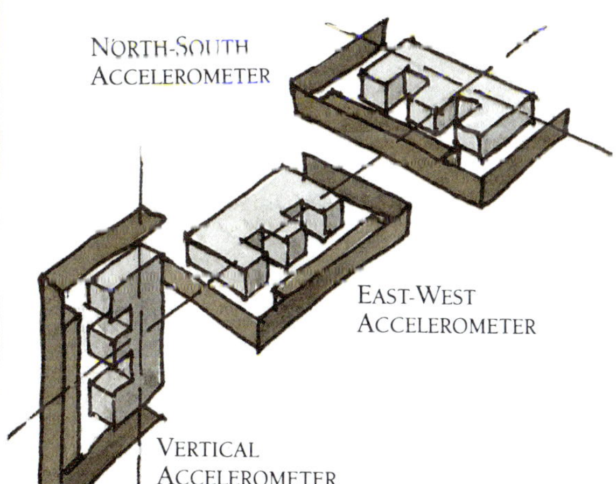

INERTIAL GUIDANCE
Inertial guidance systems contain three accelerometers mounted on a stable platform. They sense vertical forces and north-south and east-west horizontal forces. In this way, the accelerometers can detect all the movements of the aircraft. Their signals go to a computer that calculates the aircraft's current altitude and latitude and longitude to keep it on course.

BREATH TESTER

Several sensors are designed to detect the presence of specific substances. A breath tester detects and measures the concentration of alcohol in the breath, which is an accurate indication of the amount of alcohol in the blood. Breath testers use either a fuel cell (shown here) or infra-red rays, which are absorbed by alcohol vapor. Testing drivers with a breath tester enables police to check alcohol levels in a matter of seconds.

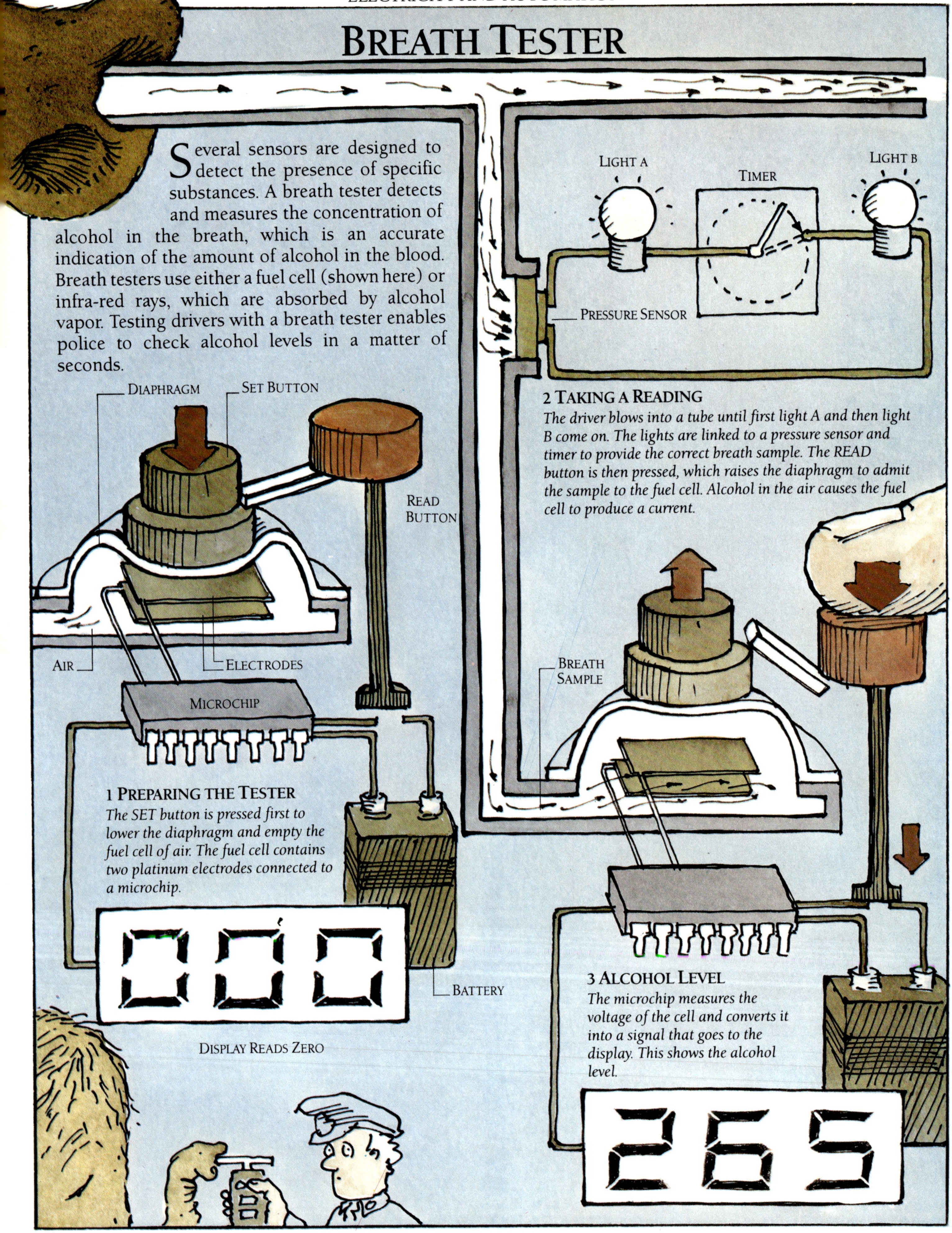

1 PREPARING THE TESTER

The SET button is pressed first to lower the diaphragm and empty the fuel cell of air. The fuel cell contains two platinum electrodes connected to a microchip.

2 TAKING A READING

The driver blows into a tube until first light A and then light B come on. The lights are linked to a pressure sensor and timer to provide the correct breath sample. The READ button is then pressed, which raises the diaphragm to admit the sample to the fuel cell. Alcohol in the air causes the fuel cell to produce a current.

3 ALCOHOL LEVEL

The microchip measures the voltage of the cell and converts it into a signal that goes to the display. This shows the alcohol level.

SMOKE DETECTOR

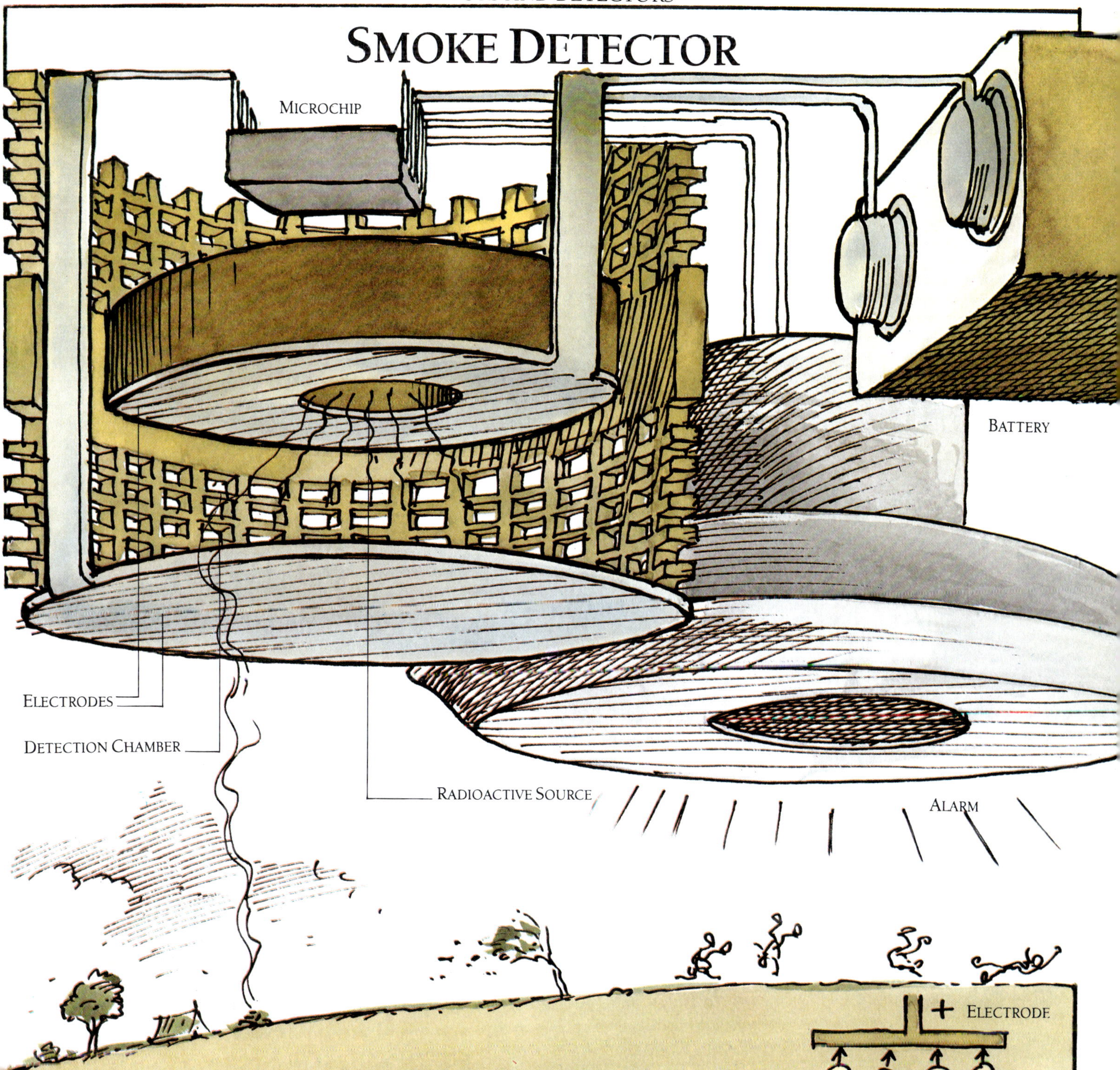

Smoke detectors can sense the small particles of smoke that rise from a smoldering object, and raise the alarm before fire breaks out. They work in two ways. Optical detectors use a light beam and light sensor that react to anything obscuring the beam. Ionizing detectors of the kind shown here are electrical sensors that can detect smaller particles than their optical equivalents.

The ionizing smoke detector contains a chamber in which a low electric current flows through the air. Smoke particles entering the chamber increase its electrical resistance so that less current flows. A microchip responds to the drop in current by switching on an alarm.

IONIZING RAYS

Rays from the radioactive source ionize the atoms in the air of the detection chamber, giving them positive and negative electric charges. The charged atoms or ions carry an electric current between the charged electrodes. Smoke particles entering the chamber attract the ions and reduce the current.

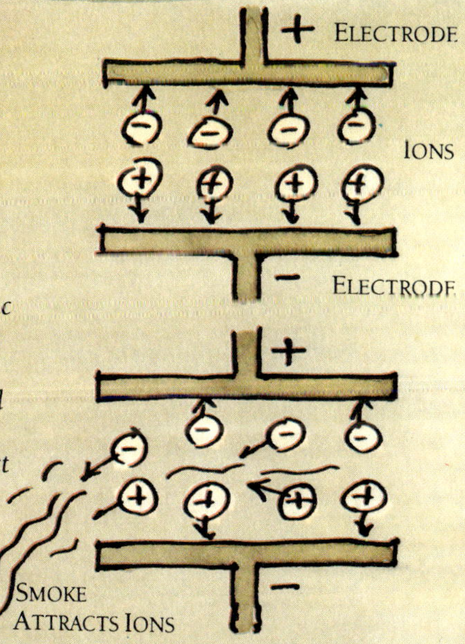

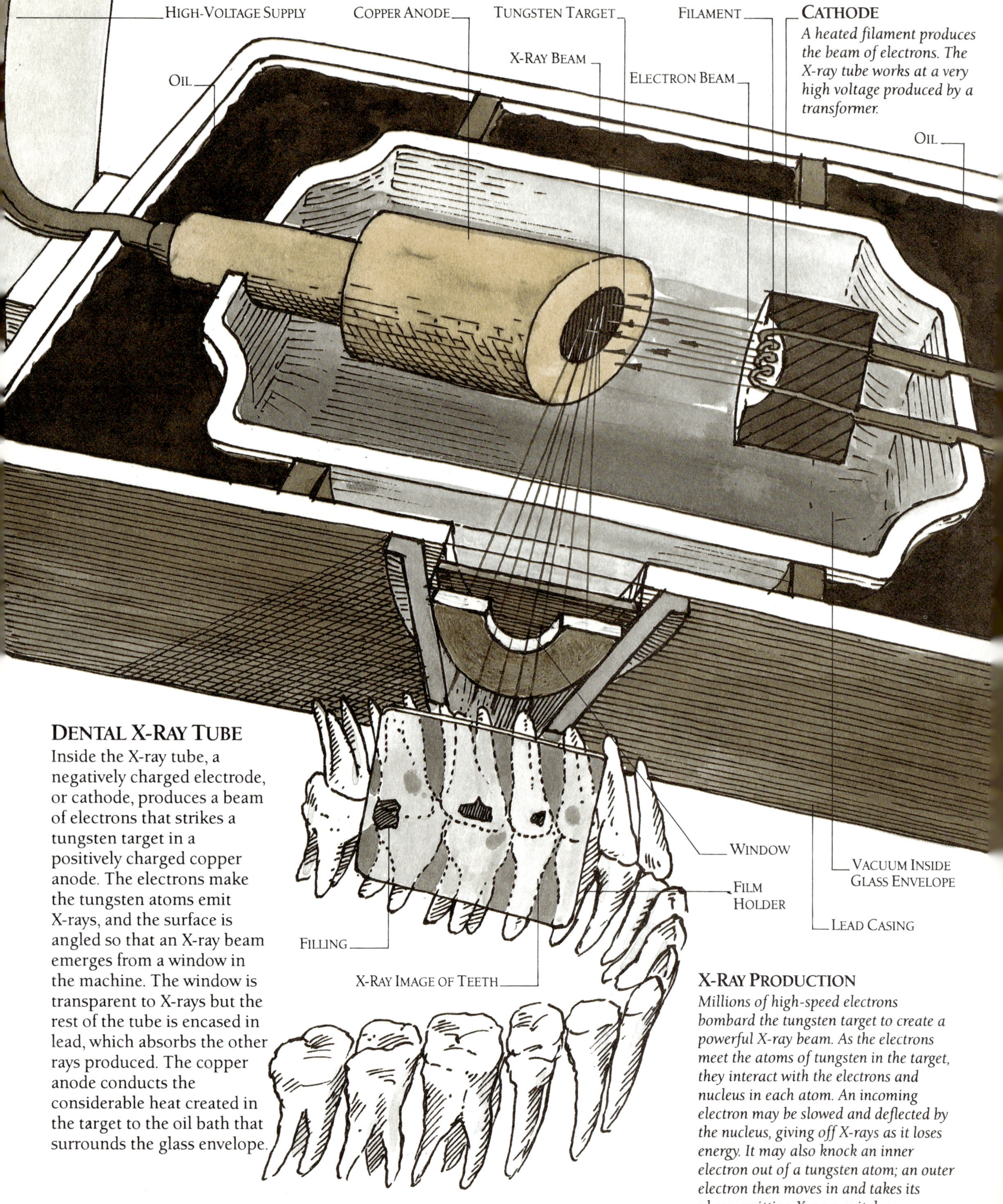

CATHODE

A heated filament produces the beam of electrons. The X-ray tube works at a very high voltage produced by a transformer.

DENTAL X-RAY TUBE

Inside the X-ray tube, a negatively charged electrode, or cathode, produces a beam of electrons that strikes a tungsten target in a positively charged copper anode. The electrons make the tungsten atoms emit X-rays, and the surface is angled so that an X-ray beam emerges from a window in the machine. The window is transparent to X-rays but the rest of the tube is encased in lead, which absorbs the other rays produced. The copper anode conducts the considerable heat created in the target to the oil bath that surrounds the glass envelope.

X-RAY PRODUCTION

Millions of high-speed electrons bombard the tungsten target to create a powerful X-ray beam. As the electrons meet the atoms of tungsten in the target, they interact with the electrons and nucleus in each atom. An incoming electron may be slowed and deflected by the nucleus, giving off X-rays as it loses energy. It may also knock an inner electron out of a tungsten atom; an outer electron then moves in and takes its place, emitting X-rays as it does so.

X-RAYS

Most of us are familiar with X-rays from pictures that the dentist takes to examine our teeth. The X-ray machine produces a beam of invisible rays. These penetrate the teeth and strike a piece of photographic film mounted in a holder clenched between the teeth. The dentist develops the film and sees a picture that shows the interior of the teeth and also any defects that need attention.

X-rays are used to look inside many things in the same kind of way. They are electromagnetic rays similar to light rays (see p.257) but with greater energy. They easily penetrate materials made of light atoms, which include the atoms in flesh. Heavier atoms, such as those of most metals, absorb them. Teeth and bones contain some calcium, which is a metal, and so the teeth and metal fillings inside them show up.

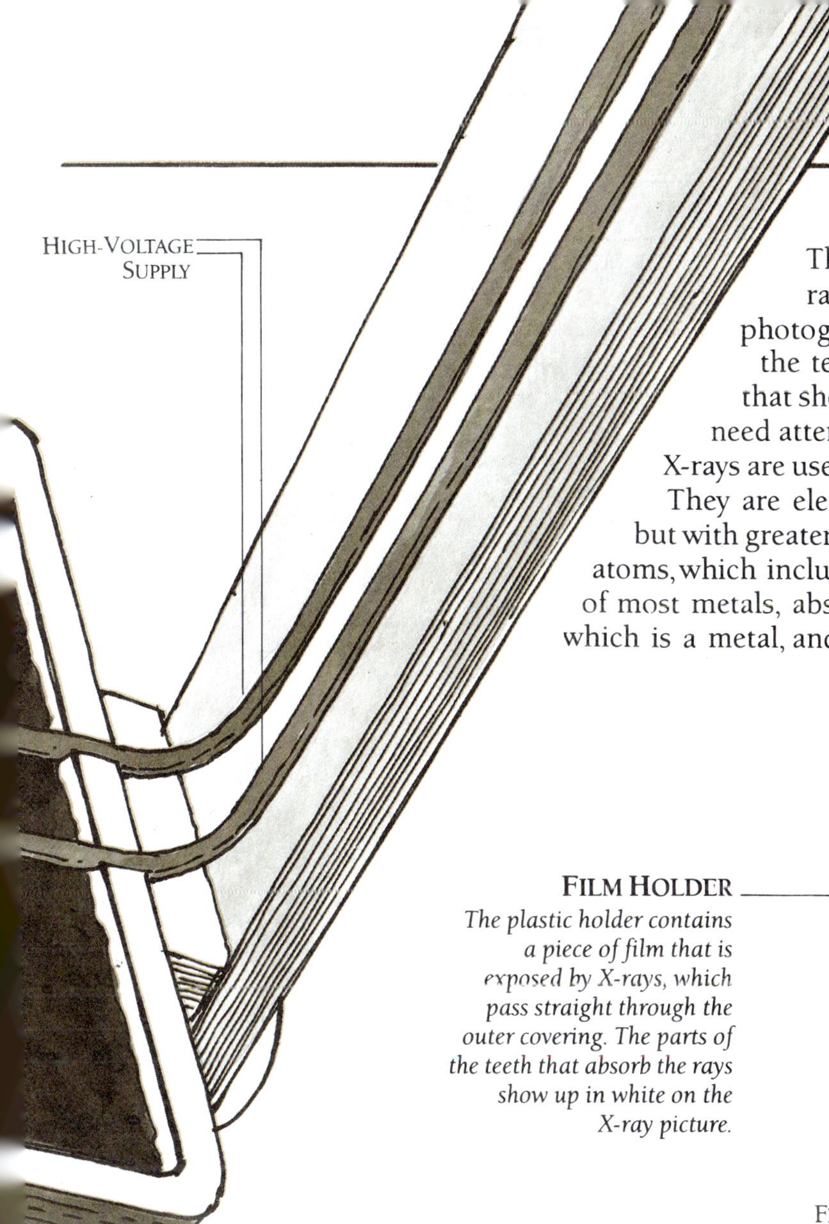

FILM HOLDER

The plastic holder contains a piece of film that is exposed by X-rays, which pass straight through the outer covering. The parts of the teeth that absorb the rays show up in white on the X-ray picture.

BAGGAGE SCANNER

Airport security requires both rapid and effective searching of passenger's baggage by a scanner. This uses X-rays to penetrate the baggage and show up metal objects inside. A very sensitive detector is used so that a low dose of X-rays is given, thereby avoiding exposure of films in the baggage.

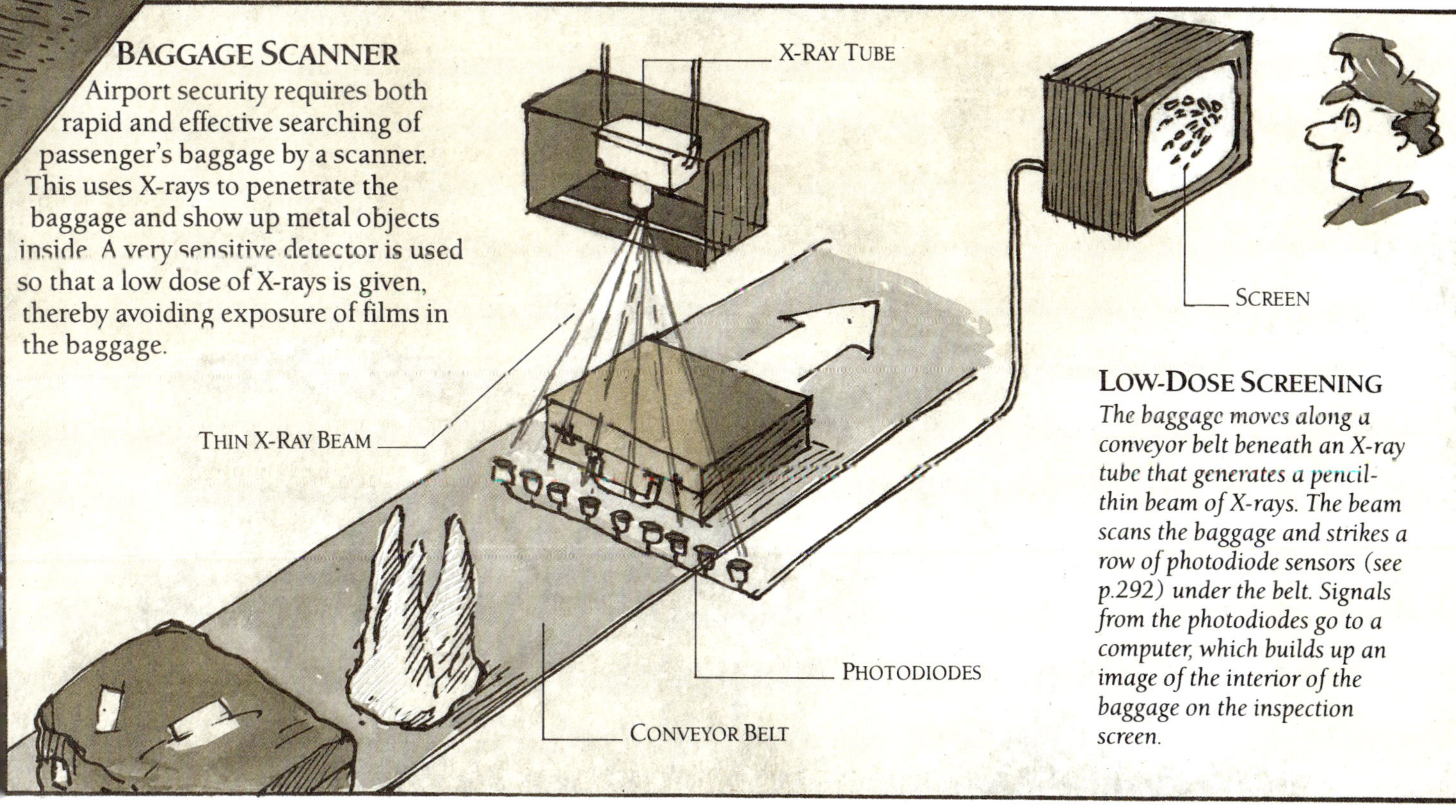

LOW-DOSE SCREENING

The baggage moves along a conveyor belt beneath an X-ray tube that generates a pencil-thin beam of X-rays. The beam scans the baggage and strikes a row of photodiode sensors (see p.292) under the belt. Signals from the photodiodes go to a computer, which builds up an image of the interior of the baggage on the inspection screen.

SONAR

Sonar, which stands for SOund Navigation And Ranging, is a sensing system that detects objects with sound waves. It is mainly used underwater, where other kinds of waves and rays do not travel so well. Ships employ sonar to measure the depth of water, to find shoals of fish and to detect wrecks. A transducer emits a pulse of sound, which travels down through the water and is reflected back. The transducer picks up this echo, and the sonar converts the time it takes the sound to return into a value for the object's distance.

ECHO SOUNDING

It takes 1 second for an echo to return from an object 2,500 feet (750 meters) deep. The returning echoes of sound produce an electric signal that goes to a screen display. The time differences of the echoes show on the display as points of light in different positions. In this way, a profile of the water beneath the ship is seen complete with depth scale, giving the location of the bottom and shoals of fish.

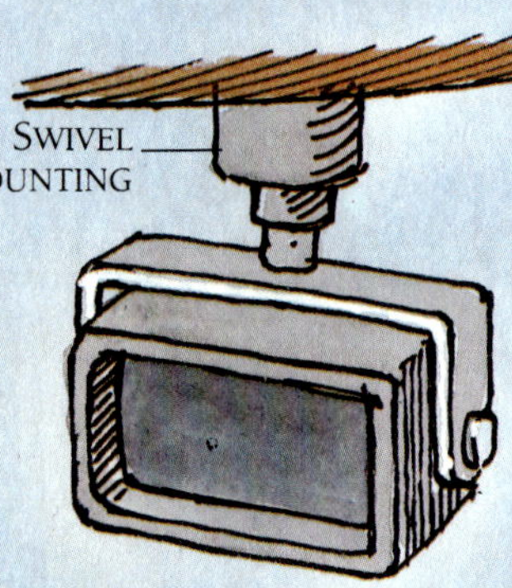

TRANSDUCER

A transducer is a device that converts one kind of signal into another. In sonar, a transducer on the hull of a ship converts an electric pulse into a pulse of sound; it then converts the returning sound waves back into an electric signal. It is similar to a combined loudspeaker and microphone.

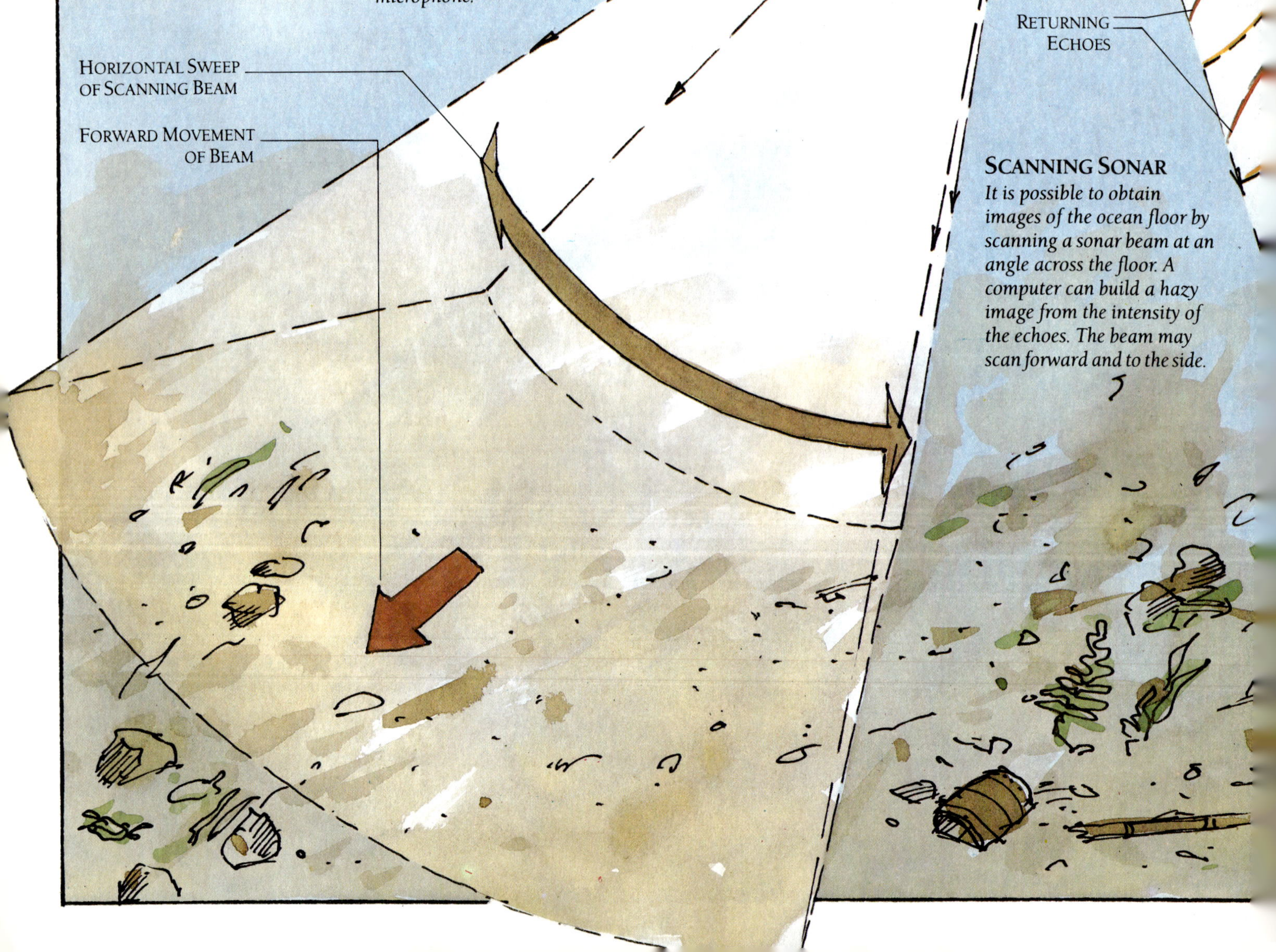

SCANNING SONAR

It is possible to obtain images of the ocean floor by scanning a sonar beam at an angle across the floor. A computer can build a hazy image from the intensity of the echoes. The beam may scan forward and to the side.

ULTRASOUND SCANNER

The principles of sonar are put to important use in the ultrasound scanner. This machine can produce an image of an unborn child inside its mother. Pulses of sound from a probe scan across the interior of the body. A computer uses the returning echoes to build up a cross-section image of the mother and baby.

The scanner produces pulses of ultrasound, which is sound with a frequency range that lies above the limit of human hearing. Ultrasound is used, not to spare the ears of doctor, mother and baby, but because it has a shorter wavelength and so enables the computer to produce more detail in the image.

1 **PROBE** *emits ultrasound pulse.* 2 **ECHO** *returns from womb.* 3 **ECHO** *returns from baby.*

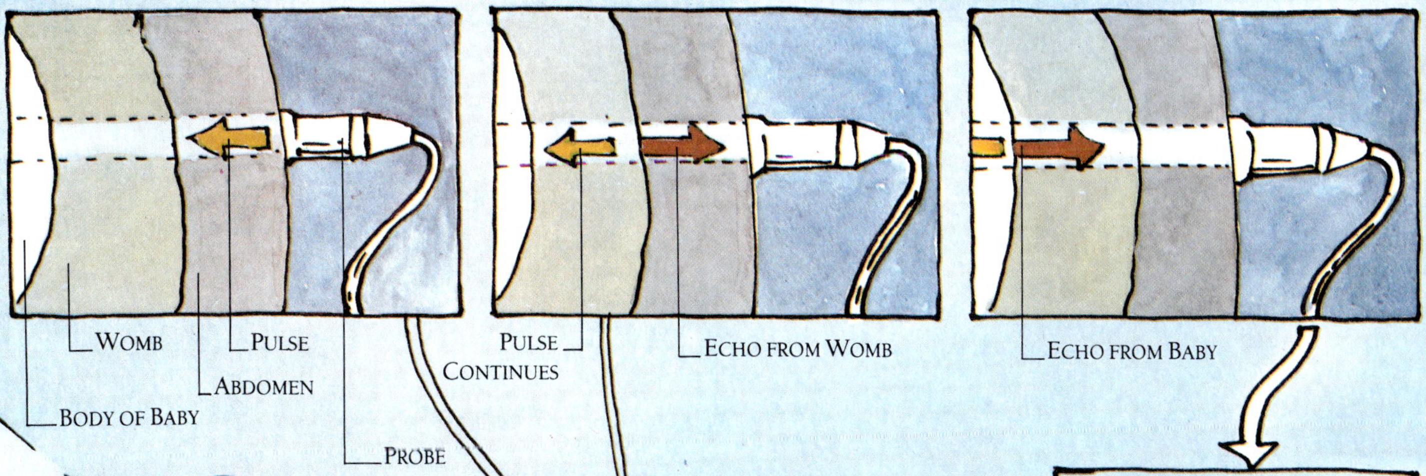

COMPUTER
A computer receives electrical signals from the probe as the echoes return. The computer plots points of light on the screen that show echoes from various depths. As the ultrasound beam scans across, the points build into lines that form an image.

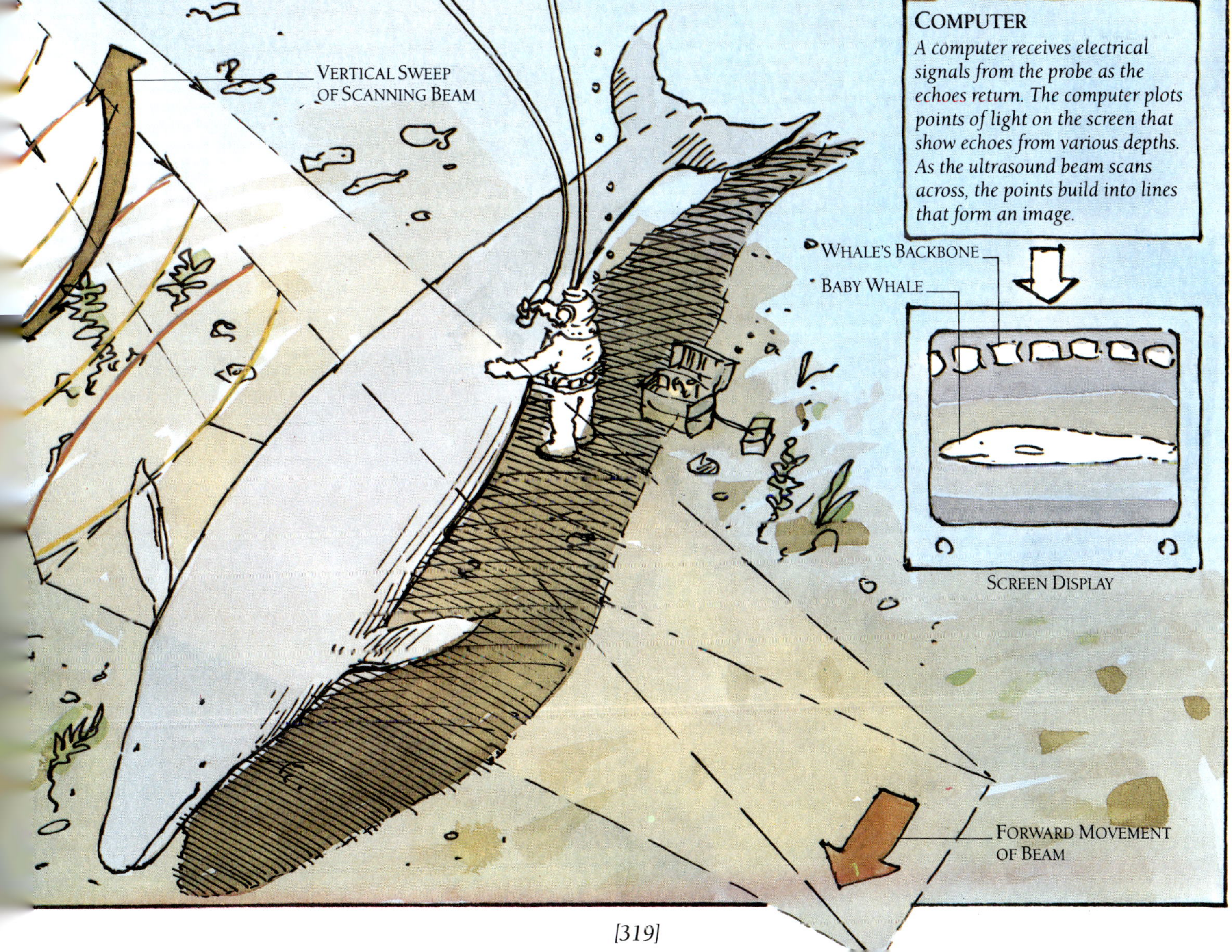

RADAR

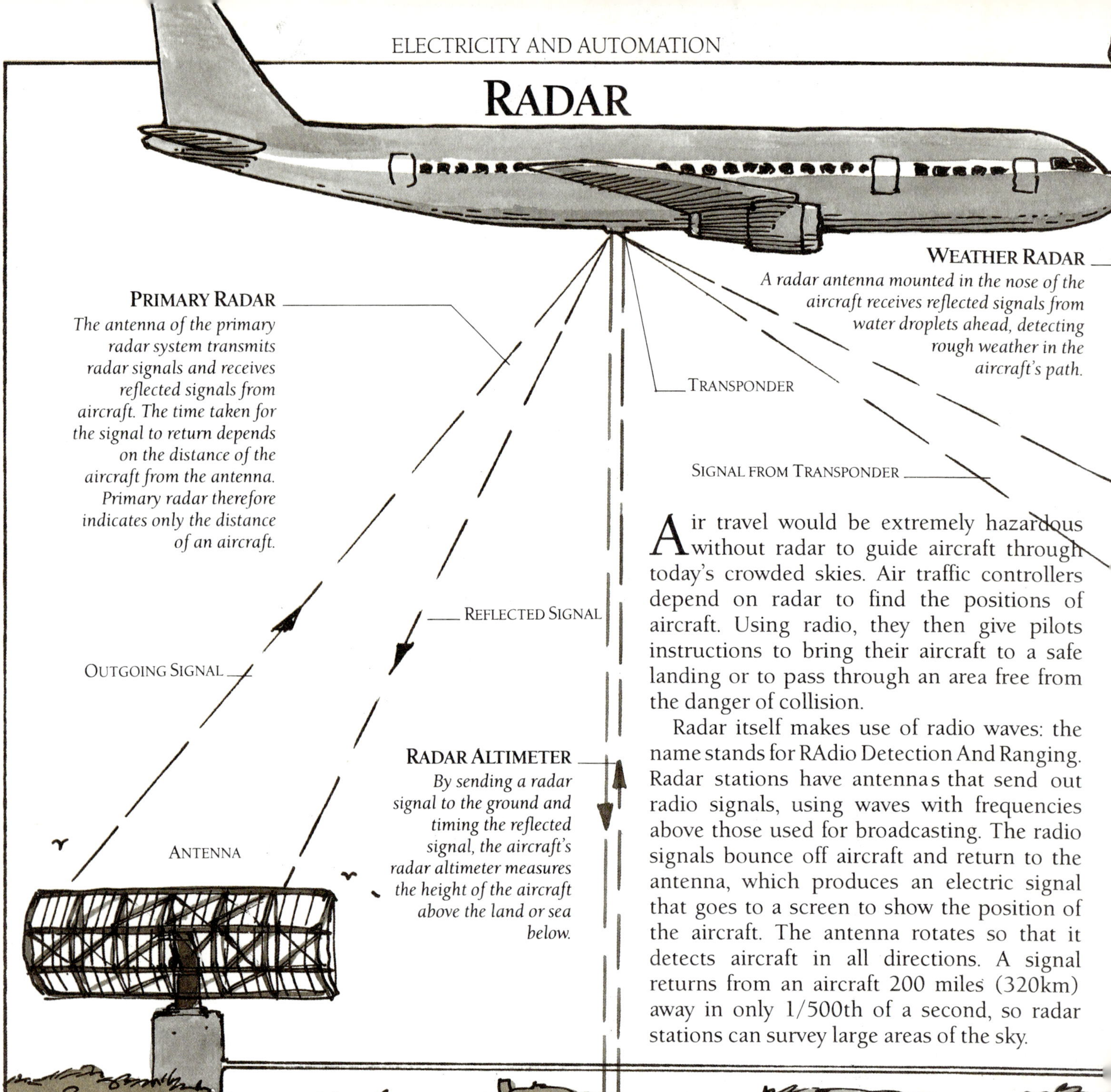

PRIMARY RADAR
The antenna of the primary radar system transmits radar signals and receives reflected signals from aircraft. The time taken for the signal to return depends on the distance of the aircraft from the antenna. Primary radar therefore indicates only the distance of an aircraft.

WEATHER RADAR
A radar antenna mounted in the nose of the aircraft receives reflected signals from water droplets ahead, detecting rough weather in the aircraft's path.

RADAR ALTIMETER
By sending a radar signal to the ground and timing the reflected signal, the aircraft's radar altimeter measures the height of the aircraft above the land or sea below.

Air travel would be extremely hazardous without radar to guide aircraft through today's crowded skies. Air traffic controllers depend on radar to find the positions of aircraft. Using radio, they then give pilots instructions to bring their aircraft to a safe landing or to pass through an area free from the danger of collision.

Radar itself makes use of radio waves: the name stands for RAdio Detection And Ranging. Radar stations have antennas that send out radio signals, using waves with frequencies above those used for broadcasting. The radio signals bounce off aircraft and return to the antenna, which produces an electric signal that goes to a screen to show the position of the aircraft. The antenna rotates so that it detects aircraft in all directions. A signal returns from an aircraft 200 miles (320km) away in only 1/500th of a second, so radar stations can survey large areas of the sky.

RADAR SPEED TRAP

A radar signal fired at a moving vehicle can be used to measure its speed. The frequency of the returning signal increases if the vehicle is approaching and decreases if it is departing. The change of frequency depends on the speed, and a radar speed trap measures this change to display the speed of the vehicle.

The frequency of a signal is the rate at which the waves of energy pass a point. If the vehicle is approaching the speed trap, it travels into the radio waves and reflects them more often to increase the frequency. If the vehicle is moving away, it takes longer for each wave to meet the vehicle and the frequency of the reflected signal decreases.

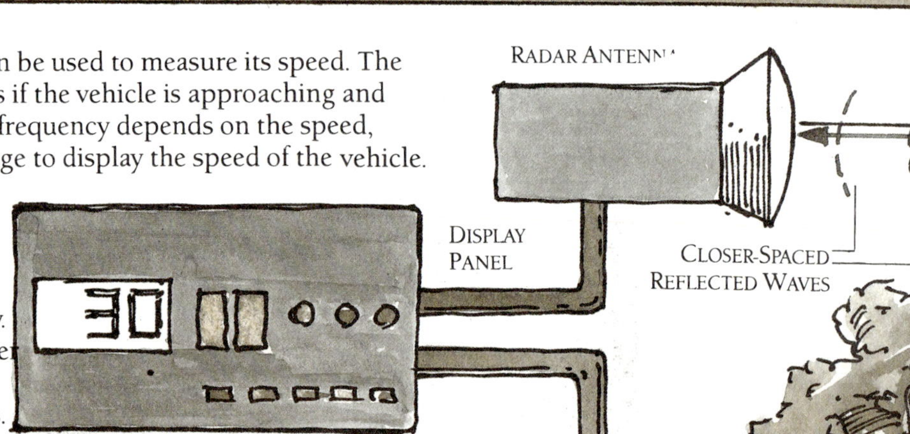

Radar Display

In a radar display (*below*), the positions of aircraft within range of the radar station appear on a screen marked with a map of the area. As the primary antenna rotates, the positions of aircraft returning radar signals light up. The computer displays the information from secondary radar beside the position of the aircraft. This information gives the aircraft's flight number (in this case TW754), its destination (LL or London) and its current height (300 or 30,000 feet). In this way, the screen displays the information that the air traffic controller requires.

Secondary Radar

The antenna of the secondary radar system sends signals to transponders on aircraft. In reply, each transponder sends back a signal giving the aircraft's height and identity.

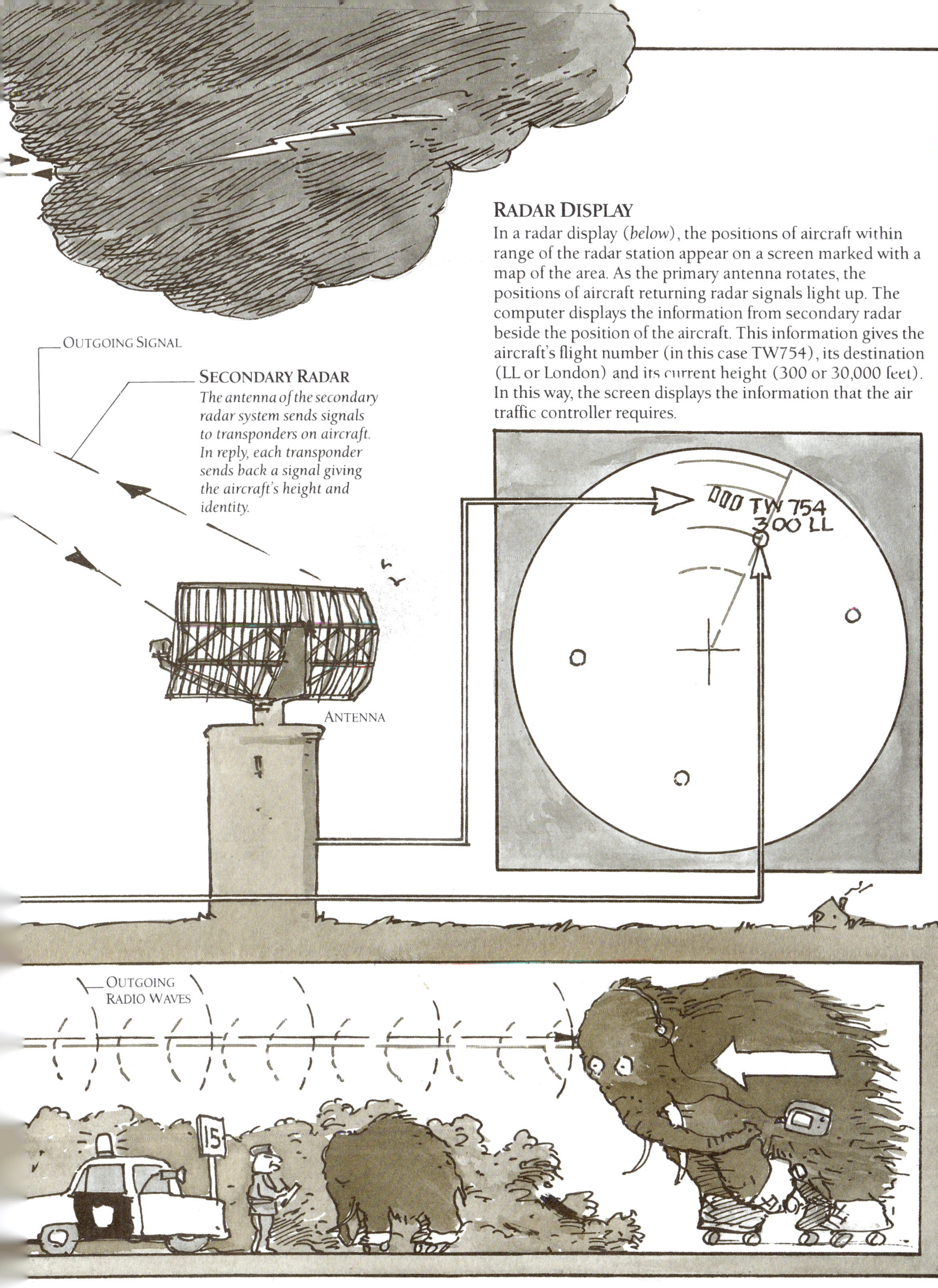

METAL DETECTOR

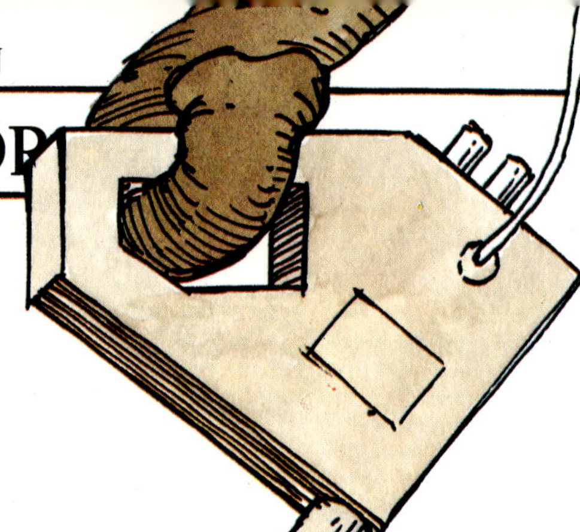

The technology that enables us to discover buried treasure also tests coins in ticket machines or vending machines, invisibly frisks people at airports and controls traffic lights. All these machines are basically metal detectors, and they work by electromagnetic induction (see pp.304-5).

When a piece of metal passes through a magnetic field or the field passes through the metal, the field produces electric eddy currents that circulate in the metal. The eddy currents in turn produce their own magnetic field, and metal detectors work by detecting this field.

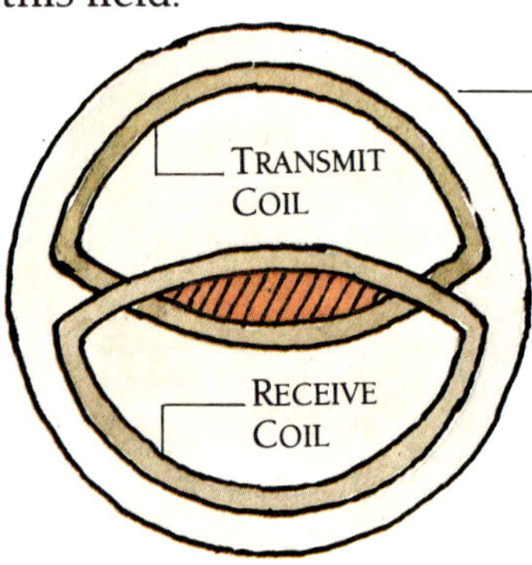

DETECTOR COILS
The transmit and receive coils overlap so that each induces a current in the other. Normally the two currents cancel out, but the magnetic field of a metal object distorts this balance and a low current appears in the receive coil.

DETECTOR HEAD
In the metal detector head, one coil usually transmits a magnetic field and another coil picks up the magnetic field produced by a metal object below. The receive coil produces an electric signal that goes to a light, earphones or a meter to indicate a find.

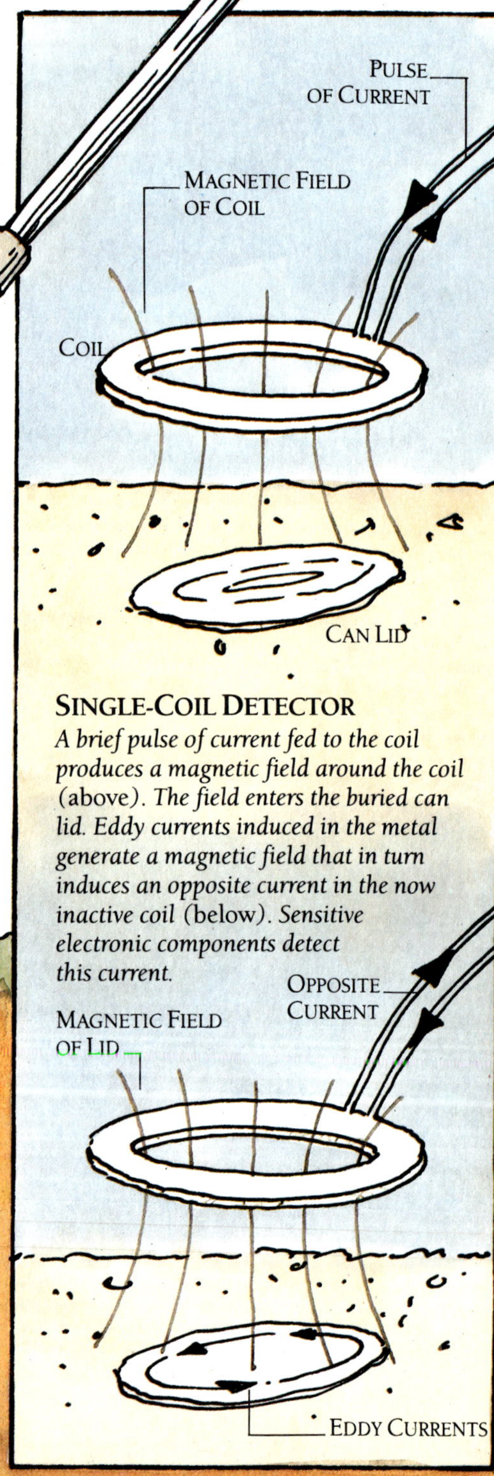

SINGLE-COIL DETECTOR
A brief pulse of current fed to the coil produces a magnetic field around the coil (above). The field enters the buried can lid. Eddy currents induced in the metal generate a magnetic field that in turn induces an opposite current in the now inactive coil (below). Sensitive electronic components detect this current.

COIN TESTER

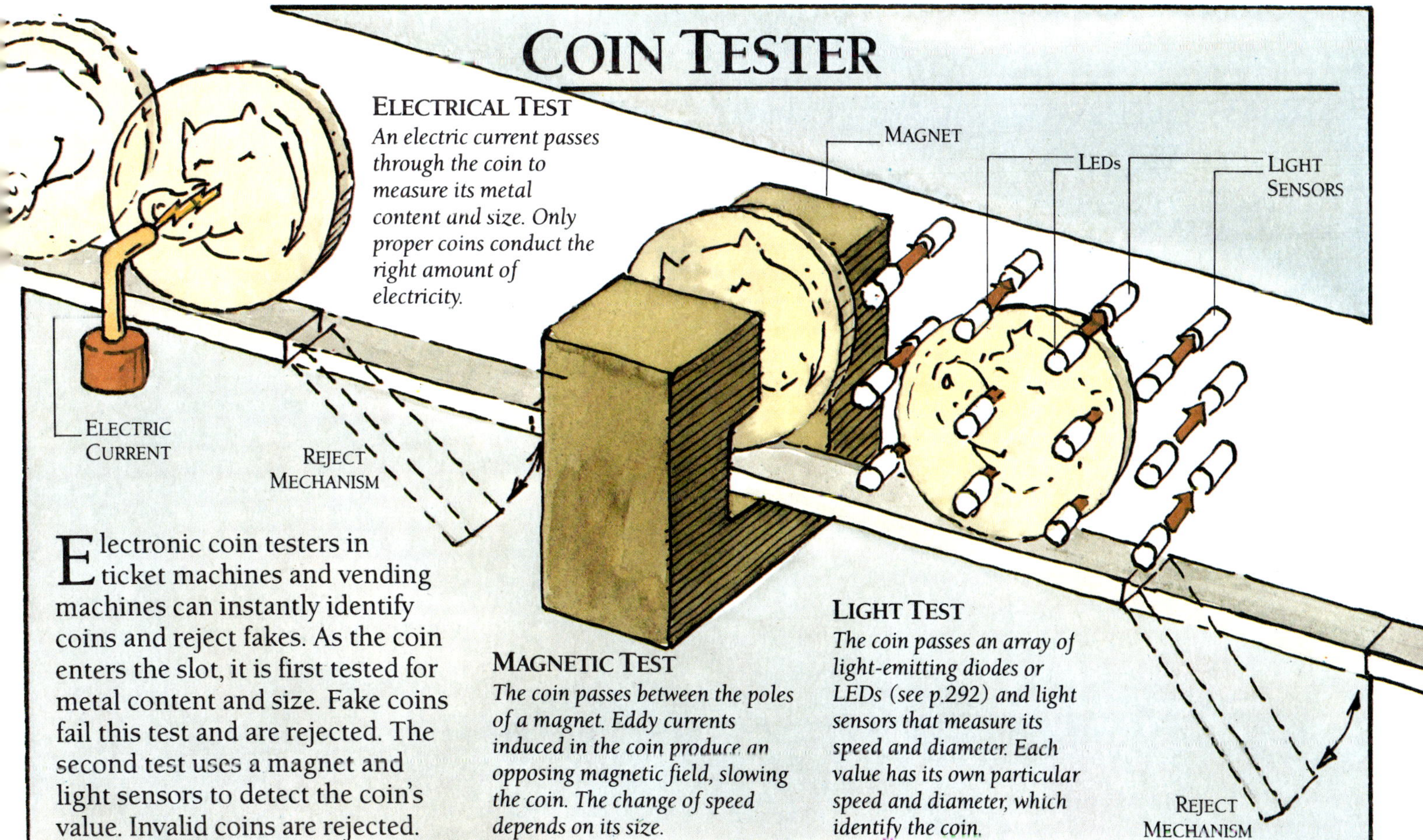

ELECTRICAL TEST

An electric current passes through the coin to measure its metal content and size. Only proper coins conduct the right amount of electricity.

Electronic coin testers in ticket machines and vending machines can instantly identify coins and reject fakes. As the coin enters the slot, it is first tested for metal content and size. Fake coins fail this test and are rejected. The second test uses a magnet and light sensors to detect the coin's value. Invalid coins are rejected.

MAGNETIC TEST

The coin passes between the poles of a magnet. Eddy currents induced in the coin produce an opposing magnetic field, slowing the coin. The change of speed depends on its size.

LIGHT TEST

The coin passes an array of light-emitting diodes or LEDs (see p.292) and light sensors that measure its speed and diameter. Each value has its own particular speed and diameter, which identify the coin.

AIRPORT DETECTOR

The gateways of metal detectors in airports contain coils similar in principle to the coils in a treasure finder. A receiver detects distortions of the transmitted field caused by metal possessions on the person passing through the gateway. The coils are shielded on the outside so that people passing nearby do not trigger the detector.

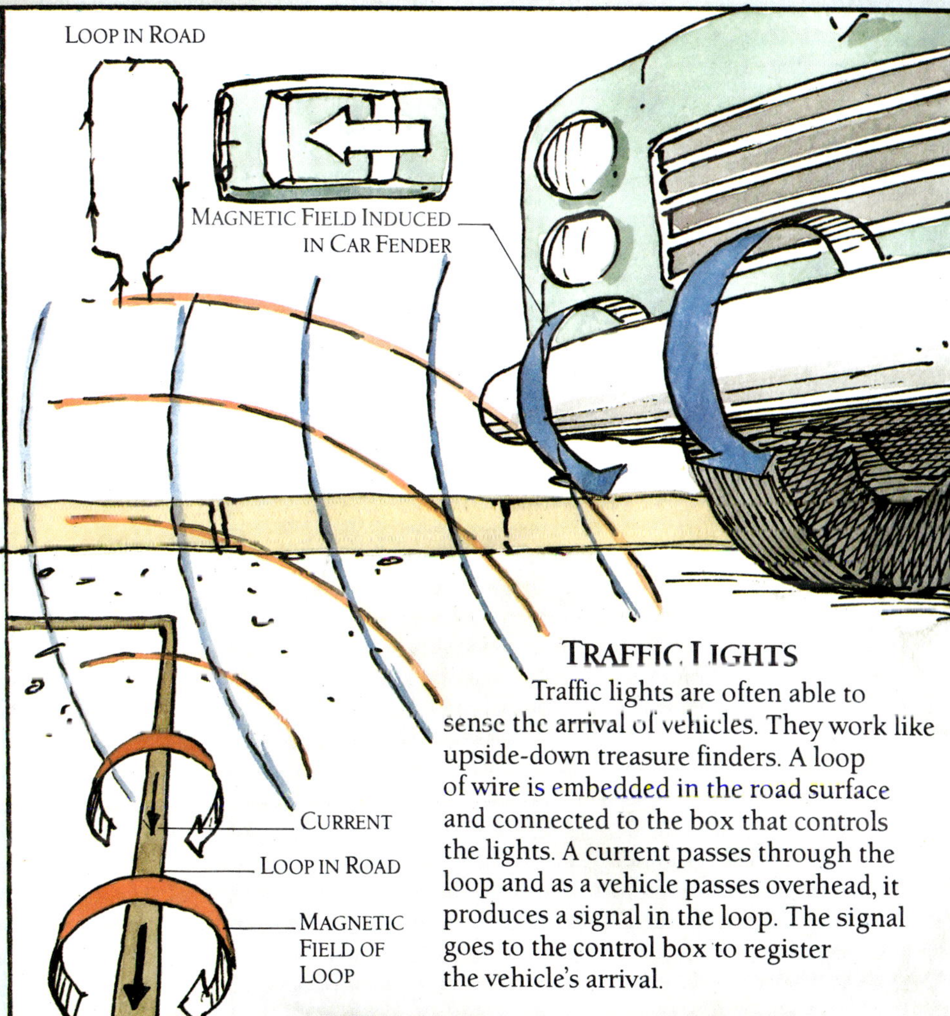

TRAFFIC LIGHTS

Traffic lights are often able to sense the arrival of vehicles. They work like upside-down treasure finders. A loop of wire is embedded in the road surface and connected to the box that controls the lights. A current passes through the loop and as a vehicle passes overhead, it produces a signal in the loop. The signal goes to the control box to register the vehicle's arrival.

BODY SCANNER

Doctors can see into any part of the body with the aid of body scanners. These can produce images of any internal organ, locating defects and diseases. Some scanners work by passing X-rays or gamma rays through the body. The most useful scanner works by nuclear magnetic resonance (NMR). When a patient lies inside an NMR scanner, the body is bombarded first by a strong magnetic field and then by pulses of radio waves. Unlike X-rays and gamma rays, these are harmless. The nuclei in the atoms of the body produce magnetic signals that are picked up by detectors, and a computer forms an image from the signals.

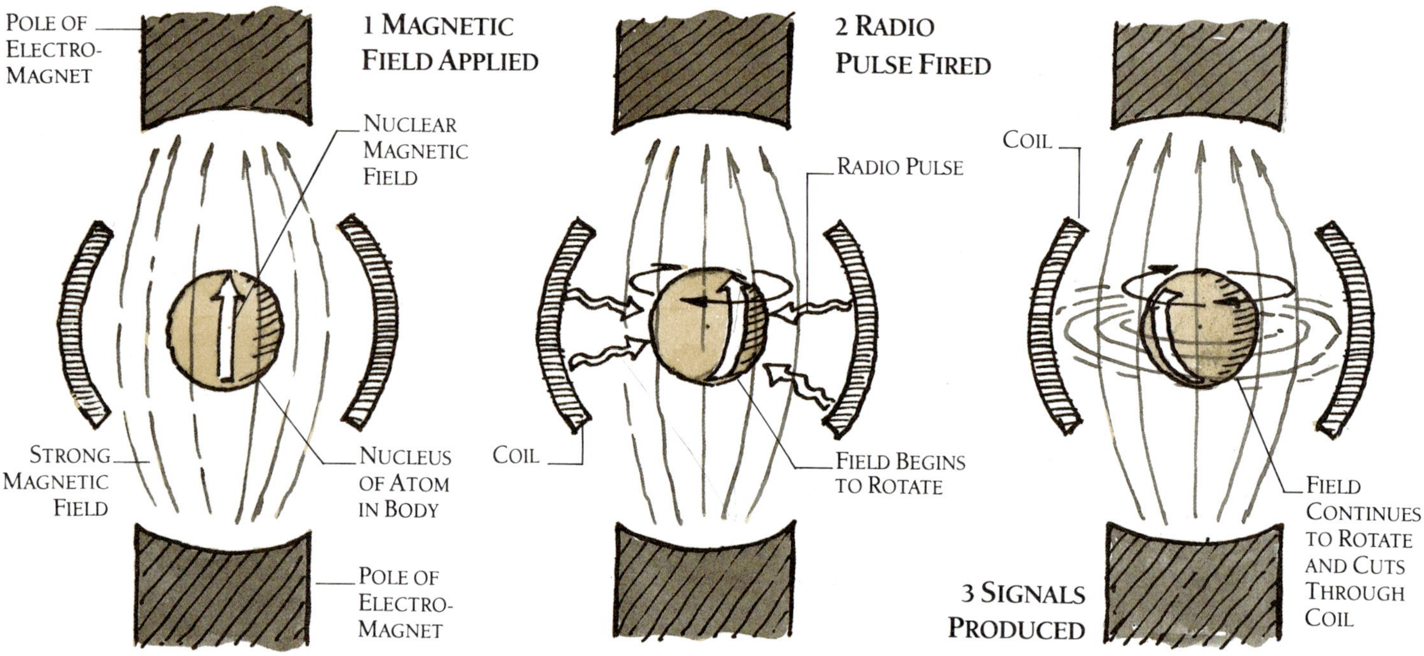

HOW THE SCANNER WORKS

Atomic nuclei each have their own tiny magnetic field. The scanner first applies a strong magnetic field (1) to line up the nuclear fields of the body's atoms. It then sends out a pulse of radio waves (2) which makes the nuclear fields rotate. The rotating fields induce electric signals in the scanner (3) which are built up into an image.

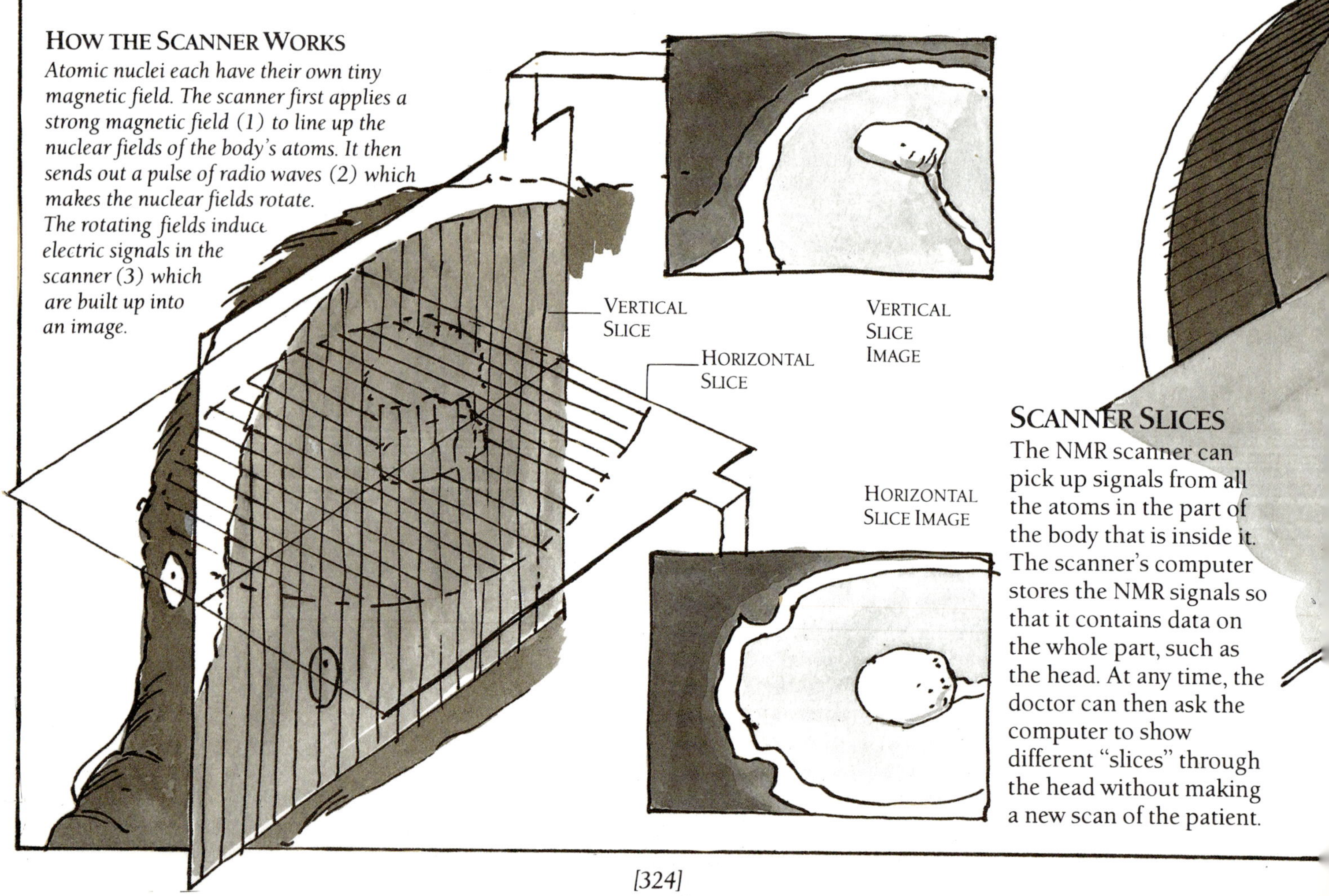

SCANNER SLICES

The NMR scanner can pick up signals from all the atoms in the part of the body that is inside it. The scanner's computer stores the NMR signals so that it contains data on the whole part, such as the head. At any time, the doctor can then ask the computer to show different "slices" through the head without making a new scan of the patient.

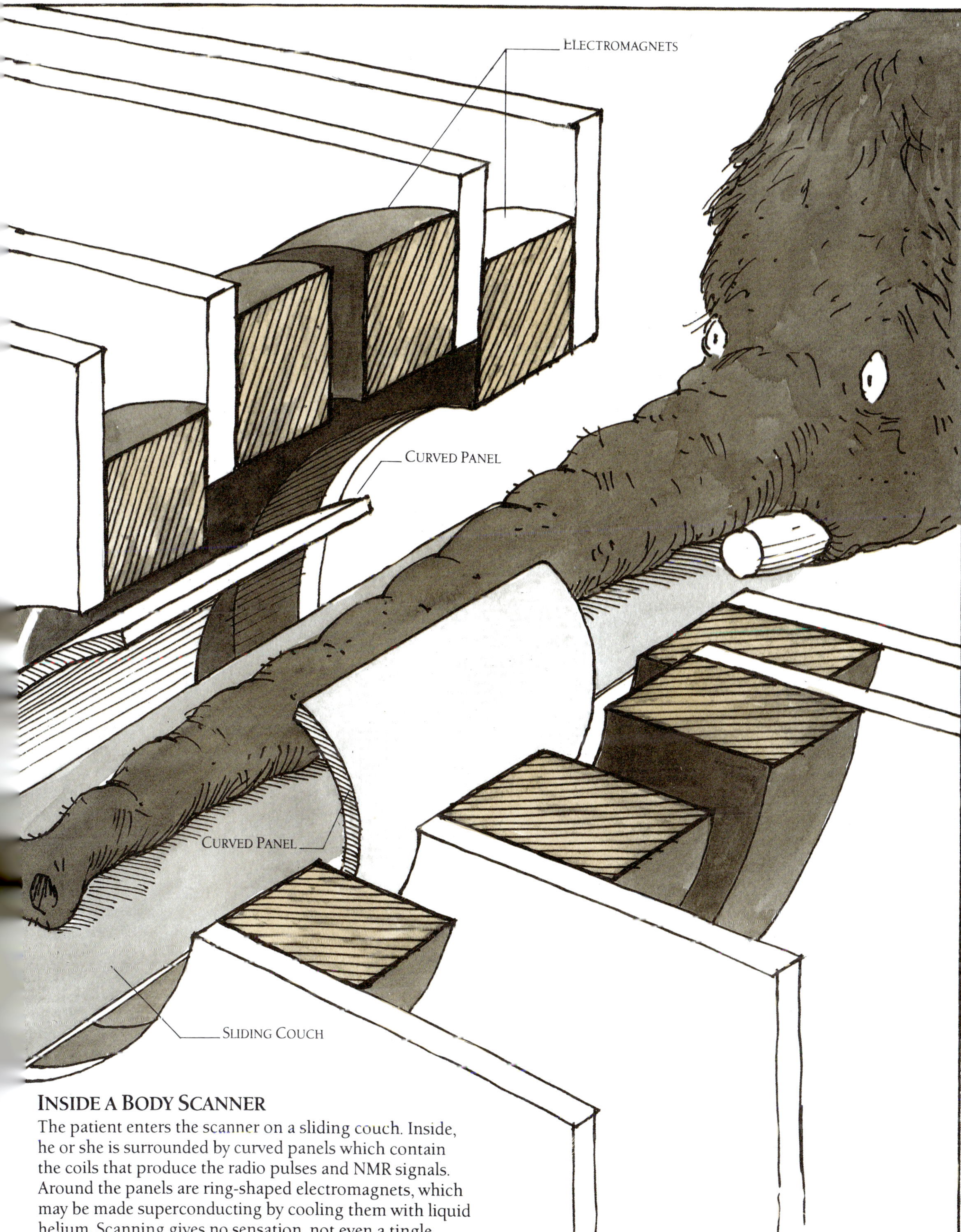

INSIDE A BODY SCANNER
The patient enters the scanner on a sliding couch. Inside, he or she is surrounded by curved panels which contain the coils that produce the radio pulses and NMR signals. Around the panels are ring-shaped electromagnets, which may be made superconducting by cooling them with liquid helium. Scanning gives no sensation, not even a tingle.

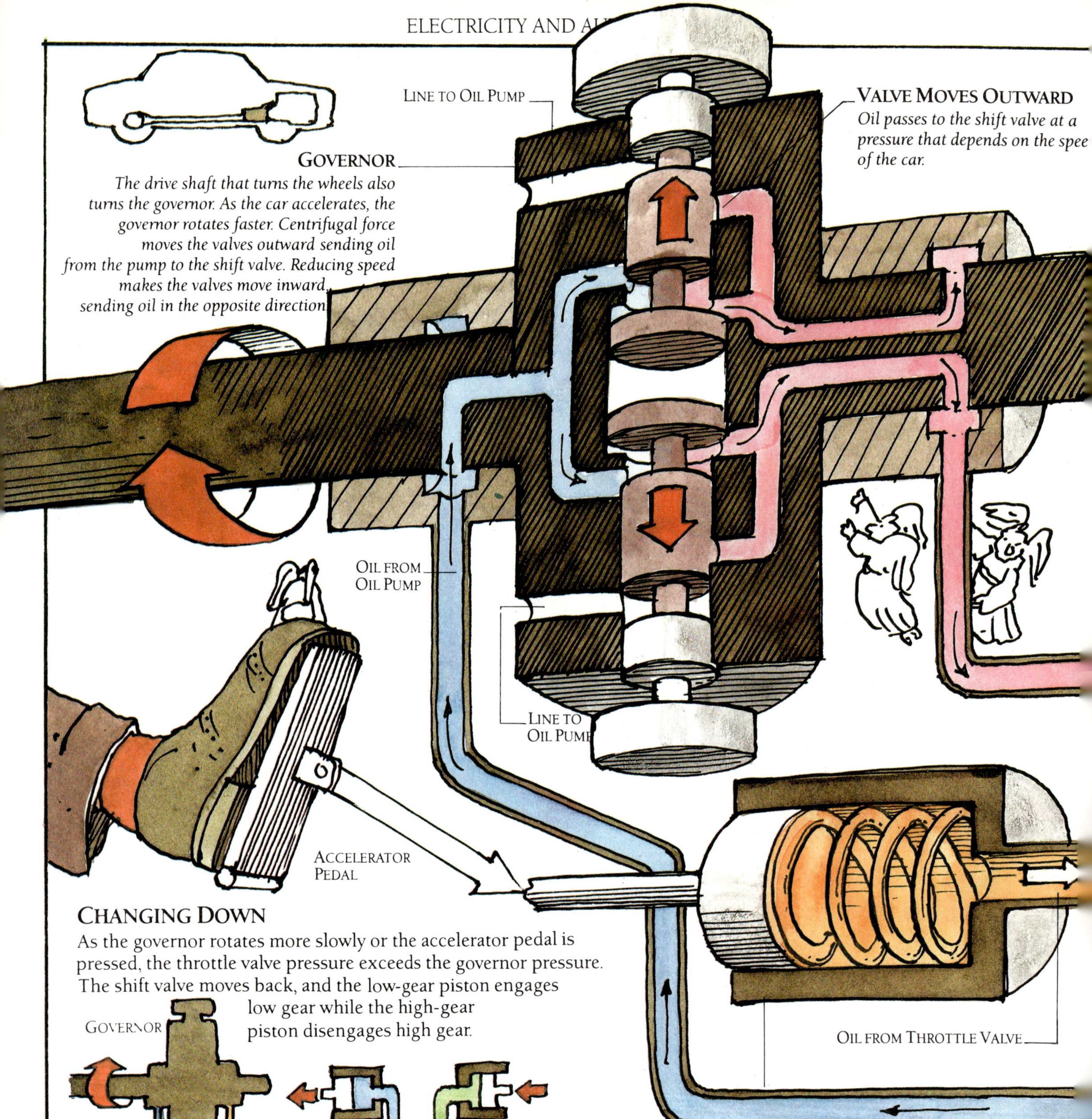

GOVERNOR

The drive shaft that turns the wheels also turns the governor. As the car accelerates, the governor rotates faster. Centrifugal force moves the valves outward sending oil from the pump to the shift valve. Reducing speed makes the valves move inward, sending oil in the opposite direction.

VALVE MOVES OUTWARD

Oil passes to the shift valve at a pressure that depends on the spee of the car.

CHANGING DOWN

As the governor rotates more slowly or the accelerator pedal is pressed, the throttle valve pressure exceeds the governor pressure. The shift valve moves back, and the low-gear piston engages low gear while the high-gear piston disengages high gear.

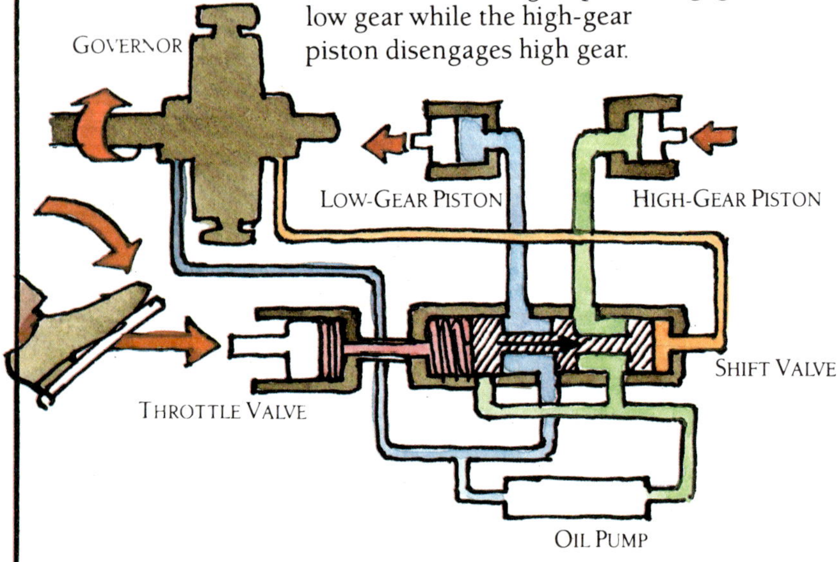

THROTTLE VALVE

The accelerator pedal moves the piston, increasing oil pressure in the valve. A spring returns the pedal, decreasing the oil pressure.

AUTOMATIC TRANSMISSION

Automatic transmission makes driving easy because there is no gear lever and clutch pedal to operate. The mechanism responds to the speed of the car, and automatically changes to a higher or lower gear as the car's speed rises and falls. It can also sense the position of the accelerator pedal.

The control system works by oil pressure. Each gear change is controlled by a shift valve. A governor linked to the wheels and a throttle valve operated by the pedal supply oil at different pressures to the shift valve. The valve moves accordingly and routes oil to the gear change mechanisms in the transmission.

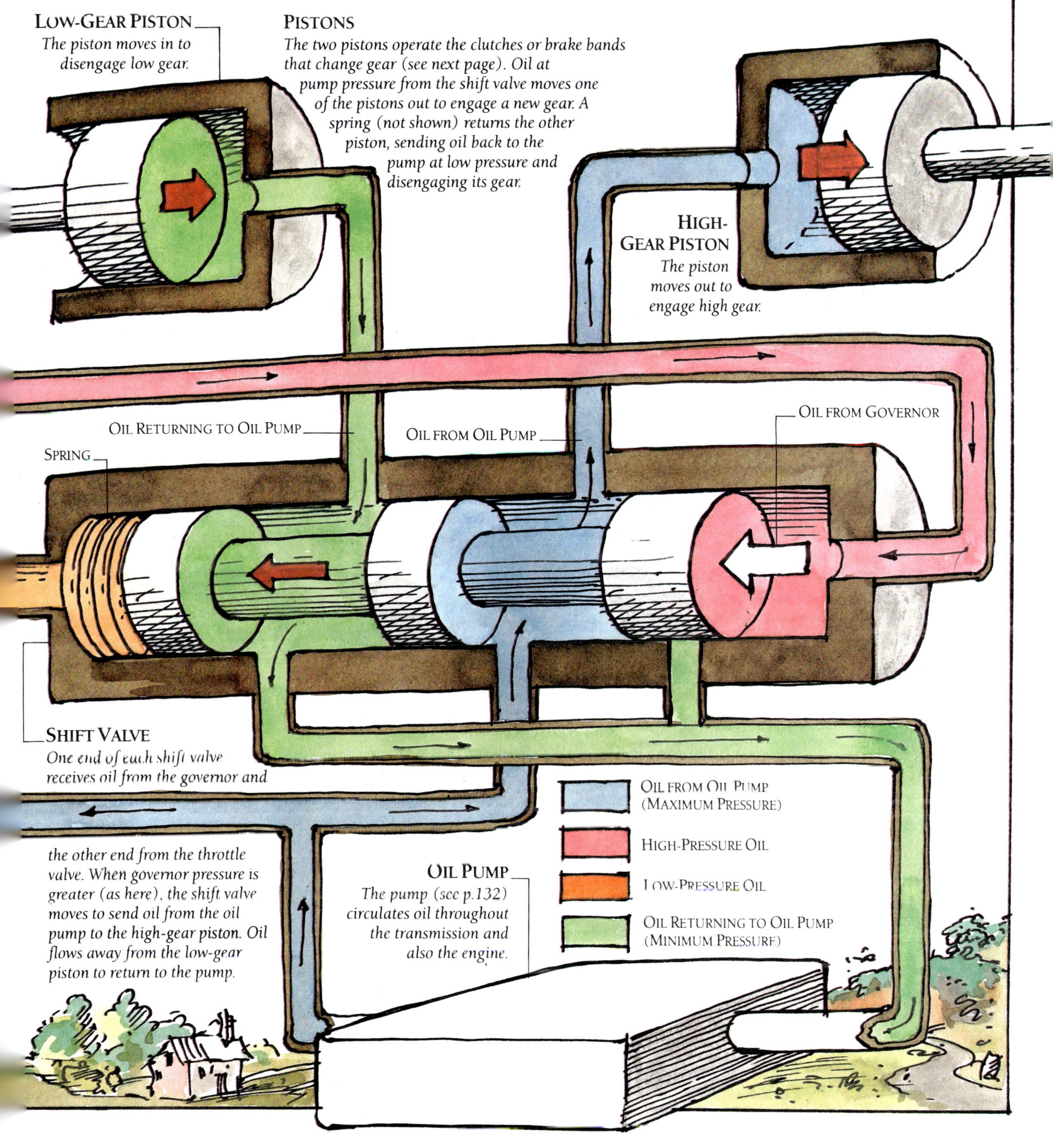

AUTOMATIC TRANSMISSION

An automatic transmission contains two main parts, the torque converter and automatic gearbox. The torque converter passes power from the engine flywheel to the gearbox. It does this progressively and smoothly so that starting and changing gear are not jerky, acting rather like the clutch in a manual gearbox (see p.88).

The automatic gearbox contains two sets of epicyclic gears (see p.43) in which gear wheels rotate at different speeds. Overall, except in top gear, the speed of the flywheel is reduced so that the car wheels turn more slowly but with more force. Reverse gear reverses the direction of the wheels.

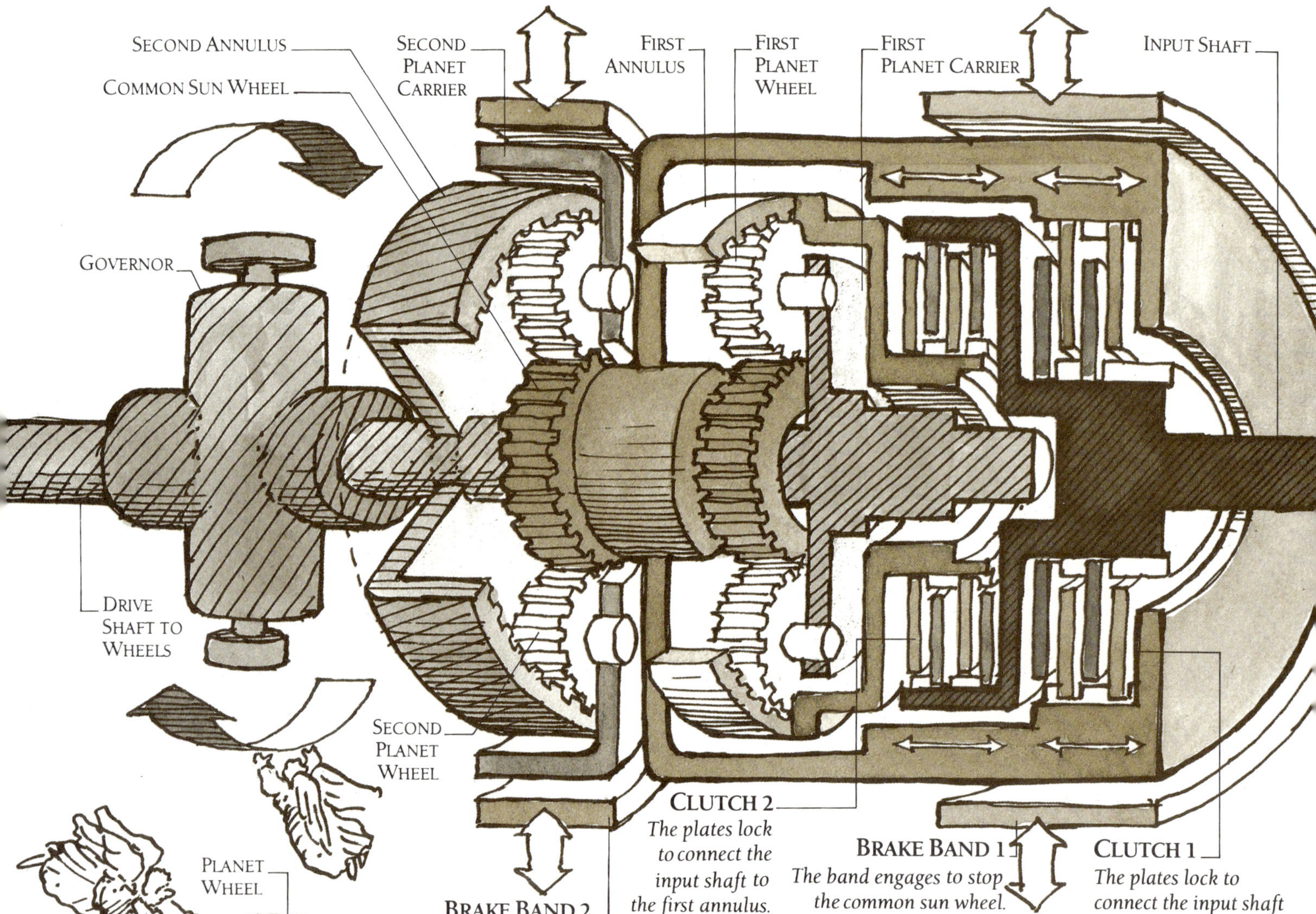

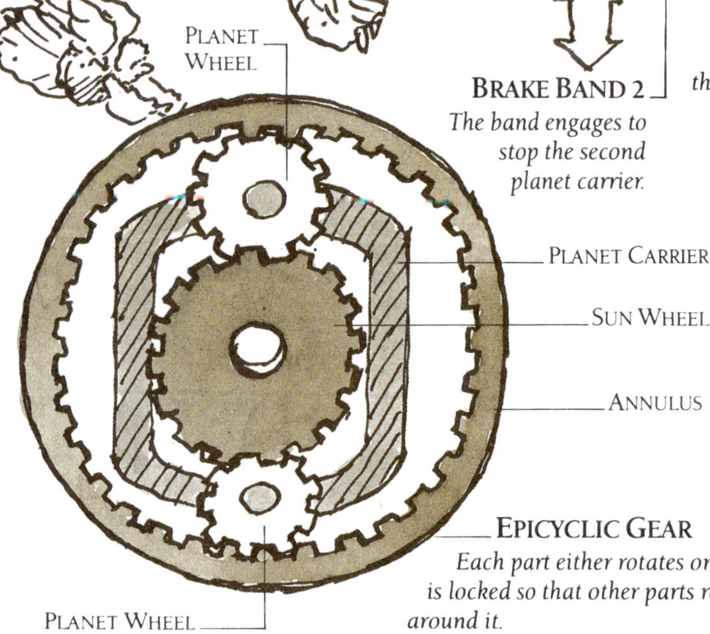

EPICYCLIC GEAR
Each part either rotates or is locked so that other parts rotate around it.

AUTOMATIC GEARBOX

This gearbox has three forward gears and one reverse gear. The various parts are controlled by multiplate clutches, which lock to transmit power, and brake bands that engage to stop a part rotating.

	Clutch 1	Clutch 2	Band 1	Band 2
First gear	Unlocked	Locked	Off	On
Second gear	Unlocked	Locked	On	Off
Third gear	Locked	Locked	Off	Off
Reverse gear	Locked	Unlocked	Off	On

CRUISE CONTROL

Many cars are fitted with cruise control, which at the press of a button automatically maintains a set speed. In this way, the driver can cruise at a speed limit or economic speed without continually checking the speedometer. The automatic system required is an example of a feedback mechanism. A sensor measures the car's speed and controls the carburetor (see p.148). It boosts fuel flow if speed begins to drop on climbing a slope, or feeds less fuel to the engine if the car begins to speed up.

The sensor may be an electromagnet on the drive shaft, which produces an electric signal related to the speed. A motor operates the carburetor. The controlling operation is best carried out by a microprocessor – the "brain" of a computer. The microprocessor continually checks the sensor signal and sends a control signal to the motor.

The advantage of a microprocessor is that it can do more than simply control speed. As it "knows" the speed and fuel flow, it can also calculate and display speed, distance and fuel consumption, and control the engine to improve consumption.

TORQUE CONVERTER

The torque converter contains three parts – an impeller turned by the engine flywheel, a turbine that turns the input shaft of the automatic gearbox, and a reactor between. The converter is filled with oil, which is moved by the impeller blades. The vanes of the reactor deflect this oil to move the turbine blades. As the impeller rotates, the speed of the turbine increases to match the impeller speed. This provides a fluid coupling between the engine and gearbox that smooths out speed changes. It also increases torque (turning force).

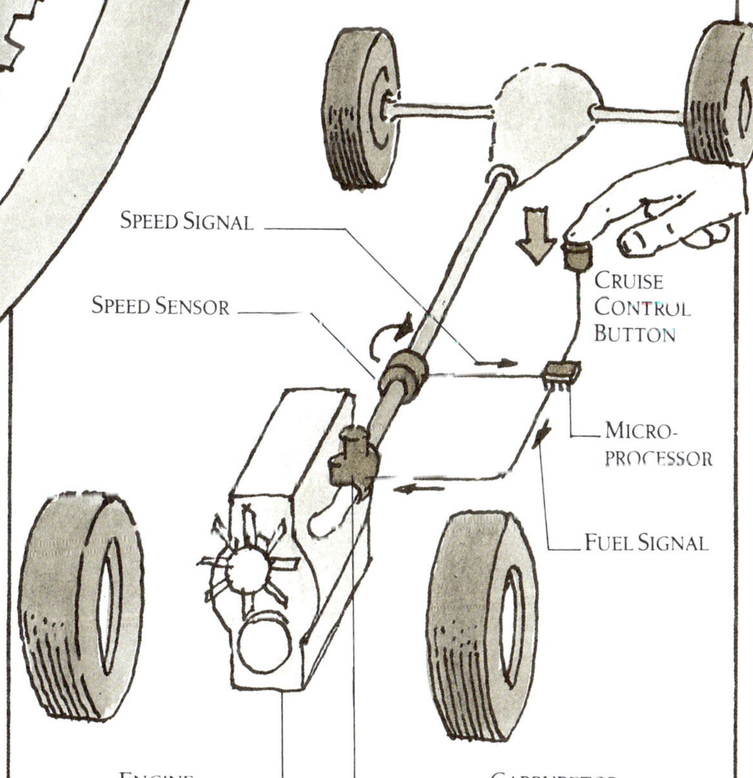

COMPUTERS

ON MAMMOTH MEMORY

When I first discovered Chip the mammoth he was using his extraordinary skill as a logger in the Great Northern Forest. His owner would simply tap on Chip's tusks for the total number of trees he wanted the creature to gather, then tug his tail. If ten taps were applied to his tusks, Chip would not stop working until he had collected ten trees.

I suggested to his owner that it might be to everyone's advantage if Chip's remarkable memory were adapted to other uses. I offered to train him myself; by tapping two numbers – one on each tusk – and then tugging his tail, I planned to teach him the art of multiplication. We worked together, night and day, for almost a year, and Chip's instruction was almost complete when suddenly I received an urgent request for his services.

MACHINES WITH MEMORIES

Computers and calculators are a revolutionary development in the history of technology. They are fundamentally different from all other machines because they have a memory. This memory stores instructions and information.

In a calculator, the instructions are the various methods of arithmetic. These are permanently remembered by the machine and cannot be altered or added to. The information consists of the numbers keyed in.

A calculator requires an input unit to feed in numbers, a processing unit to make the calculation, and an output unit to display the result. In using its extraordinary memory, the mammoth becomes a calculator. Tusk-tapping inputs the number while tail-tugging activates addition. The output is the log display. A calculator also needs a memory unit to store the arithmetic instructions for the processing unit, and to hold the temporary results that occur during calculation. The mammoth's remarkable brain contains both memory and processing units.

ANATOMY OF A CALCULATOR

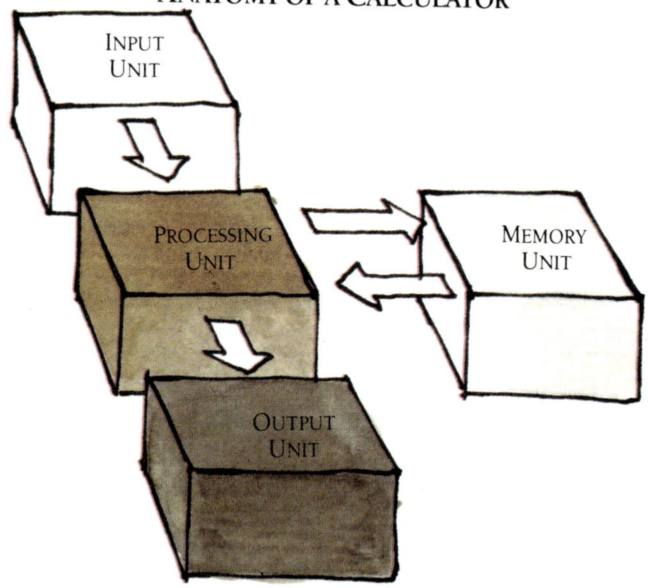

We arrived at a local restaurant where a dispute had arisen over a check. It appeared that the waiter had served seven fixed-price meals to one particular table but had incorrectly calculated the total. Here indeed was the chance I had been waiting for during months of laborious training.

The diners looked on sceptically as I requested one of their number to tap the fixed price on one of Chip's tusks; the restaurant owner tapped the number of meals on the other and then tugged his tail. Chip stood for a few moments as if confused, but then without further hesitation set off for some nearby woods.

As I had anticipated, the animal gave exactly the right answer, but to my great dismay he did it in tree trunks, thereby leveling the restaurant and three adjoining structures. After much deliberation, I reluctantly parted with him because of the dangers of dealing with large numbers.

PROGRAM POWER

A computer contains the same basic four elements as a calculator. It differs in that its memory can be given a different set of instructions, called a computer program, for different tasks. A program can turn a computer into, for example, a game player, a word processor, a paintbox or a musical instrument. It instructs the processing unit how to perform the various tasks, and stores scores, words, pictures or music.

Computer programs consist of long sequences of instructions that individually are very simple. During its training, the mammoth receives instructions just like this. It is taught how to distinguish two kinds of tusk taps and how to put these two numbers into its memory. It is also trained how to multiply, which, like a computer, it does by adding the first number to itself by the second number of times. The mammoth's program is just a few steps long and calls for some patience while the results are produced. Computers, on the other hand, can perform millions of instructions in a matter of seconds.

ANATOMY OF A COMPUTER

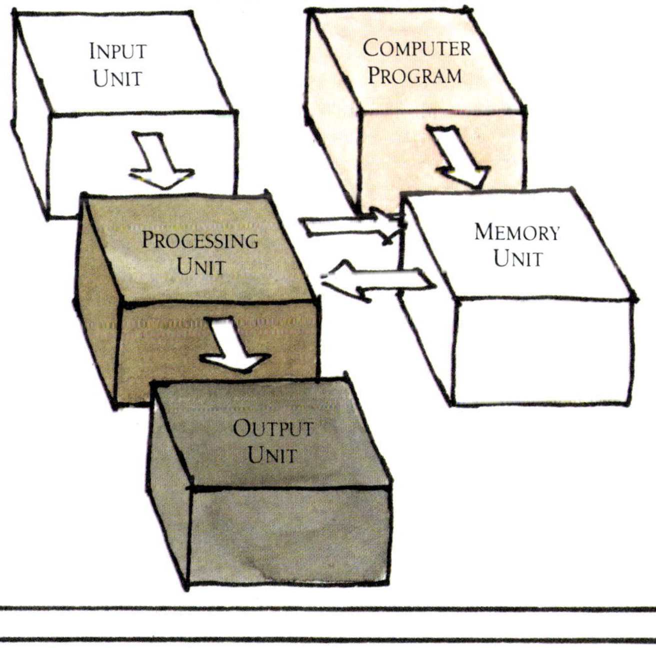

BINARY CODE

A computer appears to work with things with which we are familiar—words composed of letters or pictures made of shapes in different colors, for example. But a computer turns all of these things into sequences of code numbers. These are not the decimal numbers we use, but binary numbers. Inside calculators and computers, code signals for decimal numbers, letters, screen positions, colors and so on flash to and fro between the various components. These signals are made up of numbers in binary code.

	8	4	2	1
0	0	0	0	0
1	0	0	0	1
2	0	0	1	0
3	0	0	1	1
4	0	1	0	0
5	0	1	0	1
6	0	1	1	0
7	0	1	1	1
8	1	0	0	0
9	1	0	0	1
10	1	0	1	0

BINARY NUMBERS

Computers and calculators use binary code because it is the simplest number system. Its two digits—0 and 1—compare with our ten digits (0 to 9).

The table on the left shows how binary numbers relate to decimal numbers. Reading from right to left, the ones and zeros in each column indicate whether or not the number contains 1, 2, 4, 8 and so on, doubling each time. 0101, for example, is $0 \times 8 + 1 \times 4 + 0 \times 2 + 1 \times 1 = 5$. Each binary digit (0 or 1) is called a bit.

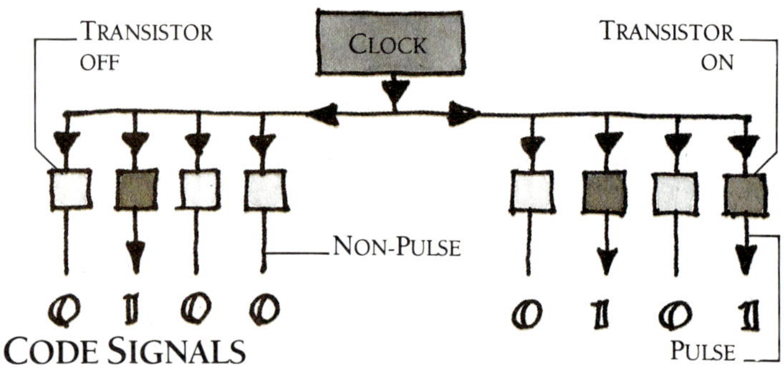

CODE SIGNALS

A binary code signal (*above*) is a sequence of electrical pulses traveling along wires. A device called a clock sends out regular pulses, and components such as transistors switch on and off to pass or block the pulses. One (1) represents a pulse, zero (0) a non-pulse.

WORKING WITH ADDITION

As computers and calculators work, they make calculations in binary arithmetic at lightning speed. Components called adders do the calculations, which all break down into sequences of additions. Subtraction, for example, is done by steps that involve adding and inverting numbers (changing each 1 to 0 and vice-versa).

BINARY ARITHMETIC

There are only four basic rules:

A *0 + 0 = 0 and carry 0*
B *0 + 1 = 1 and carry 0*
C *1 + 0 = 1 and carry 0*
D *1 + 1 = 0 and carry 1*

	B	D	A	C
5	0	1	0	1
+ 4	0	1	0	0
9	1	0	0	1
		1		

LOGIC GATES

Computers and calculators contain components called logic gates that are linked together to do electronic calculations. Three principal kinds of gates are shown here, each inside the symbol used in computer circuits.

The gates are composed of transistors (see pp.238-9) that process an input signal, changing it to an output signal. A pulse (1) entering the center of a transistor switches it off so that a clock pulse is blocked, giving a 0.

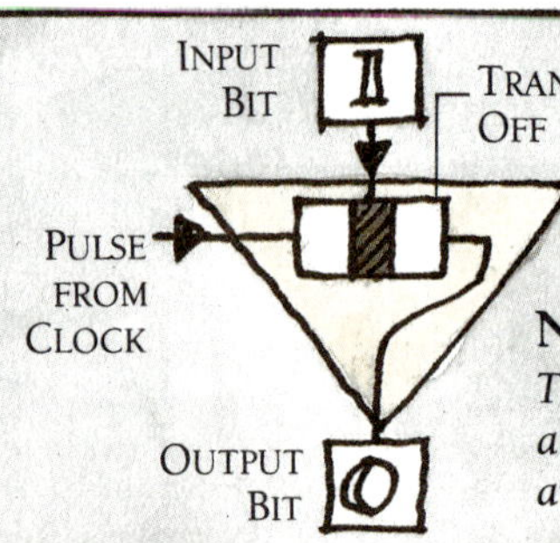

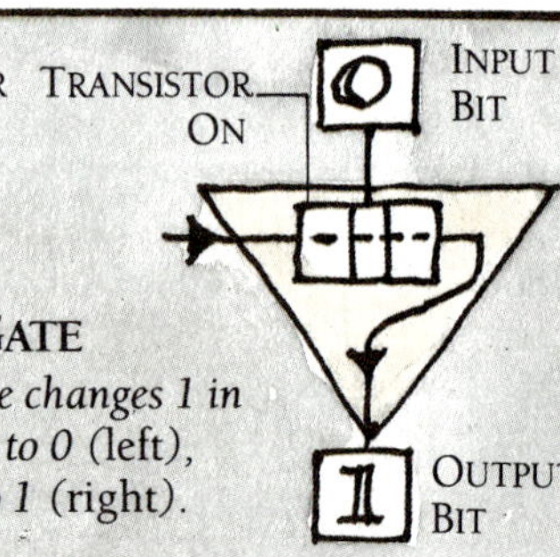

NOT GATE

This gate changes 1 in a signal to 0 (left), and 0 to 1 (right).

OR GATE

This gate (right) processes two-bit signals. If the first OR second bit is 1, then it outputs 1.

INPUT		OUTPUT
0	0	0
1	0	1
0	1	1
1	1	1

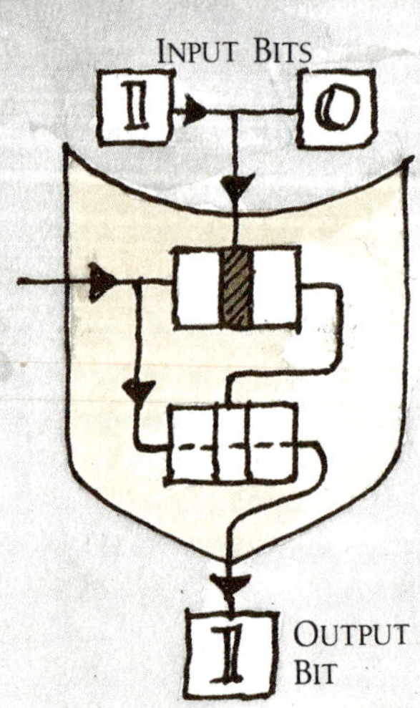

AND GATE

This gate (right) processes two-bit signals. If the first AND second bits are 1, then it outputs 1.

INPUT		OUTPUT
0	0	0
1	0	0
0	1	0
1	1	1

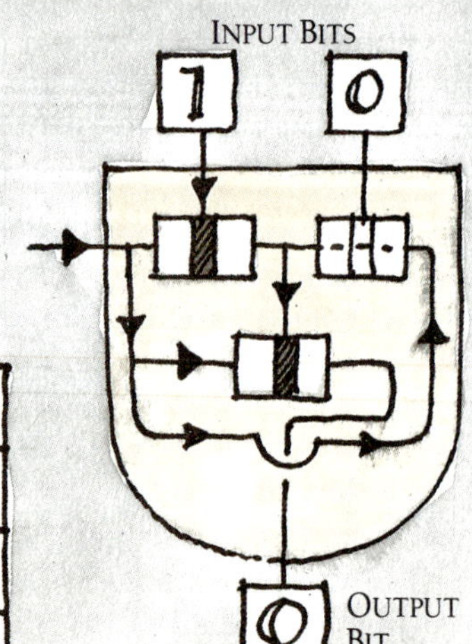

THE HALF ADDER

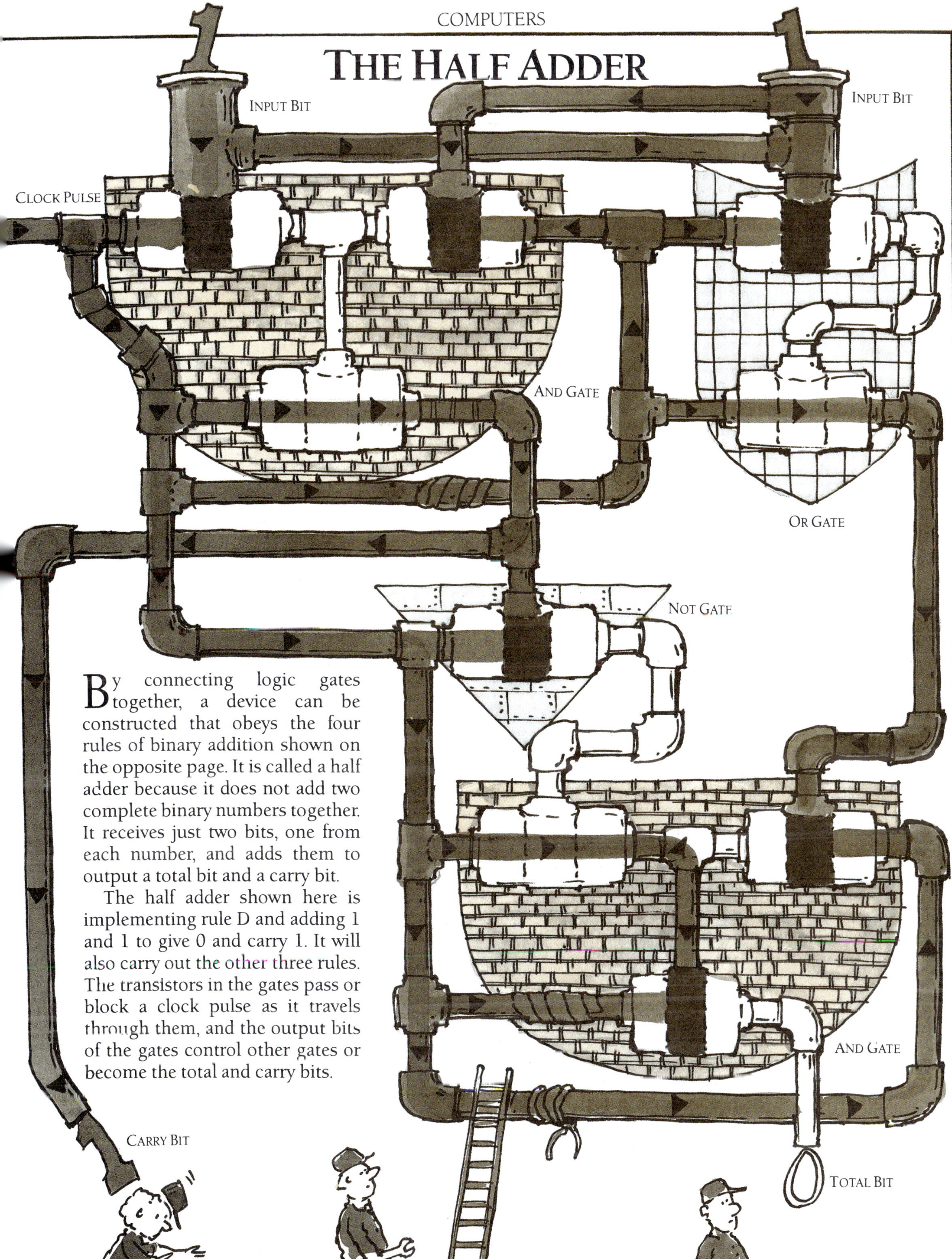

By connecting logic gates together, a device can be constructed that obeys the four rules of binary addition shown on the opposite page. It is called a half adder because it does not add two complete binary numbers together. It receives just two bits, one from each number, and adds them to output a total bit and a carry bit.

The half adder shown here is implementing rule D and adding 1 and 1 to give 0 and carry 1. It will also carry out the other three rules. The transistors in the gates pass or block a clock pulse as it travels through them, and the output bits of the gates control other gates or become the total and carry bits.

INPUT NUMBERS
0
1
0
0
0
1
0
1
FOURTH STAGE
THIRD STAGE
HALF ADDER
HALF ADDER
OR GATE
HALF ADDER
HALF ADDER

THE FULL ADDER

A cascade of half adders (see p.333), known as a full adder, can add two complete binary numbers together. A full adder consists of a first stage with one half adder to add the first bit of each number. This is followed by stages each composed of a pair of half adders to add the subsequent bits. The stages are linked by OR gates that deal with the carry bits.

The full adder shown here is adding 4 and 5 to get 9. It has four stages because 9 is a 4-bit number. In a computer, a full adder has extra stages in order to add 8-bit or 16-bit numbers. Being able to handle longer numbers in one operation makes a computer faster and more powerful, which is why 16-bit computers are superior to 8-bit computers.

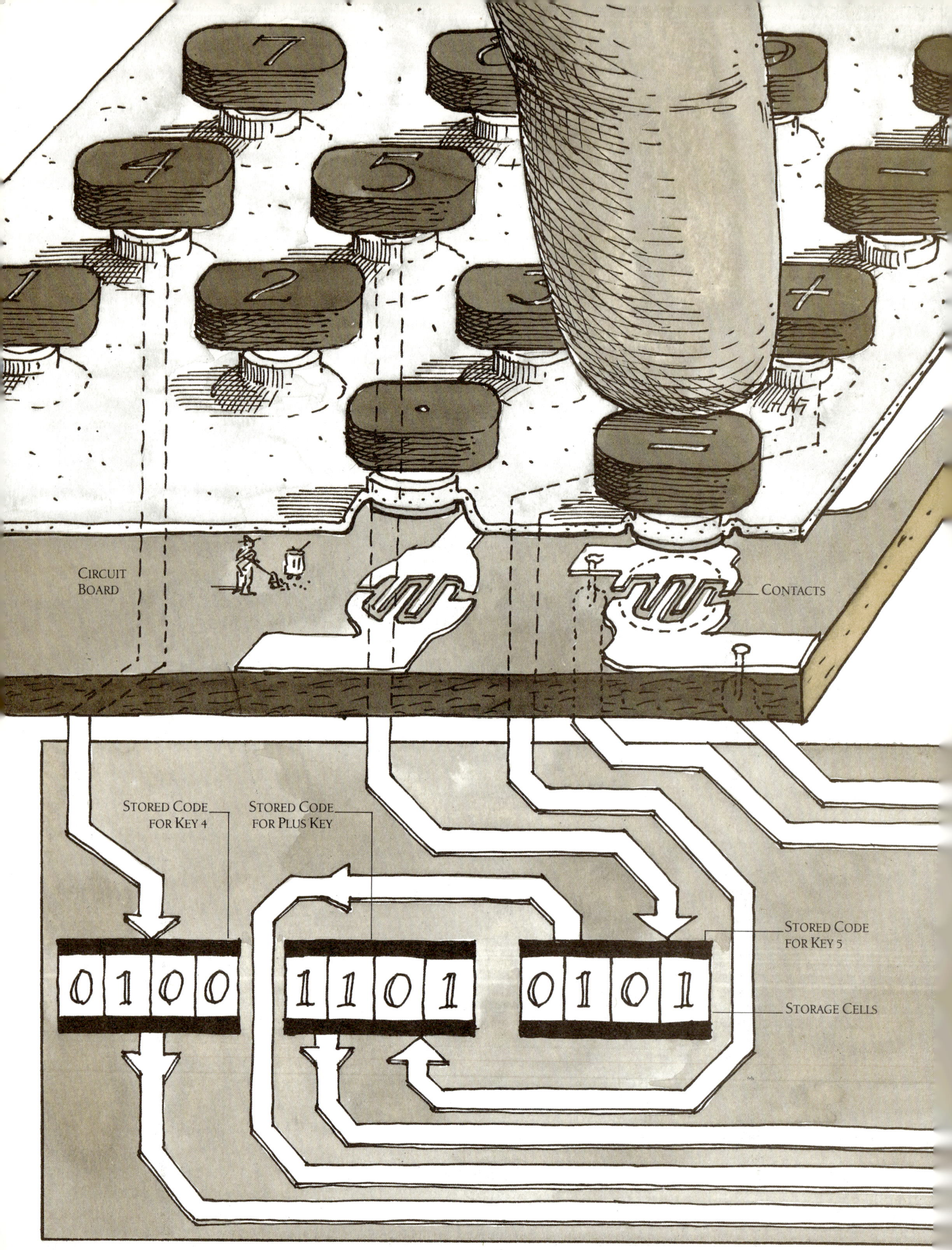
CIRCUIT BOARD
CONTACTS
STORED CODE FOR KEY 4
STORED CODE FOR PLUS KEY
STORED CODE FOR KEY 5
STORAGE CELLS
0100
1101
0101

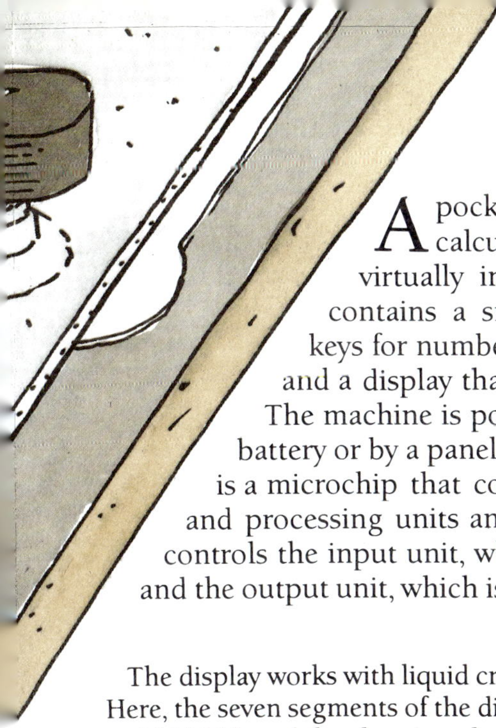

THE CALCULATOR

A pocket electronic calculator can perform virtually instant arithmetic. It contains a small keyboard with keys for numbers and operations, and a display that shows the result. The machine is powered by a small battery or by a panel of solar cells. Inside is a microchip that contains the memory and processing units and which also controls the input unit, which is the keyboard, and the output unit, which is the display.

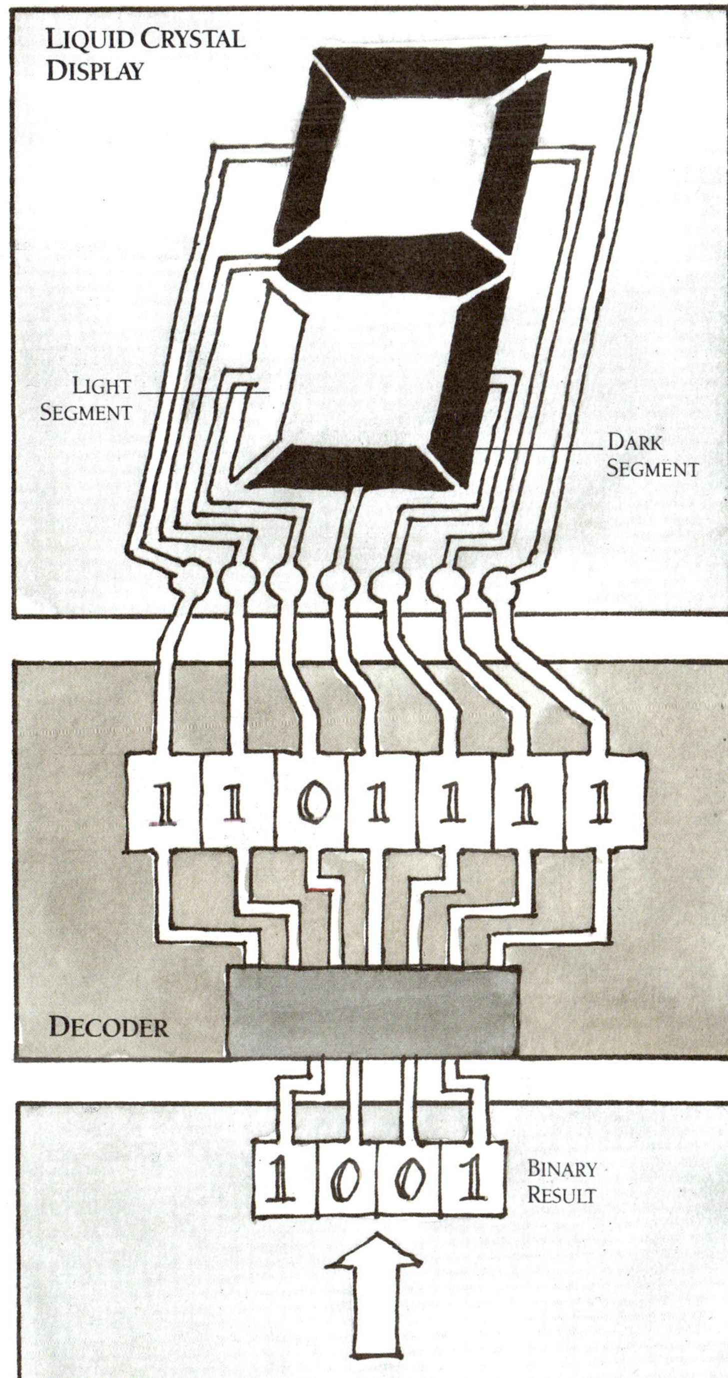

DISPLAY

The display works with liquid crystals (see pp.204-5). Here, the seven segments of the display receive the code 1101111, and the electric pulse of each 1 bit makes a segment darken. The pattern produces the figure 9.

KEYBOARD

Beneath the keys is a printed circuit board containing a set of contacts for each key. Pressing a key closes the contacts and sends a signal along a pair of lines in the circuit board to the processing unit, which stores the binary code for that key in the memory. The processing unit also sends the code to the display. Each key is connected by a different pair of lines to the processing unit, which repeatedly checks the lines to find out when a pair is linked by a key.

MEMORY UNIT

Storage cells in the memory unit (see p.341) hold the binary codes for the keys that have been pressed. Here, the memory is storing 0100 for key 4, and 0101 for key 5. The number codes, together with the operation code for the plus key, are held in temporary cells until the processing unit requires them. However, a calculator often has additional memory keys that enable an extra section of memory to store a number for use in several calculations. This number can be recalled from the memory at the press of a key.

PROCESSING UNIT

When the equals key is pressed, it sends a signal to the processing unit. This takes the operation code (addition) and the two numbers being held in the memory unit, and performs the operation on the two numbers. A full adder does the addition and the result (1001) goes to the decoder in the calculator's microchip.

The decoder (*above*) contains a set of logic gates that change the bits in the result into a sequence of 7-bit binary codes. These are the codes for the decimal numbers in the result. In this case, 1001 is converted into the 7-bit code 1101111. This code is then sent to the liquid crystal display.

THE MICROCOMPUTER

Computers come in a variety of guises, from the simple microcomputer shown here to machines as diverse as cash dispensers, robots, flight simulators and supermarket checkout counters. However, they still contain the same four basic units as a calculator – only the purposes of the input and output vary greatly.

An ordinary microcomputer has a wide range of uses because many kinds of units can be connected to its input and output ports. The keyboard is the principal input unit. This works in much the same way as a calculator keyboard, except that combinations of two or even three keys can be pressed. Also, programs can give the keys uses other than simply entering letters and numbers. A mouse is another important input unit (see pp.346-7). The case also houses the microprocessor and memory chips and it may also include a disk drive (see pp.302-3), though a drive is often connected to it. A monitor and a printer (see pp.348-9) are the principal output units.

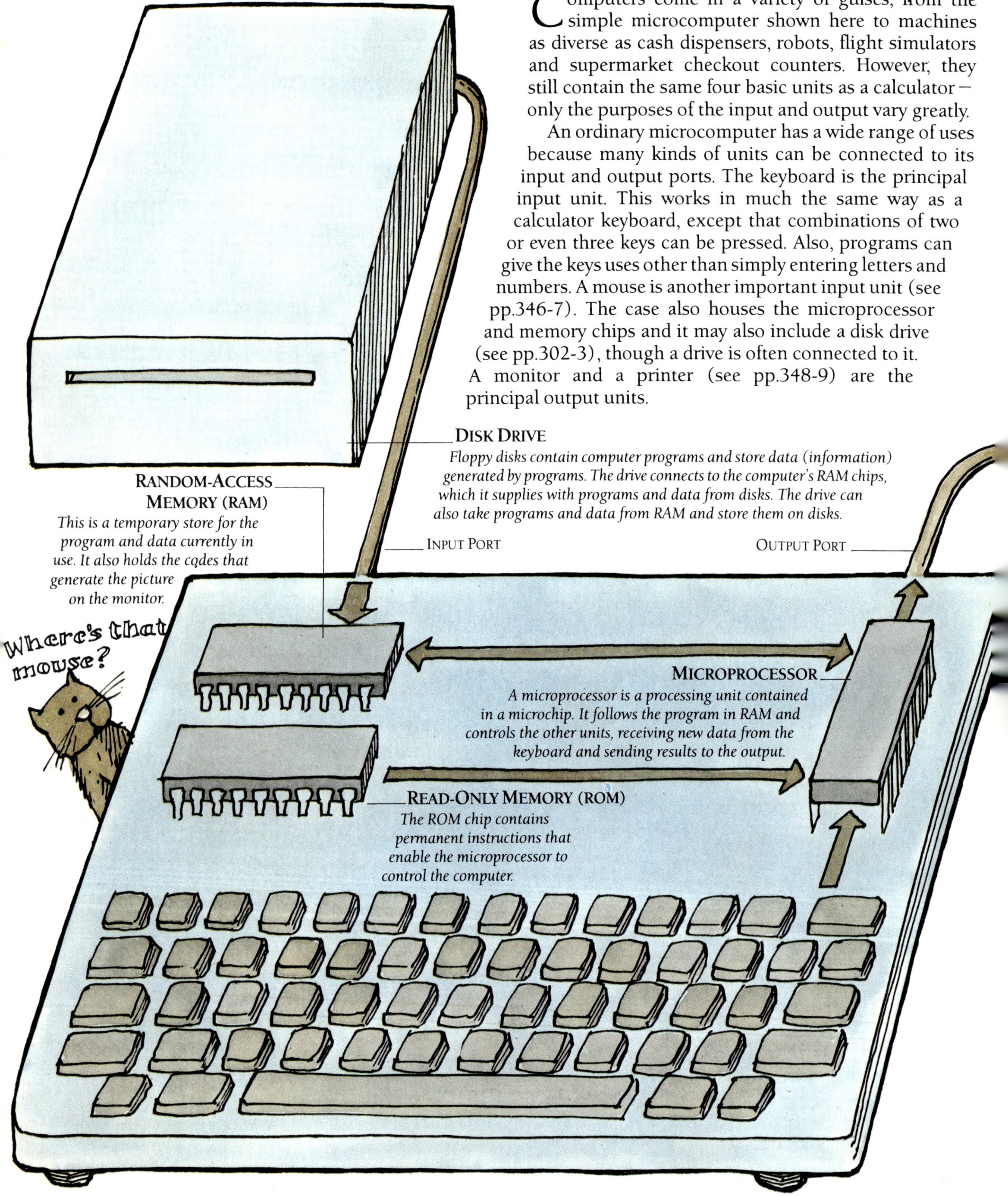

DISK DRIVE
Floppy disks contain computer programs and store data (information) generated by programs. The drive connects to the computer's RAM chips, which it supplies with programs and data from disks. The drive can also take programs and data from RAM and store them on disks.

RANDOM-ACCESS MEMORY (RAM)
This is a temporary store for the program and data currently in use. It also holds the codes that generate the picture on the monitor.

MICROPROCESSOR
A microprocessor is a processing unit contained in a microchip. It follows the program in RAM and controls the other units, receiving new data from the keyboard and sending results to the output.

READ-ONLY MEMORY (ROM)
The ROM chip contains permanent instructions that enable the microprocessor to control the computer.

HUNT THE MAMMOTH

The computer is here playing a game program called *Hunt The Mammoth*, which is on the disk in the drive. Pressing keys on the keyboard moves boy to left or right and releases the lasso. Mammoths and mooses move across the top of the screen, and the aim is to capture mammoths. A score of 12 mammoths has been made in 59 seconds.

PAINTING BY NUMBERS

The screen of a computer works in the same way as a television set (see pp.262-3). The picture consists of a grid of small picture cells called pixels, each of which has horizontal and vertical position codes and a color code. The microprocessor generates the picture as a series of codes, lighting up each pixel in a certain color. The enlarged view (right) shows the pixels in the face on the screen, together with their position and color codes.

COLOR CODES

- 0 WHITE
- 1 BLACK
- 2 RED
- 3 GREEN
- 4 BLUE
- 5 CYAN
- 6 YELLOW
- 7 MAGENTA

VERTICAL POSITION CODES	HORIZONTAL POSITION CODES							
	157	158	159	160	161	162	163	164
110	0	0	0	0	0	0	0	0
111	4	4	0	0	0	4	4	0
112	0	0	0	2	0	0	0	0
113	0	0	0	0	2	0	0	0
114	0	0	0	0	2	0	0	0
115	0	0	0	0	0	2	0	0
116	0	0	0	2	2	2	0	0
117	0	0	0	0	0	0	0	0

COMPUTER MEMORY

Although computers "think" by using on-off pulses of electric current, they have to change the pulses into some other system of on-off code in order to store or "remember" them. Computer memories do this in several different ways. Magnetic disks use patterns of magnetism in which the direction of the magnetism is either north or south. Memories made of microchips (see pp.342-3) may use on-off static electric charges, or connections and disconnections within the chip. Optical memory uses on-off light beams, compact disks (see pp.248-9), for example storing immense amounts of information.

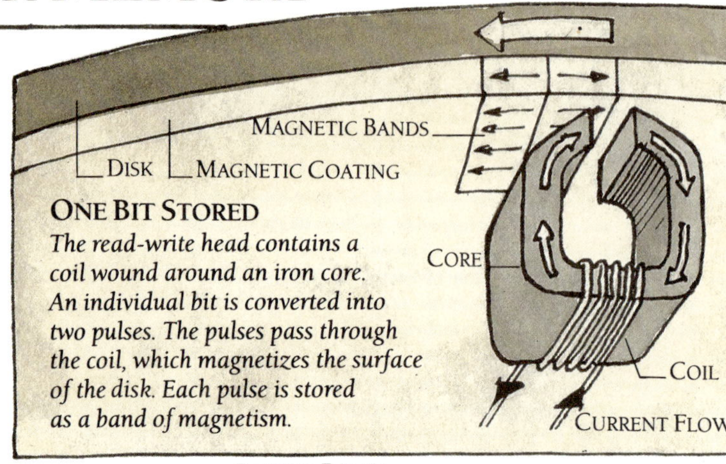

ONE BIT STORED

The read-write head contains a coil wound around an iron core. An individual bit is converted into two pulses. The pulses pass through the coil, which magnetizes the surface of the disk. Each pulse is stored as a band of magnetism.

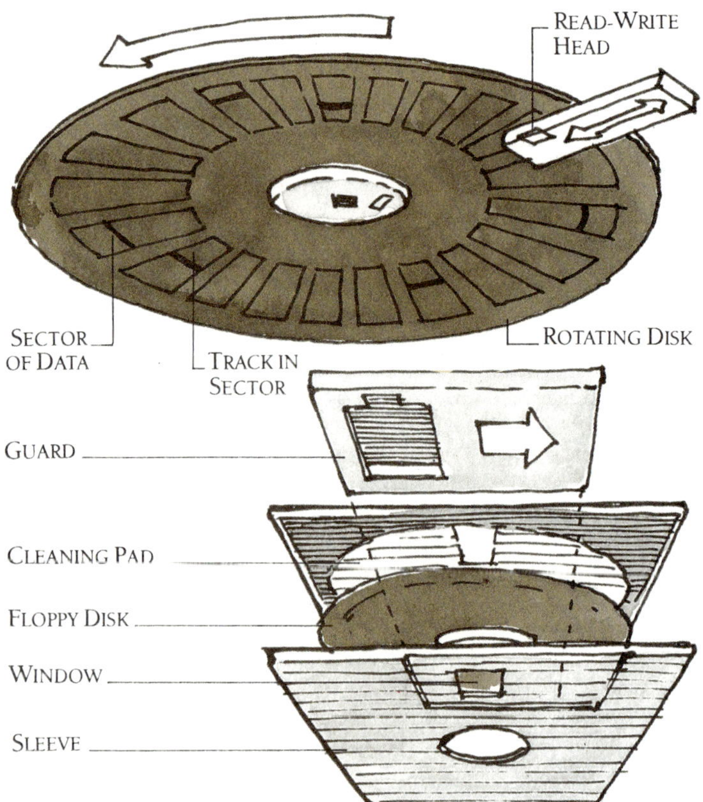

TWO BITS STORED

When a bit is stored, the first band always reverses the direction of magnetization of the previous band to indicate a new bit. The second band has the same direction if the bit is a 0, or the reversed direction if it is a 1.

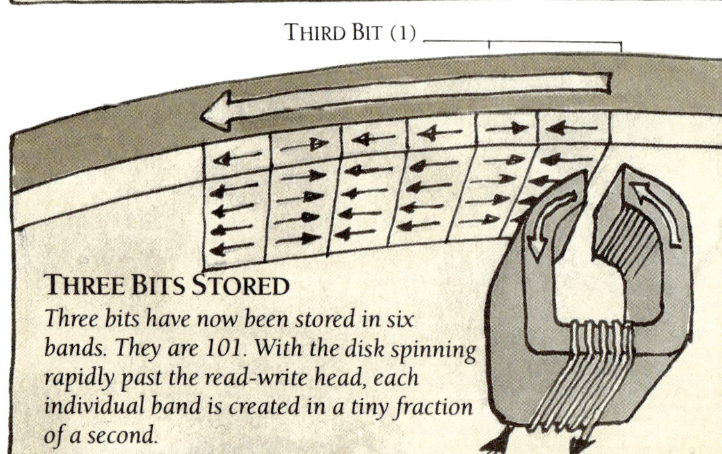

THREE BITS STORED

Three bits have now been stored in six bands. They are 101. With the disk spinning rapidly past the read-write head, each individual band is created in a tiny fraction of a second.

MAGNETIC DISKS

A floppy disk (*above*) is a flexible magnetic disk. It is used in a disk drive (see p.303) to store or retrieve programs and data. A hard disk drive (*below*) contains several hard magnetic disks permanently enclosed to exclude dust. Hard disks have a much greater memory capacity. Memory capacity is measured in kilobytes (K) or megabytes (M); a single byte is equal to 8 bits of binary code.

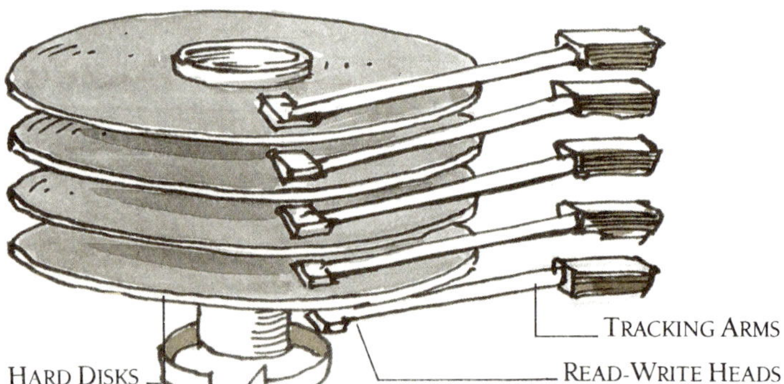

FOUR BITS STORED

Four bits – 1011 – have now been written onto the disk. To read the bits, the disk is spun past the head, inducing an electric signal in the coil that goes to the computer.

Random-Access Memory (RAM)

Although RAM chips are extremely small, they contain thousands of storage cells. Each cell stores one bit of binary code, and the cells are arranged in groups of eight so that each group stores one byte. Each byte is stored at a particular location and is identified by an "address" number. RAM chips store data only as long as the current stays on.

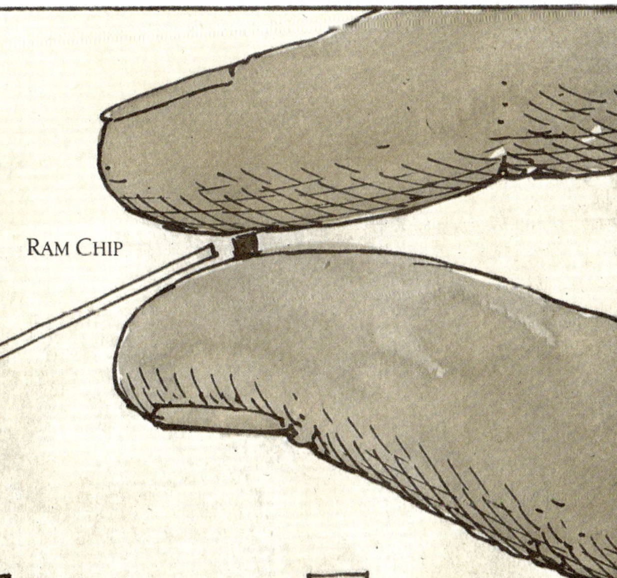

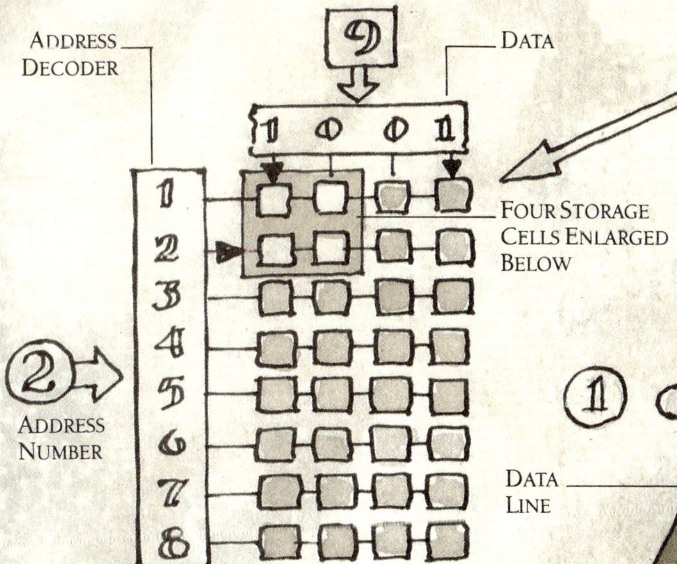

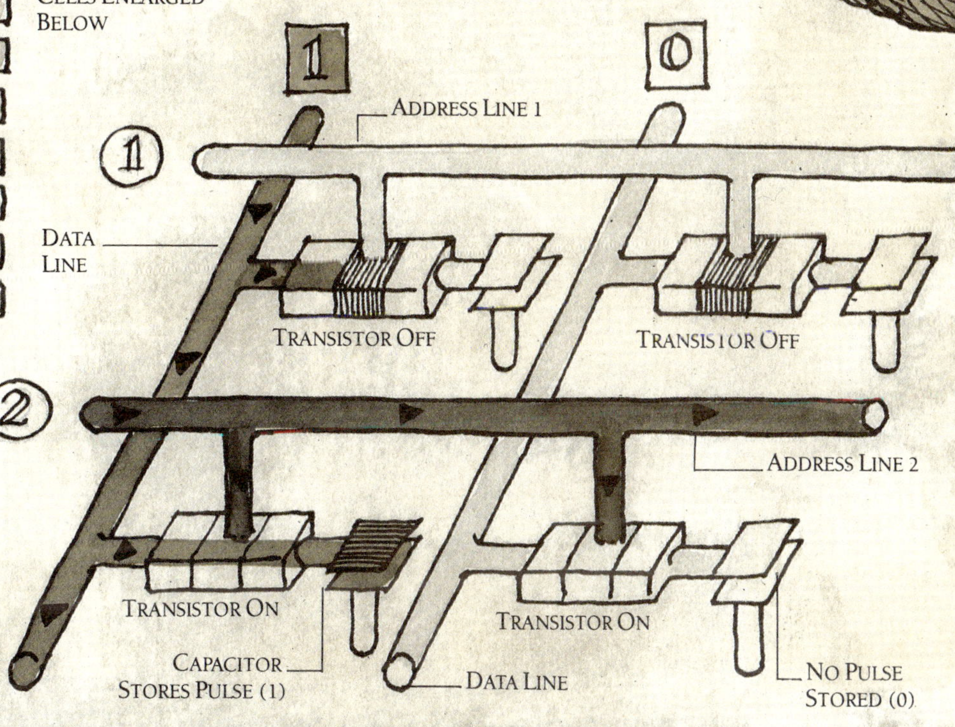

Inside RAM

Each storage cell is connected to an address line and a data line. To store a number, a pulse is first sent to all the cells along a single address line, switching on the transistor in each cell. The on-off pulses for the number then go along the data lines. Each pulse (1 bit) charges a capacitor connected to the transistor. Here 9 (1001) is stored at address 2. To read the data, the cells discharge and send pulses back along the data lines.

Read-Only Memory (ROM)

A ROM chip is a permanent memory which stores data even when the current is off. Like a RAM chip, it contains a grid of cells connected by address lines and data lines. But where these lines meet, the ROM chip is quite different. In a cell that stores a 1 bit, the two lines are connected by a diode (see p.292). In a cell that stores a 0 bit, there is no connection. The memory is read by sending a pulse to the address line, which diverts through the diodes to the data lines.

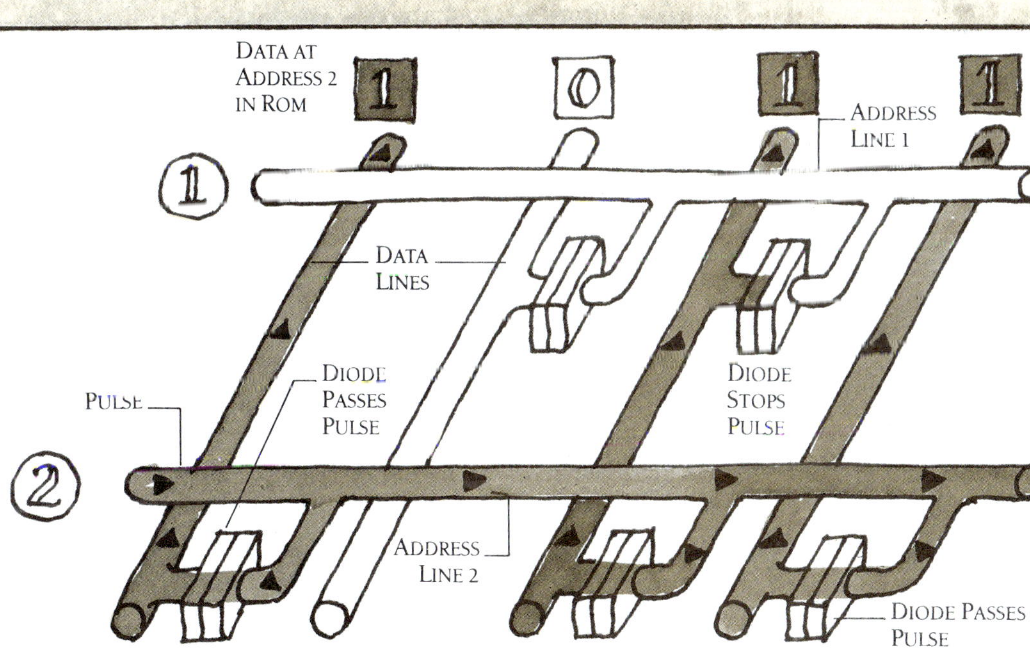

MICROCHIP

A microchip or integrated circuit contains many thousands of electronic components squeezed into a thin sliver of silicon less than 0.4 inch (1cm) square. These components are connected together to form devices such as logic gates (see p.332) and memory cells (see p.340).

The miniature components within a chip are not individually made and assembled. Instead, the components and all their connections are built up in layers of material in complex miniature patterns. These are made with masks produced by reducing large patterns photographically. These two pages illustrate the manufacture of one transistor component in a microchip.

SILICON CYLINDER

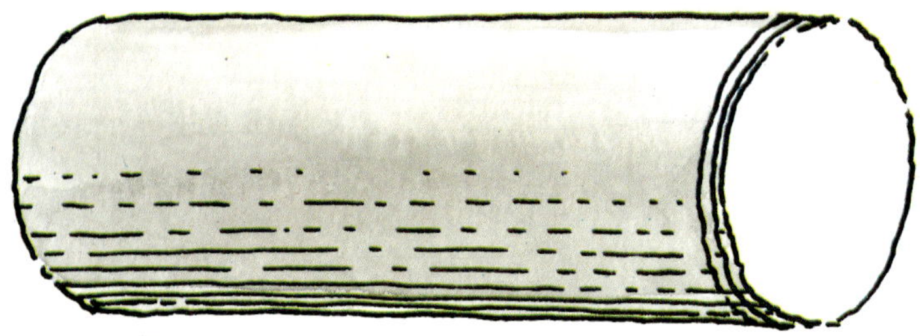

CHIP CHOP

A microchip is made mostly of *p*-type silicon (see p.291). A cylinder of silicon is first produced, and this is then sliced into wafers about 0.01 inch (0.25mm) thick. Each wafer is then treated to make several hundred microchips. The wafers are then tested and chopped up into individual chips. These are inspected under a microscope before being packaged.

MAKING A TRANSISTOR

1 FIRST MASKING

The silicon base is first coated with silicon dioxide, which does not conduct electricity, and then with a substance called photoresist. Shining ultraviolet light through a patterned mask hardens the photoresist. The unexposed parts remain soft.

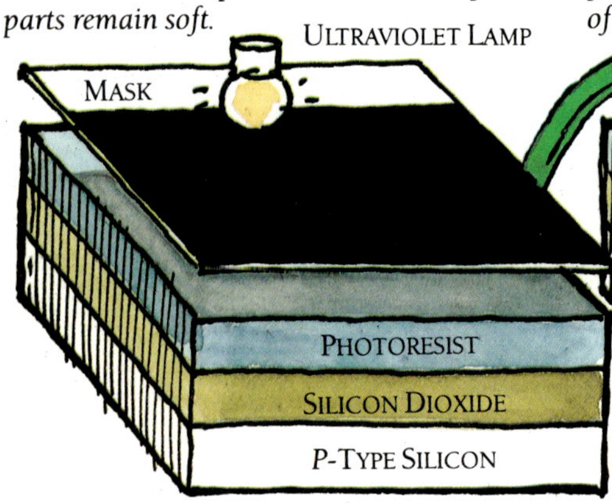

2 FIRST ETCHING

A solvent dissolves away the soft unexposed layer of photoresist, revealing a part of the silicon dioxide. This is then chemically etched to reduce its thickness. The hardened photoresist is then dissolved to leave a ridge of dioxide.

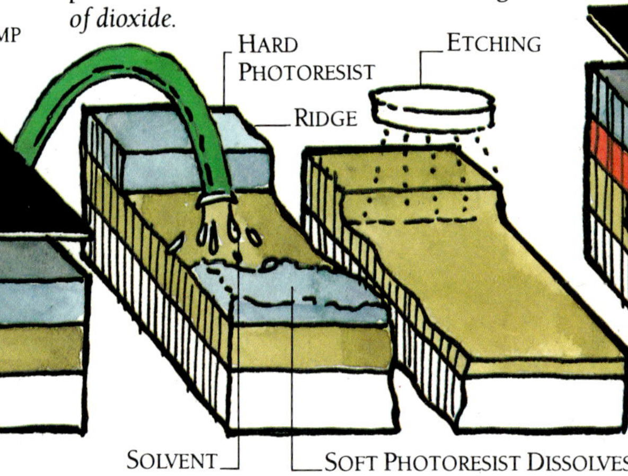

3 SECOND MASKING

Layers of polysilicon, which conducts electricity, and photoresist are applied, and then a second masking operation is carried out.

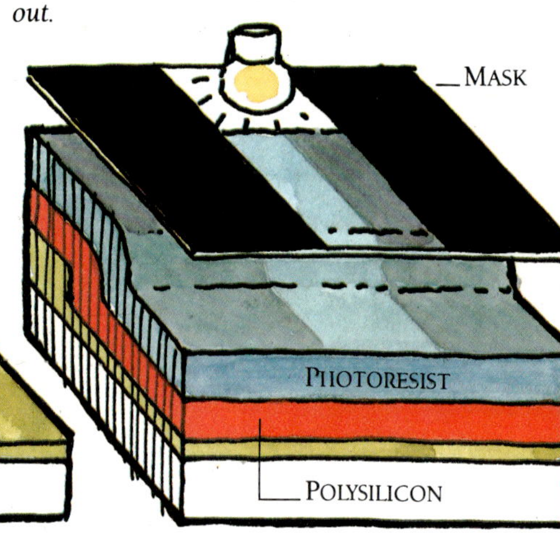

4 SECOND ETCHING

The unexposed photoresist is dissolved, and then an etching treatment removes the polysilicon and silicon dioxide beneath it. This reveals two strips of p*-type silicon.*

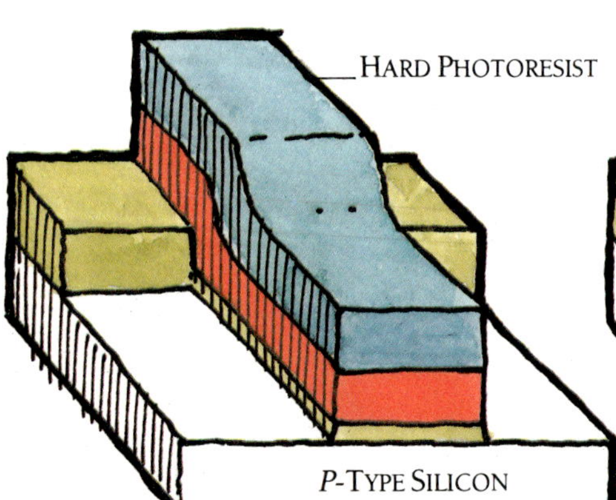

5 DOPING

The hard photoresist is removed. The layers now undergo an operation called doping which transforms the newly revealed strips of p*-type silicon into* n*-type silicon.*

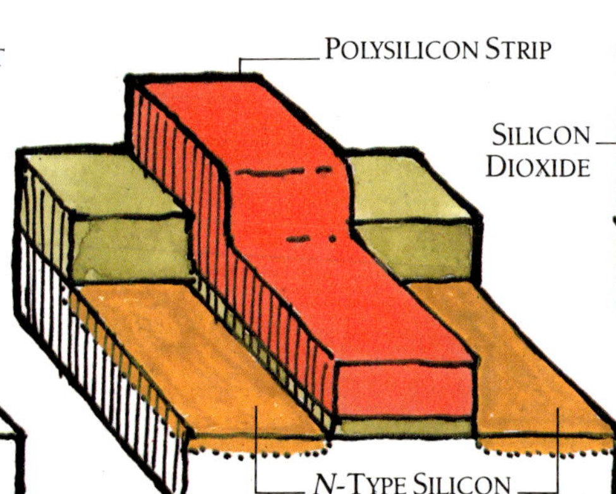

6 THIRD MASKING AND ETCHING

Layers of silicon dioxide and photoresist are added. Masking and etching creates holes through to the doped silicon and central polysilicon strip.

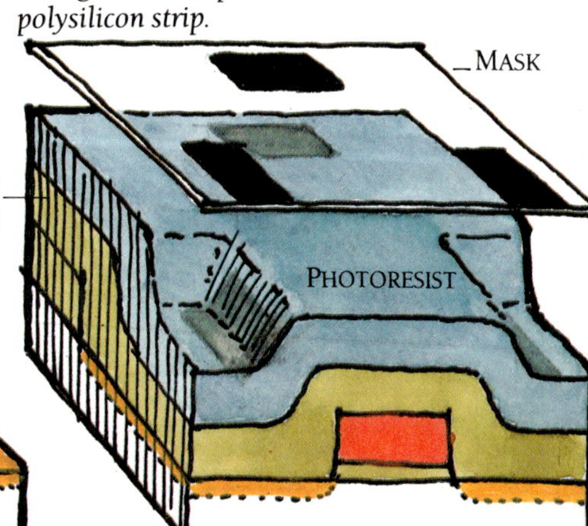

PACKAGE OF PINS

A finished microchip is dwarfed by the connections needed to fit it into a computer's circuit board. The chip is set in a plastic base and connected to two rows of pins which conduct electric signals.

7 COMPLETING THE TRANSISTOR

The photoresist is dissolved, and a final masking stage adds three strips of aluminum. These make electrical connections through the holes and complete the transistor.

In this transistor, known as an MOS type, a positive charge fed to the gate attracts electrons in the p-type silicon base. Current flows between the source and the drain, thereby switching the transistor on. A negative charge at the gate repels electrons and turns the current off.

If the transistors in a chip were really this big, then the whole chip would be the size of a city!

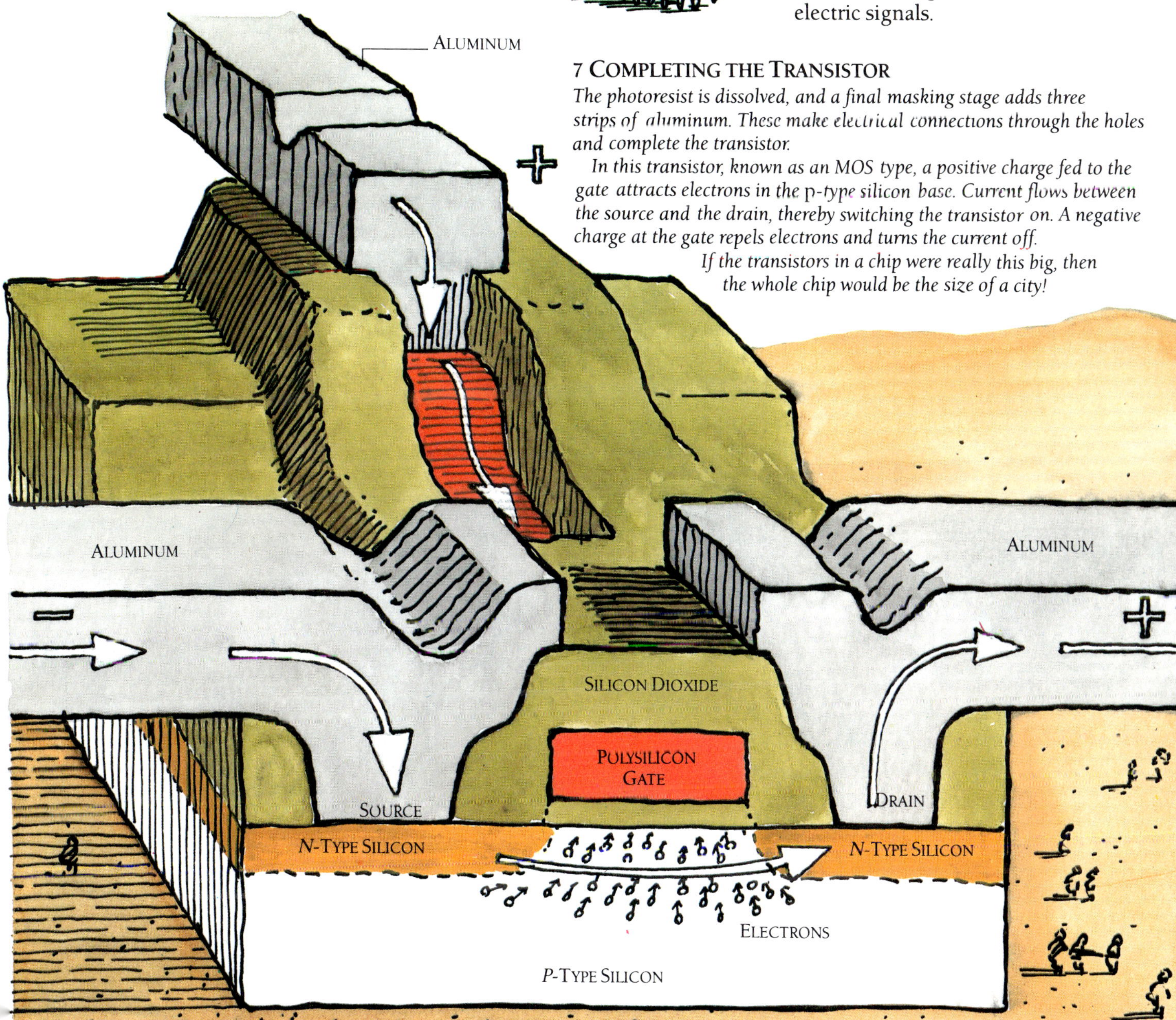

MICROPROCESSOR

The brain of a computer is its central processing unit. In the case of a microcomputer, this is a chip called the microprocessor. It is connected to the other units by buses, groups of wires along which binary code signals pass. In the microprocessor are components that carry out arithmetic and logic operations and control the computer. Here, the microprocessor is performing four operations that check the time in *Hunt the Mammoth* (see pp.338-9). They take place in a tiny fraction of a second.

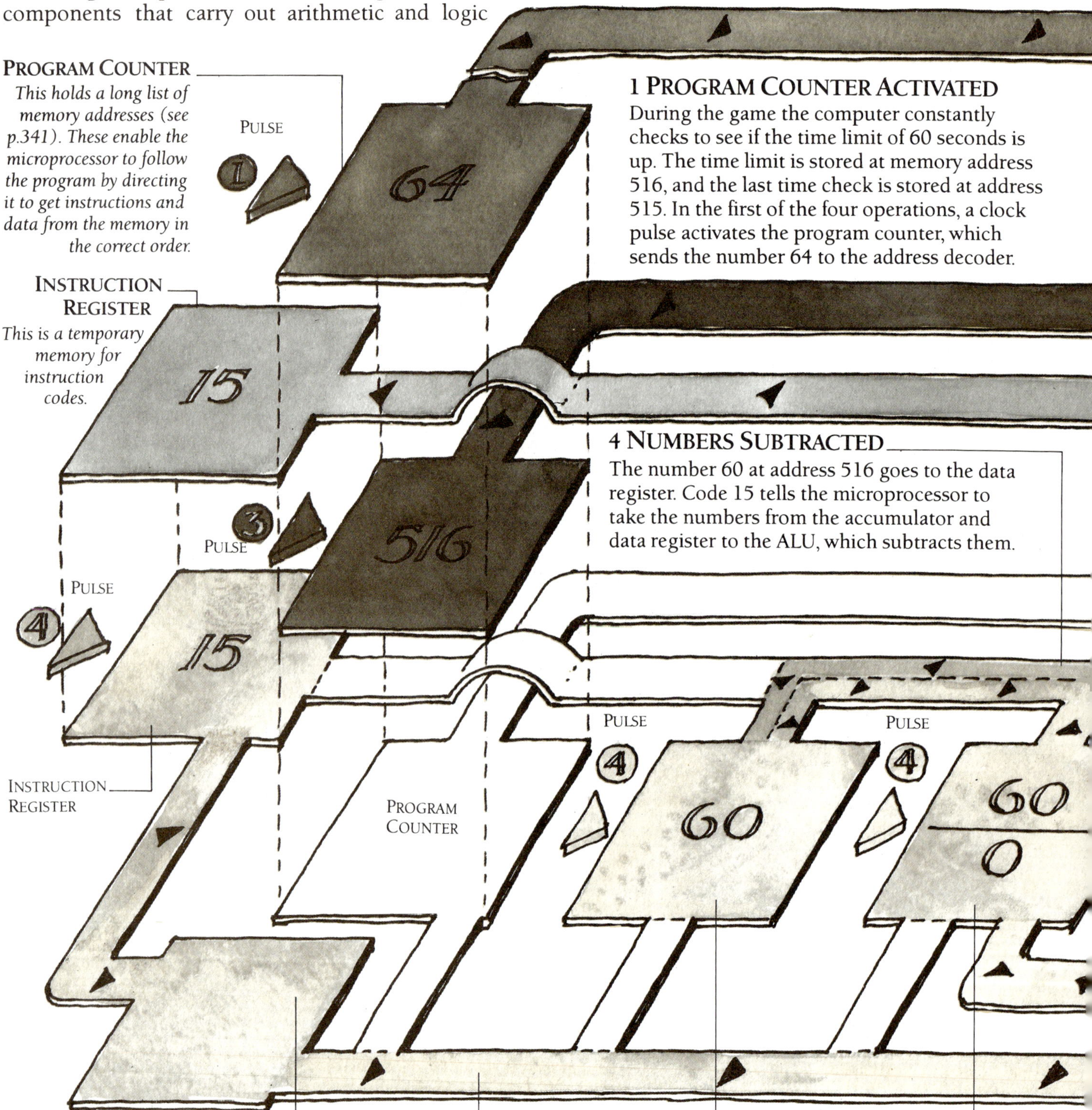

PROGRAM COUNTER
This holds a long list of memory addresses (see p.341). These enable the microprocessor to follow the program by directing it to get instructions and data from the memory in the correct order.

INSTRUCTION REGISTER
This is a temporary memory for instruction codes.

1 PROGRAM COUNTER ACTIVATED

During the game the computer constantly checks to see if the time limit of 60 seconds is up. The time limit is stored at memory address 516, and the last time check is stored at address 515. In the first of the four operations, a clock pulse activates the program counter, which sends the number 64 to the address decoder.

4 NUMBERS SUBTRACTED

The number 60 at address 516 goes to the data register. Code 15 tells the microprocessor to take the numbers from the accumulator and data register to the ALU, which subtracts them.

INSTRUCTION DECODER
This receives information from the instruction register and sends control signals to the microprocessor to perform the required instructions – in this case, subtraction.

CONTROL BUS
This carries control signals to parts of the microprocessor and other computer units.

DATA REGISTER
This is a temporary memory that holds data values. Here it is holding the time limit of 60.

ACCUMULATOR
The accumulator is a temporary memory that holds results before and after processing. Here, it contains first 60 (the time played) and then 0 (the subtraction result).

ADDRESS BUS

The address bus carries memory address numbers from the program counter to the address decoder.

ADDRESS DECODER

The address decoder receives a number from the program counter and opens the cells in the memory which have that address.

ENDING THE GAME

The data register contains a time limit. As the final second of the game ticked by, a previous addition operation (code 9 at address 63) added 1 to the last time reached (data 59 at address 515). The result, 60, is waiting in the accumulator. Subtraction (operation 4) compares the two numbers to see if they are the same. If the result sent to the accumulator after operation 4 is 0, then the time is up. The program counter diverts the computer to a sequence of operations that displays GAME OVER.

2 MEMORY ACTIVATED

A second pulse makes the memory send the number 15 at address 64 along the data bus to the instruction register.

3 PROGRAM COUNTER REACTIVATED

The next pulse makes the program counter send its next number – 516 – to the address decoder, which opens address 516.

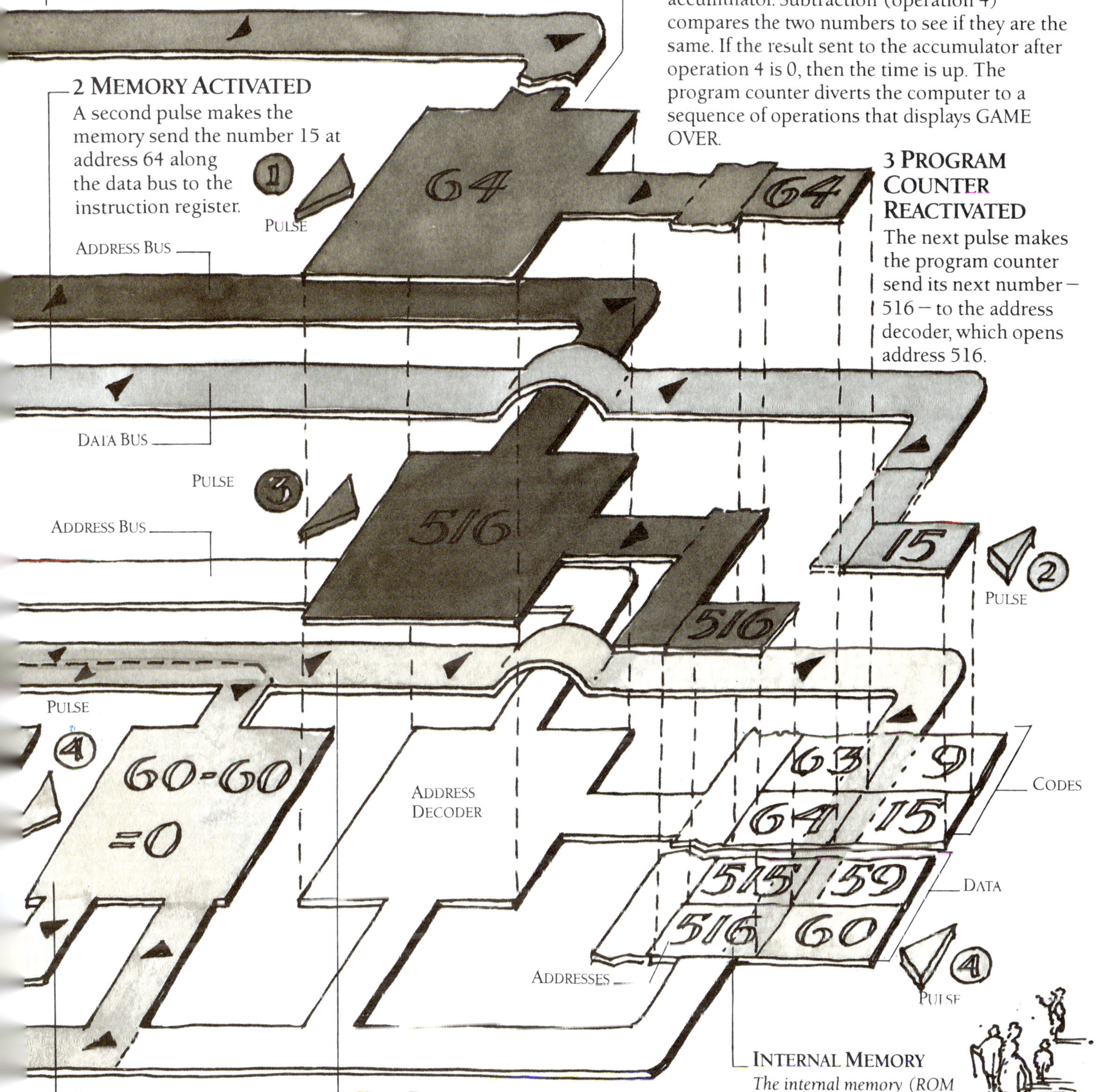

ALU

The ALU – Arithmetic and Logic Unit – carries out arithmetic and logic operations. Here it is comparing two numbers to see if they are the same.

DATA BUS

The data bus carries numbers from the memory to parts of the microprocessor and vice-versa.

INTERNAL MEMORY

The internal memory (ROM and RAM) contains instruction codes and data values at particular addresses.

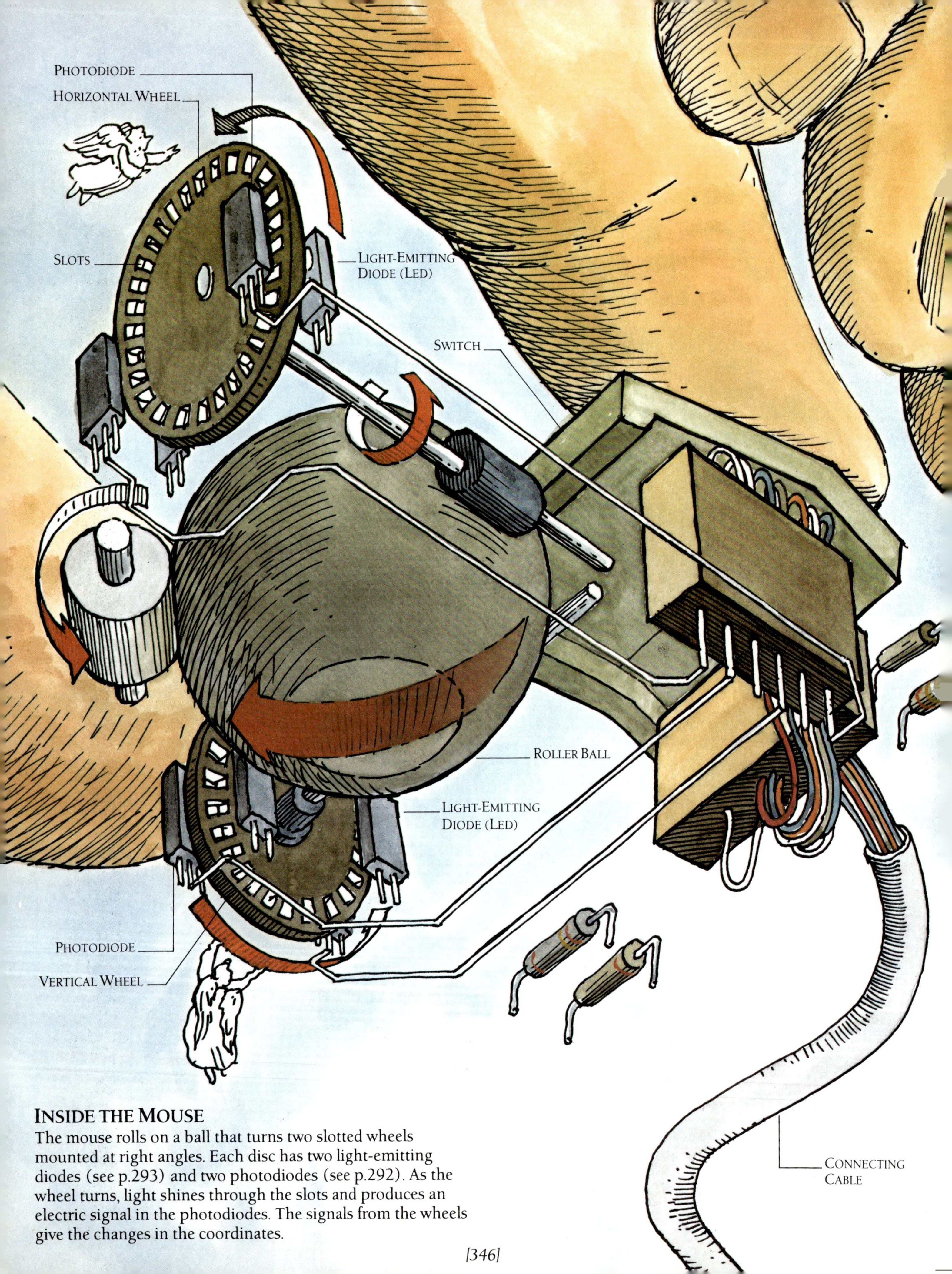

INSIDE THE MOUSE

The mouse rolls on a ball that turns two slotted wheels mounted at right angles. Each disc has two light-emitting diodes (see p.293) and two photodiodes (see p.292). As the wheel turns, light shines through the slots and produces an electric signal in the photodiodes. The signals from the wheels give the changes in the coordinates.

MOUSE

Using a keyboard to control a complex computer program is awkward and slow. The advent of the mouse and icons makes a computer much easier to use, and also gives programs much greater flexibility. The mouse is a controller that is moved over a mat or desk top. As it moves, electric pulses inform the computer of its exact change in position. The computer responds by shifting a cursor over the picture in the same direction as the mouse. The computer can be given commands by moving the mouse so that the cursor points to an icon, and then "clicking" its switch. The command represented by the selected icon is then carried out.

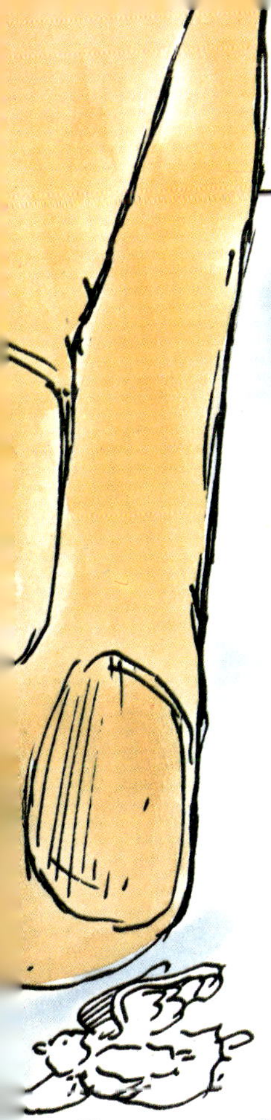

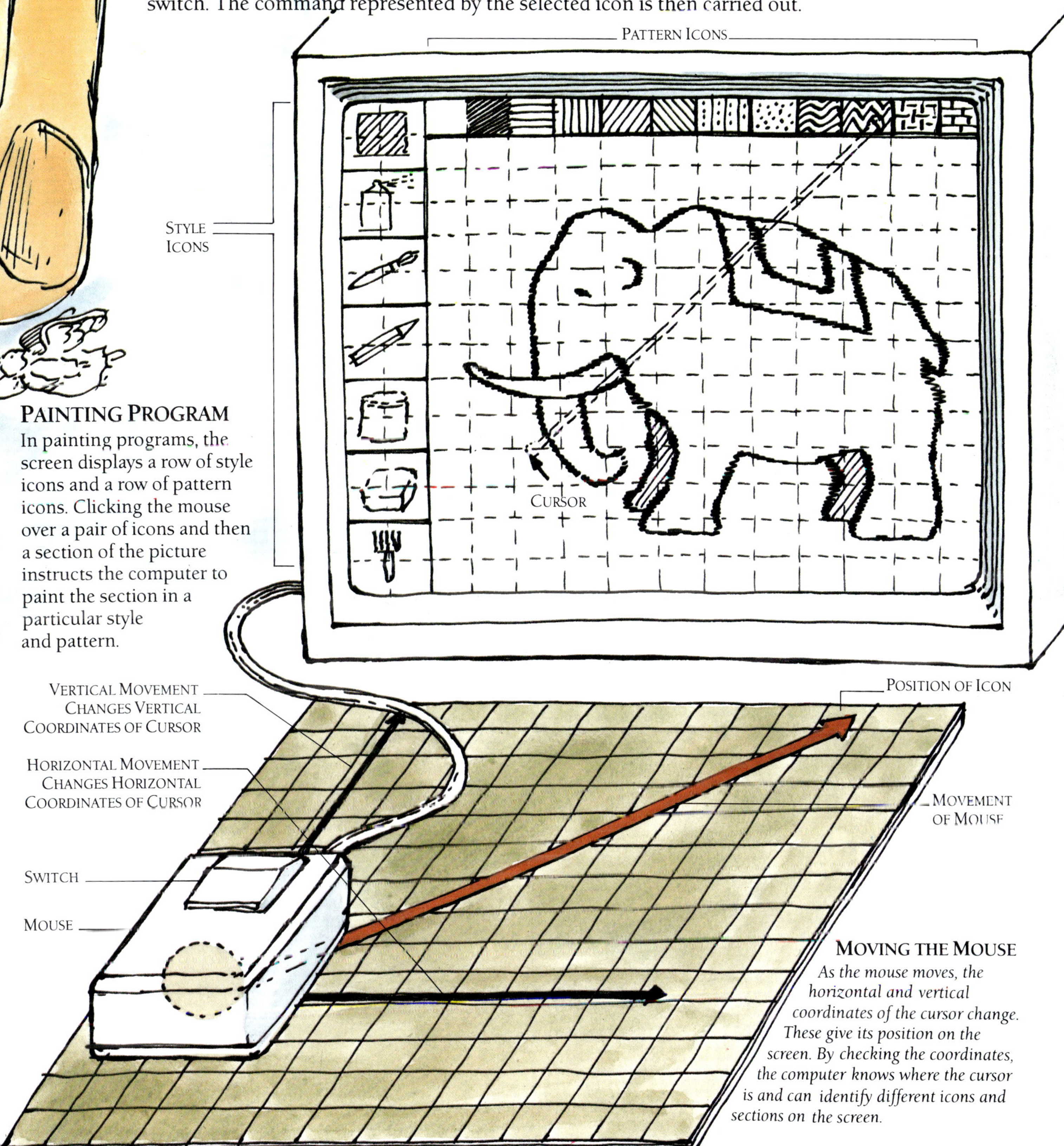

PAINTING PROGRAM

In painting programs, the screen displays a row of style icons and a row of pattern icons. Clicking the mouse over a pair of icons and then a section of the picture instructs the computer to paint the section in a particular style and pattern.

MOVING THE MOUSE

As the mouse moves, the horizontal and vertical coordinates of the cursor change. These give its position on the screen. By checking the coordinates, the computer knows where the cursor is and can identify different icons and sections on the screen.

Even though a computer is capable of storing masses of information in memory devices, one of its principal tasks is to print words and results on paper.

Several kinds of printers can be connected to a microcomputer. A dot-matrix printer (*above*) constructs patterns of closely spaced dots that merge to form letters, numbers and punctuation marks. A daisy-wheel printer contains a wheel of type-bars similar to those in a typewriter (see pp.30-1), and gives a high-quality result. Unlike daisy-wheel printers, dot-matrix printers can print *italic* and other kinds of characters simply by producing different pin patterns. Laser printers combine this sort of flexibility with extremely high-quality print.

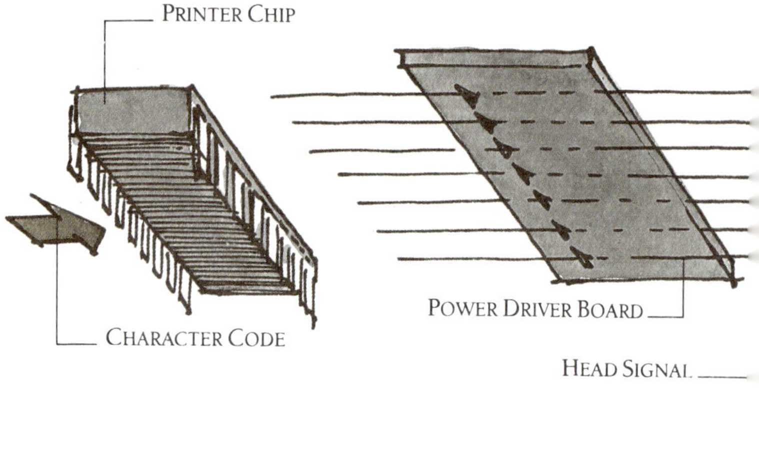

COMPUTER PRINTERS

DOT-MATRIX HEAD

The printer head of a dot-matrix printer contains a column of pins. The pins are driven by electromagnets responding to the head signals. These are binary signals that turn the electromagnets on or off.

PIN PATTERNS

The pins strike an inked ribbon that marks the paper with dots. The head signals fire the pins in different combinations so that each character is made up of several vertical dot patterns.

QUALITY PRINTING

Seven pins are shown for simplicity; good quality printers have 24 pins. To improve quality, the head may pass over the paper again and print dots that overlap with those printed on the first pass.

PRINTER CONTROL

All letters, numbers and other characters have standard codes that the computer sends to the printer. A chip in the printer (below) converts these codes into signals that drive the printer head as it moves across the paper and prints the characters. The power driver board amplifies the chip signals. Stepper motors (see p.303) move the head and paper to the right positions.

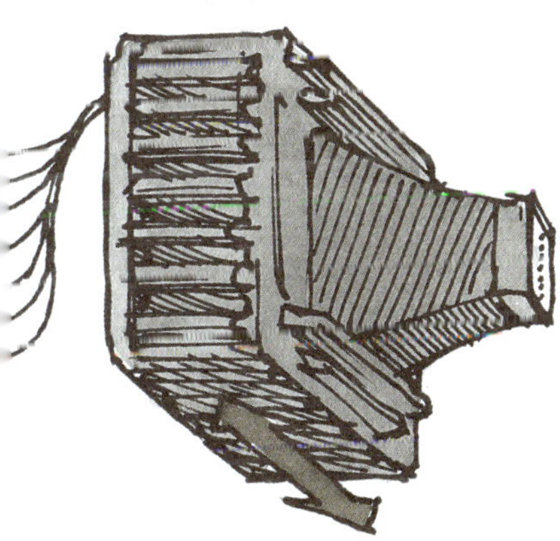

PRINTER HEAD

LASER PRINTER

Like a dot-matrix printer, a laser printer also builds characters up with dots, but the dots are so small that the printing is very detailed. A laser fires a beam of light at a spinning mirror. Another mirror and lenses then focus the moving beam onto a drum like that in a photocopier (see pp.280-1). Signals from the computer turn the beam on and off as it scans across the drum, building up an electrical image. The image is transferred onto the paper as in a photocopier.

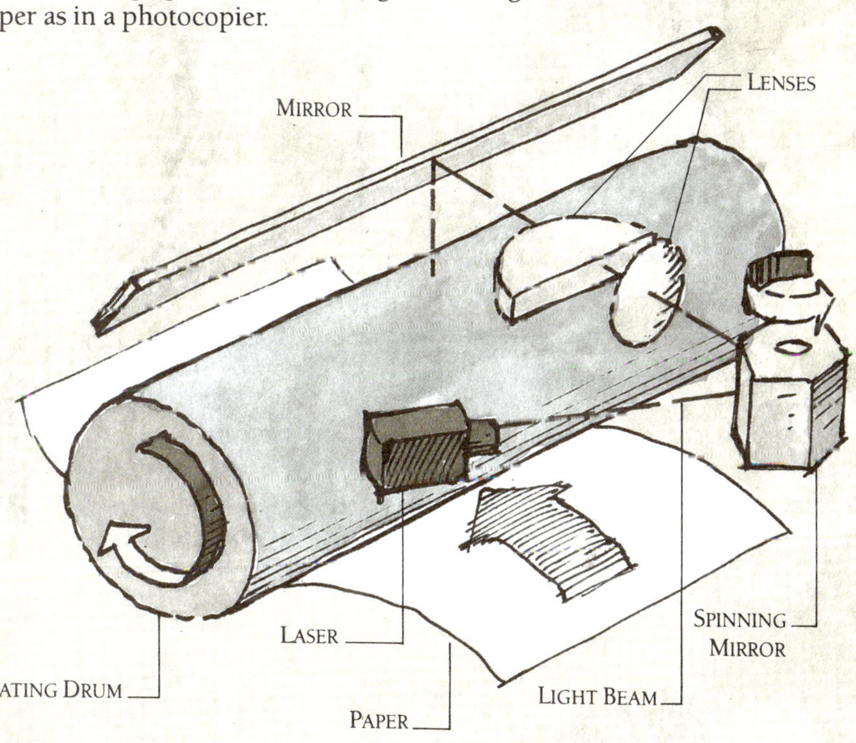

SUPERMARKET CHECKOUT

BAR CODES

A bar code is a set of binary numbers (see p.332). It consists of black bars and white spaces; a wide bar or space signifies 1 and a thin bar or space 0. The binary numbers stand for decimal numbers or letters.

There are several different kinds of bar codes. In each one, a number, letter or other character is formed by a certain number of bars and spaces. The bar code shown below uses five elements (three bars and two spaces) for numbers only.

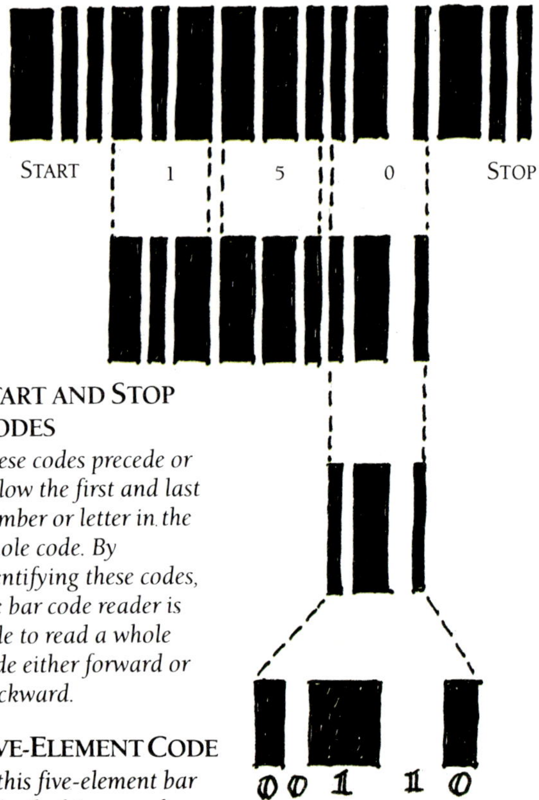

START AND STOP CODES

These codes precede or follow the first and last number or letter in the whole code. By identifying these codes, the bar code reader is able to read a whole code either forward or backward.

FIVE-ELEMENT CODE

In this five-element bar code, the binary code 00110 stands for the decimal number 0.

A supermarket checkout is a sophisticated input unit to the store's computer. Each product has a bar code that contains an identification number. The assistant moves the product over a window and an invisible infra-red laser beam scans the bar code, reading it at any angle. The number goes to the computer, which instantly shows the price of the product on the checkout display. The computer also adds up the check.

Checkout computers give the customer rapid accurate service, and they also benefit the store. The computer keeps a complete record of purchases and can work out stock levels. When the stock of a particular product falls, the computer can automatically order a new supply. The computer can also tell the store which products are selling well and which are less popular.

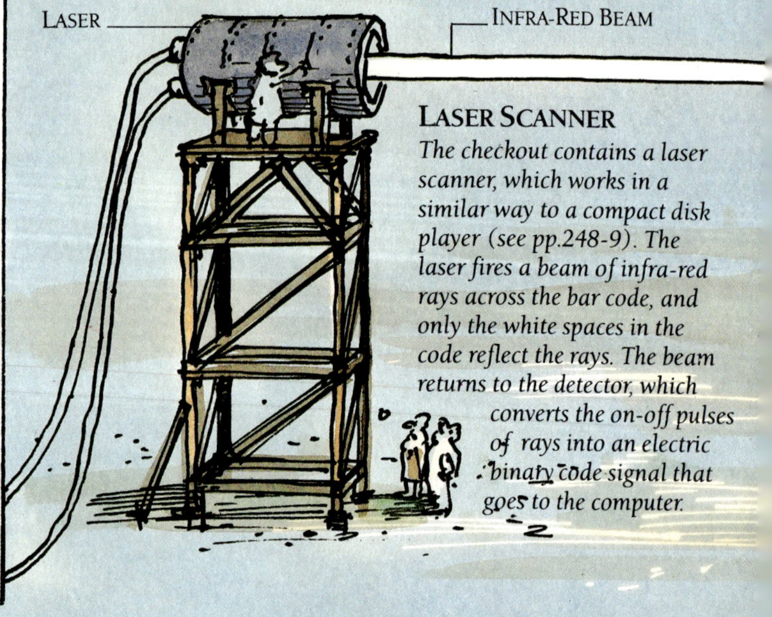

LASER SCANNER

The checkout contains a laser scanner, which works in a similar way to a compact disk player (see pp.248-9). The laser fires a beam of infra-red rays across the bar code, and only the white spaces in the code reflect the rays. The beam returns to the detector, which converts the on-off pulses of rays into an electric binary code signal that goes to the computer.

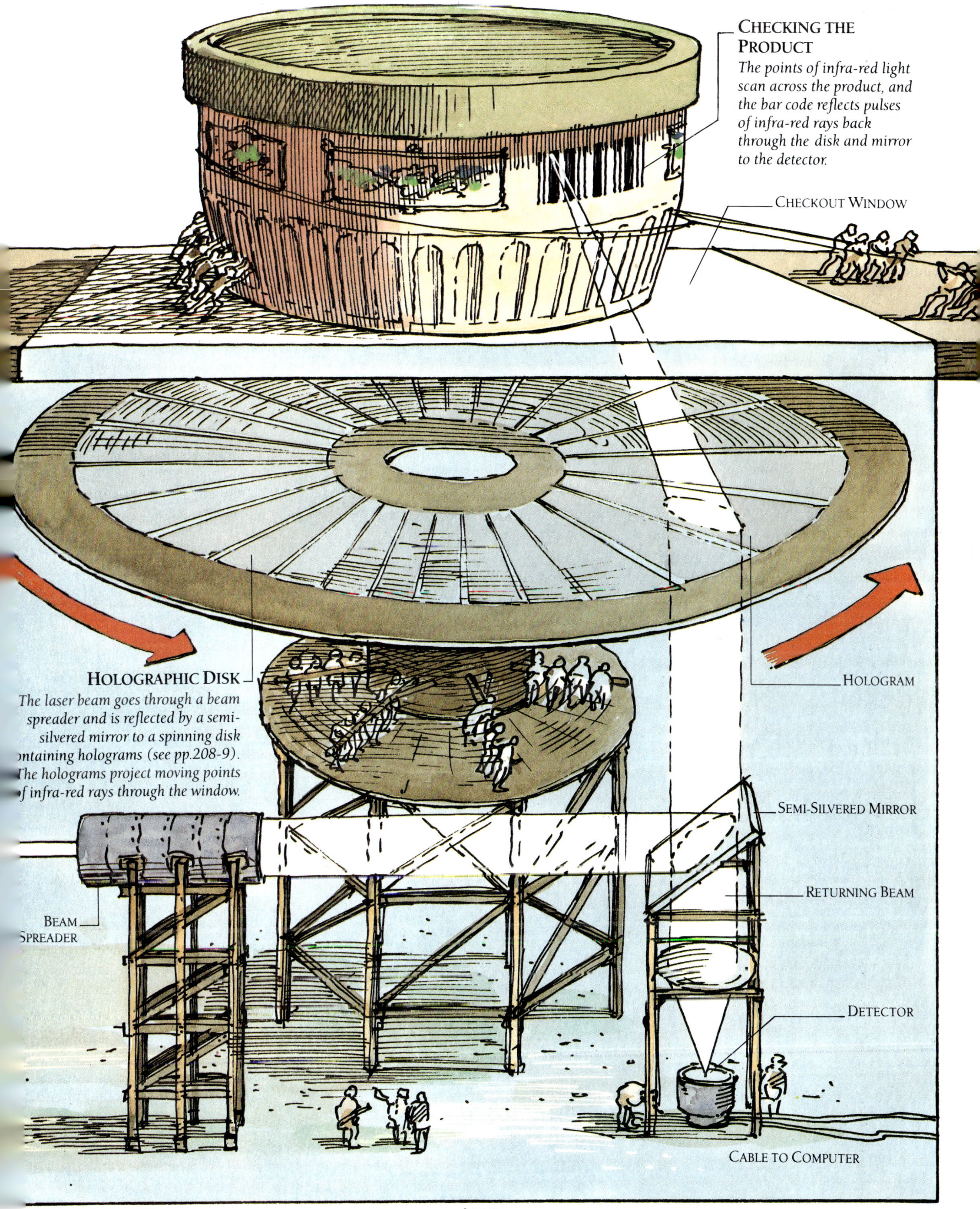

CHECKING THE PRODUCT
The points of infra-red light scan across the product, and the bar code reflects pulses of infra-red rays back through the disk and mirror to the detector.
CHECKOUT WINDOW
HOLOGRAPHIC DISK
The laser beam goes through a beam spreader and is reflected by a semi-silvered mirror to a spinning disk ntaining holograms (see pp.208-9). The holograms project moving points f infra-red rays through the window.
HOLOGRAM
SEMI-SILVERED MIRROR
RETURNING BEAM
BEAM SPREADER
DETECTOR
CABLE TO COMPUTER

SMART CARD

A smart card is an "intelligent" credit card. It contains a microchip that turns the card into a miniature computer. This card does far more than authorize the payment of cash from a cash dispenser. It stores records of all purchases made with the card and can be used to replace a check book. The card's memory can also store the holder's signature, picture and medical history, which can be instantly viewed on inserting the card into a card-reading machine.

BUILT-IN MEMORY

The microchip in a smart card contains a PROM (Programmable Read-Only Memory). This can be given data which is stored and which cannot then be changed. The chip's microprocessor controls access to the memory, which has three zones. The open zone has information, such as holder's medical history, that can be accessed by anyone. The working zone, which contains records of purchases, can only be accessed by the holder using a password. The secret zone cannot be accessed and contains the password. This is much more secure than the PIN (Personal Identification Number) recorded in the magnetic stripe on a cash card.

A PROM contains MOS transistors (see p.343) in which the gate can be given a permanent charge that switches the transistor on or off.

MICROCHIP
The microchip in a smart card contains a microprocessor to control its operation and a PROM to store information. Contacts in the card link the microprocessor to machines that transfer data to the card and display some of the information it contains.
CONNECTORS

FLIGHT SIMULATOR

One of the valuable – and potentially life-saving – uses of the computer is to simulate aircraft flight. A pilot undergoing training or retraining can practise normal flying and emergency procedures without leaving the ground. The flight simulator is not only cheaper than using a real aircraft but totally safe. Inside the simulator, the crew sits in a mock-up of the flight deck of a particular aircraft. Through the windows, they see a real airport and other computer-generated moving pictures of scenes that occur during take-off, flight and landing. Jacks beneath the simulator tilt the flight deck to simulate motion.

A powerful computer is linked to the controls on the flight deck. As the pilot operates the controls, the computer moves the picture, varies the instrument displays, sounds warnings and tilts the flight deck exactly as if the aircraft were flying through the air.

PROJECTORS
High-quality projectors throw adjacent sections of a wide color television picture generated by the computer onto a curved screen that extends around the flight deck. Half of the screen can be seen here with two of the three projectors working.

INSTRUCTOR STATION
Behind the pilot, the instructor sits at the computer console. This displays information about the performance of the simulator so that the instructor can assess the pilot. The instructor can command the computer to simulate special conditions. These include switching to night landings or foggy weather, for example, or conditions that require an emergency landing. The computer can also record a "flight" and replay it so that the pilot and instructor can go back over the training exercise.

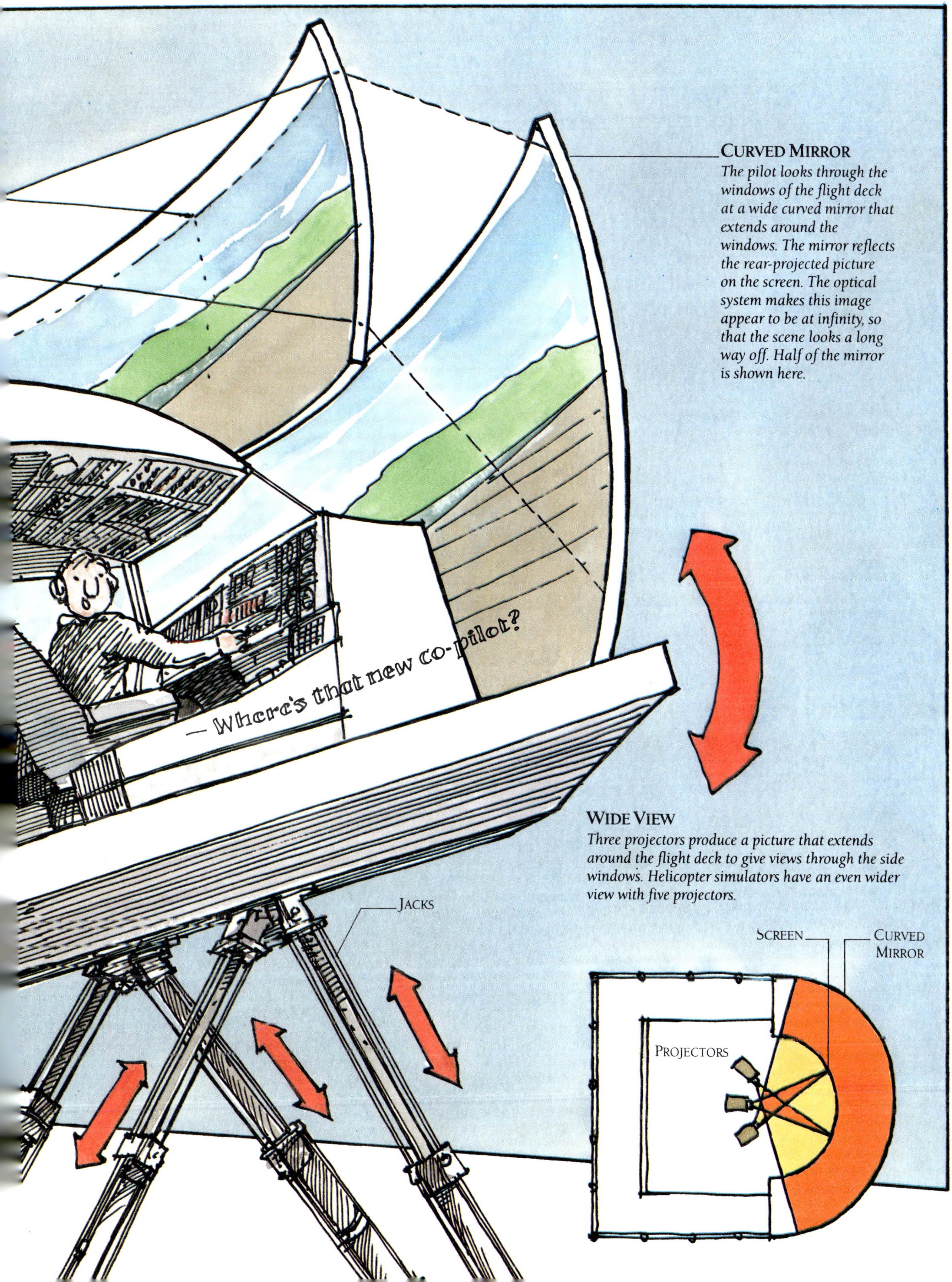

CURVED MIRROR
The pilot looks through the windows of the flight deck at a wide curved mirror that extends around the windows. The mirror reflects the rear-projected picture on the screen. The optical system makes this image appear to be at infinity, so that the scene looks a long way off. Half of the mirror is shown here.

WIDE VIEW
Three projectors produce a picture that extends around the flight deck to give views through the side windows. Helicopter simulators have an even wider view with five projectors.

ROBOTS

A robot is the ultimate of all computer-controlled machines. Its movements can be directed with immense precision, enabling it to repeat actions exactly and to relieve humans of monotonous or hazardous tasks. One of the advantages of computer control is that a robot can be "taught" to do a new job. This is often done by carrying out the new movements, which the computer then remembers and instructs the robot to copy. Computer control will eventually give robots full mobility, vision, hearing and speech.

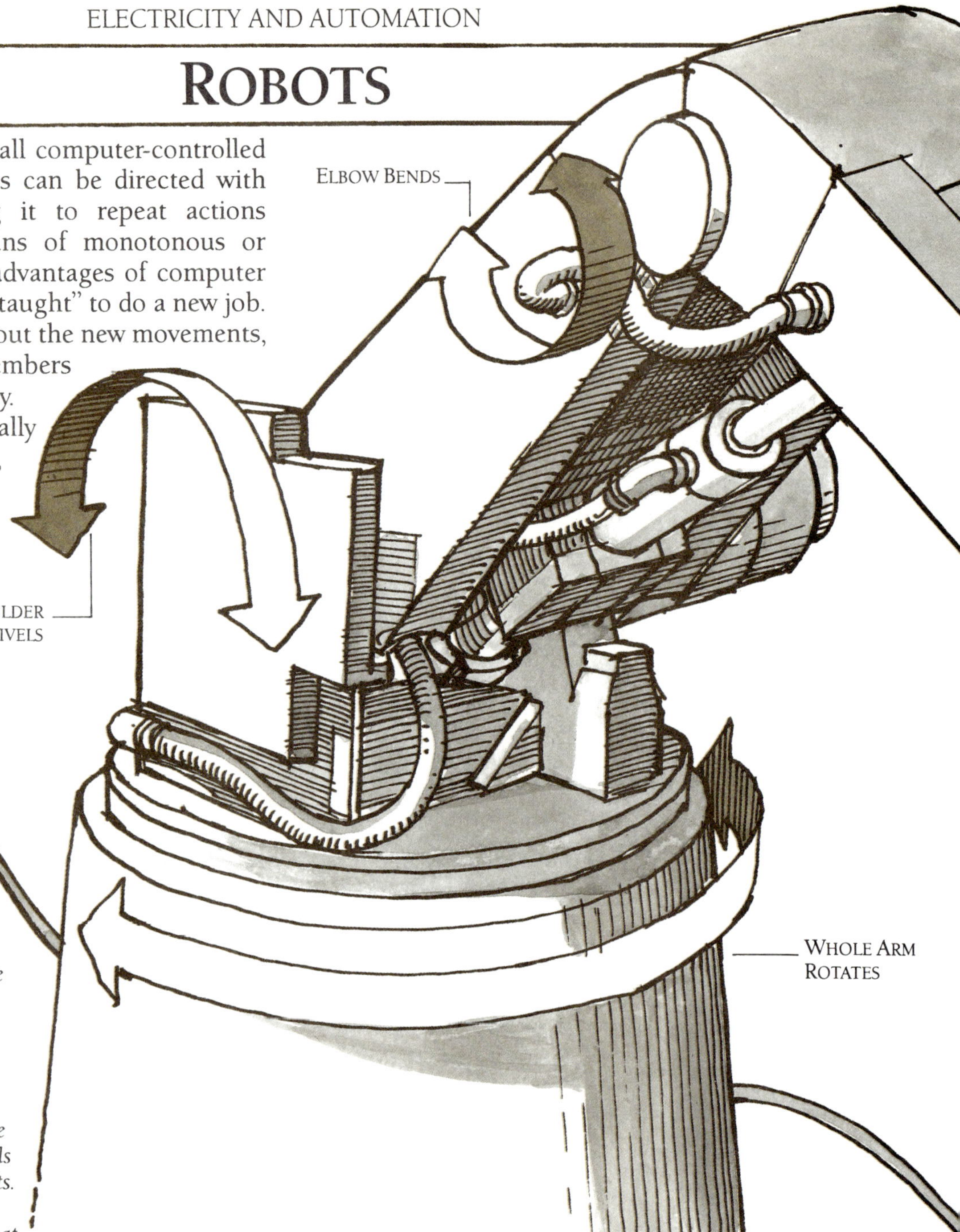

COMPUTER UNIT

A powerful computer controls the actions of the robot. Its program calculates the various angles through which each of the joints must turn. Given a position to reach, it works out the shortest way to get the robot's hand to the position and sends control signals via the power unit to the six joints. Sensors feed the angles of the joints back to the computer so that it knows when each has turned the required amount.

The computer can make the robot repeat an action over and over again by running through a sequence of angles stored in its memory.

DIRECT TEACHING

A robot may be taught an action such as paint spraying by guiding its hand through the movements. The computer stores the various angles turned in its memory.

TEACHING UNIT

A person may use a teaching unit to feed the required movements to the robot. The robot may move as it is being taught the action.

Jointed-Arm Robot

This kind of robot can perform complex actions because the various sections of the arm turn around six joints and move in the same way as the parts of a human arm. The robot's hand can move in all possible directions to any position within reach. Each of the robot's joints is driven by an electro-hydraulic system that gives the robot great strength.

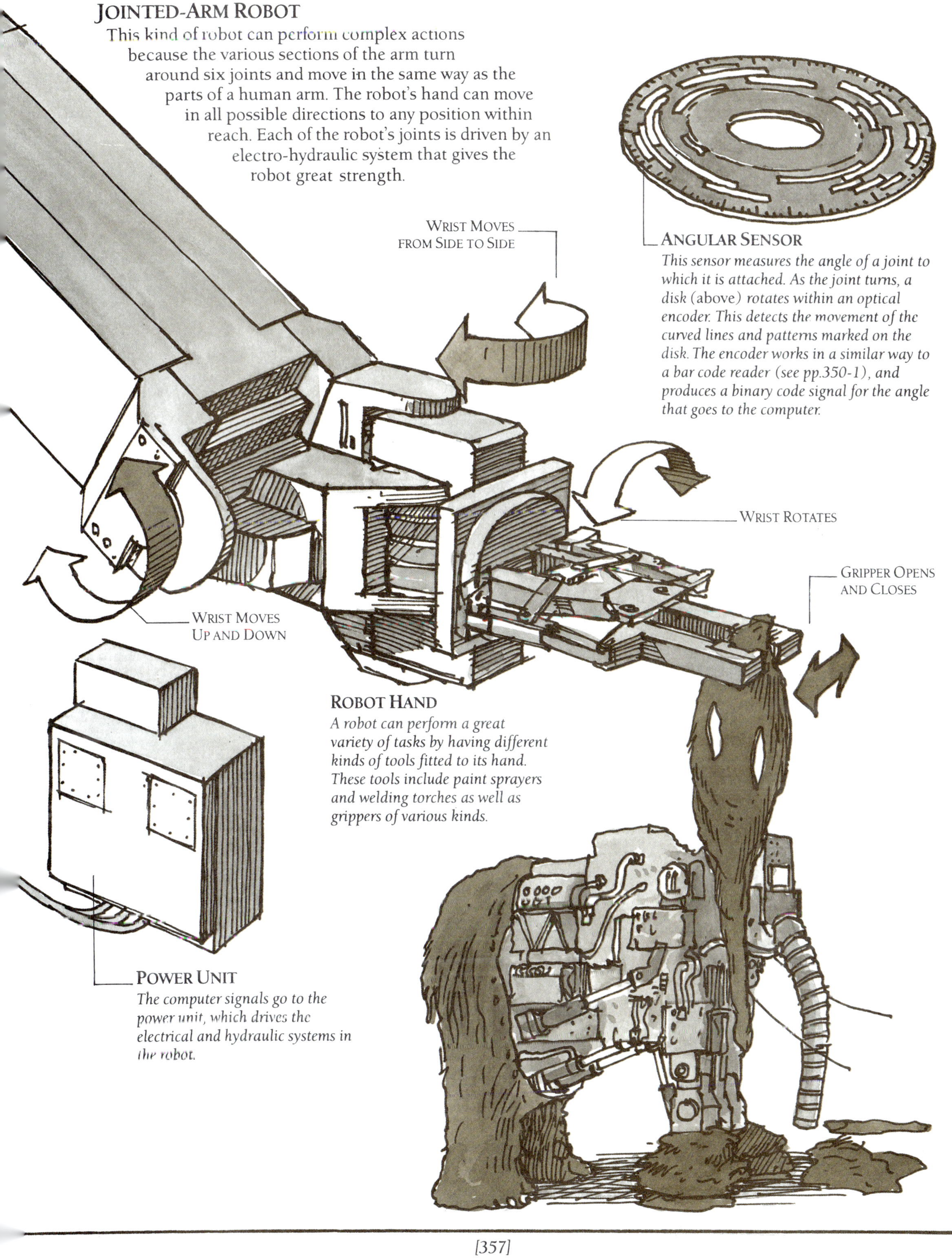

Angular Sensor

This sensor measures the angle of a joint to which it is attached. As the joint turns, a disk (above) rotates within an optical encoder. This detects the movement of the curved lines and patterns marked on the disk. The encoder works in a similar way to a bar code reader (see pp.350-1), and produces a binary code signal for the angle that goes to the computer.

Robot Hand

A robot can perform a great variety of tasks by having different kinds of tools fitted to its hand. These tools include paint sprayers and welding torches as well as grippers of various kinds.

Power Unit

The computer signals go to the power unit, which drives the electrical and hydraulic systems in the robot.

EUREKA!

THE INVENTION OF MACHINES

THE INCLINED PLANE

People have to eat to live and, necessity being ever the mother of invention, the very first machines to be invented were the tools used by prehistoric people in hunting and gathering their food. Stones crudely chipped to form tools date back about a million years, and stone axes and spearheads litter archeological sites down to the dawn of civilization.

In cutting tools, the inclined plane became the first principle of technology to be put to work. On a larger scale, it may have enabled people to build at least one of the Seven Wonders of the World – the Great Pyramid. This was constructed in Egypt in 2600BC using high earth ramps to raise great stone blocks into position.

THE PLOW

The plow was invented in the Middle East in about 3500BC. At first, it was little more than a digging stick drawn by a person or an ox, but this primitive plow enabled people to dig deeper than before. Plants could put down stronger roots in plowed soil, increasing crop yields and enabling farmers to produce a surplus of food. The plow thus freed some people from the necessity of growing their food.

LOCKS

Locks existed in ancient Egypt, and they made use of pins in the same way as the cylinder lock. The application of the inclined plane to the key, made by Linus Yale in the United States in 1848, is one of those fundamental inventions that long outlive their maker, and the cylinder lock is still often called a Yale lock. The lever lock dates from 1778 and was invented by a British engineer Robert Barron. The design resulted from a need to prevent burglars taking wax impressions inside locks and then making keys from them, and it too proved to be a fundamental advance.

THE ZIPPER

The zipper took quite a time to make its mark. It was invented by Whitcomb Judson in 1891, not in its present form as a clothes fastener, but as a device to do up boots. It did not take off until 1918, when the US Navy realized that Judson's invention would make an ideal fastener for flying suits. The name zipper, coined in 1926, clinched its success.

THE CAN OPENER

Methods of preserving food in sealed containers were invented in the early 1800s, at first using glass jars and then tin cans. The cans were ideal for transporting food, but opening them could be a problem. At first, a hammer and chisel – a crude use of the inclined plane – had to serve. Claw-like devices and levered blades were then devised to open cans, not without some danger to the user. The safe and simple can openers that we have today were not invented until the 1930s, more than a century after the appearance of the tin can

LEVERS

Levers also originated in ancient times in devices such as hoes, oars and slings. People realized intuitively that levers could aid their muscle power, but it took a genius to explain how levers work. The genius was the Ancient Greek scientist Archimedes (287-212BC), who first defined the principle of levers. He illustrated it with the famous adage "Give me a place to stand and I will move the Earth" – meaning that if he had a lever sufficiently long, he could shift the Earth by his own efforts.

The formulation of the principle of levers was a landmark in the development of science and technology. Archimedes' insight explained not only levers, because the same principle lies behind the inclined plane, gears and belts, pulleys and screws. Furthermore, Archimedes showed that by making observations and experiments, it was possible to deduce the basic principles that explain why things work.

WEIGHING MACHINES

The first device to make precise use of levers was invented long before Archimedes' time. This was the balance or scales used for weighing, which dates back from 3500BC. It may seem odd that a precision instrument was required so long ago: however, what had to be weighed was no ordinary material – it was gold. Gold dust was used as currency in the ancient civilizations of the Middle East, and amounts of it had to be weighed very precisely in order to assess their value.

Keyboard Machines

The piano was invented in Italy in 1709 by Bartolommeo Cristofori, who sought a way of varying the volume of a keyboard by using levers to strike the strings with different amounts of force. His success is reflected in the instrument's name: the pianoforte or "soft-loud". The lever system was later improved to increase the response of the piano, resulting in the highly expressive instrument that affected the whole course of music.

The first practical typewriter was invented in the United States in 1867 by Christopher Scholes, and it was taken up by the Remington company. Unlike the piano, the typewriter gives little variety in performance and is consequently well on its way to obsolescence, ousted by computer printers capable of many kinds of typefaces and type sizes.

The Parking Meter

The mechanical parking meter and other kinds of coin-in-the-slot machines involving lever action have their ancestor in a fascinating device invented by the Greek scientist Hero, who lived in Alexandria in the first century AD. Hero is justly renowned for inventing the first engine (see p.364) but he also built many ingenious devices that employed levers and other mechanical parts. Among them was a machine that delivered a cupful of holy water on inserting a coin. The falling coin tripped a lever, which raised a valve that allowed the holy water to flow.

The Wheel and Axle

The development of mechanical power has its origins in the wheel and axle. The first machines to make use of this device may well have been the windlass and the winch. The Greek physician Hippocrates, who was born in 460BC, employed a windlass to stretch the limbs of his patients, a treatment uncomfortably like the rack of medieval torture chambers. Winches have been used to draw water from wells for many centuries.

The Waterwheel and Windmill

The waterwheel dates back to the first century BC. The windmill, basically the same machine but driven by air instead of water, followed some seven centuries later.

Turbines

Modern turbines are a product of the Industrial Revolution, when the demand for power soared as factories developed. Engineers investigated blade design, seeking to maximize energy output. The Francis turbine, invented by James Francis in 1850 and now common in power-stations, was literally a product of lateral thinking because Francis made the water flow inward instead of outward.

Gears and Belts

Belts are simple devices, seen in the chains of buckets that lifted water in ancient times. The basic forms of gears were known by the first century AD. An extraordinary early application of gears is the Antikythera mechanism, a mechanical calendar made in Greece in about 100BC and recovered from a wreck sunk off the Greek island of Antikythera. This machine had 25 bronze gear wheels forming a complex train of gears that could move pointers to indicate the future positions of the Sun and Moon as well as the times when certain stars would rise or set.

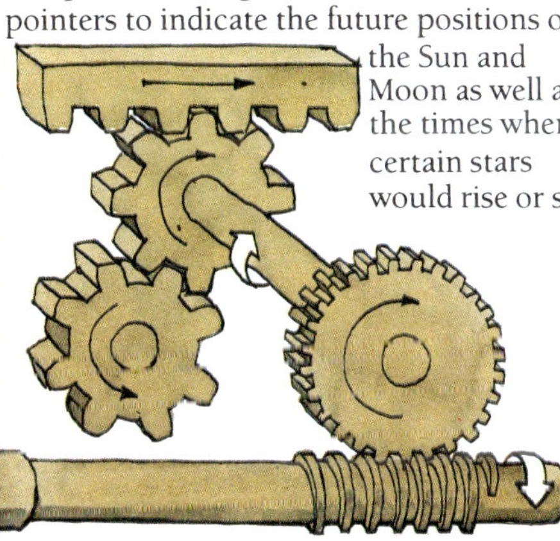

Clocks

A rack-and-pinion gear was used in a water clock built by the Greek inventor Ctesibius in about 250BC. The water clock was an ancient device in which water dropped at a constant rate into a container, the level of the water indicating the time. Ctesibius improved it by having a float raise a rack that turned a pinion connected to a pointer on a drum. The pointer turned to indicate the time in the same way as the hour hand of a mechanical clock.

The oldest surviving mechanical clocks date from the late 1300s. Gears transmitted the constant movement of a regulator to the hands or to a bell. A good regulator appeared only with the discovery of the pendulum in 1581 by the great Italian scientist Galileo, who timed a swinging chandelier with his pulse and realized that the time taken for each swing was always constant. Even so, it took nearly a century for the first pendulum clocks to appear.

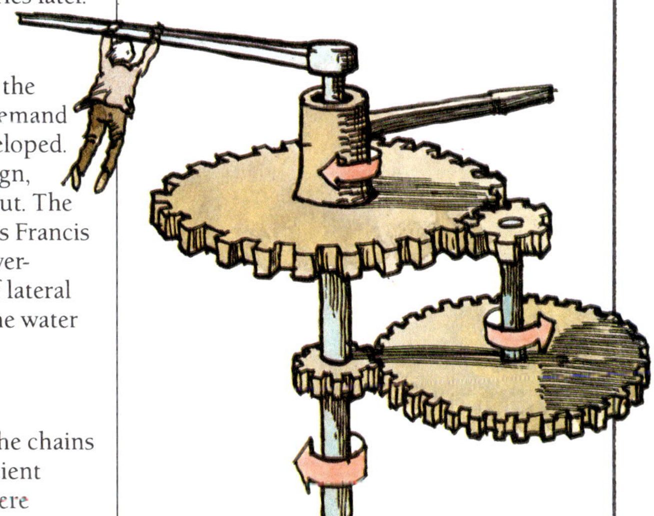

The Epicyclic Gear

The sun-and-planet or epicyclic gear is of much more recent origin than other types. It was invented in 1781 by the great British engineer James Watt, who is best known for improving the steam engine. Watt needed a device to turn the reciprocating motion of the piston of his steam engine into rotary motion, but he could not use the crank because someone else had patent protection on it. Watt's alternative was the epicyclic gear, now found in salad spinners, automatic transmission and many other devices.

The Differential

This first appeared in the "south-pointing carriage" invented in China in the third century AD. The two-wheeled carriage was surmounted by a figure that always pointed south, no matter how the carriage turned as it moved. The figure was set to south, and then a differential driven by the wheels turned the figure in the opposite direction to the carriage so that it still pointed south.

Such a machine must have appeared magical to the people of the time. However, calculations show that the mechanism could not have been sufficiently precise for the figure to point south for long. Within 3 miles (5km) it could well have been pointing north instead!

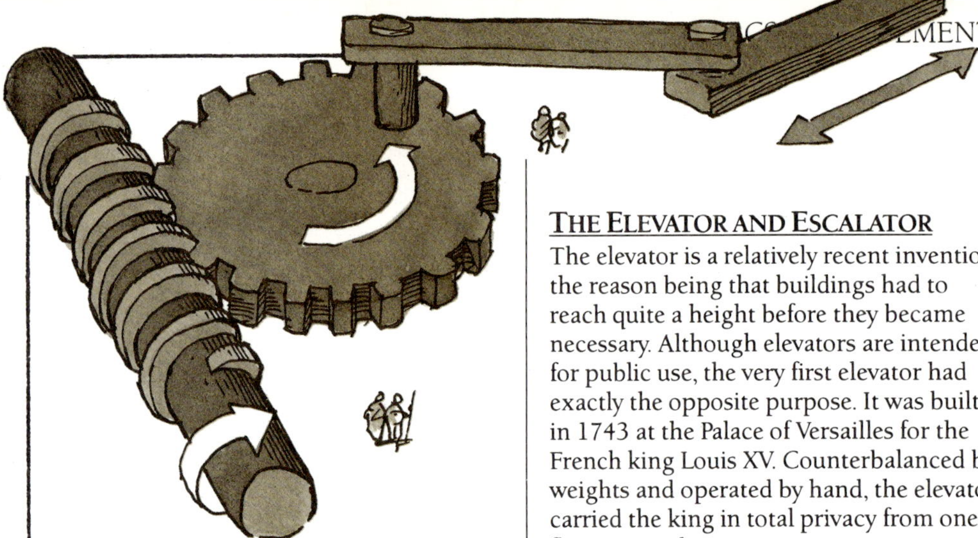

CAMS AND CRANKS

Cams and cranks are old devices too – the cam appearing in the drop hammer and the crank in a winding handle. Their application in the sewing machine was developed during the early 1800s, the first successful sewing machine being produced in the United States by Isaac Singer in 1851. The four-stroke internal combustion engine, which similarly depends on the controlling movement of cams and cranks, was first put to use in the motor car by Karl Benz in 1885. These two machines are still with us in their basic form today, along with their inventors' names.

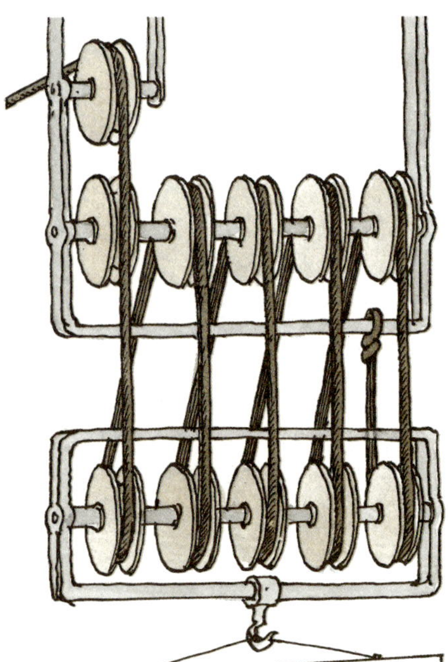

PULLEYS

Simple cranes using single pulley wheels were invented some 3,000 years ago, and compound pulleys with several wheels date back to about the 400s BC. Archimedes is said to have invented a compound pulley that was able to haul a ship ashore. The shadoof, a counter-weighted lifting machine, is also of ancient origin.

THE ELEVATOR AND ESCALATOR

The elevator is a relatively recent invention, the reason being that buildings had to reach quite a height before they became necessary. Although elevators are intended for public use, the very first elevator had exactly the opposite purpose. It was built in 1743 at the Palace of Versailles for the French king Louis XV. Counterbalanced by weights and operated by hand, the elevator carried the king in total privacy from one floor to another.

The modern safety elevator is the invention of the American engineer Elisha Otis, who dramatically demonstrated its effectiveness in 1854. He ordered the rope of the elevator carrying himself to be cut. The emergency braking system was automatically activated, and the elevator did not fall.

Escalators date from the 1890s. The first models were basically moving belts.

SCREWS

The screw is yet another machine associated with Archimedes, for the earliest known is the water-lifting auger know as Archimedes' screw. However, it may well have been invented before his time. The screw press, which contains the form of screw used in nuts and bolts, was first described by Hero of Alexandria.

Metal screws were used as a superior alternative to nails in 1556, when the German mining engineer Agricola described how to screw leather to wood to make durable bellows. The screwdriver however did not follow until 1780.

THE COMBINE HARVESTER

The combine harvester is the most important invention in farming since the plow, and a modern harvester makes use of several augers that work in exactly the same was as Archimedes' screw. The first combine harvester was built in 1835 by combining a horse-drawn reaper and a threshing machine. It took a century to develop the harvester into an effective self-powered machine.

THE MICROMETER

This important device based on the screw was invented in 1772 by James Watt. Watt's micrometer worked in much the same way as the modern micrometer, and was accurate to one-thousandth of an inch (a fiftieth of a millimeter).

ROTATING WHEELS

Ancient peoples could easily move heavy loads by rolling them on logs, and one would expect that the wheel developed in this way. But this is not the case. Unlike a roller, a wheel requires an axle on which to turn and so the potter's wheel was the first true wheel. It was invented in the Middle East in about 3500BC. From the potter's wheel, the wheel was soon developed for transport.

THE BICYCLE

The first bicycles were pushed along by the feet and not pedaled. They were novelties rather than a serious means of transport, and were known as hobby-horses. Kirkpatrick Macmillan, a blacksmith, invented the pedal-operated bicycle in Britain in 1839. Raising the feet from the ground to turn the pedals required the rider to make use of precession to balance.

GYROCOMPASS

The inherent stability or gyroscopic inertia of devices such as spinning tops has been known for centuries, but the development of the gyroscope in machines is more recent. Its most important application, the gyrocompass, was invented by Elmer Sperry and first demonstrated on an American ship in 1911.

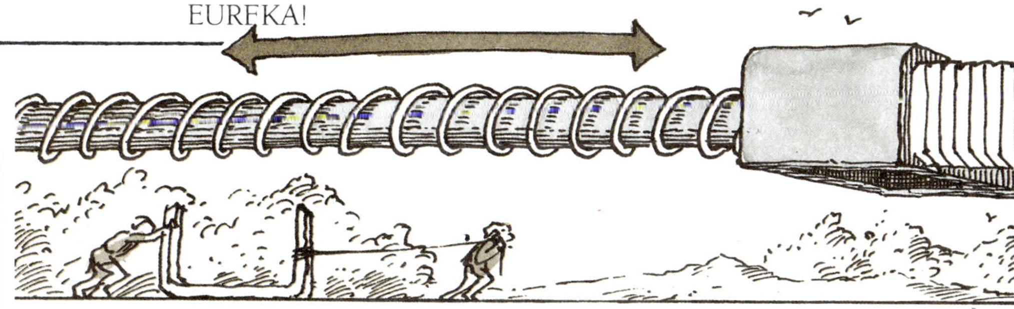

SPRINGS

Springs are also of ancient origin, being used in primitive locks. Metal springs date from the 1500s, when leaf springs were invented to provide a primitive suspension for road carriages. Springs did not become common until two centuries later, when coil springs were invented.

MEASURING MACHINES

The principle behind the spring – that the extension of a spring is proportional to the force acting on it – was discovered by the English scientist Robert Hooke in 1678 and it is consequently known as Hooke's Law. Hooke also invented the spiral spring known as a hairspring, which is used as a regulator in mechanical watches and which made portable timepieces possible.

FRICTION

People have been making use of friction ever since they first set foot on the ground, and the first friction devices to pound grain into flour date back to the beginnings of civilization.

THE PARACHUTE

This was one of several inventions that were forecast by Leonardo da Vinci, who drew one in 1485. Understandably, neither Leonardo nor anyone else was very keen to try out the idea in practice. However, there was little need for parachutes until the first balloons took to the air three centuries later. The first parachute descent proper took place in 1797, when the French balloonist Andre Garnerin successfully dropped 2,230 feet (680 meters). Early parachutes were fashioned like huge parasols and similarly named, being proof against a *chute* or fall rather than the Sun.

DRILLING MACHINES

Drilling, which is basically pounding or grinding, is a surprisingly old activity. The Chinese drilled oil wells some hundreds of yards or meters deep as early as the third century BC. They dropped a metal drilling tool into the hole to break up the rock. The first modern oil well, drilled by Edwin Drake in Pennsylvania in 1859, was drilled in the same way.

BEARINGS

Devices to reduce friction are of ancient origin, the first being log rollers which were placed under an object that was to be moved. To work effectively a wheel needs bearings on its axle. These were invented in France and Germany in about 1000BC. The bearings were made of wood and then greased to improve speed and lengthen their life. Modern bearings date back to the late 1700s. They made the development of machines during the Industrial Revolution all the more effective.

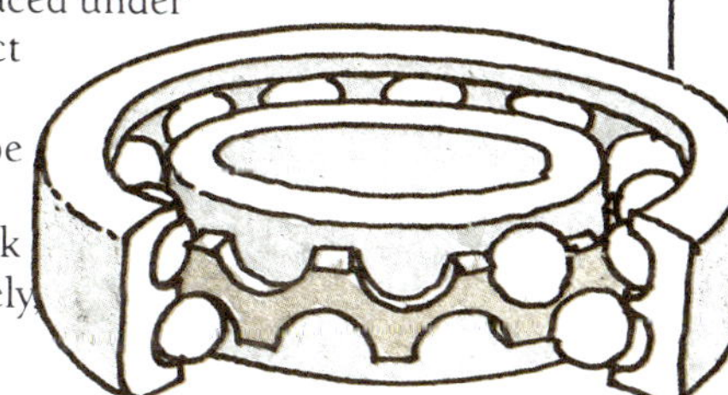

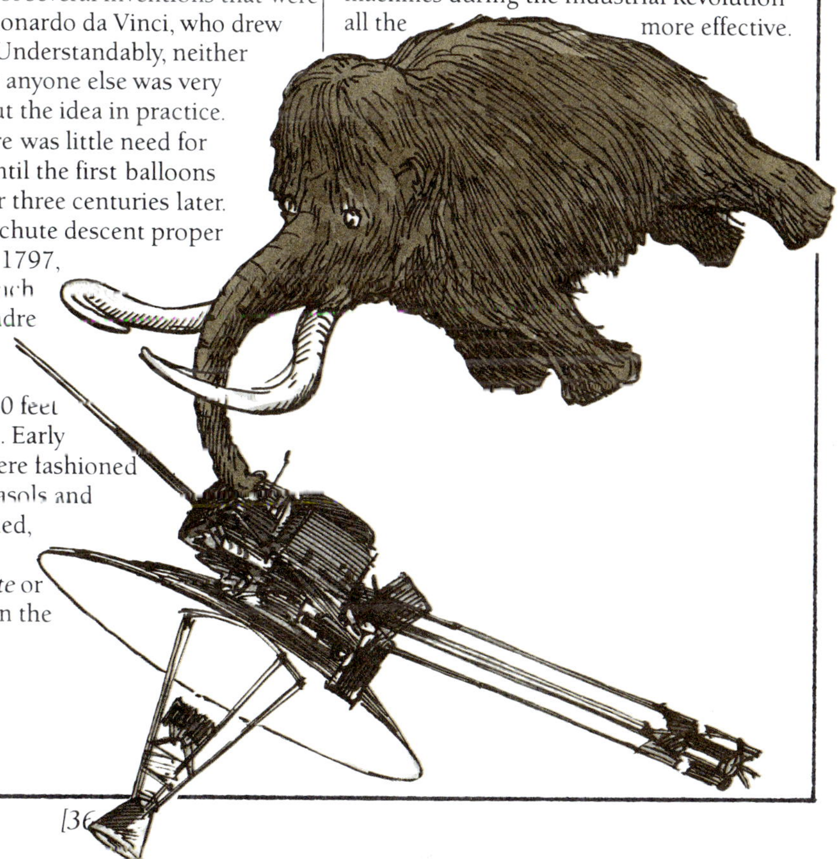

FLOATING

The first form of transport to progress under its own power was the raft. In prehistoric times, people must have hitched rides on uprooted trees that happened to be floating down rivers. Rafts borne on ocean currents probably carried people across the world's oceans long before recorded history.

The earliest known hollow boats date back to about 8000BC. These were canoes dug out of tree trunks, which were paddled through the water.

The principle of flotation, which explains how things float, was one of the many achievements of Archimedes, the great scientist who lived in Sicily (then a Greek colony) in the 200s BC. He is reputed to have made this discovery in his bath and then ran naked into the street shouting the classic inventor's cry of *Eureka*, which means "I have found it".

Although the principle that Archimedes put forward explained that an iron boat could float, nobody really believed this and all boats and ships were made of wood until just over two centuries ago. The development of the iron ship coincided with the development of a powerful steam engine, which drove paddle wheels in boats.

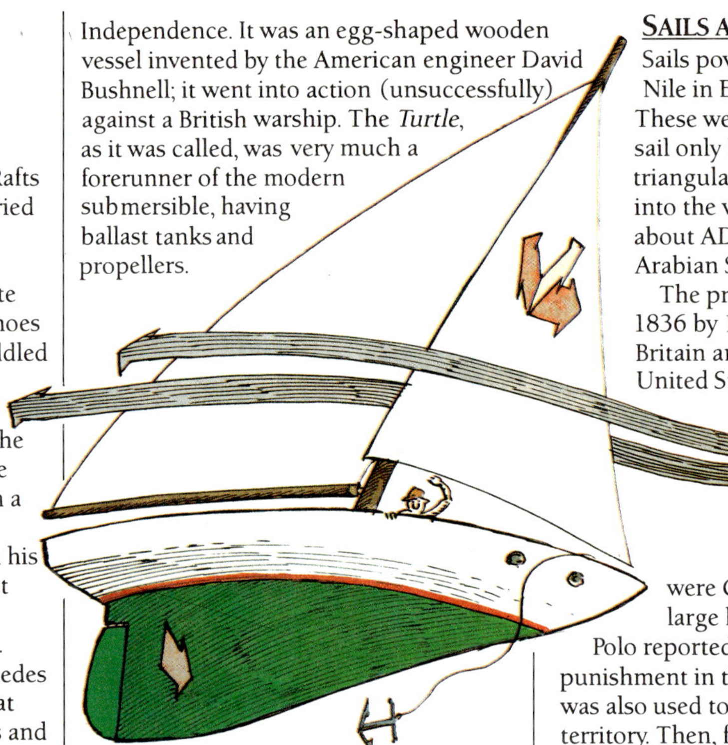

SUBMARINES

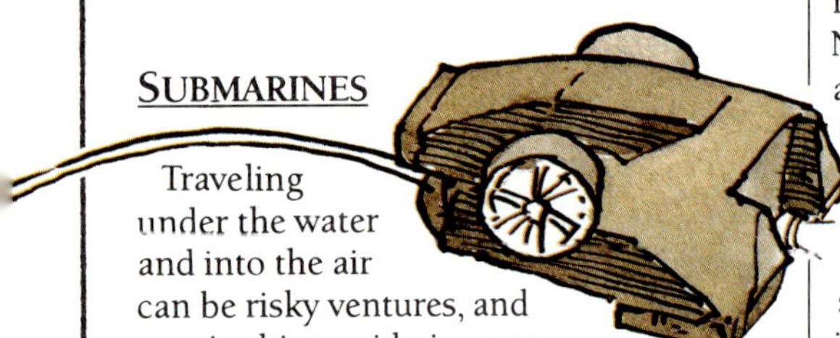

Traveling under the water and into the air can be risky ventures, and required intrepid pioneers. Understandably perhaps, the inventors of both the first submarine and balloon persuaded other people to try out their craft.

The first proper submarine took to the water in 1776 during the American War of Independence. It was an egg-shaped wooden vessel invented by the American engineer David Bushnell; it went into action (unsuccessfully) against a British warship. The *Turtle*, as it was called, was very much a forerunner of the modern submersible, having ballast tanks and propellers.

BALLOONS

The first balloon to carry passengers was a hot-air balloon invented by the Montgolfier brothers in France in 1783. It made its first flight in November of that year and flew 5 miles (8km). A few days later the first gas-filled balloon took to the Parisian skies, piloted by its inventor Jacques Charles. It contained hydrogen, which also lifted the first airship into the air in 1852. This machine, which was steam-powered, was invented by the French engineer Henri Giffard.

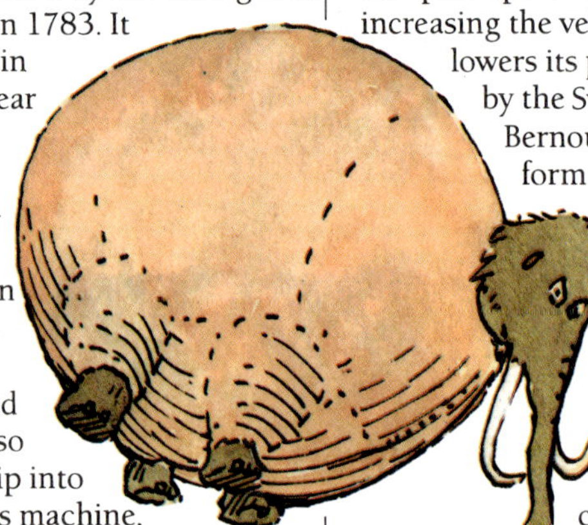

SAILS AND PROPELLERS

Sails powered boats along the River Nile in Egypt as long ago as 4000BC. These were square sails, which could sail only before the wind. The triangular sail, which is able to sail into the wind, first appeared in about AD 300 in boats on the Arabian Sea.

The propeller was invented in 1836 by Francis Pettit Smith in Britain and John Ericsson in the United States. It first powered a sea-going ship, appropriately called the *Archimedes*, in 1839.

FLYING

The first people to fly were Chinese criminals lifted by large kites. The explorer Marco Polo reported the use of such kites for punishment in the 1200s, but kite flying was also used to look out over enemy territory. Then, five centuries later, balloons began to carry people aloft.

THE AIRFOIL

The principle behind the airfoil – that increasing the velocity of a gas or liquid lowers its pressure – was discovered by the Swiss scientist Daniel Bernoulli in 1738, and the basic form of the winged aircraft was developed during the 1800s. Its design was due to the British engineer Sir George Cayley, who flew the first glider in 1849. This machine carried a child. Four years later, Cayley's coachman (against his will) became the first adult to fly a winged aircraft. On landing, he immediately resigned!

POWERED FLIGHT

The invention of powered flight is indelibly associated with the Wright brothers, the American engineers who flew the first powered airplane at Kitty Hawk in North Carolina in 1903. Unlike all modern aircraft, the wings of the Wrights' flying machine did not have ailerons. This development occurred in 1908 in aircraft built by the British engineer Henry Farman.

The Helicopter

Like the Wrights' powered airplane, the development of the helicopter was contingent upon the invention of a light but powerful engine – the gasoline engine. The very first helicopter, built by Paul Cornu, whirled unsteadily into the air in France in 1907. The development of a reliable helicopter took about thirty years.

The Hydrofoil

The first use of the principle of the airfoil was not in air but in water. In Britain in 1861, Thomas Moy tested wings by fixing them beneath a boat and found that the wings raised the hull above the water. Thus the hydrofoil was born before the airplane. The production of a practical hydrofoil took place in Italy, where it was developed by Enrico Forlanini during the first decade of this century.

Pressure Power

The achievements of Archimedes inspired generations of inventors and engineers. First was Ctesibius, who lived at Alexandria in Egypt at about the same time. Ctesibius was renowned for his self-powered devices, notably the first organ. Water was a convenient source of power and in this instrument, he used the pressure of water to drive air into the pipes: the resulting music was ear-splitting.

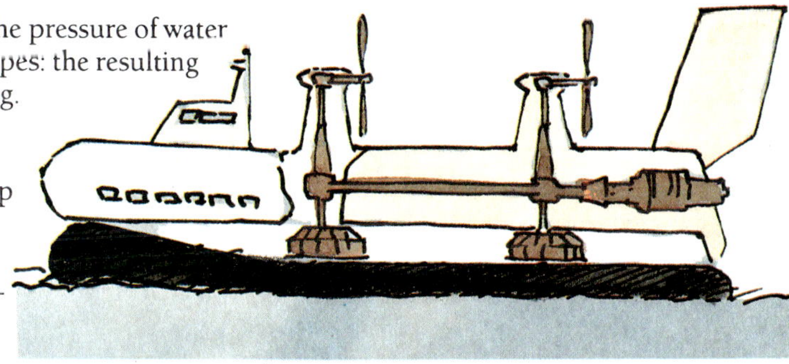

Pumps and Jets

The water-lifting pump is another invention credited to Ctesibius. However, the slow development of pumps able to produce a continuous jet of water enabled the Great Fire to destroy much of London in 1666. The first real fire engine did not appear until 1721. The pump was invented by the British engineer Richard Newsham. It was a reciprocating pump with two pistons

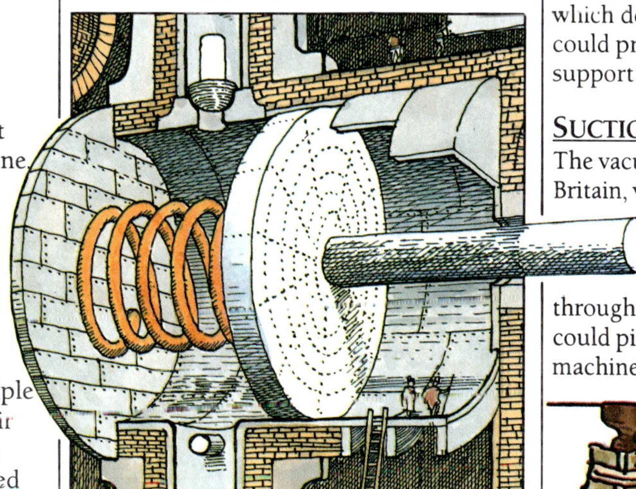

driven alternately up and down by hand. The fire engine was reputed to produce a jet of water nearly 160 feet (50m) high and to be strong enough to smash a window.

Portable fire extinguishers were developed in the nineteenth century, powered at first by compressed air and then by carbon dioxide.

Hydraulics and Pneumatics

An understanding of pressure in both air and water came with the work of the French scientist Blaise Pascal. In the mid-1600s, he discovered the principle that governs the action of pressure on a surface. Pascal's principle explains both hydraulics and pneumatics. One of its latest consequences is the hovercraft, which was invented in 1955 by the British engineer Christopher Cockerell. This machine began life as a pair of tin cans linked up to a vacuum cleaner, which demonstrated that an air cushion could produce sufficient pressure to support a hovercraft.

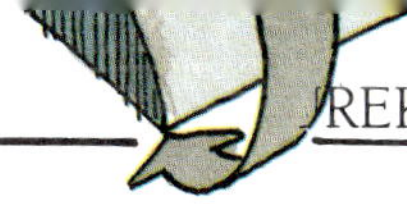

Suction Machines

The vacuum cleaner also began life in Britain, where it was invented by Hubert Booth in 1901. Again, a simple demonstration sufficed to prove its viability: Booth sucked air through a handkerchief to show how it could pick up dirt. However, a practical machine was developed in the United States in 1908 by William

Hoover, and it is his name that has always been associated with the vacuum cleaner. Its distant relative, the aqualung, is also firmly associated with its inventor, the French oceanographer Jacques Cousteau. The aqualung was developed during World War II, and Cousteau subsequently used it to pioneer exploration of the sea bed.

THE FLUSH TOILET

The water closet, the first of many euphemisms (though more accurate than most) for the flush toilet, dates back to 1589, when it was invented by Sir John Harington, a British nobleman who was a godson of Queen Elizabeth I. The tank in Harington's invention worked with a valve that released the flow of water. Harington recommended that it be flushed once or preferably twice a day.

Harington's important contribution to the history of technology was centuries ahead of its time, and the water closet did not attain its present form until the late 1800s. The use of a siphon, which does away with valves that can leak, dates from that period.

EXPLOITING HEAT

Harnessing heat was the first technological achievement to be made. The discovery of fire, which happened in Africa about a million years ago, provided heat for cooking and warmth. Millennia were to pass before heat was to be turned to the much more advanced uses such as smelting metals and providing motive power.

IRON- AND STEEL-MAKING

Iron-making dates from 1500BC when the Hittites, in what is now Turkey, built furnaces to smelt iron ore with charcoal and so produce the metal itself. The process did not develop further until 1709, when the British iron maker Abraham Darby substituted coke for charcoal and added limestone. His furnace needed a powerful blast of air to burn the coke, but it could make iron in large quantities – a factor which helped to bring about the Industrial Revolution.

The large-scale production of steel from

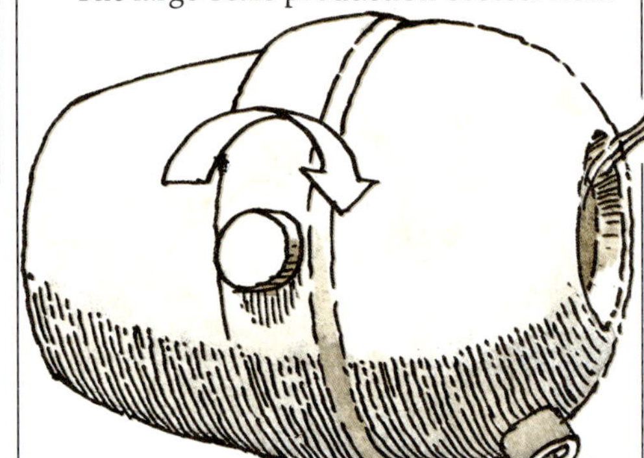

iron made in blast furnaces did not follow until the 1850s, when the steel converter was invented independently by William Kelly in the United States and Henry Bessemer in Britain. Air was blown through the molten iron to form steel, a process leading to the use of oxygen today.

THE REFRIGERATOR AND VACUUM FLASK

Although preserving food by keeping it in an ice-filled pit is an art some 4,000 years old, the first machine capable of reducing temperatures was not built until 1851. James Harrison, an Australian printer, noticed when cleaning type with ether that the type became very cold as the ether evaporated. Using this idea, he built an ether refrigerator. However, it was not very successful, being unable to compete with ice imported all the way from America.

The first practical refrigerator, which used ammonia as the refrigerant, was made by the German scientist Karl von Linde in 1876. He was able to produce liquid oxygen with it, but such a cold liquid was difficult to keep. James Dewar, a British scientist, developed the vacuum flask in 1892 to store liquid oxygen, but it has since found far wider use in storing hot drinks.

STEAM POWER

The use of heat to provide motive power came in a brilliant invention by the Greek engineer Hero. He built the first steam engine, a little device that spouted jets of steam and whirled around rather like a lawn sprinkler. Hero's engine was of no practical use and the steam engine vanished until the 1700s, when it was developed in Britain, notably by James Watt. The steam turbine was invented by another Briton, Charles Parsons, in 1884.

GASOLINE, DIESEL AND JET ENGINES

The gasoline engine followed the development of oil drilling in the mid-1800s and also the invention of a four-stroke engine running on gas at about the same time. The first two-stroke engine was invented in 1878, but it was a gasoline-powered four-stroke engine that came to power the horseless carriage. A practical gasoline engine was principally the work of the German engineer Gottlieb Daimler who developed it in 1883, fitting it first to a boat and then in 1885 to a bicycle. However, it was another German, Karl Benz, who built the first practical automobile in 1885.

The diesel engine was perfected by Rudolf Diesel in 1897, a year before the invention of the carburetor.

The gasoline engine spurred the invention of the airplane while the jet engine, being cheaper and faster, has brought us mass worldwide air travel. The jet engine was invented by the British engineer Frank Whittle in 1930.

THERMOMETERS

The measurement of temperature is associated with several famous names. The first thermometer was invented in 1593 by the great Italian scientist Galileo, who is better known for his discoveries in astronomy. The instrument used the expansion and contraction of a volume of gas, and was very inaccurate as well as bulky.

The first thermometer to use mercury was invented by the German physicist Gabriel Fahrenheit in 1714, and he also devised a temperature scale that bears his name.

GUNPOWDER AND ROCKETS

Heat also became a source of power in gunpowder, the first explosive, which appeared in China about a thousand years ago. Gunpowder had other uses too, and by the 1200s, rockets fueled by gunpowder were being fired in China.

The first person to propose that rockets be used for spaceflight was not an engineer, but a Russian schoolteacher named Konstantin Tsiolkovsky. At the turn of this century, he realized that rockets powered by liquid-fuel engines working in several separate stages would be required to provide the immense power necessary to carry people into space. However, Tsiolkovsky was a visionary and did not build rockets himself.

Liquid-fuel rocket engines were pioneered by the American engineer Robert Goddard, who first launched one in 1926. The first space rocket was built by the Russian engineer Sergei Korolev. It used liquid fuel to launch the first satellite, Sputnik 1, in October 1957, just a century after Tsiolkovsky was born.

NUCLEAR POWER

The basis of nuclear power was discovered in 1905 by the great German scientist Albert Einstein. In his special theory of relativity, Einstein explained that a little mass could theoretically be converted into a lot of energy.

Nuclear fission is the practical application of Einstein's theory, and was first achieved in the laboratory of the Italian scientist Enrico Fermi in 1934. Fermi did not realize at the time that fission had in fact occurred, and it was not until 1939 that other nuclear scientists were able to confirm that fission was possible, and with it the release of enormous amounts of energy. This information was kept secret as World War II loomed, and Fermi and the other scientists arrived in the United States. Here, at the prompting of Einstein, a crash program to build a nuclear reactor went ahead in the fear that Germany might build a fission bomb first. Fermi constructed the first experimental nuclear reactor in 1942.

The first atomic bomb was tested in the United States in July 1945, shortly after the defeat of Germany, and it was Japan that suffered the first, and so far the only, use of nuclear weapons. The hydrogen bomb was first tested by the United States in 1952. The first nuclear reactor to produce electricity was built in Russia in 1954.

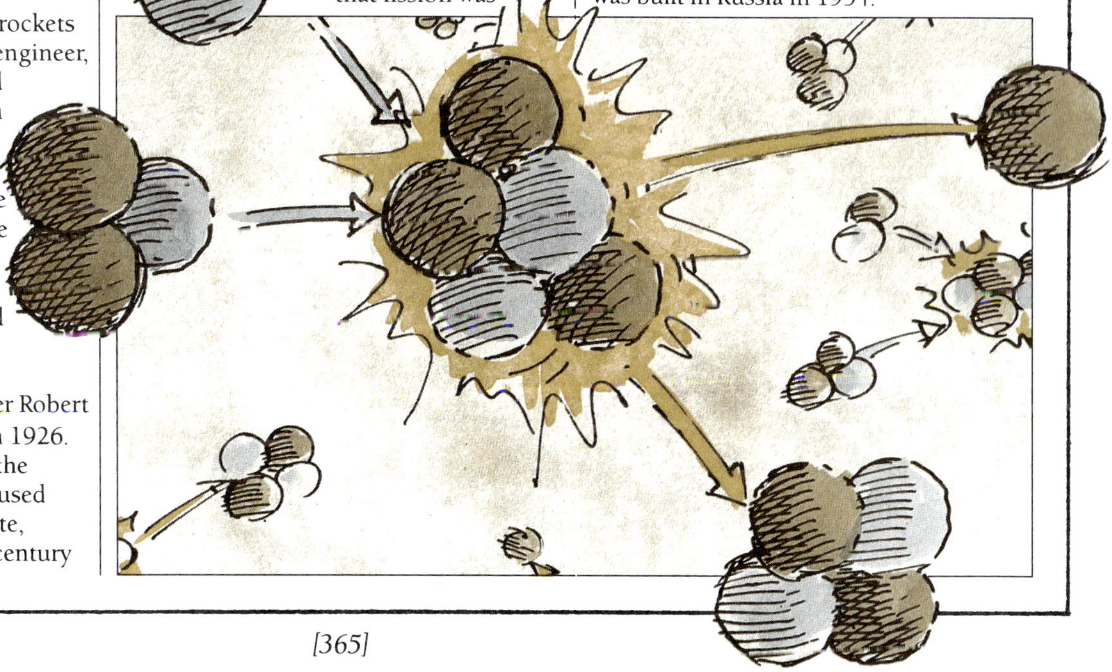

LIGHT AND IMAGES

People must have begun observing how light behaves thousands of years ago. They could see where it came from, and they could see that it was reflected by bright smooth surfaces and cast a shadow when something got in its way. The Greek philosopher Euclid was certainly familiar with the basic principles of optics around 300BC, and Alhazen, the famous Arab scholar, wrote an important treatise on the subject in the 900s AD. But no-one knew anything about the nature of light until 1666 when Isaac Newton discovered the color spectrum and 1678 when the Dutchman Christiaan Huygens suggested that light is composed of waves. Until then, Newton's assertion that light is made up of particles or "corpuscles" was regarded as more convincing.

ELECTRIC LIGHTS

The American inventor Thomas Edison is usually credited with inventing the electric light. In reality, however, he was preceded by a British competitor, Joseph Wilson Swan. Swan's filament lamp, not unlike Edison's, had been unveiled nearly a year before Edison's in December 1878. Incandescent lamps are relatively inefficient compared with fluorescent lamps, which give off little heat. Henri Becquerel, the discoverer of radioactivity, made the first experiments with fluorescence in 1859, but it was not until 1934 that the American physicist Arthur H. Compton developed the first fluorescent lamp for general use in homes and offices.

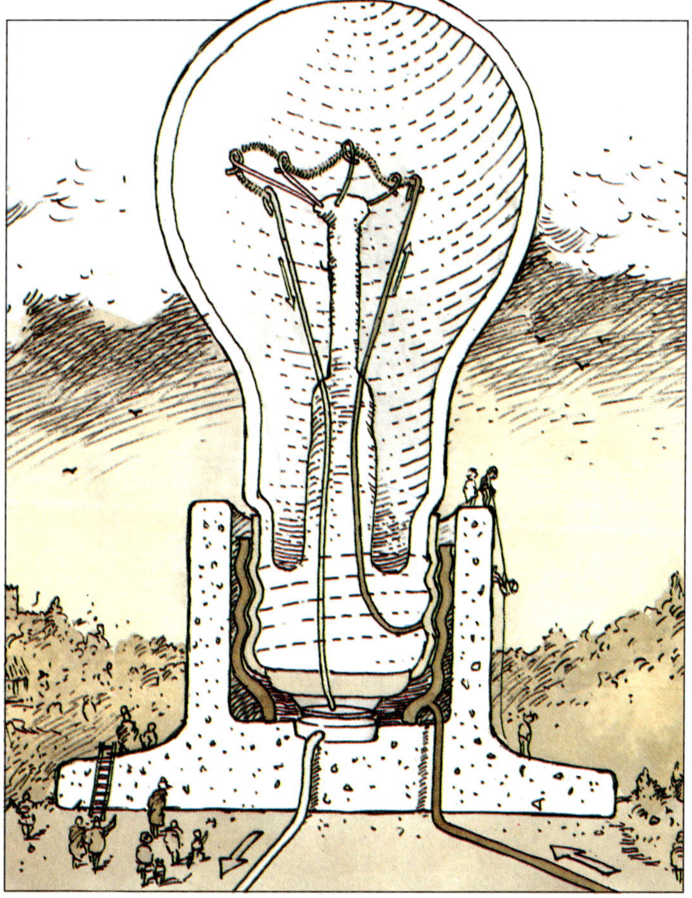

MIRRORS

"Natural" mirrors made of polished obsidian (a natural glass) were in use in Turkey 7,500 years ago, but the earliest manufactured mirrors were highly polished copper, bronze and brass. Pliny the Elder mentions glass backed with tin or silver in the first century AD, but silvering did not come into widespread use until the Venetians found a way of doing it in the thirteenth century. A German chemist, Justus von Liebig, invented the modern silvering process in 1835. Modern applications of reflecting surfaces include the periscope and the endoscope. Periscopes were developed for use in submarines in France in 1854. The flexible endoscope using glass fibers with special coatings to reflect images around corners came into use in 1958.

LENSES

The Roman Emperor Nero (AD 37 -68) was one of the first people to use a lens (although he may not have realized it) when he watched performances in the arena through a fragment of emerald which just happened to be of the right shape to benefit his poor eyesight. Spherical lenses used as burning glasses were certainly known by the 900s, when Alhazen described how they work. The first lenses to come into general use were convex lenses in eyeglasses, sometime around 1287 in Italy. The zoom lens, giving a correctly focused image at a range of focal lengths, was developed for use in the movie industry in the 1930s.

TELESCOPES

Lenses had been in use for centuries before Hans Lippershey, a Dutch eyeglass maker, happened upon the marvelous invention of the telescope. In 1608, he looked at a nearby church steeple through two lenses placed one in front of the other and found that it was magnified. The working telescopes that followed Lippershey's discovery all suffered from poor image quality caused by the refraction of light through the glass lenses. Isaac Newton solved this problem in 1668 by making a reflecting telescope that worked with mirrors rather than lenses. Binoculars are essentially two telescopes arranged side by side. They first appeared in a Paris opera house in 1823; although it is not known who invented them, they rapidly became popular for use both indoors and outside.

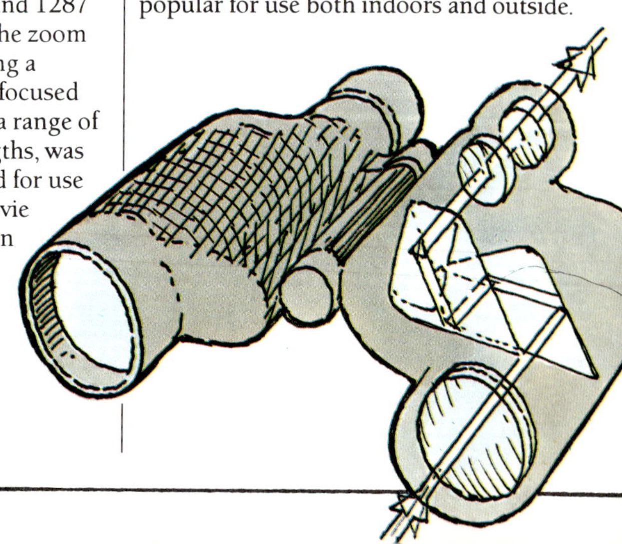

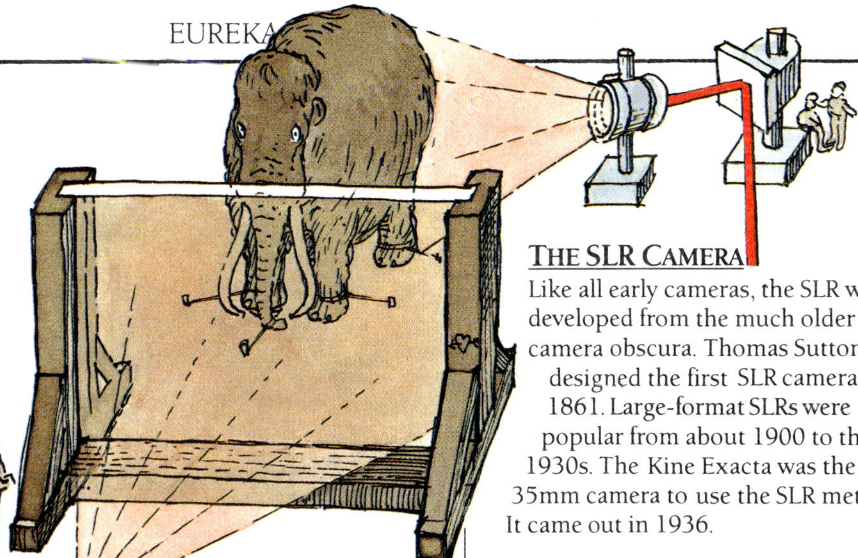

MICROSCOPES

Magnifying glasses have been used as long as convex eyeglass lenses have existed, and and they were developed into excellent single-lens microscopes by a Dutch merchant, Anton van Leeuwenhoek, in the mid-1600s. By using a tiny bead-like lens, he was able to obtain magnifications of up to 200 times.

The origin of the compound microscope, which has two lenses, is shrouded in some mystery. Another Dutch spectacle maker, Zacharias Janssen, has been credited with inventing the compound microscope in 1590. However, it seems unlikely that this would have preceded the discovery of the telescope, and Janssen's son is thought to have made up the story. Galileo is believed to have experimented with lenses for microscopy, but the biographers of the Dutch-born scientist Cornelius Drebbel insist that he built the first compound microscope in 1619. The first electron microscope was built over three centuries later in Germany in 1928.

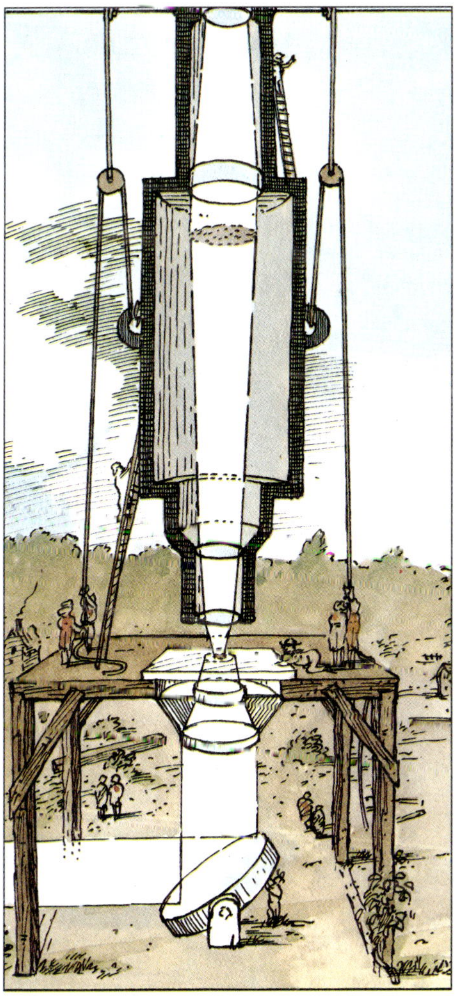

LASER AND HOLOGRAMS

The first laser was built in 1960 by Theo Maiman of the Hughes Laboratory. At the time it was scorned as the invention with no use. After this shaky start, the laser has become one of the most powerful and adaptable tools at our disposal.

A phenomenon which is only just now beginning to be explored – the hologram – depends for its existence on the laser. Dennis Gabor invented the hologram in 1947, but could not put his idea into practice until he had a coherent light source, in other words a beam of light of a single wavelength. This the laser gave him.

PHOTOGRAPHY

Photography was invented when Joseph Niepce, a Frenchman, found a way of fixing an image created in a camera obscura ("dark room" or box), a device which had been used for many years previously as a drawing aid for artists. He took his first photograph in 1826. Niepce's young partner, Louis Daguerre, invented a new process in 1837. Eventually it reduced exposure times to under half a minute, making portrait photography hugely popular. But modern photography is based on two breakthroughs made by the British inventor William Fox-Talbot in 1839: the negative-positive method of print-making, which allows many copies to be made of one exposure, and development of the latent image, leading ultimately to split-second exposure times. The first color photograph – of a tartan ribbon – was made by the British physicist James Clerk Maxwell in 1861.

THE SLR CAMERA

Like all early cameras, the SLR was developed from the much older camera obscura. Thomas Sutton designed the first SLR camera in 1861. Large-format SLRs were popular from about 1900 to the 1930s. The Kine Exacta was the first 35mm camera to use the SLR method. It came out in 1936.

INSTANT PICTURES

The American inventor Edwin Land became so absorbed in his studies of polarized light and photography that he dropped out of his regular courses at Harvard University. It proved to have been the right decision, however, because in 1947 he succeeded in producing the world's first instant picture camera, the Polaroid. Color pictures were introduced in 1963 and in 1972 a completely new design appeared, the SX-70. The SX-70 automatically ejected each photograph after exposure. It then developed as if by magic, without any need for peeling apart or timing. These had both been features of the original camera.

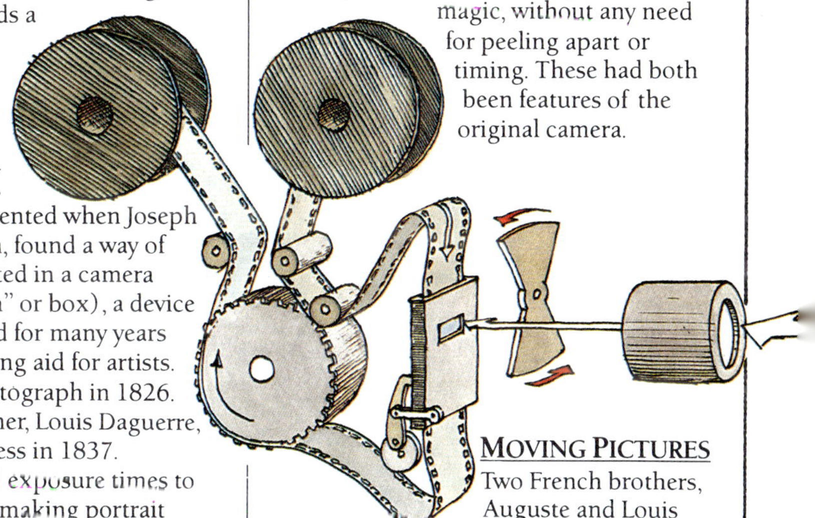

MOVING PICTURES

Two French brothers, Auguste and Louis Lumiere, invented the first practical movie camera and projector in 1895. At the first showings of their films, people fainted in the audience as a train appeared to come steaming straight out of the screen into the auditorium. Despite this impressive demonstration of its power, the brothers remained strangely unaware of their invention's enormous potential. When someone offered them a large sum of money for it, Auguste thought he was doing the eager buyer a great favor when he rejected his offer. How wrong he was!

PRINTING

A basic form of printing was practised by the Romans in the third century. About the same time Egyptian clothmakers used figures cut in blocks of wood to put marks and patterns on textiles. Block printing of books developed in isolation in both Europe and China. The Chinese produced the first block-printed book in 868 and were also the first to invent movable type in 1041. Unlike blocks, these could be used in the printing of any book, not just one, and were the vital element in Gutenberg's invention four centuries later. The Chinese made type out of baked clay. It soon became clear that only metal type could withstand repeated use. These were first made in Korea in the early fifteenth century. The letterpress method of printing is still in use today.

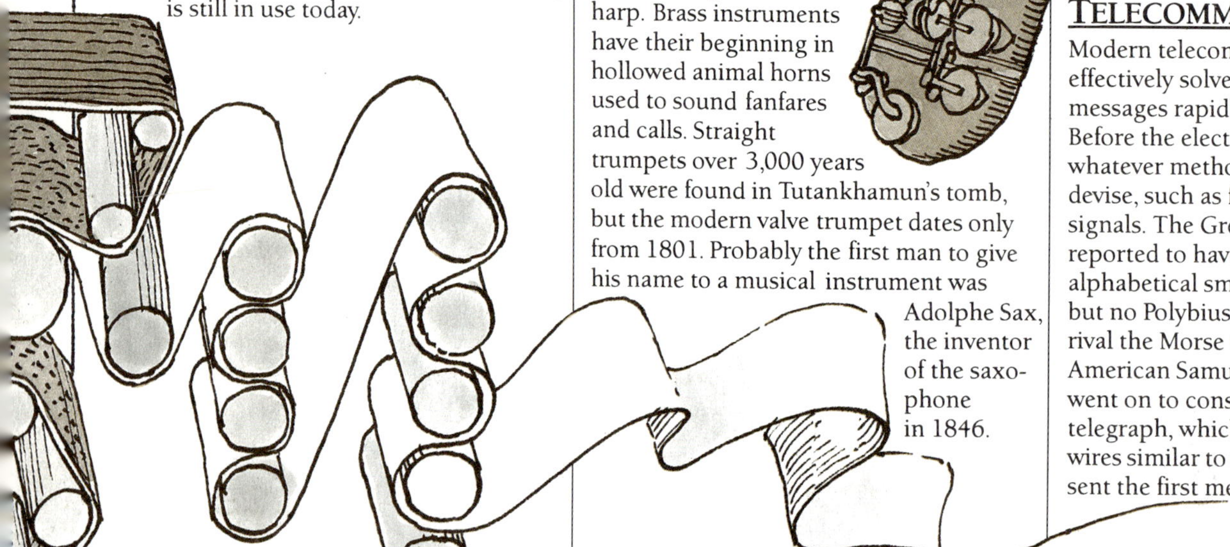

PAPER

Before paper was invented, people wrote on anything they could lay their hands on: silk and bamboo in China, palm leaves in India, clay tablets in Babylon and wax tablets in Greece. Between 3000 and 2000BC the Egyptians started using papyrus, a type of sedge which they dried into strips and then glued together in two layers to form a sheet. Paper was invented in China by Tsai Lun in AD 105. In 751, the Arabs captured some Chinese papermakers at Samarkand and so the invention set out on its four-hundred-year journey to the West. Today paper is made chiefly from fibres produced by trees.

THE PRINTING PRESS

The printing press was invented by Johan Gutenberg in Germany about 1450. It was one of a number of elements in the printing process (including movable metal type) which Gutenberg was the first to perfect. The press itself was adapted from existing screw presses used in trades like book binding and was so efficient that no significant changes were necessary until automation was introduced in the nineteenth century.

SOUND AND MUSIC

If archeological discoveries are anything to go by, the first musical instruments were hollow bones used as whistles in prehistoric times. Pottery drums have been found dating back 6,000 years. After the drum came the lyre, a stringed instrument played 4,500 years ago in the ancient city of Ur; it later developed into the harp. Brass instruments have their beginning in hollowed animal horns used to sound fanfares and calls. Straight trumpets over 3,000 years old were found in Tutankhamun's tomb, but the modern valve trumpet dates only from 1801. Probably the first man to give his name to a musical instrument was Adolphe Sax, the inventor of the saxophone in 1846.

THE TAPE RECORDER

Danish telephone engineer, Valdemar Poulsen, invented magnetic recording in 1898. Strictly speaking, Poulsen's "telegraphone" was not a tape recorder as it used wire rather than tape. The American film producer Louis Blattner made the first *tape* recorder in 1920 and AEG of Berlin produced the first *plastic* tape recorder in 1935.

THE RECORD PLAYER

The problems of recording sound and playing it back were solved by one of the greatest inventors of all time, the American Thomas Edison. Using a tinfoil cylinder as his "record" he recorded and then reproduced the nursery rhyme *Mary Had a Little Lamb* on December 6 1877. He called his invention a phonograph. Ten years later Émile Berliner, a German immigrant then living in Washington, invented the flat disk record player or gramophone.

TELECOMMUNICATIONS

Modern telecommunications have effectively solved the problem of sending messages rapidly over immense distances. Before the electronic age people had to use whatever methods their ingenuity could devise, such as flashing mirrors and smoke signals. The Greek historian Polybius is reported to have devised a system of alphabetical smoke signals in the 100s BC, but no Polybius Code is known today to rival the Morse Code, invented by the American Samuel Morse in 1838. Morse went on to construct the first electric telegraph, which carried his code over wires similar to telephone wires. In 1844, he sent the first message "What hath God wrought?"

THE TELEPHONE

During the great days of early sound engineering in the last century, inventors were often at a loss to think of things to say during their experiments and demonstrations. Alexander Graham Bell, however, had no difficulty when he used his newly invented telephone for the first time

in 1876: "Mr Watson. Come at once. I want you." were his first words. He had spilled acid from a battery down his pants and needed help from his assistant urgently. In inventing the telephone, Bell had also invented two important devices – the microphone and the loudspeaker.

RADIO

The introduction of the telegraph in 1844 solved the problem of sending messages using electricity. But the new machine had one big drawback: it depended on a physical wire link. Other scientists immediately began working on wire-less communications. A breakthrough came in 1888 when the German scientist Heinrich Hertz discovered the existence of radio waves. Seven years later, Guglielmo Marconi, the 21-year-old son of a wealthy Italian landowner, made the first successful transmission using radio waves. In 1901 he

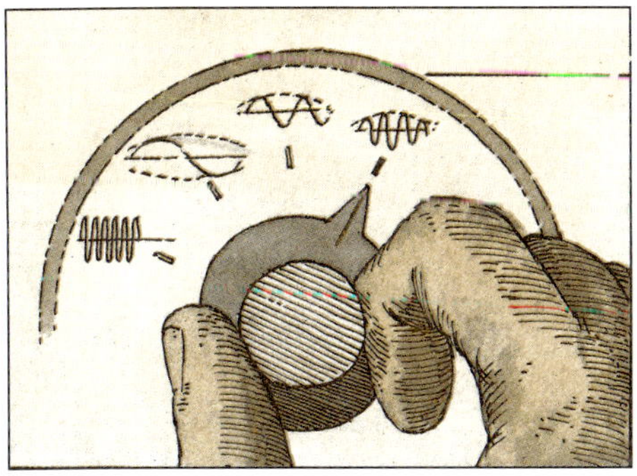

created an even bigger sensation when his radio sent a signal all the way across the Atlantic. Broadcasting began in 1906, when the Canadian inventor Reginald Fessenden first transmitted sound. However the invention of the electronic valve or tube in the same year by the American Lee de Forest was the major factor in the development broadcasting.

TELEVISION

Considering that television is the most powerful tool of mass communication known to man, it was conceived in remarkably humble circumstances. John Logie Baird was a British amateur scientist who sold shoe polish and razor blades to finance his spare-time research. In 1925 after years of work he successfully

transmitted the first television picture in his attic workshop, using a boy from the office downstairs as his subject. Because Baird's system was mechanical and gave low picture quality, it was only a matter of time before someone came along with a superior electronic product. That someone was Vladimir Zworykin, a Russian immigrant to America, who built the first electronic television in 1929. The world's first public broadcast was in 1936.

COMMUNICATIONS SATELLITES

The US government was responsible for developing the idea of communications satellites in the 1950s. In July 1962 the American Telephone and Telegraph Company launched Telstar, the first communications satellite to transmit telephone and television signals. It could only operate for a few hours each day, because its low orbit took it out of range of its transmitting and receiving stations for most of the time. Early Bird, launched in 1965, was the first satellite to solve this problem by keeping exact pace with the rotation of the Earth, maintaining an apparently stationary position.

RADIO TELESCOPE

The inventor of the radio telescope, and so of radio astronomy, was the amateur American astronomer Grote Reber. He built his first receiving dish in 1937, having heard about Karl Jansky's 1931 discovery that the Earth is constantly being bombarded with cosmic radio waves. Reber set out to focus these waves with his dish and thereby map where they came from. In 1942 he made the first radio map of the Milky Way galaxy.

SPACE PROBES

The first successful space probe was the Russian Luna 3, which sent back the first picture of the Moon's unseen far side in 1959. Probing the planets became a reality in December 1962 when the US spacecraft Mariner 2 reached Venus after a 180 million-mile (290 million-kilometer) journey lasting nearly four months.

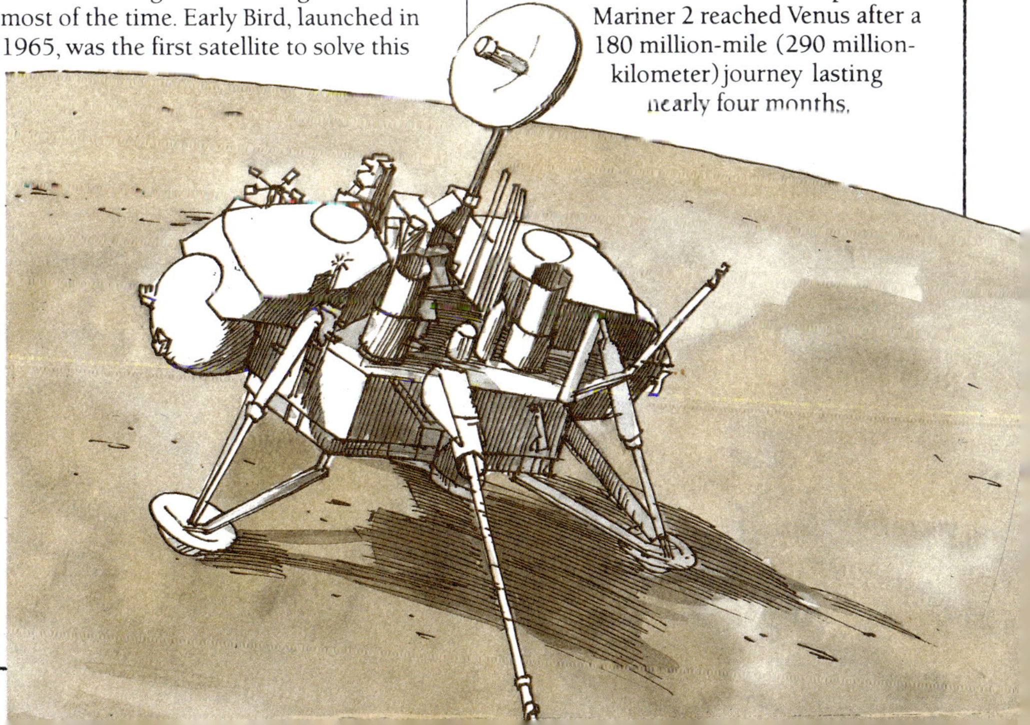

ELECTRICITY

In about 600BC the Greek philosopher Thales noticed that amber rubbed with wool somehow acquires the power to attract light objects such as straw and feathers. Over 2,000 years later in 1600, William Gilbert, physician to Queen Elizabeth I of England, called this power electricity after the Greek word for amber. It was not until the 1700s that scientists began to learn more about the nature of electricity, and one of the pioneers in the field was Benjamin Franklin, who was an intrepid investigator. In 1752, Franklin daringly flew a kite in a thunderstorm to prove that lightning is electrical in nature. This famous experiment, in which he was lucky not to have been killed, led Franklin to invent the lightning conductor. Franklin also postulated that electricity consists of two varieties of "fluid", one positive and one negative. We now know that the fluid is a stream of negative electrons, which were discovered by the British scientist J.J. Thompson in 1897.

THE BATTERY

In 1780 an Italian anatomist, Luigi Galvani, noticed that the severed leg of a dead frog could be made to twitch when touched by pieces of metal. Galvani concluded rightly that electricity was producing the reaction, but it was another Italian, Alessandro Volta, who found that the electricity came not from the frog, as Galvani had thought, but from the metals. Eventually Volta found that copper and zinc together produce a strong charge and that if he built a pile of metal disks, alternately copper and zinc separated by pads soaked in salty water, he could produce a continuous electric current. Perfected in 1800, the Voltaic pile, as it is called, was the first electric battery. Since then, a great range of different types of battery has been developed.

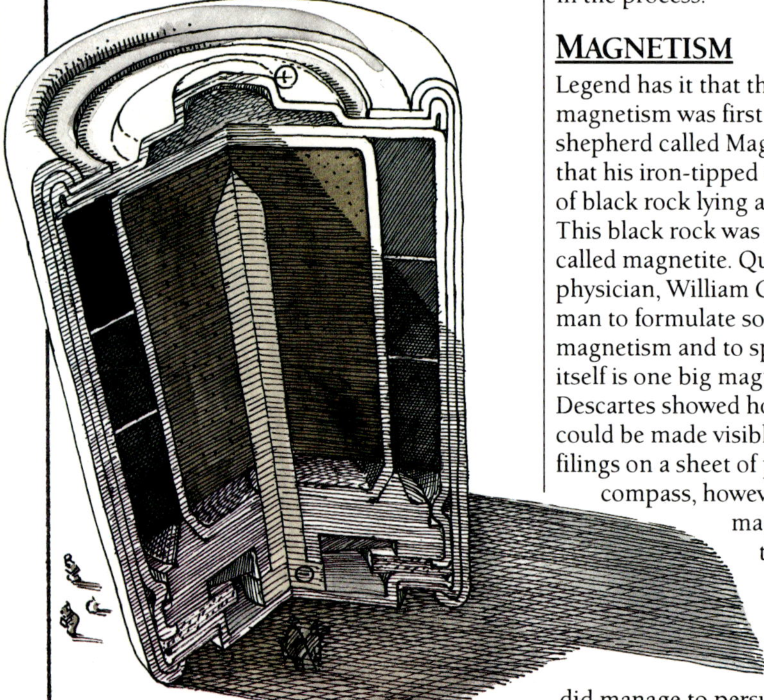

THE PHOTOCOPIER

In the 1930s Chester Carlson was working for the patents department of a large electronics firm in New York. He was happy enough in his work except for one thing – the time and expense involved in getting patents copied. Eventually he became so frustrated that he decided to invent a whole new process himself. The result was the first xerographic copy, taken on October 22 1938. Dispensing with the messy wet chemicals used in existing copiers, Carlson had invented a dry process based on the ability of an electrostatically charged plate to attract powder in the image of the original document. Several years later the rights to the process were acquired by a small family firm which later grew into the mighty Xerox corporation, making Chester Carlson a very wealthy man in the process.

MAGNETISM

Legend has it that the phenomenon of magnetism was first observed by a Greek shepherd called Magnes when he noticed that his iron-tipped crook picked up pieces of black rock lying around on the ground. This black rock was a kind of iron ore called magnetite. Queen Elizabeth I's physician, William Gilbert, was the first man to formulate some of the basic laws of magnetism and to speculate that the Earth itself is one big magnet. In 1644 René Descartes showed how magnetic fields could be made visible by scattering iron filings on a sheet of paper. Apart from the compass, however, no practical use for magnets was found until the invention of the electric motor – although Franz Anton Mesmer, the original mesmerizer, did manage to persuade eighteenth-century Parisians for a few years that magnetism was a cure for certain illnesses.

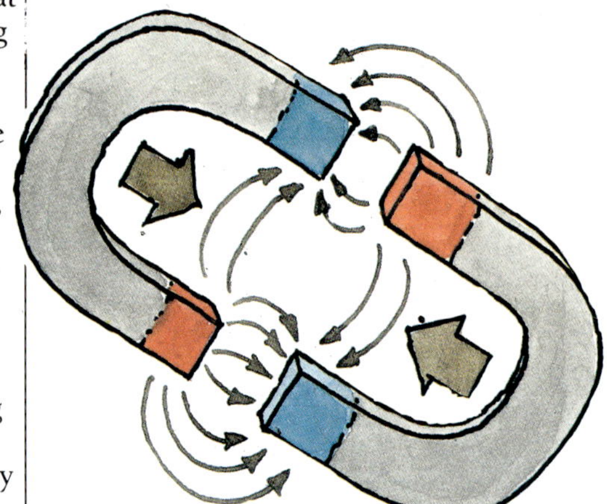

MAGNETS

The earliest magnets were made from naturally occurring magnetic rock called magnetite. Later when magnetite's directional properties were recognized, the name lodestone, meaning leading stone, was coined and it was used to make magnetic compasses. Magnets did not really come into their own until 1820 when the Danish physicist Hans Oersted made his sensational discovery of the link between magnetism and electricity. This event changed the course of human history by making possible the great electrical inventions of the nineteenth century such as the motor, the dynamo and, in the field of telecommunications, the telegraph.

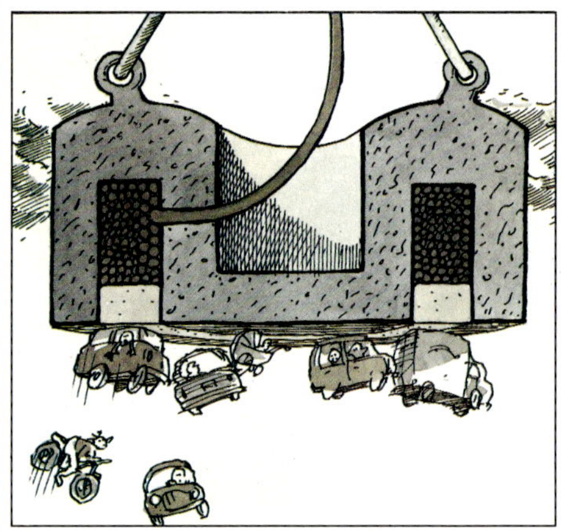

ELECTROMAGNETS

The electromagnet was one of the discoveries made possible by Oersted's great discovery. Shortly after it was announced, a French scientist, André-Marie Ampère, proved that wires could be made to behave exactly like magnets when a current was passed through them and that the polarity of the magnetism depended on the direction of the current. So the electromagnet – a magnet whose field is produced by an electric current – was born. Later the American inventor Joseph Henry found that wrapping several layers of insulated wire around a big piece of iron produced a vastly increased magnetic field. In 1829 he built the first heavy-duty working electromagnet, capable of lifting one ton.

MAGNETIC COMPASS

Chinese historians date the discovery of the magnetic compass to 2634BC. Whether or not this is true, the Chinese certainly seem to have been the first people to discover that magnetism could be useful in navigation, and by the third century AD magnetic compasses were in common use in the Far East. The Chinese were not noted navigators and it was left to the maritime nations of Europe to perfect the device. As with other inventions, the Arabs may have been responsible for transmitting the idea from East to West. By the eleventh century the Vikings were using compasses on their raids in northern Europe. More recent is a variation of the compass which measures the vertical angle that the Earth's magnetic field makes at its surface.

THE ELECTRIC MOTOR

In 1821, following Oersted's discovery the previous year, the British scientist Michael Faraday set out to show that just as a wire carrying electric current could cause a magnetized compass needle to move, so in reverse a magnet could cause a current-carrying wire to move. Suspending a piece of wire above a bowl of mercury in which he had fixed a magnet upright, Faraday connected the wire to a battery and sure enough it began to rotate. He had shown that electrical energy could be converted into mechanical energy, the principle behind the electric motor. The American scientist Joseph Henry built the first motor capable of work in 1830; by 1840 electric motors were powering machinery.

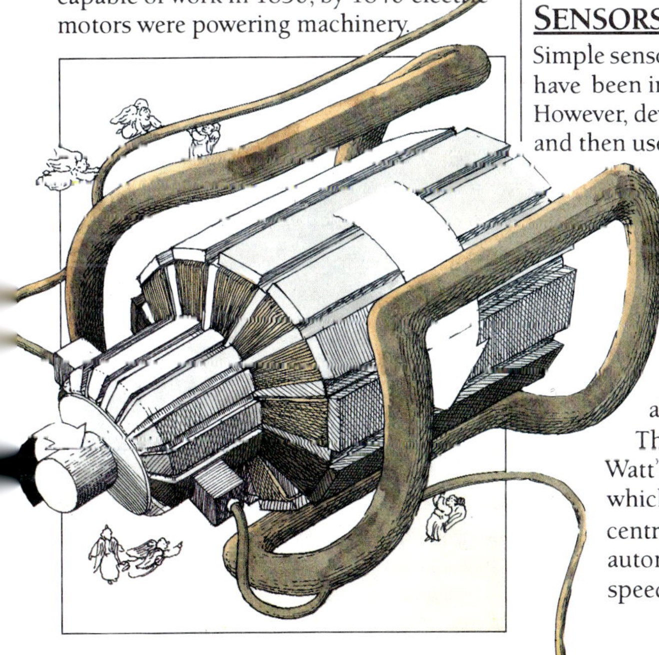

DOMESTIC ELECTRICITY SUPPLY

In the winter of 1880 a British industrialist, W.G. Armstrong, built a small hydroelectric station in the grounds of his country mansion in Northumberland to power its new electric lighting. It was the first domestic electricity supply anywhere in the world. The following winter the town council of Godalming in Surrey built the first power station to provide electric power for both private homes and public street-lighting. Take-up however was disappointingly slow and the station had to be closed a few years later. A few months later in January 1881 Thomas Edison's Electric Light Company installed a similar station at Holborn Viaduct in London. Unlike the Godalming scheme, this venture was highly popular and proved to be a roaring success.

SENSORS AND DETECTORS

Simple sensors triggered by movement have been in existence since ancient times. However, devices that can sense movement and then use this information to control machinery are more recent. Two important early examples were invented in the eighteenth century. The first was the windmill fantail, invented by Edmund Lee in 1745, which ensured that a windmill's sails always pointed into the wind. The second was James Watt's centrifugal governor, which ingeniously used centrifugal force to automatically regulate the speed of a steam engine.

THE SEISMOGRAPH

Historically, the Chinese have kept fuller records relating to earthquakes than any other country, so it is appropriate that they should also have produced the first seismograph. Invented by a mathematician, astronomer and geographer called Chang Heng (AD 78-139) it consisted of eight carefully balanced bronze balls arranged in a circle around a compass. Whenever the instrument picked up tremors from an earthquake, one of the balls would roll off, indicating which direction the vibrations had come from. The first seismograph to make use of currents produced by electromagnetism was invented by the Russian physicist Prince Boris Golitsyn in 1905.

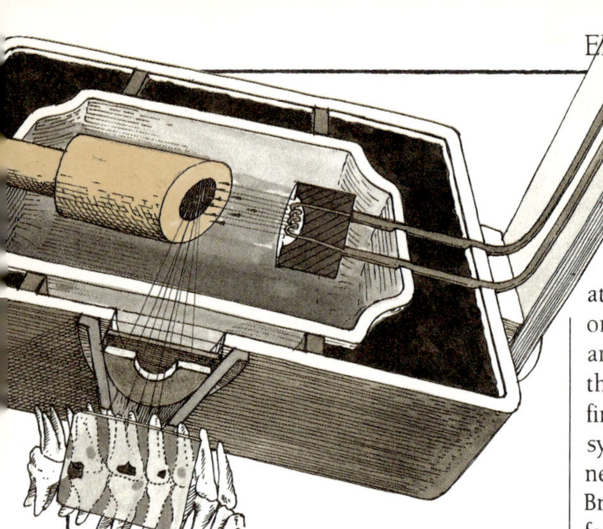

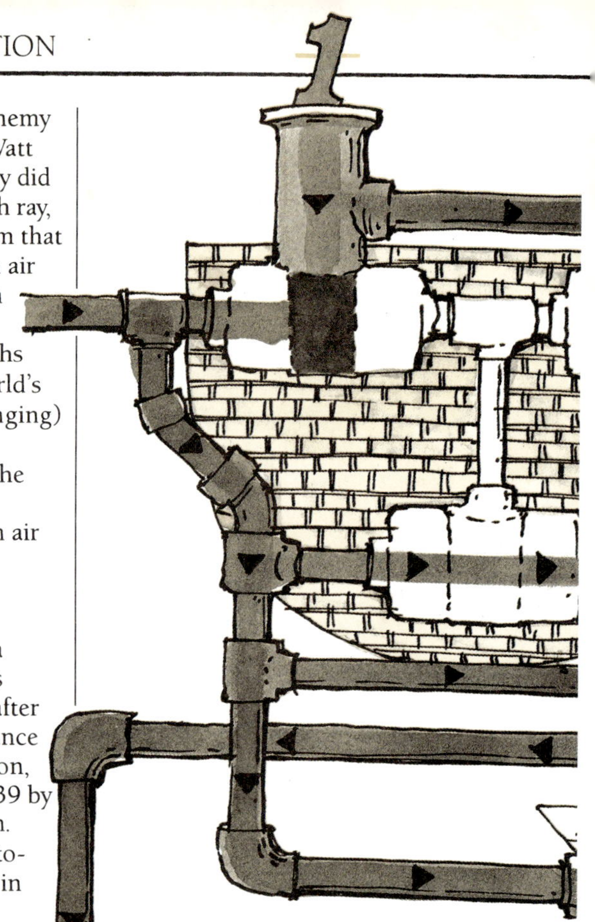

X-RAYS

In 1895 Wilhelm Röntgen, head of the physics department at Würzburg University in Germany, was amazed to see chemicals glowing on the other side of his laboratory while conducting experiments using a cathode ray tube enclosed in a container. After investigating he found that the cathode ray tube was causing the glow, but not the cathode rays, because they could not penetrate the container. Quite by chance he had discovered some completely unknown type of rays which he accordingly named X-rays. Before long he also discovered that photographic plates are sensitive to the rays even though the rays are invisible. This meant that it was possible to take photographs of objects not normally visible to the human eye, a discovery that revolutionized medical diagnosis. The X-ray measurement unit is named after him.

SONAR

In the early days of World War I, German U-boat submarines inflicted such heavy losses on Allied shipping that it became a matter of urgent priority to find some kind of effective submarine detection system. After experimenting with passive detectors, Paul Langevin, a French scientist, developed a much more sophisticated system using ultrasonic pulses generated by piezoelectricity. These found submarines even when their engines were not running by using echoes that bounced off their hulls to pinpoint each target's location.

RADAR

In 1935 the British Government asked a leading scientist, Robert Watson-Watt, about the possibility of producing a "death ray" to knock enemy aircraft out of the sky. Watt replied that the technology did not exist to produce a death ray, but that he could build a system that would give advance warning of an air attack. The details were written down on half a sheet of paper. Within the amazingly short space of a few months they had been developed into the world's first radar (RAdio Detection And Ranging) system. Within three years a whole network of radar stations protected the British coast and gave the British air force a decisive edge over the German air force in the Battle of Britain in 1940.

AUTOMATIC TRANSMISSION

Although automatic transmission is a sophisticated device, the first one was actually produced in 1896, not long after the first car. But it bore little resemblance to the first fully automatic transmission, the Hydramatic drive, invented in 1939 by American engineer Earl A. Thompson. Following Thompson's invention, automatic transmission became standard in American cars. The first model to be fitted with the new device was a 1940 Oldsmobile.

COMPUTERS

The history of computers is inextricably linked with that of its predecessor, the calculator. The first aid to counting, apart perhaps from counting on the fingers, was the abacus, which was developed in Babylonia about 5,000 years ago. The Roman abacus used *calculi* (pebbles) to represent numbers, from which our word calculation derives. Still in use, the abacus is now a frame with a set of wires carrying beads. Each wire represents the place of a digit in a number – the units, tens, hundreds and so on. Numbers are fed in by moving beads up the wire.

Interestingly, the abacus works in the same way as its electronic counterpart – the full adder in a calculator or computer. To add two numbers, beads representing both numbers are moved along the wires. If ten beads get moved up one wire, then they are moved back (to represent zero) and a single bead moved up the next wire instead. This "carrying-over" operation is exactly the same as the one performed by a full adder with binary numbers, except that one device uses beads and the other pulses electricity.

BINARY CODE

The idea that number systems do not necessarily have to be based on 10 is not a recent one. Gottfried Leibniz, working in the 1600s, developed theories of logic and binary numbers. A century later, George Boole, a British mathematician, devised a branch of logic that is still applied to binary systems in computing.

CALCULATORS

The principle of the abacus was first translated into a mechanical calculator in 1642 by the great French scientist Blaise

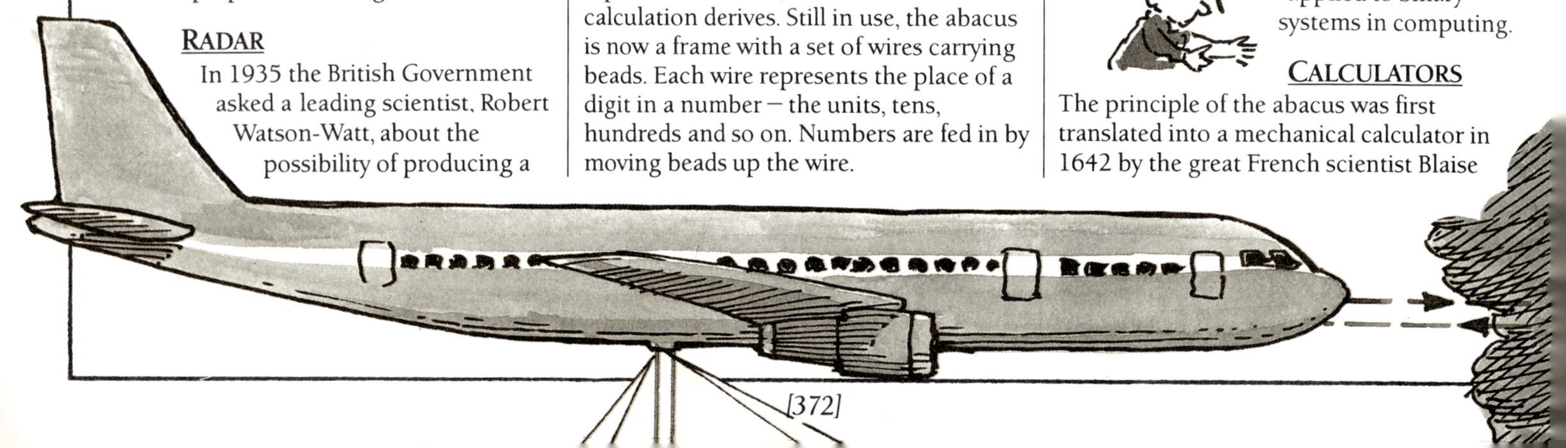

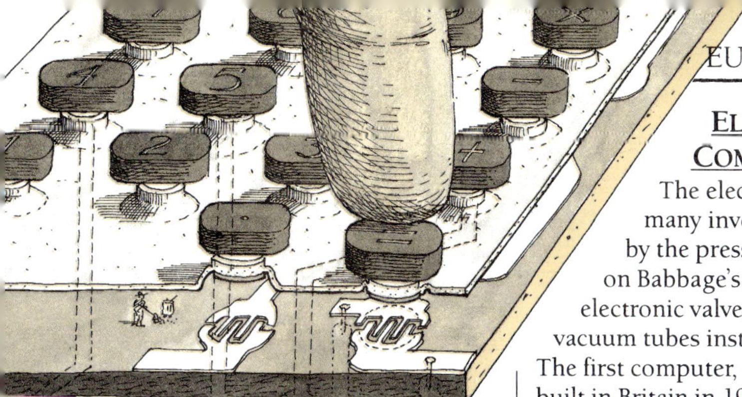

Pascal, at the tender age of 19. Numbers were fed into his machine by turning dials similar to those on a telephone, and the result appeared in a window. Pascal's calculator contained interlocking cogs, in which one cog turning ten digits tripped the next cog to turn one digit. It added or subtracted numbers with total accuracy, but was a financial flop.

In 1694, Gottfried Liebniz in Germany improved the mechanical calculator so that it could also multiply and divide. He devised a way of making it automatically perform repeated additions or subtractions. Mechanical calculators then advanced rapidly over the following two centuries, until they were gradually displaced by much faster electronic calculators from the late 1950s onward.

Mechanical Computers

Early calculators only did arithmetic; unlike a computer, they could not store results and they could not be given instructions to perform different tasks. However, the idea that such a machine could be built occurred to the British inventor Charles Babbage. He conceived the computer in 1833, a daring insight which earned him the title of "father of the computer". Babbage at that time was working on a machine called the Difference Engine capable of accurate calculation of logarithms (calculation tables) using a system called the method of differences. He hit upon a successor that could be programmed to perform different kinds of calculations and that could store results. The machine was called the Analytical Engine; it was to be given instructions on punched cards like those used in automatic looms, and it was said that the Analytical Engine would "weave algebraic patterns as the loom weaves flowers and leaves"

Sadly, this was not to be. The construction of Babbage's mechanical computer went far beyond the engineering of the day, and it was never completed. Computation needs the rapid execution of many separate steps, a requirement is very difficult to meet with large numbers of moving parts.

Electronic Computers

The electronic computer, like many inventions, was ushered in by the pressure of war. It was built on Babbage's principles but used electronic valves (see below) or vacuum tubes instead of cogs and levers. The first computer, called Colossus, was built in Britain in 1943 to break German codes, and may well have affected the outcome of World War II.

Colossus was in fact only used for code-cracking, and the first general-purpose computer was ENIAC, an American machine completed in 1946. ENIAC was hot and huge, containing 19,000 valves. Computers only shrank with the invention of the transistor and the microchip.

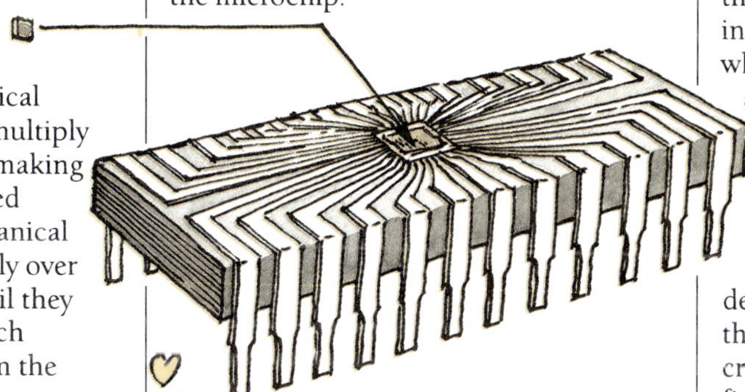

Diodes, Transistors and Microchips

Electronics really goes back to the turn of the century, when the first devices that could produce and process electric signals were invented. These were electronic valves or vacuum tubes in which a beam of electrons produced by a glowing filament carried a current between electrodes. The diode (two-electrode) valve came first, invented by the British scientist John Ambrose Fleming in 1904, followed in America by Lee de Forest's three-electrode triode valve in 1906. The diode changed alternating current to a direct-current signal, and the triode amplified a signal. These valves were crucial to the development of radio and television, and to sound recording.

However, valves were large and unreliable as the filament sooner or later burned out, making the development of small electronic machines impossible. The solution was found in 1948 by three American scientists – William Shockley, John Bardeen and Walter Brattain – working at the Bell Telephone Laboratories. Their research led to two crucial discoveries in the field of electronics – the semiconductor diode and transistor. Pieces of semiconductor replaced the filament and electrodes, enabling electronic components to be made small and fully reliable.

The next important development was to fabricate several components in a single piece of semiconductor – the integrated circuit. This was invented by the American engineer Jack Kilby in 1958 and it led to the microchip, into which many thousands of components are packed. The first microprocessor was produced in 1970.

Robots

The term robot, which is actually a Czech word meaning "labor", was first applied to machines in the 1920s. However, robots that move themselves are much older than this. They reached the height of perfection in the clockwork automata of the 1700s, which performed complex actions for the amusement of their wealthy owners. One, for example, could write a whole sentence. These early robots were entirely driven by complex gears and levers. Electronically controlled robots are a twentieth-century development. Although they are now found throughout modern factories, they are a far cry from the classic robots of science fiction that can see, listen, talk and think. These will eventually step out into the world only with major advances in computers.

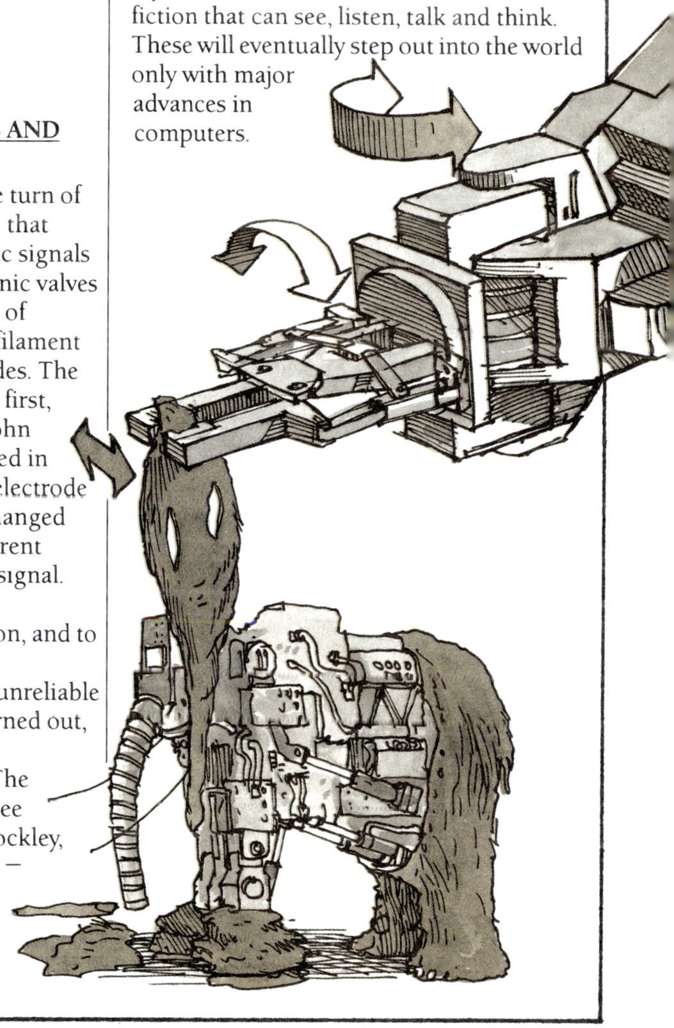

Technical Terms

A.C.
See *ALTERNATING CURRENT.*

Action and Reaction
Two forces that act whenever an object is moved. The moving force is called the action, and the object pushes back with a force called the reaction. Action and reaction are always equally strong, and they always push in opposite directions. They also occur when a liquid or gas is made to move or when they themselves make an object move.

Additive Color Mixing
Combining light sources of the three primary colors of light (red, green and blue) to produce all other colors.

Airfoil
The curved surface of a wing that produces lift as the wing moves through the air.

Alternating Current (A.C.)
Electric current in which the flow of current constantly reverses direction. In the U.S., the electricity supply alternates at a frequency of 60 times per second, or 60 hertz.

Ampère (Amp)
The unit of measurement for electric currents. A 1-amp current flows through a circuit if the resistance is 1 ohm and the voltage 1 volt.

Amplitude
The amount of energy in a ray or wave. It is equal to the change in energy (for example, pressure in a sound wave) that takes place as one complete wave passes.

Analog
A word applied to a system of recording or transmitting sound or pictures. In an analog system, the changing energy in the sound or light is converted to energy variations in another medium, for example magnetism in a tape.

Anode
An electrode with a positive charge.

Antenna
The part of a radio transmitter or receiver that sends out or picks up radio waves.

Armature
A part of an electric machine which moves in response to a current or signal, or which moves to produce a current or signal.

Atoms
The tiny particles of which the chemical elements that make up all substances are composed. An atom measures about 500-billionths of an inch (a hundred-millionth of a centimeter) in size, and consists of a central nucleus surrounded by electrons.

Auger
A large screw that rotates inside a pipe to transport water or loose materials, or a screw that is used to drill holes.

Axle
The shaft on which a wheel turns. The axle may be fixed to the wheel so that the wheel turns when the axle rotates, or alternatively, the wheel may spin freely on the axle.

Balance
A weighing machine, or the part of a mechanical watch that makes the watch keep time.

Binary Code
A code in which data is converted into binary numbers. Binary code is used in computers and other digital machines and systems, such as compact disk players and telephone networks.

Binary Number
A number in the binary number system. This number system uses only two digits (0 and 1).

Bit
Short for binary digit. In any binary number, a 0 or a 1. A 16-bit machine, for example, works with binary code made up of binary numbers each containing 16 bits (2 bytes).

Boom
The arm of a crane or excavator that raises the load.

Byte
A binary number containing eight bits, which represent decimal numbers between 0 (00000000) and 255 (11111111). Memory capacity is often measured in K or kilobytes (1024 bytes), and M or megabytes (1024 kilobytes).

Cam
A non-circular wheel which rotates in contact with a part called a follower. Together, the cam and follower are used to convert rotary motion into reciprocating motion.

Canard
A small wing placed in front of a main wing. Hydrofoils may also have canard foils, which are small forward foils. Canards aid stability in flight.

Capacitor
An electrical component that stores electric charge. Also called a condenser.

Carrier Wave
A radio wave that is broadcast at a particular frequency or wavelength and which is modulated to carry a sound or picture signal.

Cathode
An electrode with a negative charge.

Cell
A single device that produces electric current. A battery may contain several cells connected together, and a solar panel may contain several solar cells. Also a unit of memory that stores one bit of binary code.

Centrifugal
A word applied to any rotating device or part which moves away from the center of rotation.

Chip
See *MICROCHIP.*

Circuit
A source of electric current and a set of electrical devices or components that are connected together by wires so that current flows through them. A circuit board contains a printed metal pattern to conduct current to components fixed to the board.

Clock
In a calculator or computer, a device that produces regular electric pulses which synchronize the operations of the components.

Cog
A toothed gear wheel or a tooth on such a wheel.

Concave
A word applied to a surface that curves inward at the center.

Condenser
In heat, a device which cools a gas or vapor so that it changes into a liquid. In electricity, a component (also called a capacitor) that stores electric charge.

CONVEX
A word applied to a surface that curves outward at its center.

COUNTERWEIGHT
A weight that is fixed to one part of a machine to balance the weight of a load elsewhere in the machine.

CRANK
A wheel or rotating shaft to which a pivoted connecting rod is attached. As the crank turns, the rod moves to and fro; alternatively, the rod's movement may turn the crank. In a car engine crankshaft, a number of cranks are linked together and turned by rods connected to the pistons.

A winding handle is also a form of crank.

DAMPER
A part of a machine that absorbs vibration or prevents sudden movement. In a piano, the mechanism that stops the piano wires sounding.

DATA
Information, such as words or numbers, that a computer stores or requires in order to perform an operation. The data is processed by the computer and held in its memory in the form of numbers in binary code.

D.C.
See *DIRECT CURRENT*.

DENSITY
The weight of any amount of a solid, liquid or gas relative to its volume. Every pure substance has a particular density. Provided that two substances do not mix, the one with the lesser density will always float on top of the other. Wood floats on water because it has a lesser density than water.

DIFFRACTION
The bending of rays or waves that occurs as they pass through an opening or around an edge. The angle of bending depends on the wavelength.

DIGIT
A single numeral in a number, for example 2 or 7 in 27. The decimal number system uses ten different digits (0 to 9), the binary number system two different digits (0 and 1).

DIGITAL
A word applied to a system of recording or transmitting sound or pictures. In digital systems, the changing energy in the sound or light is continually measured. The measurements are then recorded or transmitted as a rapid sequence of code numbers.

DIODE
An electronic component through which current can flow in only one direction. A photodiode is sensitive to light or other rays, and a light-emitting diode (LED) emits light or other rays when a current flows through it.

DIRECT CURRENT (D.C.)
Electric current that always flows in one direction.

DRAG
The force with which air or water resists the motion of an object such as a car, boat or aircraft. Drag is also called air resistance or water resistance.

ECCENTRIC
A word applied to any object, often a wheel, that rotates about a point other than its center. An eccentric pin is an off-center projection on a wheel. It slides in a slot on an arm so that as the wheel rotates, it drives the arm to and fro.

EFFORT
The force that is applied to a machine to produce an action.

ELASTICITY
The ability of certain materials to regain their former shape and dimensions when forces cease to act on them.

ELECTRIC CHARGE
The electrical property produced by the addition (negative charge) or removal (positive charge) of electrons. The charge on the electron is the fundamental unit of electricity.

ELECTRIC CURRENT
The continual flow of electrons through a wire or other electrical conductor.

ELECTRIC FIELD
The region around an electric charge. One field affects another so that a negative charge and positive charge attract each other, and two negative charges or two positive charges repel each other.

ELECTRIC SIGNAL
A varying flow of current that is produced by changing other forms of energy into electricity. Two examples are the sound signal from a microphone and the data used by a computer in code form. The electric signals may be in analog or digital form.

ELECTRODE
Part of an electrical device or machine that either produces electrons (cathode) or receives electrons (anode).

ELECTROLYTE
A solution, paste, or molten substance that conducts electric current between electrodes.

ELECTROMAGNET
A device that uses an electric current to produce a magnetic field.

ELECTROMAGNETIC WAVES
The family of rays and waves that includes radio waves; microwaves, infra-red rays, light rays, ultraviolet rays, X-rays and gamma rays. All consist of vibrating electric and magnetic fields and travel at 186,000 miles per second (300,000 kilometers per second) which is the speed of light. All the rays and waves differ only in their wavelength or frequency. Except for gamma rays, all electromagnetic waves are generated by accelerating electrons.

ELECTROMAGNETISM
The relationship between electricity and magnetism; either can be used to produce the other.

ELECTRON
The smallest particle in an atom. An electron is about 100,000 times smaller than an atom, and has a negative electric charge. Electrons surround the central nucleus of the atom. They may be freed from atoms to flow through a conductor in an electric current, or to move through a vacuum in an electron beam. Electrons also move to produce a charge of static electricity.

ELECTROSTATIC
A word applied to a device that works by the production of an electric charge.

ELEMENT
A substance containing only one kind of atom. Some elements, such as hydrogen, nitrogen, oxygen and chlorine, are gases at normal temperatures. Others, such as iodine, sulfur and most metals, including iron, aluminum, copper, silver and gold, are solids. Only two, bromine and mercury, are liquids.

Just over 100 elements are known, including several artificial elements such as plutonium. All other substances are compounds of two or more elements.

ENERGY
The capacity to do work. Every action that occurs requires energy and, except in

nuclear reactions, converts one form of energy into another. Forms of energy include movement, heat, light and other electromagnetic waves, sound and electricity. There are also stored or potential forms of energy, such as chemical energy, that are available for conversion into other forms.

ESCAPEMENT
The part of a mechanical clock or watch that connects the train of gear wheels which moves the hands to the pendulum or to the balance, which controls the hands' speed.

EVAPORATION
The process by which a liquid turns into a vapor at a temperature below its boiling point. Evaporation occurs if the pressure of the vapor above the liquid is low enough for molecules to escape from the liquid into the vapor.

FIBER OPTICS
Devices that send images or light signals along glass fibers (optical fibers).

FISSION
A nuclear reaction in which the nuclei of atoms split apart to produce energy.

FLUORESCENT
A word often applied to something that glows with light. A fluorescent object, such as a screen, changes an invisible electron beam or ultraviolet rays into visible light.

FOCUS
A point at which rays or waves meet. With lenses, a sharp image forms at the focus of the lens. The focus of a telescope is the position at which an image is produced.

FORCE
The push or pull that makes something move, slows it down or stops it, or the pressure that something exerts on an object.

When a force acts on an object, it may be split into two smaller component forces acting at different angles. One of these component forces may move the object forward in one direction, while the other component may support its weight or overcome a separate force acting in another direction.

FREQUENCY
The rate at which waves of energy pass in sound waves and electromagnetic waves such as radio waves and light rays. Also the rate at which an alternating current changes direction, flowing forward and then backward. Frequency is measured in hertz (Hz), which is the number of waves or forward-backward cycles per second.

FRICTION
A force that appears when a solid object rubs against another, or when it moves through a liquid or gas. Friction always opposes movement, and it disappears when movement ceases.

FULCRUM
The pivot on which a device such as a lever is supported so that it can balance, tilt or swing.

FUSION
A nuclear reaction in which the nuclei of atoms combine to produce energy.

GAMMA RAYS
Invisible high-energy electromagnetic waves with wavelengths shorter than about a hundred-billionth of a meter. Gamma rays are emitted by the nuclei of atoms.

GAS TURBINE
A heat engine in which fuel burns to heat air and the hot air and waste gases drive a turbine. The jet engine is a gas turbine. Helicopters may have gas turbines in which the turbine drives the rotor.

GEAR
Two toothed wheels that intermesh either directly or through a chain so that one wheel turns to drive the other. A screw called a worm or a toothed shaft called a rack may replace one of the wheels.

In a moving machine such as a car or bicycle, a gear is also a combination of gear wheels that produces a certain speed. Top gear gives a high speed, and low gear a slow speed.

GRAVITY
The force that gives everything weight and pulls objects toward the ground. The normal pressure of the air or water is caused by gravity.

HAIRSPRING
A flat spiral spring in which one end is fixed and the other end can move.

HARMONICS
A set of accompanying waves that occurs with a main or fundamental wave. The frequencies of the harmonics are multiples of the frequency of the fundamental wave.

HEAT ENGINE
An engine in which heat is converted into movement by the expansion of a gas, which is either steam or the products of burning a fuel. There are two main kinds: external and internal combustion engines.

In an external combustion engine, the source of heat that raises the temperature of the gas is outside the engine, as in the boiler of a steam engine. In an internal combustion engine, fuel burns inside the engine. Gasoline and diesel engines, jet engines and rocket engines are all internal combustion engines.

HEAT EXCHANGER
A device in which heat is taken from a hot liquid or gas in order to warm a cool one. Inside a heat exchanger, pipes containing the hot fluid generally pass through the cool fluid.

HELICAL
A word applied to any device in the shape of a helix, such as a coil spring or a corkscrew.

HOLE
A space in an atom produced by the removal of an electron. As an electron comes from another atom to fill the hole, the hole "transfers" to the other atom.

HOLOGRAM
An image formed by laser light that has depth like a real object, or the photographic film or plate that produces the image.

HOLOGRAPHY
The production of holograms.

IMAGE
A picture of an object or scene formed by an optical instrument. A real image can form on a screen or other surface. A virtual image can only be seen in a lens, mirror or other instrument, or a hologram. Images can be recorded by photography, printing, video recording and holography.

INCLINED PLANE
A sloping surface. An inclined plane can be used to alter the effort and distance involved in doing work, such as raising loads.

INDUCTION
The production of magnetism or an electric current in a material by a magnetic field.

INERTIA
The resistance of a moving object to a change in its speed or direction, and the resistance of a stationary object to being moved.

INFRA-RED RAYS
Invisible electromagnetic waves with wavelengths longer than light rays and

ranging from a millionth to a thousandth of a meter. They include heat rays.

INTERFERENCE
The effects produced when two waves or rays meet. The combined wave has a different frequency or amplitude, giving color effects in light, for example.

INTERNAL COMBUSTION ENGINE
See *HEAT ENGINE.*

ION
An atom that has lost or gained one or more electrons and has an electric charge.

JACK
A device that raises a heavy object a short distance, with reduced effort.

LASER
A device that produces a narrow beam of very bright light or infra-red rays, in which all the waves have exactly the same frequency, are in phase and move exactly together. Laser stands for Light Amplification by Stimulated Emission of Radiation.

LENS
A device that bends light rays to form an image.

LEVER
A rod that tilts about a pivot to produce a useful movement.

LIFT
The upward force produced by an aircraft wing and helicopter rotor, and by the foils of a hydrofoil.

LIGHT RAYS
Visible electromagnetic waves ranging from 4 to 8 ten-millionths of a meter in wavelength, and respectively from blue to red in color.

LINEAR MOTION
Movement in a straight line.

LOAD
The weight of an object that is moved by a machine, or the resistance to movement that a machine has to overcome.

MAGNETIC FIELD
The region around a magnet or an electric current that attracts or repels other magnets.

MAIN SUPPLY
The supply of electricity to the home. It is alternating current at a voltage of about 110 volts and a frequency of 60 hertz.

MASS
The amount of substance that an object possesses. Mass is not the same as weight, which is the force that gravity exerts on an object to pull it to the ground. A floating object loses weight, but its mass remains the same.

MEMORY
The part of a computer or computing system that stores programs and data.

MICROCHIP
An electronic component containing many miniature circuits that can process or store electric signals. Also called a chip or integrated circuit.

MICROCOMPUTER
A small computer that can be placed on a desk or carried about.

MICROWAVES
Radio waves with very short wavelengths ranging from a millimeter to 30 centimeters.

MIRROR
A smooth surface that reflects light rays striking it. A semi-silvered mirror partly reflects and partly passes light.

MODERATOR
A substance used in a nuclear reactor to slow the fast-moving neutrons produced by fission of uranium fuel. Fast-moving neutrons do not cause further fission, and must be slowed to promote fission in the fuel.

MODULATION
Superimposing one kind of wave on another so that the first wave changes the second, often varying its amplitude (AM) or frequency (FM).

MOLECULES
The minute particles of which all materials – solid, liquids and gases – are composed. Each material has its own kind of molecules, which each consist of a particular combination of atoms. Water, for example, contains molecules each made of two hydrogen atoms fixed to an oxygen atom. In crystals, the atoms connect together in a regular network rather than forming separate molecules.

NEGATIVE
In photography, an image in which the brightness is reversed so that black becomes white and vice-versa; in a color negative, colors are reversed so that primary colors become secondary colors and vice-versa – blue becomes yellow, for example. In electricity, the charge on an electron is considered to be negative, so anything that stores or emits electrons is also negative. In waves, a minimum or opposite value of energy is considered to be negative.

***N*-TYPE SEMICONDUCTOR**
A kind of semiconductor that has been treated to produce electrons. It tends to lose these electrons and thus gain a positive charge.

NEUTRON
One of two kinds of particles that make up the nucleus of an atom. The other kind is the proton. A neutron has almost the same mass as a proton but no electrical charge. All nuclei contain neutrons except the very lightest, which is the common form of hydrogen. Deuterium and tritium, which are other forms or isotopes of hydrogen, do contain neutrons.

NUCLEUS (pl. NUCLEI)
The central part of an atom, composed mainly of two smaller particles called protons and neutrons that are held together with great force. The nucleus is about 10,000 times smaller than the whole atom. It is surrounded by electrons.

OPTICAL FIBER
See *FIBER OPTICS.*

OSCILLATOR
A device which produces sound waves or an electric signal of regular frequency.

PAWL
A pivoted arm that engages with the teeth of a ratchet.

PENDULUM
A rod or cord with a heavy weight called a bob at the lower end. The pendulum pivots at the upper end and swings to and fro. The time of each swing depends only on the length of the pendulum – not on the weight of the bob.

PINION
The smaller of two gear wheels, or a gear wheel that drives or is driven by a toothed rack.

PLANET WHEEL
A gear wheel that moves around another gear wheel, the sun wheel, as it turns.

POSITIVE
In photography, an image which looks like the original scene. In electricity, anything that receives electrons or from which electrons have been removed.

PRECESSION
A movement of a rotating wheel in response to a force on its axle. Precession makes the wheel move at right angles to the direction of this force.

PRESSURE
The force with which a liquid or a gas pushes against its container or any surface inside the liquid or gas. Units of pressure measure the force acting on a unit of surface area.

PRIMARY COLOR
A color that cannot be formed by mixing other colors. All other colors can be made by combining two or three primary colors.

PRISM
A glass block with flat sides in which light rays are reflected from the inner surfaces.

PROPELLANT
The liquid in a spray can or aerosol can which produces pressure that creates the spray, or the fuel of a rocket engine.

PROTON
One of two kinds of particles that make up the nucleus of an atom. The other kind is the neutron. A proton has almost 2,000 times the mass of an electron and has a positive electric charge. The number of protons in the nucleus defines the identity of an element. Hydrogen, for example, has one proton per nucleus, while oxygen has eight.

***P*-TYPE SEMICONDUCTOR**
A kind of semiconductor that has been treated to produce holes (spaces for electrons). It tends to gain electrons and thus acquire a negative charge.

PULLEY
A wheel over which a rope, chain or belt passes.

PULSE
A short burst of electric current.

RACK
A toothed shaft that intermeshes with a pinion.

RADIATION
The electromagnetic rays that come from any source of heat, or the rays and streams of particles that come from nuclear reactions and radioactive materials. Heat rays are harmless (unless they burn), but nuclear radiation can be highly damaging to living cells.

RADIATOR
The part of a car engine which removes heat from the cooling water that circulates through the engine; also a heater that warms a room by radiating (emitting) heat rays.

RADIOACTIVITY
The production of radiation by materials containing atoms with unstable nuclei, such as nuclear fallout and the waste from nuclear reactors.

RADIO WAVES
Invisible electromagnetic waves with wavelengths ranging from a millimeter to several kilometers. Radio waves used for radar have wavelengths of several millimeters or centimeters, shorter than the waves used for broadcasting sound radio and television.

RAM
In mechanical machines, such as an excavator, a device that exerts a strong pulling or pushing force. In computers, random-access memory – a temporary memory for programs and data.

RATCHET
A device which allows movement in one direction but not in the other. A ratchet has a toothed shaft or wheel on which a pawl rests. The pawl is pivoted so that it can move over the teeth of the ratchet in one direction. If the pawl or ratchet moves in the reverse direction, the pawl engages the teeth of the ratchet to prevent movement. A pawl may also move to and fro to turn a ratchet wheel in one direction.

RAY
An electromagnetic wave with a short wavelength.

REACTION
The equal and opposing force that always accompanies the action of a force (see *ACTION AND REACTION*). Also, in chemistry, the process by which one or more substances change to become different substances.

Chemical reactions often involve the production or consumption of heat. In chemical reactions, the atoms involved recombine in different configurations but do not themselves change. In nuclear reactions, the central nuclei of the atoms do change, producing new elements and emitting energy in the form of heat or radiation.

REAL IMAGE
See *IMAGE*.

RECIPROCATING MOTION
Movement in which an object moves repeatedly forward and backward.

REFLECTION
The reversal of direction that occurs when a wave or ray bounces off a surface. Internal reflection occurs if light rays reflect from the inner surface of a transparent material.

REFRACTION
The bending of a wave or ray that occurs as it passes from one medium or substance into another, for example from air into glass.

RESISTANCE
In mechanical machines, a force that slows the movement of an object, such as air resistance and water resistance, and the resistance of a material to cutting or breaking. In electricity, the property of an object, measured in ohms, that obstructs the flow of electrons through it.

RESONANCE
The production of vibrations or sound at a certain natural frequency in an object when it is struck by external vibrations or sound waves.

REVOLUTION
One complete turn of a rotating object.

ROM
In computers, read-only memory – a permanent memory for programs and data.

ROTARY MOTION
Movement in which an object spins around.

SCALE
A set of units or an indicator marked with units for measuring. A weighing machine is also known as a scale or scales.

SCANNING
The conversion of an image into a sequence of signals. Scanning splits up the image into a series of lines and converts the various levels of brightness and colours in each line into electric signals.

SCREW
A shaft with a helical thread or groove that turns either to move itself, or to move an object or material surrounding it.

SECONDARY COLOR
A color formed by mixing two primary colors.

SEMICONDUCTOR
A substance, such as silicon, whose electrical properties can be precisely controlled to regulate the flow of electrons and handle electric signals.

SHAFT
A bar or rod that moves or turns to transmit motion in a machine. Also a deep hole, as in an elevator shaft.

SOLAR CELL
A device that converts light into electricity.

SOUND WAVE
Waves of pressure that travel through air and other materials. At frequencies from about 20 hertz up to 20,000 hertz, we can hear these waves as sound.

SPEED
The rate at which something moves. Also a combination of gear wheels.

SPROCKET
A toothed wheel over which a chain passes.

STATIC ELECTRICITY
Electric charge produced by the movement of electrons into or out of an object.

STEREOPHONIC SOUND
Sound reproduced by two loudspeakers or earphones in which the sound sources, such as voices or instruments, are in different positions.

STEREOSCOPIC IMAGE
An image with depth. This kind of image is formed by a pair of images of an object or scene seen separately by both eyes.

SUBTRACTIVE COLOR MIXING
Combining dyes or pigments of the three secondary colors of light (yellow, cyan and magenta) to produce all other colors. These colors mix by absorbing primary colors from the light illuminating the dyes or pigments.

SUN WHEEL
A gear wheel around which a planet wheel rotates.

SUPERCONDUCTIVITY
The removal of electrical resistance in a conductor by cooling it. The conductor can then pass a very large electric current and generate a strong magnetic field.

SUPERSONIC
Faster than the speed of sound, which is about 760mph (1200km/h) at sea level.

TENSION
The force produced in a bar or a rope or string when it is stretched.

TERMINAL
The part of an electric machine to which a wire is connected to take or supply electric current. Also a unit for use by an operator that is connected to a computer.

THREAD
The helical groove around a screw or inside a nut.

THRUST
A force that moves something forwards.

THRUSTER
A propeller used for maneuvering a ship or submersible; also a small rocket engine or gas jet used for maneuvering a spacecraft.

TRANSFORMER
A device that increases or decreases the voltage of an electric current.

TRANSISTOR
An electronic component made of sections of *n*-type and *p*-type semiconductor that switches a current on or off, or amplifies the current. A controlling signal goes to the central section (the base or gate), which controls the flow of current through two outer sections (the emitter or source, and the collector or drain).

TURBINE
A machine with blades that are turned by the movement of a liquid or gas such as air, steam or water. The turbine may also turn to move the liquid or gas.

ULTRAVIOLET LIGHT
Invisible electromagnetic waves with a wavelength less than that of light and ranging from 5 billionths to 4 ten-millionths of a meter.

VALVE
A device that opens or closes to control the flow of a liquid or gas through a pipe. Valves often work one way and seal a container so that a liquid or gas can only enter it and not escape.

VAPOR
See *EVAPORATION*.

VIRTUAL IMAGE
See *IMAGE*.

VOLTAGE
The force, measured in volts, with which a source of electric current or charge moves electrons.

WATT
The unit of power. One watt is produced when a current of one amp from a source of one volt flows for one second.

WAVE
A flow of energy in which the level of energy regularly increases and decreases, like the height of a passing water wave. One complete wave is the amount of flow between one maximum of energy and the next. This distance is the wavelength.

WAVELENGTH
See *WAVE*.

WEDGE
A part of a machine with a sloping side that moves to exert force.

WEIGHT
The force with which gravity pulls on an object.

WHEEL
Any circular rotating part in a machine.

WHEEL AND AXLE
A class of rotating machines or devices in which effort applied to one part produces a useful movement at another part.

WINCH
A drum around which a rope is wound to pull, lift or lower an object.

WORM
A screw that intermeshes with a gear wheel.

X-RAYS
Invisible electromagnetic waves with wavelengths shorter than light and ranging from 5 billionths to 6 million-millionths of a meter.

INDEX

A

B

C

ACKNOWLEDGMENTS

The editorial and design staff for *THE WAY THINGS WORK* would like to extend their special thanks to the following for their help in the preparation of this book:

Hilary Bird (indexing); Andrew Duncan (research); Kathy Gill (proofreading); Fred Ford and Mike Pilley of Radius (artwork services); Lesley Goode (typing); Ray Owen (artwork services); Judy Sandeman and Jeanette Graham (production); Rupert Wheeler (technical assistance).